MECHANICAL MEASUREMENTS
Fifth Edition

MECHANICAL MEASUREMENTS
Fifth Edition

Thomas G. Beckwith
University of Pittsburgh, Emeritus

Roy D. Marangoni
University of Pittsburgh

John H. Lienhard V
Massachusetts Institute of Technology

Addison-Wesley Publishing Company
Reading, Massachusetts ▪ Menlo Park, California ▪ New York
Don Mills, Ontario ▪ Wokingham, England ▪ Amsterdam ▪ Bonn
Sydney ▪ Singapore ▪ Tokyo ▪ Madrid ▪ San Juan ▪ Milan ▪ Paris

The photograph on the cover is reproduced courtesy of AEROMETRICS.

Figure 16.18b is adapted from *LakeShore: Measurement and Control Technologies,* courtesy of LakeShore Cryotronics, Inc. Figure 16.18a is adapted from *Analog Devices: Linear Products Databook,* 1990/91, courtesy of Analog Devices, Inc. Figure 6.16 is adapted from *Pressure Sensors,* BR121/D REV 5, courtesy of Motorola Inc. Material on page 302 adapted from *Kistler Piezo-Instrumentation General Catalog,* 1st edition, 1989, courtesy of Kistler Instrument Corporation. Material on page 294 adapted from *General Purpose Linear Devices Datebook,* 1989, courtesy of National Semiconductor. Material on page 98 from *Engineering and Scientific Graphs for Publication,* courtesy of The American Society of Mechanical Engineers. All material reprinted with permission.

Library of Congress Cataloging-in-Publication Data
Beckwith, T. G. (Thomas G.)
   Mechanical measurements / Thomas G. Beckwith, Roy D. Marangoni,
   John H. Lienhard. — 5th ed.
      p.   cm.
   Includes bibliographical references and index.
   ISBN 0-201-56947-7
   1. Engineering instruments.   2. Measuring instruments.
   I. Marangoni, Roy D.   II. Lienhard, John H., 1961–   .   III. Title.
   TA165.B38   1993
   681'.2—dc20   92–23515
                                                                       CIP

2 3 4 5 6 7 8 9 10–DO–959493

PREFACE

After more than 30 years, the basic purpose of this book can still be expressed by the following three paragraphs extracted from the Preface of the first edition of *Mechanical Measurements,* published in 1961.

> Experimental development has become a very important aspect of mechanical design procedure. In years past the necessity for "ironing out the bugs" was looked upon as an unfortunate turn of events, casting serious doubts on the abilities of a design staff. With the ever-increasing complexity and speed of machinery, a changed design philosophy has been forced on both the engineering profession and industrial management alike. An experimental development period is now looked upon, not as a problem to avoid, but as an *integral phase* of the whole design procedure. Evidence supporting this contention is provided by the continuing growth of research and development companies, subsidiaries, teams, and armed services R & D programs.
>
> At the same time, it should not be construed that the experimental development (design) approach reduces the responsibilities attending the preliminary planning phases of a new device or process. In fact, knowledge gained through experimental programs continually strengthens and supports the theoretical phases of design.
>
> *Measurement* and the correct interpretation thereof are necessary parts of any engineering research and development program. Naturally, the measurements must supply reliable information and their meanings must be correctly comprehended and interpreted. *It is the primary purpose of this book to supply a basis for such measurements.*

With the fifth edition of *Mechanical Measurements* considerable changes and improvements have occurred. Foremost a new co-author, Dr. John H. Lienhard, has been added. In addition, more than half of the chapters in the book have been substantially revised. Some specific changes are as follows:

- The uncertainty material in Chapter 3 has been fully revised and follows the ASME standards.

- New material on discrete sampling and discrete Fourier transforms has been added in Chapter 4 and also in Chapter 8.

- New material on semiconductor sensors has been added in Chapter 6 and also in Chapter 16.

- The sections on filters and op-amps in Chapter 7 have been revised and updated.

- Laser-based velocity and displacement measurement has been introduced in Chapter 15 and also in Chapters 11 and 17.

- New material on sound intensity and noise measurement has been added in Chapter 18.

- Recent changes in measurement standards are incorporated in Chapter 2.

- The remainder of the material has been substantially updated, and approximately 40 percent of the problems have been replaced or revised.

The authors do not suggest that the sequence of materials as presented need be strictly adhered to. Wide flexibility of course contents should be possible, with text assignments tailored to fit a variety of basic requirements or intents. For example, the authors have found that, if desired, Chapters 1 and 2 can simply be made a reading assignment. Greater or lesser emphasis may be placed on certain chapters as the instructor wishes. Should a course consist of a lecture/recitation section plus a laboratory, available laboratory equipment may also dictate areas to be emphasized. Quite generally, as a text, the book can easily accommodate a two-semester sequence.

Acknowledgments

The authors would like to thank Senior Editor Eileen Bernadette Moran, Editorial Assistant Dana Goldberg, and the entire staff at Addison-Wesley for their energetic assistance in assembling the fifth edition. Roy Marangoni would like to express his gratitude to Dr. Joel E. Peterson for the material on digital counters and digital recording, which was adapted from his lecture and laboratory notes at the University of Pittsburgh.

John Lienhard would like to express his gratitude to Thomas Beckwith and Roy Marangoni for inviting his participation in this book—it is a great pleasure to work with you both. Making an adequate contribution to this already outstanding book has been an exceptional challenge. Thanks also go to my mentors in measurement: Millard F. Beatty, Charles W. Van Atta, Kenneth N. Helland, Jon Haugdahl, Frank A. McClintock, C. Forbes Dewey, and Ernest Rabinowicz. I would like to thank Lucille Blake for her help in preparing and handling the manuscript. Finally, I would like to thank David N. Wormley and the Department of Mechanical Engineering for supporting me in this undertaking, and I would like to acknowledge summer support from the Bernard M. Gordon Engineering Curriculum Development Fund.

CONTENTS

Contents

Important ↗

Solve and interpret the solutions to diff eq

Contents

Contents

13 Measurement of Force and Torque *537*

14 Measurement of Pressure *571*

Appendixes

Answers to Selected Problems

Index

PART I

Fundamentals of Mechanical Measurement

CHAPTER 1

The Process of
Measurement:
An Overview

1.1   Introduction

It has been said: "Whatever exists, exists in some amount." The determination of the amount is what measurement is all about. If those things that exist are related to the practice of mechanical engineering, then the determination of their amounts constitutes the subject of *mechanical measurements.**

The process or the act of measurement consists of obtaining a quantitative comparison between a predefined *standard* and a *measurand*. The word *measurand* is used to designate the particular physical parameter being observed and quantified; that is, the input quantity to the measuring process. The act of measurement produces a *result* (see Fig. 1.1).

The standard of comparison must be of the same character as the measurand, and usually, but not always, is prescribed and defined by a legal or recognized agency or organization—e.g., the National Institute of Standards and Technology (NIST), formerly called the National Bureau of Standards (NBS), the International Organization for Standardization (ISO), or the American National Standards Institute (ANSI). The meter, for example, is a clearly defined standard of length.

---

* *Mechanical measurements* are not necessarily accomplished by mechanical means: Rather, it is to the measured quantity itself that the term *mechanical* is directed. The phrase *measurement of mechanical quantities,* or *of parameters,* would perhaps express more completely the meaning intended. In the interest of brevity, however, the subject is simply called *mechanical measurements.*

Figure 1.1   Fundamental measuring process

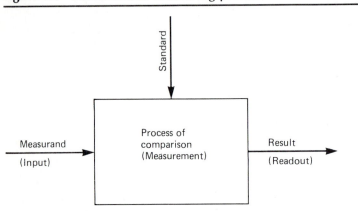

Such quantities as temperature, strain, and the parameters associated with fluid flow, acoustics, and motion, in addition to the fundamental quantities of mass, length, time, and so on, are typical of those within the scope of mechanical measurements. Unavoidably, the measurement of mechanical quantities also involves consideration of things electrical, since it is often convenient or necessary to change, or *transduce,* a mechanical measurand into a corresponding electrical quantity.

1.2   The Significance of Mechanical Measurement

Measurement provides quantitative information on the actual state of physical variables and processes that otherwise could only be estimated. As such, measurement is both the vehicle for new understanding of the physical world and the ultimate test of any theory or design. Measurement is the fundamental basis for all research, design, and development, and its role is prominent in many engineering activities.

All mechanical design of any complexity involves three elements: experience, the rational element, and the experimental element. The element of experience is based on previous exposure to similar systems and on an engineer's common sense. The rational element relies on quantitative engineering principles, the laws of physics, and so on. The experimental element is based on measurement—that is, on measurement of the various quantities pertaining to the operation and performance of the device or process being developed. Measurement provides a comparison between what was intended and what was actually achieved.

Measurement is also a fundamental element of any control process. The concept of control *requires* the measured discrepancy between the actual and the desired performances. The controlling portion of the system must know the magnitude and direction of the difference in order to react intelligently.

In addition, many daily operations require measurement for proper performance. An example is in the modern central power station. Temperatures, flows, pressures, and vibrational amplitudes must be constantly monitored by measurement to ensure proper performance of the system. Moreover, measurement is vital to commerce. Costs are established on the basis of *amounts* of materials, power, expenditure of time and labor, and other constraints.

To be useful, measurement must be reliable. Having incorrect information is potentially more damaging than having no information. The situation, of course, raises the question of the accuracy or *uncertainty* of a measurement. Arnold O. Beckman, founder of Beckman Instruments, has stated, "One thing you learn in science is that there is no *perfect* answer, no *perfect* measure."* It is quite important that engineers interpreting the results of measurement have some basis for evaluating the likely uncertainty. Engineers should *never* simply read a scale or printout and blindly accept the numbers. They must carefully place realistic tolerances on each of the measured values, and not only should have a doubting mind but also should attempt to quantify their doubts. We will discuss uncertainty in more detail in Section 1.8 and as the subject of Chapter 3.

1.3 Fundamental Methods of Measurement

There are two basic methods of measurement: (1) *direct comparison* with either a primary or a secondary standard and (2) *indirect comparison* through the use of a calibrated system.

1.3.1   Direct Comparison

How would you measure the length of a bar of steel? If you were to be satisfied with a measurement to, let us say, $\frac{1}{8}$ in. (approximately 3 mm), you would probably use a steel tape measure. You would compare the length of the bar with a *standard,* and would find that the bar is so many inches long because that many inch-units on your standard are the same length as the bar. Thus you would have determined the length by *direct comparison.* The standard that you have used is called a secondary standard. No doubt you could trace its ancestry back through no more than four generations to the primary length standard, which is related to the speed of light (Section 2.5).

Although to measure by direct comparison is to strip the measurement process to its barest essentials, the method is not always adequate. The human

---

* Emphasis added by the authors.

senses are not equipped to make direct comparisons of all quantities with equal facility. In many cases they are not sensitive enough. We can make direct comparisons of small distances using a steel rule, with a preciseness of about 1 mm (approximately 0.04 in.). Often we require greater accuracy. Then we must call for additional assistance from some more complex form of measuring system. Measurement by direct comparison is thus less common than is measurement by *indirect comparison.*

1.3.2   Using a Calibrated System

Indirect comparison makes use of some form of transducing device coupled to a chain of connecting apparatus, which we shall call, in toto, the *measuring system.* This chain of devices converts the basic form of input into an analogous form, which it then processes and presents at the output as a known function of the original input. Such a conversion is often necessary so that the desired information will be intelligible. The human senses are simply not designed to detect the strain in a machine member, for instance. Assistance is required from a system that senses, converts, and finally presents an analogous output in the form of a displacement on a scale or chart or as a digital readout.

Processing of the analogous signal may take many forms. Often it is necessary to increase an amplitude or a power through some form of amplification. Or in another case it may be necessary to extract the desired information from a mass of extraneous input by a process of filtering. A remote reading or recording may be needed, such as ground recording of a temperature or pressure in a missile in flight. In this case the pressure or temperature measurement must be combined with a radio-frequency signal for transmission to the ground.

In each of the various cases requiring amplification, or filtering, or remote recording, electrical methods suggest themselves. In fact, the majority of transducers in use, *particularly for dynamic mechanical measurements,* convert the mechanical input into an analogous electrical form for processing.

1.4   The Generalized Measuring System

Most measuring systems fall within the framework of a general arrangement consisting of three phases or stages:

Stage 1   A detection-transduction, or *sensor-transducer,* stage

Stage 2   An intermediate stage, which we shall call the *signal-conditioning* stage

Stage 3   A terminating, or *readout-recording,* stage

Each stage consists of a distinct component or group of components that performs required and definite steps in the measurement. These are called

Figure 1.2 Block diagram of the generalized measuring system

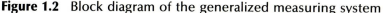
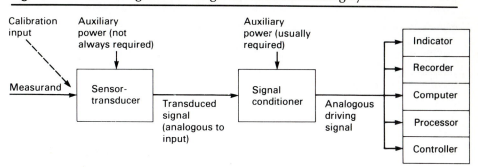

basic elements; their scope is determined by their function rather than by their construction. Figure 1.2 and Table 1.1 outline the significance of each of these stages.

1.4.1 First, or Sensor-Transducer, Stage

The primary function of the first stage is to detect or to sense the measurand. At the same time, ideally, this stage should be insensitive to every other possible input. For instance, if it is a pressure pickup, it should be insensitive to, say, acceleration; if it is a strain gage, it should be insensitive to temperature; if a linear accelerometer, it should be insensitive to angular acceleration; and so on. Unfortunately, it is rare indeed to find a detecting device that is completely selective. Unwanted sensitivity is a measuring error, called *noise* when it varies rapidly and *drift* when it varies very slowly.

Frequently one finds more than a single transduction (change in signal character) in the first stage, particularly if the first-stage output is electrical (see Section 6.3).

1.4.2 Second, or Signal-Conditioning, Stage

The purpose of the second stage of the general system is to modify the transduced information so that it is acceptable to the third, or terminating, stage. In addition, it may perform one or more basic operations, such as selective filtering to remove noise, integration, differentiation, or telemetering, as may be required.

Probably the most common function of the second stage is to increase either amplitude or power of the signal, or both, to the level required to drive the final terminating device. In addition, it must be designed for proper matching characteristics between the first and second and between the second and third stages.

Table 1.1 Stages of the general measurement system

Stage 1: Sensor-Transducer	Stage 2: Signal Conditioning	Stage 3: Readout-Recording
Senses desired input to exclusion of all others and provides analogous output	Modifies transduced signal into form usable by final stage. Usually increases amplitude and/or power, depending on requirement. May also selectively filter unwanted components or convert signal into pulsed form	Provides an indication or recording in form that can be evaluated by an unaided human sense or by a controller. Records data digitally on a computer
Types and Examples	*Types and Examples*	*Types and Examples*
Mechanical: Contacting spindle, spring-mass, elastic devices (e.g., Bourdon tube for pressure, proving ring for force), gyro	*Mechanical*: Gearing, cranks, slides, connecting links, cams, etc.	*Indicators Displacement types*: Moving pointer and scale, moving scale and index, light beam and scale, electron beam and scale (oscilloscope), liquid column
Hydraulic-pneumatic: Buoyant float, orifice, venturi, vane, propeller	*Hydraulic-pneumatic*: Piping, valving, dashpots, plenum chambers	*Digital types*: Direct alphanumeric readout
Optical: Photographic film, photoelectric diodes and transistors, photomultiplier tubes, holographic plates	*Optical*: Mirrors, lenses, optical filters, optical fibers, spatial filters (pinhole, slit)	*Recorders*: Digital printing, inked pen and chart, direct photography, magnetic recording (hard disk or tape)
Electrical: Contacts, resistance, capacitance, inductance, piezoelectric crystals and polymers, thermocouple, semiconductor junction, etc.	*Electrical*: Amplifying or attenuating systems, bridges, filters, telemetering systems, various special-purpose integrated-circuit devices	*Processors and Computers*: Various types of computing systems, either special-purpose or general, used to feed readout/recording devices and/or controlling systems
		Controllers: All types

1.4.3   Third, or Terminating Readout, Stage

The third stage provides the information sought in a form comprehensible to one of the human senses or to a controller. If the output is intended for immediate human recognition, it is, with rare exception, presented in one of the following forms:

1. As a *relative displacement,* such as movement of an indicating hand, or displacement of oscilloscope trace or oscillograph stylus

2. In *digital* form, as presented by a counter such as an automobile odometer, or by a liquid crystal display (LCD) or light-emitting diode (LED) display as on a digital voltmeter

To illustrate a very simple measuring system, let us consider the familiar tire gage used for checking automobile tire pressure. Such a device is shown in Fig. 1.3(a). It consists of a cylinder and piston, a spring resisting the piston movement, and a stem with scale divisions. As the air pressure bears against the piston, the resulting force compresses the spring until the spring and air

Figure 1.3 (a) Gage for measuring pressure in automobile tires. (b) Block diagram of tire-gage functions. In this example the spring serves as a secondary transducer (see Section 6.3).

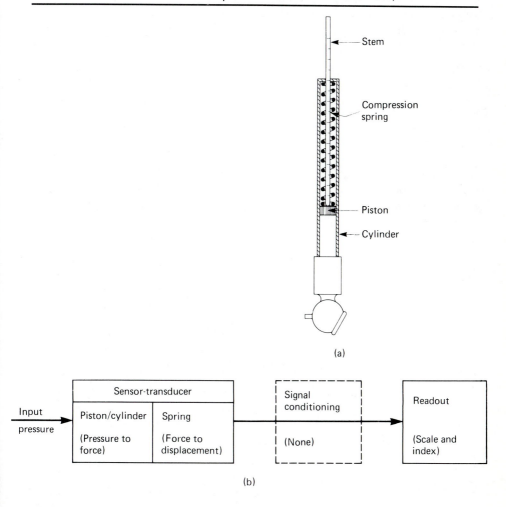

Figure 1.4 Block diagram of a relatively complex measuring system

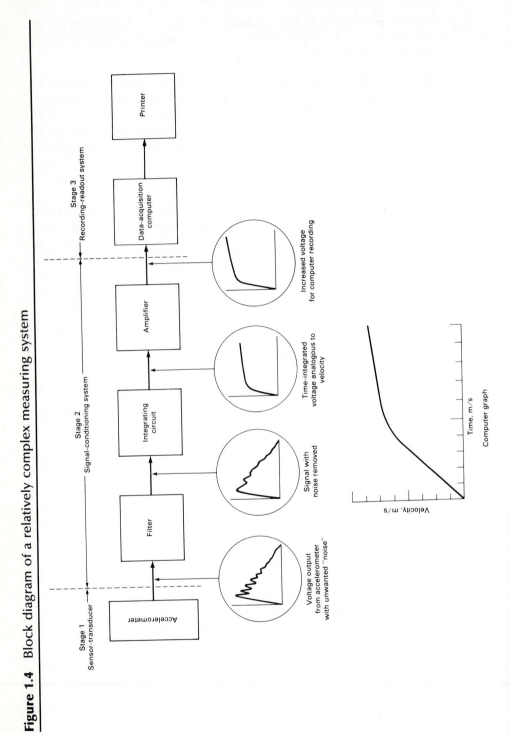

forces balance. The calibrated stem, which remains in place after the spring returns the piston, indicates the applied pressure.

The piston-cylinder combination constitutes a force-summing apparatus, sensing and transducing pressure to force. As a secondary transducer (see Section 6.3), the spring converts the force to a displacement. Finally, the transduced input is transferred *without* signal conditioning to the scale and index for readout [see Fig. 1.3(b)].

As an example of a more complex system, let us say that a velocity is to be measured, as shown in Fig. 1.4. The *first-stage* device, the accelerometer, provides a voltage analogous to acceleration.* In addition to a voltage amplifier, the *second* stage may also include a filter that selectively attenuates unwanted high-frequency noise components. It may also integrate the analog signal with respect to time, thereby providing a velocity–time relation, rather than an acceleration–time signal. Finally, the signal voltage will probably need to be increased to the level necessary to be sensed by the *third,* or *recording and readout, stage,* which may consist of a data-acquisition computer (Chapter 8) and printer. The final record will then be in the form of a computer-generated graph; with the proper calibration, an accurate velocity-versus-time measurement should be the result.

1.5   Types of Input Quantities

1.5.1   Time Dependence

Mechanical quantities, in addition to their inherent defining characteristics, also have distinctive time-amplitude properties, which may be classified as follows:

1. Static—constant in time
2. Dynamic—varying in time
   a. Steady-state periodic
   b. Nonrepetitive or transient
      i. Single pulse or aperiodic
      ii. Continuing or random

Of course, the unchanging, static measurand is the most easily measured. If the system is terminated by some form of meter-type indicator, the meter's pointer has no difficulty in eventually reaching a definite indication. The rapidly changing, dynamic measurand presents the real measurement challenge.

---

* Although the accelerometer may be susceptible to an analysis of "stages" within itself, we shall forgo such an analysis in this example.

Two general forms of dynamic input are possible: steady-state periodic input and transient input. The steady-state periodic quantity is one whose magnitude has a definite repeating time cycle, whereas the time variation of a transient quantity does not repeat. "Sixty-cycle" line voltage is an example of a steady-state periodic signal. So also are many mechanical vibrations, after a balance has been reached between a constant input exciting energy and energy dissipated by damping.

An example of a pulsed transient quantity is the acceleration–time relationship accompanying an isolated mechanical impact. The magnitude is temporary, being completed in a matter of milliseconds, with the portions of interest existing perhaps for only a few microseconds. The presence of extremely high rates of change, or wavefronts, can place severe demands on the measuring system. The nature of these inputs is discussed in detail in Chapter 4, and the response of the measuring system to such inputs is covered in Chapter 5.

1.5.2   Analog and Digital Signals

Most measurands of interest vary with time in a continuous manner over a range of magnitudes. For instance, the speed of an automobile, as it starts from rest, has some magnitude at every instant during its motion. A sensor that responds to velocity will produce an output signal having a time variation analogous to the time change in the auto's speed. We refer to such a signal as an *analog* signal because it is *analogous* to a continuous physical process. An analog signal has a value at every instant in time, and it usually varies smoothly in magnitude.

Some quantities, however, may change in a stepwise manner between two distinct magnitudes: a high and low voltage or on and off, for instance. The revolutions of a shaft could be counted with a cam-actuated electrical switch that is open or closed, depending on the position of the cam. If the switch controls current from a battery, current either flows with a given magnitude or does not flow. The current flow varies discretely between two values, which we could represent as single digits: 1 (flowing) and 0 (not flowing). The amplitude of such a signal may thus be called *digital*.

Many electronic circuits store numbers as sets of digits—strings of 1s and 0s—with each string held in a separate memory register. When digital circuits, such as those in computers, are used to record an analog signal, they do so only at discrete points in time because they have only a fixed number of memory registers. The analog signal, which has a value at every instant of time, becomes a *digital signal*. A digital signal is a set of discrete numbers, each corresponding to the value of the analog signal at a single specific instant of time. Clearly, the digital signal contains no information about the value of the analog signal at times other than sample times.

Mechanical quantities—such as temperatures, fluid-flow rates, pressure, stress, and strain—normally behave timewise in an analog manner. However, distinct advantages are often obtained in converting an analog signal to an equivalent digital signal for the purposes of signal conditioning and/or readout. Noise problems are reduced or sometimes eliminated altogether, and data transmission is simpler. Computers are designed to process digital information, and direct numerical display or recording is more easily accomplished by manipulating digital quantities. Digital techniques are discussed at length in Chapter 8.

1.6 Measurement Standards

As stated earlier, measurement is a process of comparison. Therefore, regardless of our measurement method, we must employ a basis of comparison—*standardized units*. The standards must be precisely defined, and, because different systems of units exist, the method of conversion from system to system must be mutually agreed upon. Chapter 2, "Standards and Dimensional Units of Measurement," provides a detailed discussion of this subject.

Most importantly, a relationship between the standards and the readout scale of each measuring system must be established through a process known as *calibration*.

1.7 Calibration

At some point during the preparation of a measuring system, *known* magnitudes of the input quantity must be fed into the sensor-transducer, and the system's output behavior must be observed. Such a comparison allows the magnitude of the output to be correctly interpreted in terms of the magnitude of the input. This *calibration* procedure establishes the correct output scale for the measuring system.

By performing such a test on an instrument, we both calibrate its scale and prove its ability to measure reliably. In this sense, we sometimes speak of *proving* an instrument. Of course, if the calibration is to be meaningful, the known input must itself be derived from a defined standard.

If the output is exactly proportional to the input (output = constant × input), then a single simultaneous observation of input and output will suffice to fix the constant of proportionality. This is called *single-point calibration*. More often, however, *multipoint calibrations* are used, wherein a number of different input values are applied. Multipoint calibration works when the output is not simply proportional, and, more generally, improves the accuracy of the calibration.

If a measuring system will be used to detect a time-varying input, then the calibration should ideally be made using a time-dependent input standard. Such *dynamic* calibration can be difficult, however, and a *static* calibration, using a constant input signal, is frequently substituted. Naturally, this procedure is not optimal; the more nearly the calibration standard corresponds to the measurand in all its characteristics, the better the resulting measurements.

Occasionally, the nature of the system or one of its components makes the introduction of a sample of the basic input quantity difficult or impossible. One of the important characteristics of the bonded resistance-type strain gage is the fact that, through quality control at the time of manufacture, *spot* calibration may be applied to a complete lot of gages. As a result, an indirect calibration of a strain-measuring system may be provided through the gage factor supplied by the manufacturer. Instead of attempting to apply a known unit strain to the gage installed on the test structure—which, if possible, would often result in an ambiguous situation—a resistance change is substituted. Through the predetermined gage factor, the system's strain response may thereby be obtained (see Section 12.4).

1.8 Uncertainty: Accuracy of Results

Error may be defined as the difference between the *measured* result and the *true* value of the quantity being measured (see Section 3.1). We do not know the true value; hence, we do not know the error. We can discuss an error and can estimate the size of an error, but we can never know its actual magnitude. If we estimate a likely upper bound on the error, that bound is called the *uncertainty*. We estimate, with some level of odds, that the error will be no larger than the uncertainty. There are two basic types of error (remember, we can discuss it without ever knowing its magnitude): *bias,* or *systematic, error* and *precision,* or *random, error.*

Should an unscrupulous butcher place a ball of putty under the scale pan, the scale readouts would be consistently in error. The scale would indicate a weight of product too great by the weight of the putty. This *zero offset* represents one type of systematic error.

Shrink rules are used to make patterns for the casting of metals. Cast steel shrinks in cooling by about 2%; hence the patterns used for preparing the molds are oversized by the proper percentage amounts. The pattern maker uses a shrink rule on which the dimensional units are increased by that amount. Should a pattern maker's shrink rule for cast steel be inadvertently used for ordinary length measurements, the readouts would be consistently undersized by $\frac{1}{50}$ in one (that is, by 2%). This is an example of *scale error.*

In each of the foregoing examples the errors are constant and of a systematic nature. Such errors *cannot* be uncovered by statistical analysis; however, they may be estimated by methods that we discuss in Chapter 3.

An inexpensive frequency counter may use the 60-Hz power-line frequency as a comparison standard. Power-line frequency is held very close to the 60-Hz standard. Although it does wander slowly above and below the average value, over a period of time—say, a day—the *average* is very close to 60 Hz. The wandering is random and the moment-to-moment error in the frequency meter readout (from this source) is called *precision,* or *random, error.*

Randomness may also be introduced by variations in the measurand itself. If a number of hardness readings is made on a given sample of steel, a range of readings will be obtained. An average hardness may be calculated and presented as the actual hardness. Single readings will deviate from the average, some higher and some lower. Of course, the primary reason for the variations in this case is the nonhomogeneity of the crystalline structure of the test specimen. The deviations will be random and are due to variations in the measurand. Random error may be estimated by statistical methods and is considered in more detail in Chapter 3.

1.9   Reporting Results

When experimental setups are made and time and effort are expended to obtain results, it normally follows that some form of written record or report is to be made. The purpose of such a record will determine its form. In fact, in some cases, several versions will be prepared. Reports may be categorized as follows:

1. Executive summary
2. Laboratory note or technical memo
3. Progress report
4. Full technical report
5. Technical paper

Very briefly, an executive summary is directed at a busy overseer who wants only the key features of the work: what was done and what was concluded, outlined in a few paragraphs. A laboratory note is written to be read by someone thoroughly familiar with the project, such as an immediate supervisor or the experimentalist himself. A full report tells the complete story to one who is interested in the subject but who has not been in direct touch with the specific work—perhaps top officials of a large company or a review committee of a sponsoring agency. A progress report is just that—one of possibly several interim reports describing the current status of an ongoing project, which will eventually be incorporated in a full report. Ordinarily, a technical paper is a brief summary of a project, the extent of which must be tailored to fit either a time allotment at a meeting or space in a publication.

Several factors are common to all the various forms. With each type, the first priority is *to make sure* that the *problem or project* that has been tackled *is clearly stated.* There is nothing quite so frustrating as reading details in a technical report while never being certain of the raison d'être. It is extremely important to make certain that the reader is quickly clued in on the *why* before one attempts to explain the *how* and the results. A clearly stated objective can be considered the most important part of the report. The entire report should be written in simple language. A rule stated by Samuel Clemens is not inappropriate: "Omit unnecessary adverbs and adjectives."

1.9.1   Laboratory Note or Technical Memo

The laboratory note is written for a very limited audience, possibly even only as a memory jogger for the experimenter or, perhaps more often, for the information of an immediate supervisor who is thoroughly familiar with the work. In some cases, a single page may be sufficient, including a sentence or two stating the problem, a block diagram of the experimental setup, and some data presented either in tabular form or as a plotted diagram. Any pertinent observations not directly evident from the data should also be included. Sufficient information should be included so that the experimenter can mentally reconstruct the situation and results 1 year or even 5 years hence. A date and signature should always be included and, if there is a possibility of important developments stemming from the work, a second witnessing signature should be included and dated.

1.9.2   Full Report

The full report must relate all the facts pertinent to the project. It is even more important in this case to make the purpose of the project completely clear, for the report will be read by persons not closely associated with the work. The full report should also include enough detail to allow another professional to repeat the measurements and calculations.

One format that has much merit is to make the report proper—the main body—short and to the point, relegating the supporting materials, data, detailed descriptions of equipment, review of literature, sample calculations, and so on, to appendixes. Frequent reference to these materials can be made throughout the report proper, but the option to peruse the details is left to the reader. This scheme also provides a good basis for the technical paper, should it be planned.

1.9.3   Technical Paper

A primary purpose of a technical paper is to make known (to advertise) the work of the writer. For this reason, two particularly important portions of the writing are the *problem statement* and the *results*. Adequately done, these two items will attract the attention of other workers interested in the particular field who can then make direct contact with the writer(s) for additional details and discussion.

Space, number of words, limits on illustrations, and perhaps time are all factors making the preparation of a technical paper particularly challenging. Once the problem statement and the primary results have been adequately

Figure 1.5 Conceptual organization of this book

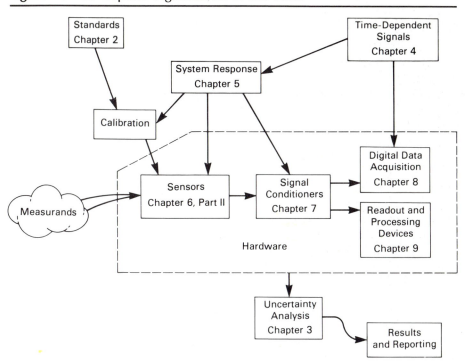

established, the remaining available space may be used to summarize procedures, test setups, and the like.

1.10 Final Remarks

An attempt has been made in this chapter to provide an overall preview of the problems of mechanical measurement. In conformance with Section 1.9, we have tried to state the problem as fully as possible in only a few pages. In the remainder of the book, we will expand on the topics introduced in this chapter. Figure 1.5 illustrates the interrelation of these topics and their organization within this book.

Suggested Readings

Dally, J. W., W. F. Riley, and K. G. McConnell. *Instrumentation for Engineering Measurements*. New York: John Wiley, 1984.

Ibrahim, K. F. *Instruments and Automatic Test Equipment: An Introductory Testbook.* New York: John Wiley, 1988.

Jones, B. E. *Instrument Science and Technology,* 3 vol. New York: Adam Hilger, 1982–85.

Morrison, R. *Instrumentation Fundamentals and Applications.* New York: John Wiley, 1984.

Sydenham, P. H. *Handbook of Measurement Science.* 2 vol. New York: John Wiley, 1983.

Tse, F., and I. Morse. *Measurement and Instrumentation in the Laboratory.* New York: Marcel Dekker, 1986.

Wolf, S., and R. F. M. Smith. *Student Reference Manual for Electronic Instrumentation Laboratories.* Englewood Cliffs, N. J.: Prentice Hall, 1990.

Problems

1.1 From the list of Suggested Readings at the end of this chapter, select one book, and write a short executive summary of a chapter discussing pressure, temperature, force, strain or motion measurement.

1.2 Consider a mercury-in-glass thermometer as a temperature-measuring system. Discuss the various stages of this measuring system in detail.

1.3 For the thermometer of Problem 1.2, specify how practical single point calibration may be obtained.

1.4 Set up test procedures you would use to estimate, with the aid only of your present judgment and experience, the magnitudes of the common quantities listed.

    a. Distance between the centerlines of two holes in a machined part

    b. Weight of two small objects of different densities

Figure 1.6   Impact test frame for Problem 1.5

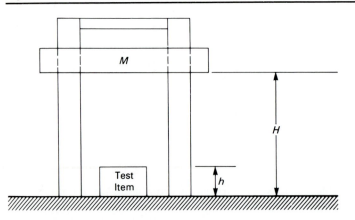

     c.   Time intervals
     d.   Temperature of water
     e.   Frequency of pure tones

1.5   Consider the impact frame shown in Figure 1.6. Mass M is raised to an initial height H and released from rest. Discuss how you would measure the mass velocity just prior to impact with the test item in order to account for friction between mass M and the guide rails.

CHAPTER 2

||||||||||||||||||||||||||||||||||||||||||||||||||||||||||

Standards and Dimensional Units of Measurement

2.1 Introduction

The basis of measurement was outlined in Section 1.3; it is the comparison between a measurand and a suitable *standard*. In this chapter we will take a closer look at the establishment and use of standards.

The term *dimension* connotes the defining characteristics of an entity (measurand), and the *dimensional unit* is the basis for quantification of the entity. For example, *length* is a dimension, whereas *centimeter* is a unit of length; *time* is a dimension, and the *second* is a unit of time. A dimension is unique; however, a particular dimension—say, length—may be measured in various units, such as feet, meters, inches, or miles. *Systems of units* must be established and agreed to; that is, the systems must be standardized. Because there are various systems, there must also be agreement on the basis for *conversion* from system to system.

It is clear, then, that standards of measurement apply to units, to systems of units, and to unitary conversion between such systems. In the following sections we will discuss those standards, systems of units, and problems of conversion that are basic to mechanical measurement.

2.2 Establishment of Dimensional Standards

Certain dimensions are considered to be *fundamental*—length, mass, time, temperature, electrical current, amount of substance, and luminous intensity. Others are *supplementary*; still others are *derived*. Of the fundamental dimensions, the first four that we listed are of special interest.

In general terms, standards are ubiquitous. There are standards governing food preparation, marketing, professional behavior, and so on. Many are established and governed by either federal or state laws.

So that we may avoid chaos, it is especially important that the basic standards carry the authority not only of federal, but also of international, laws. To begin our discussion of standards, let us review the background for such authority.

2.3 Historical Background of Measurement Standards in the United States

The legal authority to control measurement standards in the United States was assigned by the U.S. Constitution. Quoting from Article 1, Section 8, Paragraph 5, of the U.S. Constitution: "The congress shall have power to . . . fix the standard of weights and measures." Although Congress was given the power, considerable time elapsed before anything was done about it. In 1832, the Treasury Department introduced a uniform system of weights and measures to assist the customs service; in 1836 these standards were approved by Congress [1]. In 1866, the Revised Statutes of the United States, Section 3569, added the stipulation that "It shall be lawful throughout the United States of America to employ the weights and measures of the metric system . . . ." This simply makes it clear that the metric system *may* be used. In addition, this act established the following relation for conversion:

$$1 \text{ meter (m)} = 39.37 \text{ inches (in.) (exactly)}.$$

An international convention held in Paris in 1875 resulted in an agreement signed for the United States by the U.S. ambassador to France. The following is quoted therefrom: "The high contracting parties engage to establish and maintain, at their common expense, a scientific and permanent international bureau of weights and measures, the location of which shall be Paris" [2, 3]. Although this established a central bureau of standards, which was set up at Sèvres, a suburb of Paris, it did not, of course, bind the United States to make use of or adopt such standards.

On April 5, 1893, in the absence of further congressional action, Superintendent Mendenhall of the Coast and Geodetic Survey issued the following order [2, 4]:

> The Office of Weights and Measures with the approval of the Secretary of the Treasury, will in the future regard the international prototype meter and the kilogram as *fundamental standards*, and the customary units, the yard and pound, will be derived therefrom in accordance with the Act of July 28, 1866.

The Mendenhall Order turned out to be a very important action. First, it recognized the meter and the kilogram as being fundamental units on which all other units of length and mass should be based. Second, it tied together the metric and English systems of length and mass in a definite relationship, thereby making possible international exchange on an exact basis.

In response to requests from scientific and industrial sources, and to a great degree influenced by the establishment of like institutions in Great Britain and Germany,* Congress on March 3, 1901, passed an act providing that "The office of Standard Weights and Measures shall hereafter be known as 'The National Bureau of Standards' " [5]. Expanded functions of the new bureau were set forth and included development of standards, research basic to standards, and the calibration of standards and devices. The National Bureau of Standards (NBS) was formally established in July 1910, and its functions were considerably expanded by an amendment passed in 1950. In 1988, Congress changed the name of the bureau to "The National Institute of Standards and Technology" (NIST) [6].

Commercial standards are largely regulated by state laws; to maintain uniformity, regular meetings (National Conferences on Weights and Measures) are held by officials of NIST and officers of state governments. Essentially all state standards of weights and measures are in accordance with the Conference's standards and codes. International uniformity is maintained through regularly scheduled meetings (held at about 6-year intervals), called the *General Conference on Weights and Measures* and attended by representatives from most of the industrial countries of the world. In addition, numerous interim meetings are held to consider solutions to more specific problems, for later action by the General Conference.

2.4 The SI System

2.4.1 International Actions

The Eleventh General Conference on Weights and Measures (1960) adopted the International System of Units (SI) and established the rules for the various units. In June 1972 the International Standards Organization (ISO), an assembly of which the United States is a member, approved the International Standard 1000, called *SI Units and Recommendations for the Use of Their Multiples and of Certain Other Units*. The system is often popularly referred to as the *metric system*. Therein are three classes of measurement units: (1) *base units*, (2) *supplementary units*, and (3) *derived units*. The seven base units and the two supplementary units are listed in Table 2.1. In addition, various units derived from the base units are listed in the Standard. Certain of the derived units are assigned special names; others are not. For example, force $(m \cdot kg/s^2)$ is given the special name *newton*, whereas area is simply meters

---

* The National Physical Laboratory, Teddington, Middlesex, and Physikalisch-Technische Reichsanstalt, Braunschweig.

Table 2.1 Base and supplementary units

Quantity	Name and Symbol of Unit
	Base Units
Length	meter (m)
Mass	kilogram (kg)
Time	second (s)
Electric current	ampere (A)
Thermodynamic temperature	kelvin (K)
Amount of substance	mole (mol)
Luminous intensity	candela (cd)
	Supplementary Units
Plane angle	radian (rad)
Solid angle	steradian (sr)

squared (m^2). Work and energy ($m^2 \cdot kg/s^2$) are expressed in *joules*. The term *hertz* is used for frequency (s^{-1}) and the term *pascal* is used for pressure or stress (N/m^2). Derived units carrying special names are listed in Table 2.2, and selected derived units without special names are given in Table 2.3. Note that, whereas those assigned words originating from proper names are not capitalized, the corresponding abbreviations are capitalized. *It should be clear that* all *the various derived units may be expressed in terms of the base units.* In certain instances when a unit balance is attempted for a given equation, it may be desirable, or necessary, to convert all variables to base units.

Table 2.2 Derived units and their assigned special symbols

Quantity	Unit	Symbol	Formula
Electrical capacitance	farad	F	C/V
Electrical conductance	siemens	S	A/V
Electrical inductance	henry	H	Wb/A
Electric potential (voltage)	volt	V	W/A
Electrical resistance	ohm	Ω	V/A
Energy (work, quantity of heat)	joule	J	$N \cdot m$
Force	newton	N	$kg \cdot m/s^2$
Frequency	hertz	Hz	1/s
Magnetic flux	weber	Wb	$V \cdot s$
Magnetic flux density	tesla	T	Wb/m^2
Power	watt	W	J/s
Pressure, stress	pascal	Pa	N/m^2
Quantity of electrical charge	coulomb	C	$A \cdot s$

Table 2.3 Common derived units that do not have
assigned special symbols*

Quantity	Formula
Acceleration	m/s²
Angular acceleration	rad/s²
Angular velocity	rad/s
Area	m²
Density (mass)	kg/m³
Density (energy)	J/m³
Heat flux	W/m²
Moment of force	N · m
Velocity	m/s
Viscosity (absolute)	Pa · s
Volume	m³

* Also see Appendix A.

Table 2.4 Multiplying factors

Multiple and Submultiple	Prefix	Symbol	Pronunciation
10^{18}	exa	E	ĕks′ à
10^{15}	peta	P	pĕt′ à
10^{12}	tera	T	tĕr′ à
10^{9}	giga	G	jĭ′ gà
10^{6}	mega	M	mĕg′ à
10^{3}	kilo	k	kĭl′ô
10^{2}	hecto	h	hĕk′tô
10	deka	da	dĕk′ à
10^{-1}	deci	d	dĕs′ĭ
10^{-2}	centi	c	sĕn′tĭ
10^{-3}	milli	m	mĭl′ĭ
10^{-6}	micro	μ	mī′krō
10^{-9}	nano	n	năn′ō
10^{-12}	pico	p	pē′ cō
10^{-15}	femto	f	fĕm′tō
10^{-18}	atto	a	ăt′ tō

To accommodate the writing of very large or very small values, certain multiplying factors are provided (Table 2.4). For example, 2,500,000 Hz may be written as 2.5 MHz or 0.000 000 000 005 farad as 5 pF.

2.4.2 Domestic Actions

In May 1965 the United States announced its intention of adopting the metric system. In 1968, passage of Public Law 90-472 authorized the secretary to make a "U.S. Metric Study," to be reported by August 1971. After prolonged debates, studies, and public pronouncements of 10 year conversion plans, on December 23, 1975, the 94th Congress approved Public Law 94-168, called the *Metric Conversion Act* of 1975. Its stated purpose was as follows: "To declare a national policy of coordinating the increasing use of the metric system in the United States, and to establish a United States Metric Board to coordinate the voluntary conversion to the metric system." Note especially that the conversion was to be *voluntary* and that no time limit was set. The Act made clear that, in using the term *metric,* the SI system of units was intended.

In 1981, the U.S. Metric Board reported to Congress that it lacked the clear Congressional mandate needed to bring about an effective national conversion to the metric system; funding for the Board was eliminated after fiscal year 1982 [7].

As the only developed nation in the world that does not officially use the metric system, the United States increasingly finds its international industrial competitiveness threatened. Congress has at last recognized this danger, and in the *Trade and Competitiveness Act of 1988* it amended the earlier *Metric Conversion Act* to provide strong incentives for industrial conversion to SI units. The amended act declares that the metric system is "the preferred system of weights and measures for United States trade and commerce." It further requires that all federal agencies use the metric system in procurement, grants, and business-related activities; this requirement is to be met by the end of fiscal year 1992, except in cases where conversion will harm international competitiveness.

Metrication in the United States has progressed, especially in the automotive industry and certain parts of the food and drink industries. Classroom use has increased to the point that most engineering courses rely primarily on SI units. Throughout this book, we shall use both the SI system and the English Engineering system,* with the hope of encouraging the complete conversion to SI units.†

2.5 The Standard of Length

The meter was originally intended to be one ten-millionth of the earth's quadrant. In 1889 the First General Conference on Weights and Measures defined the meter as the length of the International Prototype Meter, the distance between two finely scribed lines on a platinum-iridium bar when subject to certain specified conditions. On October 14, 1960, the Eleventh General Conference on Weights and Measures adopted a new definition of the

---

* This term may appear incongruous given that the United Kingdom has adopted the SI system. However, its usage is so well established that it has outlived its origins.

† Attention is directed to reference [8], which is an excellent guide for applying the metric system.

meter as 1,650,763.73 wavelengths in vacuum of the radiation corresponding to the transition between the levels $2p_{10}$ and $5d_5$ of the krypton-86 atom. The National Bureau of Standards of the United States adopted this standard, and the inch became 41,929.398 54 wavelengths of the krypton light.

As it turned out, the wavelength of krypton light could be determined only to about 4 parts per billion, limiting the accuracy of the meter to a similar level. During the 1960s and early 1970s, laser-based measurements of frequency and wavelength evolved to such accuracy that the uncertainty of the meter became the limiting uncertainty in determining the speed of light [9, 10]. This limitation was of serious concern in both atomic and cosmological physics, and on October 20th, 1983, the Seventeenth General Conference on Weights and Measures redefined the meter directly in terms of the speed of light:

> The meter is the length of the path traveled by light in vacuum during a time interval of 1/299,792,458 of a second.

This definition has the profound effect of *defining* the speed of light to be 299,792,458 m/s, which had been the accepted experimental value since 1975 [11].

The relation between the meter and the inch as specified by the Mendenhall Order (1 m = 39.37 in.) results in

$$1 \text{ in.} = 2.540\,005\,08 \text{ cm (approximately).}$$

The convenient relation

$$1 \text{ in.} = 2.54 \text{ cm (exactly)}$$

had been used in industry and engineering for years and, through adoption of the SI system, became official for all length conversions. The difference between these two standards may be written as

$$2.540\,005\,08/2.54 - 1 = 0.000\,002,$$

or 0.0002%, which is about 1/8 in. per mile.

We gain an idea of the significance of the difference by considering the following situations. The work of the United States Coast and Geodetic Survey is based on the 39.37 in./m relation and a coordinate system with origin located in Kansas [12]. Changing the metric relation from 39.37 in./m (exactly) to 2.54 cm/in. (exactly) would cause discrepancies of almost 16 ft at a distance of 1500 mi. One can only imagine the confusion over property lines if such a change were made. On the other hand, gage blocks, which are the manufacturing secondary standards, are established on the basis of 2.54 cm/in. If they were measured on the basis of 39.37 in./m, errors of 2 μin./in. would be found. This is of the same order of magnitude as the *tolerance* of the better-quality blocks (see Section 11.3).

This problem might be serious were it not for the fact that geodetic distances and small mechanical displacements seldom need to be compared. Nevertheless, in 1959, the U.S. government defined separately the *U.S. survey foot* (12/39.37 m) and the *international foot* (12 × 2.54 cm). The survey foot is still used with U.S. geodetic data and U.S. statute miles [13].

2.6 The Standard of Mass

The *kilogram* is defined by the mass of the International Prototype Kilogram, a platinum-iridium mass kept at the International Bureau of Weights and Measures near Paris. Of the basic standards, this remains the only one established by a prototype (*the* original model or pattern, *the* unique example to which all like are referred for comparison). Even this standard, however, may someday follow the prototype meter bar into the museum and be replaced by a standard available to any laboratory. Reference [14] alludes to an X-ray interferometry measurement of Avogadro's number, enabling macroscopic mass to be tied directly to the atomic-mass unit.

Secondary standards of known relative masses are maintained by each of the primary industrial countries of the world. In the United States, the basic unit of mass is the "United States National Prototype Kilogram No. 20," carefully maintained by the National Institute of Standards and Technology.

The Mendenhall Order of 1893 included the following relationship:

$$1 \text{ lb avoirdupois} = 453.592\,427\,7 \text{ g}.$$

For 58 years, following the founding of the National Bureau of Standards, this relationship applied. However, on July 1, 1959, the equivalent was altered to

$$1 \text{ lb avoirdupois} = 453.592\,37 \text{ g}.$$

This change was brought about by the desire to unify the equivalencies used by Australia, Canada, New Zealand, South Africa, the United Kingdom, and the United States.

2.7 Time and Frequency Standards

Until 1956, the second was defined as 1/86,400 of the average period of revolution of the earth on its axis. Although this seems to be a relatively simple and straightforward definition, problems remained. There is a gradual slowing of the earth's rotation (about 0.001 seconds per century) [15], and, in addition, the rotation is irregular.

Therefore, in 1956 an improved standard was agreed on; the second was defined as 1/31,556,925.9747 of the time required by the earth to orbit the sun in the year 1900. This is called the *ephemeris second*. Although the unit is defined with a high degree of exactness, implementation of the definition is dependent on astronomical observation, which is incapable of realizing the implied precision.

In the 1950s, atomic research led to the observation that oscillations associated with certain atomic transitions may be measured with great repeatability. One, the hyperfine transition of the cesium atom, was related to the ephemeris second with an estimated accuracy of two parts in 10^9. On October 13, 1967, in Paris, the Thirteenth General Conference on Weights and Measures officially adopted the unit of time of the International System of Units as the second, defined in the following terms: "The second is the duration of 9,192,631,770 periods of the radiation corresponding to the transition between the two hyperfine levels of the fundamental state of the atom of cesium 133" [16].

An atomic-beam apparatus [17, 18], commonly called an atomic "clock," is used to produce the frequency of transition. Heated metallic cesium is caused to emit a beam of atoms that is separated into two beams of differing energies depending on the alignment of nuclei and electrons. When an oscillating electromagnetic field having a frequency characteristic of the particular transition is applied, the frequency agreement is detected and appropriately indicated. The frequency of the "master" oscillator may then be used as a standard to which the outputs of other oscillators can be compared. The best cesium standards reproduce the second to an accuracy of a few parts in 10^{14}.

2.8 Temperature Standards

In 1927 the national laboratories of the United States, Great Britain, and Germany proposed a temperature standard that became known as the *International Temperature Scale of 1927 (ITS-27)*. This standard, adopted by 31 nations, conformed as closely as possible to the thermodynamic scale proposed by Lord Kelvin in 1854. It was based on six fixed-temperature points dependent on physical properties of certain materials, including the ice and steam points of water. Revisions have been made by succeeding conferences, notably in 1948, 1968, and 1990. Currently, the International Temperature Scale of 1990 (ITS-90), adopted by the International Committee on Weights and Measures and authorized by the Eighteenth General Conference, is in effect [19].

The basic unit of temperature, the kelvin (K), is defined as the fraction 1/273.16 of the thermodynamic temperature of the triple point of water, the temperature at which the solid, liquid, and vapor phases of water exist in equilibrium. The degree Celsius is defined by the relationship

$$t = T - 273.15,$$

where t and T represent temperatures in *degrees Celsius* and in *kelvins*, respectively.

In reality, two temperature scales are defined, a *thermodynamic* scale and a *practical* scale. The latter is the usual basis for measurement. The thermodynamic scale is defined in terms of entropy and the properties of heat engines [20], but it can be implemented, for example, by using the expansion of an ideal gas. Unfortunately, this scale is inconvenient for most physical measurements, and a corresponding scale is needed that is easier to realize. Thus, the thermodynamic scale is normally approximated using a so-called practical scale. The practical scale is described in terms of certain temperature-dependent physical properties, such as electrical resistance, which are used to interpolate between the established temperatures of specific states of matter. Using the interpolation formulae and fixed points of the practical scale, one can calibrate any other temperature-measuring device. The International Temperature Scale of 1990 (ITS-90) is a practical scale, and the following discussion focuses on it.

The temperature standard relies on a set of fixed-reference temperatures. The fixed points correspond to specific thermodynamic states of materials that are accurately reproducible. Zero degrees Celsius is the temperature of equilibrium between pure ice and air-saturated pure water at normal atmospheric pressure. However, a more precise datum, independent of both ambient pressure and possible contaminants, is the *triple point* temperature of water, at which the solid, liquid, and vapor phases of water can exist in equilibrium. The value 0.0100°C is assigned to this temperature. A relatively simple apparatus can be used to reproduce this temperature fixed-point [21].

The standard defines numerous other fixed points, some of which are shown in Figure 2.1 and Table 2.5. Between these fixed points, elaborate interpolation equations are specified by ITS-90 for use with the various interpolation standards. From 0.65 K to 5.0 K, the standard is based on measurement of the vapor pressure of helium and corresponding equations describing the vapor-pressure/temperature relations of helium (Section 16.3). From 3.0 K to 24.5561 K (the triple point of neon), a constant-volume helium gas thermometer is used (Section 16.3). Over the broad range from the triple point of hydrogen (13.8033 K) to the normal freezing point of silver (1234.93 K), the standard is defined by means of a platinum resistance thermometer (Section 16.4.1). Complex equations expressing resistance as a function of temperature are prescribed, along with calibration procedures for each of several temperature subranges. To calibrate, the resistance of a platinum sensor is measured at several fixed-point temperatures within a given subrange, and these measurements are used to determine unknown constants in the temperature-resistance equations.

Finally, above the melting point of silver, temperatures are defined by measurement of the thermal energy emitted by a blackbody cavity in vacuum and use of the Planck radiation law:

$$\frac{E_\lambda(T)}{E_\lambda(T_{\text{ref}})} = \frac{\exp\left(\dfrac{C_2}{\lambda T_{\text{ref}}}\right) - 1}{\exp\left(\dfrac{C_2}{\lambda T}\right) - 1},$$

Figure 2.1 Some fixed points established by ITS-90. Some others, in
kelvins, are as follow: triple point of hydrogen, 13.8033; triple
point of neon, 24.5561; triple point of oxygen, 54.3584; triple
point of mercury, 234.3156; melting point of gallium, 302.9146;
freezing point of indium, 429.7485

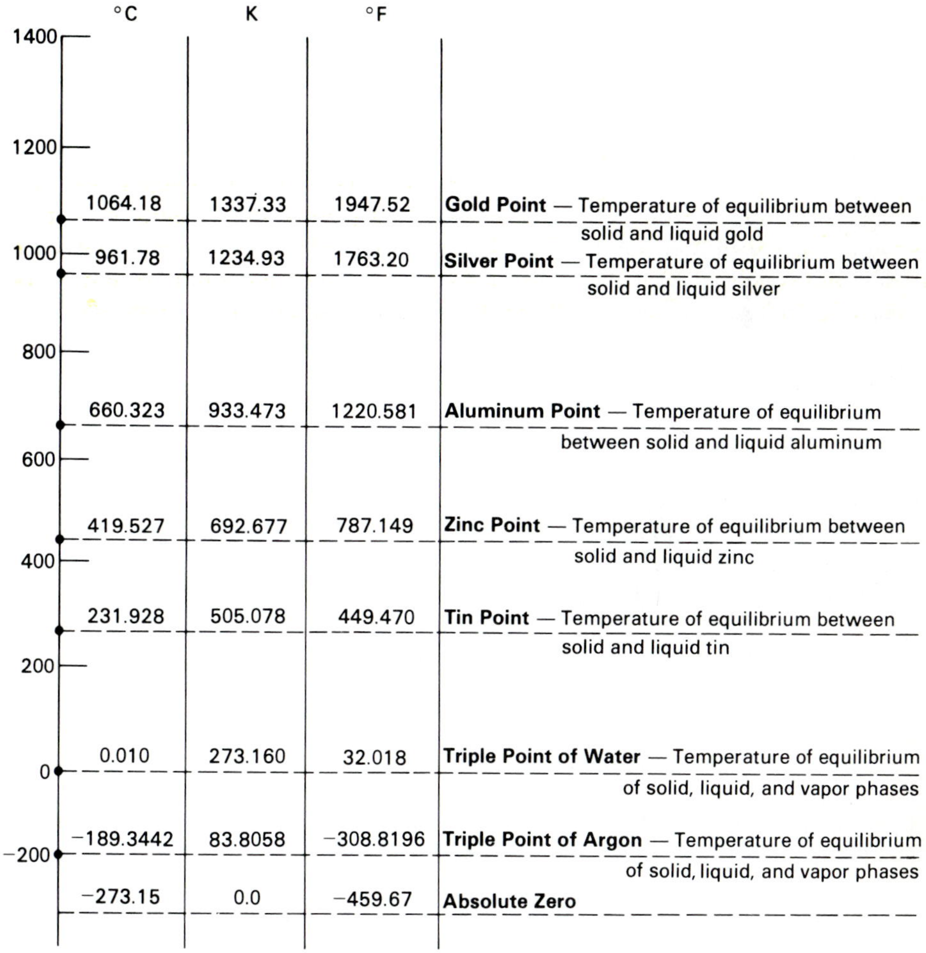

°C	K	°F	
1064.18	1337.33	1947.52	**Gold Point** — Temperature of equilibrium between solid and liquid gold
961.78	1234.93	1763.20	**Silver Point** — Temperature of equilibrium between solid and liquid silver
660.323	933.473	1220.581	**Aluminum Point** — Temperature of equilibrium between solid and liquid aluminum
419.527	692.677	787.149	**Zinc Point** — Temperature of equilibrium between solid and liquid zinc
231.928	505.078	449.470	**Tin Point** — Temperature of equilibrium between solid and liquid tin
0.010	273.160	32.018	**Triple Point of Water** — Temperature of equilibrium of solid, liquid, and vapor phases
−189.3442	83.8058	−308.8196	**Triple Point of Argon** — Temperature of equilibrium of solid, liquid, and vapor phases
−273.15	0.0	−459.67	**Absolute Zero**

where

$E_\lambda(T)$, $E_\lambda(T_{ref})$ = the radiant energy emitted by the blackbody per unit
time, per unit area, and per unit wavelength
at a wavelength λ, and temperature T or T_{ref}
respectively,

Table 2.5 Defining fixed points of the ITS-90

	Assigned Values of Temperature	
Equilibrium State	**K**	**°C**
Vapor pressure of helium*	3 to 5	−270.15 to −268.15
Triple point† of hydrogen	13.8033	−259.3467
Vapor pressure of hydrogen	≈17	≈ −256.15
Vapor pressure of hydrogen	≈20.3	≈ −252.85
Triple point of neon	24.5561	−248.5939
Triple point of oxygen	54.3584	−218.7916
Triple point of argon	83.8058	−189.3442
Triple point of mercury	234.3156	−38.8344
Triple point of water	273.16	0.01
Melting point‡ of gallium	302.9146	29.7646
Freezing point‡ of indium	429.7485	156.5985
Freezing point of tin	505.078	231.928
Freezing point of zinc	692.677	419.527
Freezing point of aluminum	933.473	660.323
Freezing point of silver	1234.93	961.78
Freezing point of gold	1337.33	1064.18
Freezing point of copper	1357.77	1084.62

* Temperature is calculated by substituting measured vapor pressure into an equation of state.

† Equilibrium between solid, liquid, and vapor phases.

‡ Melting and freezing point temperatures correspond to standard atmospheric pressure (101,325 N/m^2).

T_{ref} = the freezing point temperature of either silver (1234.93 K), gold (1337.33 K), or copper (1357.77 K),

$C_2 = 0.014388$ m · K.

The radiant energy is typically measured by optical pyrometry (Section 16.8). The unknown temperature is then calculated by comparing the emission of a source at the unknown temperature to that from a source at the reference temperature.

The International Temperature Scale thus establishes means of determining any temperature from 0.65 K to more than 4000 K. In actual applications, the standardized pyrometer, the standardized resistance thermometer, or the standardized gas thermometers are used as secondary standards for calibration of working instruments (Section 16.12). Apart from any uncertainties introduced in the calibration procedures, the major uncertainties in ITS-90 arise in realizing the fixed points. At a 1σ level (Sections 3.5–3.6), the uncertainties in the fixed-point temperatures are $\pm 0.5 - 1.5\,\text{mK}$ for temperatures up to the melting point of gallium, increasing to $\pm 60\,\text{mK}$ at the freezing point of copper [22].

2.9 Electrical Standards

In the SI system, all electrical units originate from the definition of the ampere. One ampere is the current that produces a magnetic force of $2 \times 10^{-7}\,\text{N/m}$ on a pair of thin parallel wires carrying that current and separated by 1 m. The force on an appropriate pair of conductors can be measured directly, using a *current balance* [23]. The current may be then calculated from the relations of electromagnetic theory and the assigned value of the permeability of free space, μ_0. Note that force itself is derived from the standards of mass, length, and time.

The remaining electrical units, such as volts and ohms, can all be derived from the value of the ampere and the mechanical units of mass, length, and time, again using the results of electromagnetic theory [24]. One additional electrical constant, the permittivity of free space, ε_0, was traditionally derived first in order to determine the remaining units. However, when the speed of light, c, was defined in 1983 (Section 2.5), the value of ε_0 was fixed as well, since $\mu_0\varepsilon_0 c^2 \equiv 1$. Thus other electrical units now enjoy the same stature as the ampere, in that they may also be deduced directly from theoretical relations and the mechanical standards, without first calculating ε_0. The ampere need not be the starting point, although the SI system still treats it as such.

The volt and the ohm are related to the ampere through Ohm's law ($V = IR$). If standard values are available for any two of these, the third is easily obtained. Standard voltage sources and standard resistors are the most common secondary standards used for the calibration of working instruments.

The classical voltage source is a *standard cell*, an electrochemical cell of high stability that maintains its voltage to an accuracy of a few parts in 10^{10}. However, a quantum mechanical process has recently emerged as a more promising secondary (and possibly even primary) voltage standard. The superconducting Josephson junction effect, discovered in 1962, produces a voltage stable and reproducible to a few parts in 10^{13} [25]. The ultimate accuracy of the Josephson junction voltage is limited primarily by the accuracy of the time standard.

The *standard resistor* is normally a specially alloyed wire held in an oil bath to stabilize its temperature. These standard resistors have an accuracy of a few parts in 10^7. Like the Josephson junction effect, the quantum Hall effect, discovered in 1980, provides an alternative, quantum-mechanical standard. The quantized Hall resistance allows the ohm to be determined directly from fundamental physical constants to a resolution of a few parts in 10^8 [25].

2.10   Conversion Between Systems of Units

Over the centuries, various systems of units have evolved. Five systems are listed in Table 2.6. To be acceptable, each system of units must be compatible with the physical laws of the universe. If compatible with the laws of nature, the values expressed in one system must be convertible to equally legitimate values in any of the other systems.

Although physicists may argue over the nature of gravitation, in the practice of engineering it is usually described by Newton's laws of motion. Newton's second law may be expressed in various ways, one of which is

A particle acted upon by an external force will be accelerated in proportion to the force magnitude and in inverse proportion to the mass of the particle; the direction of the acceleration will coincide with the line-of-action of the force.

Algebraically,

$$F = ma, \tag{2.1}$$

where

$$F = \text{the magnitude of the applied force,}$$

$$m = \text{the mass of particle,}^* \text{ and}$$

$$a = \text{the resulting acceleration.}$$

From experiment [26], we know that near the earth's surface a body acted on solely by gravitational attraction accelerates at a rate of about 32.2 ft/s² (9.81 m/s²).† In this situation, the acting force is *weight,* which may be expressed in pounds-force (lbf), dynes, etc., depending on the particular system of units that is used, and magnitude of mass may be expressed variously as slugs, pounds-mass (lbm), kilograms, etc. In any case, whichever system is used, a consistent, compatible balance of units must be maintained. Newton's inertial law is of particular interest in this regard because it demands a careful distinction between the units of force and mass. In the United States, it has long been the habit to use the abbreviation lb as the unit for both mass and force, except when a distinction is absolutely required; then, the abbreviations lbm and lbf are used. The movement toward use of the metric system in the United States, with promotion of the SI system of units, should help eliminate this confusion.

Table 2.6 lists the basic units for five different systems. For example, the SI system (Système International d'Unites, or International System of Units)

---

* *Caution:* Particular note should be made of the use throughout this text of the symbols "m" and "w." The symbol m is used to represent the magnitude of the dimension, *mass,* and carries the units of *kilogram* (*kg*) or *pound-mass* (lbm). Weight, *w,* which is a force, carries the units *pounds-force* (lbf) or newtons (N). Note should also be made of the use of the symbol "m" to denote the unit, meter. Context should always make clear the intent.

† The standard gravitational body force ("acceleration due to gravity") is taken as 32.174 ft/s², or 9.80665 m/s². Of course, the actual value depends on the specific locality.

Table 2.6 Systems of units*

	System				
Quantity	SI (MKS) (mass, length, time)	Absolute Metric (CGS) (mass, length, time)	English Engineering (force, mass, length, time)	Absolute English (mass, length, time)	Technical English (force, length, time)
Length	meter (m)	centimeter (cm)	foot (ft)	foot (ft)	foot (ft)
Time	second (s)	second (s)	second (s)	second (s)	second (s)
Mass	kilogram (kg)	gram (g)	pound-mass (lbm)	pound-mass (lbm)	slug†
Force	newton (N)†	dyne†	pound-force (lbf)	poundal†	pound-force (lbf)
Energy	joule (J = N-m)	erg = dyne-cm	foot-(pound-force)	foot-poundal	foot-(pound-force)
Power			= energy/second		
Dimensional constant, g_c	$1\dfrac{\text{kg} \cdot \text{m}}{\text{N} \cdot \text{s}^2}$	$1\dfrac{\text{g} \cdot \text{cm}}{\text{dyne} \cdot \text{s}^2}$	$32.17\dfrac{\text{lbm} \cdot \text{ft}}{\text{lbf} \cdot \text{s}^2}$	$1\dfrac{\text{lbm} \cdot \text{ft}}{\text{poundal} \cdot \text{s}^2}$	$1\dfrac{\text{slug} \cdot \text{ft}}{\text{lbf} \cdot \text{s}^2}$

*Four dimensions are involved in each system. For the English Engineering system, all four dimensions are assigned. This requires that the dimensional constant carry a value of 32.17. For the other systems that are listed, the numerical value of the constant is unity. In each case, the constant carries the units necessary to balance the inertial equation.

† Derived units are underscored.

assigns the units of kilograms, meters (or metres), and seconds to the dimensions mass, length, and time, respectively. The unit of force, the newton, is a derived unit. Correspondingly, the English Engineering system assigns pounds-force, pounds-mass, feet, and seconds for force, mass, length, and time, respectively. In each case, when the assigned units are applied to Eq. (2.1), a question of compatibility arises. To provide a balance of units, we must introduce a factor called the *dimensional constant* g_c, modifying Eq. (2.1) as follows:

$$F/a = W/g = m/g_c \quad \text{or} \quad g_c = ma/F. \qquad (2.2)$$

If we select the English system as an example and assume that 1 lbf acts on 1 lbm, which we know results in an acceleration of 32.2 ft/s², then we find that

$$g_c = (1 \text{ lbm})(32.2 \text{ ft/s}^2)/(1 \text{ lbf})$$
$$= (32.2)(\text{lbm} \cdot \text{ft/lbf} \cdot \text{s}^2).$$

In like manner, we may determine values and units for g_c for the other systems listed in Table 2.6. Note that the other systems are "consistent" in the sense that g_c is unity.

To help reinforce the concept of the factor g_c, we consider Example 2.1 below.

In the past, physicists have been partial to the Absolute Metric or cgs (centimeter-gram-second) system, whereas engineers have used either the English Engineering system or the Technical English system. Throughout this book, we shall use both the SI and the English Engineering systems. All systems other than SI, however, are regarded as obsolete.

Example 2.1

Water of density ρ and absolute viscosity μ flows with velocity V through a pipe of diameter D. Calculate the Reynolds number, Re, from the data supplied, using (a) the English Engineering system of units and (b) the SI system. Before making the numerical calculations, check for balance of units.

Referring to Section 15.2, we see that Re $= D\rho V/\mu$ and, as discussed in that section, its value is unitless and hence is independent of the system of units used; thus, we should obtain the same numerical answers for both parts (a) and (b).

Data (see Appendix A for conversion factors):

$$D = 8.00 \text{ in.} = 0.667 \text{ ft} = 0.2032 \text{ m}$$
$$\rho = 62.3 \text{ lbm/ft}^3 = 997.95 \text{ kg/m}^3 \text{ (see Appendix D)}$$
$$V = 4.00 \text{ ft/s} = 1.219 \text{ m/s}$$
$$\mu = 2.02 \times 10^{-5} \text{ lbf} \cdot \text{s/ft}^2$$
$$= 9.6718 \times 10^{-4} \text{ N} \cdot \text{s/m}^2 \text{ (see Appendix D)}$$

Solution.

a. If we enter the units for each of the separate quantities appearing in the equation for Re, we have

$$(\text{ft})(\text{lbm/ft}^3)(\text{ft/s})(\text{ft}^2/\text{lbf} \cdot \text{s})(1/g_c);$$

or, entering the units for g_c,

$$(\text{ft})(\text{lbm/ft}^3)(\text{ft/s})(\text{ft}^2/\text{lbf} \cdot \text{s})(\text{lbf} \cdot \text{s}^2/\text{lbm} \cdot \text{ft}).$$

We see that the various units cancel, confirming the statement that the Reynolds number is unitless.

In magnitude,

$$\text{Re} = (\tfrac{2}{3})(62.3)(4.00)/(2.02 \times 10^{-5})(32.2) \approx 255{,}000.$$

b. In terms of SI units, we have

$$(\text{m})(\text{kg/m}^3)(\text{m/s})(\text{m}^2/\text{N} \cdot \text{s})(1/g_c)$$

or, when the units for g_c are entered,

$$(\text{m})(\text{kg/m}^3)(\text{m/s})(\text{m}^2/\text{N} \cdot \text{s})(\text{N} \cdot \text{s}^2/\text{kg} \cdot \text{m}).$$

Again, we see that the units cancel.

In magnitude, using SI units,

$$\text{Re} = (0.2032)(997.95)(1.219)/(9.6718 \times 10^{-4})(1) \approx 256{,}000.$$

Note: The lack of exact numerical agreement in the final numbers is due to the combination of inexact conversions plus a rounding variation (see the next section).

2.11 Significant Digits, Rounding, and Truncation

In this section we will discuss subjects that may be considered conventions rather than standards. We will concern ourselves with the application of simple common sense to the arithmetical manipulation of numbers.

Before Amédée Mannheim invented the basic slide rule in 1859, most arithmetic calculations were made with pencil and paper. The slide rule soon became the engineer's trademark. It made multiplication or division to at least three decimal places fast and easy. Then came the scientific calculator, yielding readouts to nine or more decimal places. Desktop computers now make such calculations precise to an almost unlimited number of decimal places, often yielding misleading impressions of true values.

Let us demonstrate the problem by reviewing a bit of the arithmetic in the example in the previous section. Consider the value of ρ, the density of the water. It is given as 62.3 lbm/ft^3. Presumably, we are saying that the value is precise to the nearest tenth. Now, to convert lbm/ft^3 to kg/m^3, we check

Appendix A and find kg/m^3 = 16.01846 × lbm/ft^3. Carrying out the multiplication with our trusty pocket calculator, we may read 997.95006 kg/m^3. Intuitively, we know we have not increased the precision of the value by the simple act of multiplication. The number of digits in our answer should be reduced; that is, the number should be *rounded*. The next section contains rules for extracting the true significance from such strings of digits.

2.11.1   Definitions

- *Number*   A series of digits that, along with their decimal places, combine to evaluate a quantity. (See Appendix C for a more detailed discussion of numbers.)
- *Whole number*   A number with no fractional part—for example, 63,120.
- *Mixed number*   A number containing a fractional part—for example, 6.312 or 0.087.
- *Result*   Desired numerical objective, obtained either experimentally or by calculation. At this point we are primarily interested in the numerical manipulation of the inputted numbers.
- *Significant digits*   Digits that are meaningful in assigning a true or realistic value to a result.
- *Empty digits*   Digits that have no meaning in assigning value to a result.
- *Exact counts*   Numbers that, by their very nature, consist entirely of significant digits. Examples are 365 days in a year, π = circumference of a circle divided by its diameter, the Naperian base e, and the like.
- *Truncating*   Simplification of a number by arbitrarily cutting off, or removing right-hand digits. Ordinarily, truncation is applied only to mixed numbers. Many computer subroutines are based on approximate evaluations of infinite series. For obvious practical reasons, the number of terms is limited. Beyond a certain point, the terms are truncated, or discarded. For example,

$$3.14159265\ldots \quad \text{may be truncated to} \quad 3.1415.$$

- *Rounding of approximate numbers*   The discarding of insignificant digits in a number, discarding of digits on the right of the decimal point for a mixed number, or replacement of right-hand, nonzero digits to the left of a decimal point with zeros in a whole number. There are rather definite conventions for rounding [8, 27]:

  1. Leave unchanged the last digit of the retained portion if the first digit of the discard is smaller than 5.
  2. Increase the last digit of the retained portion by 1 if the first digit of the discard is greater than 5.
  3. If the first digit of the discard is exactly 5, add 1 to the last digit to be retained if it is an odd number; leave it unchanged if it is even.

For example,

3.14159265 . . .	rounded to four decimal places is	3.1416
	rounded to five decimal places is	3.14159
	rounded to six decimal places is	3.141593
86628535	rounded to seven significant digits is	86628530
	rounded to two significant digits is	87000000

(*Note:* The zeros are significant only for reserving decimal places.)

2.11.2   Numerical Manipulation and Significant Digits

Before proceeding, let us make clear that, in this discussion, we are not questioning *uncertainties of measurement*. Measurement uncertainties are another matter and are dealt with in detail in Chapter 3. At this point, our questions revolve around *manipulation* of the numbers.

Judgment plays an important part in the manipulation of significant and empty digits; there are, however, commonly accepted conventions that may be stated, as follows:

- *Addition or subtraction*   When numbers are added (or subtracted), the number of significant digits in the answer shall not be greater than the number of significant digits contained in the least precise number.
- *Multiplication or division*   The product (or quotient) shall contain no more significant digits than are contained in the number with the fewest significant digits used in the multiplication or division.
- *Numbers that are exact counts*   The digits of an exact count or in a truncated exact count shall all be considered to be significant.

2.11.3   Significant Zero

Of course, zeros are significant when they are used to indicate a specific magnitude, as in "204 miles per hour." However, if we write "frequency = 1700 Hz," do we mean

1. f = 1700 Hz (rounded from, e.g., 1699.8);

or do we mean

2. f = 1700 Hz (rounded from 1696);

or do we mean

3. f = 1700 Hz (rounded from 1667)?

In the first case, both zeros are significant; in the second, the leftmost zero is significant. In the third case, neither zero is significant. A better way to make clear the exact meaning would be to write the three results as follows:

1. $f = 1.700 \times 10^3 \, \text{Hz}$
2. $f = 1.70 \times 10^3 \, \text{Hz}$
3. $f = 1.7 \times 10^3 \, \text{Hz}$

2.11.4   Series Arithmetic Operations

Clearly, there can be no absolute rules for rounding. Common sense must always be applied. Many engineering calculations involve a series of operations—additions, subtractions, multiplications, and divisions—leading to a final numerical answer. For instance, see Example 2.1. In many cases, strict adherence to the rules we have given may easily result in loss of all sense of variations. A particularly vexing problem sometimes exists when a calculation involves the difference between two very nearly equal numbers.

A practical approach that is often used is to carry one or two empty digits in each number, then to apply the rounding rules to the final result. In this case it is logical that *the number of significant digits in the answer should never exceed the number of significant digits in the least precise component of the calculation.*

2.12   Summary

All measurements are based on defined standards, most of which have been established by international agreement and U.S. laws. A measurement consists of quantifying the dimensional magnitude of an unknown relative to that established by the standard.

1. The SI system of units employs seven base units (meter, kilogram, second, ampere, kelvin, mole, and candela), two supplementary units (radian, steradian), a number of derived units, and various multiplying factors (Section 2.4).

2. The standard of length is the meter, defined as the distance traveled by light in 1/299,792,458 of a second (Section 2.5).

3. The standard of mass is the kilogram, defined by the International Prototype Kilogram (Section 2.6).

4. The standard of time is the second, defined as 9,192,631,770 periods of hyperfine-transition radiation from a cesium atom (Section 2.7).

5. The operational (or "practical") temperature standard is the International Temperature Scale of 1990, defined with respect to the thermodynamic temperatures of specific states of matter (Section 2.8).

6. SI electrical units are derived from the ampere. The ampere is defined by the mechanical force present in a particular type of electrical circuit (Section 2.9).

7. Although SI is the preferred system of measurement, other systems are still commonly used. The most important of these is the English Engineering system (Section 2.10). Conversion factors for changing English Engineering units to SI units are listed in Appendix A.

8. When manipulating numbers, the precision of the original data must be preserved. Procedures to avoid degrading (or "improving") the precision of the data are discussed in Section 2.11.

Suggested Readings

ASME, *Orientation and Guide for Use of SI Units*. 9th ed. New York, 1982.

ASME, PTC 19.12-1958. *Measurement of Time*. New York, 1958.

ASME, PTC 19.16-1965. *Density Determination of Solids and Liquids*. New York, 1965.

Belecki, N. B., B. L. Dunfee and O. Petersons. *The National Measurement System for Electricity*, National Bureau of Standards Interagency Report, NBSIR 75-935, 1978.

Blair, B. E., and A. H. Morgan. *Precision Measurement and Calibration*. (Selected NBS papers on frequency and time), NBS Special Publication 300, vol. 5. Washington, D.C.: U.S. Government Printing Office, 1972.

Chiswell, B., and E. C. M. Grigg. *S. I. Units: An Introduction*. New York: John Wiley, 1971.

Hermach, F. L., and R. F. Dziuba. *Precision Measurement and Calibration*. (Selected NBS papers on electricity-low frequency), NBS Special Publication 300, vol. 3. Washington, D.C.: U.S. Government Printing Office, 1968.

International Standard ISO 1000, SI Units and Recommendations for Their Use, 1973-02-01, New York: ANSI.

The International System of Units. NIST Special Publication 330, 1991 ed. Washington, D.C.: U.S. Department of Commerce, 1991.

Kamas, G., and S. L. Howe. *Time and Frequency Users' Manual*. NBS Special Publication 559. Washington, D.C.: U.S. Government Printing Office, 1979.

Mechtly, E. A. *The International System of Units, Physical Constants and Conversion Factors (NASA)*. Washington, D.C.: U.S. Government Printing Office, 1964.

Metric Conversion Act of 1975, Public Law 94-168, 94th Congress. H.R.8674, December 23, 1975. *Amended by* Public Law 100-418, 100th Congress, H.R.4848, August 23, 1988.

Metric Practice Guide, E 380-86, Philadelphia: ASTM, 1986.

Thomas, H. P. The international temperature scale of 1990 (ITS-90). *Metrologia, 27*: 3–10, 107, 1990.

Problems

2.1 Determine the speed of light in vacuum in:
   a.   miles/hour;
   b.   foot/s.

2.2 Convert the following temperatures to equivalent temperatures in K:
   a.   100°C
   b.   100°F
   c.   595°R

2.3 Determine the following temperatures in °R:
   a.   Freezing point of tin
   b.   Freezing point of aluminum
   c.   Freezing point of copper

2.4 Calculate the force (lbf) necessary to accelerate a weight of 0.5 lbf at 5 ft/s^2.

2.5 Determine the force (N) necessary to accelerate a mass of 200 g at 25 cm/s^2. What is the magnitude of this force in units of lbf?

2.6 Prepare a list of the best secondary standards that are available to you presently for calibration sources of
   a.   Length;
   b.   Time;
   c.   Mass;
   d.   Temperature;
   e.   Pressure.

2.7 Determine by calculation the relation for converting
   a.   Pressure in lbf/in^2 units to N/m^2 units;
   b.   Viscosity in lbf · s/ft^2 units to kg/m · s units;
   c.   Specific heat in kJ/kg · °C units to Btu/lbm · °F;
   d.   Dynamic viscosity in poise (1 dyne · s/cm^2) units to lbm/h · ft;
   e.   Heat flux in W/cm^2 units to Btu/h · ft^2.

2.8 Determine the factor for converting volume flow rate in cm^3/s units to gal/min.

2.9 Express the universal gas constant of 1545 ft · lbf/lbm · mol · °R in SI units.

References

1. Judson, L. V. *Weights and Measures Standards of the United States: A Brief History*, Gaithersburg, Md.: National Bureau of Standards, Special Publication 447, March 1976.

2. *Units of Weight and Measure*. Gaithersburg, Md.: National Bureau of Standards Misc. Pub. 214, July 1955.

3. Terrien, J. Scientific metrology on the international plane and the Bureau International des Poids et Mesures. *Metrologia* 1 (2): 15, January 1965.

4. *U.S. Coast and Geodetic Survey*, Bull. 26, April 1893.

5. Cochrane, R. D. *Measures for Progress, A History of the National Bureau of Standards*. Washington, D.C.: U.S. Dept. of Commerce, 1966, p. 47.

6. United States Congress, *Omnibus Trade and Competitiveness Act of 1988.*

7. *America and the Metric System: A Capsule History*. Commerce Metric Program. Washington, D.C.: U.S. Department of Commerce Technology Administration.

8. American Society for Testing and Materials (ASTM), *Standard for Metric Practice,* E380-86.

9. Terrien, J. International agreement on the value of the velocity of light. *Metrologia,* 10: 9, 1974.

10. Svenson, K. M., et al. Speed of light from direct frequency and wavelength measurements of the methane-stabilized laser. *Phys. Rev. Letters* 29 (19): 1346–49, 1972.

11. Documents concerning the new definition of the metre. *Metrologia* 19: 163–177, 1984.

12. Silsbee, F. B. Fundamental units and standards. *Instruments* 26: 1520, October 1953.

13. "Units and Systems of Weights and Measures: Their Origin, Development, and Present Status," National Bureau of Standards, Letter Circular 1035, November 1985.

14. Morrison, P. Book review, *Sci. Am.* 233 (4): 132, 1975.

15. Clemence, G. M. Time and its measurement. *Am. Scientist* 40 (2): 260, April 1952.

16. *National Bureau of Standards Tech. News Bull.* 52 (1): 10, January 1968.

17. "Frequency and Time Standards," Application Note 52, Hewlett-Packard Co., Palo Alto, Calif., 1965.

18. Kamas, G., and S. L. Howe (eds.) *Time and Frequency User's Manual,* National Bureau of Standards Special Publ. 559. Washington, D.C.: U.S. Government Printing Office, 1979.

19. Thomas, H. P. The international temperature scale of 1990 (ITS-90). *Metrologia* 27: 3–10; 107, 1990.

20. Bejan, A. *Advanced Engineering Thermodynamics.* New York: John Wiley, 1988.

21. Mangum, B. W., and G. T. Furukawa. "Guidelines for Realizing the International Temperature Scale of 1990 (ITS-90)," National Institute of Standards and Technology, Technical Note 1265, August 1990.

22. Rusby, R. L. et al. Thermodynamic basis of the ITS-90. *Metrologia,* 28: 9–18, 1991.

23. Driscoll, R. L., and R. D. Cutkoskey. Measurement of current with the National Bureau of Standards current balance. *National Bureau of Standards J. Res.* 60, April 1958.

24. Belecki, N. B., B. L. Dunfee and O. Petersons. *The National Measurement System for Electricity,* National Bureau of Standards Interagency Report, NBSIR 75-935, 1978.

25. Kose, V., and W. Wogar. Fundamental constants and standards. *Metrologia* 22: 177–185, 1986.

26. Cook, A. H. The absolute determination of the acceleration due to gravity. *Metrologia,* 1 (3): 84, 1965.

27. Natrella, M. G. *Experimental Statistics,* National Bureau of Standards Handbook 91. Washington, D.C.: U.S. Government Printing Office, 1963.

CHAPTER 3

|||||||||||||||||||||||||||||||||||||||||||||||||||||||||||||||

Assessing and Presenting Experimental Data

3.1   Introduction

"How good are the data?" is the first question put to any experimentalist who draws a conclusion from a set of measurements. The data may become the foundation of a new theory or the undoing of an existing one. They may form a critical test of a structural member in an aircraft wing that must never fail during operation. Before a data set can be used in an engineering or scientific application, its quality must be established.

The answer to the question revolves around the meaning we assign to the word "good." Our first temptation may be to call the data "good" if they agree well with a theoretically derived result. Theory, however, is simply a model intended to mimic the behavior of the real system being studied; there is no guarantee that it actually does represent the physical system well. The accuracy of even the most fundamental theory, such as Newton's laws, is limited both by the accuracy of the data from which the theory was developed and by the accuracy of the data and assumptions used when calculating with it. Thus, measurements should not be compared to a theory in order to assess their quality. What we are really after is the *actual value* of the physical quantity being measured, and that is the standard against which data should be tested. The *error* of a measurement is thus defined as the difference between the measured value and the true physical value of the quantity. The original question could be more clearly phrased as, "What is the error of the data?"

The definition of error is helpful, but it suffers from one major flaw: The error cannot be calculated exactly unless we know the true value of the quantity being measured! Obviously, we can never know the true value of a physical quantity without first measuring it, and, because some error is present in every measurement, the true value is something we can never know exactly. Hence, we can never know the error exactly, either.

The definition of error is not as circular as it seems, however, because we can usually estimate the likelihood that a measuring error exceeds some specific value. For example, 95% of the readings from one particular flowmeter will have an error of less than 1 L/s. Thus, we can say with 95% certainty (19 times out of 20) that a reading taken from that meter has an error of 1 L/s or less, or, equivalently, that the reading has an *uncertainty* of 1 L/s at odds of 19 to 1. A theoretical result that disagrees with the reading by more than 1 L/s shows a measurable inaccuracy; a theory within 1 L/s is supported by the reading at that level of odds.

Error or uncertainty may be estimated with statistical tools when a large number of measurements are taken. However, the experimentalist must also bring to bear his or her own knowledge of how the instruments perform and of how well they are calibrated in order to establish the possible errors and their probable magnitudes. This chapter describes how to estimate the uncertainty in a measurement and how to present the corresponding experimental data in an easily interpreted way.

3.2 Common Types of Error

We have defined the error in measuring a quantity x as the difference between the measured value, x_m, and the true value, x_{true}:

$$\text{Error} = \varepsilon \equiv x_m - x_{\text{true}}. \tag{3.1}$$

A primary objective in designing and executing an experiment is to *minimize* the error. However, after the experiment is completed, we must turn our attention to estimating a *bound* on ε with some level of certainty. This bound is typically of the form

$$-u \leq \varepsilon \leq +u \qquad (n:1), \tag{3.2}$$

where u is the *uncertainty* estimated at odds of $n:1$. In other words, only one measurement in n will have an error whose magnitude is greater than u. This bound is equivalent to saying that

$$x_m - u \leq x_{\text{true}} \leq x_m + u \qquad (n:1). \tag{3.3}$$

We would, of course, give higher odds that the true value would lie within a wider interval and lower odds that it would lie within a narrower interval.

The first step in bounding a measurement's error is to identify its possible causes. The specific causes of error will vary from experiment to experiment, and even a single experiment may include a dozen sources of inaccuracy. But, in spite of this diversity, most errors can be placed into one of two general classes [1]: *bias errors* and *precision errors*.

Bias errors, also referred to as *systematic errors,* are those that occur the same way each time a measurement is made. For example, if the scale on an instrument consistently reads 5% high, then the entire set of measurements will be biased by +5% above the true value. Alternatively, the scale may have a fixed offset error, so that the indicated value for every reading of x is higher than the true value by an amount x_{offset}.

Precision errors, also called *random errors,* are different for each successive measurement but have an average value of zero. For example, mechanical friction or vibration may cause the reading of a measuring mechanism to fluctuate about the true value, sometimes reading high and sometimes reading low. This lack of mechanical precision will cause sequential readings of the same quantity to differ slightly, creating a distribution of values surrounding the true value.

If enough readings are taken, the distribution of precision errors will become apparent. The successive readings will generally cluster about a central value and will extend over a limited interval surrounding that central value. In this situation, we may use statistical analysis to estimate the likely size of the error or, equivalently, the likely range of x in which the true value lies.

In contrast, bias errors cannot be treated using statistical techniques, because such errors are fixed and do not show a distribution. However, bias error can be estimated by comparison of the instrument to a more accurate standard, from our knowledge of how the instrument was calibrated, or from our experience with instruments of that particular type.

Figure 3.1 Bias and precision errors: (a) bias error larger than the typical precision error, (b) typical precision error larger than the bias error.

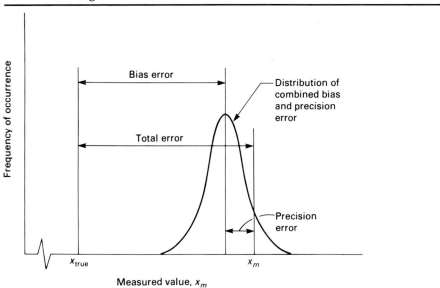

(a)

continued

Figure 3.1 *continued*

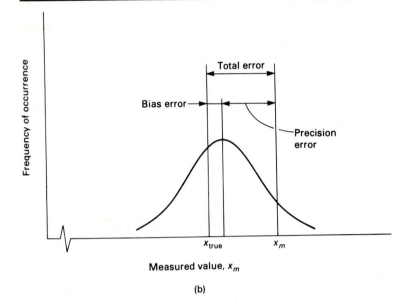

(b)

In practice, bias and precision errors occur simultaneously. The combined effect on repeated measurements of x is shown in Fig. 3.1(a) and (b). In Fig. 3.1(a), the bias error is larger than the typical precision error. In Fig. 3.1(b), the typical precision error exceeds the bias error. In other situations, bias and precision errors may be of the same size. The *total error* in a particular measured x_m is the sum of the bias and precision errors for that measurement.

3.2.1 Classification of Errors

A full classification of all possible errors as either bias or precision error would be convenient but is nearly impossible to make, since categories of error overlap and are at times ambiguous. Some errors behave as bias error in one situation and as precision error in other situations; some errors do not fit neatly into either category. However, for purposes of discussion, typical errors may be roughly sorted as follows:

1. Bias or systematic error
    a. Calibration errors
    b. Certain consistently recurring human errors
    c. Certain errors caused by defective equipment
    d. Loading errors
    e. Limitations of system resolution

2.   Precision or random error
   a.   Certain human errors
   b.   Errors caused by disturbances to the equipment
   c.   Errors caused by fluctuating experimental conditions
   d.   Errors derived from insufficient measuring-system sensitivity

3.   Illegitimate error
   a.   Blunders and mistakes during an experiment
   b.   Computational errors after an experiment

4.   Errors that are sometimes bias error and sometimes precision error
   a.   Errors from instrument backlash, friction, and hysteresis
   b.   Errors from calibration drift and variation in test or environmental conditions
   c.   Errors resulting from variations in procedure or definition among experimenters

The most common form of bias error is error in calibration. These errors occur when an instrument's scale has not been adjusted to read the measured value properly. As mentioned in Section 1.8, typical calibration errors may be *zero-offset* errors, which cause all readings to be offset by a constant amount x_{offset}, or *scale errors* in the slope of the output relative to the input, which cause all readings to err by a fixed percentage. Figure 3.2 illustrates these types of error.

Calibration procedures normally attempt to identify and eliminate these errors by "proving" the measuring system's readout scales through a comparison with a standard (Section 1.7). Of course, the standards themselves also have uncertainties, albeit smaller ones.* The impreciseness of any calibration procedure guarantees that some calibration-related bias error is present in all measuring systems.

Human errors may well be systematic, as when an individual experimenter consistently tends to read high or to "jump the gun" when synchronized readings are to be taken.

The equipment itself may introduce built-in errors resulting from incorrect design, fabrication, or maintenance. Such errors result from defective mechanical or electrical components, incorrect scale graduations, and so forth. Errors of this type are often consistent in sign and magnitude, and because of their consistency they may sometimes be corrected by calibration. When the input is time-varying, however, introducing a correction is more complicated. For example, distortion caused by poor frequency response cannot be corrected by the usual "static calibration," one based on a signal that is constant in time

---

* The uniqueness of certain primary standards makes them exceptions to this statement. In particular, the mass standard (the International Prototype Kilogram) has *by definition* a mass of exactly 1 kg. In the sense of practical applications, however, uncertainty will nevertheless occur: Even primary standards require the use of ancillary apparatus, which necessarily introduces some uncertainty.

Figure 3.2   Calibration errors. For ideal response, $x_{measured} = x_{true}$. Actual
response may include zero-offset error (x_{offset}) and scale
error ($\beta \neq 1$) so that $x_{measured} = \beta x_{true} + x_{offset}$.

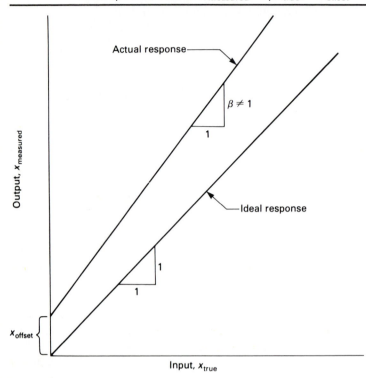

(see Section 5.20). Such frequency-response errors arise in connection with
seismic motion detectors, as discussed in Chapter 17.

   Loading error is of particular importance. It refers to the influence of the
measurement procedure on the system being tested. *The measuring process
inevitably alters the characteristics of both the source of the measured quantity
and the measuring system itself; thus, the measured value will always differ by
some amount from the quantity whose measurement is sought.* For example, the
sound-pressure level sensed by a microphone is not the same as the
sound-pressure level that would exist at that location if the microphone were
not present. Minimizing the influence of the measuring instrument on a
measured variable is a major objective in designing any experiment.

   Precision errors are also of several typical forms. The experimenter may be
inconsistent in estimating successive readings from his or her instruments.
Precision errors in the instrumentation itself may arise from outside disturbances to the measuring system, such as temperature variations or mechanical
vibrations. The measuring system may also include poorly controlled processes
that lead to random variations in the system output.

Variations in the actual quantity being measured may also appear as precision error in the results. Sometimes these variations are a result of poor experimental design, as when a system designed to run at a constant speed instead has a varying speed. Sometimes the variations are an inherent feature of the process under study, as when manufacturing variations create a distribution in the operating lifetimes of a group of light bulbs. Strictly speaking, variation in the measured quantity is not a measurement "error"; however, it is possible to apply the same statistical techniques to variations in the measured variable and to treat them as if they were errors. In particular, if you wish to find the mean value of the measured quantity, its variations may be averaged (together with the precision errors of the equipment), and the mean value may be calculated along with its uncertainty.

Illegitimate errors are errors that would not be expected to exist. These include outright mistakes (which can be eliminated through exercise of care or repetition of the measurement), such as incorrectly writing down a number, failing to turn on an instrument, or miscalculating during data reduction. Sometimes a statistical analysis will reveal such data as being extremely unlikely to have arisen from precision error.

Backlash and *mechanical friction* are important sources of variation in measuring systems. For example, friction may cause a mechanical element, such as a galvanometer needle, to lag behind advances in its intended position, thus reading low while the measured variable is increased and reading high while the measured variable is decreased. Such *hysteresis* is illustrated in Fig. 3.3. Since this error depends on how a sequence of measurements is taken, it may behave as either a bias error or a precision error. One means of detecting—and often correcting—this type of error is to make measurements while first increasing and then decreasing the measured quantity, an approach sometimes called the *method of symmetry*.

Drift of an instrument's calibration may occur if the response of an instrument varies in time. Often drift results from the sensitivity of an instrument to temperature or humidity fluctuations. If changes in environmental conditions occur between the time an instrument is calibrated and the time it is used, a bias error may appear in the readings. Conversely, if the test duration is long, environmental conditions may fluctuate throughout the test, causing different calibration errors for each successive measurement. In this case, the fluctuations create a precision error.

When an experiment is repeated using different equipment or by different experimenters, the bias errors of successive experiments are unrelated. If enough different experiments are performed, the bias errors are effectively *randomized*, and they become another form of precision error in the set of all experiments. For example, the speed of light in vacuum has been measured by many experimenters, each using different techniques and apparatus to obtain what is supposedly a unique quantity. Each experiment included its own bias and precision errors, but taken together their results show a distribution about a mean value, which may be estimated statistically (Fig. 3.4).

Figure 3.3 Hysteresis error

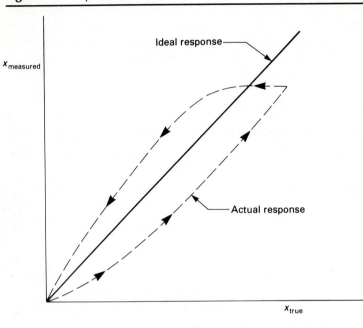

Figure 3.4 Measured values of the speed of light, 1947–1967
                (Data from Froome and Essen [2])

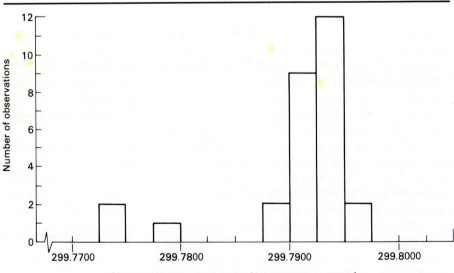

In contrast, the randomness of precision error may sometimes work against itself. For example, computer signal-processing techniques can extract desired information from a very noisy signal, as when photographs from satellites are enhanced to reveal planetary topography. In such cases, the systematic nature of the desired information enables it to be separated from the completely random overlying noise.

3.2.2 Terms Used in Rating Instrument Performance

The following terms are often employed to describe the quality of an instrument's readings. They are related to the expected errors of the instrument.

Accuracy The difference between the measured and true values. Typically, a manufacturer will specify a *maximum* error as the accuracy; manufacturers often neglect to report the odds that an error will not exceed this maximum value.

Precision The difference between the instrument's reported values during repeated measurements of the same quantity. Typically, this value is determined by statistical analysis of repeated measurements.

Resolution The smallest increment of change in the measured value that can be determined from the instrument's readout scale. The resolution is often on the same order as the precision; sometimes it is smaller.

Sensitivity The change of an instrument or transducer's output per unit change in the measured quantity. A *more sensitive* instrument's reading changes significantly in response to *smaller* changes in the measured quantity. Typically, an instrument with higher sensitivity will also have finer resolution, better precision, and higher accuracy.

Reading error refers to error introduced when reading a number from the display scale of an instrument. This type of error may sometimes be a bias error caused by truncation or rounding of the actual value to one within the resolution of the display. Reading error will also include error from inadequate instrument sensitivity if the instrument does not respond to the smallest fluctuations of the measured quantity. For example, a digital display may truncate an actual value of 10.4 to a displayed value of 10. The reading error of the digital display is thus $\pm\frac{1}{2}$ of the last digit read.* This is a bias error in the sense that 10.4 will always be displayed as 10. When many different values are to be read, the error may be thought of as a precision error if the many values have no particular relation to one another.

For a needle display on a galvanometer, one may be able to estimate to $\pm\frac{1}{2}$ or even $\pm\frac{1}{5}$ of the finest graduation. Depending on the particular experimenter, such error may be either bias or precision error, as discussed previously.

---

* It may be shown that 95% of the readings (19 out of 20) will differ from the value displayed by less than $\pm0.5(19/20) = \pm0.475 \approx \pm0.5$. The uncertainty is ±0.5 at $19:1$ odds.

3.3   Introduction to Uncertainty

When estimating uncertainty, we are usually concerned with two types of error, precision and bias error, and with two classes of experiments, *single-sample* experiments and *repeat-sample* experiments.

A *sample*, in this sense, refers to an individual measurement of a specific quantity. When we measure the strain in a structural member several times under identical conditions, we have repeatedly sampled that particular strain. With such repeat sampling, we can, for example, statistically estimate the distribution of precision errors in the strain measurement. If we measure that strain only once, we have instead a single sample of the strain, and our result does not reveal the distribution of precision error. In that case, we must resort to other means for estimating the precision error in our result.

Much of the remainder of this chapter is devoted to methods of estimating bias and precision error. Procedures for statistical analysis of precision error in repeat-sampled data are described in Sections 3.4–3.7. The estimation of bias uncertainty is covered in Section 3.9. The estimation of precision uncertainty for single-sample experiments is also considered in Section 3.9. Section 3.10 describes how uncertainty in measured variables leads to uncertainty in results calculated from those variables. Examples of uncertainty analysis are given in Section 3.11.

After determining the individual bias and precision uncertainty in a measurement of x, we must combine them to obtain the total uncertainty in our result for x. If the bias uncertainty is B_x and the precision uncertainty is P_x,* then the two may be combined in a root-mean-square sense as

$$U_x = (B_x^2 + P_x^2)^{1/2} \qquad\qquad (3.4)$$

to yield the total uncertainty, U_x.

The justification for combining the two uncertainty estimates this way is largely empirical [1]. However, the underlying assumption is that B_x and P_x are associated with independent sources of error, so that the errors are unlikely to have their maximum values simultaneously. When B_x and P_x are each estimates for 95% coverage, then U_x is also a 95% coverage; under the same conditions, it turns out [1] that simply adding B_x and P_x yields an uncertainty that covers roughly 99% of the data.

Example 3.1
A brass rod is held under a fixed tensile load and the axial strain in the rod is determined using a strain gage. Thirty results are obtained under fixed test conditions, yielding an average strain of $\varepsilon = 520\,\mu$-strain

---

* The precision uncertainty is evaluated in Section 3.6. In terms defined later, it has the value $P_x = tS_x/\sqrt{n}$, where t is the t-statistic, S_x is the sample standard deviation, and n is the sample size.

(520 ppm). Statistical analysis of the distribution of measurements gives a precision uncertainty of $P_\varepsilon = 21\,\mu$-strain with 95% confidence. The bias uncertainty is estimated to be $B_\varepsilon = 29\,\mu$-strain with odds of 19:1 (95% confidence). What is the total uncertainty of the strain?

Solution. The total uncertainty for 95% coverage is

$$U_\varepsilon = (B_\varepsilon^2 + P_\varepsilon^2)^{1/2} = 36\,\mu\text{-strain} \qquad (95\%).$$

In other words, with odds of 19:1, the true strain lies in the interval $520 \pm 36\,\mu$-strain:

$$484\,\mu\text{-strain} \le \varepsilon \le 556\,\mu\text{-strain}.$$

3.4 Estimation of Precision Uncertainty

Two fundamental concepts form the basis for analyzing precision errors. The first is that of a *distribution* of error. The distribution characterizes the probability that an error of a given size will occur. The second concept is that of a *population* from which *samples* are drawn. Usually, we have only a limited set of observations, our sample, from which to infer the characteristics of the larger population.

Statistical analysis of error usually assumes a model for the distribution of errors in a population, generally the *Gaussian,* or *normal,* distribution. Using this assumed distribution, we may estimate the probable difference between, say, the average value of a small sample and the true mean value of the larger population. This probable difference, or *confidence interval,* provides an estimate of the precision uncertainty associated with our measured sample.

This section and the next four examine some basic probability distributions, the characteristics of a population that satisfies a Gaussian distribution, and the accuracy with which the statistics of samples represent an underlying population. The *t*-distribution is introduced for treating small samples, and the χ^2-distribution is introduced for other statistical purposes. These sections should provide you with sufficient background to estimate the precision uncertainty in elementary engineering experiments.

3.4.1 Sample versus Population

Manufacturing variations in a production lot of marbles will create a distribution of diameters. To estimate the mean diameter, we may take a handful of marbles, measure them, and average the result (Figure 3.5). The handful is our *sample,* drawn from the production lot, which is our *population.*

No two samples from the same population will yield precisely the same average value; however, each should approximate the average of the population to some level of uncertainty. The difference between the sample

Figure 3.5   Sample taken from a population

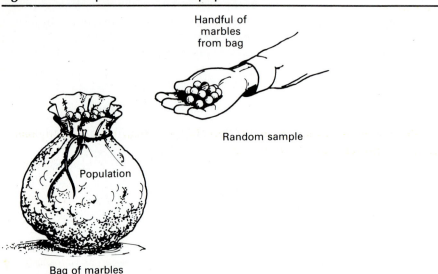

characteristics and those of the population will decrease as the sample is made larger.

Because our handful of marbles includes a number of members of the population, it is a repeat sample. By contrast, if we had drawn only one marble, we would obtain a single sample, which would give no direct evidence of the distribution of marble diameters.

Experimental errors can also be viewed in terms of population and sample. If we measure the diameter of a single marble repeatedly, the set of measured diameters gives a sample of the precision error in measuring the diameter. In this case, we could measure the diameter as many times as we liked, and each measurement would include a slightly different precision error. Thus, the population of precision errors is theoretically infinite. (Note, however, that this particular repeat sampling of the precision error is performed on a single sample from the marble population.)

The discussion of this and subsequent sections applies to both of the following two classes of sampling:

1. A sample of size n is drawn from a finite population of size p. The sample is used to estimate properties of the population. Additional data cannot be added to the population; for example, we assume that no more marbles can be added to the particular production lot from the same source. Further, the sample size is assumed to be small compared to the population size: $n \ll p$.

2. A *finite* number of items, n, is randomly drawn from what is assumed to be a population of indefinite size. The properties of the *assumed* population are inferred from the sample.

An important qualification underlies this discussion: The sample must be *randomly* selected from the population. If we select only the largest marbles from the bag, our sample will not accurately represent the whole population of marbles. Randomness can be accomplished, for example, by numbering the members of a population and then using a table of random numbers (Appendix F) to select a sample from it.

3.4.2 Probability Distributions

Probability is an expression of the likelihood of a particular event taking place, measured with reference to *all* possible events. Specifically, suppose that one of n equally likely cases will occur and that m of these cases correspond to an event A. The probability that event A will occur is m/n.

A penny is tossed. The total number of possible outcomes is two—heads and tails. If we choose heads (or tails) as event A, then the probability of A is 1 in 2, or 50%.

A slightly more complex example is that of throwing a pair of dice. One possible outcome yields a sum of 2, six outcomes yield a sum of 7, and the remaining outcomes are as distributed in Fig. 3.6. The bar chart used here is termed a *histogram*. If we divide the ordinate of the chart by the total number of possible outcomes (36), we obtain a graph of the *probability distribution*. For example, the probability of rolling a 7 is 6 in 36, or 16.6%.

Other distributions that will be considered in the following sections are these:

1. The *Gaussian*, or *normal*, *probability distribution*   When examining experimental data, this distribution is undoubtedly the first that is considered. The Gaussian distribution describes the population of possible

Figure 3.6   Distribution of results for a pair of thrown dice

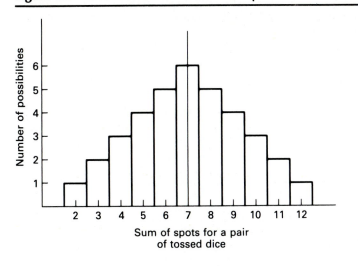

Sum of spots for a pair
of tossed dice

errors in a measurement when many independent sources of error contribute simultaneously to the total precision error in a measurement. These sources of error must be unrelated, random, and of roughly the same size.* Although we will emphasize this particular distribution, you must keep in mind that data do not always abide by the normal distribution. For tabulation and calculation, the Gaussian distribution is recast in a standard form, sometimes called the *z-distribution* [see Eq. (3.11) and Table 3.2].

2. *Student's t-distribution*   This distribution is used in predicting the mean value of a Gaussian population when only a small sample of data is available [see Eq. (3.22) and Table 3.6].

3. The χ^2-*distribution*   This distribution helps in predicting the width of a population's distribution, in comparing the uniformity of samples, and in checking the *goodness of fit* for assumed distributions (see Section 3.13 and Table 3.5).

3.5 Theory Based on the Population

From a practical standpoint, we are limited to samples from which to extract statistical information. In most cases, it is either impractical or impossible to manipulate the entire population. Nevertheless, some useful and important results can be established at the outset by considering the properties of the entire population.

Consider an infinite population of data, each datum representing a measurement of a single quantity, and assume that each datum, *x*, differs in magnitude from the rest only as a result of precision error. Effectively, each time *x* is measured, a different precision error occurs and a different member of the population is randomly selected. Some measurements are larger and some are smaller. The probability of obtaining a specific value of *x* depends on the magnitude of *x*, and the probability distribution of *x* values is described by a *probability density function, f(x)* (Fig. 3.7).

Since the population is infinite, the probability density function, or PDF, is a continuous curve, unlike the previous bar-graph histogram describing rolled dice. Consequently the ordinate of the PDF must be carefully interpreted: It represents the probability of occurrence per unit change of *x*. The probability of measuring a given *x* is *not* $f(x)$ itself; instead, the probability of measuring an *x* in the interval $\Delta x = x_2 - x_1$ is the *area* under the PDF curve between

---

* Note that the distribution of *precision errors* accompanying *experimental data* and the distribution of *dimensional variations* in *manufacturing operations* are very similar. For example, a drawing may specify tolerances within which the diameter of a shaft is allowed to vary about a nominal value. The variations that actually exist within a production lot are often found to be distributed in a normal, or Gaussian, manner. Quality control of machining operations is based on essentially the same theories as applied to distribution of experimental error.

Figure 3.7 Normal distribution curve. More precise data are represented by the dashed curve than by the solid curve.

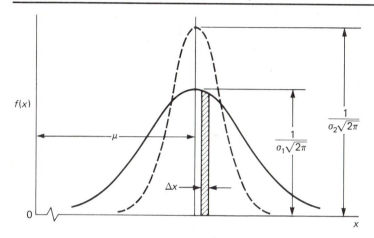

x_1 and x_2:

$$\text{Probability}_{(x_1 \to x_2)} = \int_{x_1}^{x_2} f(x)\, dx. \tag{3.5}$$

In any measurement, some value of x will be observed, so the total area under the PDF curve is unity, i.e., the probability of measuring some x value is 100%.

PDFs come in a variety of shapes, which are determined by the nature of the data considered. Precision error in experimental data is often distributed according to the familiar bell-shaped curve given by the *Gaussian*, or *normal*, *distribution* (Fig. 3.7). Most of the remaining statistical discussion in this chapter is based on that premise. For an infinite population, the mathematical expression for the Gaussian probability density function is

$$f(x) = \frac{1}{\sigma\sqrt{2\pi}} \exp\left[-\frac{(x - \mu)^2}{2\sigma^2} \right], \tag{3.6}$$

where

x = the magnitude of a particular measurement,

μ = the mean value of the entire population,

σ = the standard deviation of the entire population.

The mean value, μ, is that which would be obtained if every x in the population could be averaged together; for an infinite population, such averaging is clearly impossible, and thus μ remains unknown. Since we assume that the various x's differ as a result of precision error, μ also represents the true value of the quantity we are attempting to measure, and the average sought amounts to averaging out all the precision errors.

If a large number of measurements are taken with equal care, then the arithmetic average of these n measurements,

$$\bar{x} = \frac{x_1 + x_2 + \cdots + x_n}{n}$$

$$= \sum_{i=1}^{n} \frac{x_i}{n}, \tag{3.7}$$

can be shown to be *the most probable single value for the quantity μ.* Averaging a large sample allows us to estimate the true value.

The amount by which a single measurement is in error is termed the *deviation d*:

$$d = x - \mu. \tag{3.8}$$

The mean squared deviation, σ^2, is approximated by averaging the squared deviations of a very large sample:*

$$\sigma \approx \sqrt{\frac{d_1^2 + d_2^2 + \cdots + d_n^2}{n}}. \tag{3.9}$$

The quantity, σ, is called the *standard deviation* of the population; it characterizes both the typical deviation of measurements from the mean value and the width of Gaussian distribution (Fig. 3.7). When σ is smaller, the data are more precise. The standard deviation is a very important parameter in characterizing both populations and samples.

The actual deviation or error for a particular measurement is, of course, never known. However, the likely size of the error, or uncertainty, can be estimated with various levels of confidence by using our knowledge of the distribution of the population. For example, if a population has a Gaussian distribution, then the probability that a single measurement will have a deviation greater than $\pm\sigma$ is 31.7%, or about one chance in three. For a single measurement, then, we can be 68% confident that the deviation is less than $\pm\sigma$. A deviation greater than $\pm1.96\sigma$ has a probability of 5.0% (1 in 20); greater than $\pm3\sigma$, about 0.27% (1 in 370); and greater than $\pm4\sigma$, about 0.0063% (1 in 16,000). The 50% probability level, $\pm0.6754\sigma$, is sometimes termed "probable error," in the sense that one measurement in two will exceed it; however, the most popular uncertainty estimate is that for 95% confidence, $\pm1.96\sigma$. Table 3.1 summarizes the various levels of probability for the normal distribution.

---

* Formally, the mean and the standard deviation may be calculated as integrals of the PDF: $\mu = \int_{-\infty}^{+\infty} xf(x)\,dx$; $\sigma^2 = \int_{-\infty}^{+\infty} (x - \mu)^2 f(x)\,dx$. This amounts to summing the probable contribution to the observed value of each x or $(x - \mu)^2$.

Table 3.1 Summary of probability estimates based on the normal distribution

Common Name for "Error" Level	Error Level in Terms of σ	% Certainty That Deviation of x from Mean is Smaller	Odds That Deviation of x is Greater
"Probable error" (or mean deviation)	$\pm 0.6754\sigma$	50.0	1 in 2
Standard deviation	$\pm\sigma$	68.3	abt. 1 in 3
"90%" error	$\pm 1.6449\sigma$	90.0	1 in 10
"Two-Sigma error"	$\pm 1.96\sigma$	95.0	1 in 20
"Three-Sigma error"	$\pm 3\sigma$	99.7	1 in 370
"Maximum error"*	$\pm 3.29\sigma$	99.9+	1 in 1,000
"Four-Sigma Error"	$\pm 4\sigma$	99.994	1 in 16,000

* Some regard the 95% error as "maximum." In either case, of course, it is a practical maximum that is being considered. The actual maximum error is theoretically infinite.

Figure 3.8 Standard normal distribution curve. Note that a and b are inflection points.

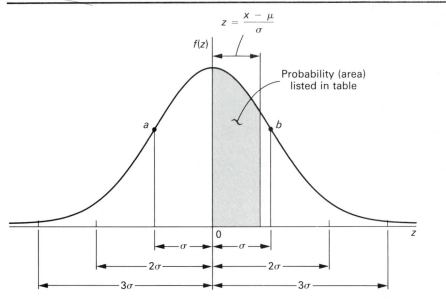

Table 3.2 Areas under the standard normal curve

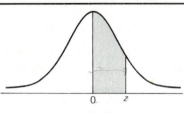

				Second Decimal Place in z					
z	**0.00**	**0.01**	**0.02**	**0.03**	**0.04**	**0.05**	**0.06**	**0.07**	**0.08**	**0.09**
0.0	0.0000	0.0040	0.0080	0.0120	0.0160	0.0199	0.0239	0.0279	0.0319	0.0359
0.1	0.0398	0.0438	0.0478	0.0517	0.0557	0.0596	0.0636	0.0675	0.0714	0.0753
0.2	0.0793	0.0832	0.0871	0.0910	0.0948	0.0987	0.1026	0.1064	0.1103	0.1141
0.3	0.1179	0.1217	0.1255	0.1293	0.1331	0.1368	0.1406	0.1443	0.1480	0.1517
0.4	0.1554	0.1591	0.1628	0.1664	0.1700	0.1736	0.1772	0.1808	0.1844	0.1879
0.5	0.1915	0.1950	0.1985	0.2019	0.2054	0.2088	0.2123	0.2157	0.2190	0.2224
0.6	0.2257	0.2291	0.2324	0.2357	0.2389	0.2422	0.2454	0.2486	0.2517	0.2549
0.7	0.2580	0.2611	0.2642	0.2673	0.2704	0.2734	0.2764	0.2794	0.2823	0.2852
0.8	0.2881	0.2910	0.2939	0.2967	0.2995	0.3023	0.3051	0.3078	0.3106	0.3133
0.9	0.3159	0.3186	0.3212	0.3238	0.3264	0.3289	0.3315	0.3340	0.3365	0.3389
1.0	0.3413	0.3438	0.3461	0.3485	0.3508	0.3531	0.3554	0.3577	0.3599	0.3621
1.1	0.3643	0.3665	0.3686	0.3708	0.3729	0.3749	0.3770	0.3790	0.3810	0.3830
1.2	0.3849	0.3869	0.3888	0.3907	0.3925	0.3944	0.3962	0.3980	0.3997	0.4015
1.3	0.4032	0.4049	0.4066	0.4082	0.4099	0.4115	0.4131	0.4147	0.4162	0.4177
1.4	0.4192	0.4207	0.4222	0.4236	0.4251	0.4265	0.4279	0.4292	0.4306	0.4319
1.5	0.4332	0.4345	0.4357	0.4370	0.4382	0.4394	0.4406	0.4418	0.4429	0.4441
1.6	0.4452	0.4463	0.4474	0.4484	0.4495	0.4505	0.4515	0.4525	0.4535	0.4545
1.7	0.4554	0.4564	0.4573	0.4582	0.4591	0.4599	0.4608	0.4616	0.4625	0.4633
1.8	0.4641	0.4649	0.4656	0.4664	0.4671	0.4678	0.4686	0.4693	0.4699	0.4706
1.9	0.4713	0.4719	0.4726	0.4732	0.4738	0.4744	0.4750	0.4756	0.4761	0.4767
2.0	0.4772	0.4778	0.4783	0.4788	0.4793	0.4798	0.4803	0.4808	0.4812	0.4817
2.1	0.4821	0.4826	0.4830	0.4834	0.4838	0.4842	0.4846	0.4850	0.4854	0.4857
2.2	0.4861	0.4864	0.4868	0.4871	0.4875	0.4878	0.4881	0.4884	0.4887	0.4890
2.3	0.4893	0.4896	0.4898	0.4901	0.4904	0.4906	0.4909	0.4911	0.4913	0.4916
2.4	0.4918	0.4920	0.4922	0.4925	0.4927	0.4929	0.4931	0.4932	0.4934	0.4936
2.5	0.4938	0.4940	0.4941	0.4943	0.4945	0.4946	0.4948	0.4949	0.4951	0.4952
2.6	0.4953	0.4955	0.4956	0.4957	0.4959	0.4960	0.4961	0.4962	0.4963	0.4964
2.7	0.4965	0.4966	0.4967	0.4968	0.4969	0.4970	0.4971	0.4972	0.4973	0.4974
2.8	0.4974	0.4975	0.4976	0.4977	0.4977	0.4978	0.4979	0.4979	0.4980	0.4981
2.9	0.4981	0.4982	0.4982	0.4983	0.4984	0.4984	0.4985	0.4985	0.4986	0.4986
3.0	0.4987	0.4987	0.4987	0.4988	0.4988	0.4989	0.4989	0.4989	0.4990	0.4990
3.1	0.4990	0.4991	0.4991	0.4991	0.4992	0.4992	0.4992	0.4992	0.4993	0.4993
3.2	0.4993	0.4993	0.4994	0.4994	0.4994	0.4994	0.4994	0.4995	0.4995	0.4995
3.3	0.4995	0.4995	0.4995	0.4996	0.4996	0.4996	0.4996	0.4996	0.4996	0.4997
3.4	0.4997	0.4997	0.4997	0.4997	0.4997	0.4997	0.4997	0.4997	0.4997	0.4998
3.5	0.4998	0.4998	0.4998	0.4998	0.4998	0.4998	0.4998	0.4998	0.4998	0.4998
3.6	0.4998	0.4998	0.4999	0.4999	0.4999	0.4999	0.4999	0.4999	0.4999	0.4999
3.7	0.4999	0.4999	0.4999	0.4999	0.4999	0.4999	0.4999	0.4999	0.4999	0.4999
3.8	0.4999	0.4999	0.4999	0.4999	0.4999	0.4999	0.4999	0.4999	0.4999	0.4999
3.9	0.5000*									

* For $z \geq 3.90$, the areas are 0.5000 to four decimal places.

One common criterion for discarding a data point as illegitimate is that the data point exceeds the 3σ level: since the probability of an error larger than this is 1 in 370, such values are unlikely in modest-sized data sets. Such data are sometimes called *outliers*.

For purposes of tabulation, the Gaussian PDF may be transformed by introducing the variable z:

$$z = \frac{x - \mu}{\sigma}. \tag{3.10}$$

Equation (3.6) is now

$$f(z) = \frac{1}{\sigma\sqrt{2\pi}} e^{-z^2/2} \tag{3.11}$$

which is the *standard* curve shown at the top of Table 3.2 and in Fig. 3.8. The table lists the *areas* under the curve between 0 and various values of z. Since the curve is symmetric about zero, the tabulation lists values for only half the curve. Bear in mind that the total area beneath the curve is equal to unity. This tabulation is sometimes called the *z-distribution*.

The following examples illustrate the nature and use of the tabulated data.

Example 3.2
a.   What is the area under the curve between $z = -1.43$ and $z = 1.43$?
b.   What is the significance of this area?

Solution.

a.   From Table 3.2, read 0.4236. This represents half the area sought. Therefore, the total area is $2 \times 0.4236 = 0.8472$.

b.   The significance is that for data following the normal distribution, 84.72% of the population lies within the range $-1.43 < z < 1.43$.

Example 3.3
What range of x will contain 90% of the data?

Solution. We need to find z such that $90\%/2 = 45\%$ of the data lie between zero and $+z$; the other 45% will lie between $-z$ and zero. Entering Table 3.2, we find $z_{0.45} \approx 1.645$ (by interpolation). Hence, since $z = (x - \mu)/\sigma$, 90% of the population should fall within the range

$$(\mu - z_{0.45}\sigma) < x < (\mu + z_{0.45}\sigma) \tag{3.12}$$

or

$$(\mu - 1.645\sigma) < x < (\mu + 1.645\sigma)$$

3.6 Theory Based on the Sample

In any real-life situation, we deal with samples from a population and not the population itself. Typically, our objective is to use average values from the sample to estimate the mean or standard deviation of the population. Thus, we would calculate the sample mean

$$\bar{x} = \sum_{i=1}^{n} \frac{x_i}{n} = \frac{x_1 + x_2 + \cdots + x_n}{n} \tag{3.13}$$

as an approximation to the population mean μ and the sample standard deviation

$$S_x = \sqrt{\frac{(x_1 - \bar{x})^2 + (x_2 - \bar{x})^2 + \cdots + (x_n - \bar{x})^2}{n - 1}} = \sqrt{\frac{(\sum_{i=1}^{n} x_i^2) - n\bar{x}^2}{n - 1}} \tag{3.14}$$

as an approximation to the population standard deviation σ. Here n is the number of data in the sample. The denominator of the standard deviation, $(n - 1)$, is called the number of *degrees of freedom*.*

The difference between population and sample values is emphasized by the use of different symbols for each:

	For Population	For Sample
Mean	μ or μ_x	$\bar{x}$
Standard Deviation	σ or σ_x	S_x

Naturally, we'd like to have some assurance that the sample mean and standard deviation accurately approximate the corresponding values for the population; more specifically, we'd like to have an estimate of the *uncertainty* in approximating μ and σ by $\bar{x}$ and S_x.

A second objective is often to infer the probability distribution of the population from that of the sample. As it turns out, these two objectives can be accomplished independently for large samples. Conversely, for small samples ($n < 30$), knowledge of the distribution is assumed in estimating the uncertainty of $\bar{x}$.

3.6.1 An Example of Sampling

During a 12-hour test of a steam generator, the inlet pressure is to be held constant at 4.00 MPa. For proper performance, the pressure should not deviate

---

* The basis for the number of degrees of freedom is the number of *independent* discrete data that are being evaluated. In computing the sample average, $\bar{x}$, all n data are independent. However, the standard deviation uses the result of the previous calculation for $\bar{x}$, which is not independent of the remaining data. Thus, the number of degrees of freedom is reduced by one in calculating S_x: we divide by $(n - 1)$ rather than n. In other instances, the data may have been used to calculate two or three necessary quantities (see Section 3.13 and Appendix G); the number of degrees of freedom is then reduced by two or three. The number of degrees of freedom is often denoted as v.

Table 3.3 Results of a 12-hour pressure test

Pressure p, in MPa	Number of Results, m
3.970	1
3.980	3
3.990	12
4.000	25
4.010	33
4.020	17
4.030	6
4.040	2
4.050	1

from this value by more than about 1%. The inlet pressure was measured 100 times during the test. Various factors caused the readings to fluctuate, and the resulting data are listed in Table 3.3. The resolution of the digital pressure gauge used was 0.001 MPa. The number of results, m, is the number of readings falling in an interval of ± 0.005 MPa centered about the listed pressure.

A first step in assessing the dispersion of the pressure readings might be to plot a histogram of the readings, as in Fig. 3.9.* A clear central tendency is

Figure 3.9 Histogram of the pressure data

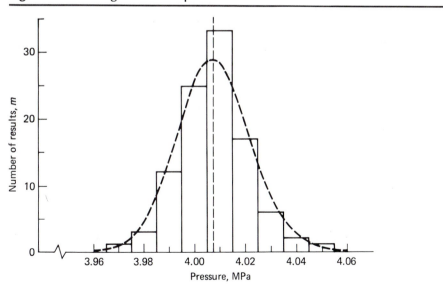

Table 3.4 Calculation of sample mean and standard deviation

Pressure p, in MPa	Number of Results, m	Deviation, d	d^2
3.970	1	−0.038	144.4×10^{-5}
3.980	3	−0.028	78.4
3.990	12	−0.018	32.4
4.000	25	−0.008	6.4
4.010	33	0.002	0.4
4.020	17	0.012	14.4
4.030	6	0.022	48.4
4.040	2	0.032	102.4
4.050	1	0.042	176.4
$\sum p = 400.77$	$n = \sum m = 100$		$\sum d^2 = 1858 \times 10^{-5}$

$\bar{p} = 400.77/100 = 4.008$ MPa

$S_p = \sqrt{1858 \times 10^{-5}/99} = 0.014$ MPa

apparent, as is the approximate width of the distribution. To quantify these values, we can compute $\bar{p}$ and S_p (Table 3.4) to obtain:

$$\bar{p} = 4.008 \text{ MPa}$$

$$S_p = 0.014 \text{ MPa}$$

Is the distribution of readings Gaussian? A simple test is just to substitute $\bar{p}$ and S_p for μ and σ in Eq. (3.6) and plot the resultant curve over the histogram, as in Fig. 3.9. (The vertical scale for the distribution has been arbitrarily increased, for purpose of comparison). An eyeball comparison indicates an approximate fit, albeit not a perfect one. How good must the fit be in order that we can claim a Gaussian distribution and apply Gaussian results? *Goodness of fit* is a legitimate concern, which is addressed further in Section 3.7.

Assuming that the population of pressure readings is in fact Gaussian distributed, with $\bar{p} \approx \mu$ and $S_p \approx \sigma$, the results of the previous section can be used to estimate the interval containing 95% of the pressure readings: $\mu \pm 1.96\sigma \approx \bar{p} \pm 1.96\,S_p = 4.008 \pm 0.027$ MPa. One objective of the pressure test was to verify that the pressure did not deviate from 4.00 MPa by more than $1\% = 0.04$ MPa. In terms of the standard deviation, 0.04 MPa $\approx 2.86\sigma$. For a Gaussian distribution, the probability of a 2.86σ fluctuation is about one chance in 240.

Two final comments should be made. First, these data do not separate measurement error from actual variations in the pressure; however, the actual pressure fluctuations are unlikely to be larger than the combined variation from measurement error and real fluctuations. Second, the analysis tells us nothing about possible bias errors in the readings.

3.6.2 Confidence Intervals for Large Samples

Last section's example showed how we can assess the dispersion of sample values about the mean value of the sample. However, it did not yield an estimate for the uncertainty in using $\bar{x}$ as an approximation to the true mean μ. In fact, we obtained from that sample only a single estimate for the mean value. If the pressure test were repeated and another 100 points acquired, the new mean value would differ somewhat from the first mean value. If we repeated the test many times, we would obtain a set of samples for the mean pressure.

These samples of the mean would also show a dispersion about a central value. A profound theorem of statistics shows that if n for each sample is very large the distribution of the mean values is *Gaussian* and that Gaussian distribution has a standard deviation

$$\sigma_{\bar{x}} = \sigma/\sqrt{n}. \tag{3.15}$$

This theorem (the *central limit theorem*) applies for very large samples even if the distribution of the underlying population is *not* Gaussian [3].

Armed with this important result, we can attack the uncertainty in our estimate that $\bar{x} \approx \mu$. Since $\bar{x}$ is Gaussian distributed, with standard deviation $\sigma_{\bar{x}}$, it follows that

$$z = \frac{\bar{x} - \mu}{\sigma_{\bar{x}}} = \frac{\bar{x} - \mu}{\sigma/\sqrt{n}}, \quad \text{(Mean of Means)} \tag{3.16}$$

where z is the same z-distribution given in Table 3.2. Hence, following Example 3.3, we can assert that $c\%$ of all readings of $\bar{x}$ will lie in the interval

$$\mu \pm z_{c/2} \frac{\sigma}{\sqrt{n}} \tag{3.17}$$

In other words, with $c\%$ confidence, the true mean value, μ, lies in the following interval about any single reading of $\bar{x}$:

$$\bar{x} - z_{c/2} \frac{\sigma}{\sqrt{n}} < \mu < \bar{x} + z_{c/2} \frac{\sigma}{\sqrt{n}}. \tag{3.18}$$

How sure are we of the mean $\bar{x}$

This interval is termed a $c\%$ *confidence interval*.

The confidence interval suffers from only one limitation: σ is usually unknown. However, a reasonable approximation to σ is S_x when n is *large*. Thus, for large samples, standard practice is to set $\sigma \approx S_x$ in estimating the confidence interval for $\bar{x}$:

$$\bar{x} - z_{c/2} \frac{S_x}{\sqrt{n}} < \mu < \bar{x} + z_{c/2} \frac{S_x}{\sqrt{n}}. \tag{3.19}$$

Often, the *standard error of the sample mean*, $S_{\bar{x}}$, is introduced in this context:

$$S_{\bar{x}} = \frac{S_x}{\sqrt{n}} \tag{3.20}$$

Example 3.4

Determine a 99% confidence interval for the mean pressure calculated in
Section 3.6.1.

Solution. First evaluate $z_{c/2} = z_{0.495} = 2.575$ from Table 3.2. Then, with
$\sigma \approx S_p$,

$$\bar{p} - z_{0.495}\frac{S_x}{\sqrt{n}} < \mu_p < \bar{p} - z_{0.495}\frac{S_x}{\sqrt{n}}$$

or

$$\mu_p = 4.008 \pm 2.575\frac{0.014}{\sqrt{100}}\text{MPa} = 4.008 \pm 0.0036\,\text{MPa} \qquad (99\%).$$

Note that this interval is much narrower than the dispersion of the data
itself, because we are focusing on the accuracy of the estimate of the
population mean.

Equation (3.19) is appropriate when we want the uncertainty in using the
sample mean, $\bar{x}$, as an estimate for the population mean, μ. The approach of
Table 3.1 and Section 3.6.1 is appropriate when we desire an estimate of the
width of the population distribution or the likelihood of observing a value that
deviates from the mean by some particular amount. For example, in Section
3.6.1 we used Table 3.1 to estimate the interval containing 95% of all
individual readings of p; in the last example, we estimated an interval
containing 99% of all measurements of $\bar{p}$. The $c\%$ confidence interval for the
mean value is narrower than that of the data by a factor of $1/\sqrt{n}$, because n
observations have been used to average out the random deviations of
individual measurements.

Figure 3.10 Probability distributions for the χ^2-statistic

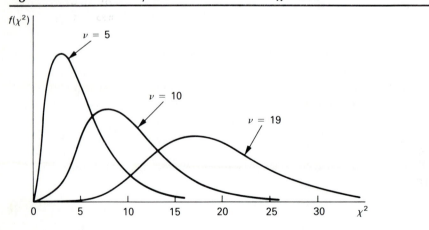

Table 3.5 The χ^2-distribution (values of $\chi^2_{\alpha,\nu}$)

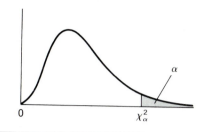

ν	$\chi^2_{0.995}$	$\chi^2_{0.99}$	$\chi^2_{0.975}$	$\chi^2_{0.95}$	$\chi^2_{0.05}$	$\chi^2_{0.025}$	$\chi^2_{0.01}$	$\chi^2_{0.005}$	ν
1	0.000	0.000	0.001	0.004	3.841	5.024	6.635	7.879	1
2	0.010	0.020	0.051	0.103	5.991	7.378	9.210	10.597	2
3	0.072	0.115	0.216	0.352	7.815	9.348	11.345	12.838	3
4	0.207	0.297	0.484	0.711	9.488	11.143	13.277	14.860	4
5	0.412	0.554	0.831	1.145	11.070	12.832	15.086	16.750	5
6	0.676	0.872	1.237	1.635	12.592	14.449	16.812	18.548	6
7	0.989	1.239	1.690	2.167	14.067	16.013	18.475	20.278	7
8	1.344	1.646	2.180	2.733	15.507	17.535	20.090	21.955	8
9	1.735	2.088	2.700	3.325	16.919	19.023	21.666	23.589	9
10	2.156	2.558	3.247	3.940	18.307	20.483	23.209	25.188	10
11	2.603	3.053	3.816	4.575	19.675	21.920	24.725	26.757	11
12	3.074	3.571	4.404	5.226	21.026	23.337	26.217	28.300	12
13	3.565	4.107	5.009	5.892	22.362	24.736	27.688	29.819	13
14	4.075	4.660	5.629	6.571	23.685	26.119	29.141	31.319	14
15	4.601	5.229	6.262	7.261	24.996	27.488	30.578	32.801	15
16	5.142	5.812	6.908	7.962	26.296	28.845	32.000	34.267	16
17	5.697	6.408	7.564	8.672	27.587	30.191	33.409	35.718	17
18	6.265	7.015	8.231	9.390	28.869	31.526	34.805	37.156	18
19	6.844	7.633	8.907	10.117	30.144	32.852	36.191	38.582	19
20	7.434	8.260	9.591	10.851	31.410	34.170	37.566	39.997	20
21	8.034	8.897	10.283	11.591	32.671	35.479	38.932	41.401	21
22	8.643	9.542	10.982	12.338	33.924	36.781	40.289	42.796	22
23	9.260	10.196	11.689	13.091	35.172	38.076	41.638	44.181	23
24	9.886	10.856	12.401	13.848	36.415	39.364	42.980	45.558	24
25	10.520	11.524	13.120	14.611	37.652	40.646	44.314	46.928	25
26	11.160	12.198	13.844	15.379	38.885	41.923	45.642	48.290	26
27	11.808	12.879	14.573	16.151	40.113	43.194	46.963	49.645	27
28	12.461	13.565	15.308	16.928	41.337	44.461	48.278	50.993	28
29	13.121	14.256	16.047	17.708	42.557	45.722	49.588	52.336	29
30	13.787	14.953	16.791	18.493	43.773	46.979	50.892	53.672	30

Confidence intervals may also be formed for the standard deviation. However, in this case we must *assume that the population is normally distributed*. If the standard deviation is calculated for many samples of size n from a Gaussian population, the results are distributed according to the so-called χ^2-distribution. The distribution is shown in Fig. 3.10 and Table 3.5; its value depends on the number of degrees of freedom of the sample, $v = n - 1$. Note that this distribution is not symmetrical like the Gaussian distribution.

Additional analysis shows that the squared standard deviation, σ^2, lies in the interval

$$\frac{(n-1)S_x^2}{\chi_{\alpha/2}^2} < \sigma^2 < \frac{(n-1)S_x^2}{\chi_{1-\alpha/2}^2} \quad (c\%) \tag{3.21}$$

with a confidence of $c\% = (1 - \alpha)$. Further uses of the χ^2-distribution are discussed in Section 3.13.

3.6.3 Confidence Intervals for Small Samples

Equations (3.18) and (3.19) provide confidence intervals for the sample mean, $\bar{x}$, when σ is known or can be approximated by S_x. The condition $\sigma \approx S_x$ will generally apply when the sample is "large," which, from a practical viewpoint, means $n \geq 30$. Unfortunately, in many engineering experiments, n is substantially less than 30, and the preceding intervals are really not much help.

An amateur statistician, writing under the pseudonym Student, addressed this matter by considering the distribution of a quantity t,

$$t = \frac{\bar{x} - \mu}{S_x / \sqrt{n}}, \tag{3.22}$$

which replaces σ in Eq. (3.16) by S_x. Student calculated the probability distribution of the t-statistic under the assumption that *the underlying population satisfies the Gaussian distribution*. This PDF, $f(t)$, is shown in Fig. 3.11. Note that the distribution depends on the number of samples taken, through the degrees of freedom, $v = n - 1$.

The t-distribution is qualitatively similar to the z-distribution. The PDF is symmetric about $t = 0$, and the total area beneath the distribution is again unity. Moreover, since the t-distribution is a PDF, the probability that t will lie in a given interval $t_2 - t_1$ is equal to the area beneath the curve between t_2 and t_1. As n (or v) becomes large, the t-distribution approaches the standard Gaussian PDF; for $n > 30$, the two distributions are identical.

By analogy to the previous treatment of the z-distribution, the area beneath the t-distribution is tabulated in Table 3.6. However, in this case, the area α between t and $t \to \infty$ is listed; selected areas are given for a range of sample sizes. Thus the area α corresponds to the probability, for a sample size of $n = v + 1$, that t will have a value *greater* than that given in the table [Fig. 3.12(a)]. Since the distribution is symmetric, α is also the probability that t has

Figure 3.11 Probability distribution for the *t*-statistic

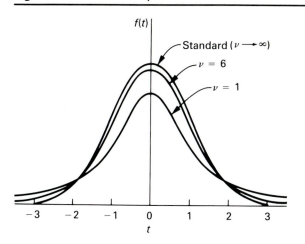

a value less than the negative of the tabulated value [Fig. 3.12(b)]. Conversely, we can assert with a confidence of $c\% = (1 - \alpha)$ that the actual value of t does *not* fall in the shaded area (i.e., if $\alpha = 0.05$, then $c = 0.95 = 95\%$).

Often, we want a two-sided confidence interval for the mean $\bar{x}$ of a small sample, so that both upper and lower bounds on the mean are stated. This interval follows directly from the preceding intervals; with a confidence of $c\%$,

Figure 3.12 Confidence intervals for the *t*-statistic: (a) one-sided, right (b) one-sided, left (c) two-sided. With a confidence of $c\% = (1 - \alpha)$, the value of t lies in the unshaded interval.

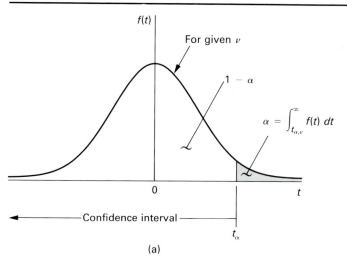

(a)

continued

Figure 3.12 *continued*

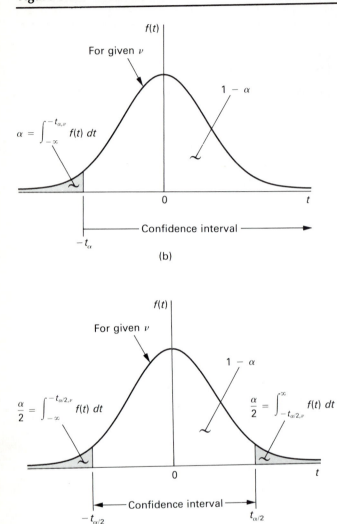

(b)

(c)

the true mean value lies in the interval

$$\bar{x} - t_{\alpha/2,v}\frac{S_x}{\sqrt{n}} < \mu < \bar{x} + t_{\alpha/2,v}\frac{S_x}{\sqrt{n}} \qquad (c\%) \qquad (3.23)$$

where $\alpha = 1 - c$ and $v = n - 1$ [Fig. 3.12(c)]. This equation is the small-sample analog of Eq. (3.19). Sometimes α is referred to as the *level of significance*.

Table 3.6 Student's t-distribution (values of $t_{\alpha,v}$)

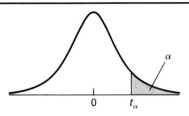

v	$t_{0.10}$	$t_{0.05}$	$t_{0.025}$	$t_{0.01}$	$t_{0.005}$	v
1	3.078	6.314	12.706	31.821	63.657	1
2	1.886	2.920	4.303	6.965	9.925	2
3	1.638	2.353	3.182	4.541	5.841	3
4	1.533	2.132	2.776	3.747	4.604	4
5	1.476	2.015	2.571	3.365	4.032	5
6	1.440	1.943	2.447	3.143	3.707	6
7	1.415	1.895	2.365	2.998	3.499	7
8	1.397	1.860	2.306	2.896	3.355	8
9	1.383	1.833	2.262	2.821	3.250	9
10	1.372	1.812	2.228	2.764	3.169	10
11	1.363	1.796	2.201	2.718	3.106	11
12	1.356	1.782	2.179	2.681	3.055	12
13	1.350	1.771	2.160	2.650	3.012	13
14	1.345	1.761	2.145	2.624	2.977	14
15	1.341	1.753	2.131	2.602	2.947	15
16	1.337	1.746	2.120	2.583	2.921	16
17	1.333	1.740	2.110	2.567	2.898	17
18	1.330	1.734	2.101	2.552	2.878	18
19	1.328	1.729	2.093	2.539	2.861	19
20	1.325	1.725	2.086	2.528	2.845	20
21	1.323	1.721	2.080	2.518	2.831	21
22	1.321	1.717	2.074	2.508	2.819	22
23	1.319	1.714	2.069	2.500	2.807	23
24	1.318	1.711	2.064	2.492	2.797	24
25	1.316	1.798	2.060	2.485	2.787	25
26	1.315	1.706	2.056	2.479	2.779	26
27	1.314	1.703	2.052	2.473	2.771	27
28	1.313	1.701	2.048	2.467	2.763	28
29	1.311	1.699	2.045	2.462	2.756	29
∞	1.282	1.645	1.960	2.326	2.576	∞

This confidence interval defines the precision uncertainty in the value $\bar{x}$. From Eq. (3.23), the precision uncertainty in $\bar{x}$ is

$$P_x = t_{\alpha/2,v}\frac{S_x}{\sqrt{n}} \quad (c\%) \tag{3.24}$$

at a confidence level of $c\%$. This precision uncertainty is that needed for Eq. (3.4).

Example 3.5
Twelve values in a sample have an average of $\bar{x}$ and a standard deviation of S_x. What is the 95% confidence interval for the true mean value, μ?

Solution. The required level of significance is $\alpha = 1 - 0.95 = 0.05$ and the degree of freedom is $v = 11$. The necessary value of t is $t_{\alpha/2,v} = t_{0.025,11} = 2.201$ (from Table 3.6). Hence, the 95% confidence interval is

$$\bar{x} - 2.201\frac{S_x}{\sqrt{12}} < \mu < \bar{x} + 2.201\frac{S_x}{\sqrt{12}} \quad (95\%).$$

Example 3.6
A simple postal scale of the equal-arm balance type, is supplied with $\frac{1}{2}$-, 1-, 2-, and 4-oz machined brass weights. For a quality check, the manufacturer randomly selects a sample of 14 of the 1-oz weights and weighs them on a precision scale. The results, in ounces, are as follows:

1.08	1.03	0.96	0.95	1.04
1.01	0.98	0.99	1.05	1.08
0.97	1.00	0.98	1.01	

Question: Based on this sample and the assumption that the parent population is normally distributed, what is the 95% confidence interval for the *population mean?*

Solution. Using the t-test, we first calculate the sample mean and standard deviation, with $n = 14$. They are respectively,

$$\bar{x} = 1.009 \text{ oz} \qquad S_x = 0.04178.$$

From Table 3.6, for $v = n - 1 = 13$, we find $t_{0.025,13} = 2.160$.
Calculating the *two-sided* confidence limits, we have

$$\frac{\pm t_{0.025,13}S_x}{n^{1/2}} = \frac{\pm 2.160(0.04178)}{(14)^{1/2}} = \pm 0.02412.$$

Hence we may write: $\mu = 1.009 \pm 0.024 \text{ oz}$, with a confidence of 95%.

If, for some reason, we wanted the one-sided limits (upper or lower), at 95% confidence, the only change would be to use $t_{0.05,13}$ instead of $t_{0.025,13}$. Then our calculations would be as follows. First, checking Table 3.6 for $v = 13$ we find $t_{0.05,13} = 1.771$. Then,

$$\frac{t_{0.05,13} S_x}{n^{1/2}} = \frac{1.771(0.04178)}{(14)^{1/2}} = 0.01978$$

and $\mu \leq 1.029$ oz or $\mu \geq 0.990$ oz at 95% confidence.

3.6.4 The t-Test Comparison of Sample Means

If we wish to compare two samples solely on the basis of their means, we can use a form of Eq. (3.22) [4]:

$$t = \frac{\bar{x}_1 - \bar{x}_2}{\sqrt{(S_1^2/n_1) + (S_2^2/n_2)}}, \tag{3.25}$$

where $\bar{x}_1$, S_1, n_1 and $\bar{x}_2$, S_2, n_2 are the means, standard deviations and the sizes of the two respective samples. This value of t is compared to the interval $\pm t_{\alpha/2, v}$ found in Table 3.6, in which α is for an arbitrarily chosen confidence level $(1 - \alpha)$. The degree of freedom v may be approximated by the following expression:

$$v = \frac{[(S_1^2/n_1) + (S_2^2/n_2)]^2}{\dfrac{(S_1^2/n_1)^2}{n_1 - 1} + \dfrac{(S_2^2/n_2)^2}{n_2 - 1}}, \tag{3.25a}$$

where v is rounded down to the nearest integer [4]. If the value t falls inside the interval $\pm t_{\alpha/2, v}$, then we can conclude that the two means $\bar{x}_1$ and $\bar{x}_2$ are not significantly different at the chosen level of confidence.

Example 3.7
An apartment manager wishes to compare the lifetimes, under similar conditions, of two major brands of light bulbs. In the following sample data, the lifetime is in months.

Bulb A   7.2, 7.6, 6.9, 8.2, 7.3, 7.8, 6.6, 6.9, 5.5, 7.4, 5.7, 6.2

Bulb B   7.5, 8.7, 7.7, 7.5, 6.7, 11.2, 7.0, 10.7, 7.0, 8.6, 6.1, 6.3,

7.8, 8.7, 6.1

Solution. When we calculate the means and standard deviations of each sample, we find

Bulb A	Bulb B
$\bar{x}_A = 6.94$ mo	$\bar{x}_B = 7.84$ mo
$S_A = 0.82$ mo	$S_B = 1.53$ mo
$n_A = 12$	$n_B = 15$

From Eq. (3.25a) we determine the degrees of freedom

$$\nu \approx \frac{[(0.82)^2/12 + (1.53)^2/15]^2}{\dfrac{[(0.82)^2/12]^2}{12 - 1} + \dfrac{[(1.53)^2/15]^2}{15 - 1}}$$

$$\approx 22 \quad \text{(rounded down)},$$

and from Eq. (3.25) we calculate the test statistic

$$t = \frac{6.94 - 7.84}{\sqrt{(0.82)^2/12 + (1.53)^2/15}} = -1.954.$$

For a confidence level $(1 - \alpha) = 0.95$ we find the critical values of t from Table 3.6 to be

$$\pm t_{0.05/2,22} = \pm t_{0.025,22} = \pm 2.074.$$

Since the value of t falls within the region, we conclude that there is not a significant difference in the lifetimes of bulbs A and B at a 95% confidence level.

3.7 Goodness of Fit

As stated previously, distribution of experimental data often approximately abides by the Gaussian form expressed by Eq. (3.6). One must keep in mind, however, that this assumption is not always justified. For example, fatigue strength data for some metals approximate the so-called Weibull distribution; other distributions are described in the Suggested Readings at the end of the chapter. Since a given set of data may or may not abide by the assumed distribution and since, at best, the degree of adherence can be only approximate, some estimate of *goodness of fit* should be made before critical decisions are based on statistical error calculations. In the following paragraphs we discuss tests of fit that may be applied to the common Gaussian distribution as defined by Eq. (3.6).

At the outset we advise the reader that there is no absolute check in the sense of producing some perfect and indisputable figure of merit. At best, a qualifying *confidence level* must be applied, with the final acceptance or rejection left to the judgment of the experimenter.

The simplest method is simply to plot a histogram and to "eyeball" the result: yes, the distribution appears to approximate a bell shaped one; or no, it does not. This approach can easily result in misleading conclusions. The appearance of the histogram can sometimes be altered quite radically simply by readjusting the number of class intervals.

A second method, which is relatively easy and much more effective, is to make a graphical check using a *normal probability plot.* This technique requires a special graph paper* available from most bookstores that deal in technical supplies. One axis of the graph represents the cumulative probability (in percent) of the summed data frequencies. The other must be scaled to accommodate the range of data values in the sample. The more nearly the data plots as a straight line and the more nearly the mean corresponds to the 50% point, the better the fit to normal distribution. The final determination is subjective; it depends on the judgment of the experimenter. Considerable deviation from a straight line should raise serious doubts as to the value of any Gaussian-based calculations, particularly the significance of the calculated standard deviation. The following example demonstrates the procedure.

Example 3.8

We will illustrate the graphical method by using the data of pressures given in Table 3.3.

Solution. In treating these data we will arbitrarily center our class intervals, or bins, on the mean and will assume eight intervals, each $0.75S_p$ in width. Using these ground rules we prepare Table 3.7.

Table 3.7   Pressure data of Table 3.3 arranged for a normal probability plot

A	B	C	D
3.967			
	1	1	1.01
3.977			
	3	4	4.04
3.987			
	12	16	16.16
3.998			
	25	41	41.41
4.008 (mean)			
	33	74	74.75
4.018			
	17	91	91.92
4.029			
	6	97	97.98
4.039			
	2	99	99.00
4.049			
	1*	100	100.00

A = Limits on class intervals, arbitrarily taken as $0.75 \times S_p =$ 0.01027

B = Number of data items falling within respective class intervals

C = Cumulative number of data items

D = Cumulative number of data items in percent

* A rule of thumb often used is to discard arbitrarily any data falling outside a $\pm 3S_x$ limit. Theoretically, discarding out-of-tolerance items could make a readjustment of the mean and the standard deviation necessary. In this particular case, the changes would be so slight as to make the additional work unprofitable.

* See Ref. [5], page 25, for directions for preparing one's own normal probability paper.

Figure 3.13 Normal probability plot of data listed in Table 3.7

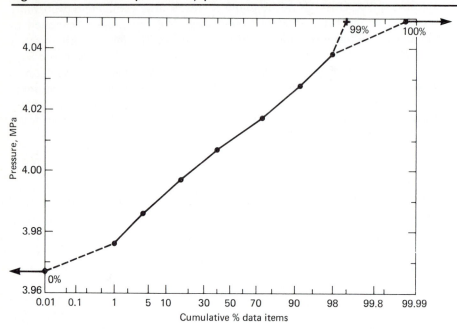

Figure 3.14 Graphical effects of data skew and offset
                as displayed on a normal probability plot

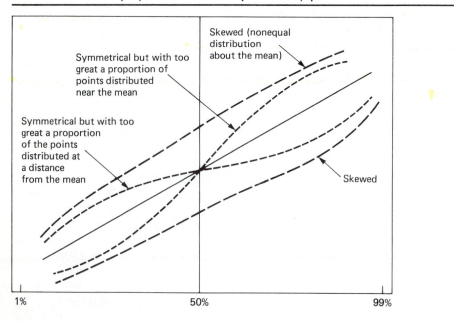

The ordinate of the graph is in terms of the upper limit of each interval. This quantity correlates with the cumulative values, which are plotted as the abscissa. Data from column A are plotted versus the percentages in column D, yielding Fig. 3.13. To plot either 0 or 100% is impossible. For this reason, and also because either absence or presence of even one extra data point in the extreme intervals unduly distorts the plot, the two endpoints are generally given little consideration in making a final judgment. On the basis of Fig. 3.13 we can say that the pressure data show a reasonably good Gaussian distribution. Figure 3.14 illustrates the general discrepancies that may be discerned from a non-straight-line normal probability plot and their causes.

Another common test for goodness-of-fit is based on the χ^2-distribution. This test and some other applications of the χ^2-distribution are considered in Section 3.13.

3.8 Statistical Analysis by Computer

Statistical analysis can often involve very large sets of data or require the application of a broad range of statistical tests. Consequently, a number of statistical software packages have been developed to assist in data analysis. Commercially marketed versions include Minitab* and SPSS.† Packages such as these are available for use on machines ranging from mainframes to personal computers.

The statistical methods of the preceding sections were developed in the last century to reduce the laborious calculations that would otherwise be required in drawing statistical conclusions from samples. These methods are possible largely because the population's probability distribution has been assumed known. However, the digital computer makes detailed statistical computations easier. As a result, some current statistical research is directed toward using computer methods to relax the assumptions associated with classical statistics. For example, can we determine small-sample confidence intervals *without* assuming that the population is Gaussian distributed?

3.9 Bias and Single-Sample Uncertainty

Precision error in repeat-sampled data reveals its own distribution, enabling us to bound its magnitude using statistical methods. Bias error, by virtue of its systematic nature, provides no direct evidence of either its magnitude or its presence. The only direct method for uncovering bias error in a measurement

---

* Minitab is registered to Minitab, Inc., 3081 Enterprise Drive, State College, PA 16801.

† SPSS is a registered trademark of SPSS, Inc., 444 N. Michigan Avenue, Chicago, IL, 60611.

is by comparison with measurements made using a separate, and presumably more accurate, apparatus.

Unfortunately, a second set of apparatus is seldom used owing to cost and time constraints. Instead, we rely on knowledge of our own equipment to make estimates for the likely sizes of bias errors. Estimation of bias uncertainty relies heavily on experience and on an understanding of calibration accuracy and dimensional tolerances. Even with such experience and understanding, unexpected sources of bias error can be overlooked. Diligence, persistence, and careful examination of one's results are essential in identifying and eliminating such errors.

Estimates of bias uncertainty should be accompanied by odds or a confidence level [6]. Unlike statistical confidence levels, odds for bias uncertainty cannot be rigorously determined. The level of confidence assigned is a product of our knowledge of the system, reflecting our assessment of the fraction of bias errors likely to land within the uncertainty interval. Sometimes the term *coverage* is used in place of confidence to reflect the empirical nature of these estimates in contrast to those derived using statistical methods.

We have previously discussed general sources of bias error. Let's look at a few in more detail.

Data reduction often requires knowledge of physical properties, system dimensions, or electrical characteristics. For example, a flowmeter measurement may depend on the density of water and a tube diameter; an amplifier gain may depend on the value of a resistor. Differences between the assumed values of these components and their actual values can systematically shift all data taken, creating a bias error.

To find water's density, the temperature is specified and a tabulated handbook value is taken, giving, for example, $998\,kg/m^3$. Inaccuracy or ambiguity in the temperature may cause a bias uncertainty in density of $2\,kg/m^3$, for instance. This uncertainty may cover 95% of the temperature range we expect to occur in the system. This inaccuracy will *systematically* affect all data reduced using the particular value of density.

Similarly, the nominal diameter of a pipe may differ from the production diameter by a percent or so; and a manufacturer's rated value for a resistors can vary substantially from the resistor's actual resistance. For carbon resistors, the manufacturing tolerance may be ±5% or ±10%; for higher-quality, metal-film resistors, the tolerance may be ±0.01%. Typically, these tolerances might represent 95% coverage—that is, the variation of 95% of all resistors. Potential sources of error such as these remain unchanged for each measurement made with the system.

If the uncertainty related to manufacturing tolerance is unacceptably large, taking our own measurement of the specific part can usually reduce the uncertainty substantially. The uncertainty in a resistor's value can be reduced to the accuracy of the ohmmeter measuring it; or a pipe's diameter can be measured to the accuracy of a set of calipers. Physical property data can also

Figure 3.15   Block diagram showing calibration procedure

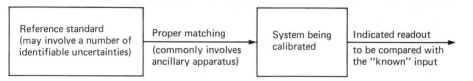

be measured, if need be, although it is more common to trust the carefully determined handbook values.

Calibration uncertainty is another very common source of bias uncertainty. Calibration requires a reference or standard against which system response can be compared. The reference may be *fixed* or *one-valued,* such as the triple point of water or the other triple points and melting points used to define the practical temperature scale (Section 2.8). Alternatively, the standard may be capable of supplying a range of inputs comparable to the range of the system, as do various commercially available voltage references. Naturally, the uncertainty of the standard should be considerably less than that of the system being calibrated. A rule of thumb is that the uncertainty of the standard should be no more than one-tenth that of the system being calibrated.

Figure 3.15 shows a typical calibration arrangement. Normally, the indicated readout is compared to the reference standard and a relation between the two is determined. Sometimes the readout scale can be adjusted until agreement with the standard is obtained; sometimes a line fit is used to relate the readout to the standard's value. In either case, additional uncertainty appears in the comparison and adjustment process.

Instrument manufacturers often supply calibration data with their products, which can assist in estimating the uncertainty of the instrument. For example, a particular position transducer is rated at 0.8 V output per millimeter of sensor displacement. The manufacturer has not specified odds or uncertainty directly. However, we might assume that the uncertainty is roughly 0.05 V/mm, since this is the apparent resolution of the calibration. The coverage is also unknown, but our experience using the device may suggest that 90 to 95% coverage is a reasonable assumption. If necessary, we could reduce the uncertainty by conducting our own calibration.

Examples of estimating bias uncertainty are given in Section 3.11.

3.9.1   Single-Sample Precision Uncertainty

When only one, two, or three repeat observations are made, the confidence intervals calculated statistically can be quite large. In that circumstance, you may determine a narrower range for the mean value by treating precision errors like bias errors and estimating a standard deviation based on your knowledge of the instruments. For example, random variations in test conditions may cause a DMM reading to fluctuate; but if the reading is made only once, the random variation simply produces an overall range of uncertainty for the true value of this variable. The uncertainty (at 19:1 odds) is twice the standard deviation of the test condition. In other words, $\pm 1.96\sigma \approx \pm 2\sigma$ will cover 95% (or 19:20) of the readings made. If, on the other hand, the same measurement is made several times, the random variations can be averaged out, and statistics can be used to place a narrower bound on the mean value.

We can construct the single-sample estimate a little more formally. We begin by estimating σ as a value (σ_e, say) that is based on our knowledge of the experimental system. Thus, we are assuming that σ of the population is *known*, and, with Eq. (3.18), the precision uncertainty can be estimated as

$$P_x \approx z_{c/2} \frac{\sigma_e}{\sqrt{n}}. \tag{3.26}$$

With a single reading, $n = 1$, so that no averaging is performed to reduce the uncertainty. Taking a 95% confidence level, $z_{0.95/2} = 1.96$ and

$$P_x \approx 1.96\sigma_e \quad (95\%). \tag{3.27}$$

The potential precision error underlying a single measurement of a random variable can usually be estimated from your knowledge of how finely an instrument will resolve, of how precisely an instrument may repeat a reading, or of how much the test condition fluctuates. Often, these estimates can be made in advance of performing the experiment, in order to gauge the expected uncertainty in the result.

Section 3.11 includes an example of single-sample uncertainty analysis.

3.10 Propagation of Uncertainty

Often several quantities are measured, and the results of those measurements are used to calculate a desired result. For example, experimental values of density are usually determined by dividing the measured mass of a sample by the measured volume of that sample. Each measurement includes some uncertainty, and these uncertainties will create an uncertainty in the calculated result. What is that uncertainty?

Finding the uncertainty in a result due to uncertainties in the independent variables is called finding the *propagation of uncertainty*. For uncertainties in the independent variables, the procedure rests on a statistical theorem that is exact for a linear function y of several independent variables x_i with standard deviations σ_i; the theorem states that the standard deviation of y is

$$\sigma_y = \sqrt{\left(\frac{\partial y}{\partial x_1}\sigma_1\right)^2 + \left(\frac{\partial y}{\partial x_2}\sigma_2\right)^2 + \cdots + \left(\frac{\partial y}{\partial x_n}\sigma_n\right)^2}. \tag{3.28}$$

Likewise, a calculated result y is a function of several independent measured variables, $\{x_1, x_2, \ldots, x_n\}$; for example, density is a function of mass and volume. Each measured value has some uncertainty, $\{u_1, u_2, \ldots, u_n\}$, and these uncertainties lead to an uncertainty in y, which we call u_y. To estimate u_y, we assume that each uncertainty is small enough that a first-order Taylor expansion of $y(x_1, x_2, \ldots, x_n)$ provides a reasonable approximation:

$$y(x_1 + u_1, x_2 + u_2, \ldots, x_n + u_n)$$

$$\approx y(x_1, x_2, \ldots, x_n) + \frac{\partial y}{\partial x_1} u_1 + \frac{\partial y}{\partial x_2} u_2 + \cdots + \frac{\partial y}{\partial x_n} u_n. \quad \textbf{(3.29)}$$

Under this approximation, y is a linear function of the independent variables. Now we can apply the theorem, assuming that uncertainties will behave much like standard deviations:

$$u_y = \sqrt{\left(\frac{\partial y}{\partial x_1} u_1\right)^2 + \left(\frac{\partial y}{\partial x_2} u_2\right)^2 + \cdots + \left(\frac{\partial y}{\partial x_n} u_n\right)^2} \quad (n:1). \quad \textbf{(3.30)}$$

Here, all uncertainties must have the same odds and must be independent of each other. This approach was established by S. J. Kline and F. A. McClintock in 1953 [6].

The uncertainties, u_i, may be either bias uncertainties or precision uncertainties. Normally, the bias uncertainties and precision uncertainties in y are propagated separately. The overall uncertainty, U_y, is then calculated by combining B_y and P_y using Eq. (3.4). Section 3.11.2 illustrates this procedure.

Equation (3.30) can be simplified when y has certain common functional forms, as illustrated in the following examples.

Example 3.9

Suppose that y has the form

$$y = Ax_1 + Bx_2$$

and that the uncertainties in x_1 and x_2 are known with odds of $n:1$. What is the uncertainty in y?

Solution.

$$\frac{\partial y}{\partial x_1} = A$$

$$\frac{\partial y}{\partial x_2} = B$$

Using Eq. (3.30),

$$u_y = \sqrt{(Au_1)^2 + (Bu_2)^2} \quad (n:1). \quad \textbf{(3.31)}$$

For additive functions, the *absolute* uncertainties in each term are combined in an rms sense.

Example 3.10

Suppose that y has the form

$$y = A \frac{x_1^m x_2^n}{x_3^k}$$

and that the uncertainties in x_1, x_2, and x_3 are known with odds of $n:1$. What is the uncertainty in y?

Solution.

$$\frac{\partial y}{\partial x_1} = mA \frac{x_1^{(m-1)} x_2^n}{x_3^k}$$

$$\frac{\partial y}{\partial x_2} = nA \frac{x_1^m x_2^{(n-1)}}{x_3^k}$$

$$\frac{\partial y}{\partial x_3} = -kA \frac{x_1^m x_2^n}{x_3^{(k+1)}}$$

Using Eq. (3.30),

$$u_y = \sqrt{\left(mA \frac{x_1^{(m-1)} x_2^n}{x_3^k} u_1 \right)^2 + \left(nA \frac{x_1^m x_2^{(n-1)}}{x_3^k} u_2 \right)^2 + \left(-kA \frac{x_1^m x_2^n}{x_3^{(k+1)}} u_3 \right)^2}$$

$$(n:1),$$

so that, for this case,

$$\frac{u_y}{y} = \sqrt{\left(m \frac{u_1}{x_1} \right)^2 + \left(n \frac{u_2}{x_2} \right)^2 + \left(k \frac{u_3}{x_3} \right)^2} \qquad (n:1). \qquad (3.32)$$

For multiplicative functions, the *fractional* uncertainties are combined in an rms sense. Note carefully the weighting factors, m, n, and k in Eq. (3.32) and their sources.

Normally, each source of error is independent of the other sources. The errors will not all be of the same sign, nor will they all take on their maximum values simultaneously. For that reason, Eq. (3.30) combines the uncertainties in a root-mean-square sense. In some situations, however, various sources of uncertainty are not independent. Dependent errors should be added together, before combining them in the root-mean-square sense with other independent sources of error.

3.11 Examples of Uncertainty Analysis

3.11.1 Rating Resistors

Example 3.11

Carbon resistors are painted with color-coded bands that specify their nominal resistance. The actual resistance of each resistor varies randomly about the nominal value, owing to manufacturing variations. The percentage variation in the resistance of the population of resistors is referred to as the *tolerance* of the resistors. For commercial carbon resistors, this variation is 5, 10, or 20%.

A lab technician has just received a box of 2000 resistors. As a result of a production error, the color-coded bands have not been painted on this lot. To determine the nominal resistance and tolerance, the technician selects ten resistors and measures their resistances with a digital multimeter. His results are as tabulated.

Number	Resistance (kΩ)
1	18.12
2	17.95
3	18.17
4	18.45
5	16.24
6	17.82
7	16.28
8	16.32
9	17.91
10	15.98

What is the nominal value of the resistors? What is the uncertainty in that value? Consider both precision and bias uncertainty. Can you estimate the tolerance?

Solution. The precision error in the resistors can be averaged to find a 95% confidence interval. From the tabulated data,

$$\bar{R} = 17.32 \text{ k}\Omega$$

$$S_R = 0.982 \text{ k}\Omega$$

The mean resistance, $\bar{R} = 17.32$ kΩ, is clearly the apparent nominal value of the resistors.

To find the uncertainty in this mean value, both the precision and the bias uncertainties must be estimated. Consider the precision uncertainty first. From Table 3.6, with $\nu = 10 - 1 = 9$ and $\alpha = (1 - 0.95)/2 = 0.025$,

$$t_{\alpha,\nu} = t_{0.025,9} = 2.262$$

Applying Eq. (3.23), the (unbiased) population mean, μ_R, is in the range

$$\mu_R = \bar{R} \pm t_{\alpha,\nu}\frac{S_R}{\sqrt{n}}$$

$$= 17.32 \pm 2.262\frac{0.982}{\sqrt{10}} \text{ k}\Omega$$

$$= 17.32 \pm 0.70 \text{ k}\Omega \qquad (95\%)$$

However, this answer accounts for *only* precision error, specifically, $P_R = 0.70 \text{ k}\Omega$. What is the *bias* uncertainty in this result?

The manual for the DMM describes its calibration; possible bias uncertainty (from temperature drift, connecting-lead resistances, and other sources) is rated as

$$\pm (0.5\% \text{ of reading} + 0.05\% \text{ of full scale} + 0.2 \,\Omega)$$

The confidence is not given, but we shall assume it to be 95%. The full-scale reading of the meter is $20 \text{ k}\Omega$, and after evaluating the terms and summing, the meter's bias uncertainty can be estimated as

$$B_R = \pm 96.8 \,\Omega = \pm 0.10 \text{ k}\Omega \qquad (95\%)$$

Notice that the reading error in the DMM scale is only $0.005 \text{ k}\Omega$, which is *much* lower than the actual uncertainty in the DMM reading! This DMM has relatively high resolution and precision but much lower accuracy.

The total uncertainty in the mean of the population is, from Eq. (3.4),

$$U_R = (B_R^2 + P_R^2)^{1/2}$$

$$= [(0.10)^2 + (0.70)^2]^{1/2} \text{ k}\Omega$$

$$= 0.71 \text{ k}\Omega \quad (95\%)$$

The uncertainty of the nominal value is $U_R = 0.71 \text{ k}\Omega$ (95%), or about $\pm 4\%$.

The precision uncertainty in the mean is the major source of uncertainty. On the other hand, if a sample of 1000 resistors were used, the precision uncertainty would be reduced by a factor of ten (why?), and the bias uncertainty would be dominant.

The tolerance of the resistors remains to be found. What we'd like is an estimate of the percentage deviation from the nominal value which includes, say, 95% of the resistors. One approach is to note that 95% of a Gaussian population lies within $\pm 1.96\sigma$ of the population mean μ (see Table 3.1). On that basis, we could approximate $\sigma_R \approx S_R$ and $\mu_R \approx \bar{R}$, so that

$$\text{Tolerance } \% = \frac{1.96\sigma_R}{\mu_R} \approx \frac{1.96 S_R}{\bar{R}} = \frac{1.96 \cdot 0.982}{17.32} = 0.111,$$

i.e., a tolerance of 11% (or about 10%, since that's the nearest production tolerance).

In a manufacturing situation, engineers are usually more interested in estimating an *interval* that is $c\%$ certain to contain *at least* some percentage b of the population. For example, the manufacturer might wish to report, with $c = 95\%$ confidence, that $b = 95\%$ of resistors will have resistances within some specific range of resistances. As it turns out, the approach used in the preceding paragraph is a very poor way to estimate such *tolerance limits*, because it ignores the inaccuracy of S_R and $\bar{R}$ as estimates of the population's σ_R and μ_R. Although 95% of a Gaussian population lies in the interval $\mu_R \pm 1.96\sigma_R$, that is not true of the interval $\bar{R} \pm 1.96 S_R$. For example, our estimate of the mean has a 4% precision uncertainty; this means that the interval likely to contain 95% of the population should be broadened by something like an additional $\pm 4\%$ of $\bar{R}$ beyond $\bar{R} \pm 1.96 S_R$. A proper estimate of tolerance must allow for this uncertainty as well as that in σ_R. More advanced statistical methods [3] show that, at a confidence of 95%, the 95% tolerance interval for the population is almost *twice* as large as that estimated above (i.e., the interval that is 95% certain to contain at least $\mu \pm 1.96\sigma$ turns out to be $\bar{R} \pm 3.532 S_R$).

After this extended discussion, it may interest you to learn that the resistors actually tested were nominally $18\,k\Omega$ with a tolerance of 10%.

3.11.2 Expected Uncertainty for Flowmeter Calibration

Example 3.12
Obstruction meters such as venturis and orifice plates are commonly used to measure the steady flow rates of fluids (see Section 15.3). Tables of empirical calibration coefficients, K, are available for use in theoretically based relationships such as Eq. (15.5e):

$$Q = KA_2 \sqrt{2g_c(P_1 - P_2)/\rho} \tag{3.33}$$

This technique provides a means of measuring flow rate in terms of the pressure drop across the obstruction. The published coefficients will yield

Figure 3.16 Setup for calibrating an orifice

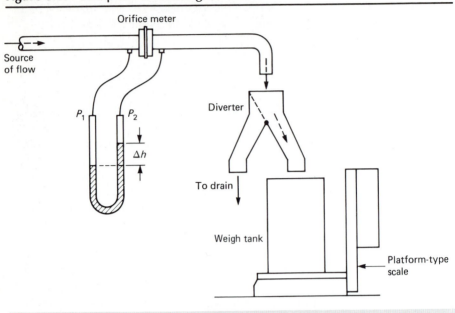

approximate flow rates, but accurate measurement requires careful experimental determination of the coefficient for each specific installation.

Figure 3.16 shows a proposed arrangement for calibration of a thin-plate orifice meter. In this case, calibration consists of experimentally determining the coefficient K in Eq. (3.33) by collecting the flowing fluid (water, in this case) in a weigh tank for some period of time. During the calibration period, the flow rate is held as constant as possible, and the pressure difference, $\Delta P = P_1 - P_2$, is recorded. The flow rate, Q, is the measured weight, W, divided by the liquid density, ρ, and the elapsed time, t ($Q = W/\rho t$); the area, A_2, is $\pi D^2/4$, for D the orifice diameter.

Substituting these values into Eq. 3.33, we may solve for the calibration constant, K:

$$K = \frac{4W}{\pi D^2 t} \sqrt{\frac{1}{2\rho\Delta P}} \qquad (3.34)$$

By inserting the observed values of W, t, and ΔP, along with a measured value of D and tabulated data for ρ, the experimental value of K is obtained.

Before undertaking this experiment, we'd like to estimate the expected accuracy of the result. We could, of course, make the following calculations *after* the experiment, but by doing it ahead of time, we can identify those parts of the experiment that contribute most of the uncertainty and, if necessary, improve them.

Solution. Both bias and precision error should be considered. However, since we have no samples for statistical analysis, we can only estimate the expected size of potential precision errors, using estimates for the standard deviations. Thus this analysis is a *single-sample* uncertainty estimate. To proceed, we first estimate the single-sample precision and bias uncertainty in each measured variable and then propagate these uncertainties into K.

The weight, W, is measured with a platform scale. What bias uncertainty exists in the scale measurement? Has the scale's calibration been checked? How recently and against what standard? (See Section 13.1 for further discussion of this point.) Presumably, we have made some sort of check, at a minimum several point calibrations using reliable proof weights. If not, the user's manual should include such data. Let us say that, in our judgment, an uncertainty of $\pm 1\%$ is justified, with a confidence of about 95%. In practice, the bias uncertainty is undoubtedly dependent to some degree on the magnitude of the weight measured relative to the scale range; this effect could also be taken into account if warranted.

Precision uncertainty in the weight measurement will be caused by reading error in the scale, and possibly by hysteresis, friction, or backlash in the scale mechanism. The size, or standard deviation, of the reading error will depend on how finely the scale is graduated; perhaps 0.1% error covers one standard deviation in the scale reading (so that $\pm 0.1\%$ covers 68% of the reading errors). Hysteresis, friction, and backlash should be negligible if the scale is in good condition; however, if these effects are observed, their errors should also be taken into account.

The diameter of the orifice must be determined. Assume that it is a sharp-edged orifice and that we have checked it with an inside micrometer. Did we check the micrometer against gauge blocks, or are we accepting its scale as is? How experienced are we in using a micrometer? We intend to measure the orifice only once, using this value in all subsequent applications of the meter, so the diameter's uncertainty all appears as bias in the results. After these considerations, we estimate a 95% bias uncertainty of ± 0.008 cm in the nominal 4-cm diameter ($\pm 0.2\%$ uncertainty).

How accurate is the determination of the time period, t? If a hand-operated diverter is used, precisely when did the flow start and stop? That is, how well is the time period *defined*? If a stopwatch is used, how good is the synchronization between the diverter and watch actions?

Note that these time uncertainties are essentially *precision* errors, which are likely to vary from run to run. If we carefully make a series of repeated runs of this particular procedure, we could accumulate enough data to perform a precision uncertainty analysis, were such accuracy required. In this case, we simply estimate the likely standard deviation to be ±3 s; if the total time period is 5 min, the standard deviation of t is ±1%.

Bias error in the time determination may arise if the watch is fast or slow (probably a *very* small error in 5 min) or if we systematically stop or start the watch too soon. Without other information, we'll assume that bias in the time reading is negligible compared to the precision error already discussed.

The density will be read from a handbook table at the temperature of the experiments. Between 0°C and 38°C, the density of water decreases by about 0.7%. Temperature may vary slightly between each experiment, leading to a precision error if only one value of density is used. However, if we use a thermocouple to measure the correct temperature for each experiment, precision error will still arise from the reading error of the temperature measurement. If standard deviation in temperature is ±0.1°C, then the corresponding density variation is only 0.002%.

On the other hand, the bias error in the temperature reading may be higher, perhaps ±1°C, and the resulting bias uncertainty in the density used would then be ±0.02% (again at 95%). If we don't bother to measure the temperature, the bias uncertainty would be larger still, maybe 0.2% if we just assume a standard room temperature of 20°C.

The value of ΔP is measured using a manometer (See Section 14.5). The dominant uncertainty is that in reading the difference in the heights of the manometer columns, essentially a precision uncertainty resulting from reading error. This uncertainty has relatively constant size, independent of the magnitude of ΔP, so that a percentage uncertainty is a bit misleading. At small ΔP, it may be 3%, but at high ΔP it may only be 0.1%. To keep the uncertainty analysis simple, let's take a representative value of ±1% for the standard deviation in the pressure. Bias uncertainty in the manometer turns out to be substantially smaller, about 0.1% at about 95% confidence.

Summarizing our results gives the following.

Variable	Bias Uncertainty, B_x/x (95%)	Standard Deviation, σ_x/x
Weight, W	1%	0.1%
Diameter, D	0.2%	≈0
Time, t	≈0	1.0%
Density, ρ	0.02%*, 0.2%†	0.002%
Pressure, ΔP	0.1%	1%

* If temperature is measured.

† If temperature is not measured.

Now we can apply Eq. (3.32) to K:

$$\frac{u_K}{K} = \left[\left(\frac{u_W}{W}\right)^2 + \left(2\frac{u_D}{D}\right)^2 + \left(\frac{u_t}{t}\right)^2 + \left(\frac{1}{2}\frac{u_\rho}{\rho}\right)^2 + \left(\frac{1}{2}\frac{u_{\Delta P}}{\Delta P}\right)^2\right]^{1/2}$$

First, we calculate the bias uncertainty in K:

$$\frac{B_K}{K} = \left[\left(\frac{B_W}{W}\right)^2 + \left(2\frac{B_D}{D}\right)^2 + \left(\frac{B_t}{t}\right)^2 + \left(\frac{1}{2}\frac{B_\rho}{\rho}\right)^2 + \left(\frac{1}{2}\frac{B_{\Delta P}}{\Delta P}\right)^2\right]^{1/2}$$

$$= \left[(1)^2 + (2 \times 0.2)^2 + (0)^2 + \left(\frac{1}{2} \times 0.02\right)^2 + \left(\frac{1}{2} \times 0.1\right)^2\right]^{1/2}$$

$$= 1.08\% \qquad (95\%)$$

Likewise, the standard deviation of K is

$$\frac{\sigma_K}{K} = \left[\left(\frac{\sigma_W}{W}\right)^2 + \left(2\frac{\sigma_D}{D}\right)^2 + \left(\frac{\sigma_t}{t}\right)^2 + \left(\frac{1}{2}\frac{\sigma_\rho}{\rho}\right)^2 + \left(\frac{1}{2}\frac{\sigma_{\Delta P}}{\Delta P}\right)^2\right]^{1/2}$$

$$= \left[(0.1)^2 + (2 \times 0)^2 + (1)^2 + \left(\frac{1}{2} \times 0.002\right)^2 + \left(\frac{1}{2} \times 1\right)^2\right]^{1/2}$$

$$= 1.12\%$$

Our estimate for the single-sample precision uncertainty in K is, from Eq. (3.27),

$$\frac{P_K}{K} = \frac{1.96\sigma_K}{K} = 1.96(0.0112) = 2.20\% \qquad (95\%).$$

From Eq. (3.4), the total uncertainty in K is

$$\frac{U_K}{K} = \left[\left(\frac{B_K}{K}\right)^2 + \left(\frac{P_K}{K}\right)^2\right]^{1/2}$$

$$= [(0.0108)^2 + (0.0220)^2]^{1/2}$$

$$= 2.45\% \qquad (95\%)$$

Inspection of these results quickly reveals the parameters having the greatest contribution to the uncertainty. Improvement of the timing and weighing procedures would improve the results the most. Improvement of the pressure and diameter measurements would contribute significantly less improvement.

Most of the total uncertainty is caused by precision uncertainty. We can reduce that uncertainty considerably by repeating the calibration experiment several times and averaging the results. Since P_K will decrease as $\sqrt{n}$, taking $n = 4$ experiments will reduce P_K to about 1.0%.

Note that the density contributes almost nothing to the total uncertainty. Even if we don't bother to measure the temperature, the contribution of density uncertainty remains negligible; i.e., $(\frac{1}{2} \cdot 0.2)^2 \ll (1)^2$. We conclude that careful temperature measurement, in this case, would be a waste of effort.

Are the various uncertainty estimates simply good guesses? To a degree, they are, but dismissing them as nothing *but* guesses would be flippant. The specific considerations leading to each estimate were not arbitrary; when properly made, such "guesses" have a strong foundation in the actual performance of the equipment and the method of taking the data. Even the estimated confidence percentages (usually 95%) are a quantitative assessment of our expectation for the variability of the data, although they *are* essentially just educated guesses. But if we admit to guessing, why not simply guess the overall uncertainty and skip all the intermediate steps? In answer, the detailed analysis provides a means for evaluating the relative effect of each identifiable source of error, thereby separating the more important ones from the less important ones. Furthermore, one can evaluate the uncertainties of each of the individual variables with considerably more assurance than one could judge the total.

3.12 Minimizing Error in Designing Experiments

The best time to minimize experimental error is in the design stage, when an experimental procedure is being developed. First and foremost, one should perform a single-sample uncertainty analysis of the proposed experimental arrangement *prior* to beginning construction, in order to determine whether the expected uncertainty is acceptably small and to identify the major sources of uncertainty. Some general precautions to observe when designing an experiment are as follows:

1. Avoid approaches that require two large numbers to be measured in order to determine the small difference between them. For example, large uncertainty is likely when measuring $\delta = (x_1 - x_2)$ if $\delta \ll x_1$, unless x_1 can be measured with great accuracy.

2. Design experiments or sensors that amplify the signal strength in order to improve sensitivity. For example, a thermopile uses several thermocouples to resolve a single temperature, and a strain gauge uses many loops of wire to measure a single strain.

3. Build "null designs," in which the output is measured as a change from zero rather than as a change in a nonzero value. This reduces both bias and precision error. Such designs often make the output proportional to the difference of two sensors. An excellent example of this approach is the Wheatstone bridge circuit (See Sections 7.8, 7.9).

4. Avoid experiments in which large correction factors must be applied as part of the data-reduction procedure.

5. Attempt to minimize the influence of the measuring system on the measured variable.

6. Calibrate entire systems, rather than individual components, in order to minimize calibration-related bias errors.

3.13   The Chi-Squared (χ^2) Distribution

The χ^2 distribution (Fig. 3.10, Table 3.5) is the basis of another *goodness-of-fit* test. It may be used to test the probability that a sample of data is or is not normally distributed. In addition the test may be used for obtaining a measure of agreement between non-Gaussian samples, either of like or of unlike sizes.

The χ^2-distribution can be defined as

$$\chi^2 = \sum_{i=1}^{N} \frac{(O_i - E_i)^2}{E_i}, \qquad (3.35)$$

where

O_i = the number of observed data in a bin,

E_i = the number of expected data in a bin,

N = the number of data bins, or "class intervals."

The theoretical basis for this relationship must be left to more specialized texts.

The χ^2-distribution takes on various shapes depending on the relative values of the variables (Fig. 3.10). Table 3.5 shows the probabilities for values of χ^2 for various degrees of freedom, v.

From Fig. 3.10 note the following characteristics of the χ^2-distribution:

1. As with the normal and t-distributions, the total area under the curve is unity.

2. The curve starts at $\chi^2 = 0$.

3. The curve is not symmetrical; however, as v (hence the number of data) increases, the shape becomes similar to that of the normal distribution curve.

Implementation of the method requires considerable data manipulation and, as with other methods, a judgment on the part of the experimenter. In addition, the method does not lend itself well to small samples of data. The usual practice is to divide the test data into a reasonable number of *class intervals*, or *bins*, to determine the number of observations O_i in each interval, and then to compare these numbers with an *expected* number E_i of data items. The expected numbers are based on a "standard," the source of which depends on the objective of the test. If the test is to determine the normalcy of test data, then the standard is the normal probability distribution. On the other hand, the standard may simply be another set of data that, in terms of an objective, is considered satisfactory; for example, how well do test data fit a standard norm?

Definite limitations apply:

1. The original, *experimentally determined* values of O_i and E_i must be numerical counts. They are *integer frequencies*; fractional events do not occur.

2. Frequency values for O_i and E_i in each bin should be equal to or greater than to unity. There should be no unoccupied bins.

3. The use of χ^2 is usually questioned if 20% of the values in either the O_i or the E_i cells or bins have counts less than 5. Often the cells or bins can be redefined to avoid the problem.

After the value of χ^2 has been calculated, how is the result interpreted? The curve accompanying Table 3.5 defines the significance level α. We see that α and $(1 - \alpha)$ have the same meanings as for the t-distribution (Section 3.6.3). Figure 3.17 is a plot of χ^2 versus ν for two arbitrary values of α—namely, $\alpha = 0.05$ and $\alpha = 0.95$. (Inspection of the values in Table 3.5 indicates a basis for the selections.) Figure 3.17 is divided into three areas, designated I, II, and III. As indicated on the figure, values falling in area I indicate extremely good fits. [Note that, from Eq. (3.35), $\chi^2 = 0$ for an exact fit.] In this area the fit is so unusually good as to cast suspicion on the randomness of the data.

Figure 3.17 χ^2 versus ν plots for $\alpha = 0.05$ and $\alpha = 0.95$

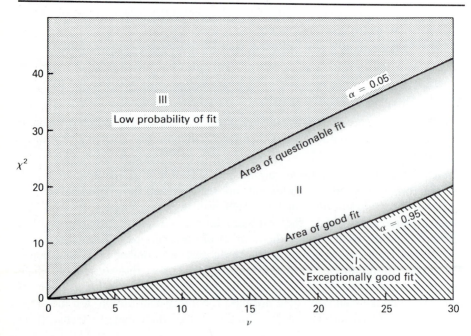

Area III includes large values of χ^2, corresponding to large discrepancies between O_i and E_i. When χ^2 falls in this region, lack of fit is indicated. This leaves area II, the region in which judgment must be applied. If the calculated value of χ^2 falls in this area, especially if the value tends toward region I, then there is reasonable evidence that samples being tested are comparable. Values tending toward region III indicate a questionable fit.

In the following sections we will consider three applications of the χ^2-distribution.

1. A test for a sample's goodness of fit to the normal distribution—that is, how well does a given sample adhere to normalcy? (Note that the χ^2-test is best for large samples; in contrast, the t-test is applied to small samples.)

2. A simple application making a comparison between two samples from related Gaussian or non-Gaussian data.

3. A test of the comparison between two samples of unequal size from a normally distributed population, or from related non-Gaussian data.

3.13.1 Testing for Sample Normalcy

The χ^2 distribution may be used to judge how well a given sample matches a normal distribution. The following example demonstrates this application.

Example 3.13

Let us apply the χ^2-test to the pressure data listed in Table 3.7.

Solution. Column B lists the frequencies of occurrence in each of the corresponding pressure intervals (column A). These are the *observed* data, O_i. To calculate χ^2 we must now determine the values for the *expected* frequencies E_i.

At this point we observe several important aspects of the data. First, one data point falls outside the usual 3σ limit (i.e., it is an outlier); thus we will ignore that point. Moreover, the total number of intervals in which the observed data O_i has a frequency count less than 5, is 3. Since 8 is the total number of intervals being compared, 20% of 8 equals 1.6 (rounded to 2). According to the restrictions for the usage of χ^2 stated in this section, we have too many intervals with frequency counts that are too small. This problem can be corrected by redefining the intervals, as shown in Table 3.8, so that $N = 4$.

Inasmuch as we now have 99 instead of 100 test points, we will recalculate the mean and the sample standard deviation, yielding

$$\bar{p} = 4.007 \, \text{MPa} \approx \mu$$

and

$$S_p = 0.013 \, \text{MPa} \approx \sigma.$$

Table 3.8 Revised intervals for Example 3.13

Interval Limits	O_i	E_i	$(O_i - E_i)^2 / E_i$
3.967			
	4	6	0.67
3.987			
	37	43	0.84
4.007			
	50	43	1.14
4.027			
	8	6	0.67
4.047			
			$\chi^2 = 3.32$

As indicated, at this point these are also our best estimates of population values.

Using the estimates of the population mean and standard deviation, we can now calculate z, which guides us to a value in Table 3.2 from which we can determine normal distributions corresponding to these parameters. This provides our best estimate of E_i for each interval. Note that we have rounded the values for E_i to avoid nonintegral frequencies.

To assign a number to the degrees of freedom we must subtract 3 from the number of bins. This is because we have already manipulated the data three times—in creating the bins and in calculating the mean and the standard deviation. Hence $v = 4 - 3 = 1$. Entering Table 3.5 for $\chi^2 = 3.32$ and $v = 1$, we see that α falls in region II in Fig. 3.17, tending toward region I. Hence we can judge that the data show a reasonable probability of being normally distributed.

3.13.2 Significance of Differences Between Samples from the Same Population

Quite often the χ^2-test is applied to make a comparison between two samples from related *non-Gaussian* data. In this section we consider a simple application where both samples have the same size. In the following section the two samples are of different sizes.

Example 3.14

A hardware manufacturer produces 10-mm dowel pins and knows that there will be variation in the diameters of the pins. She also knows that as production progresses, tools will dull and variations in pin sizes will result. Periodically, she must compare samples from production with a standard. From the results she will know when tools must be changed and equipment refurbished.

Table 3.9 Data for Example 3.14

Ranges of Diameters	Expected Frequency E_i (Standard)	Observed Frequency O_i (Test)	$\dfrac{(O_i - E_i)^2}{E_i}$
9.800–9.899	8	6	0.50
9.900–9.999	10	6	1.60
10.000–10.099	19	20	0.05
10.100–10.199	8	9	0.00
10.200–10.299	6	10	2.67
Total	51	51	$\chi^2 = 4.82$

Solution. The E_i column in Table 3.9 lists acceptable frequencies in each of five class intervals. These data are taken as a *standard*. The O_i column lists the frequencies of a sample. From these data the value of χ^2 is found to be 4.82. This statistic will be used to determine how well or how poorly the test sample compares with the standard. If the samples are similar, the value of χ^2 should be a small number. In this case, $v = 5 - 1 = 4$ because there are five bins and the data have already been used for separation into bins. From Table 3.5, for $v = 4$ and for a significance level $\alpha = 0.05$, we find that $\chi^2 = 9.488$. Since our calculated value is less than this result, we can conclude that the two samples are comparable. Hence we may conclude that, on the basis of the sample, the production remains within tolerances.

3.13.3 Comparison of Samples of Different Sizes

Example 3.15
Machinery is being designed for bagging granular polystyrene. During the developmental stage, tests are run to study the uniformity with which the bags are filled. During test runs, bags are individually weighed, with a resolution of 0.1 lbm, using carefully calibrated weighing scales. A particular run of 110 bags is selected to provide what is considered as a satisfactory "standard" against which succeeding runs may be compared. As design alterations are made in the equipment, new runs are made, and the results are compared with the standard. Table 3.10 shows the weight distributions for the standard of 110 bags and a test run of 52 bags. How well does the distribution of test weights compare with the standard?

Solution. Since the simple is of a different size than the standard, we will normalize the standard by proportioning the total of 110 data points to an

Table 3.10   Data for Example 3.15

	Standard		Sample	
Weight per Bag	*No. of Bags*	*Gross Weight*	*No. of Bags*	*Gross Weight*
40.3	3	120.9	2	80.6
40.4	4	161.6	3	121.2
40.5	4	162.0	3	121.5
40.6	5	203.0	4	162.4
40.7	10	407.0	4	162.8
40.8	19	775.2	5	204.0
40.9	16	654.4	4	163.6
41.0	12	492.0	4	164.0
41.1	10	411.0	3	123.3
41.2	8	329.6	5	206.0
41.3	6	247.8	3	123.9
41.4	4	165.6	4	165.6
41.5	4	166.0	3	124.5
41.6	3	124.8	3	124.8
41.7	2	83.4	2	83.4
Totals:	110	4504.3	52	2131.6
Avg. wt./bag:		40.948		40.992

"equivalent," 52. Column C in Table 3.11 lists the fractional proportions falling in each weight interval. Column E lists the proportionate number of bags (rounded), based on 52.

We have used the data in segregating the total into the intervals; hence, $v = 8 - 1 = 7$. Referring to Table 3.5 and Figure 3.17 we see that for seven degrees of freedom the probability is relatively high that the sample is a poor or questionable representation of the standard.

3.14   Graphical Presentation of Data

According to the American Standards Association [7]:

> When used to present facts, interpretations of facts, or theoretical relationships, a graph usually serves to communicate knowledge from the author to his readers, and to help them visualize the features that he considers important.
>
> A graph should be used when it will convey information and portray significant features more efficiently than words or tabulations.

A graph is nearly always the most effective format for conveying the interrelation of experimental variables. Graphs are invaluable in constructing acceptable curve fits of experimental data and in identifying outliers or

Table 3.11 Grouped data for Example 3.15

| A | Standard | | | Sample | |
| | B | C | D | E | F |
Weight Ranges in Each Cell	No. of Observed Values in Each Range	Proportion of Total*	No. of Observed Values in Each Range (O_i)	No. of Expected Values in Each Range (E_i)†	x^2
40.25–40.45	7	0.0636	5	3	1.33
40.45–40.65	9	0.0818	7	4	2.25
40.65–40.85	29	0.2636	9	14	1.79
40.85–41.05	28	0.2545	8	13	1.92
41.05–41.25	18	0.1636	8	9	0.11
41.25–41.45	10	0.0909	7	5	0.80
41.45–41.65	7	0.0636	6	3	3.00
41.65–41.85	2	0.0182	2	1	1.00
Totals:	110	1.0000	52	52	12.20

* Value in column C = value in column B ÷ 110.
† Value in column E = value in column C × 52.

Table 3.12   Atmospheric pressure during Hurricane Bob

Time of Day	Pressure (mbar)
10:00 A.M.	1009.0
11:30	984.2
1:00 P.M.	999.8
2:15	989.0
3:40	977.1
4:40	981.2
5:40	990.0

erroneous measurements. Graphs are also useful in testing theoretical calculations against real experimental results and in identifying the conditions under
which a theoretical model fails.

The clarity imparted by graphing data can be substantial. Table 3.12 shows
the atmospheric pressure measured in Cambridge, Massachusetts, during
Hurricane Bob on August 19, 1991. In tabular form, the trend is unclear. If,
on the other hand, the data are graphed (Fig. 3.18), the progress of the storm
is apparent. Atmospheric pressure declined steadily until about 4:00 P.M. and
then began to increase. The 11:30 A.M. reading lies well below the trend
defined by the other data, and it can probably be assumed to be in error (some
checking of the data reductions verified this conclusion). A faired curve has

Figure 3.18   Atmospheric pressure during Hurricane Bob

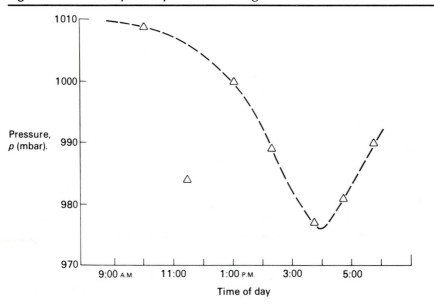

been empirically sketched through the remaining data.* This curve may be used to estimate the pressure at times other than those measured.

3.14.1 General Rules for Making Graphs

By observing the following guidelines, you can help to ensure that your graph will be easy for your readers to understand. Figure 3.19(a) and (b) illustrates some of these points [7].

1. The graph should be designed to require minimum effort from the reader in understanding and interpreting the information it conveys.
2. The axes should have clear labels that name the quantity plotted, its units, and its symbol if one is in use.
3. Axes should be clearly numbered and should have tick marks for significant numerical divisions. Typically, ticks should appear in increments of 1, 2, or 5 units of measurement multiplied or divided by factors of ten $(1, 10, 100, \ldots)$. Not every tick needs to be numbered; in fact, using too

Figure 3.19 (a) A poor graph

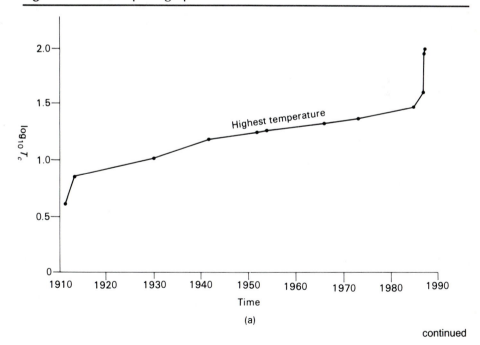

continued

* *Faired* in this sense means smooth and without irregularity: a draftsman's curve through the data.

Figure 3.19   (b) Graph of Figure 3.19(a) improved by following graphing guidelines numbers 2, 3, 5, 8, 10, and 12

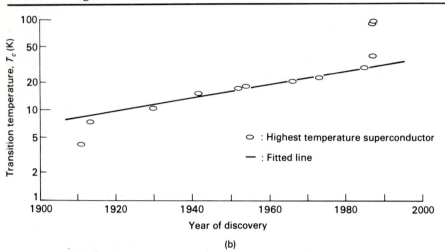

(b)

many numbers will just clutter the axes. Tick marks should be directed toward the *interior* of the figure.

4. Use scientific notation to avoid placing too many digits on the graph. For example, use 50×10^3 rather than 50,000. A particular power of ten need appear only once along each axis; avoid confusing labels such as "Pressure, Pa $\times 10^5$."

5. When plotting on semilog or full-log coordinates, use real logarithmic axes; do *not* plot the logarithm itself (e.g., plot 50, not 1.70).* Logarithmic scales should have tick marks at powers of ten and intermediate values, such as 10, 20, 50, 100, 200, ...

6. The axes should usually include zero; if you wish to focus on a smaller range of data, include zero and break the axis, as shown in Fig. 3.4.

7. The choice of scales and proportions should be commensurate with the relative importance of the variations shown in the results. If variations by increments of ten are significant, the graph should not be scaled to emphasize variations by increments of one.

8. Use symbols such as $\bigcirc$, $\square$, $\triangle$, and $\diamondsuit$ for data points. Do *not* use dots (·) for data. Open symbols should be used before filled symbols. You may place a legend defining symbols on the graph (if space permits) or in the figure caption.

9. Place error bars on data points to indicate the estimated uncertainty of the measurement or else use symbols that are the same size as the range of uncertainty.

---

* An exception is made when the unit *decibels* is plotted.

10. When several curves are plotted on one graph, different lines (solid, dashed, dash-dot, . . .) should be used for each if the curves are closely spaced. The graph should include labels or a legend identifying each curve. Avoid using colors to differentiate curves, since colors are usually lost when the graph is photocopied. Theoretical curves should be plotted as lines, without showing calculated points. Curves fitted to data do not need to pass through every measurement like a dot-to-dot cartoon; however, if a data point lies far from the fitted curve, a discrepancy may be indicated [as for the first and the last three points in Fig. 3.19(b)].

11. Lettering on the graph should be held to the minimum necessary for clarity. Too much text (or too much data) creates crowding and confusion.

12. Labels on the axes and curves should be oriented to be read from the bottom or from the right. Avoid forcing the reader to rotate the figure in order to read it.

13. The graph should have a descriptive but concise title. The title should appear as a caption to the figure rather than on the graph itself.

Good graphing software can help produce graphs that adhere to these guidelines. However, some graphing packages violate even the simplest of these rules. Discretion is advised!

3.14.2 Choosing Coordinates and Producing Straight Lines

The first step in making any graph is to decide which variables to plot and on what scale to plot them. Four basic graphical scales occur frequently in engineering work (Fig. 3.20). *Linear* coordinates have a linear variation of both the x and y scales. If a variable changes by several orders of magnitude or is exponentially related to another variable, then a logarithmically scaled axis may be preferable. Graphs having one logarithmically scaled axis and one linearly scaled axis are called *semilogarithmic* (or *semilog*). Those for which both axes are logarithmic are called *full logarithmic* (*full log,* or *log-log*). When a quantity varies with an angle, *polar coordinates* provide a physically suggestive format for the data.

The choice of which scaling to use is normally guided by your expectations for the physical behavior of the system being studied. You may also attempt to deduce the right scaling by studying a test graph made on linear coordinates. Often, the objective in selecting a scaling is to find coordinates in which the plotted data fall on a straight line, because straight lines are the easiest curves to fit.

Figure 3.21(a) shows a set of data that represent the cooling of a warm metal slug suddenly immersed in cold liquid. The difference between the slug's temperature and the liquid temperature was recorded at several times after the slug was submerged. The graph has the form of an exponential decay; indeed,

Figure 3.20   (a) Linear coordinates

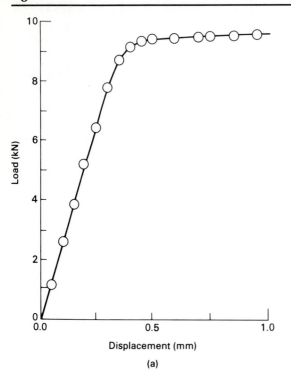

(a)

Figure 3.20   (b) Semilogarithmic coordinates

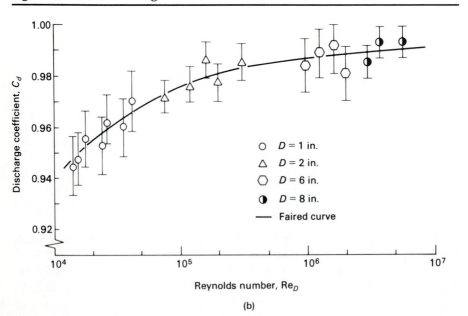

(b)

continued

Figure 3.20 (c) Full logarithmic coordinates

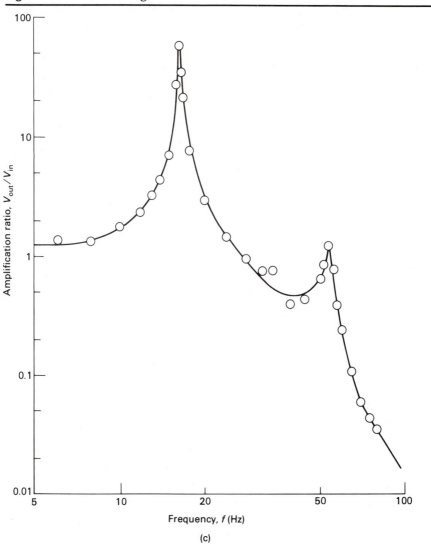

(c)

continued

heat transfer theory suggests that the cooling curve should have the form

$$\Delta T = \Delta T_0 \exp\left(-\frac{t}{\tau}\right) \tag{3.36}$$

where ΔT is the measured temperature difference at any time, ΔT_0 is the temperature difference before the slug is immersed, and τ is a time constant for the cooling. You may want to find an experimental value for τ, so that you can use Eq. (3.36) to estimate ΔT at values of t where you have no

Figure 3.20 (d) Polar coordinates

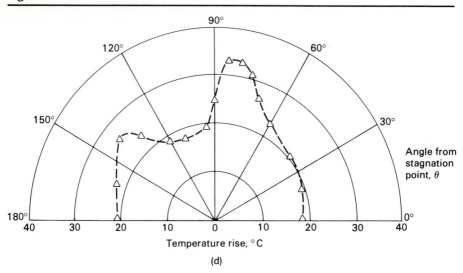

(d)

Figure 3.21 Cooling data (a) linear coordinates

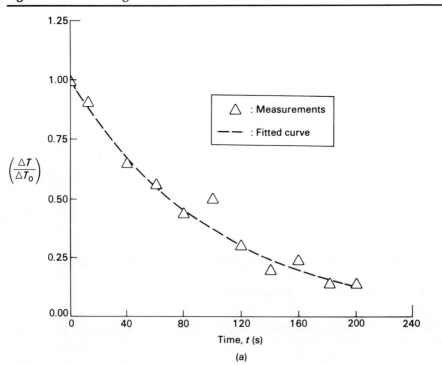

(a)

continued

Figure 3.21 (b) Cooling data, semilogarithmic coordinates. (Note logarithmic variation of $\Delta T/\Delta T_0$)

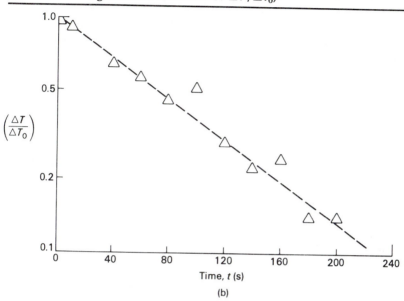

(b)

measurements. That task is not straightforward using the linear scaling of Fig. 3.21(a).

Instead, you could plot $\log \Delta T/\Delta T_0$ as a function of t. Then, the relationship between ΔT and t is

$$\log\left(\frac{\Delta T}{\Delta T_0}\right) = -\frac{0.4343t}{\tau}, \tag{3.37}$$

which is the equation of a line with slope $-0.4340/\tau$ and intercept zero.* The graph is most easily made using semilogarithmic coordinates [Figure 3.21(b)], and τ can be calculated from the slope of the line:

$$-0.4343/\tau = \frac{\log(\Delta T(t_1)/\Delta T_0) - \log(\Delta T(t_2)/\Delta T_0)}{t_1 - t_2}, \tag{3.37a}$$

resulting in $\tau = 98\,\mathrm{s}$. Note that while $\Delta T/\Delta T_0$ is plotted on the logarithmic coordinates, $\log(\Delta T/\Delta T_0)$ must be used in calculating the slope.

Semilog paper, full-log paper, and polar graph paper are available from most university bookstores or drafting suppliers. Many computer spreadsheet and plotting programs can also generate these coordinate systems. Moreover, plotting software can expedite experimentation with different scalings and coordinates, so that the most informative ones can be easily identified.

---

* Base 10 logarithms are standard in graphical work; $\log_{10} e = 0.4340$.

Figure 3.22   Plot of $y = 1.0 + (2.5/x)$ as (a) y versus
x and (b) y versus $(1/x)$

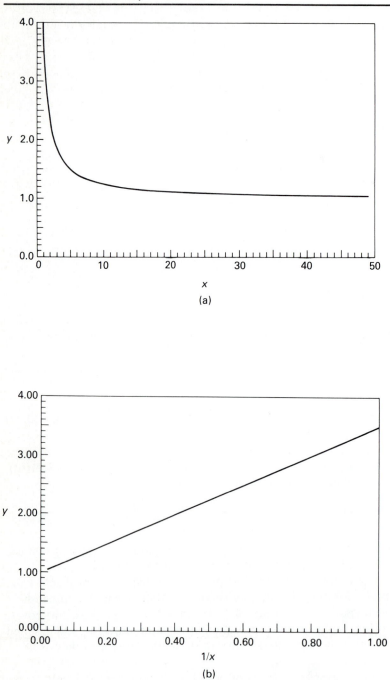

Table 3.13 Straight-line transformations: $y = f(x) \rightarrow Y = A + BX$

f(x)	Variables to Be Plotted		Straight-Line Intercept, A	Slope, B
	Y	X		
$y = a + b/x$	y	$1/x$	a	b
$y = 1/(a + bx)$ or $1/y = a + bx$	$1/y$	x	a	b
$y = x/(a + bx)$ or $x/y = a + bx$	x/y	x	a	b
$y = ab^x$	$\log y$	x	$\log a$	$\log b$
$y = ac^{bx}$	$\log y$	x	$\log a$	$b \log c$
$y = ax^b$	$\log y$	$\log x$	$\log a$	b
$y = a + bx^n$, if n is known	y	x^n	a	b

Adapted from Ref. [8].

Logarithmic scaling is only one approach to creating straight-line representations of data. For example, the function

$$y = a + \frac{b}{x} \tag{3.38}$$

does not give a straight-line variation of y with x [Fig. 3.22(a)]. However, it does give a straight-line variation of y with $1/x$. The solution is to plot y as a function of $1/x$ rather than of x itself; then a can be determined as the intercept of resultant line and b as its slope [Fig. 3.22(b)].

Table 3.13 offers a guide to straight-line transformations in which a function $y = f(x)$ is transformed to

$$Y = A + BX \tag{3.39}$$

by plotting an appropriate pair of modified variables, Y and X.

3.15 Line Fitting and the Method of Least Squares

Once the data are in a straight-line form, the correct line must be passed through them and its slope and intercept determined. The simplest approach is just to draw what *appears* to be a good straight line through the data. When this approach is used, the probable tendency is to draw a line that minimizes

Figure 3.23 Bias and precision error in line fitting

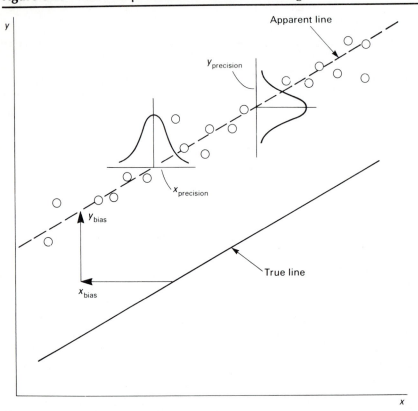

the total deviation of all points from the line. Results obtained this way are often acceptably accurate, particularly if the data set includes only a few points or if both x and y have significant error.

The data plotted may include both bias and precision errors. Bias errors will tend to shift the entire data set away from the true line or, perhaps, to change its slope. Precision errors will cause the data to scatter about the true line. Either y or x may include both precision and bias error (Fig. 3.23). The objective of a curve fit is to average out the *precision* errors by drawing a curve that follows the apparent central tendency of the scattered data. Curve fitting, like statistical analysis of precision error, does nothing to uncover or reduce bias error. Bias errors in fitted curves must be identified by other methods, such as comparison to independent data sets.

3.15.1 Least Squares for Line Fits

When the precision error in y is substantially greater than that in x, the *method of least squares*, or linear regression, enables us to calculate a line, $y = a + bx$, through the data [3]. The method identifies the slope b and intercept a

that minimize the sum of the squared deviations of the data from the fitted line, S^2:

$$S^2 = \sum_{i=1}^{n} [y_i - y(x_i)]^2 \tag{3.40}$$

Here, for the various measured values of x_i, y_i is the experimentally determined ordinate and $y(x_i) = a + bx_i$ is the corresponding value calculated from the fitted line; n is the number of experimental observations used. The result is:

$$a = \frac{\sum y_i \sum x_i^2 - \sum x_i \sum x_i y_i}{n \sum x_i^2 - (\sum x_i)^2} \tag{3.41}$$

$$b = \frac{n \sum x_i y_i - \sum x_i \sum y_i}{n \sum x_i^2 - (\sum x_i)^2} \tag{3.42}$$

Most scientific calculators incorporate programs for calculating least squares lines, a great convenience to the experimentalist. Consequently, least squares has become increasingly popular as *the* method for fitting lines. But to some degree this predominance has also promoted misuse of the method. The least squares method addresses *only* the precision error in y_i; poor results are obtained if x_i also includes large precision error. Least squares assumes, in effect, that the experimental x_i are *error-free*.

To indicate the reliability of the fit, most pocket calculators and software packages return the *correlation coefficient*, r, along with the least squares results:

$$r^2 = \frac{\text{explained squared variation about } y_m}{\text{total squared variation about } y_m}, \tag{3.43}$$

where y_m is the mean of the measured y_i:

$$y_m = \frac{1}{n} \sum_{i=1}^{n} y_i \tag{3.44}$$

The *explained* squared variation results from the straight-line change of y with x:

$$\sum_{i=1}^{n} (y(x_i) - y_m)^2$$

The *total* squared variation also includes the precision error:

$$\sum_{i=1}^{n} (y_i - y_m)^2 = \cdots = S^2 + \sum_{i=1}^{n} (y(x_i) - y_m)^2$$

Thus,

$$r^2 = \frac{\sum (y(x_i) - y_m)^2}{S^2 + \sum (y(x_i) - y_m)^2} \tag{3.45}$$

If the sum of squared deviations S^2 is assumed to result only from precision error, then a "perfect fit" occurs when $S^2 \to 0$ and $r^2 \to 1$. Hence, the nearer r is to ± 1, the "better" the fit.

Unfortunately, when the data look basically linear, one usually obtains $|r| > 0.9$; the correlation coefficient is not a very sensitive indicator of the precision of the data. It turns out that $(1 - r^2)^{1/2}$ is a better indicator of the fit's quality (Appendix G, [9]); the closer it is to zero, the lower the precision error in the data. The quantity $(1 - r^2)^{1/2}$ is roughly the ratio of the vertical standard deviation of the data about the line to the total vertical variation of the data.

When both y and x have significant precision error, least squares should not be used; this case is usually identifiable by the highly scattered appearance of the data. In addition to eyeball estimates, various other semiempirical line-fitting procedures are available in that situation, as discussed in Ref. [10]. Conversely, if both y and x have precision errors, but these errors are *small* relative to the *overall* variation of the data, then least squares results may still be acceptably accurate. In this situation, the data will still appear to fall on a straight line.

3.15.2   Uncertainty in Line Fits

Sometimes the real issue is to estimate an uncertainty for the slope or intercept of a fitted line. Eyeball estimates of the uncertainty are often acceptably accurate; for example, you may be able to vary the slope of a hand-fitted line to determine what range of slopes will still fit the data with 95% confidence. Similar estimates can be applied to finding the intercept. This approach is best either when the sample is small or when both y and x have large precision errors. Statistical confidence intervals may also be derived, as discussed in Appendix G.

Another major concern in line or curve fitting is that of identifying outliers. Points well beyond the trend of the remaining data can often be identified by eye. When a particular point is in doubt, exclude it temporarily, and fit a new line through the remaining data. If that line shows a much better fit, the point can probably be dropped. Again, within the restrictive assumptions of least squares, statistical tests can be applied; essentially, these consist of estimating a 3σ band around the fitted curve (Appendix G). Sometimes, however, data deviate from a fitted line because the actual relationship is *not* a straight line. One must always avoid forcing nonlinear data to fit an assumed straight-line form, since valuable information can be lost in the process.

3.15.3   Software for Curve-fitting

Line fitting is actually a special case of the method of least squares; least squares can be generalized to fit polynomials of any order. Many other methods of curve fitting have been developed, some considerably more sophisticated than least squares. Good discussions of such methods can be found in most texts on numerical analysis. (Reference [11] gives an introduction to these issues.)

One example of higher-order curve fitting can be found in Table 16.6 of this book, where high-order polynomials have been used to fit thermocouple voltages as functions of temperature.

Example 3.16

A cantilever beam deflects downward when a mass is attached to its free end. The deflection, δ (m), is a function of the *beam stiffness, K* (N/m), the applied mass, M (kg), and the gravitational body force, $g = 9.807$ m/s:

$$K\delta = Mg$$

To determine the stiffness of a small cantilevered steel beam, a student places various masses on the end of the beam and measures the corresponding deflections. The deflections are measured using a scale (a ruler) marked in 1 mm increments. Each mass is measured in a balance. His results are as follow:

Mass (g)	Deflection (mm)
0	0
50.15	0.6
99.90	1.8
150.05	3.0
200.05	3.6
250.20	4.8
299.95	6.0
350.05	6.2
401.00	7.5

The estimated precision uncertainty in the measured mass, largely from reading error, is ± 0.05 g (95%) and the bias uncertainty, largely from calibration uncertainty, is ± 0.1 g (95%). Thus, the overall uncertainty in the mass is $U_M = 0.11$ g (95%), corresponding to about 0.05% of a typical load.

For the deflection, reading error is the most likely cause of precision uncertainty; the student estimates this uncertainty as ± 0.2 mm (95%). The bias uncertainty in the ruler, from manufacturing error, is estimated at ± 0.1 mm (95%). The overall uncertainty in the deflection is typically 5–10% of the measured value.

What is the stiffness of the beam, and what is its uncertainty?

Solution. The stiffness can be found by taking a least squares fit through the data, using the deflection as the y variable, since it has a much greater precision uncertainty than does the mass. Setting $y = \delta$ and $x = M$, we calculate the required sums (perhaps using a calculator subroutine that processes the entered data):

$$n = 9$$

$$\sum x = 1801 \text{ g}$$

$$\sum x^2 = 5.109 \times 10^5 \text{ g}^2$$

$$\sum y = 33.50 \text{ mm}$$

$$\sum y^2 = 179.3 \text{ mm}^2$$

$$\sum xy = 9959 \text{ g} \cdot \text{mm}$$

The least squares results are then

$$y = a + bx \qquad \left(\text{or } \delta = 0 + \frac{g}{K} M \right)$$

$$a = -0.0755 \text{ mm}$$

$$b = \frac{g}{K} = 0.0190 \text{ mm/g}$$

$$r = 0.995886$$

The data and the line fit are shown in Fig. 3.24. The experimental stiffness of the beam is

$$K = \frac{g}{b} = \frac{9.807}{0.0190} = 516 \text{ N/m}.$$

From the figure, these data do appear to fall on a straight line. The correlation coefficient, r, is nearly unity, but a better test is to consider $(1 - r^2)^{1/2} = 0.0906$. This value indicates that the vertical standard deviation of the data (from precision error) is only about 9% of the total vertical variation caused by the straight-line relationship between y and x.

What is the uncertainty in the stiffness? The answer is equivalent to the uncertainty in the line's slope. One way to estimate this is just to vary the fitted line by eye to see what range of slopes fit the data with 95% certainty. These bounds are shown in Fig. 3.24, giving a variation in b of about 10%. Thus, the 95% uncertainty in K is also about ±10%. A more precise statistical interval is calculated in Appendix G, but the result is essentially the same.

Figure 3.24 Beam deflection for various masses

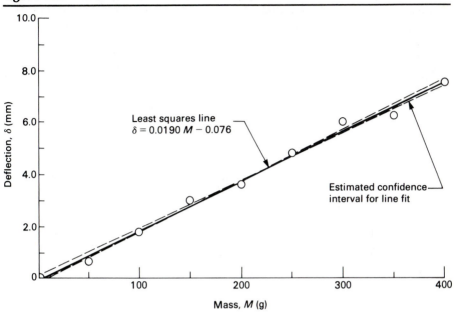

3.16 Summary

Every measurement includes some level of error, and this error can never be known exactly. However, a probable bound on the error can usually be estimated. This bound is called *uncertainty*. Uncertainty should always be accompanied by the odds (or confidence percentage) that a particular error will fall within this bound. When presenting data either graphically or numerically, the uncertainty should also be shown.

1. Errors can usually be classified as either *bias* error or *precision* error. Bias (or systematic) errors occur same way for each measurement made. Precision (or random) errors vary in size and sign with a zero average value. Bias uncertainty must be estimated from our knowledge of the measuring equipment or by comparison to other, more accurate systems. Precision uncertainty can be estimated statistically (Sections 3.1, 3.2).

2. The total uncertainty in a measurement includes both bias and precision error:

$$U_x = (B_x^2 + P_x^2)^{1/2}$$

The bias and precision uncertainty should have the same confidence level, typically 95% (Sections 3.3, 3.11).

3. Random variables, such as precision error, may be characterized in terms

of a *population* and its *distribution*. The most common distribution for precision error is the *Gaussian* or *normal* distribution (Sections 3.4, 3.5).

4. Properties of a population are estimated by taking a *sample* from it. The sample mean, $\bar{x}$, and sample standard deviation, S_x, are used to estimate the population mean and population standard deviation. The precision uncertainty in $\bar{x}$ is

$$P_x = t_{\alpha/2, v} \frac{S_x}{\sqrt{n}} \quad (c\%)$$

with $\alpha = 1 - c$ and $v = n - 1$. (Sections 3.6, 3.11).

5. Bias uncertainty and single-sample precision uncertainty are estimated from our knowledge of the measuring system. (Sections 3.9, 3.11).

6. When experimental data are used to compute a final result, the uncertainty of the data must be *propagated* to determine the uncertainty in the result:

$$u_y = \sqrt{\left(\frac{\partial y}{\partial x_1} u_1\right)^2 + \cdots + \left(\frac{\partial y}{\partial x_n} u_n\right)^2}$$

Here u_i is either a bias uncertainty, B_i, or a precision uncertainty, P_i (Sections 3.10, 3.11).

7. The accuracy of experiments can be improved *before* they are conducted by identifying and eliminating major sources of uncertainty (Sections 3.11.2, 3.12).

8. The accuracy with which a set of data fit the Gaussian distribution can be checked using either a normal probability plot (Section 3.7) or the χ^2 distribution (Section 3.13). The χ^2 distribution may also be used to compare the distribution of one sample to the distribution of another (Section 3.13).

9. Graphical presentation of data is often the most effective way to convey your results and conclusions. Applying a few simple techniques when making your graphs can help ensure that your readers will fully and easily understand your work (Section 3.14).

10. Line fits enable your results to take an analytical form. Line fits also help to average out precision errors in data. The method of least squares will yield good line-fits when: (a) the y precision error is much larger than the x precision error or (b) both x and y precision errors are small compared to the overall variation of the data. Line fits do *not* compensate for bias error (Section 3.15).

Suggested Readings

ANSI/ASME PTC 19.1-1985. ASME Performance Test Codes, Supplement on Instruments and Apparatus, Part 1, Measurement Uncertainty. New York, 1985.
Note: The above source contains a multitude of valuable references.

Barker, T. B. *Quality by Experimental Design.* New York: Marcel Dekker, 1985.

Bartee, E. M. *Engineering Experimental Design Fundamentals.* Englewood Cliffs, N.J.: Prentice Hall, 1968.

Collection of papers related to engineering measurement uncertainties. *Trans. of ASME, Jour. of Fluids Engrg.* 107, June 1985.

Note: The above source contains a multitude of valuable references.

Haugen, E. B. *Probabilistic Approaches to Design.* New York: John Wiley, 1968.

Kline, S. J., and F. A. McClintock. Describing uncertainties in single-sample experiments. *Mech. Eng.* 75: 3–8, January 1953.

Ku, H. H. *Precision Measurement and Calibration.* (Selected papers of statistical concepts), NBS Spcl. Publication 300, vol. 1. Washington, D.C.: U.S. Government Printing Office, 1969.

Lipson, Charles, and N. J. Seth. *Statistical Design and Analysis of Engineering Experiments.* New York: McGraw-Hill, 1973.

Miller, I. R., J. E. Freund, and R. Johnson. *Probability and Statistics for Engineers.* 4th ed. Englewood Cliffs, N.J.: Prentice Hall, 1990.

Moffat, R. J. Contributions to the theory of single-sample uncertainty analysis. *Trans. of ASME, Jour. of Fluids Engrg.* 104, June 1982.

Natrella, M. G. *Experimental Statistics.* National Bureau of Standards Handbook 91, U.S. Government Printing Office, 1963.

Neville, A. M., and J. B. Kennedy. *Basic Statistical Methods for Engineers and Scientists.* Scranton, Penn.: International, 1964.

Schenck, J., Jr. *Theories of Engineering Experimentation.* 2nd ed. New York: McGraw-Hill, 1968.

Tufte, E. R. *The Visual Display of Quantitative Information.* Chesire, Conn: Graphics Press, 1983.

Weiss, N. A., and M. J. Hassett. *Introductory Statistics.* 2nd ed. Reading, Mass.: Addison-Wesley, 1987.

Problems

Note: Unless otherwise specified, all uncertainties are assumed to represent 95% coverage (19:1 odds).

3.1   For a very large set of data, the measured mean is found to be 200 with a standard deviation of 20. Assuming the data to be normally distributed, determine the range within which 60% of the data are expected to fall.

3.2   From long-term plant-maintenance data, it is observed that pressure downstream from a boiler in normal operation has a mean value of 303 psi with a standard deviation of 33 psi. What is the probability that the pressure will exceed 350 psi during any one measurement in normal operation?

Note: The following information should be helpful in solving Problems 3.3–3.8.
    For pure electrical elements in series:
      Resistances add directly.
      Reciprocals of capacitances add to yield the reciprocal of the overall
      capacitance.
      Inductances add directly.
    For pure electrical elements in parallel:
      Reciprocals of resistances add to yield the reciprocal of the overall
      resistance.
      Capacitances add directly.
      Reciprocals of inductances add to yield the reciprocal of the overall
      inductance.

3.3   **a.**   A 68-kΩ resistor is paralleled with a 12-kΩ resistor. Each resistor has a $\pm 10\%$ tolerance. What will be the nominal resistance and the uncertainty of the combination?

      b.   If the values remain the same except that the tolerance on the 68-kΩ resistor is dropped to $\pm 5\%$, what will be the uncertainty of the combination?

3.4   Five 100-Ω resistors, each having a 5% tolerance, are connected in series. What is the overall nominal resistance and tolerance of the combination?

3.5   Three 1000-Ω resistors are connected in parallel. Each resistor has a $\pm 5\%$ tolerance. What is the overall nominal resistance and what is the best estimate of the tolerance of the combination?

3.6   A 47-Ω resistor is connected in series with a parallel combination of a 100-Ω resistor and a 180-Ω resistor. What is the overall resistance of the array and what is the best estimate of its tolerance?

      a.   For individual tolerances equal to 1%

      b.   For tolerances of the 47-Ω resistors equal to 10% and for the 180-Ω resistors equal to 5%

3.7   A capacitor of 0.05 $\mu F \pm 10\%$ is parallel with a capacitor of 0.1 $\mu F \pm 10\%$.

      a.   What are the resulting nominal capacitance and uncertainty?

      b.   What would be the nominal capacitance and uncertainty if the two elements were connected in series?

3.8   Two inductances are connected in parallel. Their values are 0.5 mH and 1.0 mH. Each carries a tolerance of $\pm 20\%$. Assuming no mutual inductance, what is the nominal inductance and uncertainty of the combination?

3.9   Power can be measured as $I^2 R$ or VI. Using the data of Problem 3.17, which method will give the most accurate measurement?

3.10 The volume of a cylinder is to be determined from measurements of the diameter and length. If the length and diameter are measured at four different locations by means of a micrometer with an uncertainty of 0.5% of reading, determine the uncertainty in the measurement.

Diameter	Length (in.)
3.9920	4.4940
3.9892	4.4991
3.9961	4.5110
3.9995	4.5221

3.11 A tube of circular section has a nominal length of 52 cm ± 0.5 cm, an outside diameter of 20 cm ± 0.04 cm, and an inside diameter of 15 cm ± 0.08 cm. determine the uncertainty in calculated volume.

3.12 A cantilever beam of circular section has a length of 6 ft and a diameter of $2\frac{1}{2}$ in. A concentrated load of 350 lbf is applied at the beam end, perpendicular to the length of the beam. If the uncertainty in the length is ±1.5 in., in the diameter is ±0.08 in., and in the force is ±5 lbf, what is the uncertainty in the calculated maximum bending stress?

3.13 If it is determined that the overall uncertainty in the maximum bending stress for Problem 3.12 may be as great as, but no greater than, 6%, what maximum uncertainty may be tolerated in the diameter measurement if the other uncertainties remain unchanged?

3.14 It is desired to compare the design of two bolts based on their tensile strength capabilities. The following lists the results of the sample testing.

Group	Failure Load	Std. Deviation	Number of Tests
A	30 kN	2 kN	21
B	34 kN	6 kN	9

Is there a difference between the two samples at the 95% confidence level?

3.15 From a sample of 150 marbles having mean diameter of 10 mm and a standard deviation of ±3.4 mm, how many marbles would you expect to find in the range from 10 to 15 mm?

3.16 During laboratory testing of a thin-wall pressure vessel, the cylinder diameter and thickness were measured at ten different locations; the resulting data were as follows:

$$\overline{D} = 10.25 \text{ in.} \qquad S_D = 0.25 \text{ in.}$$

$$\overline{t} = 0.25 \text{ in.} \qquad S_t = 0.05 \text{ in.}$$

If the pressure inside the vessel is measured to be 100 psi with an estimated uncertainty of ±10 psi, determine the best estimate of the tangential or hoop stress in the vessel. (*Note: $\sigma_\theta = PD/2t$.*)

3.17 In order to determine the power dissipated across a resistor, the current flow and resistance values are measured separately. If $I = 3.2$ A and $R = 1000\ \Omega$ are measured values, determine the uncertainty if the following instruments are used:

Instrument	Resolution	Uncertainty (% of reading)
Voltmeter	1.0 mV	0.5%
Ohmmeter	1.0 Ω	0.1%
Ammeter	0.1 A	0.5%

3.18 A total of 120 hardness measurements are performed on a large slab of steel. If, using the Rockwell C scale, the mean of the measurements is 39 and the standard deviation is 4.0, how many of the measurements can be expected to fall between the hardness readings of 35 and 45?

3.19 In order to determine whether the use of a rubber backing material between a concrete compression sample and the platen of a testing machine affects the compressive strength, six samples with packing and six without were tested. The strengths are listed below. Determine if the packing material has any effect. Use a 99% confidence level.

	Tensile Strength MN/m²	
Sample No.	With Packing	Without Packing
1	2.48	2.18
2	2.76	2.48
3	2.96	2.38
4	2.72	2.00
5	2.62	2.10
6	2.65	2.28

3.20 Results from a chemical analysis for the carbon content of two materials is as follows:

	Carbon Content, %					
Material A	93.52	92.81	94.32	93.77	93.57	93.12
Material B	92.38	93.21	92.55	92.05	92.54	

Determine if there is a significant difference in carbon content at the 99% confidence level.

3.21 Figure 3.9 shows a histogram based on the values listed in Table 3.3. As suggested in Section 3.10, prepare histograms representing the data, based on (a) seven bins, (b) eight bins, and (c) ten bins.

3.22 The manufacturer of inexpensive outdoor thermometers checks a sample of ten against a 68°F standard. The following results were obtained:

      68.5   67.5   67   69   68   67   67.5   69   69.

Using the Student's t-test, calculate the range within which the population mean may be expected to exist with a confidence level of 95%.

3.23 Spacer blocks are manufactured in quantity to a nominal dimension of 125 mm. A sample of 12 blocks was selected and the following measurements were made.

            1.28   1.32   1.29   1.23
            1.26   1.26   1.20   1.29
            1.24   1.23   1.26   1.22

Using the Student's t-test, determine the upper and lower tolerance values within which the population mean may be expected to fall with a significance level of 10%.

3.24 In a laboratory it is suspected that the results from two different viscometers do not agree. Ten fluid samples were tested using apparatus A and corresponding samples were tested using apparatus B. The results are as follows:

Viscosity (Dimensionless)

Sample No.	Using Apparatus A	Using Apparatus B
1	72	73
2	43	45
3	54	56
4	75	75
5	50	53
6	48	50
7	73	72
8	55	54
9	48	48
10	50	52

Determine whether there is a significant difference in the two systems at the 99% confidence level.

3.25 Consider the equation

$$y = 1.0 - 0.2x + 0.01x^2 \qquad (0 \le x \le 3).$$

Determine the maximum uncertainty in y for $\pm 2\%$ uncertainty in the variable x.

3.26 For the following data determine the equation for $y = y(x)$ by graphical analysis.

x	0	0.43	0.76	1.21	2.60	3.5
y	1.00	1.54	3.61	5.25	10.0	13.50

3.27 For the following data, determine the equation $y = y(x)$ by graphical analysis.

x	1.21	1.35	2.75	5.1	8.1
y	12.0	18.2	88.0	325.0	800

3.28 For the following data, determine the equation $y = y(x)$ by graphical analysis.

x	0	0.43	2.6	2.9	4.3
y	94	71	26	19.5	11.5

3.29 The influence of the size of the test specimen on the tensile strength of an epoxy resin was determined by casting seven samples of each size and testing them accordingly. The experimental data are as follows:

Specimen Strengths (kN/m²)

Sample of Small Specimens	Sample of Large Specimens
3475	1813
4326	3145
2262	4140
7415	6867
3418	3842
4404	3984
3788	3053

Determine whether there is a significant difference between the two samples at the 95% confidence level.

3.30 In 200 tosses of a coin, 116 heads and 84 tails were observed. Determine if the coin is fair using a confidence level of 95% or a significance level of 5%.

3.31 A random number table of 100 digits showed the following distribution of the digits 0, 1, 2 . . . 9. Determine if the distribution of the digits differs significantly from the expected distribution at the 1% significance level.

Digit	0	1	2	3	4	5	6	7	8	9
Observed frequency	7	12	12	7	6	8	14	12	8	14

3.32 A quality control engineer wants to determine if the diameters of ball bearings produced by a machine are normally distributed. From a random sample of 300 bearings, he determines that the sample mean is 10.00 mm with a sample standard deviation of ±0.10 mm. Moreover, he obtains the following frequency distribution for the diameters.

Diameter, mm	Observed Frequency
Under 9.80	8
9.80–Under 9.90	42
9.90–Under 10.00	107
10.00–Under 10.10	97
10.10–Under 10.20	38
10.20 and Over	8

Are the bearing diameters normally distributed at the 5% significance level?

3.33 Using the data of Problem 3.32, construct a normal probability plot. What conclusions can you determine from this graphical representation regarding the normalcy of the data?

3.34 A sample of 100 test specimens of a steel alloy provides the following breaking strengths. Determine if the data are normally distributed at the 1% significance level if the mean breaking strength is 67.45 ksi and the standard deviation is 2.92 ksi.

Breaking Strength, ksi	Observed Frequency
59.5–62.5	5
62.5–65.5	18
65.5–68.5	42
68.5–71.5	27
71.5–74.5	8

3.35 A company subcontracts the mass production of a die casting of fixed design. Four primary types of defects have been identified and records have been kept providing a "standard" against which defect distribution for batches may be judged. For a given batch of 2243 castings the following data apply. Do the batch data vary significantly from the standard?

Defect Identification	Distribution Percent	Results for Batch 2073
Type A	7.2	125
Type B	4.6	60
Type C	1.9	75
Type D	0.9	31
Nondefective	85.4	1952
	100.0	2243

3.36 A system is calibrated statically. The accompanying table lists the results.

Input	Output (Increasing Input)	Output (Decreasing Input)
0.12	1.6	2.2
0.17	2.7	2.3
0.27	3.7	3.2
0.32	3.9	4.2
0.38	4.3	4.8
0.46	5.6	5.2
0.53	6.7	6.5
0.64	7.4	7.4

a. Plot output versus input.
b. Calculate the best straight-line fit, first for the increasing output, then for the decreasing output and finally for the combined data.
c. What is the maximum deviation in each case?
d. If it is assumed that zero input should yield zero output, what is the zero offset (bias) that should be assigned?

3.37 The following data describe the temperature distribution along a length of heated pipe. Determine the best straight-line fit to the data.

Temperature, °C	Distance from a Datum, cm
100	11.0
200	19.0
300	29.0
400	39.0
500	50.5

3.38 The force-deflection data for a spring are tabulated below. Determine a least squares fit.

Deflection, in.	Force, lbf
0.10	9
0.20	19
0.30	22
0.40	40
0.50	52
0.60	59

3.39 Solve Problem 3.38 adding one more set of data—namely, zero deflection under zero load.

3.40 The data in the accompanying tabulation (from several sources) provide the resistivity of platinum at various temperatures.
- **a.** Make a linear plot of the data points.
- **b.** Determine the constants for a linear least squares fit of the entire data set. Plot the fitted line on your graph of the data.
- **c.** Because the resistivity is not a perfectly linear function of temperature, a more accurate fit can be obtained by limiting the range of temperature considered. Obtain the constants for a linear least squares fit over the range of 0°C to 1000°C only. Plot the result on your graph.

Temperature, °C	Resistivity, $\Omega \cdot$ cm
0	10.96
20	10.72
100	14.1
100	14.85
200	17.9
400	25.4
400	26.0
800	40.3
1000	47.0
1200	52.7
1400	58.0
1600	63.0

3.41 In constructing a spring-mass system, a deflection constant of 50 lbf/in. is required. Four springs are available, two having deflection constants of 25.0 lbf/in. with an uncertainty (tolerance) of ±2.0 lbf/in. and two having deflection constants of 100 lbf/in. with uncertainties of ±4.0 lbf/in. What combinations can be used for a system deflection constant of 50 lbf/in.? What will be the uncertainty in each case?

3.42 Show that $y = a + bx^n$ will plot as a straight line on linear graph paper when y is plotted as the ordinate and x^n is plotted as the abscissa. Show that the intercept is equal to a and the slope is equal to b.

3.43 Show that if $1/y$ versus $1/x$ is plotted on linear paper, the function $y = x/(ax + b)$ (which may also be written $1/y = a + b/x$) will yield a straight line, with a as the intercept and b as the slope.

3.44 Show that $y = ac^{bx}$ will plot as a straight line on linear paper when $\log y$ is plotted as the ordinate and x is plotted as the abscissa and that the intercept is equal to $\log a$ and the slope is equal to $b \log c$. Note that with the slope known, b and c may be found by simultaneous solution of the slope equation and the original equation written for a selected (x_i, y_i) point.

3.45 Select a range for x and make an x versus y plot of $y = 12x^{2/3}$ on linear graph paper. Now transform the data to $\log y$ and $\log x$ and plot on linear paper. The second set of data should plot as a straight line with a slope of $\frac{2}{3}$ and an intercept of 12.

3.46 From 1960 to 1983, the standard meter was defined as 1,650,763.73 wavelengths of the light emitted during the transition between the $2p_{10}$ and the $5d_5$ levels of the krypton-86 atom. That emission line is slightly asymmetric, and its wavelength has a total uncertainty of about $\pm 2.4 \times 10^{-5}$ Å (68%).

By the early 1970s, laser technology permitted highly-precise determination of the speed of light, c, by using the relation $c = \lambda f$ and measured values of laser wavelength, λ, and frequency, f. Laser frequency measurements had at that time reached relative uncertainties of $u_f/f = 6 \times 10^{-10}$ (68%).

a. What was the uncertainty (95%) in the measured speed of light at that time? What factor limited the accuracy of this measurement?

b. In 1983, the meter was redefined as "the distance travelled by light in vacuum during a time interval of 1/299792458 of a second." How did this affect the uncertainty in the speed of light (in meters per second)?

References

1. ANSI/ASME 19.1–1985. ASME Performance Test Codes. Supplement on Instruments and Apparatus, Part 1, *Measurement Uncertainty*. New York, 1985.

2. Froome, K. D., and L. Essen. *The Velocity of Light and Radio Waves*. New York: Academic Press, 1969.

3. Miller, I. R., J. E. Freund, and R. Johnson. *Probability and Statistics for Engineers*. 4th ed. Englewood Cliffs, N.J.: Prentice Hall, 1990.

4. Weiss, N. A., and M. J. Hassett. *Introductory Statistics*. 2nd ed. Reading, Mass.: Addison-Wesley, 1987.

5. Schenk, J., Jr. *Theories of Engineering Experimentation*. 2nd ed. New York: McGraw-Hill, 1968.

6. Kline, S. J., and F. A. McClintock, Describing uncertainties in single-sample experiments. *Mech. Engr.* 75: 3–8, January 1953.

7. *Engineering and Scientific Graphs for Publications,* American Standards Association. New York: American Society of Mechanical Engineers, July 1947.

8. Natrella, M. G. Experimental Statistics. *National Bureau of Standards Handbook 91*. Washington, D.C.: U.S. Government Printing Office, 1963.

9. McClintock, F. A. *Statistical Estimation: Linear Regression and the Single Variable,* Research Memo 274, Fatigue and Plasticity Laboratory. Cambridge: Massachusetts Institute of Technology, February 14, 1987.

10. Rabinowicz, E. *Introduction to Experimentation*. Reading, Mass.: Addison-Wesley, 1970.

11. Hornbeck, R. W. *Numerical Methods*. New York: Quantum Publishers, 1975.

CHAPTER 4

|||||||||||||||||||||||||||||||||||||||||||||||||||||||||||||||

The Analog Measurand: Time-Dependent Characteristics

4.1 Introduction

A parameter common to all of measurement is *time*: All measurands have time-related characteristics. As time progresses, the magnitude of the measurand either changes or does not change. The time variation of any change is often fully as important as is any particular amplitude.

In this chapter we will discuss those quantities necessary to define and describe various time-related characteristics of measurands. As in Chapter 1, we classify time-related measurands as either

1. Static—constant in time
2. Dynamic—varying in time
   a. Steady-state periodic
   b. Nonrepetitive or transient
      i. Single-pulse or aperiodic
      ii. Continuing or random

4.2 Simple Harmonic Relations

A function is said to be a *simple harmonic* function of a variable when its second derivative is proportional to the function but of opposite sign. More often than not, the independent variable is time t, although any two variables may be related harmonically.

Some of the most common harmonic functions in mechanical engineering relate displacement and time. In electrical engineering, many of the variable quantities in alternating-current (AC) circuitry are harmonic functions of time. The harmonic relation is quite basic to dynamic functions, and most quantities that are time functions may be expressed harmonically.

In its most elementary form, *simple harmonic motion* is defined by the relation

$$s = s_0 \sin \omega t, \tag{4.1}$$

in which

s = instantaneous displacement from equilibrium,

s_0 = amplitude, or maximum displacement from equilibrium,

ω = circular frequency (rad/s),

t = any time interval measured from the instant when $t = 0$ s.

A small-amplitude pendulum, a mass on a beam, a weight suspended by a rubber band—all vibrate with simple harmonic motion, or very nearly so.

By differentiation, the following relations may be derived from Eq. (4.1):

$$v = \frac{ds}{dt} = s_0 \omega \cos \omega t \tag{4.2}$$

and

$$v_0 = s_0 \omega. \tag{4.2a}$$

Also,

$$a = \frac{dv}{dt} = -s_0 \omega^2 \sin \omega t \tag{4.3}$$

$$= -s\omega^2. \tag{4.3a}$$

In addition,

$$a_0 = -s_0 \omega^2. \tag{4.3b}$$

In the preceding equations,

v = velocity,

v_0 = maximum velocity or velocity amplitude,

a = acceleration,

a_0 = maximum acceleration or acceleration amplitude.

Equation (4.3a) satisfies the description of simple harmonic motion given in the first paragraph of this section: The acceleration a is proportional to the displacement s but is of opposite sign. The proportionality factor is ω^2.

4.3 Circular and Cyclic Frequency

The frequency with which a process repeats itself is called *cyclic frequency, f,* and is typically measured in cycles per second, or *hertz*: 1 Hz = 1 cycle/s. However, the idea of *circular frequency, ω,* is also useful in studying cyclic relations. Circular frequency has units of radians per second (rad/s). The connection between the two frequencies is conveniently illustrated by the well-known *Scotch-yoke* mechanism.

Figure 4.1(a) shows the elements of the Scotch yoke, consisting of a crank, *OA*, with a slider block driving the yoke-piston combination. If we measure the piston displacement from its midstroke position, the displacement amplitude will be ±*OA*. If the crank turns at ω radians per second, then the crank angle θ may be written as ω*t*. This, of course, is convenient because it introduces time *t* into the relationship, which is not directly apparent in the term θ. Piston displacement may now be written as

$$s = s_0 \sin \omega t,$$

which is the same as Eq. (4.1). One cycle takes place when the crank turns through 2π rad, and, if *f* is the frequency in hertz, then

$$\omega = 2\pi f. \qquad (4.4)$$

Figure 4.1 (a) The Scotch-yoke mechanism provides a simple harmonic motion to the piston, (b) a spring-mass system that moves with simple harmonic motion

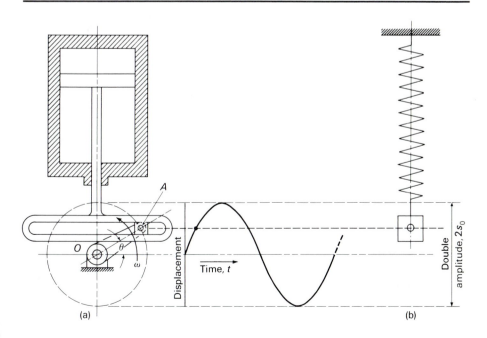

Figure 4.2   Motions that are out of phase. The dashed curve lags
the solid curve by a phase angle ϕ

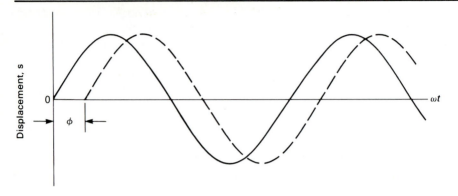

Thus the displacement may instead be expressed in terms of cyclic frequency:

$$s = s_0 \sin 2\pi ft. \tag{4.5}$$

Either displacement equation shows that the yoke-piston combination
moves in simple harmonic motion. Many other mechanical and electrical
systems display simple harmonic relationships. The spring-mass system shown
in Fig. 4.1(b) is such an example. If its amplitude and natural frequency just
happened to match the amplitude and frequency of the Scotch-yoke mechanism, then the mass and the piston could be made to move up and down in
perfect synchronization.

To put it another way, *for every simple harmonic relationship, an
analogous Scotch-yoke mechanism may be devised or imagined.* The crank
length OA will represent the vector amplitude, and the angular velocity ω of
the crank, in radians per second, will correspond to the circular frequency of
the harmonic relation. If the mass and piston have the same frequencies and
simultaneously reach corresponding extremes of displacement, their motions
are said to be *in phase.* When they both have the same frequency but do not
oscillate together, the time difference (lag or lead) between their motions may
be expressed by an angle referred to as the *phase angle, ϕ* (Fig. 4.2).

4.4   Complex Relations

Most complex dynamic mechanical signals, steady-state or transient, whether
they are functions such as pressure, displacement, strain, or something else,
may be expressed as a combination of simple harmonic components. Each
component will have its own amplitude and frequency and will be combined in
various phase relations with the other components. A general mathematical

statement of this may be written as follows:

$$y(t) = \frac{A_0}{2} + \sum_{n=1}^{\infty} (A_n \cos n\omega t \pm B_n \sin n\omega t), \qquad (4.6)$$

in which

A_0, A_n, and B_n = amplitude-determining constants called *harmonic coefficients*,

n = integers from 1 to ∞, called *harmonic orders*.

When n is unity, the corresponding sine and cosine terms are said to be *fundamentals*. For $n = 2, 3, 4$, etc., the corresponding terms are referred to as second, third, fourth *harmonics*, and so on. Equation (4.6) is sometimes called a *Fourier series* for $y(t)$.

The variable part of Eq. (4.6) may be written in terms of either the sine or the cosine alone, by introducing a phase angle. Conversion is made according to the following rules:

Case I: For $y = A \cos x + B \sin x$,

$$y = C \cos(-x + \phi_2) \qquad (4.6a)$$

or

$$y = C \sin(x + \phi_1). \qquad (4.6b)$$

Case II: For $y = A \cos x - B \sin x$,

$$y = C \cos(x + \phi_2) \qquad (4.6c)$$

or

$$y = C \sin(-x + \phi_1). \qquad (4.6d)$$

In both cases, $C = \sqrt{A^2 + B^2}$; and ϕ_1 and ϕ_2 are *positive acute angles*, such that

$$\phi_1 = \tan^{-1} \frac{|A|}{|B|}$$

and

$$\phi_2 = \tan^{-1} \frac{|B|}{|A|}.$$

Note that in calculating ϕ_1 and ϕ_2, A and B are taken as absolute values.

Although Eq. (4.6) indicates that all harmonics may be present in defining the signal-time relation, such relations usually include only a limited number of harmonics. In fact, all measuring systems have some upper and some lower frequency limits beyond which further harmonics will be attenuated. In other words, no measuring system can respond to an infinite frequency range.

Figure 4.3   Examples of two-component waveforms with second-harmonic
component of various relative amplitudes

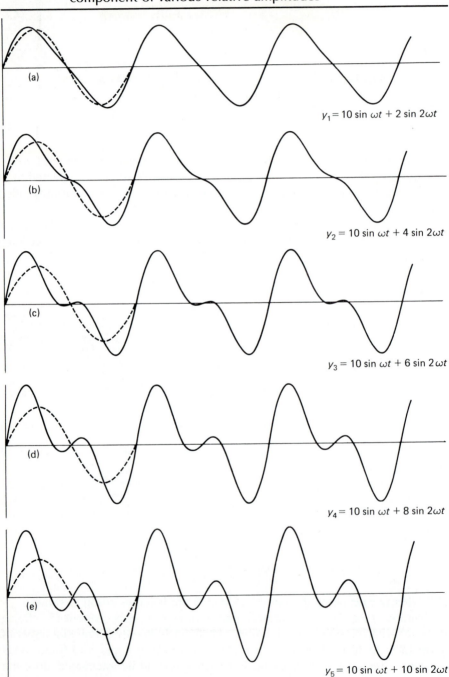

(a)

$y_1 = 10 \sin \omega t + 2 \sin 2\omega t$

(b)

$y_2 = 10 \sin \omega t + 4 \sin 2\omega t$

(c)

$y_3 = 10 \sin \omega t + 6 \sin 2\omega t$

(d)

$y_4 = 10 \sin \omega t + 8 \sin 2\omega t$

(e)

$y_5 = 10 \sin \omega t + 10 \sin 2\omega t$

Although it would be utterly impossible to catalog all the many possible harmonic combinations, it is nevertheless useful to consider the effects of a few variables such as relative amplitudes, harmonic orders n, and phase relations ϕ. Therefore, Figs. 4.3 through 4.7 are presented for two-component relations, in each case showing the effect of only one variable on the overall waveform. Figure 4.3 shows the effect of relative amplitudes; Fig. 4.4 shows the effect of relative frequencies; Fig. 4.5 shows the effect of various phase relations; Fig. 4.6 shows the appearance of the waveform for two components having considerably different frequencies; and Fig. 4.7 shows the effect of two frequencies that are very nearly the same.

Example 4.1

As an example of a relation made up of harmonics, let us analyze a relatively simple pressure-time function consisting of two harmonic terms:

$$P = 100 \sin 80t + 50 \cos\left(160t - \frac{\pi}{4}\right). \tag{4.7}$$

Solution. Inspection of the equation shows that the circular frequency of the fundamental has a value of 80 rad/s, or $80/2\pi = 12.7$ Hz. The period for the pressure variation is therefore $1/12.7 = 0.0788$ s. The second term has a frequency twice that of the fundamental, as indicated by its circular frequency of 160 rad/s. It also lags the fundamental by one-eighth cycle, or $\pi/4$ rad. In addition, the equation indicates that the amplitude of the fundamental, which is 100, is twice that of the second harmonic, which is 50. A plot of the relation is shown in Fig. 4.8.

Example 4.2

As another example, let us analyze an acceleration-time relation that is expressed by the equation

$$a = 3800 \sin 2450t + 1750 \cos\left(7350t - \frac{\pi}{3}\right) + 800 \sin (36,750t). \tag{4.8}$$

where a = angular acceleration (rad/s^2) and t = time (s).

Solution. The relation consists of three harmonic components having circular frequencies in the ratio 1 to 3 to 15. Hence the components may be referred to as the fundamental, the third harmonic, and the fifteenth harmonic. Corresponding frequencies are 390, 1170, and 5850 Hz.

Figure 4.4 Examples of two-component waveforms with second term of various relative frequencies

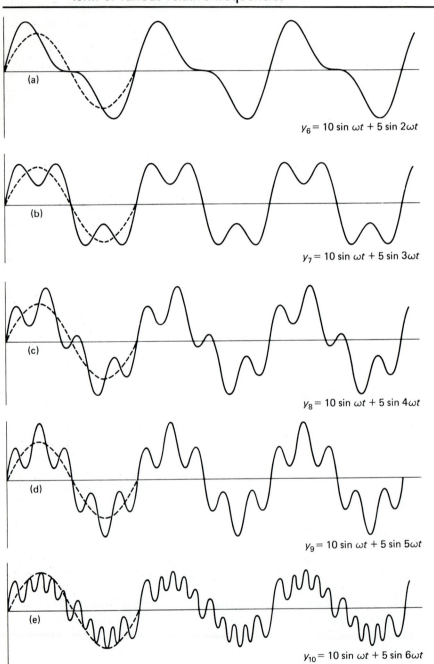

$y_6 = 10 \sin \omega t + 5 \sin 2\omega t$

$y_7 = 10 \sin \omega t + 5 \sin 3\omega t$

$y_8 = 10 \sin \omega t + 5 \sin 4\omega t$

$y_9 = 10 \sin \omega t + 5 \sin 5\omega t$

$y_{10} = 10 \sin \omega t + 5 \sin 6\omega t$

Figure 4.5   Examples of two-component waveforms with the second
harmonic having various degrees of phase shift

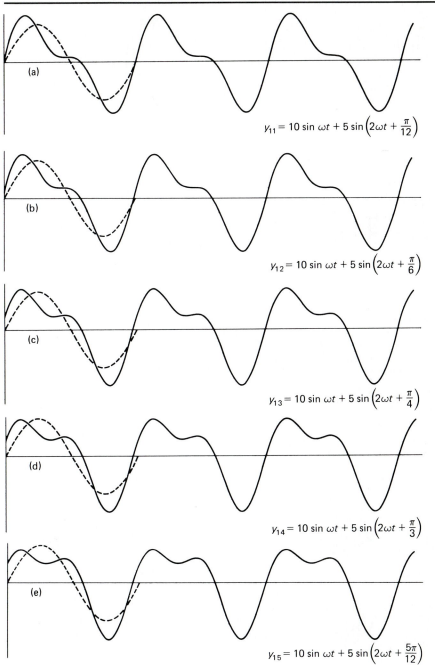

$$y_{11} = 10 \sin \omega t + 5 \sin \left(2\omega t + \frac{\pi}{12}\right)$$

$$y_{12} = 10 \sin \omega t + 5 \sin \left(2\omega t + \frac{\pi}{6}\right)$$

$$y_{13} = 10 \sin \omega t + 5 \sin \left(2\omega t + \frac{\pi}{4}\right)$$

$$y_{14} = 10 \sin \omega t + 5 \sin \left(2\omega t + \frac{\pi}{3}\right)$$

$$y_{15} = 10 \sin \omega t + 5 \sin \left(2\omega t + \frac{5\pi}{12}\right)$$

Figure 4.6   Examples of waveforms with the two components
having considerably different frequencies

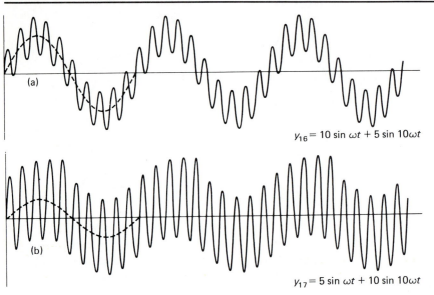

(a)

$y_{16} = 10 \sin \omega t + 5 \sin 10\omega t$

(b)

$y_{17} = 5 \sin \omega t + 10 \sin 10\omega t$

Figure 4.7   Examples of waveforms with two components having
frequencies that are very nearly the same

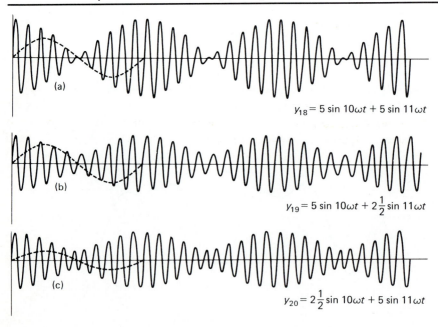

(a)

$y_{18} = 5 \sin 10\omega t + 5 \sin 11\omega t$

(b)

$y_{19} = 5 \sin 10\omega t + 2\frac{1}{2} \sin 11\omega t$

(c)

$y_{20} = 2\frac{1}{2} \sin 10\omega t + 5 \sin 11\omega t$

Figure 4.8 Pressure-time relation, $P = 100 \sin(80t) + 50 \cos(160t - \pi/4)$

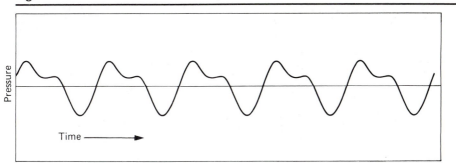

4.4.1 Beat Frequency and Heterodyning

The situation shown in Fig. 4.7(a) is the basis of an important method of frequency measurement. Here two waves of equal amplitude and *nearly* equal frequency have been added. If one wave has a cyclic frequency of f_0 and the second wave has a frequency of $f_1 = f_0 + \Delta f$, then the resultant wave is

$$y = A \sin(2\pi f_0 t) + A \sin[2\pi (f_0 + \Delta f)t]$$
$$= 2A \cos\left(2\pi \frac{\Delta f}{2} t\right) \sin\left(2\pi \frac{f_0 + f_1}{2} t\right). \tag{4.9}$$

This wave undergoes slow "beats" where the amplitude rises and falls with a cyclic frequency of $\Delta f/2$.

Such addition of waves occurs when a tuning fork is used to tune a musical instrument. The tuning fork and the musical instrument produce nearly equal tones, and, when the two sound waves are heard together, a lower beat frequency is also heard. The instrument is adjusted until the beat frequency is zero, so that the instrument's frequency is identical to that of the tuning fork.

When the difference frequency, Δf, is much smaller than f_0, addition of waves allows us to measure Δf with less uncertainty than if we measured f_0 and f_1 separately and subtracted them. This technique for frequency measurement is called *heterodyning* (Section 10.8). It is very important in laser-doppler velocity measurements (Section 15.11) and in radio applications.

Example 4.3

Helium-neon laser light has a frequency of 473.8 THz (473.8×10^{12} Hz). A helium-neon laser beam is reflected from a moving target. This creates a doppler shift in the beam, which increases its frequency by 3 MHz

$(3 \times 10^6 \, \text{Hz}).$* If the reflected beam is "added" to an unshifted beam of equal intensity, what is the resulting beat frequency?

Solution. From Eq. (4.9), with $f_0 = 473.8 \, \text{THz}$ and $f_1 = f_0 + \Delta f = 473.8 \, \text{THz} + 3 \, \text{MHz}$,

$$y = 2A \cos\left(2\pi \frac{\Delta f}{2} t\right) \sin\left(2\pi \frac{f_0 + f_1}{2} t\right)$$

$$= 2A \cos[2\pi(1.5 \times 10^6)t] \sin[2\pi(473.8 \times 10^{12})t].$$

The beats occur with a frequency of $1.5 \times 10^6 \, \text{Hz} = 1.5 \, \text{MHz}$. Note that the beat frequency is more than a 100 million times smaller than the frequency of the original light. To find the difference between f_0 and f_1 directly by subtraction, we would have needed to measure each frequency to an accuracy of better than 1 part in 100 million! As a result, the beats provide a much easier way to determine Δf.

4.4.2   Special Waveforms

A number of frequently used special waveforms may be written as infinite trigonometric series. Several of these are shown in Fig. 4.9. Table 4.1 lists the corresponding equations.

Both the square wave and the sawtooth wave are useful in checking the response of dynamic measuring systems. In addition, the skewed sawtooth form, Fig. 4.9(c), is of the form required for the voltage-time relation necessary for driving the horizontal sweep of a cathode-ray oscilloscope. All these forms may be obtained as voltage-time relations from electronic signal generators.

For each case shown in Fig. 4.9, all the terms in the infinite series are necessary if the precise waveform indicated is to be obtained. Of course, with increasing harmonic order, their effect on the whole sum becomes smaller and smaller.

As an example, consider the square wave shown in Fig. 4.9(a). The complete series includes all the terms indicated in the relation

$$y = \frac{4A}{\pi}\left(\sin \omega t + \frac{1}{3}\sin 3\omega t + \frac{1}{5}\sin 5\omega t + \cdots\right).$$

By plotting only the first three terms, which include the fifth harmonic, the waveform shown in Fig. 4.10(a) is obtained. Figure 4.10(b) shows the results of plotting terms through and including the ninth harmonic, and Fig. 4.10(c) shows the form for the terms including the fifteenth harmonic. As more and

---

* A *doppler shift* is an apparent change in the frequency of a light or sound wave that occurs when the wave source and receiver are in motion relative to one another. One typical example is the change in pitch of a passing train's whistle.

Figure 4.9 Various special waveforms of harmonic nature. In each case, the ordinate is y and the abscissa is ωt.

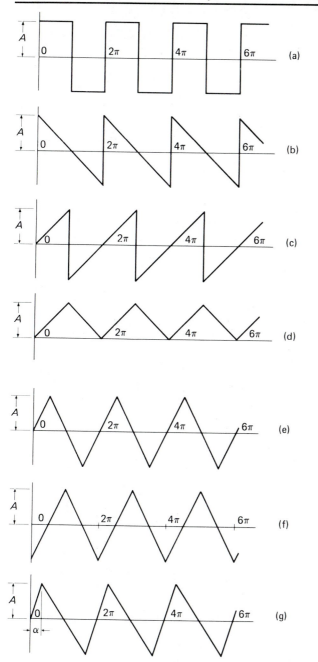

Table 4.1　Equations for special periodic waveforms shown in Fig. 4.9

Figure	Equation*
4.9(a)	$y = \dfrac{4A}{\pi}\left(\sin \omega t + \dfrac{1}{3}\sin 3\omega t + \dfrac{1}{5}\sin 5\omega t + \cdots\right) = \dfrac{4A}{\pi}\sum_{n=1}^{\infty}\left[\dfrac{1}{2n-1}\sin(2n-1)\omega t\right]$
4.9(b)	$y = \dfrac{2A}{\pi}\left(\sin \omega t + \dfrac{1}{2}\sin 2\omega t + \dfrac{1}{3}\sin 3\omega t + \cdots\right) = \dfrac{2A}{\pi}\sum_{n=1}^{\infty}\left[\dfrac{1}{n}\sin n\omega t\right]$
4.9(c)	$y = \dfrac{2A}{\pi}\left(\sin \omega t - \dfrac{1}{2}\sin 2\omega t + \dfrac{1}{3}\sin 3\omega t - \dfrac{1}{4}\sin 4\omega t + \cdots\right) = \dfrac{2A}{\pi}\sum_{n=1}^{\infty}\left[\dfrac{(-1)^{n+1}}{n}\sin n\omega t\right]$
4.9(d)	$y = \dfrac{A}{2} - \dfrac{4A}{(\pi)^2}\left(\cos \omega t + \dfrac{1}{(3)^2}\cos 3\omega t + \dfrac{1}{(5)^2}\cos 5\omega t + \cdots\right) = \dfrac{A}{2} - \dfrac{4A}{\pi^2}\sum_{n=1}^{\infty}\left[\dfrac{1}{(2n-1)^2}\cos(2n-1)\omega t\right]$
4.9(e)	$y = \dfrac{8A}{(\pi)^2}\left(\sin \omega t - \dfrac{1}{(3)^2}\sin 3\omega t + \dfrac{1}{(5)^2}\sin 5\omega t - \cdots\right) = \dfrac{8A}{(\pi)^2}\sum_{n=1}^{\infty}\left[\dfrac{(-1)^{n+1}}{(2n-1)^2}\sin(2n-1)\omega t\right]$
4.9(f)	$y = -\dfrac{8A}{(\pi)^2}\left(\cos \omega t + \dfrac{1}{(3)^2}\cos 3\omega t + \dfrac{1}{(5)^2}\cos 5\omega t + \cdots\right) = \dfrac{8A}{(\pi)^2}\sum_{n=1}^{\infty}\left[\dfrac{1}{(2n-1)^2}\cos n\omega t\right]$
4.9(g)	$y = \dfrac{2A}{\alpha(\pi - \alpha)}\left(\sin \alpha \sin \omega t + \dfrac{1}{(2)^2}\sin 2\alpha \sin 2\omega t + \dfrac{1}{(3)^2}\sin 3\alpha \sin 3\omega t + \cdots\right) = \dfrac{2A}{\alpha(\pi - \alpha)}\sum_{n=1}^{\infty}\left[\dfrac{1}{n^2}\sin n\alpha \sin n\omega t\right]$

* n as used in these equations does not necessarily represent the harmonic order.

Figure 4.10 Plot of square-wave function: (a) plot of first three terms only (includes the fifth harmonic), (b) plot of the first five terms (includes the ninth harmonic), (c) plot of the first eight terms (includes the fifteenth harmonic)

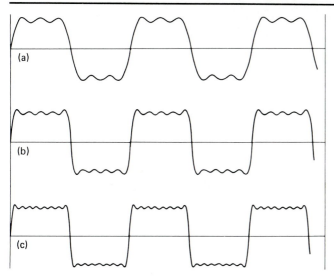

Figure 4.11 (a) Acceleration-time relationship resulting from shock-test, (b) considering the nonrepeating function as one real cycle of a periodic relationship

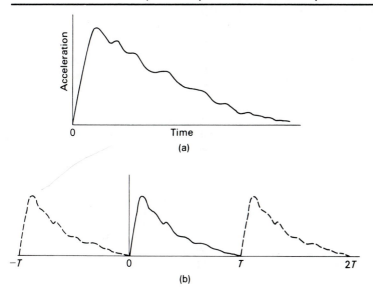

more terms are added, the waveform gradually approaches the square wave, which results from the infinite series.

The analytical calculation of the equations for Fig. 4.9(a) and (d) is performed in Appendix B.

4.4.3   Nonperiodic or Transient Waveforms

In the foregoing examples of special waveforms, various combinations of harmonic components were used. In each case the result was a periodic relation repeating indefinitely in every detail. Many mechanical inputs are not repetitive—for example, consider the acceleration-time relation resulting from an impact test [Fig. 4.11(a)]. Although such a relation is transient, it may be thought of as one cycle of a periodic relation in which all other cycles are fictitious [Fig. 4.11(b)]. On this basis, nonperiodic functions may be analyzed in exactly the same manner as periodic functions. If the nonperiodic waveform is sampled for a time period T, then the fundamental frequency of the fictitious periodic wave is $f = 1/T$ (cyclic) or $\omega = 2\pi/T$ (circular).

4.5   Frequency Spectrum

Figures 4.3 through 4.7 are plotted using time as the independent variable. This is the most common and familiar form. The waveform is displayed as it would appear on the face of an ordinary cathode-ray oscilloscope or on the paper of a strip-chart recorder. A second type of plot is the *frequency spectrum,* in which frequency is the independent variable and the amplitude of each frequency component is displayed as the ordinate. For example, the frequency spectra for the plots of Fig. 4.3(a) and (d) are shown in Fig. 4.12. Spectra corresponding to Fig. 4.4(a) and (c) are shown in Fig. 4.13, and the frequency spectrum for the square wave is shown in Fig. 14.14. Figures 4.12 through 4.14 use circular frequency ω; cyclic frequency f is often used instead.

The frequency spectrum is useful because it allows us to identify at a

Figure 4.12   (a) Frequency spectrum corresponding to Fig. 4.3(a), (b) frequency spectrum corresponding to Fig. 4.3(d)

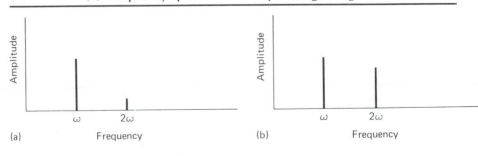

Figure 4.13 (a) Frequency spectrum corresponding to Fig. 4.4(a), (b) frequency spectrum corresponding to Fig. 4.4(c)

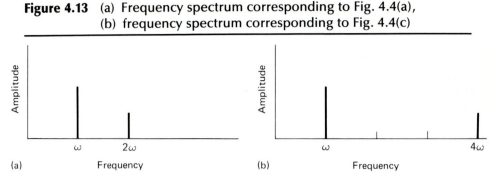

(a)

(b)

glance the frequencies present in a signal. For example, if the waveform results from a vibration test of a structure, we could use the frequency spectrum to identify the structure's natural frequencies.

The application of frequency spectrum plots has increased greatly since the development of the *spectrum analyzer* and *fast Fourier transform*. The spectrum analyzer is an electronic device that displays the frequency spectrum on a cathode-ray screen (Sections 9.12 and 18.5.4). The fast Fourier transform is a computer algorithm that calculates the frequency spectrum from computer-acquired data (see Section 4.6.1).

4.6 Harmonic, or Fourier, Analysis

In the preceding sections, we saw how known combinations of waves could be summed to produce more complex waveforms. In an experiment, the task is reversed: We measure the complex waveform and seek to determine which frequencies are present in it! The process of determining the frequency

Figure 4.14 Frequency spectrum for the square wave shown in Fig. 4.9(a)

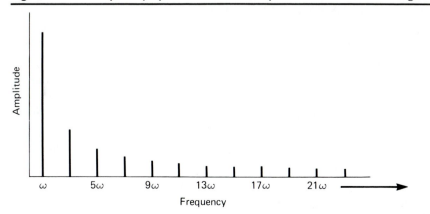

spectrum of a known waveform is called *harmonic analysis,* or *Fourier analysis.**

Fourier analysis a branch of classical mathematics on which entire textbooks have been written. The basic equations are derived in Appendix B, and more detailed discussions are available in the Suggested Readings for this chapter. The key to harmonic analysis is that the harmonic coefficients in Eq. (4.6) are integrals of the waveform $y(t)$:

$$A_n = \frac{\omega}{\pi} \int_0^{2\pi/\omega} y(t) \cos(\omega n t)\, dt \qquad n = 0, 1, 2, \ldots, \tag{4.10}$$

$$B_n = \frac{\omega}{\pi} \int_0^{2\pi/\omega} y(t) \sin(\omega n t)\, dt \qquad n = 1, 2, \ldots. \tag{4.11}$$

These relations are reciprocal to Eq. (4.6). When the harmonic coefficients are already known, Eq. (4.6) can be summed to obtain $y(t)$. Conversely, when $y(t)$ is known (as from an experiment), the integrals can be evaluated to determine the harmonic coefficients.

Experimentally, the waveform is usually measured only for a finite time period T. It turns out to be more convenient to write the integrals in terms of this time period, rather than the fundamental circular frequency, ω. Since $\omega = 2\pi/T$, the integrals are just

$$A_n = \frac{2}{T} \int_0^T y(t) \cos\left(\frac{2\pi}{T} nt\right) dt \qquad n = 0, 1, 2, \ldots, \tag{4.12}$$

$$B_n = \frac{2}{T} \int_0^T y(t) \sin\left(\frac{2\pi}{T} nt\right) dt \qquad n = 1, 2, \ldots. \tag{4.13}$$

Practical harmonic analysis usually falls into one of the following four categories:

1. The waveform $y(t)$ is known mathematical function. In this case, the integrals (4.12) and (4.13) can be evaluated analytically. These calculations are illustrated in Appendix B for two cases.
2. The waveform $y(t)$ is an analog signal from a transducer. In this case, the waveform may be processed with an electronic spectrum analyzer to obtain the signal's spectrum (Sections 9.12, 18.5, and 18.6).
3. Alternatively, the analog waveform may be recorded by a digital computer, as discussed in Chapter 8. The computer will store $y(t)$ only at a series of discrete points in time. Integrals (4.12) and (4.13) are replaced by sums and evaluated, as discussed in the next section.
4. The waveform is known graphically, for instance, from a strip-chart recorder or the screen of an oscilloscope. In this case, $y(t)$ may be read

---

* The term *spectral analysis* is also used.

from the graph at a discrete series of points, and the integrals may again be
evaluated as sums.*

4.6.1   The Discrete Fourier Transform

The case when $y(t)$ is known only at discrete points in time is very important in
practice because of the wide use of computers and microprocessors for recording
signals. Normally, a computer will read and store signal input at time intervals
of Δt (Fig. 4.15). The computer records a total of N points over the time
period $T = N\,\Delta t$.† Therefore, in the computer's memory, the analog signal
$y(t)$ has been reduced to a series of points measured at times $t = \Delta t$,
$2\,\Delta t, \ldots, N\,\Delta t$, specifically, $y(\Delta t), y(2\,\Delta t), \ldots, y(N\,\Delta t)$. We can write this
series more compactly as $y(t_r)$ by setting $t_r = r\,\Delta t$ for $r = 1, 2, \ldots, N$.

To perform a Fourier analysis of a discrete time signal like this, the
integrals in Eqs. (4.12) and (4.13) must be replaced by sums. Likewise, the
continuous time t is replaced by the discrete time $t_r = r\,\Delta t$, and the period T is
replaced by $N\,\Delta t$. Making these substitutions in Eq. (4.12), we get

$$
A_n = \frac{2}{N\,\Delta t} \sum_{r=1}^{N} y(t_r) \cos\left(\frac{2\pi}{N\,\Delta t} nr\,\Delta t\right) \Delta t
$$

$$
= \frac{2}{N} \sum_{r=1}^{N} y(r\,\Delta t) \cos\left(\frac{2\pi rn}{N}\right).
$$

Figure 4.15   Discrete sampling of a continuous analog signal. The value of
the signal is recorded at intervals Δt apart for a period T.

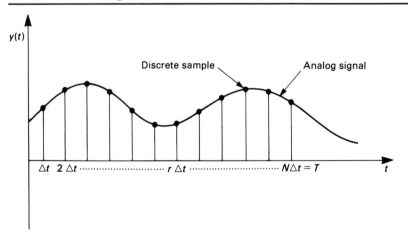

---

* An example of this approach is given in [1], Section 4.7.
† Assume that the point at $t = 0$ is not recorded and that N is even.

Thus, the harmonic coefficients of a discretely sampled waveform are

$$A_n = \frac{2}{N} \sum_{r=1}^{N} y(r\,\Delta t) \cos\left(\frac{2\pi rn}{N}\right) \qquad n = 0, 1, \ldots, \frac{N}{2} \tag{4.14}$$

$$B_n = \frac{2}{N} \sum_{r=1}^{N} y(r\,\Delta t) \sin\left(\frac{2\pi rn}{N}\right) \qquad n = 1, 2, \ldots, \frac{N}{2} - 1 \tag{4.15}$$

for N an even number. The corresponding expression for the discrete waveform $y(t_r)$ is

$$y(t_r) = \frac{A_0}{2} + \sum_{n=1}^{N/2-1} \left[A_n \cos\left(\frac{2\pi rn}{N}\right) + B_n \sin\left(\frac{2\pi rn}{N}\right) \right]$$

$$+ \frac{A_{N/2}}{2} \cos(\pi r). \tag{4.16}$$

Equations (4.14) and (4.15) are called the *discrete Fourier transform* (DFT) of $y(t_r)$ [2]. Equation (4.16) is called the *discrete Fourier series.*

In practice, the discrete sample is taken by an *analog-to-digital converter* (Section 8.12) connected to a computer or a microprocessor-driven electronic spectrum analyzer (Section 18.16). The computer or microprocessor evaluates the sums, Eqs. (4.14) and (4.15), often by using the fast Fourier transform algorithm.* The result, which approximates the spectrum of the original analog signal, is then displayed.

Like the ordinary Fourier series [Eq. (4.6)], the discrete Fourier series expresses $y(t)$ as a sum of frequency components, since

$$\frac{2\pi rn}{N} = 2\pi\left(\frac{n}{N\,\Delta t}\right) r\,\Delta t = 2\pi(n\,\Delta f)t_r$$

for a fundamental cyclic frequency

$$\Delta f \equiv \frac{1}{N\,\Delta t} \tag{4.17}$$

and harmonic orders $n = 1, \ldots, N/2$. In other words,

$$y(t) = \frac{A_0}{2} + \sum_{n=1}^{N/2-1} [A_n \cos(2\pi n\,\Delta f\,t) + B_n \sin(2\pi n\,\Delta f\,t)]$$

$$+ \frac{A_{N/2}}{2} \cos\left(2\pi \frac{N\,\Delta f}{2} t\right). \tag{4.18}$$

Note that the DFT yields *only* harmonic components up to $n = N/2$, whereas

---

* The *fast fourier transform*, or *FFT*, algorithm is special factorization of these sums that applies when N is a power of 2 ($N = 2^m$). The number of calculations normally required to evaluate these sums is proportional to N^2; when the FFT algorithm is used, the number is proportional to $N \log_2 N$. Thus, the FFT requires less computer work when N is large.

the ordinary Fourier series [Eq. (4.6)] may have an infinite number of frequency components. This very important fact is a consequence of the discrete sampling process itself.

4.6.2 Frequencies in Discretely Sampled Signals: Aliasing and Frequency Resolution

When an analog waveform is recorded by discrete sampling, some care is needed to ensure that the waveform is accurately recorded. The two sampling parameters that we can control are the *sample rate*, $f_s = 1/\Delta t$, which is the frequency with which samples are recorded, and the number of points recorded, N. Typically, the software controlling the data-acquisition computer will request values of f_s and N as input.

Figure 4.16 shows two examples of sampling a particular waveform. In Fig. 4.16(a), the sample rate is low (Δt is large), and as a result the high frequencies of the original waveform are not well resolved by the discrete samples—the signal seen in the discrete sample (the dashed curve) does not show the sharp peaks of the original waveform. The total time period of sampling is also fairly short (N is small), and thus the low frequencies of the signal are missed as well; it isn't clear how often the signal repeats itself. In Fig. 4.16(b), the sample rate and the number of points are each increased, improving the resolution of both high and low frequencies. These figures illustrate the importance of sample rate and total sampling time period in determining how well a discrete sample represents the original waveform.

What is the minimum sample rate needed to resolve a particular frequency? Consider the cases shown in Fig. 4.17, where a signal of frequency f is sampled at increasing rates. In (a), when the waveform is sampled at a frequency of $f_s = f$, the discretely sampled signal appears to be constant! No frequency is seen. In (b), the waveform is sampled at a higher rate, between f and $2f$; the discrete signal now appears to be a wave, but it has a frequency *lower* than f. In (c), the waveform is sampled at a rate $f_s = 2f$, and the discrete sample appears to be a wave of the correct frequency, f. Unfortunately, if the sampling begins a quarter-cycle later at this same rate [(d)], then the signal again appears to be constant. Only when the sample rate is increased *above 2f,* as in (e), do we always obtain the correct signal frequency with the discrete sample.

The highest frequency resolved at a given sampling frequency is determined by the *Nyquist frequency,*

$$f_{\mathrm{Nyq}} = \frac{f_s}{2}. \qquad\qquad (4.19)$$

Signals with frequencies lower than $f_{\mathrm{Nyq}} = f_s/2$ are accurately sampled. Signals with frequencies greater than f_{Nyq} are not accurately sampled; the frequencies above the Nyquist frequency incorrectly appear as lower frequencies in the discrete sample. The phenomenon of a discretely sampled signal taking on a

Figure 4.16 Effect of sample rate and number of samples taken:
(a) undersampled: both sample rate and number of points
are too low, (b) resolution of waveform is improved by
raising the sample rate and number of points

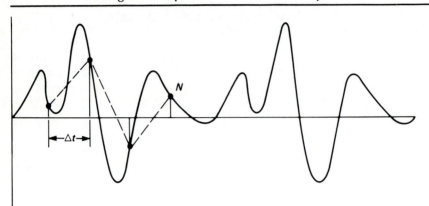

(a)

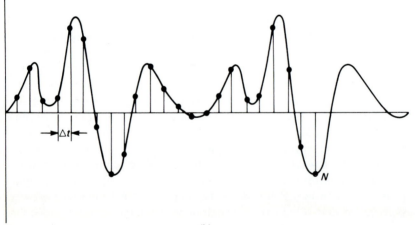

(b)

different frequency, as in Fig. 4.17(b), is called *aliasing*.* Aliasing occurs
whenever the Nyquist frequency falls below the signal frequency. Furthermore, the *phase ambiguity* shown by Fig. 4.17(c) and (d) prohibits sampling at
the Nyquist frequency itself. To prevent these problems, the sampling
frequency should always be chosen to be *more than twice* a signal's highest
frequency.

---

* An incorrectly sampled signal takes on a new identity, or *alias*.

   The Nyquist frequency tells us how to resolve correctly the highest
frequencies of a signal. It also tells us why the DFT contains only a finite
number of frequency components: Frequencies higher than the Nyquist are too
fast to be resolved with samples taken at the given sample rate.
   In a similar fashion, we can determine the lowest nonzero frequency in the
DFT and justify the discrete spacing of frequencies. If a waveform is to be
resolved by discrete sampling, one or more *full periods* of that waveform must

Figure 4.17   Effect of varying the sample rate, f_s, on the apparent
                  signal obtained by discrete sampling

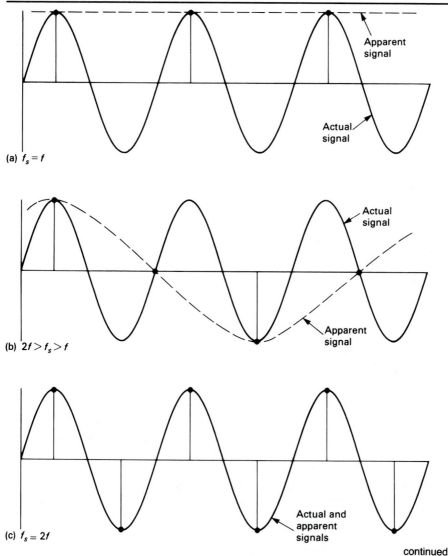

(a) $f_s = f$

(b) $2f > f_s > f$

(c) $f_s = 2f$

continued

Figure 4.17 *continued*

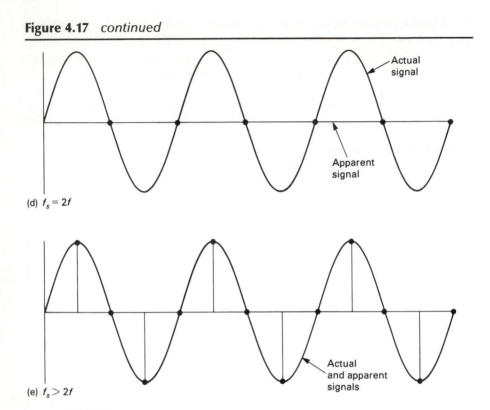

(d) $f_s = 2f$

(e) $f_s > 2f$

be present in the discrete sampling period, as shown for one wave in Fig. 4.18(a). Since the period of sampling is $T = N\,\Delta t = N/f_s$, the frequency of this wave is:

$$f_{\text{lowest}} = \frac{1}{T} = \frac{1}{N\,\Delta t} = \frac{f_s}{N} \equiv \Delta f \qquad (4.20)$$

Thus the fundamental frequency of the DFT, Δf, is also that of the lowest-frequency full wave that fits within the sampling period. No lower frequency (other than $f = 0$) is resolved.

The next-lowest frequency is that for which two full waves fit in the sampling period [Fig. 4.18(b)]. Since two periods of the wave equal the sampling period, the wave's frequency is

$$f_2 = \frac{2}{T} = 2\frac{f_s}{N} = 2\,\Delta f.$$

We can continue adding full waves to show that the only frequencies resolved by the DFT are

$$0, \Delta f, 2\,\Delta f, \dots, n\,\Delta f, \dots, \frac{N}{2}\,\Delta f = f_{\text{Nyq}}.$$

Figure 4.18 Resolving low frequencies: (a) one full wave in the sampling period, (b) two full waves in the sampling period

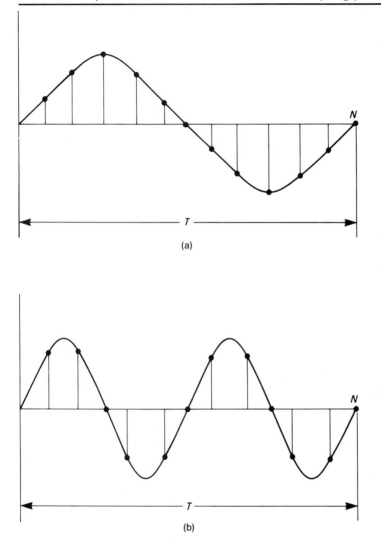

(a)

(b)

Note that although the Nyquist frequency itself *is* present in the DFT, it may not correctly represent the underlying signal, owing to phase ambiguity.

The frequencies of the DFT are spaced in increments of Δf, and thus Δf is sometimes called the *frequency resolution* of the DFT. If an analog signal contains a frequency f_0 that lies between two resolved frequencies, say $n \, \Delta f < f_0 < (n + 1) \, \Delta f$, then this frequency component will "leak" to the adjacent frequencies of the DFT. The adjacent frequencies can each show some contribution from f_0. As a result, each frequency component observed in the DFT has an uncertainty in frequency of approximately $\pm\Delta f/2$ (95%) relative to the frequencies actually present in the original analog signal. We can reduce leakage and sharpen the peaks in the frequency spectrum by decreasing Δf.

This discussion leads us to the following steps for accurate discrete sampling:

1. First, estimate the highest frequency in the signal and choose the Nyquist frequency to be greater than it. In other words, make the sample rate, f_s, *greater than twice* the highest frequency in the signal.

2. If limitations in the sample rate force you to pick a Nyquist frequency less than the highest frequency in the signal, then use a low-pass filter (Sections 7.20–7.22) to block frequencies greater than the Nyquist frequency while sampling.

3. After the sample rate is chosen, estimate the lowest frequency in the signal or estimate the frequency resolution needed to accurately resolve the frequency components in the signal. Then choose the number of points in the sample, N, to yield the desired $\Delta f = f_s/N$ at the previously determined value of the sampling frequency, f_s.

Figure 4.19 Experimental setup in which pure tones from three sound sources are mixed, detected by a microphone, displayed on an oscilloscope, and recorded by a computer

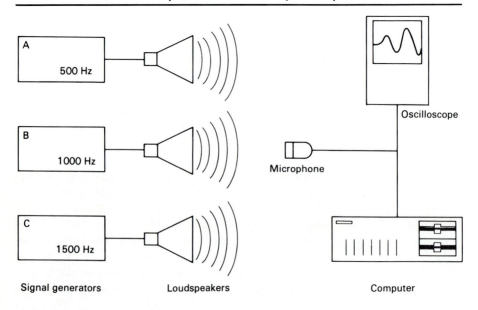

Signal generators   Loudspeakers   Computer

4.6.3 An Example of Discrete Fourier Analysis

Figure 4.19 illustrates a simple experiment. Each of three signal generators was set to produce sine waves and connected to a loudspeaker. Generator A was set to 500 Hz, generator B to 1000 Hz, and generator C to 1500 Hz. A microphone, which converts sound pressure to voltage, was used to measure the sound level. The sound level from each speaker was individually adjusted so that the microphone voltage displayed on the oscilloscope was 50 mV in amplitude. Then all three sources were run simultaneously, so that the three waveforms were mixed, and the voltage signal produced was recorded by the computer. Note that the signal generators were not in phase. The data acquired are shown in Table 4.2.

The computer sampled the microphone voltage at a rate of $f_s = 9000\,\text{Hz}$, corresponding to a Nyquist frequency of

$$f_{\text{Nyq}} = 4500\,\text{Hz}$$

The lowest frequency in the signal was known to be 500 Hz, so 18 points were used in the DFT ($N = 18$) to yield a frequency resolution of

$$\Delta f = \frac{f_s}{N} = 500\,\text{Hz}$$

Table 4.2 Experimental data for microphone voltage

Time (ms)	Voltage (mV)
0.11	10
0.22	30
0.33	70
0.44	65
0.56	15
0.67	−40
0.78	−50
0.89	15
1.00	100
1.11	135
1.22	90
1.33	−40
1.44	−130
1.56	−140
1.67	−110
1.78	−50
1.89	−10
2.00	0

Table 4.3 Calculated harmonic coefficients for the data of Table 4.2

Harmonic Order, n	Frequency, $n \, \Delta f$, Hz	A_n, mV	B_n, mV	C_n, mV
0	0	−1.04	0.00	1.04
1	500	−29.11	46.83	55.14
2	1000	49.09	52.70	72.02
3	1500	−21.96	−47.41	52.25
4	2000	2.78	2.23	3.56
5	2500	1.97	−0.470	2.03
6	3000	−0.96	−0.51	1.09
7	3500	0.15	2.93	2.93
8	4000	−0.37	−1.11	1.17
9	4500	−1.08	0.00	1.08

Figure 4.20 Frequency spectrum for data of Table 4.2

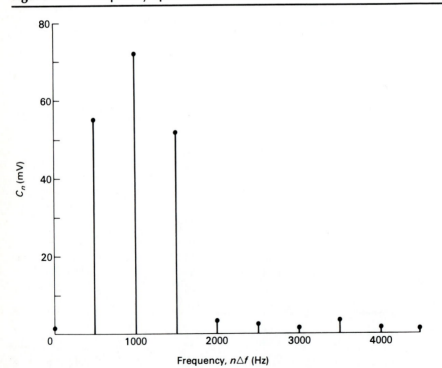

Thus, the samples were taken at intervals of 0.11 ms ($\Delta t = 1/f_s = 0.11$ ms), covering a period of $N \Delta t = 2.00$ ms.

The test data were analyzed using the computer's fast Fourier transform program. The resulting harmonic coefficients are listed in Table 4.3. Only harmonic components up to $n = N/2 = 9$ can be determined with the DFT (or FFT), and the last of these is the Nyquist frequency component, which suffers from phase ambiguity.

The first, second, and third harmonics were originally set to amplitudes of 50 mV. Since each wave is phase-shifted [recall Eqs. (4.6a–d)], we consider the amplitude of the sum of sine and cosine waves at each frequency, $C_n = \sqrt{A_n^2 + B_n^2}$. This frequency spectrum is shown in Fig. 4.20. For the first and third harmonic, the calculated amplitudes are within 10% of what we thought they should be. The second harmonic is 40% higher than we thought. Apart from the lack of precision of the data and any approximation introduced in the DFT calculation, the test condition itself may have contributed to this discrepancy. The test was run in a conventional laboratory environment, and sound reflections from the walls and ceiling may have also contributed to some distortion of the original signals. Ideally, the test should have been run in an anechoic chamber.

The sum of the harmonic components [Eq. (4.18)] is plotted together with the data in Fig. 4.21. Visually, the computed curve appears to fit the data perfectly. The DFT calculation reconstructs the original signal to within the accuracy of the original data.

Figure 4.21 Comparison of Fourier series with actual test data

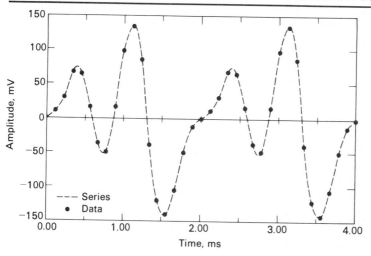

4.7   Amplitudes of Waveforms

The magnitude of a waveform can be described in several ways. The simplest waveform is a sine or cosine wave:

$$V(t) = V_a \sin 2\pi ft.$$

The *amplitude* of this waveform is V_a. The *peak-to-peak amplitude* is $2V_a$.

On the other hand, we may want a time-average value of this wave. If we simply average it over one period, however, we obtain an uninformative result:

$$\bar{V} = \frac{1}{T} \int_0^T V_a \sin 2\pi ft \, dt = -\frac{V_a}{2\pi} (\cos 2\pi - \cos 0) = 0.$$

The net area beneath one period of a sine wave is zero. Thus it is more useful to work with a *root-mean-square* (rms) value:

$$V_{\text{rms}} = \sqrt{\frac{1}{T} \int_0^T V^2(t) \, dt} = \sqrt{\frac{1}{T} \int_0^T V_a^2 \sin^2 2\pi ft \, dt} = \cdots = \frac{V_a}{\sqrt{2}}. \quad (4.21)$$

For more complex waveforms, the frequency spectrum provides a complete description of the amplitude of each individual frequency component in the signal. However, the spectrum can be cumbersome to use, and a single time-average value is often more convenient to work with. For this reason, the rms amplitude is generally applied to complex waveforms as well. Meters that measure the rms value of a waveform are discussed in Sections 9.3–9.5.

4.8   Summary

Mechanical and electrical measuring systems often produce time-varying output signals. We have seen that even fairly complex signals may be broken down and analyzed as a mixture of harmonic components, each having different frequency and amplitude. While reading the following chapters, keep in mind that all dynamic inputs, those whose magnitudes vary with time, are in reality only combinations of simple sinusoidal building blocks.

1.   Simple harmonic motion, such as $s = s_0 \sin \omega t$, is the most basic form of time-dependent behavior. The frequency of such motion can be described by either the cyclic frequency, f, or the circular frequency, $\omega = 2\pi f$. The relationship of these two frequencies may be visualized in terms of the Scotch-yoke mechanism (Sections 4.2–4.3).

2.   More complex waveforms can be represented by a sum of simple harmonic components having different amplitudes and frequencies. Even fairly sharp waveforms, such as the square wave, can be described by such sums. These sums are called Fourier series (Section 4.4, 4.4.2).

3.   When a nonperiodic waveform is recorded for a time interval T, the recorded portion may be viewed as a periodic waveform of period T and frequency $f = 1/T$ (Section 4.4.3).

4. When two waves of nearly equal frequency are added, the resulting waveform undergoes periodic beats at a frequency of one-half the frequency difference of the original waves (Section 4.4.1).

5. The frequencies present in a complex waveform may be described using the frequency spectrum, which shows the amplitude of each frequency component present (Section 4.5).

6. The process of determining the frequency spectrum of a complex signal is called harmonic analysis or Fourier analysis. Several methods of harmonic analysis are available, depending on the nature of the signal being studied (Section 4.6, Appendix B).

7. When a signal is recorded by a computer, only discrete points are stored. The discrete Fourier transform, or DFT, may be used to find the frequency spectrum of the recorded data (Section 4.6.1, Appendix B).

8. Accurate discrete sampling can be ensured by selecting appropriate values of the sampling frequency, f_s, and the number of sample points recorded, N. The correct values are determined by: (a) the Nyquist frequency, $f_{Nyq} = f_s/2$, which must be greater than the highest frequency in the signal; and (b) the frequency resolution, $\Delta f = f_s/N$, which is both the lowest frequency in the discrete signal and the spacing of frequencies in the signal's DFT (Section 4.6.2).

9. The average amplitude of a complex signal is often described using the root-mean-square, or rms, value (Section 4.7).

Suggested Readings

Churchill, R. V., and J. W. Brown. *Fourier Series and Boundary Value Problems.* 3rd ed. New York: McGraw-Hill, 1978.

Greenberg, M. D. *Foundations of Applied Mathematics.* Englewood Cliffs, N.J.: Prentice Hall, 1978.

Oppenheim, A. V., A. S. Willsky, and I. T. Young. *Signals and Systems.* Englewood Cliffs, N.J.: Prentice Hall, 1983.

Problems

4.1 The following expression represents the displacement of a point as a function of time:

$$y(t) = 100 + 95 \sin 15t + 55 \cos 15t.$$

    a.   What is the fundamental frequency in hertz?
    b.   Rewrite the equation in terms of cosines only.

4.2 Rewrite each of the following expressions in the form of Eq. (4.6).
    a. $y = 3.2 \cos(0.2t - 0.3) + \sin(0.2t + 0.4)$
    b. $y = 12 \sin(t - 0.4)$

Figure 4.22   Oscilloscope trace for Problem 4.8

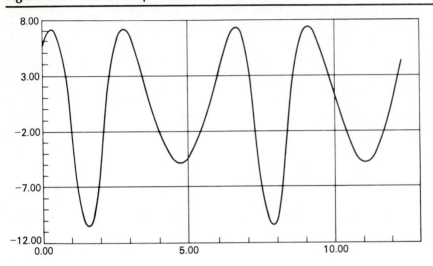

4.3   Construct a frequency spectrum for Fig. 4.9(a).

4.4   Construct a frequency spectrum for Fig. 4.9(c).

4.5   Construct a frequency spectrum for Fig. 4.9(e).

4.6   Construct a frequency spectrum for Fig. 4.9(f).

4.7   Construct a frequency spectrum for Fig. 4.9(g).

4.8   Figure 4.22 represents a trace from an oscilloscope where the ordinate is in volts and the abscissa is in milliseconds. Determine its discrete Fourier transform and sketch its frequency spectrum. Use a sampling frequency of 2000 Hz. Check for periodicity in choosing your data window.

Figure 4.23   Pressure-time record for Problem 4.9

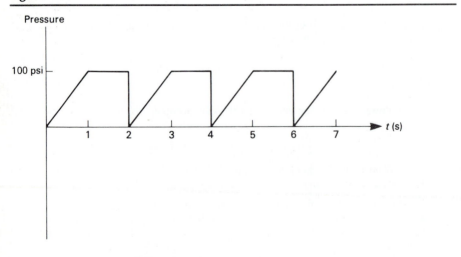

4.9   Consider a pressure-time record as shown in Fig. 4.23. Determine its frequency spectrum.

4.10 If the signal of Problem 4.9 were to be sampled digitally for its discrete Fourier transform, what sampling frequency would you recommend?

4.11 Solve Problem 4.8 using a sampling frequency of 10 Hz and compare your results with those of Problem 4.8.

4.12 Using the data files in Table 4.4, if t is in milliseconds and $f(t)$ is in volts, determine the discrete Fourier transform for each set of digital data.
   a.   Use $f_1(t)$.
   b.   Use $f_2(t)$.
   c.   Use $f_3(t)$.
   d.   Use $f_4(t)$.
   e.   Use $f_5(t)$.
   f.   Use $f_6(t)$.

4.13 A 500 Hz sine wave is sampled at a frequency of 4096 Hz. A total of 2048 points are taken.
   a.   What is the Nyquist frequency?
   b.   What is the frequency resolution?
   c.   The student making the measurement suspects that the sampled waveform contains several harmonics of 500 Hz. Which of these can be accurately measured? What happens to the others?

4.14 A 150 Hz cosine wave is sampled at a rate of 200 Hz.
   a.   Draw the wave and show the temporal locations at which it is measured.
   b.   What apparent frequency is measured?
   c.   Describe the relation of the measured frequency to aliasing. Give a numerical justification for your answer.

4.15 a.   Suppose that a 500 Hz sinusoidal signal is sampled at 750 Hz. Draw the discrete time signal found and determine the apparent frequency of the signal.
   b.   If a 200 Hz component were present in the signal of part a, would it be detected? Explain.
   c.   If a 375 Hz component were present in the signal of part a, would it be detected? Explain.

4.16 A engineer is studying the vibrational spectrum of a large diesel engine. Her modeling estimates suggest that a strong resonance is likely at 250 Hz, and that weaker frequencies of up to 2000 Hz may be excited also. She has placed an accelerometer on the machine to measure the vibration spectrum. She samples the accelerometer output voltage using her computer's analog-to-digital converter board.
   a.   What is the minimum sample rate she should use?
   b.   To reliably test her model of the machine's vibration, she must resolve the peak resonant frequency to ± 1 Hz. How can she achieve this level of resolution?

4.17 A temperature measuring circuit responds fully to frequencies below 8.3 kHz; above this frequency, the circuit attenuates the signal. This circuit is to be used to measure a temperature signal with an unknown frequency spectrum. Accuracy of

Table 4.4 Data for Problem 4.12

t	$f_1(t)$	$f_2(t)$	$f_3(t)$	$f_4(t)$	$f_5(t)$	$f_6(t)$
0	5.4	3.76	10.2	4	10.1	−9.4
1	4.74	3.88	10	2.58	9.34	−6.8
2	3.01	4.19	9.83	0.99	8.32	−3.5
3	0.8	4.54	9.57	1.54	7.05	−0.2
4	−1.2	4.67	9.26	4.4	5.57	2.52
5	−2.4	4.46	8.92	8.02	3.91	4.11
6	−2.7	4.06	8.56	10.7	2.14	4.58
7	−2.4	3.73	8.21	11.6	0.3	4.17
8	−1.8	3.6	7.89	11.2	−1.6	3.3
9	−1.6	3.55	7.62	10	−3.4	2.47
10	−1.8	3.35	7.43	8.52	−5.1	2.03
11	−2.4	2.89	7.34	7.19	−6.6	2.11
12	−2.7	2.36	7.36	6.66	−7.9	2.57
13	−2.4	2.01	7.5	7.3	−9	3.05
14	−1.2	1.98	7.74	8.45	−9.9	3.11
15	0.8	2.14	8.07	8.46	−10	2.36
16	3.01	2.24	8.48	5.85	−11	0.64
17	4.74	2.16	8.94	0.9	−11	−1.9
18	5.4	2.06	9.41	−4	−10	−4.9
19	4.74	2.16	9.87	−6	−9.3	−7.6
20	3.01	2.57	10.3	−4.5	−8.3	−9.3
21	0.8	3.13	10.6	−1.5	−7.1	−9.5
22	−1.2	3.56	10.9	−0.9	−5.6	−7.9
23	−2.4	3.73	11.1	−4.6	−3.9	−4.5
24	−2.7	3.76	11.2	−11	−2.1	0.06
25	−2.4	3.88	11.3	−15	−0.3	4.93
26	−1.8	4.19	11.2	−15	1.55	9.15
27	−1.6	4.54	11.2	−10	3.35	11.8
28	−1.8	4.67	11.1	−5.1	5.05	12.4
29	−2.4	4.46	10.9	−3.7	6.59	10.7
30	−2.7	4.06	10.8	−6.7	7.94	7.1
31	−2.4	3.73	10.7	−11	9.04	2.29
32	−1.2	3.6	10.6	−12	9.87	−2.7
33	0.8	3.55	10.5	−8.5	10.4	−7
34	3.01	3.35	10.4	−2.4	10.6	−9.7
35	4.74	2.89	10.3	2.57	10.5	−11
36	5.4	2.36	10.2	4	10.1	−9.4

±1 Hz is desired in the frequency components. If no frequency components above 8.3 kHz are present in the circuit's output, what sample rate and number of samples should be used?

4.18 A cantilever beam of stiffness K supports a large mass M on its free end. The vibrational frequency of the beam approximately equal to

$$f = \frac{1}{2\pi} \sqrt{\frac{K}{M}}$$

In an experiment, this frequency was measured by attaching a strain gage to the beam to produce a waveform corresponding to the oscillating motion. The waveform was then discretely sampled using the lab computer, and the frequency of motion was obtained from an FFT of the waveform.

For one particular case, the mass at the end of the beam was measured to be 60.10 g, to an uncertainty of ±0.11 g (95%). The waveform was sampled at a rate of 128 Hz for 128 points, and a peak frequency of 10 Hz was returned by the FFT calculation.

a. Calculate the beam stiffness and its uncertainty (95%).

b. What is the primary source of uncertainty in this result? What is the best way to reduce the uncertainty of the result?

c. The experimenter's disk is nearly full, so he does not wish to increase the number of points sampled when he repeats the experiment. If the sample rate can be adjusted in increments of 1 Hz, what sample rate would allow the lowest uncertainty with the same number of points? Estimate the uncertainty in stiffness for that sample rate.

References

1. Beckwith, T. G., and R. D. Marangoni. *Mechanical Measurements.* 4th ed. Reading, Mass.: Addison-Wesley, 1990.

2. Oppenheim, A. V., A. S. Willsky, and I. T. Young. *Signals and Systems.* Englewood Cliffs, N.J.: Prentice Hall, 1983.

CHAPTER 5

||||||||||||||||||||||||||||||||||||||||||||||||||||||||||

The Response of Measuring Systems

5.1 Introduction

Quite simply, *response* is a measure of a system's fidelity to purpose. It may be defined as an evaluation of the system's ability to faithfully sense, transmit, and present all the pertinent information included in the measurand and to exclude all else.

We would like to know if the output information truly represents the input. If the input information is in the form of a sine wave, a square wave, or a sawtooth wave, does the output appear as a sine wave, a square wave, or a sawtooth wave, as the case may be? Is each of the harmonic components in a complex wave treated equally, or are some attenuated, completely ignored, or perhaps shifted timewise relative to the others? These questions are answered by the response characteristics of the particular system—that is, (1) amplitude response, (2) frequency response, (3) phase response, and (4) slew rate.

5.2 Amplitude Response

Amplitude response is governed by the system's ability to treat all input amplitudes uniformly. If an input of 5 units is fed into a system and an output of 25 indicator divisions is obtained, we can generally expect that an input of 10 units will result in an output of 50 divisions. Although this is the most common case, other special nonlinear responses are also occasionally required. Whatever the arrangement, whether it be linear, exponential, or some other amplitude function, discrepancy between design expectations in this respect and actual performance results in poor amplitude response.

Of course no system exists that is capable of responding faithfully over an

Figure 5.1 Gain versus input voltage for amplifier section of a
commercially available strain measuring system for a
frequency of 1 kHz (Gain = output voltage/input voltage)

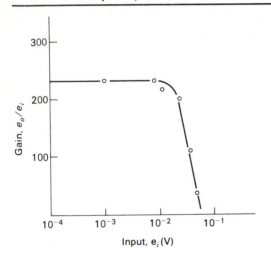

unlimited range of amplitudes. All systems can be overdriven. Figure 5.1
shows the amplitude response of a voltage amplifier suitable for connecting a
strain-gage bridge to an oscilloscope. The usable range of the amplifier is
restricted to the horizontal portion of the curve. The plot shows that for inputs
above about 0.01 V the amplifier becomes overloaded and the amplification
ceases to be linear.

5.3 Frequency Response

Good frequency response is obtained when a system reacts to all *frequency
components* in the same way. If a 100-Hz sine wave with an input amplitude of
5 units is fed into a system and a peak-to-peak output of $2\frac{1}{2}$ in. results on an
oscilloscope screen, we can expect that a 500-Hz sine-wave input of the *same*
amplitude would also result in a $2\frac{1}{2}$-in. peak-to-peak output. Changing the
frequency of the input signal should not alter the system's output magnitude so
long as the input amplitude remains unchanged.

Yet here again there must be a limit to the range over which good
frequency response may be expected. This is true for any dynamic system,
regardless of its quality. Figure 5.2 illustrates the frequency response relations
for the same voltage amplifier used in Fig. 5.1. Frequencies above about
10 kHz are attenuated. Only inputs below this frequency limit are amplified in
the correct relative proportion.

Figure 5.2 Frequency response curve for amplifier section of a
commercially available strain-measuring system; $e_i = 10\,mV$

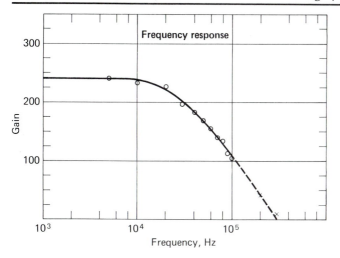

5.4   Phase Response

Amplitude and frequency responses are important for all types of input waveforms, whether simple or complex. *Phase response*, however, is of primary importance for the complex wave only.

Time is required for the transmission of a signal through any measuring system. Often, when a simple sine-wave voltage is amplified by a single stage of amplification, the output trails the input by approximately 180°, or one-half cycle (see Fig. 5.20). For two stages, the phase shift may be about 360°, and so on. The actual shift will not be an exact multiple of half-wavelengths but will depend on the equipment and the frequency. It is the frequency dependence that defines phase response.

For a single-sine-wave input, any phase shift would normally be unimportant. The output produced on the oscilloscope screen would show the true waveform, and its proper parameters could be determined. The fact that the shape being shown was actually formed on the screen a few microseconds or a few milliseconds after being generated is of no consequence.

Let us consider, however, a complex wave made up of numerous harmonics. If the phase lag is different for each frequency, then each component of the complex wave is delayed by a different amount. The harmonic components then emerge from the system in phase relations different from when they entered. The whole waveform and its amplitudes are changed, a result of poor phase response.

Figure 5.3 illustrates phase response characteristics for a typical voltage amplifier.

Figure 5.3 Phase lag versus frequency (phase response) for
the same amplifier used in Fig. 5.2

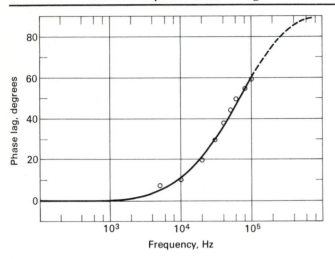

5.5 Predicting Performance for Complex Waveforms

Response characteristics of an existing system or a component of a system may
be determined experimentally by injecting as input a signal of known form,
then determining the output, and finally comparing the results (see Section
10.6). Of course, the most basic test waveform is the sine wave.

It we know the sine-wave response of a device, can we use this information
to predict how it will respond to a complex input, such as a square wave or one
of the various sawtooth waveforms? The answer is yes, as we demonstrate in
the following example.

Example 5.1
Using a computer program and information given in Figs. 5.2 and 5.3,
predict the form of amplifier output to be expected if a perfect square
wave is the input. Do this for input having fundamental frequencies of
1000 and 2000 Hz and amplitude of 1 mV.

Solution. Recall that in Table 4.1 a square wave is defined by the infinite
series

$$y = \frac{4A}{\pi} \sum_{n=1}^{\infty} \left[\frac{1}{(2n-1)} \sin(2n-1)2\pi ft \right]. \tag{5.1}$$

Figure 5.4 (a) Computer-determined response to a 1000-Hz square wave by the amplifier whose characteristics are shown in Figs. 5.1 through 5.3, (b) computer-determined response to a 2000-Hz square wave by the amplifier whose characteristics are shown in Figs. 5.1 through 5.3 (Ideal amplitude = 240 mV)

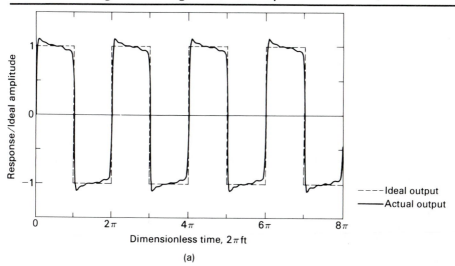

(a)

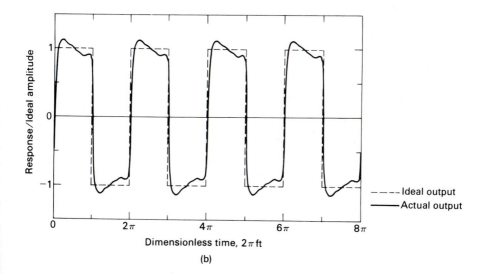

(b)

We may modify Eq. (5.1) by introducing frequency and phase distortion factors:

$$y = \frac{4A}{\pi} \sum_{n=1}^{\infty} \frac{G_n}{(2n-1)} \sin[(2n-1)2\pi ft - \phi_n], \qquad (5.1a)$$

where

G_n = an amplitude factor based on frequency,

ϕ_n = a phase distortion factor.

Magnitudes for G_n and ϕ_n for each harmonic order can be extracted from the response curves, Figs. 5.2 and 5.3. For example, if $n = 15$, then $(2n - 1)f = 29(1000\,\text{Hz}) = 29\,\text{kHz}$, and we read $G_{15} = 200$ from Fig. 5.2 and $\phi_{15} = 30°$ from Fig. 5.3. The computer code can incorporate tables of such frequency-response and phase-response data taken directly from the sine-wave response curves. The results of using these data in Eq. (5.1a) are plotted in Fig. 5.4(a) and (b).

We can make similar calculations for any waveform for which a harmonic series can be written. In particular, to investigate a measuring system's response to a waveform of interest, we can make a Fourier analysis of that waveform and investigate the system's response characteristics using sine-wave test results, as before.

5.6 Delay, Rise Time, and Slew Rate

Finally, a fourth type of response, which is actually another form of frequency response, is *delay,* or *rise time.* When a stepped or relatively instantaneous input is applied to a system, the output may lag, as shown in Fig. 5.5. The time delay after the step is applied, but before proper output magnitude is reached is known as rise time. It is a measure of the system's ability to handle transients. Sometimes rise time is defined specifically as the time, Δt, required for the system to pass from 10% to 90% of its final response. Alternatively, transient response may be characterized by the *settling time* required for the system response to remain within some small percentage of its final value.

Slew rate is the *maximum* rate of change that the system can handle. In electrical terms, it is *de/dt,* or volts per unit time (e.g., 25 V/μs). The term slew rate or slew speed is also used in other ways; for example, the slew speed of the stylus of an *xy* plotter would be expressed in terms of centimeters/second, a reference to the maximum speed with which the stylus can traverse its range of motion.

Figure 5.5   Response of a typical system to a step input,
showing rise time (Δt) and settling time

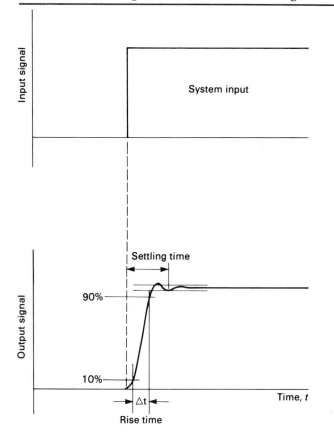

5.7   Response of Experimental System Elements

An experimentalist can usually avoid operating a measuring system under conditions when amplitude response or slew rate are limiting factors. For example, a solid-state amplifier that is overloaded typically produces a constant output of ± 5 V irrespective of the input; such a condition should be fairly obvious, and its result is clearly useless.

Frequency- and phase-response limitations are not so readily apparent. One would prefer zero phase shift and flat frequency response for all frequencies of interest. However, because experimental inputs often contain a wide range of frequencies, poor response to some subset of frequencies may go unnoticed. An evaluation procedure is generally necessary to establish a particular system's response.

We have seen that a sine-wave test can provide the frequency and phase

Figure 5.6 Response of a system element to a sine-wave input

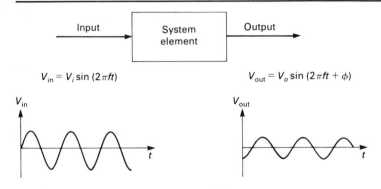

$V_{in} = V_i \sin(2\pi ft)$ $V_{out} = V_o \sin(2\pi ft + \phi)$

response frequency by frequency. The sine-wave response of a measuring system element, such as a transducer or an amplifier, provides a foundation for evaluating the performance of the overall measurement system.

Each element in the system transfers its input to an output (Fig. 5.6). The output can differ from the input in both amplitude and phase. The ratio of output amplitude to input amplitude is the gain (or amplification) for that frequency, and the variation of the gain over all frequencies is what we have called frequency response,

$$G(f) = \frac{V_o}{V_i}.$$

Similarly, we called the variation of the phase shift with frequency the phase response, $\phi(f)$. These are, of course, the functions shown in Figs. 5.2 and 5.3.*

A complete measuring system consists of a series of elements, from sensors and transducers to signal conditioners to recording and display devices. For any given frequency component, the system's gain is the product of the gains of all system elements at that frequency. Likewise, the system phase shift is the sum of every stage's phase shift. To obtain an acceptable measurement, every measuring stage must have acceptable response over the frequency range of interest. If any single stage does not respond properly, it will distort the signal and contaminate the entire measurement.

Figure 5.7 shows a typical measuring system. A periodic measurand is detected by a sensor. This measurand might be a position, x, that oscillates at a frequency f_x. The sensor produces an output voltage in millivolts, which varies with x. However, the sensor also picks up high-frequency electrical noise from nearby equipment. Thus, the output of the sensor includes both the low frequency of the signal, f_x, and the high frequency of the noise, f_n (see Fig.

---

*The reader who has studied system dynamics will recognize that $G(f)$ and $\phi(f)$ are the magnitude and phase, respectively, of the periodic transfer function, $H(j\omega)$, of a linear system [1].

Figure 5.7 Measuring system composed of several system elements;
the signal is shown as it leaves each element

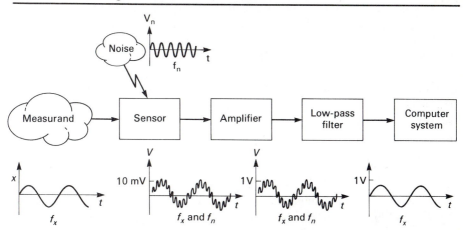

4.6). The sensor's signal is received by an amplifier (gain = 100), which raises
the signal level to a value convenient for computer recording. Apart from the
added noise, both measuring stages respond faithfully to the input signal.

Certain system elements are chosen specifically for the limitations of their
frequency-response characteristics. In the case of Fig. 5.7, the measurand
varies at a fairly low frequency, whereas the unwanted electrical noise is at a
much higher frequency. Since the experimentalist has evaluated the output
signal on an oscilloscope, she has a clear idea of which range of frequencies
characterizes the measurand and which range characterizes the noise. Thus,
she inserted a signal-conditioning *low-pass filter* after the amplifier to eliminate
the noise. This filter has flat frequency response at low frequencies (such as f_x)
and zero response at high freqencies (such as f_n), as shown in Fig. 5.8. The

Figure 5.8 Frequency response for the low-pass filter of Fig. 5.7

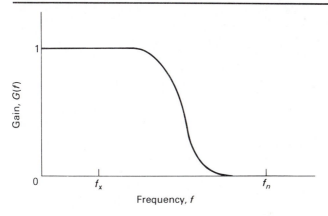

signal that the filter transmits to the computer recording system excludes the noise component. We discuss filters in more detail when we consider signal conditioning in Chapter 7 (Sections 7.20–7.22).

In contrast to the filter, a well-designed sensor or transducer should respond to all frequencies equally. Unfortunately, most actual sensors and transducers do not. Instead, such devices are characterized by an upper or lower frequency beyond which response is attenuated (much like a filter) or by a high or low frequency at which the sensor resonates with the input, producing an output that is ridiculously large. Determination of such limiting frequencies is extremely important if a dynamic measurand is to be accurately recorded.

The frequency and phase response of a system element (or of the entire chain of elements) can be determined several ways. For simple elements, we may be able to construct a physical model of the device that accurately predicts its response. For more complex systems, we may wish to test the response experimentally, for example, by using a sine-wave test. In other circumstances, we may rely on response data provided by the manufacturer. In any case, once we have determined the range of frequencies for which the system responds accurately, we will disregard frequencies outside this range, if they can be identified, and we will take precautions to prevent such frequencies from entering the measuring chain.

The remainder of this chapter is concerned with the identification of system response. Sections 5.8 through 5.19 address the modeling of frequency and phase response for simple physical systems. Section 5.20 returns to the matter of experimental determination and calibration of system response.

5.8 Simplified Physical Systems

What basic physical factors govern response? In terms of practical systems, we are confronted with two fundamental segments of construction: mechanical and electrical. The basic mechanical elements are mass, damping, and some form of equilibrium-restoring element, such as a spring. Corresponding electrical elements are inductance, resistance, and capacitance. Although many, if not most, devices and systems involve both electrical and mechanical elements, for our immediate purposes it is advantageous to consider the two separately. In the next several sections we will discuss some of the mechanical aspects; and beginning with Section 5.17, we will consider the electrical.

5.9 Mechanical Elements

A discussion of the dynamic characteristics of an elementary mechanical system necessitates a short description of the elements composing such a system.

5.9.1 Mass

It is obvious that in all cases mass will be a factor. Under certain conditions, however, the masses making up the device or system will not affect its performance. We will consider such cases in Sections 5.14 and 5.15.

By its very nature, mass must be distributed throughout some volume. In many cases, however, it is not only convenient but also correct, or nearly so, to assume that the mass of a member is concentrated at a point. Depending on the geometry of the member and its application, the point of concentration may or may not be the center of gravity. In certain cases, the center of percussion may be the location of effective concentration.

5.9.2 Spring Force

Many mechanical members deflect in direct proportion to the force exerted on them, that is, $\Delta F/\Delta\delta = k = $ a constant, where ΔF is an applied force increment and $\Delta\delta$ is the resulting deflection increment. Most coil springs, beams, and tension/compression members abide by this relationship. It may be noted that the force is opposed to the deflection, that is, the resulting force always attempts to restore equilibrium.

Torsional members commonly adhere to the relationship $\Delta T/\Delta\theta = k_t = $ a constant, where ΔT is an applied torque increment and $\Delta\theta$ is the resulting torsional deflection increment. The constants k and k_t are called *spring constants,* or *deflection constants.*

Elasticity is not always the source of the restoring force, however. In certain cases, such as for a beam balance (see Section 5.10), the restoring force may be supplied by gravity.

When the motion of a concentrated mass is constrained by an equilibrium-seeking member [Fig. 5.9(a)], simple vibration theory shows that the combina-

Figure 5.9 Elementary spring-mass systems (a) without damping, (b) with viscous damping

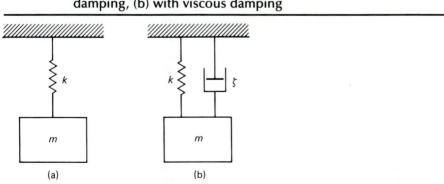

(a)   (b)

tion will have a natural frequency

$$\omega_n = \sqrt{\frac{kg_c}{m}},$$ (5.2)

where

ω_n = circular frequency in radians per second (see Section 4.3)

    $= 2\pi f,$

f = frequency of vibration in hertz,

g_c = the dimensional constant (see Table 2.6).

A system of this sort is said to have a *single degree of freedom,* that is, it is assumed to be constrained in some way to oscillate in a single mode or manner, needing only one coordinate to fully describe its motion.

5.9.3   Damping

Another factor important to the usefulness of any system of this type is damping [Fig. 5.9(b)]. Damping in this connection is usually thought of as viscous, rather than Coulomb or frictional, and may be obtained by fluids (including gases) or by electrical means.

Viscous damping is a function of velocity, and the force opposing the motion may be expressed as

$$F = -\zeta \frac{ds}{dt},$$ (5.3)

where ζ = the damping coefficient and ds/dt = the velocity. We can see that the damping coefficient is an evaluation of force per unit velocity. The negative sign indicates that the resulting force opposes the velocity. The effect of viscous damping on a freely vibrating single-degree-of-freedom system is to reduce the vibrational amplitudes with respect to time according to an exponential relation.

Damping magnitude is conveniently thought of in terms of *critical damping,* which is the minimum damping that can be used to prevent overshoot when a damped spring-mass system is deflected from equilibrium and then released (see Section 5.16.1 for further discussion). This limiting condition is shown in Fig. 5.10. The value of the critical damping coefficient, ζ_c, for a simple spring-supported mass m is expressed by the relation

$$\zeta_c = 2\sqrt{\frac{mk}{g_c}}.$$ (5.4)

Damping is often specified in terms of the dimensionless damping ratio,

$$\xi = \frac{\zeta}{\zeta_c}.$$

Figure 5.10   Time-displacement relations for damped motion: (a) for
damping greater than critical, (b) for critical
damping, (c) for damping less than critical

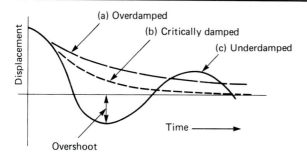

Many measuring devices or system components involve elements constrained by gravity or spring force, whose deflection is analogous to the signal
input. The ordinary balance scale is an example, as are the D'Arsonval meter
movement and the mirror-type galvanometer (Section 9.9). The same is true of
most pressure transducers, elastic-force transducers, and many other measuring devices.

If the system is a translational one, a spring-constrained mass may be
involved. Our tire gage of Section 1.4 illustrates the case in which the piston
and stem and a portion of the spring constitute the mass whose motion is
controlled by the interaction of the applied pressure and the spring force.
Other examples are the seismic-type instruments discussed in Chapter 17.

These devices depend on equilibrium for correct indication. When
equilibrium is disturbed by a change of input, the system requires time to
readjust to the new equilibrium, and a number of oscillations may take place
before the new output is correctly indicated. The rate at which the amplitude
of such oscillations decreases is a function of the system's damping. In
addition, the frequency of oscillation is a function of both damping and
sensitivity.

5.10   An Example of a Simple Mechanical System

Let us consider a symmetrical scale beam *without* damping (Fig. 5.11). For
simplification, we will assume that the masses of the scale pans and the weights
being compared are concentrated at points A and B, and that they are also
included in the moment of inertia I, which is referred to the main pivot point
O. Further, we will assume that a small difference, ΔW, exists between the
two weights being compared, and that points A, B, and O lie along a
straight line. We define sensitivity, η, as the ratio of the displacement of the

Figure 5.11 Schematic diagram of a beam balance

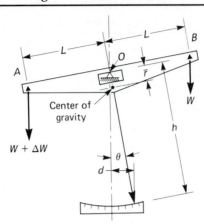

end of the pointer to the length of the pointer, h, divided by ΔW, or

$$\eta = \frac{1}{\Delta W}\frac{d}{h} = \frac{1}{\Delta W}\tan\theta. \tag{5.5}$$

The system behaves as a compound pendulum, and it can be shown that the period of oscillation will be

$$T = 2\pi\sqrt{\frac{I}{\bar{r}w_b g_c}} \tag{5.6}$$

where

I = the moment of inertia,

$\bar{r}$ = the distance between the center of gravity of the beam alone and about pivot point O,

w_b = the weight of the beam.

With the weights applied,

$$w_b\bar{r}\sin\theta = \overline{\Delta W}L\cos\theta \quad\text{or}\quad \tan\theta = \frac{L\,\overline{\Delta W}}{w_b\bar{r}}. \tag{5.7}$$

Hence, using Eq. (5.5), we find that

$$\eta = \frac{L}{w_b\bar{r}}. \tag{5.7a}$$

Combining Eqs. (5.6) and (5.7a), we have

$$\eta = \frac{L}{I}\left(\frac{T}{2\pi}\right)^2\cdot g_c = \frac{L}{I}\left(\frac{1}{2\pi f}\right)^2\cdot g_c, \tag{5.7b}$$

where f = natural frequency.

Equation (5.7b) indicates that the sensitivity is a function of T, the *period of oscillation* of the balance scale, with increased sensitivity corresponding to a long beam and low moment of inertia. In other words, the more sensitive instrument oscillates more slowly than the less sensitive instrument. This is an important observation having significant bearing on the dynamic response of most single-degree-of-freedom instruments.

5.11   The Importance of Damping

The importance of proper damping to dynamic measurement may be understood by assuming that our scale beam in the previous example is part of an instrument that is required to come to *different* equilibria as rapidly as possible. A situation of this sort exists in the application of light-beam galvanometers to recording oscillographs. The galvanometer suspension is driven by varying frequency inputs, and its ability to follow is governed by its natural period and by damping.

Suppose that our scale beam has very low damping. When a disturbing force is applied, the scale will be caused to oscillate, and the oscillation will continue for a long period of time. A final balance will be obtained only by prolonged waiting, and thus the frequency with which the weighing process may be repeated is limited.

On the other hand, suppose considerable damping is provided—well above critical. An extreme example of this would be to submerge the entire scale in a container of molasses. Balance would be approached at a very slow rate again, but in this case there would be no oscillation. Here, again, excessive time would be required before the next weighing operation could commence.

It appears, therefore, that if we were to design a beam-type scale for quickly determining magnitudes of different masses, the final form would necessarily be a result of compromise. We would like equilibrium to be reached as quickly as possible in order to *get on with the job*. It would seem that there might be an optimum value that should be used. Although this is not exactly the case because of other factors involved, damping of the order of 60% to 75% of critical is provided in many instruments of this type (see Section 5.16.2 for further clarification of this point).

Although damping will tend to decrease the frequency of oscillation, it does not change the inherent sensitivity of the device, which is related to the undamped natural frequency. However, some compromises must still be made in regard to sensitivity. Sensitivity increases in proportion to the undamped natural period, as shown in the previous section. Because a high natural period (or low natural frequency) usually corresponds to a lessened frequency response, sensitivities greater than those required by the application should be avoided in the interest of maintaining adequate response.

5.12   Dynamic Characteristics of Simplified Mechanical Systems

By making certain simplifying modeling assumptions, we may place the dynamic characteristics of *most* measuring systems in one of several categories. The basic assumptions are that any restoring element (such as a spring) is linear, that damping is viscous, and that the system may be approximated as a single-degree-of-freedom system.

5.13   Single-Degree-of-Freedom Spring-Mass Damper Systems

Figure 5.12 shows a simple single-degree-of-freedom mechanical system. It is single-degree because only one coordinate of motion is necessary to completely define the motion of the system. We will also assume a general form of excitation, $F(t)$, which may or may not be periodic. Forces acting on the mass will result from the spring, damping, and the external force, $F(t)$. Using Newton's second law we can write

$$F(t) - ks - \zeta \frac{ds}{dt} = \frac{m}{g_c} \left(\frac{d^2s}{dt^2} \right). \tag{5.8}$$

Note that the spring force will always oppose the displacement and that the damping is proportional to velocity and opposite the velocity direction. This relationship can be rearranged to read

$$\frac{1}{g_c} \left[m \left(\frac{d^2s}{dt^2} \right) \right] + \zeta \left(\frac{ds}{dt} \right) + ks = F(t). \tag{5.8a}$$

If we assume $F(t)$ to be periodic with time, we can substitute the

Figure 5.12   Mechanical model of a force-excited second-order system

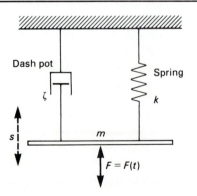

appropriate Fourier series for $F(t)$ or, in general (see Section 4.4),

$$F(t) = \frac{A_0}{2} + \sum_{n=1}^{\infty} C_n \cos(n\Omega t - \phi_n), \tag{5.9}$$

where Ω = fundamental circular forcing frequency and

$$C_n = \sqrt{A_n^2 + B_n^2} \quad \text{and} \quad \tan\phi = \frac{B_n}{A_n}. \tag{5.9a}$$

We will consider this general case in Section 5.16.3. First, however, we will consider several special cases.

5.14   The Zero-Order System

A nearly trivial case occurs if we remove the spring and damper. The voltage-dividing potentiometer (see Sections 6.6 and 7.7) is an example. In its simplest form this device is a single slide wire. Aside from the mass of the slider and any member attached to it, there is no appreciable resistance to movement. In particular, an equilibrium-seeking force is not present and the output is independent of time, that is,

$$\text{Output} = \text{constant} \times \text{input.}$$

Dynamically, the zero-order system requires no further consideration.

5.15   Characteristics of First-Order Systems

If we assume the mass, m, in Fig. 5.12 and Eq. (5.8a) to be zero, we obtain a *first-order system*. Such systems have a balance between damping forces ($\zeta\, ds/dt$), restoring forces (ks), and externally applied forces [$F(t)$]. Among instruments, the most common first-order systems are temperature-measuring devices. A temperature sensor usually responds as a first-order system because the rate of change of its temperature (dT_s/dt) is proportional to its current temperature (T_s).* Obviously, a mercury-in-glass thermometer or a thermocouple (Section 16.5) bears no physical resemblance to Fig. 5.12. However, the dynamic responses of the two systems are identical.

For the first-order system we can write

$$\zeta\frac{ds}{dt} + ks = F(t). \tag{5.10}$$

---

* For further discussion, see Section 16.10.3.

5.15.1   The Step-Forced First-Order System

Let

$$F(t) = 0, \qquad \text{for } t < 0,$$

and

$$F(t) = F_0, \qquad \text{for } t \geq 0.$$

For force equilibrium on the connecting element (which is assumed to be massless),

$$\zeta \frac{ds}{dt} + ks = F_0, \tag{5.11}$$

where

$t =$ time,

$s =$ displacement,

$\zeta =$ the damping coefficient,

$k =$ the deflection constant,

$F_0 =$ the amplitude of the constant input force.

Then

$$\int_0^t dt = \zeta \int_{s_A}^s \frac{ds}{(F_0 - ks)},$$

from which we obtain

$$\frac{F_0 - ks}{F_0 - ks_A} = e^{-kt/\zeta} = e^{-t/\tau}. \tag{5.12}$$

The units of $\tau = \zeta/k$ are seconds, and this quantity is known as the *time constant*.

Equation (5.12) can be written

$$s = s_\infty[1 - e^{-t/\tau}] + s_A e^{-t/\tau} = s_\infty + [s_A - s_\infty]e^{-t/\tau}, \tag{5.13}$$

where

$s_\infty = F_0/k =$ the final displacement of the system as $t \to \infty$,

$s_A =$ the initial displacement at $t = 0$.

We have assumed that the first-order system represents any dynamic condition wherein the elements are essentially massless, the displacement constraint is linear, and a significant viscous rate constraint is present. Generally, Eq. (5.13) can be written

$$P = P_\infty[1 - e^{-t/\tau}] + P_A e^{-t/\tau}$$

or

$$P = P_\infty + [P_A - P_\infty]e^{-t/\tau}, \tag{5.14}$$

where

P = the magnitude of any first-order process at time t,

P_∞ = the limiting magnitude of the process as $t \to \infty$,

P_A = the initial magnitude of the process at $t = 0$.

Figure 5.13 Characteristics of a first-order system subjected to a step input at $t = 0$: (a) for a progressive process, (b) for a decaying process

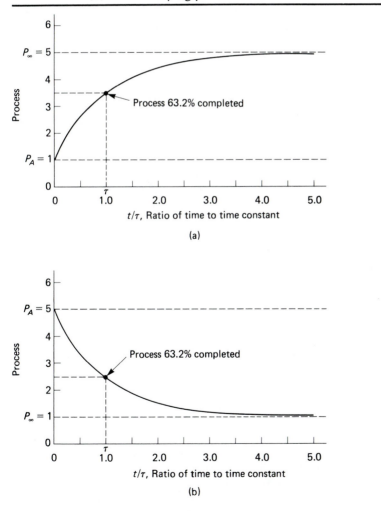

(a)

(b)

Although the basic relationship was derived in terms of a spring-dashpot arrangement, other processes that behave in an analogous manner include: (1) a heated (or cooled) bulk mass, such as a temperature sensor subjected to a step temperature change; (2) simple capacitive-resistive or inductive-resistive circuits; and (3) the decay of a radioactive source with time.

Figure 5.13 represents two different process–time conditions for the step-excited first-order system: (a) a *progressive* process, wherein the action is an increasing function of time; and (b) the *decaying* process, wherein the magnitude decreases with time.

Significance of the Time Constant, τ

If we substitute the magnitude of one time constant for t in Eq. (5.14),

$$P = P_\infty + (P_A - P_\infty)(0.368),$$

from which we see that $(1 - 0.368)$, or 63.2%, of the dynamic portion of the process will have been completed. Two time constants yield $(1 - 0.135) = 86.5\%$, three yield 95.0%, four yield 98.2%, and so on. These percentages of completed processes are important because they will always be the same regardless of the process, provided that the process is described by the conditions of the step-excited first-order system.

It is often assumed that a process is completed during a period of five time constants.

Example 5.2

Assume that a particular temperature probe approximates first-order behavior in a particular application, that it has a time constant of 6 s, and that it is suddenly subjected to a temperature step of 75°F–300°F. What temperature will be indicated 10 s after the process has been initiated?

Solution. Applying Eq. (5.14), we find that

$$P_\infty = 300°F, \qquad P_A = 75°F, \qquad t = 10\,s,$$
$$P = 300 + (75 - 300)e^{-10/6} = 257°F.$$

Example 5.3

Assume the same conditions as those of Example 5.2, but with a step of 300°F–75°F. Find the indicated temperature after 10 s.

Solution.

$$P_\infty = 75°F, \qquad P_A = 300°F, \qquad t = 10\,s,$$
$$P = 75 + (300 - 75)e^{-10/6} = 117°F.$$

5.15.2   The Harmonically Excited First-Order System

Again referring to Eq. (5.10), let us now consider the case of

$$F(t) = F_0 \cos \Omega t,$$

or

$$\zeta \frac{ds}{dt} + ks = F_0 \cos \Omega t, \qquad (5.15)$$

where

F_0 = the amplitude of the forcing function,

Ω = the circular frequency of the forcing function in radians per second.

The solution of Eq. (5.15) yields

$$s = A_1 e^{-t/\tau} + \frac{F_0/k}{\sqrt{1 + (\tau\Omega)^2}} \cos(\Omega t - \phi), \qquad (5.16)$$

where

A_1 = a constant whose value depends on the initial conditions,

τ = the time constant = $\dfrac{\zeta}{k}$,

ϕ = the phase lag = $\tan^{-1} \dfrac{\Omega\zeta}{k} = \tan^{-1} \dfrac{2\pi}{T} \tau$, \qquad (5.17)

$T = \dfrac{2\pi}{\Omega}$ = the period of excitation cycle in seconds.

We see that the first term on the right side of Eq. (5.16), the complementary function, is *transient* and after a period of several time constants becomes very small. The second term is the *steady-state* relationship and, except for the short initial period, we can write

$$s = \frac{F_0/k}{\sqrt{1 + (\tau\Omega)^2}} \cos(\Omega t - \phi) \qquad (5.18)$$

or

$$\frac{s}{s_s} = \frac{\cos(\Omega t - \phi)}{\sqrt{1 + (\tau\Omega)^2}},$$

and

$$\frac{s_d}{s_s} = \frac{1}{\sqrt{1 + (\tau\Omega)^2}}$$

$$= \frac{1}{\sqrt{1 + (2\pi\tau/T)^2}}, \qquad (5.19)$$

where

s_d = the maximum amplitude of the periodic dynamic displacement

and

$$s_s = \frac{F_0}{k}.$$

The quantity s_s is the static deflection that would occur, should the force amplitude F_0 be applied as a *static* force. The ratio s_d/s_s is often called the *amplification ratio*. For analogous situations, Eq. (5.19) may be written

$$\frac{P_d}{P_s} = \frac{1}{\sqrt{1 + (2\pi\tau/T)^2}}, \tag{5.19a}$$

where P represents the magnitude of the applicable process.

Figures 5.14 and 5.15 illustrate the relationships of the phase angle and the amplification ratio described by Eqs. (5.17) and (5.19a), respectively.

By calculating the response of the first-order system to a sinusoidal input, we have in fact determined its frequency and phase response. To show this explicitly in the notation of Section 5.7, we can rewrite Eq. (5.19a) in terms of the cyclic forcing frequency, $f = 1/T$:

$$G(f) \equiv \frac{P_d}{P_s} = \frac{1}{\sqrt{1 + (2\pi f\tau)^2}}. \tag{5.19b}$$

Likewise, from Eq. (5.17), the phase response is

$$\phi(f) = \tan^{-1}(2\pi f\tau). \tag{5.19c}$$

Note that the ideal response ($G \to 1$ and $\phi \to 0$) is obtained at frequencies small enough that $2\pi f\tau \ll 1$. From Figs. 5.14 and 5.15, we can see that this is

Figure 5.14   Phase lag versus ratio of excitation period to time constant for the harmonically excited first-order system

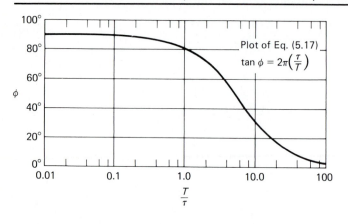

Figure 5.15 Amplitude ratio versus ratio of excitation period to time constant for the harmonically excited first-order system

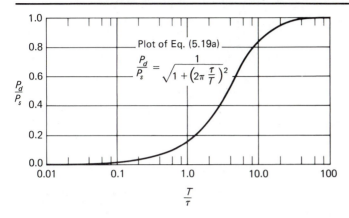

equivalent to the statement that *the frequency and phase response of a first-order system are best when the time constant of the system is small compared to the period of forcing, $\tau \ll T$, so that the system responds rapidly in comparison to the variations it is measuring.*

In most mechanical systems, a moving mass exists and cannot be ignored. In such cases, the system is of second order, and has characteristics that are discussed in subsequent sections. However, for temperature-sensing systems, first-order response is usually a good model, and we can use this model to examine the response characteristics of temperature probes.

Example 5.4

A temperature probe has a time constant of 10 s when used to measure a particular gas flow. The gas temperature varies harmonically between 75°F and 300°F with a period of 20 s (i.e., with a frequency of 0.05 Hz). What is the temperature readout in terms of the input gas temperature? What time constant should the probe have to give 99% of the correct temperature amplitude?

Solution. In this case the temperature input can be expressed as

$$T_{gas}(t) = \left(\frac{300 + 75}{2}\right) + \left(\frac{300 - 75}{2}\right)\cos\left(\frac{2\pi}{20}t\right)$$

$$= 187 + 112\cos\left(\frac{2\pi}{20}t\right).$$

From Eq. (5.19a), we find that

$$\frac{T_d}{T_s} = \frac{1}{\sqrt{1 + \left(\frac{2\pi}{20} \times 10\right)^2}},$$

$$T_d = \frac{112}{3.3} = 34°F.$$

From Eq. (5.17), we find that the phase lag for a forcing period of $T = 20\,\text{s}$ is

$$\phi = \tan^{-1} 2\pi(\tau/T) = \tan^{-1}\left(\frac{2\pi \times 10}{20}\right) = \tan^{-1} \pi = 72\tfrac{1}{2}° \quad (\text{angle})$$

or

$$\text{Time lag} = \frac{72\tfrac{1}{2}}{360} \times 20 = 4\,\text{s}.$$

A graphical representation of the situation is shown in Fig. 5.16.
   To obtain a 99% amplification ratio, we require

$$\frac{T_d}{T_s} = \frac{1}{\sqrt{1 + (2\pi\tau/20)^2}} = 0.99.$$

Solving, we find that the probe would need a time constant of $\tau = 0.45\,\text{s}$. In this case, τ is very small compared to T.

Figure 5.16 Response of temperature probe for conditions described in text

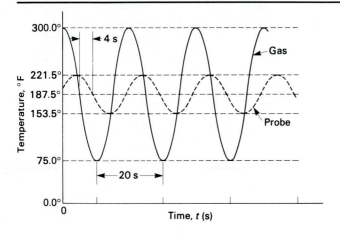

5.16 Characteristics of Second-Order Systems

Figure 5.17 illustrates the essentials of a second-order system. This arrangement approximates many actual mechanical arrangements including simple weighing systems, such as an elastic-type load cell supporting a mass, D'Arsonval meter movements, including the ordinary galvanometers, and many force-excited mechanical-vibration systems such as accelerometers.

As with the first-order system, many excitation modes are possible, ranging from the simple step and simple harmonic functions to complex periodic forms. These modes approximate many actual situations, and since all periodic inputs can be reduced to combinations of simple harmonic components (Section 4.4), the latter can give us insight into system performance when subject to most forms of dynamic input.

5.16.1 The Step-Excited Second-Order System

Referring to Fig. 5.17, we let

$$F = 0, \quad \text{when} \quad t < 0$$

and

$$F = F_0, \quad \text{when} \quad t \geq 0.$$

Application of Newton's second law yields

$$\frac{d^2s}{dt^2} + \left(\frac{\zeta g_c}{m}\right)\frac{ds}{dt} + \left(\frac{k g_c}{m}\right)s = F_0\left(\frac{g_c}{m}\right) \tag{5.20}$$

If we assume *underdamping*, that is, $(kg_c/m) > (\zeta g_c/2m)^2$ [see Eq. (5.21a)],

Figure 5.17 Schematic representation of a second-order system

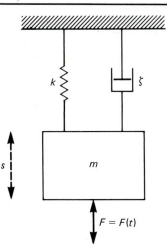

the general solution of Eq. (5.20) can be written as

$$s = e^{-(\zeta g_c/2m)t}[A \cos \omega_{nd}t + B \sin \omega_{nd}t] + \frac{F_0}{k}, \qquad (5.21)$$

where A and B are constants governed by initial conditions and

$$\omega_{nd} = \text{damped natural frequency}$$

$$= \sqrt{\frac{kg_c}{m} - \left(\frac{\zeta g_c}{2m}\right)^2}. \qquad (5.21a)$$

Note that the exponential multiplier may be written as $e^{-t/\tau}$, where the time constant $\tau = 2m/\zeta g_c$. If we let $s = 0$ and $ds/dt = 0$ at $t = 0$ and evaluate A and B, then by rearrangement and substitution of terms, we can write Eq. (5.21) as

$$\frac{s}{s_s} = 1 - e^{-\xi\omega_n t}\left[\frac{\xi}{\sqrt{1 - \xi^2}} \sin\sqrt{1 - \xi^2}\,\omega_n t + \cos\sqrt{1 - \xi^2}\,\omega_n t\right]. \qquad (5.22)$$

An alternative form is

$$\frac{s}{s_s} = 1 - e^{-\xi\omega_n t}\sqrt{\frac{1}{1 - \xi^2}} \cos(\omega_{nd}t - \beta), \qquad (5.22a)$$

$$\beta = \tan^{-1}\left[\frac{\xi}{\sqrt{1 - \xi^2}}\right], \qquad (5.22b)$$

where

$\omega_n = \sqrt{kg_c/m}$ = the undamped natural frequency in radians per second,

$\zeta_c = 2\sqrt{mk/g_c}$ = the critical damping coefficient,

ξ = the critical damping ratio, ζ/ζ_c,

$s_s = F_0/k$ = the "static" amplitude, or the amplitude that is reached as $t \to \infty$.

Here again we may introduce the general idea of an analogous process, P, or

$$\frac{P}{P_s} = \frac{s}{s_s}.$$

For the overdamped condition, $\xi = \zeta/\zeta_c > 1$, the solution of Eq. (5.20) can be written as

$$\frac{P}{P_s} = \frac{-\xi - \sqrt{\xi^2 - 1}}{2\sqrt{\xi^2 - 1}}e^{(-\xi+\sqrt{\xi^2-1})\omega_n t} + \frac{\xi - \sqrt{\xi^2 - 1}}{2\sqrt{\xi^2 - 1}}e^{(-\xi-\sqrt{\xi^2-1})\omega_n t} + 1. \qquad (5.23)$$

Figure 5.18 shows the plots for Eqs. (5.22) and (5.23) for various damping

Figure 5.18   Response of a second-order system to a step input at $t = 0$

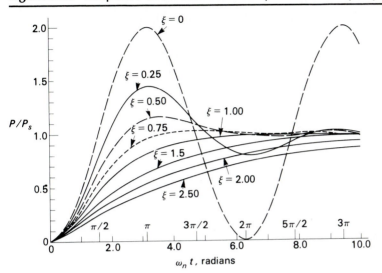

ratios.* When the system has a nonzero damping, it approaches a static condition with $P/P_s = 1$ as the transient dies out.

5.16.2   The Harmonically Excited Second-Order System

Referring to Fig. 5.17, when

$$F(t) = F_0 \cos \Omega t$$

we can write

$$\frac{m}{g_c}\frac{d^2s}{dt^2} + \zeta\frac{ds}{dt} + ks = F_0 \cos \Omega t. \tag{5.24}$$

For underdamped systems the solution becomes

$$s = e^{-(\zeta g_c/2m)t}[A \cos \omega_{nd}t + B \sin \omega_{nd}t] + \frac{(F_0/k)\cos(\Omega t - \phi)}{\sqrt{\left[1 - \dfrac{m\Omega^2}{kg_c}\right]^2 + \left(\dfrac{\zeta\Omega}{k}\right)^2}} \tag{5.25}$$

$$= e^{-t/\tau}[A \cos\sqrt{1 - \xi^2}\,\omega_n t + B \sin\sqrt{1 - \xi^2}\,\omega_n t]$$

$$+ \frac{s_s \cos(\Omega t - \phi)}{\sqrt{[1 - (\Omega/\omega_n)^2]^2 + [2\xi(\Omega/\omega_n)]^2}}, \tag{5.25a}$$

---

* Note that the cases of zero damping and critical damping require special treatment.

where A and B are constants that depend on particular initial conditions, and

Ω = the frequency of excitation (rad/s),

$$\phi = \tan^{-1}\left[\frac{2\xi\Omega/\omega_n}{1-(\Omega/\omega_n)^2}\right] = \text{the phase angle,} \qquad (5.25\text{b})$$

$s_s = F_0/k.$

We see that the first term on the right side of Eq. (5.25a) is transient and will disappear after several time constants. The second term is the steady-state relationship, for which we may write*

$$\frac{s_d}{s_s} = \frac{P_d}{P_s} = \frac{1}{\sqrt{[1-(\Omega/\omega_n)^2]^2 + [2\xi\Omega/\omega_n]^2}}$$
$$= \text{the amplification ratio,} \qquad (5.25\text{c})$$

where

s_d = the amplitude of the periodic steady-state displacement.

Figures 5.19 and 5.20 are plots of Eqns. (5.25c) and (5.25b), respectively, for various values of the damping ratio, ξ. The ratio s_s/s_d is none other than the frequency response, G, of the system; we could write it in terms of cyclic forcing frequency, f, by the substitution $\Omega = 2\pi f$.

Figure 5.19 Plot of Eq. (5.25c) illustrating the frequency response to harmonic excitation of the system shown in Fig. 5.17

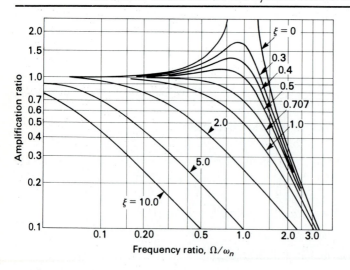

---

* Note that inasmuch as the complementary or homogeneous solution (transient) is not involved, this relationship is valid for under-, over-, and critically damped conditions.

Figure 5.20   Plot of Eq. (5.25b) illustrating the phase response
of the system shown in Fig. 5.17

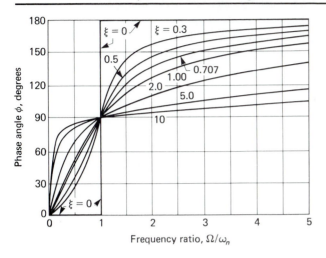

Frequency ratio, Ω/ω_n

The ratio s_d/s_s is a measure of the system response to the frequency input. Normally, we hope that this relationship is constant with frequency; that is, we would like the system to be insensitive to changes in the frequency of input $F(t)$. Inspection of Fig. 5.19 shows that the amplitude ratio is reasonably constant for only a limited frequency range and then only for certain damping ratios. We see that for a given damping ratio, ideal response ($s_d/s_s = 1$) may occur at only one or two frequencies. If the system is to be used for general dynamic measurement applications, rather definite damping must be used and an upper frequency limit must be established. Practically, if a damping ratio in the neighborhood of 65%–75% is used, then the amplitude ratio will approximate unity over a range of frequency ratios of about 0%–40%. Even for these conditions, inherent error ($s_d/s_s - 1$) exists, and a usable system can be had only through compromise.

It should be made clear that the basic reason for optimizing the damping ratio is to extend the usable range of exciting frequency Ω. Certain devices, notably piezoelectric sensors (see Section 6.14), commonly possess such high undamped natural frequencies, ω_n, that the range of normal *operating* frequency ratios, Ω/ω_n, may extend from zero to only 10% or even less. In such cases, damping ratio magnitudes are of lesser interest.

Inspection of Fig. 5.20 indicates that damping ratios of the order of 65%–75% of critical provide an approximately linear phase shift for the frequency ratio range of 0%–40%. This approach is desirable if a proper time relationship is to be maintained between the harmonic components of a complex input. (See Sections 17.6.2 and 17.7.1 for further discussion of phase relationships.)

Example 5.5
A particular pressure transducer consists of a circular steel diaphragm mounted in a housing. One side of the diaphragm is exposed to varying pressures, which cause the diaphragm to deflect. The elastic deflection of the diaphragm is sensed by a piezoelectric quartz crystal mounted within the housing, on the rear side of the diaphragm.

This transducer is effectively a second-order spring-mass system. The elastic stiffness of the steel diaphragm provides the spring force. The effective mass of the vibrating diaphragm contributes inertia. Damping in the system is slight ($\xi = 0.025$). The diaphragm has a diameter of 12 mm and a thickness of 1.75 mm.

The transducer is being considered for measuring combustion-engine cylinder pressures. Will this transducer have adequate frequency and phase response at typical engine speeds of 3000 to 6000 rpm? What is the amplification ratio at the transducer's natural frequency?

Solution. The first step is to estimate the deflection constant of the diaphragm and its effective mass while vibrating. Appropriate elasticity calculations show that $k \approx 131$ MN/m and $m \approx 3.7$ g. Thus

$$\omega_n = \sqrt{\frac{kg_c}{m}} \approx 188 \times 10^3 \, \text{rad/s},$$

or 29.9 kHz. The engine's highest operating frequency is

$$\Omega = (6000 \, \text{rpm})(2\pi \, \text{rad/rev})/(60 \, \text{s/min}) = 628 \, \text{rad/s},$$

or 100 Hz. Thus

$$\frac{\Omega}{\omega_n} = 0.0033.$$

From Figures 5.19 and 5.20, we see that the transducer will perform with negligible phase shift and amplification ratio of unity; substitution of ξ and Ω/ω_n into Eqs. (5.25b) and (5.25c) verifies this conclusion. The transducer's response characteristics are well suited to the application.

If the transducer were subjected to pressure variations near its natural frequency, however, the amplification ratio would be enormous; from Eq. (5.25c),

$$\frac{P_d}{P_s} = \frac{1}{\sqrt{0 + [2(0.025)(1)]^2}} = 20.$$

This device is clearly not designed to operate at frequencies that high.

5.16.3   General Periodic Forcing

We now return to the general case of periodic forcing as suggested in Section 5.13. For convenience we will restate Eqs. (5.8a) and (5.9)

$$\frac{m}{g_c}\left(\frac{d^2s}{dt^2}\right) + \zeta\left(\frac{ds}{dt}\right) + ks = F(t), \tag{5.26}$$

where

$$F(t) = \left(\frac{A_0}{2}\right) + \sum_{n=1}^{\infty} C_n \cos(n\Omega t - \phi_n) \tag{5.26a}$$

and

$$C_n = \sqrt{A_n^2 + B_n^2} \quad \text{and} \quad \tan\phi_n = \frac{B_n}{A_n}. \tag{5.26b}$$

By substituting Eq. (5.26a) into Eq. (5.26) we obtain

$$\frac{m}{g_c}\left(\frac{d^2s}{dt^2}\right) + \zeta\left(\frac{ds}{dt}\right) + ks = \frac{A_0}{2} + \sum_{n=1}^{\infty} C_n \cos(n\Omega t - \phi_n). \tag{5.27}$$

Although this expression appears quite formidable, we can easily recognize that it yields a combination of the solutions given by Eqs. (5.21) and (5.25a). Using the reasoning that the cosine terms on the right side of Eq. (5.27) will give results similar to those of a harmonically forced system, we can write a solution for $\xi < 1$ in the form

$$s = e^{(-\zeta g_c/2m)t}[A \cos \omega_{nd}t + B \sin \omega_{nd}t]$$

$$+ r_0 + \sum_{n=1}^{\infty} r_n \cos(n\Omega t - \phi_n - \psi_n) \tag{5.28}$$

where

$$r_0 = \frac{A_0}{2k},$$

$$r_n = \frac{C_n}{k}\left[\frac{1}{\sqrt{\left[1 - \left(\frac{n\Omega}{\omega_n}\right)^2\right]^2 + \left[2\xi\frac{n\Omega}{\omega_n}\right]^2}}\right],$$

$$\tan\psi_n = \frac{2\xi\left(\frac{n\Omega}{\omega_n}\right)}{1 - \left(\frac{n\Omega}{\omega_n}\right)^2}.$$

It helps to recall that

$$\omega_{nd} = \text{the damped natural frequency} = \sqrt{\frac{kg_c}{m} - \left(\frac{\zeta g_c}{2m}\right)^2},$$

$$\omega_n = \text{the undamped natural frequency} = \sqrt{\frac{kg_c}{m}}.$$

As we discussed previously, the first term on the right side of Eq. (5.28) is transient and dies out after several time constants. The remaining terms, then, represent the steady-state response.

Example 5.6

a. Write an expression for the steady-state response of the single-degree-of-freedom system shown in Fig. 5.17 when subjected to the sawtooth forcing function described in Table 4.1 and shown in Fig. 4.9(b).

b. Let

$$m = 1 \text{ kg, or } 2.2 \text{ lbm},$$

$$k = 1000 \text{ N/m, or } 5.71 \text{ lbf/in.},$$

$$\zeta = 31.6 \text{ N} \cdot \text{s/m, or } 0.180 \text{ lbf} \cdot \text{s/in.},$$

$$A = 10 \text{ N, or } 2.248 \text{ lbf}.$$

Using these data (the SI values), obtain computer-plotted waveforms for input frequencies of 10, 30, and 50 rad/s.

Solution.

a. In general, the steady-state response is given by

$$s = \frac{A_0}{2k} + \sum_{n=1}^{\infty} \frac{C_n}{k} \left[\frac{\cos(n\Omega t - \phi_n - \psi_n)}{\sqrt{\left[1 - \left(\frac{n\Omega}{\omega_n}\right)^2\right]^2 + \left[2\xi\frac{n\Omega}{\omega_n}\right]^2}} \right],$$

$$\phi_n = \tan^{-1}\left(\frac{B_n}{A_n}\right),$$

$$C_n = \sqrt{A_n^2 + B_n^2},$$

$$\psi_n = \tan^{-1}\left[\frac{2\xi\left(\frac{n\Omega}{\omega_n}\right)}{1 - \left(\frac{n\Omega}{\omega_n}\right)^2}\right].$$

In this case, the forcing function, $F(t)$, is [see Table 4.1, case 4.9(b)]

$$F(t) = \frac{2A}{\pi} \sum_{n=1}^{\infty} \frac{1}{n} \sin(n\Omega t),$$

so that both $A_0 = 0$ and $A_n = 0$. Thus

$$\phi_n = \tan^{-1} \infty = \frac{\pi}{2},$$

$$C_n = B_n = \frac{2A}{\pi} \frac{1}{n}.$$

Therefore,

$$s = \frac{2A}{\pi k} \sum_{n=1}^{\infty} \frac{1}{n} \left[\frac{\cos\left(n\Omega t - \frac{\pi}{2} - \psi_n\right)}{\sqrt{\left[1 - \left(\frac{n\Omega}{\omega_n}\right)^2\right]^2 + \left[2\xi \frac{n\Omega}{\omega_n}\right]^2}} \right].$$

$$\omega_n = \sqrt{\frac{kg_c}{m}} = \sqrt{1000} \approx 31.6\,\text{rad/s},$$

$$\zeta_c = 2\sqrt{\frac{mk}{g_c}} = 2\sqrt{1000} \approx 63.3\,\text{N} \cdot \text{s/m},$$

$$\xi = \frac{\zeta}{\zeta_c} = \frac{31.6}{63.3} \approx 0.50,$$

$$\frac{A}{k} = \frac{10}{1000} = 1\,\text{cm},$$

$$\psi_n = \tan^{-1} \left[\frac{2 \times 0.5 \left(\dfrac{n\Omega}{31.6}\right)}{1 - \left(\dfrac{n\Omega}{31.6}\right)^2} \right],$$

$$s = \frac{2}{\pi} \sum_{n=1}^{\infty} \frac{1}{n} \frac{\cos\left(n\Omega t - \dfrac{\pi}{2} - \psi_n\right)}{\sqrt{\left[1 - \left(\dfrac{n\Omega}{31.6}\right)^2\right]^2 + \left[\dfrac{n\Omega}{31.6}\right]^2}}$$

$$= \frac{2}{\pi} \sum_{n=1}^{\infty} \frac{1}{n} \frac{\cos\left(n\Omega t - \dfrac{\pi}{2} - \psi_n\right)}{\sqrt{\left[1 - \left(\dfrac{n\Omega}{31.6}\right)^2\right]^2 + \left(\dfrac{n\Omega}{31.6}\right)^2}}.$$

Note that s is in centimeters.

The most practical approach to obtaining numerical results for this part of the example is through the use of a computer. Computer-plotted results are shown in Fig. 5.21(a) through 5.21(c).

From the foregoing discussions it is apparent that if simple measurements are to be made as rapidly as possible or, more importantly, if the input signal is continuous and complex, rather definite limitations are imposed by the measuring system. Additional discussion of such limitations can be found in Chapters 9 and 17.

5.17   Electrical Elements

As we discussed in Section 5.8, most measurement systems are composed of a combination of mechanical and electrical elements. Very often the basic detecting element of the sensor is mechanical and its output is immediately transduced into an electrical signal by a secondary element. The signal conditioning that follows is largely by electrical means; however, termination sometimes requires conversion to something basically mechanical, such as a controller, a galvanometer-type recorder or plotter, etc. It is clear, then, that overall performance results from a combination of mechanical and electrical responses. In previous sections we have discussed the response of simple, purely mechanical systems. In succeeding sections we will look at the corresponding electrical elements.

In preparation for the discussion that follows, Table 5.1 lists some fundamental electrical quantities and defining relationships. Table 5.2 lists certain mechanical-electrical analogies. For verification of these the reader is referred to any basic electric circuits text [2] or physics text [3].

In addition, at this point it is useful to recall Kirchhoff's two laws for electrical circuits, namely,

1. The algebraic sum of all currents entering a junction point is zero, and
2. The algebraic sum of all voltage drops taken in a given direction around a closed circuit is zero.

5.18   First-Order Electrical System

Consider the circuit shown in Fig. 5.22. Assume that the capacitor carries no initial charge; then let the SPDT (single-pole, double-throw) switch be moved to contact A, thereby inserting the battery into the circuit. Now the capacitor

Figure 5.21 Computer-plotted sawtooth wave response for the second-order system specified in the text: (a) for $\Omega = 10$ rad/s, (b) for $\Omega = 30$ rad/s, (c) for $\Omega = 50$ rad/s

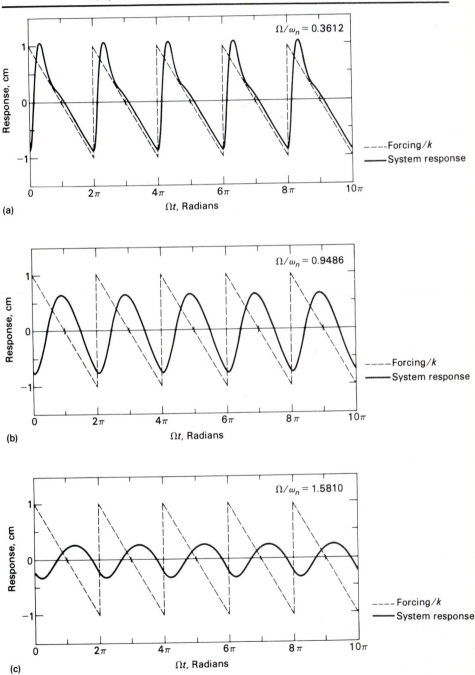

Table 5.1 Some basic electrical definitions and relationships

Symbol	Definition	Unit
E	Electric potential	Volt (V)
I	Electric current	Ampere (A)
Q	Electric charge	Coulomb (C)
R	Electrical resistance	Ohm (Ω)
L	Electrical inductance	Henry (H)
C	Electrical capacitance	Farad (F)

Some defining relationships:

For a capacitance: $I = dQ/dt = C(dE/dt)$, $E = Q/C$

For a resistance: $E = IR$ (Ohm's law)

For an inductance: $E = L(dI/dt) = L(d^2Q/dt^2)$

begins to charge; what is response of the circuit in terms of the voltage across the capacitor?

By employing Kirchhoff's law of voltages, we may write

$$IR + \frac{Q}{C} - E = 0, \tag{5.29}$$

but

$$I = \frac{dQ}{dt};$$

hence

$$\frac{dQ}{dt} + \frac{Q}{RC} - \frac{E}{R} = 0. \tag{5.29a}$$

Solving, we have

$$Q = CE(1 - e^{-t/RC}).$$

We define the circuit time constant as

$$\tau = RC.$$

The voltage drop across the capacitor is $E_c = Q/C$; hence,

$$E_c = E(1 - e^{-t/\tau}). \tag{5.30}$$

In a similar manner, if the switch contact is moved from A to B after the capacitor is charged, we may write

$$IR + \frac{Q}{C} = 0, \tag{5.31}$$

Table 5.2   Dynamically analogous mechanical and electrical system elements

Symbol	Mechanical Quantity	Symbol	Electrical Quantity
m	Mass, kg (lbm)		
I	Moment of inertia, kg · m^2 (lbm · in.^{2})	L	Inductance, H
k	Deflection constant, N/m (lbf/in.)		
k_t	Torsional deflection constant, N · m/rad (lbf · in./rad)	$1/C$	Reciprocal of capacitance, F^{-1}
ζ	Damping coefficient, N · s/m (lbf · s/in.)		
ζ_t	Torsional damping coefficient, N · m · s/m (lbf · in. · s/in.)	R	Resistance, Ω
F	Force, N (lbf)		
T	Torque, N · m (lbf · in.)	E	Voltage, V
x	Translational displacement, m (in.)		
θ	Angular displacement, rad	Q	Charge, C
dx/dt	Translational velocity, m/s (in./s)		
$d\theta/dt$	Angular velocity, rad/s (rad/s)	dQ/dt	Current, A
Ω	Forcing frequency, rad/s (rad/s)	Ω	Forcing frequency, rad/s
d^2x/dt^2	Translational acceleration, m/s^2 (in./s^2)		
$d^2\theta/dt^2$	Angular acceleration, rad/s^2 (rad/s^2)	d^2Q/dt^2	Rate of change of current, A/s

Note: The uppercase C has long been used as the symbol for capacitance. It has also been assigned as the symbol for the SI unit for electric charge, the coulomb. Likewise, Ω is the SI symbol for resistance, ohms. It is also widely used to represent an exciting frequency in radians per second. In this text we will let the symbols retain each meaning. Should the contexts not make clear the meanings intended, we will include clarifying statements.

Figure 5.22   Series resistance-capacitance circuit

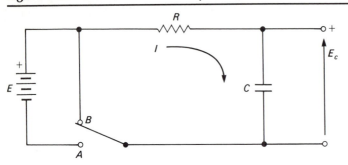

for which

$$E_c = Ee^{-t/\tau}. \tag{5.31a}$$

It is apparent, then, that Eqs. (5.14) and (5.10), which apply to a mechanical system, hold equally well for the electrical circuit discussed here. Should the battery in Fig. 5.22 be replaced with a sinusoidal voltage source, analysis would show that equations such as (5.17) and (5.19a, b, c) also apply to electrical systems. In each case, it is necessary to insert the appropriate time constant and to properly interpret the response variable.

Example 5.7

Figure 5.23(a) shows a simple circuit consisting of a capacitor, a resistor and a 5 kHz voltage source connected in series. Determine the amplitude and phase shift of the voltage appearing across the capacitor. Compare this to the voltage across the resistor.

Solution. From Kirchhoff's law of voltages, we obtain

$$R\frac{dQ}{dt} + \frac{1}{C}Q = E_0 \cos \Omega t.$$

By analogy to Section 5.15.2 [Eqs. (5.15), (5.17), and (5.18)], the steady-state solution of this equation is

$$Q = \frac{E_0 C}{\sqrt{1 + (\tau\Omega)^2}} \cos(\Omega t - \phi),$$

with

$$\phi = \tan^{-1}\left(\frac{2\pi}{T}\tau\right)$$

Figure 5.23 Resistor and capacitor in series: (a) excited by a 5000 Hz voltage source, (b) voltage across the capacitor for the circuit shown in (a). The angle ϕ is a phase *lag*.

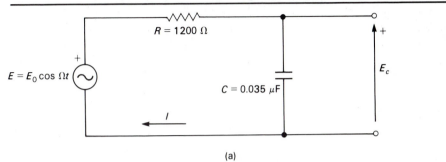

(a)

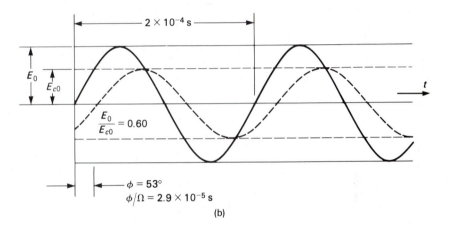

(b)

and

$$\tau = RC$$

for T the period of the exciting voltage. The voltage across the capacitor is $E_c = Q/C$. Therefore, the voltage amplitude across the capacitor is

$$\frac{E_{c_0}}{E_0} = \frac{1}{\sqrt{1 + (\tau\Omega)^2}} = \frac{1}{\sqrt{1 + (2\pi\tau/T)^2}}$$

Observe that these results are identical to those plotted in Figs. 5.14 and 5.15.

Numerically,

$$\tau = 1200(0.035 \times 10^{-6}) = 4.2 \times 10^{-5}\,\text{s},$$

$$T = \frac{1}{f} = 2 \times 10^{-4}\,\text{s},$$

so

$$\frac{E_{c_0}}{E_0} = 0.6$$

and

$$\phi = 53°.$$

Figure 5.23(b) illustrates the resulting relationship.

The resistor behaves somewhat differently. In terms of Q, the voltage across the resistor is

$$E_r = R\frac{dQ}{dt} = \frac{E_0 RC\Omega}{\sqrt{1 + (\tau\Omega)^2}}[-\sin(\Omega t - \phi)]$$

or

$$E_r = \frac{E_0\tau\Omega}{\sqrt{1 + (\tau\Omega)^2}}\cos(\Omega t + \pi/2 - \phi).$$

Thus, the resistor voltage has an amplitude

$$\frac{E_{r_0}}{E_0} = \frac{\tau\Omega}{\sqrt{1 + (\tau\Omega)^2}} = \frac{2\pi\tau/T}{\sqrt{1 + (2\pi\tau/T)^2}}$$

and a phase *lead* of $\pi/2 - \phi$. Note that the resistor voltage amplitude is small when $\tau/T \ll 1$ (at low excitation frequencies), whereas the capacitor voltage is small when $\tau/T \gg 1$ (at high excitation frequencies). The contrasting behavior of the two elements is the basis for this circuit's application to high-pass and low-pass filters, as we shall see in Chapter 7.

First-order circuits may also be constructed using an *inductor* and a resistor. Similar results can be developed, the primary difference being that time constant is instead $\tau = L/R$.

5.19 Simple Second-Order Electrical System

Figure 5.24 illustrates a circuit consisting of R, L, and C elements in series with a voltage source. Referring to Table 5.1 for the voltage drop across each

Figure 5.24 An *RLC* circuit

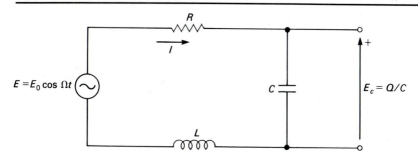

element and then applying Kirchhoff's law of voltages, we can write

$$L\left(\frac{d^2Q}{dt^2}\right) + R\left(\frac{dQ}{dt}\right) + \frac{Q}{C} = E_0 \cos \Omega t. \tag{5.32}$$

We recognize this expression as having the same form as Eq. (5.24), and thus we can quickly write a solution:

$$Q = e^{-t/\tau}[A \cos \omega_{nd}t + B \sin \omega_{nd}t] + \frac{E_0 \cos(\Omega t - \phi)}{\sqrt{[1/C - L\Omega^2]^2 + (R\Omega)^2}}. \tag{5.32a}$$

If, for example, we consider the steady-state voltage amplitude across the capacitor we may write (see Table 5.1)

$$E_c = \frac{Q}{C} = \frac{E_0}{C\sqrt{[1/C - L\Omega^2]^2 + (R\Omega)^2}}. \tag{5.32b}$$

Using the analogies in Table 5.2 (see also Section 7.11) we may write

$$\omega_n = \sqrt{\frac{1}{LC}} \tag{5.32c}$$

and

$$R_c = 2\sqrt{\frac{L}{C}}, \tag{5.32d}$$

where

ω_n = a resonance frequency corresponding to the undamped natural frequency of the mechanical system,

R_c = a critical resistance analogous to critical damping.

By algebraic manipulation, Eqs. (5.32a) and (5.32b) may now be written in the dimensionless forms, where E_c is the dynamic amplitude of the voltage across C; that is,

$$\frac{E_c}{E_0} = \frac{1}{\sqrt{[1 - \Omega/\omega_n)^2]^2 + [2(R/R_c)(\Omega/\omega_n)]^2}} \tag{5.32e}$$

and

$$\tan \phi = \frac{2(R/R_c)(\Omega/\omega_n)}{1 - (\Omega/\omega_n)^2}. \qquad (5.32f)$$

Except for the symbols we see that Eqs. (5.32e) and (5.32f) are identical to Eqs. (5.25c) and (5.25b), respectively. It follows, then, that Figs. 5.19 and 5.20 apply equally well to the electric circuit that we have just investigated.

5.20 Calibration of System Response

The final and positive proof of a measuring system's performance is direct measurement of the actual system's response to a completely defined and known input. Physical modeling of system response (as in Sections 5.8–5.19) provides important understanding of the device's essential characteristics, such as its time constant or natural frequency. However, if the device is too complex to model well or if the modeling assumptions are tenuous, greater accuracy can be obtained through an experimental test of the system's response. Such testing is always required in high-accuracy experimentation.

Testing a measuring system's response is really a process of *calibration* (Section 1.7). What output is obtained for an input of given amplitude and frequency? Can we experimentally evaluate the behavior of the system when it is confronted with an input that is rapidly changing with time? This is a considerable challenge. It is not difficult to produce relatively rapid changes in most measurands; to do so *and* to assure ourselves that we *really know* the driving function is the crux of the problem.

Calibration sources having sinusoidal time variation are undoubtedly the easiest to produce and the most used (recall Section 5.7). With electrical input signals, commercial voltage sources are easily applied to this task. Various classical complex waveforms, such as square waves or sawtooth waves, may also be employed. For example, the response of many electrical components can be judged through the use of square-wave inputs. A skilled technician can frequently pinpoint reasons for distortion by observing the tested apparatus's treatment of square-wave input.* More importantly, such tests can identify limits beyond which system performance is questionable.

For some measurands, even a sine-wave test is relatively hard to implement: How do you create a well-defined sinusoidal variation in temperature or fluid velocity? For such variables, step inputs or pulse inputs are often easier to produce, and they can also yield useful information about the system's rise time, time constant, or resonant frequencies.

In cases where any time variation of the input is too difficult or too

---

* Square-wave testing is so useful in adjusting the frequency response of hot-wire anemometer bridges (Section 15.10) that many bridges are sold with built-in square-wave generators.

Figure 5.25   Comparison of static and dynamic calibrations. In a static
calibration, $G(0)$ is measured and used to approximate
$G(f)$ at higher frequencies. In a dynamic calibration,
$G(f)$ itself is measured.

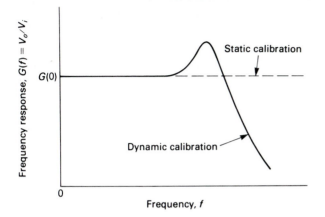

expensive to produce, constant inputs are occasionally substituted. For
example, we may measure the output of a temperature sensor at several
different constant temperatures to obtain $T_{measured}$ as a function of T_{actual}. This
tells us how the sensor responds to inputs of different amplitude. We would
then assume that the measured amplitude relationship applied for all frequencies of interest.

In this sense, we can distinguish two type of calibrations: *static calibration*
and *dynamic calibration*. In Section 5.7, we noted that the frequency response
of a system was the ratio of output to input amplitude at a given frequency:
$G(f) = V_o/V_i$. In a static calibration, we measure V_o and V_i at zero frequency
to obtain $G(0)$. We then assume that $G(f) \approx G(0)$ over the frequency range
of interest (Fig. 5.25). In a dynamic calibration, we measure V_o and V_i over a
range of frequencies to obtain $G(f)$ itself. Static calibration approximates
dynamic calibration at frequencies low enough that the device's frequency
response is flat. Static calibrations must always be used with caution and an
awareness of their inherent frequency limitations.

Calibration practices for various mechanical measurands are considered in
more detail in Part II of this book (see Sections 14.13, 15.12, 16.12, 17.11 and
18.9).

5.21   Summary

Response is a vital feature of a measuring system's ability to accurately resolve
time-varying inputs. In this chapter, we have examined various important types

of response, procedures for testing response, and simplified physical models of the response of instruments.

1. Amplitude response, frequency response, phase response, and rise time are each important aspects of measuring-system performance. Amplitude response is generally defined by the overload condition of an instrument. Frequency and phase response are considerations in determining the range of frequencies over which an instrument is accurate (Sections 5.2–5.7).

2. Each element in a measuring system must have adequate frequency and phase response for the measurand at hand. Some system elements, like filters, are selected specifically for the limitations of their frequency response (Section 5.7).

3. Often, a physical model of a measuring device can identify the important characteristics of its behavior. Mechanical models rely on the identification of masses, spring forces, and damping (Sections 5.8–5.11).

4. Selection of the correct amount of damping is vital to obtaining optimal frequency response (Sections 5.11, 5.15, 5.16, 5.18).

5. The spring-mass-damper system is a fundamental model of the dynamic response of mechanical systems (Sections 5.12–5.16). Some systems behave as if massless or first-order, especially temperature sensors (Section 5.15). More systems have mass and are second-order; examples include accelerometers, elastic diaphragm transducers, and various moving mechanisms (Section 5.16).

6. Electrical-system response is similar to mechanical-system response. Models for electrical systems yield results analogous to those for mechanical systems (Sections 5.17–5.19).

7. For a first-order system, the system's time constant, τ, is of critical importance to its response. The time constant should normally be small compared to the period of forcing, T, in order to obtain good response. First-order response to a step input is described by Eq. (5.14) and Fig. 5.13; response to harmonic forcing is described by Eqs. (5.17) and (5.19a, b, c) and Figs. 5.14 and 5.15 (Sections 5.15, 5.18).

8. For a second-order system, both the natural frequency, ω_n, and the critical damping ratio, ξ, must be considered. For most systems, good response is obtained when the natural frequency is large compared to the forcing frequency, Ω, and when the damping ratio is 65%–75%. Second-order response to a step input is described by Eqs. (5.22) and (5.23) and Fig. 5.18; response to harmonic input is described by Eqs. (5.25a, b, c) and Figs. 5.19 and 5.20 (Sections 5.16, 5.19).

9. Experimental determination of system response, or calibration, is often required. A sine-wave test is sometimes useful (Section 5.7). Other kinds of test signals, such as square-wave or step input, can also be applied. Static calibrations (for $f = 0$) are occasionally used when dynamic calibrations (for $f > 0$) are too difficult (Section 5.20).

Suggested Readings

Cannon, R. H., Jr. *Dynamics of Physical Systems*. New York: McGraw-Hill, 1967.

Carlson, A. B. and D. G. Gisser. *Electrical Engineering*. Reading, Mass.: Addison-Wesley, 1981.

Dorf, R. C. *Modern Control Systems*. Reading, Mass.: Addison-Wesley, 1992.

Keast, D. N. *Measurement of Mechanical Dynamics*. New York: McGraw-Hill, 1967.

Vierck, R. K. *Vibration Analysis*. New York: Harper & Row, 1979.

Problems

5.1 A simple U-tube monometer is shown in Fig. 14.3. Show that the period of oscillation may be approximated by

$$T = 2\pi \sqrt{\frac{L}{2g}}.$$

5.2 The simple U-tube monometer shown in Fig. 14.3 is subjected to a time-varying pressure with a cyclic circular frequency of $\sqrt{g/8L}$ rad/s. Determine the error in the pressure reading.

5.3 A temperature measuring system (assumed to be a first-order system) is excited by a 0.25 Hz harmonically varying input. If the time constant of the system is 4.0 s and the indicated amplitude is 10°F, what is the true temperature?

5.4 A mercury-in-glass thermometer initially at 25°C is suddenly immersed into a liquid that is maintained at 100°C. After a time interval of 2.0 s the thermometer reads 76°C. Assuming a first-order system, estimate the time constant of the thermometer.

5.5 A 5 kg mass is statically suspended from a load (force) cell and the load cell deflection caused by this mass is 0.01 mm. Estimate the natural frequency of the load cell.

5.6 If the load cell of Problem 5.5 is assumed to be a second-order system with negligible damping, determine the practical frequency range over which it can measure dynamic loads with an inherent error of less than 5%.

5.7 If the load cell of Problem 5.5 actually has a damping ratio of 0.707, determine the practical frequency range over which it can measure dynamic loads with an inherent error of less than 5%.

5.8 A manufacturer lists the specifications of a dynamic tension-compression load cell as follows:

Undamped natural frequency = 1000 Hz

Damping ratio = 0.707

Sensitivity = 10 mV/lbf

Stiffness = 100,000 lbf/in.

If a dynamic force as indicated in Fig. 5.26 is applied to the load cell, determine the steady-state output voltage as a function of time.

Figure 5.26 Dynamic load for Problem 5.8

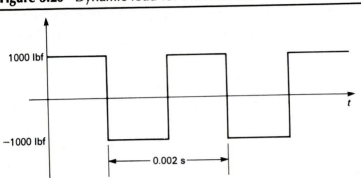

5.9   Plot the voltage output of Problem 5.8 as a function of time using at least the first five terms.

5.10   An *RC* circuit as shown in Fig. 5.22 is required to have a time constant of 1 ms. Determine three different combinations of *R* and *C* to accomplish this.

5.11   A tank containing an initial volume of water V_0 discharges water from an opening at the bottom of the tank. If the discharge rate is directly proportional to the volume of water in the tank, determine the time constant for this system.

5.12   For the tank of Problem 5.11, determine the time constant if the initial discharge rate is $\dot{Q}$ (liters/min).

5.13   A temperature sensor is expected to measure an input having frequency components as high as 50 Hz with an error no greater than 5%. What is the maximum time constant for the temperature sensor that will permit this measurement?

5.14   A thermocouple with a time constant of 0.05 s is considered to behave as a first-order system. Over what frequency range can the thermocouple measure dynamic temperature fluctuations (assumed to be harmonic) with an error less than 5%?

5.15   A 100 μF capacitor is charged to a voltage of 100 V. At time $t = 0$, it is discharged through a 1.0 MΩ resistor. Determine the time for the voltage across the capacitor to reach 10 V.

5.16   A pressure transducer behaves as a second-order system. If the undamped natural frequency is 4000 Hz and the damping is 75% of critical, determine the frequency range(s) over which the measurement error is not greater than 5%.

5.17   What will be the frequency range(s) for Problem 5.16 if the damping ratio is changed to 0.5?

5.18   Consider the pressure transducer of Problem 5.16 to be damaged such that its viscous damping ratio is unknown. When the transducer is subjected to a harmonic input of 2400 Hz, the phase angle between the output and input is measured as 45°. With this in mind, determine the error when the transducer is used to measure a harmonic pressure signal of 1800 Hz. What is the phase angle between the input and output at this frequency?

Figure 5.27 Forcing for Problem 5.20

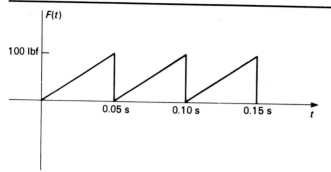

5.19 A force transducer behaves as a second-order system. If the undamped natural frequency of the transducer is 1800 Hz and its damping is 30% of critical, determine the error in the measured force for a harmonic input of 950 Hz. What is the magnitude of the phase angle?

5.20 Consider a second-order system with a damping ratio of 0.70 and a undamped natural frequency of 50 Hz. If the value of k is 100 lbf/in., determine the steady-state output if the forcing is as shown in Fig. 5.27.

5.21 Refer to Fig. 5.28.

   a.   What is the time constant if $X = 200\ \mu F$ and $R = 10,000\ \Omega$?

   b.   What is the voltage across the resistor 0.5 s after the switch is closed?

5.22 Refer to Fig. 5.28.

   a.   What is the time constant if $X = 500\ mH$ and $R = 10\ \Omega$?

   b.   What is the voltage across the resistor 0.02 s after the switch is closed?

   c.   What is the voltage across the resistor 0.05 s after the switch is closed?

Figure 5.28 Circuit for Problems 5.21 and 5.22

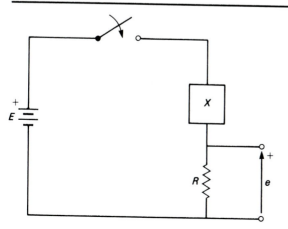

References

1. Dorf, R. C. *Modern Control Systems*. Reading, Mass.: Addison-Wesley, 1992.

2. Fitzgerald, A. E., D. E. Higginbotham, and A. Grabel. *Basic Electrical Engineering*. 3rd ed. New York: McGraw-Hill, 1967.

3. Halliday, D., R. Resnick, and J. J. Merril. *Fundamentals of Physics,* 3rd ed., extended. New York: John Wiley, 1988.

CHAPTER 6

Sensors

6.1 Introduction

In Section 1.4 we divided the general measuring system into three distinct sections: the sensor-detector stage, the signal conditioning stage, and the terminating readout stage. In this chapter and in Chapters 7 and 9 we discuss in more detail what goes on in each of these stages.

The first contact that a measuring system has with the measurand is through the input sample accepted by the detecting element of the first stage (see Section 1.4). This act is usually accompanied by the immediate transduction of the input into an analogous form.

The medium handled is information. The detector senses the information input, I_{in}, and then transduces or converts it to a more convenient form, I_{out}. The relationship may be expressed as

$$I_{out} = f(I_{in}); \tag{6.1}$$

further,

$$\text{Transfer efficiency} = \frac{I_{out}}{I_{in}}. \tag{6.1a}$$

This may not be more than unity, because the pickup cannot generate information but can only receive and process it. Obviously, as high a transfer efficiency as possible is desirable.

Sensitivity may be expressed as

$$\eta = \frac{dI_{out}}{dI_{in}}. \tag{6.1b}$$

Very often sensitivity approximates a constant; that is, the output is the linear function of the input.

211

6.2   Loading of the Signal Source

Energy will always be taken from the signal source by the measuring system, which means that the information source will always be changed by the act of measurement. This is a measurement axiom. This effect is referred to as *loading*. The smaller the load placed on the signal source by the measuring system, the better.

Of course, the problem of loading occurs not only in the first stage, but throughout the entire chain of elements. While the first-stage detector-transducer loads the input source, the second stage loads the first stage, and finally the third stage loads the second stage. In fact, the loading problem may be carried right down to the basic elements themselves.

In measuring systems made up primarily of electrical elements, the loading of the signal source is almost exclusively a function of the detector. Intermediate modifying devices and output indicators or recorders receive most of the energy necessary for their functioning from sources *other* than the signal source. A measure of the quality of the first stage, therefore, is its ability to provide a usable output without draining an undue amount of energy from the signal.

6.3   The Secondary Transducer

As an example of a system of mechanical elements only, consider the Bourdon-tube pressure gage, shown in Fig. 6.1. The primary detecting-transducing element consists of a circular tube of approximately elliptical cross section. When pressure is introduced, the section of the flattened tube tends toward a more circular form. This in turn causes the free end A to move outward and the resulting motion is transmitted by link B to sector gear C and hence to pinion D, thereby causing the indicator hand to move over the scale.

In this example, the tube serves as the primary detector-transducer, changing pressure into near linear displacement. The linkage-gear arrangement acts as a secondary transducer (linear to rotary motion) and as an amplifier, yielding a magnified output.

A modification of this basic arrangement is to replace the linkage-gear arrangement with either a differential transformer (Section 6.11) or a voltage-dividing potentiometer (Section 6.6). In either case the electrical device serves as a secondary transducer, transforming displacement to voltage.

As another example, let us analyze a simplified compression-type force-measuring *load cell* consisting of a short column or strut, with electrical resistance-type strain gages (see Section 6.7) attached (Fig. 6.2). When an applied force deflects or strains the block, the force effect is transduced to deflection (we are interested in the unit deflection in this case). The load is transduced to strain. In turn, the strain is transformed into an electrical resistance change, with the strain gages serving as secondary transducers.

Figure 6.1 Essentials of a Bourdon-tube pressure gage

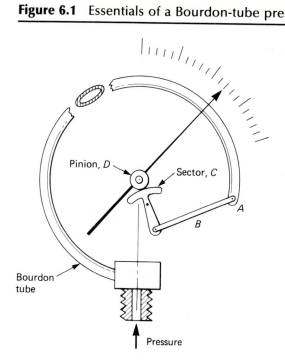

Pinion, D

Sector, C

A

B

Bourdon tube

Pressure

Figure 6.2 Schematic representation of a strain-gage load cell. The block forms the primary detector-transducer and the gages are secondary transducers

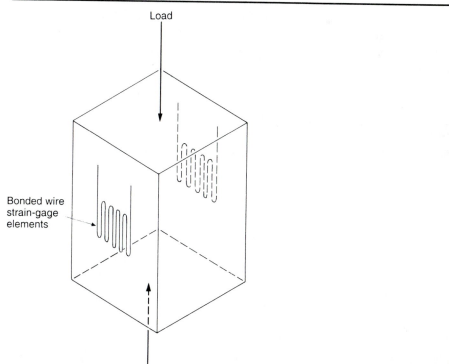

Load

Bonded wire strain-gage elements

213

6.4 Classification of First-Stage Devices

It appears, therefore, that the stage-one instrumentation may be of varying basic complexity, depending on the number of operations performed. This leads to a classification of first-stage devices as follows:

Class I. First-stage element used as detector only

Class II. First-stage elements used as detector and single transducer

Class III. First-stage elements used as detector with two transducer stages

A generalized first stage may therefore be shown schematically, as in Fig. 6.3.

Stage-one instrumentation may be very simple, consisting of no more than a mechanical spindle or contacting member used to transmit the quantity to be measured to a secondary transducer. Or it may consist of a much more complex assembly of elements. In any event the primary detector-transducer is an *integral* assembly whose function is (1) to sense selectively the quantity of interest, and (2) to process the sensed information into a form acceptable to stage-two operations. It does not present an output in immediately usable form.

More often than not the initial operation perfomed by the first-stage device is to transduce the input quantity into an analogous displacement. Without attempting to formulate a completely comprehensive list, let us consider Table 6.1 as representing the general area of the primary detector-transducer in mechanical measurements.

We make no attempt now to discuss all the many combinations of elements listed in Table 6.1. In most cases we have referred in the table to sections where thorough discussions can be found. The general nature of many of the elements is self-evident. A few are of minimal importance, included merely to round out the list. However, we can make several pertinent observations at this point.

Figure 6.3 Block diagram of a first-stage device with primary and secondary transducers

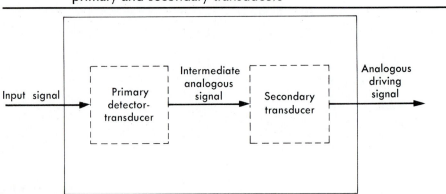

Table 6.1   Some primary detector-transducer elements
                    and operations they perform

Element	Operation
I. Mechanical	
A. Contacting spindle, pin, or finger	Displacement to displacement
B. Elastic member	
1. Load cells (Chapter 13)	
a. Tension/compression	Force to linear displacement
b. Bending	Force to linear displacement
c. Torsion	Torque to angular displacement
2. Proving ring (Chapter 13)	Force to linear displacement
3. Bourdon tube (Chapter 14)	Pressure to displacement
4. Bellows	Pressure to displacement
5. Diaphragm (Chapter 14)	Pressure to displacement
6. Helical spring	Force to linear displacement
C. Mass	
1. Seismic mass (Chapter 17)	Forcing function to relative displacement
2. Pendulum	Gravitational acceleration to frequency or period
3. Pendulum (Chapter 13)	Force to displacement
4. Liquid column (Chapter 14)	Pressure to displacement
D. Thermal (Chapter 16)	
1. Thermocouple	Temperature to electric potential
2. Bimaterial (includes mercury in glass)	Temperature to displacement
3. Thermistor	Temperature to resistance change
4. Chemical phase	Temperature to phase change
5. Pressure thermometer	Temperature to pressure
E. Hydropneumatic	
1. Static	
a. Float	Fluid level to displacement
b. Hydrometer	Specific gravity to relative displacement
2. Dynamic (Chapter 15)	
a. Orifice	Fluid velocity to pressure change
b. Venturi	Fluid velocity to pressure change
c. Pitot tube	Fluid velocity to pressure change
d. Vanes	Velocity to force
e. Turbines	Linear to angular velocity

continued

Table 6.1 *continued*

Element	Operation
II. Electrical	
A. Resistive (Sections 6.5–6.8)	
1. Contacting	Displacement to resistance change
2. Variable-length conductor	Displacement to resistance change
3. Variable-area conductor	Displacement to resistance change
4. Variable dimensions of conductor	Strain to resistance change
5. Variable resistivity of conductor	Temperature to resistance change
B. Inductive (Sections 6.10–6.12)	
1. Variable coil dimensions	Displacement to change in inductance
2. Variable air gap	Displacement to change in inductance
3. Changing core material	Displacement to change in inductance
4. Changing core positions	Displacement to change in inductance
5. Changing coil positions	Displacement to change in inductance
6. Moving coil	Velocity to change in induced voltage
7. Moving permanent magnet	Velocity to change in induced voltage
8. Moving core	Velocity to change in induced voltage
C. Capacitive (Section 6.13)	
1. Changing air gap	Displacement to change in capacitance
2. Changing plate areas	Displacement to change in capacitance
3. Changing dielectric constant	Displacement to change in capacitance
D. Piezoelectric (Section 6.14)	Displacement to voltage and/or voltage to displacement
E. Semiconductor junction (Section 6.15)	
1. Junction threshold voltage	Temperature to voltage change
2. Photodiode current	Light intensity to current
F. Photoelectric (Section 6.16)	
1. Photovoltaic	Light intensity to voltage*
2. Photoconductive	Light intensity to resistance change*
3. Photoemissive	Light intensity to current*
G. Hall Effect (Section 6.17)	Displacement to voltage

* Also sensitive to wavelength of light.

Close scrutiny of Table 6.1 reveals that, whereas many of the mechanical sensors transduce the input to displacement, many of the electrical sensors change displacement to an electrical output. This is quite fortunate, for it yields practical combinations in which the mechanical sensor serves as the primary transducer and the electrical sensor as the secondary. The two most commonly used electrical means are variable resistance and variable inductance, although others, such as photoelectric and piezoelectric effects, are also of considerable importance.

In addition to the inherent compatibility of the mechano-electric transducer combination, electrical elements have several important relative advantages:

1. Amplification or attenuation can be easily obtained.
2. Mass-inertia effects are minimized.
3. The effects of friction are minimized.
4. An output power of almost any magnitude can be provided.
5. Remote indication or recording is feasible.
6. The transducers can often be miniaturized.

Most of the remainder of this chapter is devoted to a discussion of electrical transducers. Modification of their outputs, or signal conditioning, is discussed in Chapter 7.

6.5 Variable-Resistance Transducer Elements

Resistance of an electrical conductor varies according to the following relation:

$$R = \frac{\rho L}{A}, \qquad (6.2)$$

where

R = resistance (Ω),

L = the length of the conductor (cm),

A = cross-sectional area of the conductor (cm^2),

ρ = the resistivity of material ($\Omega \cdot$ cm).

Many sensors are based on changes in the factors determining resistance. Some examples include sliding-contact devices and potentiometers, in which L changes (Section 6.6); resistance strain-gages, in which L, A, and ρ change (Section 6.7 and Chapter 12); and thermistors, photoconductive light detectors, piezoresistive strain gages, and resistance temperature detectors, in which ρ changes (Sections 6.8 and 16.4, 6.15 and 6.16, 6.15, and 16.4).

Probably the simplest mechanical-to-electrical transducer is the ordinary *switch* in which resistance is either zero or infinity. It is a yes-no, conducting-nonconducting device that can be used to operate an indicator. Here a lamp is fully as useful for readout as a meter, since only two values of quantitative information can be obtained. In its simplest form, the switch may be used as a limiting device operated by direct mechanical contact (as for limiting the travel of machine tool carriages) or it may be used as a position indicator. When actuated by a diaphragm or bellows, it becomes a pressure-limit indicator, or if controlled by a bimetal strip, it is a temperature limit indicator. It may also be combined with a proving ring to serve as either an overload warning device or a device actually limiting load-carrying, such as a safety device for a crane.

6.6   Sliding-Contact Devices

Sliding-contact resistive transducers convert a mechanical displacement input into an electrical output, either voltage or current. This is accomplished by changing the effective length L of the conductor in Eq. (6.2). Some form of electrical resistance element is used, with which a contactor or brush maintains electrical contact as it moves. In its simplest form, the device may consist of a stretched resistance wire and slider, as in Fig. 6.4. The effective resistance existing between either end of the wire and the brush thereby becomes a measure of the mechanical displacement. Devices of this type have been used for sensing relatively large displacements [1].

More commonly, the resistance element is formed by wrapping a resistance wire around a form, or *card*. The turns are spaced to prevent shorting, and the brush slides across the turns from one turn to the next. In actual practice, either the arrangement may be wound for a rectilinear movement or the resistance element may be formed into an arc and angular movement used, as shown in Fig. 6.5(a).

Sliding-contact devices are also made using conductive films as the variable-resistance elements, rather than wires [Fig. 6.5(b)]. Common examples include carbon-composition films, in which graphite or carbon particles are suspended in an epoxy or polyester binder, and ceramic-metal

Figure 6.4   Variable resistance consisting of a wire and movable contactor or brush. This is often referred to as a slide wire.

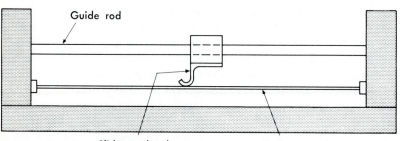

Figure 6.5   Angular-motion variable resistance, or potentiometer:
(a) wire-wound, (b) carbon-composition

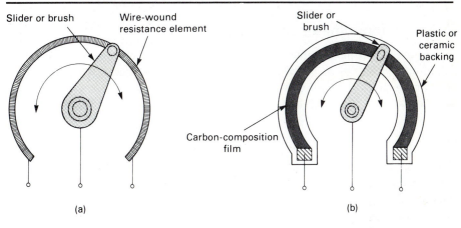

(a)   (b)

composition films (or *cermet*), in which ceramic and precious metal powders
are combined. In each case, the thin film is supported by a ceramic or plastic
backing. Conductive-film devices are less expensive than wire-wound devices.
The carbon-film devices, in particular, have outstanding wear characteristics
and long life [2], although they are more susceptible to temperature drift and
humidity effects.

These devices are commonly called *resistance potentiometers,** or simply
pots. Variations of the basic angular or rotary form are the multiturn, the
low-torque, and various nonlinear types [4]. Multiturn potentiometers are
available with various numbers of revolutions, up to 40. See also Section 7.7.

6.6.1   Potentiometer Resolution

Resistance variation available from a sliding-contact moving over a wire-
wound resistance element is not a continuous function of contact movement.
The smallest increment into which the whole may be divided determines the
resolution. In the case of wire-wound resistance, the limiting resolution equals
the reciprocal of the number of turns. If 1200 turns of wire are used and the
winding is linear, the resolution will be 1/1200, or 0.09083%. The meaning of
this quantity is apparent: No matter how refined the remainder of the system
may be, it will be impossible to divide, or resolve, the input into parts smaller
than 1/1200 of the total potentiometer range.

---

* Unfortunately, another entirely different device is also called a potentiometer. It is the
voltage-measuring instrument wherein a standard reference voltage is adjusted to counterbalance
the unknown voltage (see [3]). The two devices are different and must not be confused.

For conductive-film potentiometers, resolution is negligibly small and variation of slider contact-resistance is a more significant limitation.

6.6.2   Potentiometer Linearity

When used as a measurement transducer, a linear potentiometer is normally required. Use of the term *linear* assumes that the resistance measured between one of the ends of the element and the contactor is a direct linear function of the contactor position in relation to that end. Linearity is never completely achieved, however, and deviation limits are usually supplied by the manufacturer.

6.7   The Resistance Strain Gage

Experiment has shown that each term in Eq. (6.2) is simultaneously affected by the input strain in a resistance strain gage. The resistance element is cemented to the surface of the member to be strained, and as it elongates with application of strain (assuming a tensile strain) its cross-sectional area reduces and a longer length of smaller element results. However, simply accounting for these dimensional changes does not completely explain the behavior of the gage; there is also a change in resistivity with strain.

This device is of sufficient importance in mechanical measurements to warrant a more complete discussion than can be given at this point. Chapter 12 is devoted to the theory and use of strain gages.

6.8   Thermistors

Thermistors are thermally sensitive variable resistors made of ceramic-like semiconducting materials. Oxides of manganese, nickel, and cobalt are used in formulations having resistivity values of 100 to 450,000 $\Omega \cdot$ cm.

These devices have two basic applications: (1) as temperature-detecting elements used for the purpose of measurement or control, and (2) as electric-power-sensing devices wherein the thermistor temperature—and hence resistance—are a function of the power being dissipated by the device. The second application is particularly useful for measuring radio-frequency power.

Further discussion of thermistors is given in Section 16.4.3.

6.9   The Thermocouple

While two dissimilar metals are in contact, an electromotive force exists whose magnitude is a function of several factors, including *temperature*. Junctions of this sort, when used to measure temperature, are called *thermocouples*. Often the junction is formed by twisting and welding together two wires.

Because of its small size, its reliability, and its relatively large range of usefulness, the thermocouple is a very important primary sensing-element.

Further discussion of its application is reserved for Chapter 16 (see especially Section 16.5).

6.10   Variable-Inductance Transducer Elements

Inductive transducers are based on the voltage output of an inductor (or *coil*) whose inductance changes in response to changes in the measurand. The coil is often driven by an ac excitation, although in dynamic measurements motion of the coil relative to a permanent magnet may create sufficient voltage. A classification of inductive transducers, based on the fundamental principle used, is as follows:

1. Variable self-inductance
   a. Single coil (simple variable permeance)
   b. Two coil (or single coil with center tap) connected for inductance ratio
2. Variable mutual inductance
   a. Simple two coil
   b. Three coil (using series opposition)
3. Variable reluctance
   a. Moving iron
   b. Moving coil
   c. Moving magnet

Inductive reactance, which quantifies the ac inductive effect, may be expressed by the relation

$$X_L = 2\pi f L, \tag{6.3}$$

where

$$X_L = \text{the inductive reactance } (\Omega),$$

$$f = \text{the frequency of applied voltage (Hz)},$$

$$L = \text{inductance (H)}.$$

Inductance, L, is influenced by a number of factors, including the number of turns in the coil, the coil size, and especially the permeability of the flux path. Some coils are wound with only air as the core material. They are generally used at relatively high frequencies; however, they will occasionally be found in transducer circuitry. Often some form of magnetic material will be used in the flux path, commonly in conjunction with one or more air gaps.

An expression that may be used to estimate the inductance of a cylindrical air-core coil is as follows [5]:

$$L = \frac{d^2 n^2}{18d + 40a} \tag{6.3a}$$

where

$$L = \text{inductance } (\mu\text{H}),$$

$$d = \text{the coil diameter (in.),}$$

$$a = \text{the coil length (in.),}$$

$$n = \text{number of turns.}$$

When the flux path includes both a magnetic material (usually iron) and an air gap or gaps, the inductance may be estimated by use of the following relation [6]:

$$L = \frac{1.26n^2 \times 10^{-8}}{(h_i/\mu a_i) + (h_a/a_a)}, \tag{6.3b}$$

in which

$h_i = $ the length of the iron circuit (cm),

$h_a = $ the length of the air gaps (cm),

$a_i = $ the cross-sectional area of iron (cm^2),

$a_a = $ the cross-sectional area of the air gap (cm^2),

$\mu = $ the permeability of the magnetic material at maximum flux density

In many instances the permeability of the magnetic material is sufficiently high so that only the air gaps need be considered. In such cases, Eq. (6.3b) reduces to

$$L = 1.26n^2 \frac{a_a}{h_a} \times 10^{-8}. \tag{6.3c}$$

The total impedance of a coil may be expressed by the relation

$$Z = \sqrt{X_L^2 + R^2}, \tag{6.4}$$

in which R is the dc resistance of the coil. The higher the inductance of a coil relative to its dc resistance, the higher is said to be its quality, which is designated by the symbol $Q = X_L/R$. In most cases, high Q is desired.

Inductive transducers may be based on variation of any of the variables indicated in the foregoing equations, and most have been tried at one time or another. The following are representative.

6.10.1 Simple Self-Inductance Arrangements

When a simple single coil is used as a transducer element, the mechanical input usually changes the permeance of the flux path generated by the coil, thereby changing its inductance. The change in inductance is then measured by suitable circuitry, indicating the value of the input. The flux path may be changed by a change in air gap (Fig. 6.6); however, a change in either the amount or type of core material may also be used.

Figure 6.6 A simple self-inductance arrangement wherein a change
in the air gap changes the pickup output

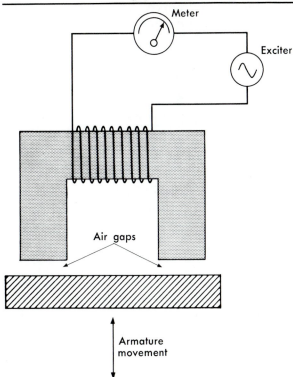

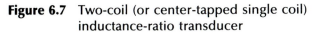

Figure 6.7 illustrates a form of *two-coil* self-inductance. (This may also be thought of as a single coil with a center tap.) Movement of the core or armature alters the relative inductance of the two coils. Devices of this type are usually incorporated in some form of inductive bridge circuit (see Section

Figure 6.7 Two-coil (or center-tapped single coil)
inductance-ratio transducer

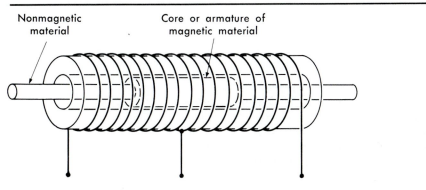

Figure 6.8 A mutual-inductance transducer. Coil *A* is the energizing coil and *B* is the pickup coil. As the armature is moved, thereby altering the air gap, the output from coil *B* is changed, and this change may be used as a measure of armature movement.

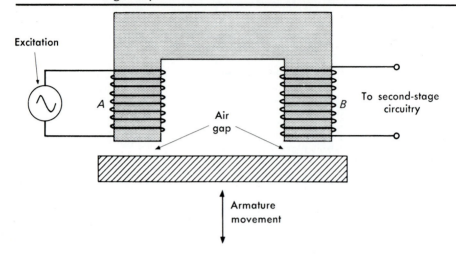

7.10) in which variation in the inductance ratio between the two coils provides the output. An application of a two-coil self-inductance used as a secondary transducer for pressure measurement is described in Section 14.8.2.

6.10.2   Two-Coil Mutual-Inductance Arrangements

Mutual-inductance arrangements using two coils are shown in Figs. 6.8 and 6.9. Figure 6.8 illustrates the manner in which these devices function. The magnetic flux from a power coil is coupled to a pickup coil, which supplies the

Figure 6.9 Two-coil inductive pickup for "an electronic micrometer"

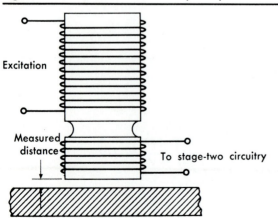

output. Input information, in the form of armature displacement, changes the coupling between the coils. In the arrangement shown, the air gaps between the core and the armature govern the degree of coupling. In other arrangements the coupling may be varied by changing the relative positions of the coils and armature, either linearly or angularly.

Figure 6.9 shows the detector portion of an *electronic micrometer* [7]. Inductive coupling between the coils, which depends on the permeance of the magnetic-flux path, is changed by the relative proximity* of a permeable material. A variation of this has been used in a transducer for measuring small inside diameters [8]. In that case the coupling is varied by relative movement between the two coils.

6.11 The Differential Transformer

Undoubtedly the most broadly used of the variable-inductance transducers is the differential transformer (Fig. 6.10), which provides an ac voltage output proportional to the displacement of a core passing through the windings. It is a mutual-inductance device making use of three coils generally arranged as shown.

Figure 6.10   The differential transformer: (a) schematic arrangement
(b) section through typical transformer

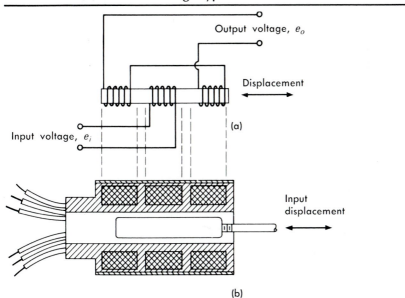

* Sensors that detect the presence or position of a nearby object are often called *proximity sensors*.

The center coil is energized from an ac power source, and the two end coils, connected in phase opposition, are used as pickup coils. This device is discussed in detail in Section 11.19.

6.12 Variable-Reluctance Transducers

In transducer practice, the term *variable reluctance* implies some form of inductance device incorporating a *permanent magnet*. In most cases these devices are limited to dynamic application, either periodic or transient, where the flux lines supplied by the magnet are cut by the turns of the coil. Some means of providing relative motion is incorporated into the device.

In its simplest form, the variable-reluctance device consists of a coil wound on a permanent-magnet core (Fig. 6.11). Any variation of the permeance of the magnetic circuit causes a change in the flux. As the flux field expands or collapses, a voltage is induced in the coil, according to Faraday's law:

$$V = -n\frac{d}{dt}\Phi \qquad (6.5)$$

where

V = induced voltage (V)

n = number of turns in coil,

Φ = magnetic flux through coil (Wb)

Practical applications of this arrangement are discussed in Sections 10.9 and 15.7.1.

Whereas the preceding arrangement depends on changing permeance,

Figure 6.11   A simple variable-reluctance pickup

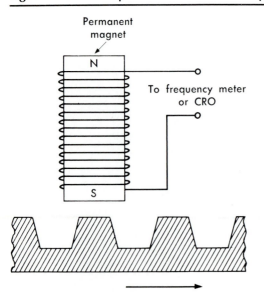

Permanent
magnet

N

To frequency meter
or CRO

S

other devices based on variable reluctance depend on relative movement between the flux field and the coil (Section 17.7.2).

6.13 Capacitive Transducers

An equation for calculating the capacitance of a set of equally spaced parallel plates is [9]

$$C = \frac{0.2249KA(N - 1)}{d},$$ (6.6)

where

C = the capacitance (pF),

K = the dielectric constant ($= 1$ for air),

A = the area of one side of one plate (in.²),

N = the number of plates,

d = the separation of plate surfaces (in.).

All the terms represented in this equation, except possibly the number of plates, have been used in transducer applications. The following are examples of each.

Changing Dielectric Constant

Figure 6.12 shows a device developed for the measurement of level in a container of liquid hydrogen [10]. The capacitance between the central rod and

Figure 6.12 Capacitance pickup for determining level of liquid hydrogen

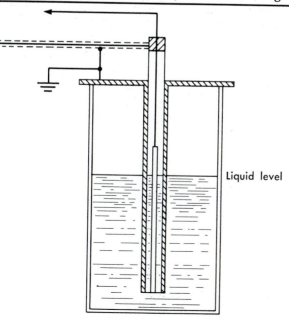

Liquid level

Figure 6.13   Section showing relative arrangement of teeth in capacitance-type torque meter

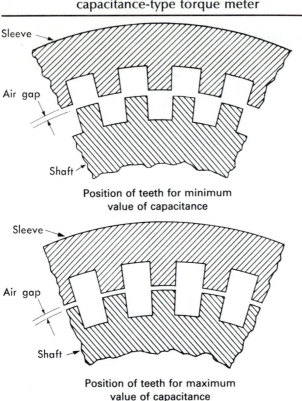

Sleeve

Air gap

Shaft

Position of teeth for minimum
value of capacitance

Sleeve

Air gap

Shaft

Position of teeth for maximum
value of capacitance

the surrounding tube varies with changing dielectric constant brought about by changing liquid level. The device readily detects liquid level even though the difference in dielectric constant between the liquid and vapor states may be as low as 0.05.

Changing Area

Capacitance change depending on changing effective area has been used for the secondary transducing element of a torque meter [11]. The device uses a sleeve with teeth or serrations cut axially, and a matching internal member or shaft with similar axially cut teeth. Figure 6.13 illustrates the arrangement. A clearance is provided between the tips of the teeth, as shown. Torque carried by an elastic member causes a shift in the relative positions of the teeth, thereby changing the effective area. The resulting capacitance change is calibrated in terms of torque.

Changing Distance

Varying the distance between the plates of a capacitor is undoubtedly the most common method for using capacitance in a pickup.

Figure 6.14 illustrates a capacitive-type pressure transducer, wherein the capacitance between the diaphragm to which the pressure is applied and the

Figure 6.14 Section through capacitance-type pressure pickup

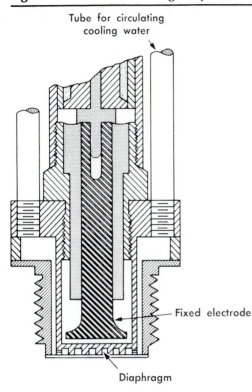

Tube for circulating
cooling water

Fixed electrode

Diaphragm

electrode foot is used as a measure of the diaphragm's relative position [12–14]. Flexing of the diaphragm under pressure alters the distance between it and the electrode.

6.14 Piezoelectric Sensors

Certain materials can generate an electrical charge when subjected to mechanical strain or, conversely, can change dimensions when subjected to voltage [Fig. 6.15(a)]. This is known as the *piezoelectric* effect*. Pierre and Jacques Curie are credited with its discovery in 1880. Notable among these materials are quartz, Rochelle salt (potassium sodium tartarate), properly polarized barium titanate, ammonium dihydrogen phosphate, certain organic polymers, and even ordinary sugar.

Of all the materials that exhibit the effect, none possesses all the desirable properties, such as stability, high output, insensitivity to temperature extremes and humidity, and the ability to be formed into any desired shape. Rochelle salt provides a very high output, but it requires protection from moisture in the air and cannot be used above about 45°C (115°F). Quartz is undoubtedly the

---

* The prefix "piezo" is derived from the Greek *piezein*, meaning "to press" or "to squeeze."

Figure 6.15 (a) Basic deformation modes for piezoelectric plates;
electrodes (+ and −) shown, (b) equivalent
circuit for a piezoelectric element

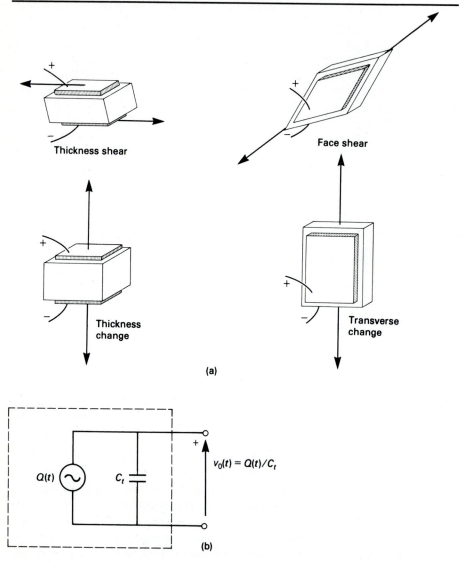

Thickness shear

Face shear

Thickness
change

Transverse
change

(a)

$v_0(t) = Q(t)/C_t$

$Q(t)$ C_t

(b)

most stable, but its output is low. Because of its stability, quartz is quite
commonly used for regulating electronic oscillators (Section 10.4). Often the
quartz is shaped into a thin disk with each face silvered for the attachment of
electrodes. The thickness of the plate is ground to the dimension that provides
a mechanical resonance frequency corresponding to the desired electrical
frequency. This crystal may then be incorporated in an appropriate electronic
circuit whose frequency it controls.

Rather than existing as a single crystal, as are many piezoelectric materials, barium titanate is polycrystalline; thus it may be formed into a variety of sizes and shapes. The piezoelectric effect is not present until the element is subjected to polarizing treatment. Although exact polarizing procedure varies with the manufacturer, the following procedure has been used [15]. The element is heated to a temperature above the Curie point of 120°C, and a high dc potential is applied across the faces of the element. The magnitude of this voltage depends on the thickness of the element and is on the order of 10,000 V/cm. The element is then cooled with the voltage applied, which results in an element that exhibits the piezoelectric effect.

Piezoelectric polymers, such as polyvinylidene fluoride [16], provide low-cost piezoelectric transducers with relatively high voltage outputs. These semi-crystalline polymers are formed into thin films, perhaps 30 μm thick, with silvered electrodes on either side and are coated onto a somewhat-thicker Mylar backing. The resulting transducer is light, flexible, and easily manipulated.

Figure 6.15(b) shows an equivalent circuit for a piezoelement, consisting of a charge generator and a shunting capacitance, C_t. (The reader should refer to Tables 5.1 and 5.2 for the relationships between charge Q and other electrical units). When mechanically strained, the piezoelement generates a charge $Q(t)$, which is temporarily stored in the element's inherent capacitance. As with all capacitors, however, the charge dissipates with time owing to leakage, a fact which makes piezodevices most valuable for dynamic measurements.

The voltage across the piezoelement at any time is simply

$$V_o(t) = \frac{Q(t)}{C_t} \qquad \textbf{(6.7)}$$

Measurement of this voltage, however, requires a very high impedance circuit to prevent charge loss. A *charge amplifier* (Section 7.18.2) is normally used for this purpose. The voltage can alternatively be expressed in terms of the stress on the piezoelement,

$$V_o = Gh\sigma, \qquad \textbf{(6.7a)}$$

where h is the thickness of the element between the electrodes and σ is the stress. G is a material constant equal to 0.055 Vm/N for quartz in compressive stress (thickness change) and 0.22 Vm/N for polyvinylidene fluoride in axial stress (transverse change).

Piezoelectric transducers are used to measure surface roughness (Section 11.20), force and torque (Section 13.6), pressure (Section 14.8.3), motion (Section 17.8), and sound (Section 18.5.1). In addition, the piezofilm transducers are used to sense thermal radiation (16.8.4): since the film expands in response to temperature change, a charge is developed when infrared radiation is absorbed (the *pyroelectric* effect). Pyroelectric transducers are common elements in household motion-detectors.

Ultrasonic transducers may use barium titanate. Such elements are found in industrial cleaning apparatus and in underwater detection systems known as *sonar*.

6.15 Semiconductor Sensors

The semiconductor revolution has profoundly influenced measurement technology. Apart from the appearance of digital voltmeters, computer data-acquisition systems, and other readout and data-processing systems, semiconductor technology has produced compact and inexpensive sensors. A principal strength of semiconductor sensors is that they take advantage of microelectronic fabrication techniques. Thus, the sensors can be quite small, mechanical structures (such as diaphragms and beams) can be etched into the device, and other electronic components (resistors, transistors, etc.) can be directly implanted with the sensor to form a transducer having onboard signal conditioning.

6.15.1 Electrical Behavior of Semiconductors

Semiconducting materials include both elements, such as silicon and germanium, and compounds, such as gallium arsenide and cadmium sulphide. Semiconductors differ from metals in that relatively few free electrons are available to carry current. Instead, when a bound electron is separated from a particular atom in the material, a positively charged *hole* is formed and will move in the direction opposite the electron. Both negatively charged electrons and positively charged holes contribute to the flow of current in a semiconductor.

The number of charge carriers (electrons or holes) in a semiconductor is a strong function of temperature, T. Typically,

$$n_c = \text{number per unit volume} \propto T^{3/2} \exp\left(\frac{-\text{constant}}{T}\right) \qquad (6.8)$$

Since the resistivity of a material is proportional to $1/n_c$, a semiconductor's resistance decreases rapidly with increasing temperature. For silicon near room temperature, the resistivity decreases by about 8%/°C [17].

Greater control over a semiconductor's electrical behavior is obtained by *doping* it with impurity atoms. These atoms may be either *electron donors* or *electron acceptors*. Electron donor atoms (such as phosphorus or arsenic) raise the number of free electrons in the material. Electron acceptor atoms (such as gallium and aluminum) hold electrons, thus raising the number of holes. Since the number of doping atoms is usually large relative to the number of free electrons in the undoped material, the dopant sets the *majority current carrier* of the material. Specifically, doping with donor atoms creates a predominance of negative charge carriers (electrons), giving an *n-type* semiconductor. Doping with acceptor atoms creates a predominance of positive charge carriers (holes) giving a *p-type* semiconductor.

Semiconductors, either doped or undoped, are useful as temperature sensors. For undoped semiconductors, the number of carriers increases rapidly with temperature [Eq. (6.8)], so that the resistance is a strongly decreasing function of temperature. Thermistors (Sections 6.8, 16.4.3) are based on this effect. Because such sensors have a negative temperature coefficient of

Figure 6.16 Cross section of a semiconductor-diaphragm absolute-pressure sensor (Motorola, not to scale). External pressure change causes diaphragm to deflect, straining the gage

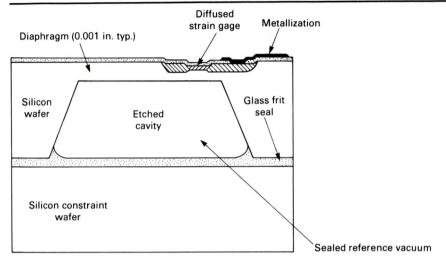

resistance, they are sometimes called *NTC* sensors. When semiconductors are heavily doped, the mobility of the carriers *decreases* with increasing temperature, so that the resistance increases with temperature; these positive-temperature-coefficient devices are called *PTC* sensors.

Semiconductors also respond to strain. For example, a *p*-type region diffused into an *n*-type base functions as a resistor whose value increases strongly when it is strained (this behavior is called *piezoresistivity*). Such resistors are the basis of semiconductor strain gages, semiconductor diaphragm pressure sensors (Figure 6.16, Section 14.8.4), and semiconductor accelerometers [18, 19].

6.15.2 pn-Junctions

Most semiconductor devices involve a *junction*, at which *n*-type and *p*-type doping meet (Fig. 6.17). Current flows easily from the *p*-type to the *n*-type material, since holes (+ charge) easily enter the *n*-type material and electrons (− charge) easily enter the *p*-type material. Current flow in the opposite direction meets much greater resistance. Thus, this junction behaves like a diode.

When a voltage is applied to the junction, the current through it varies as shown in Fig. 6.18. When V is positive, we say that the junction is *forward-biased*. The current becomes very large once the voltage reaches a threshold level. If the voltage is instead negative, the junction is *reverse-biased*, and only a very small current flows. As the reverse-bias voltage is raised, the current quickly reaches a value $-I_0$, which is nearly independent of V. The current I_0 is called the *reverse saturation current*. Typical values of I_0

Figure 6.17   (a) pn-junction with applied voltage
                  (b) circuit representation as a diode

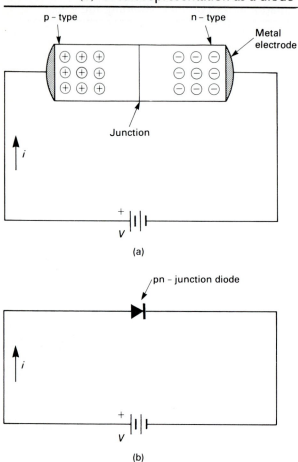

(a)

(b)

are on the order of nanoamperes for silicon and microamperes for germanium. The reverse saturation current is more nearly voltage-independent for germanium than for silicon, as a result of various secondary effects [20].

The voltage-current curve in Fig. 6.18 is described by the equation

$$i = I_0 \left[\exp\left(\frac{q}{kT} V\right) - 1 \right], \tag{6.9}$$

which shows a strong dependence of current on temperature (q is the charge of an electron and k is Boltzmann's constant). The saturation current is also a function of temperature, roughly doubling with every 10°C increase in temperature for silicon and germanium diodes near room temperature [17].

Figure 6.18 Voltage-current curve for a pn-junction. I_o is the reverse saturation current and V_t is the forward threshold voltage.

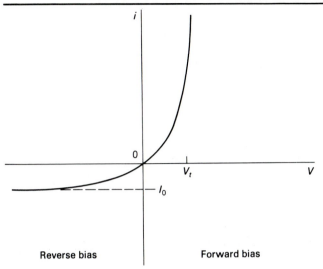

When forward-biased, the voltage drop (or threshold voltage) across a silicon pn-junction limits to about 0.7 V at room temperature. This voltage drop decreases by about 2 mV/°C as temperature increases, and the changing voltage forms the basis of some semiconductor-junction temperature sensors. Through integrated-circuit techniques, junction temperature-dependences have been applied to make linear-response, temperature-sensing chips (Section 16.6).

One disadvantage of many semiconductor-junction sensors is that the operating temperature must remain below about 150°C to prevent degradation of the junction's electrical characteristics.

6.15.3 Photodiodes

Semiconductor junctions are sensitive to light as well as heat. If a junction is formed near the surface of a semiconductor, photons reaching the junction can create new pairs of electrons and holes, which then separate and flow in opposite directions. Thus, the irradiating light produces an additional current, I_λ, at the junction:

$$i = I_0\left[\exp\left(\frac{q}{kT}V\right) - 1\right] - I_\lambda. \tag{6.10}$$

The photocurrent, I_λ, is directly proportional to the intensity, H, of the incoming light (in W/m^2):

$$I_\lambda = \text{constant} \times H,$$

Figure 6.19   Photodiode *i-V* characteristic for various
                  incident light intensities, *H*

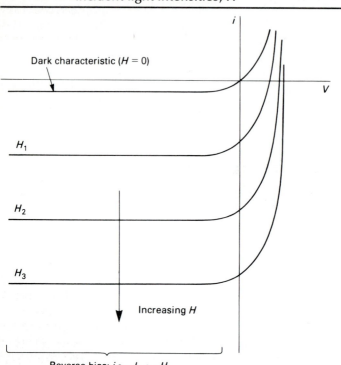

where the constant is proportional to the area of the diode exposed to light.
The voltage-current characteristics of a photodiode are shown in Fig. 6.19 [21].

   The photocurrent is typically on the order of milliamperes and thus is
much larger than the reverse saturation current $(I_\lambda \gg I_0)$. By operating a
photodiode with a reverse-bias voltage, the output current is made directly
proportional to the incident light intensity $(i \approx -I_\lambda \propto -H)$.

   Semiconductor junctions are most responsive to infrared wavelengths, but
sensitivity can extend to visible wavelengths and near-ultraviolet wavelengths
as well. Common photodiodes are usually made from inexpensive silicon
junctions, although several other semiconductors, such as germanium, are also
in use.

   The sensitivity of a photodiode is limited by its "dark current," which is
the usual junction current with no incident light [i.e., Eq. (6.9)]. Since this
current decreases with temperature, sensitivity can be improved by cooling the
diode to very low temperatures. For example, in high-performance infrared
sensing, photodiodes may be operated at liquid-nitrogen temperatures or even
liquid-helium temperatures (−198°C or −269°C, respectively) [22].

Some bulk semiconductors, without a pn-junction, also respond to light. Photons create additional electron-hole pairs in the material, thereby reducing its resistance. Common examples of such *photoconductive* materials include cadmium sulphide (CdS) and cadmium selenide (CdSe). For high-performance infrared sensing, low-temperature doped germanium may be used.

Photodetectors are discussed further in the next section.

6.16 Light-Detecting Transducers

Light-sensitive detectors, photosensors, or photocells may be categorized as either *thermal detectors* or *photon detectors*. The thermal detectors involve a temperature-sensitive element, which is heated by incident light; some examples are considered in Section 16.8.2. The photon devices respond directly to absorbed photons, either by emitting electrons from a surface (the *photoelectric effect*) or by creating additional electron-hole pairs in a semiconductor (as discussed in Section 6.15.3). Photon devices are categorized in Table 6.2.

Photoemissive detectors (Type A, Table 6.2) consist of a cathode-anode combination in an evacuated glass or quartz envelope. In the proper circuit (commonly requiring a dc source of several hundred volts), light impingement on the cathode causes electrons to be emitted. The electrons travel to the anode, thereby providing a small current. By adding several successively higher-voltage electrodes (or *dynodes*) to the envelope, substantial current amplification is obtained, producing a *photomultiplier tube,* or *PMT.* Since the invention of small semiconductor-photosensors, these devices are used only in rather specialized applications. They can outperform semiconductor sensors when very fast response and high gain are central concerns or when short wavelengths (e.g., ultraviolet) are involved [23].

Semiconductor photosensors are of several types. In general, they perform best at near-infrared wavelengths.

Photoconductive cells (Type B, Table 6.2) consist of a thin layer of material such as cadmium selenide, several of the metallic sulphides, or doped germanium coated between electrodes on the surface of a glass plate. The cell behaves as a light-controlled variable resistor whose resistance is reduced when it is exposed to a light source. In conjunction with resistance-sensitive circuitry (Sections 7.6–7.9), an output may be obtained that is a function of the intensity of the light source.

Photovoltaic cells (Type C, Table 6.2) consist of a sandwich of unlike materials, such as an iron base covered with a thin layer of iron selenide. When the cell is exposed to light, a voltage is developed across the sandwich. A distinguishing feature of this cell is that it requires no external power other than the light: It is the well-known *solar cell.*

The *photodiode* (Types D1 and D2, Table 6.2) utilizes a pn-junction and is similar to the photoconductive cell (see Section 6.15.3). Basically, it is a

Table 6.2 Photocells

Type	Symbol and Typical Circuit	Form of Output	Relative Frequency Response	Comments
A. Photoemissive or Photomultiplier	Load	Current	Extremely fast	Cathode-anode in evacuated glass or quartz envelope. PMT gain can be 10^3 to 10^8. Bulky; requires high voltage; and has given way to solid-state devices.
B. Photoconductive (or photoresistive)	AC or DC — Load	Resistance change	Slow	Light-sensitive resistor. Increased light intensity causes reduced resistance.
C. Photovoltaic (solar cell)	Out	Voltage	Fast	Typical open-circuit voltage, 0.45 V. In bright sunlight, 0.4 to 0.5 mA.
D1. Photodiode (PN junction) D2. PIN photodiode	Load	Current	Very fast	Primary disadvantage is low output current. "Dark current" very low (nanoampere range), but not zero. PIN diode has "intrinsic" layer between P and N layers that provides response over wider range of light wavelengths. PIN is faster than PN type.
E1. Phototransistor	Load	Current	Slower than photodiode	Produces much higher current for given input than photodiode does because of its amplifying ability. Base lead, if accessible, is seldom used.
E2. Photodarlington	Load	Current	Slower than phototransistor	Much more sensitive than phototransistor.

light-sensitive current source. The PIN photodiode differs from the common variety in that a layer of undoped (or *intrinsic*) semiconductor is sandwiched between the *p* and *n* layers in order to expand the range of sensitivity to longer wavelengths. Silicon photovoltaic cells are obtained by operating a photodiode in the region of Figure 6.19 where *i* is negative and *V* is positive.

Both the *phototransistor* and the *photodarlington* cells (Types E1 and E2, Table 6.2) are basically photodiodes followed by one or two stages of amplification incorporated in the same package to enhance the output.

Photosensors may be made selectively sensitive to light, not only in the visible spectrum but also in the infrared and ultraviolet ranges. Heat-seeking infrared sensors are commonly of the photoconductor type. The response of photosensors to sudden variations in light intensity is not instantaneous. It is determined both by the cell itself and by related circuitry. Rise and fall times, as determined by the cell type, may range from a few nanoseconds to several thousand milliseconds [23–26].

Applications of the photocell in mechanical measurements include simple counting, where the interruption of a beam of light is used (Fig. 10.1), strain measurement [27], dew-point controls [28], temperature measurement (Section 16.8) and edge and tension controls [29]. Two-dimensional arrays of photo-detectors are the basis of digital cameras, which reduce a visual image to a discrete array of photodetector voltages.

Special packages, optointerrupters and optoisolaters, consist of photocells combined with light-emitting diodes (LEDs), arranged so that the light from the LED impinges on the cell (see Figs. 6.20 and 6.21). The interrupter is configured so that some form of mechanical mask may be used to break the light beam between the LED and the cell, thereby providing on–off switching for counting or a variety of other purposes. The optoisolator is used to match

Figure 6.20 A photointerrupter consisting of an LED light source (often infrared) and a photocell sensor. Mechanical interruption of the light path can be used for various purposes, such as counting, triggering, and synchronization.

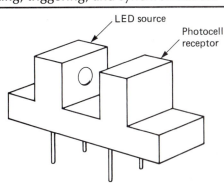

Figure 6.21   The essentials of a photoisolator, used for connecting
low-impedance current circuits to high-impedance voltage
circuits. The isolator is also useful for providing complete
electrical isolation between circuits, sometimes
imperative in health-related electronics.

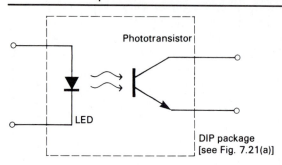

low-impedance current circuits to high-impedance voltage circuits, or vice
versa. It also provides a high-impedance isolation between circuits, an
important feature in some forms of health-related electronics [30].

6.17   Hall-Effect Sensors

The *Hall effect* is the appearance of a transverse voltage difference on a
conductor carrying a current perpendicular to a magnetic field [31]. This
voltage is directly proportional to the magnetic field strength. If the magnetic
field is made to vary with the position of a nearby object, the Hall effect can be
the basis of a proximity sensor.

In Fig. 6.22(a), a conductor carries current in the x-direction, so that
electrons flow in the $-x$ direction with a velocity $\mathbf{v}_d$. Because the electrons
carry a charge $-q$, they experience a magnetic force $\mathbf{F}_B$ in the z-direction

$$\mathbf{F}_B = -q\mathbf{v}_d \times \mathbf{B}$$

This force deflects electrons upward and so creates a negative charge along the
top of the conductor and a positive charge along the bottom [Fig. 6.22(b)].
This charge distribution in turn creates an electric field, $\mathbf{E}$, whose force in
steady-state is equal and opposite the magnetic force on the electrons:

$$\mathbf{F}_E = -q\mathbf{E} = -\mathbf{F}_B$$

From these, the magnitude of the electric field is $E = v_d B$. Since the electric
field is the gradient of voltage, the voltage difference across a conductor of
height l is

$$V_{\text{Hall}} = lE = v_d lB \qquad (6.11)$$

This is the Hall-effect voltage.

Figure 6.22 The Hall effect: (a) a conductor carries current in a perpendicular magnetic field; (b) electrons are driven upward by magnetic force, creating an opposing electric field.

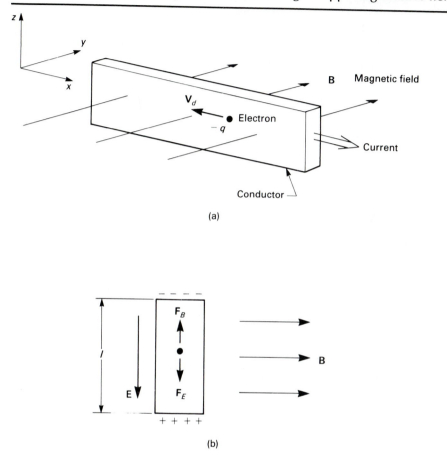

(a)

(b)

The Hall effect is present in any conductor carrying current in a magnetic field, but it is much more pronounced in semiconductors than in metals. Thus most Hall-effect transducers use a semiconducting material as the conductor, often in conjunction with an integrated-circuit signal conditioner. A permanent magnet may be built into the transducer to provide the needed magnetic field (Fig. 6.23). If a passing object, such as another magnet or a ferrous metal, alters the magnetic field, the change in the Hall-effect voltage is seen at the transducer's output terminals. Hall-effect transducers are used as position sensors, as solid-state keyboards actuators, and as current sensors [32]. Low-cost, ruggedly packaged versions are used as automotive crankshaft-timing sensors [33].

Figure 6.23 Hall-effect gear-tooth sensor [32]

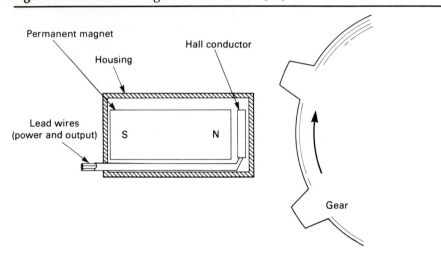

6.18   Some Design-Related Problems

Accuracy, sensitivity, dynamic response, repeatability, and the ability to reject unwanted inputs are all qualities highly desired in each component of a measuring system. Many of the parameters that combine to provide these qualities present conflicting problems and must be compromised in the final design. Uncertainty considerations in experimental design were discussed in Sections 3.11.2 and 3.12. Response requirements were presented in Chapter 5. At this point we will discuss some additional design problems.

6.18.1   Manufacturing Tolerances

Conception of a component or system on paper is a necessary and important beginning; to be useful, however, the apparatus must be produced, and no manufacturing process can reproduce *exact* length or angular dimensions. Dimensions must always be assigned with some specified or implied tolerances. How can one predict the effect of such variations on performance? The following example describes one approach to the problem.

Example 6.1

Suppose a spring scale such as that shown in Fig. 6.24 is to be designed. We will assume a force capacity of 50 N and a maximum deflection of 10 cm. This gives us a deflection constant of 5 N/cm.

Figure 6.24 Common spring-type "fish" scale

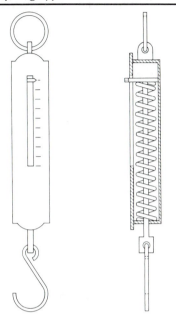

Solution. Using conventional coil-spring design relations [34], we find that a spring made with a mean coil diameter D_m of 2 cm and a steel wire diameter of 2 mm will meet stress requirements provided we ensure against overload by including appropriate deflection limits or stops. The deflection equation commonly used for coil springs is

$$K = \frac{F}{y} = \frac{E_s D_w^4}{8 D_m^3 n},$$

where

$\qquad K$ = the deflection constant (N/m),

$\qquad F$ = the design load (N),

$\qquad y$ = the corresponding deflection (m),

$\qquad E_s$ = the torsional elastic modulus (about 80×10^9 Pa for steel),

$\qquad n$ = the number of coils.

Using this relation and the design values above, we find that 40 coils are needed to provide the required deflection constant. If we apply reasonable tolerances, our specifications become

$$D_w = 2 \pm 0.01 \, \text{mm},$$

$$D_m = 2 \pm 0.05 \, \text{cm},$$

$$n = 40 \pm \tfrac{1}{3} \, \text{coils, and}$$

$$E_s = 80 \times 10^6 \pm 3.5 \times 10^6 \, \text{kPa}.$$

Let us now consider how various uncertainties (manufacturing tolerances) may affect the deflection constant, K. (We direct attention to Ref. [35] at this point.) We assume that 99% of the coils will not exceed *any* of these tolerances; in other words, we take the tolerances to represent 99% uncertainty limits for each variable. Then, using the approach described in Section 3.10, we have

$$\frac{U_K}{K} = \sqrt{\left(\frac{4 \times 0.01}{2}\right)^2 + \left(\frac{3 \times 0.05}{2}\right)^2 + \left(\frac{0.33}{40}\right)^2 + \left(\frac{3.5}{80}\right)^2}$$

$$= 0.0895 \approx 9\% \qquad (99\% \text{ certainty})$$

or

$$K = 500 \pm 45 \, \text{N/m} \qquad (99\%).$$

Note that mm, cm, and m have been used in the preceding example; hence care must be used in placing decimal points.

We see then that if we lay out the gradations corresponding to nominal values, a force of 50 N may actually be indicated as anything in the range from 45.5 to 54.5 N, depending on how the manufacturing tolerances may fall. Should this result not be satisfactory, our only recourse is (1) to provide better control of the manufacturing tolerances, or (2) to provide some means for adjusting calibration.

Various methods of calibration can be used, depending on the intended "quality" of the device. In this instance, two or three faceplates can be provided, each with gradations scaled to cover a portion of the calibration range. At the time of assembly, a simple calibration would determine the most appropriate plate to use. Section 13.4.1 presents another approach that could be applied to our particular example.

At this point it is appropriate to make an additional observation. *Weight* is basically a *force*; hence we should express the calibration in newtons rather than in kilograms. Should we wish a scale calibrated in kilograms, then, to be completely correct we should include the assumed value of gravity acceleration on the faceplate. The standard due to gravity is 9.80665 m/s², and the kilogram range corresponding to our 50 N range becomes 0 to $50/9.80665 \approx 5.1$ kg. (Use of the non-SI symbol, kgf, is discouraged.)

It should be clear that the procedures used in the preceding example are applicable to most elastic transducer configurations, as well as to many other tolerance problems.

6.18.2 Some Temperature-Related Problems

An ideal measuring system will react to the design signal only and ignore all else. Of course, this ideal is never completely fulfilled. One of the more insidious extraneous stimuli adversely affecting instrument operation is temperature. It is insidious in that it is almost impossible to maintain a constant-temperature environment for the general-purpose measuring system. The usual solution is to accept the temperature variation and to devise methods to compensate for it.

Temperature variations cause dimensional changes and changes in physical properties, both elastic and electrical, resulting in deviations (bias error) referred to as *zero shift* and *scale error* [36, 37]. Zero shift, as the name implies, results in a change in the no-input reading. Factors other than temperature may cause zero shift; however, temperature is probably the most common cause. In most applications the zero indication on the output scale would be made to correspond to the no-input condition. For example, the indicator or the spring scales referred to earlier should be set at zero when there is no weight in the pan. If the temperature changes after the scale has been set to zero, there may be a differential dimensional change between spring and scale, altering the no-load reading. This change would be referred to as *zero shift*. Zero shift is primarily a function of linear dimensional change caused by expansion or contraction with changing temperature.

Dimensional changes are expressed in terms of the coefficient of expansion by the following familiar relations:

$$\alpha = \frac{1}{\Delta T} \frac{\Delta L}{L_0} \qquad (6.12)$$

and

$$L_1 = L_0(1 + \alpha \, \Delta T), \qquad (6.13)$$

in which

α = the coefficient of linear expansion (ppm/deg temp. $\times 10^{-6}$),

L/L_0 = the unit change in length,

ΔT = the change in temperature, $T_1 - T_0$,

L_0 = the length dimension at the reference temperature T_0,

L_1 = the length dimension at any other temperature T_1.

In addition to causing zero shift, temperature changes usually affect scale calibration when resilient load-carrying members are involved. The coil and wire diameters of our spring would be altered with temperature change, and so too would the modulus of elasticity of the spring material. These variations would cause a changed spring constant, and hence changed load-deflection calibration, resulting in what is referred to as *scale error*.

The thermoelastic coefficient is defined by the relations

$$c = \frac{1}{\Delta T} \frac{\Delta E}{E_0} \tag{6.14}$$

and

$$E_1 = E_0(1 + c\,\Delta T), \tag{6.15}$$

in which

c = the temperature coefficient for the tensile modulus of elasticity (ppm/deg temp. $\times\ 10^{-6}$),

$\Delta E/E_0$ = the unit change in the tensile modulus of elasticity,

E_0 = the tensile modulus of elasticity at temperature T_0,

E_1 = the tensile modulus of elasticity at temperature T_1.

Similarly, the coefficient for torsional modulus may be written

$$m = \frac{1}{\Delta T} \frac{\Delta E_s}{E_{s_0}} \tag{6.16}$$

and

$$E_{s_1} = E_{s_0}(1 + m\,\Delta T), \tag{6.17}$$

where

m = the temperature coefficient for the torsional modulus of elasticity (ppm/deg temp. $\times\ 10^{-6}$),

$\Delta E_s/E_{s_0}$ = the unit change in the torsional modulus of elasticity,

E_{s_0} = the torsional modulus of elasticity at temperature T_0,

E_{s_1} = the torsional modulus of elasticity at temperature T_1.

Representative values of these quantities are given in Table 6.3.

The manner in which temperature changes in elastic properties affect instrument performance can be demonstrated by the following example. Assume that a restoring element in an instrument is essentially a single-leaf cantilever spring of rectangular section, for which the deflection equation at reference temperature T_0 is

$$K_0 = \frac{F}{y} = \frac{3E_0 I_0}{L_0^3} = \frac{E_0 w_0 t_0^3}{4L_0^3}, \tag{6.18}$$

Table 6.3   Temperature characteristics for some materials

Material	Tensile Modulus of Elasticity, E Pa × 10^{-10} (psi × 10^{-6})	Torsional Modulus of Elasticity, E_s Pa × 10^{-10} (psi × 10^{-6})	Coefficient of Linear Expansion, α ppm/°C (ppm/°F)	Coefficient of Tensile Modulus of Elasticity, c^* ppm/°C (ppm/°F)
High-carbon spring steel	20.7 (30)	7.93 (11.5)	11.6 (6.5)	−220 (−122)
Chrome-vanadium steel	20.7 (30)	7.93 (11.5)	12.2 (6.8)	−260 (−145)
Stainless steel Type 302	19.3 (28)	6.9 (10)	16.7 (9.3)	−439 (−244)
Spring brass	10.3 (15)	3.8 (5.5)	20.2 (11.2)	−391 (−217)
Phosphor bronze	10.3 (15)	4.3 (6.3)	17.8 (9.9)	−380 (−211)
Invar†	14.8 (21.4)	5.6 (8.1)	1.1 (0.6)	+48.1 (+27)
Isoelastic†	18.0 (26)	6.3 (9.2)	7.2 (4)	−36 to +13 (−20 to +7.3)
Aluminum	6.9 (10)	2.6 (3.8)	23 (13)	−270 to −400 (−150 to −220)

* c may be used for torsional modulus also.

† Trade names.

in which

K_0 = the deflection constant,

I_0 = the moment of inertia,

w_0 = the width of the section at reference temperature,

t_0 = the thickness of the section at reference temperature,

L_0 = the length of the beam at the reference temperature.

A second equation may be written for any other temperature, T_1, as follows:

$$K_1 = \frac{[E_0(1 + c\,\Delta T)][w_0(1 + \alpha\,\Delta T)][t_0(1 + \alpha\,\Delta T)]^3}{4[L_0(1 + \alpha\,\Delta T)]^3}. \tag{6.19}$$

Thus we have

$$\begin{array}{l}\text{Percent error in} \\ \text{deflection scale}\end{array} = \left(\frac{K_0 - K_1}{K_0}\right) \times 100$$

$$= [1 - (1 + c\,\Delta T)(1 + \alpha\,\Delta T)] \times 100,$$

which we may simplify, by expanding and neglecting the second-order term, to read

$$\text{Percent scale error} = -(c + \alpha)\,\Delta T \times 100. \tag{6.20}$$

If our spring is made of spring brass,

$$\text{Percent scale error/°F} = -(-217 + 11.2) \times 10^{-6} \times 100 = 0.021\%.$$

Hence a temperature change of +50°F would result in a scale error of about +1%. (This means that the reading is too high; our spring is too flexible, and a given load deflects the spring more than it should.)

It is interesting to note that for our example the scale error is a function of material or materials. It should be clear that we are speaking of the load-deflection relation for resilient members in this connection, and that this would not include members whose duty it is simply to transmit motion, such as the linkage in a Bourdon-tube pressure gage.

Although not a mechanical quantity, another item affected by temperature change is electrical resistance. The basic resistance equation may be written in the form

$$R = \rho\frac{L}{A}, \tag{6.21}$$

where

R = the electrical resistance (Ω),

ρ = the resistivity ($\Omega \cdot$ cm),

L = the length of the conductor (cm),

A = the cross-sectional area of the conductor (cm^2).

As temperature changes, a change in the resistance of an electrical conductor will be noted. This will be caused by two different factors: dimensional changes due to expansion or contraction and changes in the current-opposing properties of the material itself. For an unconstrained conductor, the latter is much more significant than the former, causing more than 99% of the total change for copper [37]. Therefore, in most cases it is not very important whether the dimensional effect is accounted for or not. If dimensional changes caused by temperature are ignored, change in resistivity with temperture may be expressed as

or

$$b = \frac{1}{\Delta T} \frac{\Delta \rho}{\rho_0} \tag{6.22}$$

in which

$$\rho_1 = \rho_0(1 + b \, \Delta T), \tag{6.23}$$

   b = the temperature coefficient of resistivity $[(\Omega \cdot cm)/(\Omega \cdot cm \cdot deg)]$,

   ΔT = the temperature change (deg),

   $\Delta \rho / \rho_0$ = the unit change in resistivity,

   ρ_0 = the resistivity at the reference temperature T_0 $(\Omega \cdot cm)$,

   ρ_1 = the resistivity at any temperature T_1 $(\Omega \cdot cm)$.

If we account for temperature-dimensional changes, the equation reads

$$\rho_1 = \frac{R_0 A_0}{L_0}(1 + b \, \Delta T)(1 + \alpha \, \Delta T)$$
$$= \rho_0(1 + b \, \Delta T)(1 + \alpha \, \Delta T). \tag{6.24}$$

Table 6.4 lists values of the coefficients of resistivity for selected materials.

6.18.3   Methods for Limiting Temperature Errors

Three approaches to a solution of the temperature problem in instrumentation are as follows: (1) *minimization* through careful selection of materials and operating temperature ranges, (2) *compensation* through balancing of inversely reacting elements or effects, and (3) *elimination* through temperature control. Although each situation is a problem unto itself, thereby making specific recommendations difficult, a few general remarks with regard to these possibilities may be made.

Minimization

As we pointed out earlier, temperature errors may be caused by thermal expansion in the case of simple motion-transmitting elements, by thermal expansion and modulus change in the case of calibrated resilient transducer

Table 6.4 Resistivity and temperature coefficients of
resistivity for selected materials

Material	Composition (for alloys)	Resistivity at 20°C (68°F) $\Omega \cdot cm \times 10^6$	Coefficient of Resistivity, b $\Omega/\Omega \cdot deg \times 10^6$	
			Per °C	Per °F
Aluminum	—	2.8	3900	2170
Constantan*	60% Cu, 40% Ni	44	11	6
Copper (annealed)	—	1.72	3900	2180
Iron	99.9% pure	10	5000	2800
Isoelastic*	36% Ni, 8% Cr, 4% Mn, Si, and Mo, remainder Fe	48	470	260
Manganin*	9–18% Mn, $1\frac{1}{2}$–4% Ni; remainder Cu	44	11	6
Monel*	33% Cu, 67% Ni	42	2000	1100
Nichrome*	75% Ni, 12% Fe, 11% Cr, 2% Mn	100	400	220
Nickel	—	7	6400	3550
Silver	—	1.6	4000	2250

Note: Values should be considered as quite approximate. Actual values depend on exact composition and, in certain cases, degree of cold work.

* Trade names.

elements, and by thermal expansion and resistivity change in the case of electrical resistance transducers. All these effects may be minimized by selecting materials with low-temperature coefficients in each of the respective categories. Of course, minimum temperature coefficients are not always combined with other desirable features such as high strength, low cost, corrosion resistance, and so on.

Compensation

Compensation may take a number of different forms, depending on the basic characteristics of the system. If a mechanical system is being used, a form of compensation making use of a composite construction may be employed. If the system is electrical, compensation is generally possible in the electrical circuitry.

An example of composite construction is the balance wheel in a watch or clock. As the temperature rises, the modulus of the spring material reduces and, in addition, the moment of inertia of the wheel (if of simple form) increases because of thermal expansion, both of which cause the watch to *slow down*. If we incorporate a bimetal element of appropriate characteristics in the rim of the wheel, the moment of inertia decreases with temperature enough to compensate for both expansion of the wheel spokes and change in spring modulus. (See also Section 11.6 for a discussion of temperature effect on linear measuring devices.)

Electrical circuitry may use various means for compensating temperature effects. The thermistor, discussed in some detail in Sections 6.8 and 16.4.3, is quite useful for this purpose. Most circuit elements possess the characteristic of increasing dc resistance with rising temperature. The thermistor has an opposite temperature-resistance property, along with reasonably good stability, both of which make it ideal for simple temperature-resistance compensation.

Resistance strain gages are particularly susceptible to temperature variations. The actual situation is quite complex, involving thermal-expansion characteristics of both the base material and all the gage materials (support, cement, and grid) and temperature-resistivity properties of the grid material, combined with the fact that heat is dissipated by the grid since it is a resistance device. Temperature compensation is very nicely handled, however, by pitting the temperature effect output from like gages against one another while subjecting them differentially to strain. This outcome is accomplished by use of a resistance bridge circuit arrangement, which is used extensively in strain-gage work (see Section 12.10). In addition through careful selection of grid materials, so-called self-compensating gages have been developed. (See also Section 13.5.)

Elimination

The third method—that is, eliminating the temperature problem by temperature control—really requires no discussion. Many methods are possible, extending from the careful control of large environments to the maintenance of constant temperature in small instrument enclosures. An example of the latter is the "crystal oven," often used to stabilize a frequency-determining quartz crystal.

6.19 Summary

We have in no sense exhausted the list of possible devices or principles suitable for sensing mechanical inputs. In certain instances, we discuss others elsewhere in this book, and in Table 6.1 have attempted to reference some of these. For further information on basic sensing devices, we refer you to the Suggested Readings.

1. Sensors often include both a first, detecting stage and a second, transducing stage. Each stage may convert the sensed information into a different form, often resulting a final electrical signal. Specific sensor design and selection is normally guided by the requirements of sufficient sensitivity and minimal source loading (Sections 6.1–6.4).

2. A wide range of transducers are based on changes in the resistance of a sensing element (Section 6.5–6.8). Variations in inductance (Sections 6.10–6.12) and capacitance (Section 6.13) are also used frequently.

3. Some sensing elements are self-powering. These include thermocouples (Section 6.9), which generate an electromotive force dependent upon temperature, and piezoelectric sensors (Section 6.14), which generate a charge when loaded.

4. Semiconductors devices are increasingly common among sensing elements (Section 6.15). These sensors can take advantage of microelectronic fabrication technology.

5. Light-detecting transducers may be divided into thermal and photon devices. A variety of photon devices are described in Section 6.16.

6. Hall-effect sensors are often used in position sensing and related applications (Section 6.17).

7. Manufacturing tolerance and temperature errors must be considered when designing a sensing system. Often, the magnitude of these problems can be estimated in advance, using the methods of uncertainty analysis (Section 6.18; also see Sections 3.10, 3.11.2, and 3.12). Such estimates can provide guidelines for improving transducer performance.

Suggested Readings

Alley, C. L., and K. W. Atwood. *Microelectronics.* Englewood Cliffs, N.J.: Prentice Hall, 1986.

Elion, G. R., and H. A. Elion. *Electro-Optics Handbook.* New York: Marcel Dekker, 1979.

Geddes, L. A. *Biomedical Instrumentation.* New York: John Wiley, 1968.

Hix, C. F., Jr., and R. P. Alley. *Physical Laws and Effects.* New York: John Wiley, 1958.

ISA Directory of Instrumentation. Research Triangle Park, N.C.: Instrument Society of America, yearly editions.

Jones, B. E. *Current Advances in Sensors.* New York: Adam Hilger, 1987.

Juds, S. M. *Photoelectric Sensors and Controls.* New York: Marcel Dekker, 1988.

Kannatley-Asibu, E., A. G. Ulsoy, and R. Komanduri (eds.). *Sensors and Controls for Manufacturing,* PED vol. 18. New York: ASME, 1985.

Lion, K. S. *Instrumentation in Scientific Research.* New York: McGraw-Hill, 1959.

Luxmoore, A. R. (ed.). *Optical Transducers and Techniques in Engineering Measurement.* New York: Elsevier Science Publishers, 1983.

Norton, H. N. *The Handbook of Transducers*. Englewood Cliffs, N.J.: Prentice Hall, 1989.

Nunley, W., and J. S. Bechtel. *Infrared Optoelectronics: Devices and Applications*. New York: Marcel Dekker, 1987.

Sydenham, P. H. *Transducers in Measurement and Control*. New York: Adam Hilger, 1984.

Todd, C. D. *The Potentiometer Handbook*. New York: McGraw-Hill, 1975.

Trietley, H. *Transducers in Mechanical and Electronic Design*. New York: Marcel Dekker, 1986.

Wilson, J., and J. F. B. Hawkes *Optoelectronics, an Introduction*. 2nd ed. Hemel Hempstead, U.K.: Prentice Hall International, 1989.

Wolf, S., and R. F. M. Smith. *Student Reference Manual for Electronic Instrumentation Laboratories*. Englewood Cliffs, N.J.: Prentice Hall, 1990.

Problems

6.1 Consider an inductive displacement probe having a diameter of 0.25 in. If the probe is set at a "stand-off" distance of 0.050 in. relative to a shaft, determine the probe sensitivity (mV/0.001 in. displacement) when the probe is used as shown in the circuit shown in Figure 6.25. Assume Eq. (6.3c) is valid here with $n = 100$ and that the excitation frequency is 1000 Hz.

6.2 It is desired to construct a dynamic compression force cell capable of measuring forces in the range of ±1000 N. If a quartz disk 1.0 mm thick and 10 mm in diameter is used as the sensing element, determine the force cell sensitivity (mV/N).

6.3 For a capacitive displacement transducer whose behavior can be represented by Eq. (6.6), determine an expression for the sensitivity $de_o/d(d)$ for an excitation frequency f if the transducer is used as shown in Fig. 6.26.

6.4 A proving-ring-type force transducer is a very reliable device for checking the calibration of material-testing machines. An equation for estimating the deflec-

Figure 6.25

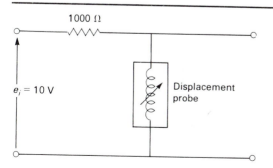

Figure 6.26 Circuit for Problems 6.3, 6.6, and 6.8

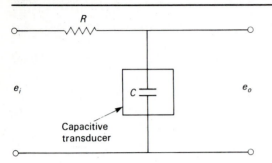

tion constant of the elemental ring, loaded in compression, is given Table 13.1. If $D = 10$ in. (25.4 cm) $\pm$ 0.010 in. (0.25 mm), $t = $ the radial thickness of the section = 0.6 in. (15.24 mm) $\pm$ 0.005 in. (0.127 mm), $w = $ the axial width of the section = 2 in. (5.08 cm) $\pm$ 0.015 in. (0.381 mm), and $E = 30 \times 10^6$ lbf/in.^{2} $(20.68 \times 10^{10}$ N/m$^2) \pm 0.5 \times 10^6$ lbf/in.^{2} $(0.34 \times 10^{10}$ N/m$^2)$, calculate the value of K and its uncertainty, using English units.

6.5 Solve Problem 6.4 using SI units.

6.6 Consider the capacitive displacement transducer in Fig. 6.26 to be governed by the following relationship:

$$C = \frac{0.225A}{d},$$

where

$$A = \text{cross-sectional area of transducer tip (in.}^2),$$

$$d = \text{air-gap distance (in.).}$$

Determine the change in voltage when the air gap changes from 0.010 in. to 0.015 in.

6.7 A capacitive displacement transducer as shown in Fig. 6.27 is constructed of two plates with area 2.0 in. separated by a distance of 0.006 in. If air is the separating medium, determine the sensitivity of the transducer in picofarads per 0.001 in. change in x.

Figure 6.27

6.8 If the transducer of Problem 6.7 is inserted into the circuit of Fig. 6.26, determine the change in output voltage when x changes from 0.010 in. to 0.015 in. Is the sensitivity constant in this range?

6.9 **a.** A commercial force sensor uses a piezoelectric quartz crystal as the sensing element. The quartz element is about 0.2 in. thick and has a cross-section of about 0.3 in by 0.3 in. The sensing element is compressed in the thickness direction when a load is applied over its cross-section. The output voltage is measured across the thickness. What is the output of the sensor in volts per newton?

    b. A polyvinylidene fluoride film is used as a piezoelectric load sensor. The film is $25\,\mu$m thick, 1 cm wide, and 2 cm in the axial direction. It is stretched in the axial direction by the load. The output voltage is measured across the thickness. What is the output in volts per newton?

6.10 The circuit of Figure 6.28 may be used to operate a photodiode. The voltage V_r is a reverse-bias voltage large enough to make diode current, i, proportional to the incident light intensity, H. Under this condition, $i/H = 1\,\mu\mathrm{A}/(\mathrm{W/m^2})$.

    a. Show that the output voltage, V_{out}, varies linearly with H.

    b. If $H = 1000\,\mathrm{W/m^2}$, $V_r = 5\,\mathrm{V}$, and an output voltage of 1 V is desired, determine an appropriate value of R_{load}.

6.11 A digital readout weighing scale with a 300 lbf (1334 N) capacity uses an aluminum cantilever beam with strain gages (see Fig. 12.28) as the force-sensing element. A lever system is used to attenuate the load by a factor of 18; i.e., a 300 lbf load on the scale exerts 300/18 lbf on the beam. The beam has an effective length $L = 7$ in., a rectangular section $w = \frac{1}{2}$ in. (12.7 mm), and $t = \frac{1}{4}$ in. (6.35 mm), oriented as shown in Fig. 12.28. The deflection constant for the beam, loaded in this manner, is $3EI/L^3$ (see Table 13.1). E is Young's modulus (10×10^6 lbf/in.^{2} or 6.89×10^{10} Pa), and I is the moment of inertia ($I = wt^3/12$). Assign tolerances and determine the uncertainty in K.

6.12 Solve Problem 6.11 using SI units.

Figure 6.28

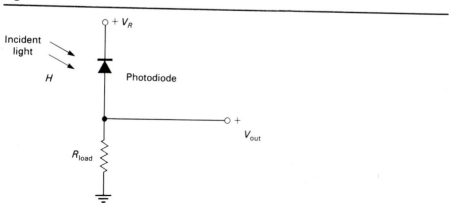

6.13 Assuming that the specifications for Example 6.1 in Section 6.18.1 are for a nominal temperature of 20°C, calculate the nominal value for the deflection constant K for temperatures of (a) 40°C and (b) −20°C. (Use values for high-carbon spring steel.)

6.14 Calculate the deflection constant for the force transducer specified in Problem 6.4 for temperatures of (a) 40°C and (b) −20°C.

6.15 Calculate the deflection constant for the force transducer specified in Problem 6.11 for temperatures of (a) 40°C and (b) −20°C.

References

1. Kneen, W. A review of the electric displacement gages used in railroad car testing. *ISA Proc.* 6: 74, 1951.

2. Michael, P. C., N. Saka, and E. Rabinowicz. Burnishing and adhesive wear of an electrically conductive polyester-carbon film. *Wear*, 132: 265–285, 1989.

3. Beckwith, T. G., and R. D. Marangoni. *Mechanical Measurements.* 4th ed. Reading, Mass.: Addison Wesley, 1990, Section 7.8.

4. Todd, C. D. *The Potentiometer Handbook.* New York: McGraw-Hill, 1975.

5. *The Radio Amateur's Handbook,* 66th ed. West Hartford, Conn.: American Radio Relay League, 1989, pp. 2–17. Revised annually.

6. Hetenyi, M. *Handbook of Experimental Stress Analysis.* New York: John Wiley, 1950, p. 239.

7. Electronic micrometer uses dual coils. *Prod. Engr.* 19: 134, Jan. 1948.

8. Brenner, A., and E. Kellogg. An electric gage for measuring the inside diameter of tubes. *NBS J. Res.* 42: 461, May 1949.

9. *The Radio Amateur's Handbook,* 66th ed. West Hartford, Conn.: American Radio Relay League, 1989, pp. 2–14. Revised annually.

10. Low-temperature liquid-level indicator for condensed gases. *NBS Tech. News Bull.* 38: 1, January 1954.

11. Heteny, M. *Handbook of Experimental Stress Analysis.* New York: John Wiley, 1950, p. 287.

12. Sihvonen, Y. T., M. Rassweiler, A. F. Welch, and J. W. Bergstrom. Recent improvements in a capacitor-type pressure transducer. *ISA J.* 2: November 1955.

13. Leggat, J. W., G. M. Rassweiler, and Y. T. Sihvonen. Engine pressure indicators, application of a capacitor type. *ISA J.* 2: August 1955.

14. Welch, Weller, Hanysz and Bergstrom. Auxiliary equipment for the capacitor-type transducer. *ISA J.* 2: December 1955.

15. Fleming, L. T. A ceramic accelerometer of wide frequency range. *ISA Proc.* 5: 62, 1950.

16. *Kynar Piezo Film Technical Manual.* Valley Forge, Penn: Pennwalt Corporation, 1987.

17. Carlson, A. B., and D. G. Gisser. *Electrical Engineering.* Reading, Mass.: Addison-Wesley, 1981, pp. 287–294.

18. *Pressure Sensors,* Catalog BR121/D. Phoenix, Ariz.: Motorola, Inc., 1991.

19. *Solid-State Sensor Handbook.* Sunnyvale, Calif.: SenSym, Inc., 1989.

20. Alley, C. L., and K. W. Atwood. *Microelectronics.* Englewood Cliffs, N.J.: Prentice Hall, 1986, pp. 56–57.

21. Wilson, J., and J. F. Hawkes. *Optoelectronics, an Introduction,* 2nd ed. Hemel Hempstead, U.K.: Prentice-Hall International, 1989, pp. 280–286.

22. *Handbook of Infrared Radiation Measurements.* Stamford, Conn.: Barnes Engineering Company, 1983, pp. 51–56.

23. *Photomultiplier Handbook.* Lancaster, Penn.: Burle Technologies, Inc., 1980.

24. McDermott, J. R. Control designer's guide to solid state photosensors. *Control Eng.* 71, October 1960.

25. Jamieson, J. A. Detectors for infrared systems. *Electronics* 33 (5): 82, 1960.

26. Bube, R. H. *Photoconductivity of Solids.* New York: John Wiley, 1960.

27. Gadd, C. W., and T. C. Van Degrift. A short-gage length extensometer and its application to the study of crankshaft stresses. *J. Appl. Mech.* 9: A.15, March 1942.

28. *Dew Point Equipment to Measure Moisture in Gases,* Bulletin GEC-588. Schenectady, N.Y.: General Electric, 1950.

29. Campbell, J. O. Special electrical applications in the steel industry. *Iron and Steel Eng.* 19: 78–89, February 1942.

30. Carr, J. J. *How to Design and Build Electronic Instrumentation.* Blue Ridge Summit, Penn.: Tab Books, 1978, p. 278.

31. Tipler, P. A. *Physics.* New York: Worth Publishers, Inc., 1976, pp. 848–850.

32. *Hall Effect Transducers: How to Apply Them as Sensors.* Freeport, Ill.: Micro Switch, A Honeywell Division, 1982.

33. Shuller, J., and A. Lee. Personal Communication to J. H. Lienhard, Chrysler Motors Corporation, February 1992.

34. Shigley, J. E. *Mechanical Engineering Design.* 2nd ed. New York: McGraw-Hill, 1972, Ch. 8.

35. Haugen, E. B. *Probabilistic Approaches to Design.* New York: John Wiley, 1968.

36. Gitlin, R. How temperature affects instrument accuracy. *Control Eng.* 2, May 1955.

37. Laws, F. A. *Electrical Measurements.* 2nd ed. New York: McGraw-Hill, 1938, p. 217.

CHAPTER 7

|||||||||||||||||||||||||||||||||||||||||||||||||||||

Signal Conditioning

7.1   Introduction

Once a mechanical quantity has been detected and possibly transduced, it is usually necessary to modify the stage-one output further before it is in satisfactory form for driving an indicator or recorder. We will now consider some of the methods used in this intermediate, signal conditioning step.

Measurement of dynamic mechanical quantities places special requirements on the elements in the signal conditioning stage. Large amplifications, as well as good transient response, are often desired, both of which are difficult to obtain by mechanical, hydraulic, or pneumatic methods. As a result, electrical or electronic elements are usually required.

An input signal is often converted by the detector-transducer to a mechanical displacement (see Table 6.1). It is then commonly fed to a secondary transducer, which converts it into a form, often electrical, that is more easily processed by the intermediate stage. In some cases, however, such a displacement is fed to mechanical intermediate elements, such as linkages, gearing, or cams; these mechanical elements present design problems of considerable magnitude, particularly if dynamic inputs are to be handled.

In the field of dynamic measurements, strictly mechanical systems are much more uncommon than they were in years past, largely because of several inherent disadvantages, which we will discuss only briefly.*

Mechanical amplification by these elements is quite limited. When amplification is required frictional forces are also amplified, resulting in considerable undesirable signal loading. These effects, coupled with backlash and elastic deformations, result in poor response. Inertial loading results in reduced frequency response and in certain cases, depending on the particular configuration of the system, phase response is also a problem.

---

* The first and second editions of this book contain a more thorough discussion of strictly mechanical signal conditioning methods and problems.

7.2   Advantages of Electrical Signal Conditioning

As we have already seen, many detector-transducer combinations provide an output in electrical form. In these cases, of course, it is convenient to perform further signal conditioning electrically. Such conditioning may typically include converting resistance changes to voltage changes, subtracting offset voltages, increasing signal voltages, or removing unwanted frequency components. In addition, in order to minimize friction, inertia, and structural flexibility requirements, we also prefer electrical methods for their ease of *power amplification*. Additional power may be fed into the system to provide a greater output power than input by the use of power amplifiers, which have no important mechanical counterpart in most instrumentation.* This technology is of particular value when recording procedures employ stylus-type recorders, mirror galvanometers, or magnetic-disk methods.

7.3   Modulated and Unmodulated Signals

Measurands may be "pure" in the sense that the analog electrical signal contains nothing other than the real-time variation of the measurand information itself. On the other hand, the signal may be "mixed" with a *carrier,* which consists of a voltage oscillation at some frequency higher than that of the signal. A common rule of thumb is that the frequency ratio should be at least ten to one. The signal is said to *modulate* the carrier. The measurand affects the carrier by varying either its amplitude or its frequency. In the former case the carrier frequency is held constant and its amplitude is varied by the measurand. This process is known as amplitude modulation, or AM [Fig. 7.1(a)]. In the latter case the carrier amplitude is held constant and its frequency is varied by the measurand. This is known as frequency modulation, or FM [Fig. 7.1(b)]. The most familiar use of AM and FM transfer of signals is in AM and FM radio broadcasting.

When modulation is used in instrumentation, amplitude modulation is the more common form. Nearly any mechanical signal from a passive pickup can be transduced into an analogous AM form. Sensors based on either inductance or capacitance *require* an ac excitation. The differential transformer (Section 6.11) is an example of the former, whereas the capacitive pickup for liquid level (Fig. 6.12) is an example of the latter. In addition, however, resistive-type sensors may also use ac excitation, as with some strain-gage circuits (e.g., Fig. 12.12).

---

* It is true that hydraulic and pneumatic systems may be set up to increase signal power; however, their use is limited to relatively slow-acting control applications, primarily in the fields of chemical processing and electric power generation. As in the case of mechanical systems, friction and inertia severely limit transient response of the type required for measurement of dynamic inputs.

Extracting the signal information from the modulated carrier is required. When AM is used, this operation may take several forms. The simplest is merely to display the entire signal using an oscilloscope or oscillograph, and then to "read" the result from the envelope of the carrier. More commonly, the mixed signal and carrier are *demodulated* by rectification and filtering, as shown in Fig. 7.1(a). FM demodulation is a more complex operation and may be accomplished through the use of frequency discrimination, ratio detection, or IC phase-locked loops. Further discussion is beyond the scope of this text.

Figure 7.1   (a) Amplitude modulation, whereby the envelope of the carrier contains the signal information, (b) frequency modulation, whereby the signal information is contained in the frequency variation of the carrier

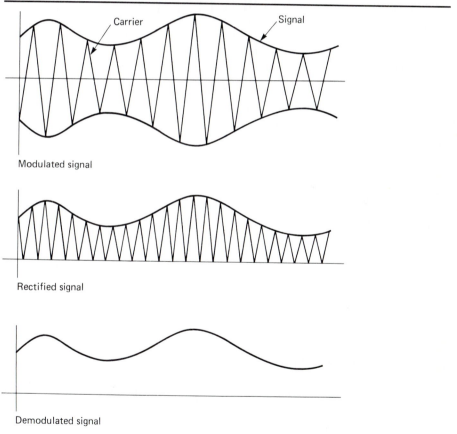

Modulated signal

Rectified signal

Demodulated signal

(a)

continued

Figure 7.1 *continued*

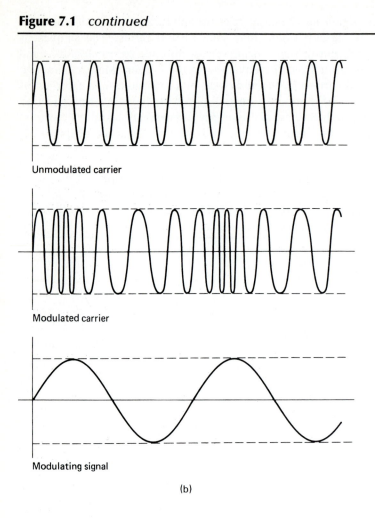

Unmodulated carrier

Modulated carrier

Modulating signal

(b)

7.4 Input Circuitry

Electrical detector-transducers are of two general types: (1) *passive, those requiring an auxiliary source of energy in order to produce* a signal, and (2) *active,* those that are self-powering. The simple bonded strain gage is an example of the former, whereas the piezoelectric accelerometer is an example of the latter.

Although it may be possible to use an active, or self-powering, detector-transducer directly with a minimum of circuitry, the passive type, in general, requires special arrangements to introduce the auxiliary energy. The particular arrangement required will depend on the operating principle involved. For example, resistive-type pickups may be powered by either an ac or a dc source, whereas capacitive and inductive types, with an exception or two, require an ac source.

Although not all-inclusive, the following list classifies the most common forms of input circuits used in transducer work: (1) simple current-sensitive circuits, (2) ballast circuits, (3) voltage-dividing circuits, (4) bridge circuits, (5) resonant circuits, (6) amplifier input circuits. Often, the input circuits will be followed by some type of filter circuit. These circuits are discussed in the following sections.

7.5   The Simple Current-Sensitive Circuit

Figure 7.2(a) illustrates a simple current-sensitive circuit in which the transducer may use any one of the various forms of variable-resistance elements. We will let the transducer resistance be kR_t, where R_t represents the maximum value of transducer resistance and k represents a percentage factor that may vary between 0.0 and 1.0 (0% and 100%), depending on the magnitude of the input signal. Should the transducer element be in the form of a sliding contact resistor, the value of k could vary through the complete range of 0% to 100%. On the other hand, if R_t represents, say, a thermistor, then k would fall within some limiting range not including 0.0%. We will let R_m represent the remaining circuit resistance, including both the meter resistance and the internal resistance of the voltage source.

If i_o is the current flowing through the circuit and hence the current indicated by the readout device, we have, using Ohm's law (Section 5.17),

$$i_o = \frac{e_i}{kR_t + R_m}.$$ (7.1)

This may be rewritten as

$$\frac{i_o}{i_{max}} = \frac{i_o R_m}{e_i} = \frac{1}{1 + \left(\dfrac{R_t}{R_m}\right)k}.$$ (7.2)

Note that maximum current flows when $k = 0$, at which point the current is $i_{max} = e_i/R_m$.

Figure 7.2(b) shows plots of Eq. (7.2) for various values of resistance ratio. The abscissa is a measure of *signal input* and the ordinate a measure of *output*. First of all, it is observed that the input–output relation is nonlinear, which of course would generally be undesirable. In addition, the higher the relative value of transducer resistance R_t to R_m, the greater will be the output variation or sensitivity. It will also be noted that the output is a function of i_{max}, which in turn is dependent on e_i. Thus careful control of the driving voltage is necessary if calibration is to be maintained.

Figure 7.2   (a) Simple current-sensitive circuit, (b) plot of Eq. (7.2),
            showing variation of current in terms of input signal
            k for a simple current-sensitive circuit

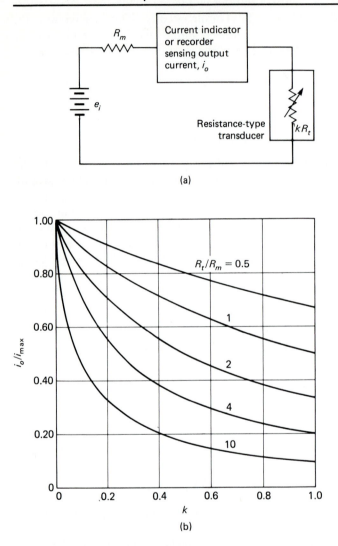

(a)

(b)

7.6   The Ballast Circuit

Now let us look at a variation of the current-sensitive circuit, often referred to
as the *ballast circuit*, shown in Fig. 7.3. Instead of a current-sensitive indicator
or recorder through which the total current flows, we shall use a voltage-
sensitive device (some form of voltmeter) placed across the transducer. The
ballast resistor R_b is inserted in much the same manner as R_m was used in the

Figure 7.3 Schematic diagram of a ballast circuit

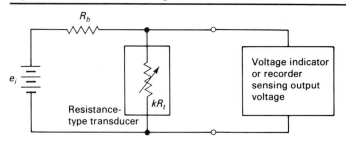

previous circuit. It will be observed that in this case, were it not for R_b, the indicator would show no change with variation in R_t; it would always indicate full source voltage. So some value of resistance R_b is necessary for the proper functioning of the circuit.

Two different situations may exist, depending on the relative impedance of the meter. First, the meter may be of high impedance, as would be the case if some form of electronic voltmeter (Section 9.4) were used; in this case any current flow through the meter may be neglected. Second, the meter may be of low impedance, so that consideration of such current flow is required.

Assuming a high-impedance meter, we have by Ohm's law

$$i = \frac{e_i}{R_b + kR_t}.$$
(7.3)

Then, if e_o = the voltage across kR_t (which is indicated or recorded by the readout device),

$$e_o = i(kR_t) = \frac{e_i kR_t}{R_b + kR_t}.$$
(7.4)

This equation may be written as

$$\frac{e_o}{e_i} = \frac{kR_t/R_b}{1 + (kR_t/R_b)}.$$
(7.5)

For a given circuit, e_o/e_i is a measure of the output, and kR_t/R_b is a measure of the input.

Defining η as the sensitivity, or the ratio of change in output to change in input, we have

$$\eta = \frac{de_o}{dk} = \frac{e_i R_b R_t}{(R_b + kR_t)^2}.$$
(7.6)

We may change R_b by inserting different values of resistance. In that case the sensitivity would be altered, which would mean that there may be some optimum value of R_b so far as sensitivity is concerned. By differentiation with respect to R_b, we should be able to determine this value:

$$\frac{d\eta}{dR_b} = \frac{e_iR_t(kR_t - R_b)}{(R_b + kR_t)^3}.$$ (7.7)

The derivative will be zero under two conditions: (1) for $R_b = \infty$, which results in minimum sensitivity, and (2) for $R_b = kR_t$, for which maximum sensitivity is obtained.

The second relation indicates that for full-range usefulness, the value R_b must be based on compromise because R_b, a constant, cannot always have the value of kR_t, a variable. However, R_b may be selected to give maximum sensitivity for a certain point in the range by setting its value to correspond to that value of kR_t.

This circuit is occasionally used for dynamic applications of resistance-type strain gages [1, 2]. In this case the change in resistance is quite small compared with the total gage resistance, and the relations above indicate that a ballast resistance equal to gage resistance is optimal.

Figure 7.4 shows the relation between input and output for a circuit of this type as given by Eq. (7.5).

It will be noted that the same disadvantages apply to this circuit as to the current-sensitive circuit discussed previously—namely: (1) a percentage variation in the supply voltage, e_i, results in a greater change in output than does a similar percentage change in k, so that very careful voltage regulation must be used; and (2) the relation between output and input is not linear.

Figure 7.4   Curves showing relation between input
and output for a ballast circuit

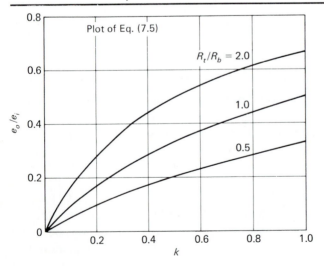

7.7 Voltage-Dividing Circuits

The *voltage divider* (Fig. 7.5) is a ubiquitous element of instrumentation circuits. Very simply, this circuit uses a pair of resistors to divide an input voltage, e_i, into a smaller output voltage, e_o. If a negligible current is drawn from the output terminals, the current through the resistors follows from Ohm's law:

$$i = \frac{e_i}{(R_1 + R_2)}$$

The output voltage measured across R_2 is then

$$e_o = iR_2 = \frac{R_2}{R_1 + R_2} e_i \qquad (7.8)$$

Voltage dividers appear throughout this chapter. The ballast circuit of the preceding section is essentially a voltage divider [see Eq. (7.4)] in which the fraction of input voltage at the output depends on the transducer resistance; bridge circuits (Section 7.9) are essentially pairs of voltage dividers; and the noninverting amplifier (Example 7.5) also incorporates a voltage divider.

7.7.1 The Voltage-Dividing Potentiometer

Figure 7.6 is a very useful voltage-divider arrangement for sliding contact resistance transducers. It is known as the *voltage-dividing potentiometer circuit*. Note that the circuit is connected not to the slider, as it would be in the ballast circuit, but across the complete resistance element. The terminating, or readout, device is connected to sense the voltage drop across the portion of resistance element R_p determined by k.

Two different situations may occur with this arrangement, depending on the relative impedance of the resistance element and the indicator-recorder. If the terminating instrument is of sufficiently high relative impedance, no

Figure 7.5 The voltage-divider circuit

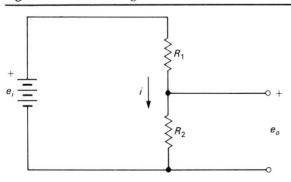

Figure 7.6   Simple voltage-dividing potentiometer circuit

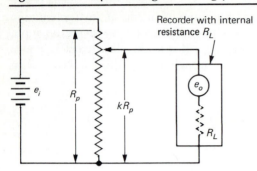

appreciable current will flow through it, and it may be considered a simple "*pressure*-measuring" device. The circuit then becomes a true voltage divider, and the indicated output voltage e_o may be determined from Eq. (7.8),

$$e_o = \frac{kR_p}{R_p} e_i = ke_i,$$

or

$$k = \frac{e_o}{e_i}. \tag{7.8a}$$

7.7.2   Loading Error

On the other hand, if the readout device draws appreciable current, a *loading error* (see Section 6.2) will result. This error may be analyzed as follows. Referring to Fig. 7.6, we find that the total resistance *seen* by the source of e_i will be

$$R = R_p(1 - k) + \frac{kR_p R_L}{kR_p + R_L}$$

and

$$i = \frac{e_i}{R} = \frac{e_i(kR_p + R_L)}{kR_p^2(1 - k) + R_p R_L}.$$

The output voltage will then be

$$e_o = e_i - iR_p(1 - k)$$

or

$$\frac{e_o}{e_i} = \frac{k}{1 + (R_p/R_L)k - (R_p/R_L)k^2}. \tag{7.9}$$

If we assume the simpler relation given by Eq. (7.8a) to hold, an error in e_o

Figure 7.7 Curves showing error caused by loading a voltage-dividing potentiometer circuit

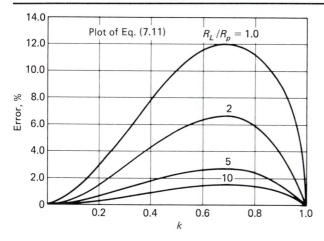

will be introduced according to the following relation:

$$\text{Error} = e_i\left[k - \frac{k}{k(1-k)(R_p/R_L)+1}\right]$$

$$= e_i\left[\frac{k^2(1-k)}{k(1-k)+(R_L/R_p)}\right]. \qquad (7.10)$$

On the basis of *full-scale output*, this relation may be written as

$$\text{Percent error} = \left[\frac{k^2(1-k)}{k(1-k)+(R_L/R_p)}\right] \times 100. \qquad (7.11)$$

Except for the endpoints ($k = 0.0$ or 1.0), where the error is zero, the error will always be on the negative side; that is, the measured value of voltage will be lower than would be the case if the system performed as a linear voltage divider. Figure 7.7 shows a plot of the variation in error with slider position for various ratios of load to potentiometer resistance. Obviously, the higher the value of load resistance compared with potentiometer resistance, the lower will be the error; thus high-input resistance is a desirable feature in voltage-reading devices.

7.7.3 Use of End Resistors

It will be observed that the nonlinearity in the relation between the potentiometer output and the input displacement k may be reduced if only a portion of the available potentiometer range is used. For example, a 1000-Ω potentiometer may be selected, but the input could be limited to only a 500-Ω

Figure 7.8 Method for improving linearity of potentiometer circuits when low-impedance indicating devices are used. Resistors R_e are termed end resistors.

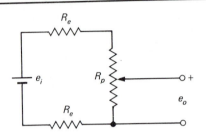

portion of the total range. This limitation would reduce the potentiometer resolution and would be generally impractical; however, it would result in a reduction in the deviation from linearity. A similar result may be obtained through use of what are known as *end resistors* (Fig. 7.8). When either an upper- or lower-end resistor, or both, are used, it is often possible to compensate for the reduced potentiometer output caused by the increased resistance by increasing the voltage input e_i by a proportional amount.

7.8 Small Changes in Transducer Resistance

Some resistance transducers show only very small changes in their resistance. For example, the resistance of a foil strain gage will normally vary by only about 0.0001% during use! The smallness of the resistance change has important ramifications for the choice of signal-conditioning circuit.

Suppose that a voltage-divider (or ballast) circuit is formed from a transducer of resistance R_2 and a second resistor R_1 [Fig. 7.9(a)]. The resistances are initially made equal, $R_1 = R_2 = R_0$, so that the initial output voltage is

$$e_o = \frac{R_2}{R_1 + R_2} e_i = \frac{R_0}{R_0 + R_0} e_i = \frac{e_i}{2}.$$

If the resistance of the transducer then increases from $R_2 = R_0$ to $R_2 = (R_0 + \Delta R)$, the output changes to

$$e_o + \Delta e_o = \frac{(R_0 + \Delta R)}{R_0 + (R_0 + \Delta R)} e_i$$

$$= \frac{R_0 + \Delta R}{2R_0 + \Delta R} e_i$$

$$= \frac{R_0}{2R_0} \left(\frac{1 + \Delta R/R_0}{1 + \Delta R/2R_0} \right) e_i$$

Figure 7.9   The use of voltage-dividers in measuring
small resistance changes

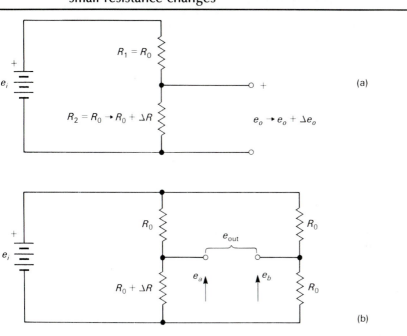

(a)

(b)

$$= \frac{e_i}{2}\left(1 + \frac{\Delta R/2R_0}{1 + \Delta R/2R_0}\right)$$

$$= \frac{e_i}{2} + \frac{e_i}{2}\left(\frac{\Delta R}{2R_0}\right)\left(\frac{1}{1 + \Delta R/2R_0}\right).$$

Assuming that the resistance change is small, so that $\Delta R/2R_0 \ll 1$, we can approximate the last factor on the right-hand side as unity; hence,

$$e_o + \Delta e_o \approx \frac{e_i}{2} + \left(\frac{e_i}{2}\right)\left(\frac{\Delta R}{2R_0}\right) \tag{7.12}$$

$$= e_o + \frac{\Delta R}{4R_0}e_i. \tag{7.12a}$$

Thus, for small resistance changes, the output voltage shows straight-line variation with ΔR.* Such variation is advantageous, because it simplifies data reduction. However, the severe disadvantages of this circuit become apparent when we look at the numerical values arising for small resistance changes.

---

* In much the same way, end resistors in the voltage-dividing potentiometer (Section 7.7.3) serve to make the transducer resistance change small relative to the other resistances, creating an approximately straight-line variation of the output.

Taking the strain gage transducer as an example, a typical resistance change might be $\Delta R = 240 \mu\Omega$ in a gage of initial resistance $R_0 = 120\ \Omega$. Hence,

$$\frac{\Delta e_o}{e_o} = \frac{(\Delta R / 4R_0)e_i}{e_i / 2} = \frac{\Delta R}{2R_0} = 10^{-6}$$

Since we measure the sum $e_o + \Delta e_o$, we will need a meter with a resolution of better than one part in a million, in order to see any change in e_o at all; this excludes many common voltmeters, which may resolve to only 0.01%, although it is within reach of the best commercial meters.

An even more important limitation is the stability of the input voltage, e_i. If e_i drifts slightly between the initial and final readings of the output (to $e_i + \Delta e_i$), then Eq. (7.12) shows that the output becomes

$$e_o + \Delta e_o \approx \frac{e_i + \Delta e_i}{2} + \left(\frac{e_i + \Delta e_i}{2} \right)\left(\frac{\Delta R}{2R_0} \right)$$

$$\approx e_o + \frac{\Delta e_i}{2} + \frac{\Delta R}{4R_0} e_i$$

Thus if e_i drifts by even 0.1% ($\Delta e_i = 0.001 e_i$), the change in Δe_o caused by voltage drift will be $0.001(e_i/2)$—a thousand times larger than the strain-induced voltage change $(\Delta R / 4R_0)e_i = 10^{-6}(e_i/2)$!

The difficulty, of course, is that we are trying to resolve a voltage change which is a tiny fraction of the total output voltage, and it illustrates an important principle in measurement: *Avoid measurements based on a small difference between large numbers.* Such measurements are limited by the accuracy with which the *large* numbers can be measured.

In this case, the solution is to design a circuit having output voltage proportional to Δe_o itself, without the large offset voltage, e_o. We can do this by introducing another voltage divider with fixed resistors R_0 [Fig. 7.9(b)] and measuring the output voltage as the difference between the midpoint voltages of the two dividers:

$$e_{\text{out}} = e_a - e_b = (e_o + \Delta e_o) - e_o = \Delta e_o = \frac{\Delta R}{4R_0} e_i. \tag{7.13}$$

The problems caused by the offset voltage, e_o, are thus eliminated.

This arrangement of two voltage dividers is, in fact, identical to the Wheatstone bridge circuit discussed in the next section; however, the Wheatstone bridge is not always restricted to small resistance changes.

7.9   Resistance Bridges

Bridge circuits are the most common method of connecting passive transducers to measuring systems. Of all the possible configurations, the Wheatstone resistance bridge devised by S. H. Christie in 1833 [3, 4] is undoubtedly used to

Figure 7.10   Simple Wheatstone bridge circuit

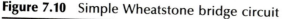

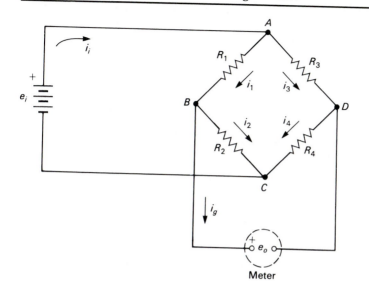

Meter

the greatest extent. Figure 7.10 shows a dc Wheatstone bridge circuit consisting of four resistor *arms* with a voltage source (battery) and a detector (meter). In applications, one or more of the arms is a resistance transducer whose resistance is to be determined. Typical resistance transducers used with a circuit of this kind include resistance thermometers, thermistors, or resistance-type strain gages.

Bridge circuits enable high-accuracy resistance measurements. These measurements are accomplished either by *balancing* the bridge—making known adjustments in one or more of the bridge arms until the voltage across the meter is zero—or by determining the magnitude of *unbalance* from the meter reading. If the circuit appears complicated to you, it may help to recognize that, when negligible current flows through the meter, the bridge is simply a pair of voltage-divider circuits (*ABC* and *ADC*) with the output taken between the midpoints of the two dividers (*B* to *D*). The great advantage of the bridge circuit is that the offset voltages of the two dividers cancel, so that the bridge output voltage can be directly related to changes in transducer resistance.

Using Fig. 7.10, we may analyze the requirements for balance. At balance, the voltage across the meter is zero and no current flows through it; hence, $i_g = 0$. In that case, we also know that $i_1 = i_2$ and $i_3 = i_4$. Since the potential across the meter is zero, $i_1 R_1 = i_3 R_3$ and $i_2 R_2 = i_4 R_4$. By eliminating i_1 and i_3

from these relations, we obtain the condition for balance, namely,

$$\frac{R_1}{R_2} = \frac{R_3}{R_4},$$

(7.14)

or

$$\frac{R_1}{R_3} = \frac{R_2}{R_4}.$$

(7.14a)

From these two equations we may formulate a statement to assist us in remembering the necessary balance relation. *In order for the Wheatstone resistance bridge to balance, the ratio of resistances of any two adjacent arms must equal the ratio of resistances of the remaining two arms, taken in the same sense.* (*Note:* "Taken in the same sense" means that if the first resistance ratio is formed from two adjacent resistances reading from left to right, the balancing ratio must also be formed by reading from left to right, etc.)

Basic bridge types are summarized in Table 7.1. When a *null* bridge is

Table 7.1 Types of electrical bridge circuits

Bridge Type	Bridge Features
Voltage-sensitive bridge	Readout instrument does not "load" bridge; that is, it requires no current; e.g., electronic voltmeter or CRO.
vs.	
Current-sensitive bridge	Readout requires current; e.g., a low-impedance indicator such as a simple galvanometer is used.
Null balance bridge	Adjustment is required to maintain balance. This becomes source of readout; e.g., manually adjusted strain indicator.
vs.	
Deflection bridge	Readout is deviation of bridge output from initial balance; e.g., as required by CRO.
AC bridge	Alternating-current voltage excitation is used.
vs.	
DC bridge	Direct-current voltage excitation is used.
Constant voltage	Voltage input to bridge remains constant; e.g., battery or voltage-regulated power supply is used.
vs.	
Constant current	Current input to bridge remains constant regardless of bridge unbalance; e.g., current-regulated power supply is used.
Resistance bridge	Bridge arms made up of "pure" resistance elements.
vs.	
Impedance bridge	Bridge arms may include reactance elements.

Figure 7.11   Arrangements used to balance dc resistance bridges

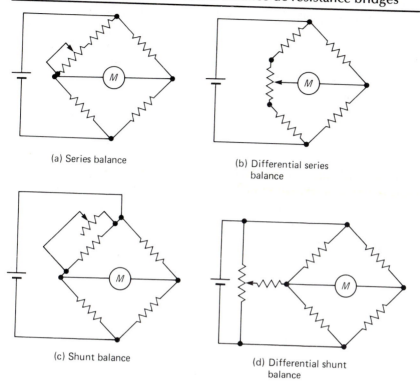

(a) Series balance

(b) Differential series balance

(c) Shunt balance

(d) Differential shunt balance

used, the resistance of one unknown arm is determined by finding values of the other arms for which the bridge is balanced. Thus, some provision must be made for adjusting the resistance of one or two arms so as to reach balance. Some balancing arrangements are shown in Fig. 7.11. An important factor in determining the type to use is bridge sensitivity. If large resistance changes are to be accommodated, large resistance adjustments must be provided; thus one of the series arrangements would be most useful and could well be the type to use for sliding-contact variable-resistance transducers or thermistors. When small resistance changes are to take place, as in the case of resistance strain gages, then the shunt balance would be used. In order to provide for a range of resistances, a bridge with both series and shunt balances might be utilized.

When the *deflection* bridge is used, bridge unbalance, as indicated by the meter reading, is the measure of input. Usually, the deflection bridge is balanced initially and later changes in transducer resistance cause the unbalance. Manufacturing variations in real resistors make it virtually impossible to obtain three resistors that will match the initial transducer resistance well enough to satisfy Eq. (7.14a) precisely; hence, provision is generally made for initial balancing by adjusting one or more arms, again using an arrangement from Figure 7.11. For static inputs, an ordinary voltmeter may be used to display the output; for dynamic signals, however, the output may be displayed by a cathode-ray oscilloscope (Section 9.6) or recorded by an oscillograph (Section 9.9), or the output may be fed to an analog-to-digital converter and a computer for display, recording, or immediate application (Section 8.12).

The output from a deflection bridge may be connected to either a high- or a low-impedance device. If the bridge is connected to a simple D'Arsonval meter or most galvanometers, the output circuit will be of low impedance, and an appreciable current (i_g) is drawn from the bridge. In most cases in which amplification is necessary, the bridge output will be connected to a high-impedance device and the bridge would supply essentially no current. Such is the case when either an oscilloscope or an electronic voltmeter is used. In the former instance the bridge is *current-sensitive;* in the latter it is *voltage-sensitive.*

7.9.1   The Voltage-Sensitive Wheatstone Bridge

Let us consider the simplest case first, in which the bridge output is connected directly to a high-impedance device, say an oscilloscope. Referring to Fig. 7.10, we see that the output voltage is the difference between the voltages at B and D

$$e_o = e_B - e_D$$

and, making use of the voltage-divider relation [Eq. (7.8)], we may write

$$e_o = e_i\left(\frac{R_2}{R_1 + R_2} - \frac{R_4}{R_3 + R_4}\right) \tag{7.15}$$

$$= e_i\left(\frac{R_2 R_3 - R_4 R_1}{(R_1 + R_2)(R_3 + R_4)}\right). \tag{7.15a}$$

We will now assume that the resistance R_2 changes by an amount ΔR_2, or

$$e_o + \Delta e_o = e_i\left[\frac{(R_2 + \Delta R_2)R_3 - R_4 R_1}{(R_1 + R_2 + \Delta R_2)(R_3 + R_4)}\right]$$

$$= e_i\left\{\frac{1 + (\Delta R_2/R_2) - (R_4 R_1/R_2 R_3)}{[(1 + (R_1/R_2) + (\Delta R_2/R_2)][1 + (R_4/R_3)]}\right\}. \tag{7.16}$$

The relation may be simplified by assuming all resistances to be initially equal (in which case $e_o = 0$). Then

$$\frac{\Delta e_o}{e_i} = \frac{\Delta R_2/R}{4 + 2(\Delta R_2/R)}. \tag{7.17}$$

Figure 7.12(a), plotted from Eq. (7.17), shows the relation for the output of a voltage-sensitive deflection bridge whose resistance arms are initially equal. Inspection of the curve indicates that this type of resistance bridge is

Figure 7.12 (a) Output from a voltage-sensitive deflection bridge whose resistance arms are initially equal, (b) output from a current-sensitive deflection bridge whose resistance arms are initially equal, plotted for different relative galvanometer resistances

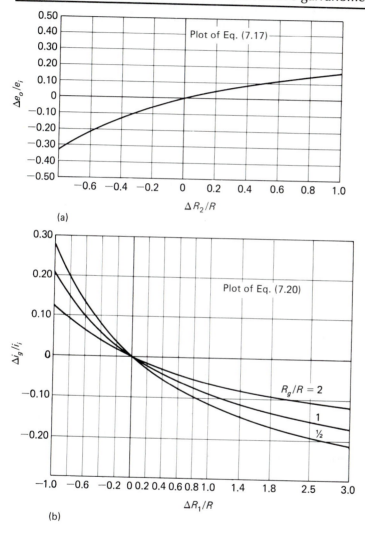

(a)

(b)

inherently nonlinear. In many cases, however, the actual resistance change is so small that the arrangement may be assumed linear. This assumption applies to most resistance strain-gage circuits. In those cases, $\Delta R_2/2R \ll 1$ and the linearized output is

$$\frac{\Delta e_o}{e_i} = \frac{\Delta R_2}{4R},$$

which is identical to Eq. (7.13).

7.9.2   The Current-Sensitive Wheatstone Bridge

When the deflection-bridge output is connected to a low-impedance device such as a galvanometer, appreciable current flows and the galvanometer resistance must be considered in the bridge equation. Galvanometer current may be expressed by the following relation [5]:

$$i_g = \frac{i_i(R_2R_3 - R_1R_4)}{R_g(R_1 + R_2 + R_3 + R_4) + (R_2 + R_4)(R_1 + R_3)}, \qquad (7.19)$$

in which

$$i_g = \text{the galvanometer current,}$$

$$i_i = \text{the input current,}$$

$$R_g = \text{the galvanometer resistance.}$$

The remaining symbols are as defined in Fig. 7.10.

If we assume that an initial bridge balance is upset by an incremental change in resistance ΔR_1 in arm R_1 and all arms are of equal initial resistance R, we may write

$$\frac{\Delta i_g}{i_i} = \frac{-\Delta R_1/R}{4[1 + (R_g/R)] + [2 + (R_g/R)](\Delta R_1/R)}. \qquad (7.20)$$

Figure 7.12(b) shows Eq. (7.20) plotted for various values of R_g/R.

7.9.3   The Constant-Current Bridge

To this point our discussion of bridge circuits has assumed a constant-voltage energizing source (a battery, for example). As the bridge resistance is changed, the total current through the bridge will, therefore, also change. In certain instances (see Section 12.9), use of a *constant-current* bridge* may be desirable [6, 7]. Such a circuit is usually obtained through the application of a commercially available *current-regulated* dc power supply,† whereby the total current flow i_i through the bridge (Fig. 7.10) is maintained at a constant value. It should be noted that such a bridge may still be either voltage-sensitive or current-sensitive, depending on the relative impedance of the readout device.

Relationships for the *voltage-sensitive constant-current* bridge may be

---

* The term "Wheatstone," as applied to bridge circuits, is commonly limited to the *constant-voltage resistance bridge*. We shall abide by this convention and avoid referring to the constant-current bridge as a Wheatstone bridge.

† Constant current is obtained by using the voltage drop across a series resistor in the supply-output line to provide a regulating feedback voltage. It may also be approximated by placing a large ballast resistor between the bridge and the voltage source; the resistor is made large enough that variations in the bridge resistors have a negligible effect on i_i.

developed as follows. Referring to Fig. 7.10, we may write

$$i_i = \frac{e_i}{R_1 + R_2} + \frac{e_i}{R_3 + R_4}, \tag{7.21}$$

or

$$e_i = i_i \left[\frac{(R_1 + R_2)(R_3 + R_4)}{R_1 + R_2 + R_3 + R_4} \right].$$

Substituting in Eq. (7.15a), we have

$$e_o = i_i \left[\frac{R_2 R_3 - R_4 R_1}{R_1 + R_2 + R_3 + R_4} \right]. \tag{7.22}$$

which is the basic equation for the voltage-sensitive constant-current bridge, provided that i_i is maintained at a constant value. If the resistance of one arm, say R_2, is changed by an amount ΔR, then

$$e_o + \Delta e_o = i_i \left[\frac{(R_2 + \Delta R)R_3 - R_4 R_1}{R_1 + (R_2 + \Delta R) + R_3 + R_4} \right]$$

and

$$\Delta e_o = i_i \left[\frac{(R_2 + \Delta R)R_3 - R_4 R_1}{R_1 + (R_2 + \Delta R) + R_3 + R_4} - \frac{R_2 R_3 - R_4 R_1}{R_1 + R_2 + R_3 + R_4} \right]. \tag{7.23}$$

For equal initial resistances ($R_1 = R_2 = R_3 = R_4 = R$),

$$\Delta e_o = i_i \left[\frac{\Delta R}{4 + (\Delta R/R)} \right]. \tag{7.24}$$

The constant-current bridge has better linearity than the constant-voltage bridge, as is apparent upon comparing Eqs. (7.17) and (7.24).

7.9.4   The AC Resistance Bridge

Resistance bridges powered by ac sources may also be used. An additional problem, however, is the necessity for providing reactance balance. In spite of the fact that the Wheatstone bridge, strictly speaking, is a resistance bridge, it is impossible to completely eliminate stray capacitances and inductances resulting from such factors as closely placed lead wires in cables to and from the transducer, and wiring and component placement in associated equipment. In any system of reasonable sensitivity, such unintentional reactive components must be accounted for before satisfactory bridge balance can be achieved.

Reactive balance can usually be accomplished by introducing an additional balance adjustment in the circuit. Figure 7.13 shows how this may be provided. Balance is accomplished by alternately adjusting the resistance and reactance

Figure 7.13   Circuit arrangement for balancing an ac bridge

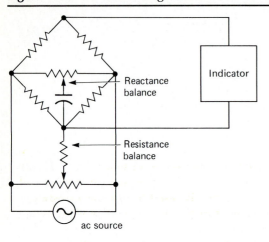

balance controls, each time reducing bridge output, until proper balance is finally achieved.

7.9.5   Compensation for Leads

Frequently a sensor and a bridge-type instrument must be separated by an appreciable distance. Wires, or leads, are used to connect the two as illustrated in Fig. 7.14(a), which shows the sensor as some type of resistance element such

Figure 7.14   (a) Simple bridge with remotely located sensor, (b) circuit
                 similar to that shown in (a), but with a compensating wire

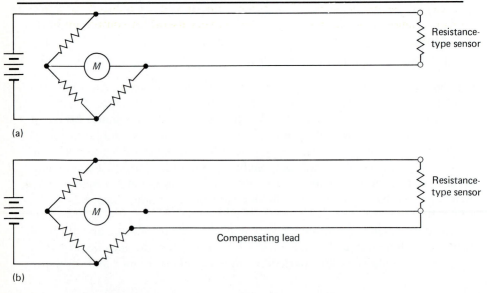

as a resistance thermometer or strain gage. In addition to the extra resistance introduced by the leads, temperature gradients along the wires may, in certain cases, cause error (see Sections 12.8.3 and 16.4.2). We can compensate for this type of error by using a three-wire circuit as illustrated in Fig. 7.14(b). Inspection shows that the additional lead serves to balance the total lead-wire lengths in the two adjacent arms, thereby eliminating any unbalance from this source.

7.9.6 Adjusting Bridge Sensitivity

Adjustable bridge sensitivity may be desired for several reasons. (1) Such adjustment may be used to attenuate inputs that are larger than desired. (2) It may be used to provide a convenient relation between system calibration and the scale of the readout instrument. (3) It may be used to provide adjustment for adapting individual transducer characteristics to precalibrated systems. (This method is used to insert the gage factor for resistance strain gages in some commercial circuits.) (4) It provides a means for controlling certain extraneous inputs such as temperature effects (see Section 13.5).

A very simple method of adjusting bridge output is to insert a variable series resistor in one or both of the input leads, as shown in Fig. 7.15. If we assume equal initial resistance R in all bridge arms, the resistance seen by the voltage source will also be R. If a series resistance is inserted as shown, then, thinking in terms of a voltage-dividing circuit, we see that the input to the bridge will be reduced by the factor

$$n = \frac{R}{R + R_s} = \frac{1}{1 + (R_s/R)}. \tag{7.25}$$

We call n the *bridge factor*. The bridge output will be reduced by a proportional amount, which makes this method very useful for controlling bridge sensitivity.

Figure 7.15 Method for adjusting bridge sensitivity through use of variable series resistance, R_s

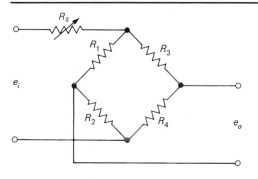

Figure 7.16 Impedance bridge arrangements

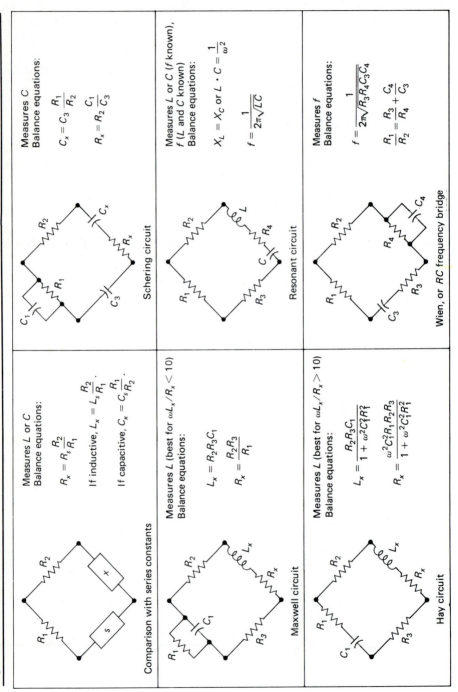

Comparison with series constants

Measures L or C
Balance equations:

$$R_x = R_s \frac{R_2}{R_1}$$

If inductive, $L_x = L_s \dfrac{R_2}{R_1}$.

If capacitive, $C_x = C_s \dfrac{R_1}{R_2}$.

Maxwell circuit

Measures L (best for $\omega L_x / R_x < 10$)
Balance equations:

$$L_x = R_2 R_3 C_1$$

$$R_x = \frac{R_2 R_3}{R_1}$$

Hay circuit

Measures L (best for $\omega L_x / R_x > 10$)
Balance equations:

$$L_x = \frac{R_2 R_3 C_1}{1 + \omega^2 C_1^2 R_1^2}$$

$$R_x = \frac{\omega^2 C_1^2 R_1 R_2 R_3}{1 + \omega^2 C_1^2 R_1^2}$$

Schering circuit

Measures C
Balance equations:

$$C_x = C_3 \frac{R_1}{R_2}$$

$$R_x = R_2 \frac{C_1}{C_3}$$

Resonant circuit

Measures L or C (f known),
f (L and C known)
Balance equations:

$$X_L = X_C \text{ or } L \cdot C = \frac{1}{\omega^2}$$

$$f = \frac{1}{2\pi\sqrt{LC}}$$

Wien, or RC frequency bridge

Measures f
Balance equations:

$$f = \frac{1}{2\pi\sqrt{R_3 R_4 C_3 C_4}}$$

$$\frac{R_1}{R_2} = \frac{R_3}{R_4} + \frac{C_4}{C_3}$$

7.10 Reactance or Impedance Bridges

Reactance or impedance bridge configurations are of the same general form as the Wheatstone bridge, except that reactive elements (capacitors and inductors) are involved in one or more of the arms. Because such elements are inherently frequency-sensitive, impedance bridges are ac-excited. Obviously the multitude of variations that are possible preclude more than a general discussion in a work of this nature; thus the reader is referred to more specialized works for detailed coverage [8].

Figure 7.16 shows several of the more common ac bridges, along with the type of element usually measured and the balance requirements.

7.11 Resonant Circuits

Capacitance-inductance combinations present varying impedance, depending on their relative values and the frequency of the applied voltage. When connected in parallel, as in Fig. 7.17(a), the inductance offers small opposition to current flow at low frequencies, whereas the capacitive reactance is low at high frequencies. At some intermediate frequency, the opposition to current flow, or impedance, of the combination is a maximum [Fig. 7.17(b)]. A similar but opposite variation in impedance is the series-connected combination.

The frequency corresponding to maximum effect, known as the *resonance frequency*, may be determined by the relation

$$f = \frac{1}{2\pi\sqrt{LC}}, \tag{7.26}$$

Figure 7.17 Parallel *LC* circuit with curve showing frequency–impedance characteristics

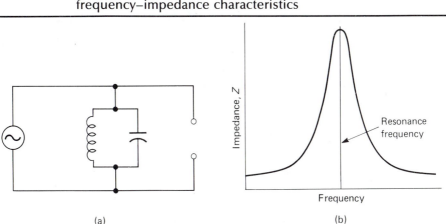

(a)   (b)

in which

$$f = \text{the frequency (Hz)},$$

$$L = \text{the inductance (H)},$$

$$C = \text{the capacitance (F)}.$$

It is evident that should, say, a capacitive transducer element be used, it could be in combination with an inductive element to form a resonant combination. Variation in capacitance caused by variation in an input signal (e.g., mechanical pressure) would then alter the resonance frequency, which could then be used as a measure of input.

7.11.1 Undesirable Resonance Conditions

On occasion, resonance conditions that occur may introduce spurious outputs. Most circuits are susceptible because they use some combination of inductance and capacitance and most are called on to handle dynamic signal inputs. In certain cases the capacitance and inductance may be not more than the stray values existing between the circuit components, including the wiring. Hence there may be resonance conditions that can result in nonlinearities at certain input or exciting frequencies.

Normally such situations are avoided in the design of commercial equipment insofar as possible. However, the instrument designer is not always in a position to predict the exact manner in which general-purpose components may be assembled or the exact nature of the input signal fed to the equipment. As a result, it is quite possible to unintentionally set up arrangements of circuit elements combined with frequency conditions that result in undesirable resonance conditions.

7.12 Electronic Amplification or Gain

The ratio of output to input for an electronic signal-conditioning device is referred to variously as gain, amplification ratio (if greater than unity), or attenuation (if less than unity). It may be defined in terms of voltages, currents, or powers, that is,

Voltage gain = voltage output/voltage input,

Current gain = current output/current input,

Power gain = power output/power input.

Another way of expressing *power gain* is through use of the *decibel*. A decibel (dB) is one-tenth of a *bel* and is based on a ratio of powers:

$$\text{Decibel (dB)} = 10 \log_{10}(P_o/P_i), \tag{7.27}$$

where P_o = the output power and P_i = the input power, both expressed in the same units.

The average human ear can just detect a loudness change from an audio amplifier when a power ratio change of one decibel is made. It has also been observed that this is nearly true regardless of the power level.

Solving Eq. (7.27) for the ratio (P_o/P_i) corresponding to one decibel yields a ratio of 1.26. In other words, for the average human ear to just detect an increase in sound output from an amplifier (feeding some form of earphone or loudspeaker), an increase of approximately 26% in power is required. Some other useful power ratios, as expressed in decibels, are

$$P_o = 2 \times P_i: \qquad 3\,\text{dB}, \qquad P_o = \frac{1}{2} \times P_i: \qquad -3\,\text{dB}.$$

$$P_o = 10 \times P_i: \qquad 10\,\text{dB}, \qquad P_o = \frac{1}{10} \times P_i: \qquad -10\,\text{dB},$$

$$P_o = 100 \times P_i: \quad 20\,\text{dB}, \qquad P_o = \frac{1}{100} \times P_i: \quad -20\,\text{dB},$$

The half-power point, $-3\,\text{dB}$, is often used in characterizing the frequency response of amplifiers and, especially, of filters.

For a pure resistance, electric power may be expressed as

$$\text{Power} = ei = e^2/R = i^2 R,$$

where e = the voltage, i = the current, and R is a pure resistance. Substituting either of the last two forms into Eq. (7.27) yields

$$\text{dB} = 20 \log_{10}\left(\frac{e_o}{e_i}\right) - 10 \log_{10}\left(\frac{R_o}{R_i}\right) \tag{7.28}$$

or

$$\text{dB} = 20 \log_{10}\left(\frac{i_o}{i_i}\right) + 10 \log_{10}\left(\frac{R_o}{R_i}\right). \tag{7.28a}$$

Should $R_i = R_o$, then the last term in each case reduces to zero. The relationship of the decibel to power and voltage ratios is illustrated in Fig. 7.18 for $R_i = R_o$.

One should remember that the decibel is fundamentally a *power ratio* and that "forgetting" the R's in the preceding equations is strictly legitimate only if the two loads, with and without amplification, are equal. Nevertheless, output/input ratios are often described using the decibel even when no load is directly involved, and one frequently sees voltage ratios expressed in decibels as

$$\text{dB} = 20 \log_{10}\left(\frac{e_o}{e_i}\right). \tag{7.28b}$$

Another common use of the decibel is in constructing a *Bode plot* of frequency

Figure 7.18   Relationship of power and voltage to the decibel

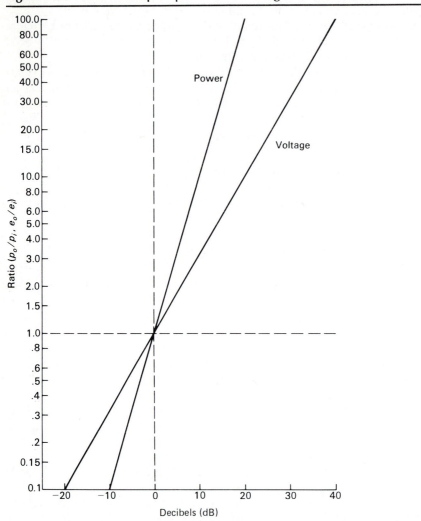

response. In such a graph, the gain in decibels is plotted against a logarithmic frequency axis, rather than showing e_o/e_i versus f on linear coordinates (compare Figs. 7.28 and 7.29).

Amplification calculations based on the decibel offer two important advantages: (1) Reasonably small numbers are involved, and (2) combining the effects of various stages of a system may be accomplished by simple addition.

Voltmeters often carry a decibel scale. When using such a scale one must always be cognizant of three important factors: (1) In reality the measurement is not in decibels, but in voltage; (2) because the decibel is a ratio, the scale must be based on some *reference voltage*; and (3) reference to Eq. (7.28) shows that the scale must assume a *reference load*.

Most voltmeter scales are based on a reference of 1 mW across 600 Ω, or

$$P = \frac{e^2}{R},$$

hence,

$$e = (PR)^{1/2} = (0.001 \times 600)^{1/2} = 0.7746 \text{ V},$$

which means that zero on the decibel scale has been arbitrarily set to correspond to 0.7746 V. In some instances the references are indicated directly on the meter face. Often the abbreviation dBm is used to indicate the above conventions. Why the 600-Ω load rather than something else? The answer is that this is a long-established industrial standard, predating the field of electronics and originated by telegraph and telephone practices.

Suppose we use a voltmeter to indicate decibels. Suppose also that the signal source impedance is R_s rather than R_r, where the latter is the reference. What correction should be applied? The following provides the proper result [9]:

$$dB_{(corrected)} = dB_{(indicated)} + 10 \log\left(\frac{R_r}{R_s}\right). \tag{7.28c}$$

Example 7.1

Suppose a reading of 50 dBm is obtained across a 16-Ω load, using a voltmeter with scale referenced to 600 Ω. What is the true dB value?

Solution.

$$dB_{(corrected)} = 50 + 10 \log\left(\frac{600}{16}\right) = 65.7.$$

As we discussed before, corrections must be made to obtain *true* dB values when load and reference conditions differ. Very conveniently, however, if we require only *differences* or *changes* in decibels, then we may not need corrections in individual readings. This situation exists if the loads remain unchanged during the actual measurements.

7.13 Electronic Amplifiers

It is not the purpose of this section, or of the book, to be concerned with electronics or electronic theory beyond the barest minimum required to make intelligent use of such equipment for the purposes of mechanical measurement.

The following discussion, therefore, is brief and is directed primarily to applications rather than to specific theory of operation.

Some form of amplification is almost always used in circuitry intended for mechanical measurement. Traditionally, the term *electronic,* as opposed to the word *electrical,* assumes that in some part of the circuit electrons are caused to flow through space in the absence of a physical conductor, thus assuming the use of vacuum tubes. With the advent of *solid-state* devices (diodes, transistors, and the like) the word electronics has taken on a broader meaning. Throughout the remainder of the book it will be understood that, unless more specific reference is made, the word electronics is being used in its broadest sense.

Electronic amplifiers are used in mechanical measurements to provide one or a combination of the following basic services: (a) voltage gain, (b) current gain (power), and (c) impedance transformations. In most cases in which mechanical or electrical transduction is used, voltage is the electrical output that is the analogous signal. Often the voltage level available from the transducer is very low; thus a voltage amplifier is used to increase the level for subsequent processing. Occasionally, the input signal must finally be used to drive a recording stylus or some control apparatus. In this case, voltage gain may not be sufficient in itself because power must be increased; hence a current or power amplifier is needed. In certain instances a transducer produces sufficient signal level but is accompanied by an unacceptably high output impedance level. This is true of most piezoelectric-type transducers. A disadvantage of high-impedance lines is their susceptibility to noise. If the signal is to be transmitted any appreciable distance (even a few inches in some cases), the noise pickup from the environment may be unacceptable. Low-impedance lines are much less prone to this problem. Hence it may well be desirable to insert an impedance transformation in the form of an amplifier that will accept a high-impedance input but produce a low-impedance output. This type of amplifier is often called a *buffer.*

There are several generalities that can be listed for the ideal (but nonexistent) electronic amplifier:

1. Infinite input impedance: no input current, hence no load on the previous stage or device.

2. Infinite gain (lower gain can be obtained by adding attenuation circuits).

3. Zero output impedance (low noise).

4. Instant response (wide frequency bandwidth).

5. Zero output for zero input.

6. Ability to ignore or reject extraneous inputs.

Although none of these aims can be completely realized, it is often possible to approach them, and their assumption simplifies circuit analysis.

7.14   Vacuum-Tube Amplifiers

Electronic amplification originated with the invention of the triode vacuum tube. Thomas Alva Edison discovered that electrons could flow from a heated cathode to an anode in an evacuated space; hence the term "Edison effect." Lee deForest is credited with showing that the flow could be *controlled* by inserting a third element, the grid, between the cathode and the anode. This resulted in the triode electron tube and, in various configurations, many with additional elements, provided the basis for electronic amplification.

Of course, vacuum tubes are little used today in instrumentation. In certain instances in which high power is required, they may still offer advantages. Most instrumentation amplification elements are now solid-state devices, in the form of either discrete circuit elements or integrated circuits.

7.15   Solid-State Amplifiers

Transistorized measurement devices are quite common. The transistor can perform most of the functions of the vacuum tube and can do so without the heated filament, high voltages, and shock-sensitive elements inherent in the construction and operation of the vacuum tube. The basic transistor is a three-element device and in this respect is similar to a triode vacuum tube. It is also adaptable to various circuit arrangements, as shown in Fig. 7.19.

Several advantages accrue from the use of transistors in measurement apparatus. The transistor, coupled with module-type integrated circuits, permits considerable weight saving and reduced size. In addition, the inherent characteristics of the transistor permit the use of relatively low powering-voltages, often conveniently supplied by batteries, thereby permitting easy field use. Vibration and shock occasionally used to make vacuum-tube "microphonics" a problem, one that is essentially eliminated in transistorized equipment. Among the disadvantages is the fact that transistors are more sensitive to voltage extremes or incorrect polarities. Proper performance is also more temperature-dependent.

Although transistors may be used as discrete, hard-wired circuit elements, they are more often incorporated with other elements into single package integrated circuits, or ICs.

7.16   Integrated Circuits*

As the name implies, integrated circuits are groups of circuit elements combined to perform specific purposes. For the most part the elements consist of transistors, diodes, resistors, and, to a lesser extent, capacitors, all

---

* See also Section 8.4.

Figure 7.19 Basic transistor circuits (bipolar junction type)

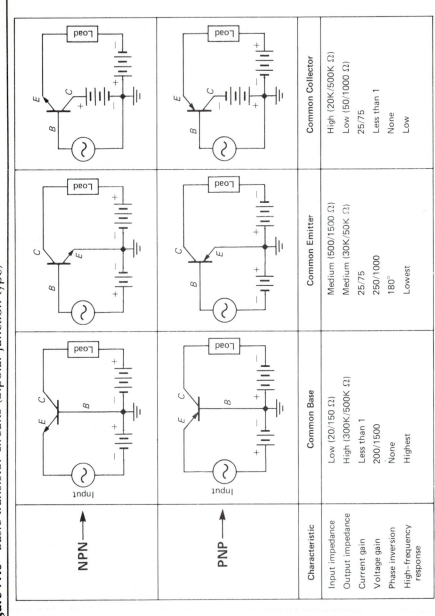

Characteristic	Common Base	Common Emitter	Common Collector
Input impedance	Low (20/150 Ω)	Medium (500/1500 Ω)	High (20K/500K Ω)
Output impedance	High (300K/500K Ω)	Medium (30K/50K Ω)	Low (50/1000 Ω)
Current gain	Less than 1	25/75	25/75
Voltage gain	200/1500	250/1000	Less than 1
Phase inversion	None	180°	None
High–frequency response	Highest	Lowest	Low

connected and packaged in convenient plug-in units. They form the building blocks used to construct more complex circuits: differential amplifiers, mixers (for combining signals), timers, filters, audio preamps, audio power amplifiers, voltage references, regulators and comparators, and many of the digital devices discussed in Chapter 8. Of particular importance to mechanical measurements is the operational amplifier, or op amp. In the following paragraphs we will discuss some of these in more detail.

7.17 Operational Amplifiers

The op amp is basically a dc differential voltage amplifier. By dc we mean that it will process input signals over a frequency range extending down to and *including* a dc voltage. As a differential amplifier it accepts two inputs and responds to the *difference* in the voltages applied to the input terminals. One of these inputs, called *noninverting*, is conventionally identified with the (+) symbol (Fig. 7.20). The other, called the *inverting* input, carries the (−) symbol. The voltage at the output terminal, e_o, is the product of the amplifier gain, G, and the voltage difference:

$$e_o = G(e_+ - e_-) \tag{7.29}$$

The output voltage is roughly limited to the power supply voltages, V_{cc} and V_{ee}, as the voltage difference increases; if the voltage difference becomes too large, the output *saturates* near one of these values and remains constant. Op-amp response is illustrated in Fig. 7.21.

The op amp's differential characteristic has great importance in instrumentation because it eliminates offset voltages and noise signals common to both input terminals. For example, nearby power lines may induce 60-cycle noise in the exterior circuitry leading to the amplifier. Such line noise is often present

Figure 7.20 Diagram showing typical operational amplifier connections

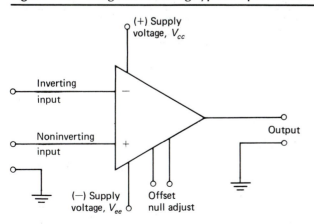

Figure 7.21   Op-amp output response

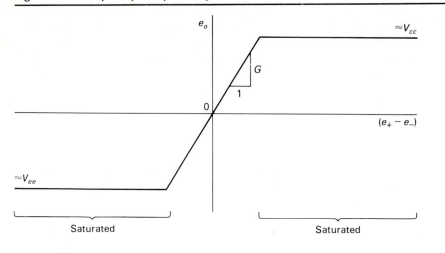

in identical form at both input terminals, and it is thus canceled by the differential amplification. This behavior is known as *common-mode rejection.* If, instead, an op amp receives the output of a voltage-sensitive Wheatstone bridge, the common offset voltages of the two voltage dividers are canceled, and only the desired difference voltage is amplified.

Figure 7.20 shows the configuration of the exterior circuitry of the op amp. Two power sources of equal magnitude but opposite polarity are generally required ($-V_{cc} = V_{ee}$). These voltages usually fall somewhere in the range of 5 to 30 V dc. Quite often, common 9 V dc transistor-radio batteries may be used. Most op amps are packaged in either the dual-in-line package (DIP) form or the one of the standard "TO" cans [Fig. 7.22(a) and (b)].

Figure 7.22   (a) Typical DIP (dual in-line package) integrated circuit,
                     (b) typical TO integrated circuit package

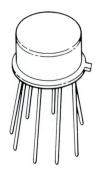
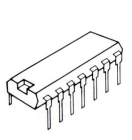

(a)   (b)

The op amp very nearly satisfies the ideal voltage amplifier characteristics of Section 7.13 for the following reasons:

1. It has very high input impedance (megaohms to teraohms).

2. It is capable of very high gain (10^5–10^6 or 100 dB–120 dB).

3. It has very low output impedance (down to a fraction of an ohm with feedback).

4. It has very fast response or high slew rate (output can change several volts per μs.)

5. It is quite effective in rejecting common-mode inputs.

One nonideal characteristic of most op amps is that they do not completely satisfy the differential amplifying property: With both inputs grounded, a residual output voltage remains. The multitude of transistors, resistors, and other elements within the op amp are never perfectly matched, so the amp output actually reaches zero at some small *nonzero* input voltage. To accommodate this *input offset voltage,* the common op amp is provided with pins marked "offset null" or "balance," which provide a means for adjusting the unwanted offset voltage toward zero (see Example 7.7).

A second limitation is that the actual common-mode rejection is finite. If the two input signals each include a common-mode voltage, e_{cm}, the op amp's actual output will be

$$e_o = G(e_+ - e_-) + G_{cm}e_{cm}.$$

The finite common-mode rejection is characterized by the *common-mode rejection ratio* in decibels:

$$\text{CMRR} = 20\log_{10}\left(\frac{G}{G_{cm}}\right). \tag{7.30}$$

Typical op amps have a CMRR of 60 to 120 dB; thus, the common-mode gain is typically 10^3 to 10^6 times smaller than the differential gain. Obviously, a high CMRR is desirable.

In addition, thermal drift can limit op-amp performance. Both internal and external circuit elements may be temperature sensitive, and the design of each circuit usually includes compensating features. A wide variety of op amps are available, and their differences largely represent attempts to improve thermal stability, CMRR, offset voltages, or frequency response. Understandably, such refinements are reflected in cost.

7.17.1   Typical Op-Amp Specifications

Op amps are often designed to optimize those aspects of performance needed for a specific application. One common general-purpose amplifier is the LF411. In comparison to more sophisticated op amps, the LF411 is quite simple; yet in a package the size of a fingernail, it incorporates 23 transistors, 11 resistors, 3 diodes, and 1 capacitor. Typical LF411 specifications are as follows:

Open-loop gain	2×10^5 (depends on frequency)
Input impedance	$10^{12}\,\Omega$
Input offset voltage	$0.8\,\text{mV}$
Input offset voltage drift	$7\,\mu\text{V}/°\text{C}$
Input offset current	$25\,\text{pA}$
Input bias current	$50\,\text{pA}$
CMRR	$100\,\text{dB}$
Maximum output current	$25\,\text{mA}$
Slew rate	$15\,\text{V}/\mu\text{s}$
Maximum power supply voltages	$\pm 18\,\text{V}$
Power supply current	$1.8\,\text{mA}$
Maximum input voltage range	$\pm 15\,\text{V}$
Maximum differential input	$\pm 30\,\text{V}$
Common-mode input voltage range	$-11.5/+14.5\,\text{V}$
Short-circuit output time	Indefinite

7.17.2   Applications of the Op Amp

Operational amplifiers may be used as the basic components of linear voltage amplifiers, differential amplifiers, integrators and differentiators, voltage comparators, function generators, filters, impedance transformers, and many other devices. They are *not* power amplifiers, nor do they have exceptionally wide bandwidth capabilities. Undistorted frequency response is typically limited to about 1 MHz when the circuit gain is low, and it decreases as the gain is raised. In general, an op amp's maximum voltage output is limited by the supply voltage.

Since the number of applications of the op amp to mechanical measurements is almost limitless, we can describe here only a few. Yet this will give the reader some idea of the tremendous versatility of the device and will suggest additional uses (see also the Suggested Readings at the end of the chapter).

One feature common to most op-amp circuits is a *negative feedback loop*. Because op-amp gain is so high, even a slight input voltage-difference will drive the amplifier to saturation. To prevent this, a connection is made between the output terminal and the inverting (−) terminal. With this connection in place, an increase in e_o will be fed back to e_-, reducing the input voltage-difference. The net effect is to produce a circuit that holds $e_- \approx e_+$, preventing saturation. Example 7.2 describes a circuit with no feedback, and several subsequent examples treat circuits having feedback loops.

Example 7.2

The *open-loop configuration** has the following characteristics:

1. No feedback loop. R_L is the load resistance powered by e_o. The circuit may be free floating or grounded.

2. Amplifier is run wide open. Any input other than zero will drive the amplifier to saturation; i.e., a very small input will drive the output to the limit permitted by the power supply.

3. It is seldom used; however, it may be employed as a voltage comparator. With different voltages applied to (+) and (−), open-loop output polarity (positive or negative saturation) will be controlled by the larger input. For sinusoidal input, a square-wave output would result.

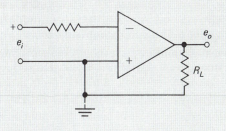

Example 7.3

The *voltage follower,* or *impedance transformer,* has a feedback loop connecting the full output to the inverting input.

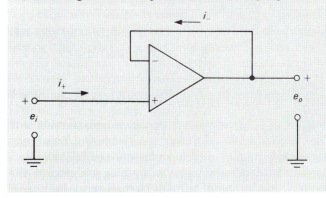

---

* It is conventional in op-amp circuit diagrams to show only those terminals that are used in the particular configuration. Power supply inputs are always required, whether shown or not. Null adjustment is often not shown, although it may be required for optimal performance (see Example 7.7).

The feedback loop prevents saturation by holding $e_- \approx e_+$. Since $e_i = e_+$ and $e_o = e_- \approx e_+$, the output voltage is equal to (follows) the input voltage: $e_o = e_i$. The circuit gain is $G = 1$.

This circuit capitalizes on the high input impedance of the op amp: Since the input impedance is so large, the input current i_+ is in nanoamps (nA) or even picoamps (pA). Source loading is minimized and can often be entirely neglected ($i_+ \approx 0$). In contrast, the output terminal can deliver up to the maximum current of the op amp. This circuit acts as an impedance transformer in that the input impedance is in gigaohms, whereas the output impedance is a fraction of an ohm.

This example demonstrates two important rules of thumb that can be applied any op-amp circuit having negative feedback:

1. The input currents, i_+ and i_-, are essentially zero: $i_+, i_- \approx 0$.
2. The input voltages, e_+ and e_-, are held equal by the negative feedback: $e_+ \approx e_-$.

Example 7.4

The *inverting amplifier* is one of the most used op-amp circuits. Feedback is provided through resistor R_2.

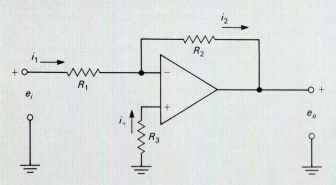

Since $i_+ \approx 0$, Ohm's law shows $e_+ = 0$. Because negative feedback is present, $e_- = e_+ = 0$. The inverting input also draws no current, so that $i_1 = i_2$. Thus we can apply Ohm's law to resistors R_1 and R_2 to find the relation between e_i and e_o:

$$i_1 = \frac{e_i - 0}{R_1} = \frac{e_i}{R_1}$$

$$i_2 = \frac{0 - e_o}{R_2} = -\frac{e_o}{R_2}$$

or

$$e_o = -\frac{R_2}{R_1} e_i$$

The output is opposite in sign from the input (inverted, or 180° out of phase), and the gain of the circuit is $G = -R_2/R_1$.

The resistor R_3 is commonly made approximately equal to the parallel value of R_1 and R_2, i.e., $R_3 \approx R_1 R_2/(R_1 + R_2)$. This choice provides nearly equal input impedances at the $(-)$ and $(+)$ terminals.

Example 7.5
The *noninverting amplifier* is as shown.

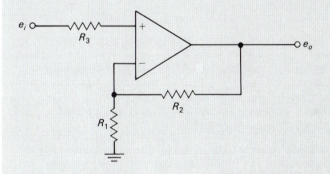

The input voltage is applied to the $(+)$ terminal $(e_i = e_+)$; because negative feedback is present, $e_- = e_+ = e_i$. The output voltage is related to the voltage at the inverting terminal by the voltage-divider relation:

$$e_- = \left(\frac{R_1}{R_1 + R_2}\right) e_o$$

Rearranging,

$$e_o = \left(\frac{R_1 + R_2}{R_1}\right) e_i$$

Thus, the output and the input are in phase and the circuit gain is $G = (R_1 + R_2)/R_1$. Resistor R_3 serves the same purpose as in the inverting amplifier.

Example 7.6

In the *differential*, or *difference*, *amplifier*:

1. If $R_1 = R_2$ and $R_3 = R_4$, then $e_o = (R_3/R_1)(e_{i_1} - e_{i_2})$.
2. The need for offset null adjustment (see Example 7.7) is minimized by making input resistances at $(-)$ and $(+)$ equal.
3. Precise resistor matching is necessary to achieve high CMRR.

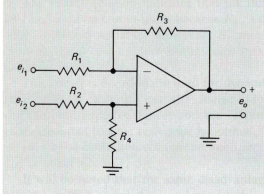

Example 7.7

An amplifier with offset null adjustment is exemplified by the accompanying diagram.

1. The circuit allows trimming to zero output with zero input.
2. Specific example shown illustrates pin numbering.
3. In this circuit, 470-kΩ resistors adjust input impedance and provide

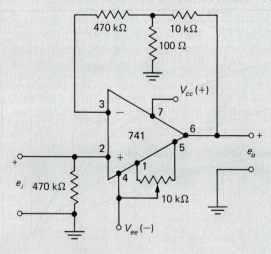

nearly equal resistances at pins 2 and 3, thereby reducing demand on null adjustment.

Example 7.8

The *voltage comparator* has the following features:

1. A small voltage difference between e_i and e_{ref} swings output to limit permitted by power supplies; e_{ref} is set to desired reference voltage. No feedback is used.

2. When $e_i > e_{ref}$, output is positively saturated; when $e_i < e_{ref}$, output is negatively saturated. This provides output indication for the size of e_i relative to e_{ref}. For example, should e_i be gradually rising, when its value reaches e_{ref} the output polarity would reverse. This could be used to trigger external action. (See Section 8.12.2 for application to analog-to-digital conversion.)

3. $R_1 \approx R_2 R_3 / (R_2 + R_3)$ to provide nearly equal impedances at $(+)$ and $(-)$.

4. Diodes serve to limit differential input.

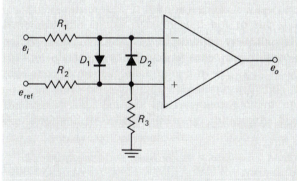

Example 7.9

The *summing amplifier* has the following characteristics:

1. $e_o = -[e_1(R_4/R_1) + e_2(R_4/R_2) + e_3(R_4/R_3)]$.

2. If $R_1 = R_2 = R_3 = R$, then $e_o = -(R_4/R)(e_1 + e_2 + e_3)$.

3. This circuit has application to digital-to-analog converters (Section 8.12.1). Also note similarity to inverting amplifier (Example 7.4).

Example 7.9 *continued*

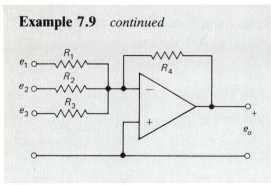

7.18   Special Amplifier Circuits

7.18.1   Instrumentation Amplifiers

In practice, transducer signals are often small voltage differences that must be accurately amplified in the presence of large common-mode signals. Simultaneously, the current drawn from the transducer must remain small to avoid loading the transducer and degrading its signal. Standard op-amp circuits, such as the differential amplifier (Example 7.6), may not provide adequate input impedance or CMRR when high-accuracy measurements are needed.

The *instrumentation amplifier* uses three op amps to remedy these problems (Fig. 7.23). The instrumentation amp is essentially a differential amplifier with a voltage follower placed at each input (this is easily seen if R_1 is temporarily removed). The voltage followers increase the $(+)$ and $(-)$ input impedances to the op-amp impedances. The addition of R_1 between the two followers has the effect of raising CMRR. Resistor matching is less critical for this circuit than for a differential op-amp circuit alone.

Instrumentation amplifiers may be built from discrete components, or they may be purchased as single integrated circuits. The typical instrumentation amp may have CMRR reaching 130 dB, input impedance of $10^9 \, \Omega$ or more, and circuit gain of up to 1000.

7.18.2   The Charge Amplifier

The *charge amplifier* is used with piezoelectric transducers (Sections 6.14, 13.6, 14.8.3, 17.8, and 18.5). These transducers are composed of a high-impedance material that generates electric charge $Q(t)$ in response to a varying load. The charge amp produces an output proportional to the charge while avoiding the potential noise difficulties of a high-impedance source. The complete circuit is shown in Fig. 7.24.

The transducer, cable, and feedback capacitances are C_t, C_c, and C_f, respectively (see Sections 5.17–5.19 for a brief review of charge and capacitance). If the large feedback resistor R_f is ignored, the output of the

Figure 7.23 An instrumentation amplifier circuit

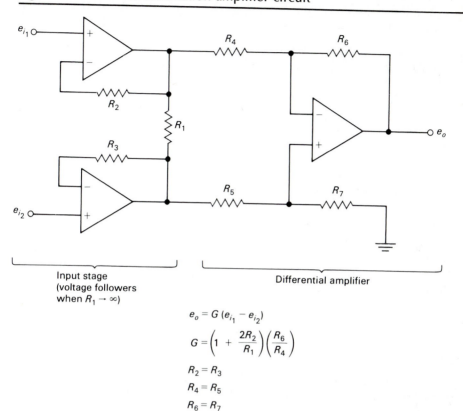

Input stage
(voltage followers
when $R_1 \to \infty$)

Differential amplifier

$$e_o = G\,(e_{i_1} - e_{i_2})$$

$$G = \left(1 + \frac{2R_2}{R_1}\right)\left(\frac{R_6}{R_4}\right)$$

$$R_2 = R_3$$
$$R_4 = R_5$$
$$R_6 = R_7$$

Figure 7.24 A charge amplifier circuit

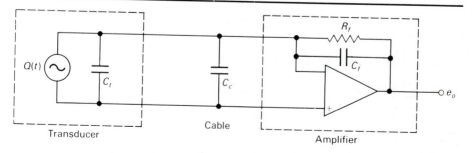

Transducer

Cable

Amplifier

circuit can be expressed as [10]

$$e_o = \frac{-Q(t)}{C_f + (C_t + C_c + C_f)/G},$$

where G is the open-loop gain of the op amp. Because op-amp gains are enormous, the second term in the denominator is usually negligible, and the effective output is just

$$e_o = -\frac{Q(t)}{C_f}.$$

Note that the charge amp's output is independent of cable and transducer capacitance.

The resistor R_f limits the response of the charge amp at frequencies below $f = 1/2\pi R_f C_f$. Such parallel resistance is often introduced to eliminate low-frequency contributions to output; however, some parallel resistance is always present, owing to the finite resistances of real capacitors.

Although the piezoelectric effect was known in the nineteenth century, it did not become technologically important until very-high-input-impedance amplifiers were developed in the 1950s and 1960s. The charge amp itself was patented by W. P. Kistler in 1950 and gained wide use following the development of MOSFET circuits and high-grade electrical insulators such as Teflon and Kapton [11].

7.18.3 Additional IC Devices

In the preceding several sections we have discussed at some length only one IC device, the operational amplifier. A multitude of additional solid-state devices are useful in mechanical measurements, including power amplifiers, solid-state relays, voltage regulators, precision voltage references, voltage comparators, voltage-controlled oscillators (VCOs), multiplexers, sample-and-hold circuits, analog-to-digital and digital-to-analog converters, precision timers, frequency-to-voltage converters, temperature transducers, various logic devices, etc. The families of microprocessor ICs form entire systems within themselves.

Discussion of some of these devices will be found in Chapter 8, "Digital Techniques in Mechanical Measurement." For the remainder, space permits us only to refer the reader to the Suggested Readings at the end of this chapter.

7.19 Shielding and Grounding

7.19.1 Shielding

Shielding applied to electrical or electronic circuitry is used for either or both of two related, but different purposes:

1. To isolate or retain electrical energy within an apparatus.

2. To isolate or protect the apparatus from outside sources of energy.

An example of the former is the shielding required by the Federal Communications Commission to minimize radio-frequency radiation from computers. In the second case, shielding may be required to protect low-level circuitry from entry of unwanted outside signals. A very common source of outside energy is the ubiquitous 60 Hz power line.

Shielding is of two basic types: (a) electrostatic and (b) electromagnetic. In each case the shielding normally consists of some form of metallic enclosure; for example, metallic braid may be used to shield signal-carrying wiring, or circuitry may be partially or entirely enclosed in metal boxes.

Only nonmagnetic metals may be used for electromagnetic shielding, whereas almost any conducting metal such as steel, aluminum, or copper may be used for electrostatic shielding. Circuits within a device often must be shielded from each other; however, connections must still be made between the subcircuits through use of special amplifiers or transformers. For example, power transformers are often provided with copper shielding between primary and secondary windings. The copper provides electrostatic shielding without hindering the transfer of electromagnetic power. Some rules for shielding are as follows [12]:

Rule 1   An electrostatic shield enclosure, to be effective, should be connected to the zero-signal reference potential of any circuitry contained within the shield.

Rule 2   The shield conductor should be connected to the zero-signal reference potential at the signal-to-earth connection.

Rule 3   The number of separate shields required in a system is equal to the number of independent signals being processed plus one for each power entrance.

7.19.2   Grounding

When low-level circuitry is employed, some form of grounding is inevitably required. *Grounding* is needed for one, or both, of two reasons: (1) to provide an electrical reference for the various sections of a device, or (2) to provide a drainage path for unwanted currents.

A *ground reference* may be either of two types, (1) earth ground, or (2) chassis ground. In the latter case, "chassis" commonly refers to the basic mounting structure (e.g., the ground plane of a circuit board) or the enclosure within which the circuitry is mounted. Conventional schematic symbols for the two are as shown in Fig. 7.25.

Figure 7.25   Conventional symbols for ground references

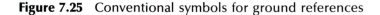

Earth   Chassis

In a text such as this, only superficial coverage of this complicated topic is possible. However, certain "rules" and observations may be listed as follows:

1. An entire system can be "grounded" and need not involve earth at all. For example, circuitry in aircraft and spacecraft are referenced to some common datum.

2. The word "circuit" need not imply wires or components. Each of the various elements in a device may, unless effectively shielded, possess capacitive paths to one or more of the others.

3. Shielding can be at any potential and still provide shielding.

4. The assumption that two nearby points are at the same potential is often invalid—not only earth points, but also points in any ground plane.

5. Potential characteristics of an element are not the same at "radio" or high frequencies as they are at "power" or low frequencies. For example, a capacitance exists between a bonded strain gage element and the structure on which it is mounted. At dc or low frequency, such capacitance may be unimportant, but at radio frequencies the capacitance may provide a ready electrical path.

6. A ground bus is protection against effects of equipment faulting, but it is not the source of zero potential for the solution of instrumentation processes.

7. All metal enclosures and housings should be earthed and bonded together, but no current should be permitted to flow in these connections.

8. Good practice suggests that it is wise to insulate an equipment rack from the obvious ties, such as building earths and conduit connections, so that the rack can be ohmically connected to a potential most favorable to the instrumentation processes.

9. Rules that are applicable at one frequency range may be inadequate at another.

10. Safety practices demanded by various civil codes can seem to be in direct conflict with good instrumentation practice.

11. Electrostatic shields are simply metallic enclosures that surround signal processes. To be effective, these shields should be tied to a zero signal potential where the signal makes its external, or ground connection.

To reiterate, shielding and grounding are very complex subjects, and often some degree of trial and error, coupled with experience, is required to find a solution.

7.20   Filters

As we have seen, time-varying measurands commonly consist of a combination of many frequency components or harmonics. In addition, unwanted inputs (noise) are often picked up, thereby resulting in distortion and masking of the

Figure 7.26   (a) Some terminology as applied to a low-pass filter
(b) band-pass filter characteristics

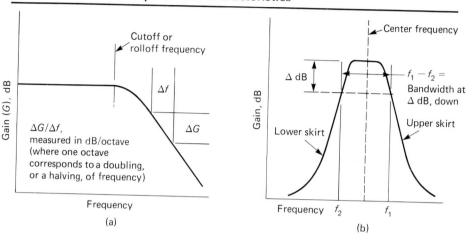

(a)

(b)

true signal. It is usually possible to use appropriate circuitry to selectively filter out some or all of the unwanted noise (see, for example, Section 5.7).

Filtering is the process of attenuating unwanted components of a measurand while permitting the desired components to pass. Filters are of two basic classes, *active* and *passive*. An active filter uses powered components, commonly configurations of op amps, whereas a passive filter is made up of some form of *RLC* arrangement. In addition, filters may be classified by the descriptive terms *high-pass*, *low-pass*, *band-pass*, and *notch* or *band-reject*. In each case, reference is to the signal frequency; for example, the high-pass filter permits components above a certain cutoff frequency to pass through. The notch filter attenuates a selected band of frequency components, whereas the band-pass filter permits only a range of components about its center frequency to pass. Figure 7.26(a) and (b) illustrate certain terms applied in filter design and use. Similar terms are applicable to the high-pass and notch filters, respectively.

7.21   Some Filter Theory

The simplest low-pass and high-pass filters are made from a single resistor and capacitor. Electrically, these passive *RC* filters are first-order systems (Section 5.18). The *RC* low-pass and high-pass filters are shown in Fig. 7.27(a) and (b), respectively.

Consider first the *RC* low-pass filter. Since a capacitor tends to block low-frequency currents and pass high-frequency currents, the basic effect of the capacitor in this filter is to short-circuit the high-frequency components of the

Figure 7.27   First-order RC filters: (a) low pass, (b) high pass

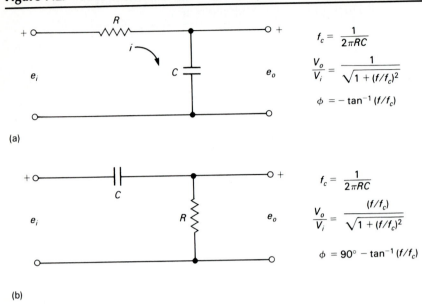

(a)

$$f_c = \frac{1}{2\pi RC}$$

$$\frac{V_o}{V_i} = \frac{1}{\sqrt{1 + (f/f_c)^2}}$$

$$\phi = -\tan^{-1}(f/f_c)$$

(b)

$$f_c = \frac{1}{2\pi RC}$$

$$\frac{V_o}{V_i} = \frac{(f/f_c)}{\sqrt{1 + (f/f_c)^2}}$$

$$\phi = 90° - \tan^{-1}(f/f_c)$$

input signal. To determine the frequency characteristics, we must find the filter output, e_o, for a harmonic input voltage, e_i:

$$e_i = V_i \sin(2\pi ft).$$

If negligible current is drawn at the output, the currents through the resistor and the capacitor are equal, so that

$$i = \frac{e_i - e_o}{R} = C\frac{d}{dt}e_o$$

or

$$\frac{d}{dt}e_o + \frac{1}{RC}e_o = \frac{1}{RC}e_i = \frac{V_i}{RC}\sin(2\pi ft). \tag{7.31}$$

Solution of this equation gives (cf. Example 5.7)

$$e_o = V_o\sin(2\pi ft + \phi) = \frac{V_i}{\sqrt{1 + (2\pi RCf)^2}}\sin(2\pi ft + \phi), \tag{7.32}$$

where the phase lag, ϕ, is

$$\phi = -\tan^{-1}(2\pi RCf). \tag{7.32a}$$

Figure 7.28 Frequency response of the *RC* low-pass
filter (linear coordinates)

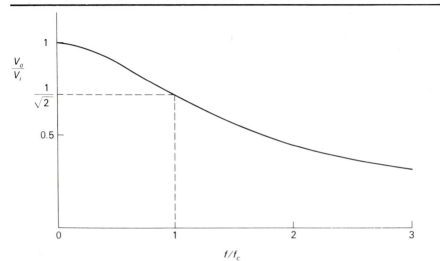

Filter performance is normally characterized by defining a *cutoff frequency*, f_c:

$$f_c \equiv \frac{1}{2\pi RC} \tag{7.33}$$

In terms of f_c, the frequency response (or gain), from Eq. (7.32), is

$$\frac{V_o}{V_i} = \frac{1}{\sqrt{1 + (f/f_c)^2}} \tag{7.33a}$$

and the phase response, from Eq. (7.32a), is

$$\phi = -\tan^{-1}\left(\frac{f}{f_c}\right) \tag{7.33b}$$

The frequency response is shown on linear coordinates in Fig. 7.28. Graphed this way, the filter response seems to change only slightly with frequency. However, the graph shows only a factor-of-three increase in frequency, and when using filters the frequencies of concern may cover a factor of 100 or more. A logarithmic graph, such as a Bode plot (Section 7.12), is needed to illustrate such variation.

A Bode plot of the low-pass filter's response is shown in Fig. 7.29. The frequency varies over several orders of magnitude, and the amplitude attenuation runs from 0 to $-40\,dB$ or more. The cutoff frequency corresponds to $-3\,dB$ response, which means that signal power is reduced by one-half at f_c (and the voltage is reduced by $1/\sqrt{2} = 0.707$). For frequencies well below f_c, the filter's response is flat and shows no signal reduction. The transition from

Figure 7.29 Frequency response of *RC* low-pass and
high-pass filters (Bode plot)

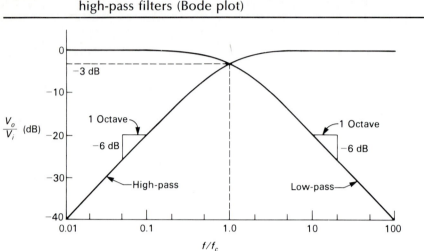

the passband to rejection band occurs gradually with increasing frequency. In
the rejection band itself, at frequencies well above f_c, the amplitude rolloff is
$-6\,$dB/octave (an octave being a factor-of-two change in frequency) or
$-20\,$dB/decade (a decade being a factor of ten).

In addition to reducing amplitude, this filter also produces an increasing
phase shift as signal frequency rises (Fig. 7.30). At the $-3\,$dB point, the
output lags the input by 45°.

The *RC* high-pass filter is obtained by interchanging the resistor and
capacitor [Fig. 7.27(b)]. Now the capacitor blocks low frequencies while

Figure 7.30 Phase response of *RC* high-pass and low-pass filters

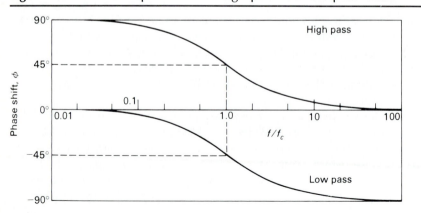

passing high frequencies. The results are quite similar:

$$e_o = V_o \sin(2\pi ft + \phi) = \frac{V_i(2\pi RCf)}{\sqrt{1 + (2\pi RCf)^2}} \sin(2\pi ft + \phi)$$

where the phase shift, ϕ, is now a *lead* ($\phi > 0$) rather than a lag ($\phi < 0$)

$$\phi = 90° - \tan^{-1}(2\pi RCf).$$

The high-pass filter's cutoff frequency is identical to the low-pass filter's:

$$f_c \equiv \frac{1}{2\pi RC}. \tag{7.34}$$

In terms of the cutoff frequency, the frequency response and phase lead are

$$\frac{V_o}{V_i} = \frac{(f/f_c)}{\sqrt{1 + (f/f_c)^2}}, \tag{7.34a}$$

$$\phi = 90° - \tan^{-1}\left(\frac{f}{f_c}\right). \tag{7.34b}$$

The $-3\,\text{dB}$ point is again f_c, and the rolloff in the rejection band is again $-6\,\text{dB/octave}$ or $-20\,\text{dB/decade}$. The high-pass frequency and phase response are shown in Figs. 7.29 and 7.30. One common use of this filter is to remove dc ($f = 0$) offsets.

First-order RC filters have a fairly slow rolloff above the cutoff frequency (not many decibels per octave), but their simplicity still gains them wide use in situations where the desired and undesired frequencies are widely separated. Similar high-pass and low-pass filters can be made using a single resistor-inductor pair. However, first-order RL filters are seldom used.

Example 7.10

A transducer responding to a 5000 Hz signal also picks up 60 Hz noise. The resulting output is

$$\{5 \sin(2\pi \cdot 60 \cdot t) + 25 \cos(2\pi \cdot 5000 \cdot t)\}\,\text{mV}.$$

To remove the 60-cycle noise, a high-pass filter with cutoff of 1000 Hz is introduced. What is the filtered output?

Solution. The amplitude and phase shift are computed separately for each component:

$$\left(\frac{V_o}{V_i}\right)_{60} = \frac{(60/1000)}{\sqrt{1 + (60/1000)^2}} = 0.060,$$

$$\left(\frac{V_o}{V_i}\right)_{5000} = \frac{(5000/1000)}{\sqrt{1 + (5000/1000)^2}} = 0.98,$$

$$\phi_{60} = 90° - \tan^{-1}\left(\frac{60}{1000}\right) = 86.6° = 1.51 \text{ rad},$$

$$\phi_{5000} = 90° - \tan^{-1}\left(\frac{5000}{1000}\right) = 11.3° = 0.197 \text{ rad}.$$

Then

$$e_0 = \{0.3\sin(2\pi \cdot 60 \cdot t + 1.51) + 24.5\cos(2\pi \cdot 5000 \cdot t + 0.197)\} \text{ mV}.$$

The noise amplitude is reduced from 20% of the signal amplitude to only 1.2%. Note that the signal itself undergoes a slight amplitude reduction as well as a phase shift. Such changes in the signal are undesirable, and they often motivate the use of more complex filters.

Three desirable elements of filter performance are:

1. Nearly flat response over the pass and rejection bands;
2. High values of rolloff for low- and high-pass filters, as measured in decibels per octave;
3. Steep skirts for band-pass and band-rejection filters.

Significant improvements in performance are obtained by using combinations of several capacitors, inductors, or resistors to produce second-order (or higher-order) electrical response. Such filters can have steeper rolloff and sharper transition from pass to rejection bands. In addition, such compound *RLC* arrangements can produce band-pass and notch filters. For example, Fig. 7.31(a) and (b) show an *RC* band-pass filter and its response:

$$\frac{V_o}{V_i} = \frac{1}{\sqrt{[1 + R_1/R_2 + C_2/C_1]^2 + [2\pi R_1 C_2 f - (1/2\pi R_2 C_1 f)]^2}}. \quad (7.35)$$

Inductors and capacitors used together allow resonant behavior, which can produce steeper filter skirts than are possible with first-order *RC* circuits. In fact, the resonant circuit of Section 7.11 is sometimes used to build very narrow band-pass filters known as *tuned filters*. Some additional *LC* designs are shown in Fig. 7.32.

Two practical issues influence the design and use of passive filters. First, the filters considered here are all designed as if negligible current is drawn from the output terminals. If several filters are placed in series, to steepen rolloff, then the current drawn by one filter can alter the performance of the filter that precedes it. To avoid this output loading, a voltage follower (Example 7.3) should be introduced as a buffer between each successive filter.

Figure 7.31 (a) A circuit for a simple band-pass filter, (b) performance characteristics of band-pass filter shown in (a)

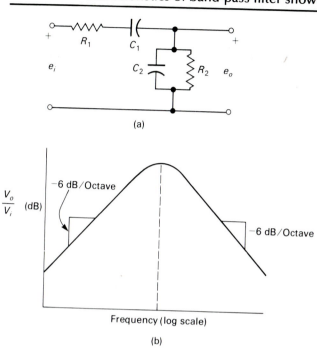

(a)

(b)

The inductors themselves are the second problem. At the frequencies encountered in mechanical measurements, which rarely exceed 100 kHz, the required inductors may be quite large and bulky. In addition, the optimal inductor values are not always easily obtained, and the inductors may have substantial internal resistances as well. As often happens in engineering, inductors can be much less satisfactory in practice than they seem on paper. The usual way of avoiding these problems is to employ an active filter, as described next.

7.22 Active Filters

Op amps can be used to construct filter circuits without inductors and without the problems of output loading. These active filters can also have very steep rolloff, arbitrarily flat passbands, and even adjustable cutoff frequencies. Active filters are a rich subject, and entire textbooks have been devoted to their design.

The basic active filter is shown in Fig. 7.33. Passive filter networks are linked to an op amp, which provides power and improves impedance characteristics. The passive network is built from resistors and capacitors only;

Figure 7.32 Examples of *LC* filter arrangements
and their output characteristics

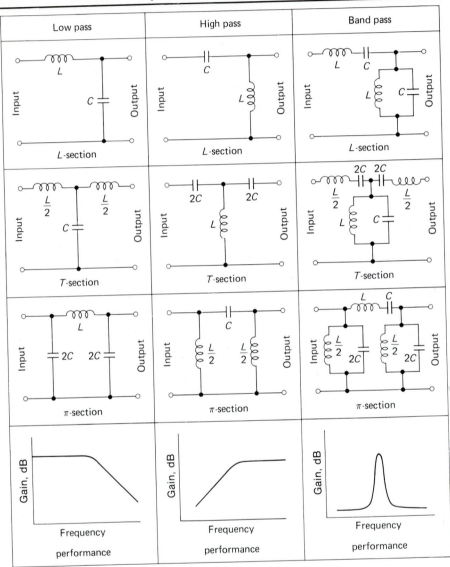

inductive characteristics are simply simulated by the circuit. Since the output impedance is generally low, these filters can deliver an output current without reduced performance. Some typical active filters are shown in Fig. 7.34.

Active filters are available with rolloffs of 80 dB/octave and more than 60 dB attenuation in the rejection band. High-order active filters are even sold as integrated circuits contained in a single DIP package. For further study, see the Suggested Readings at the end of this chapter.

Figure 7.33   Basic active filter circuit

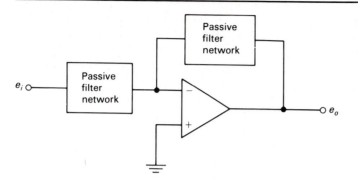

Figure 7.34   First-order active filters: (a) low pass,
(b) high pass, (c) band pass

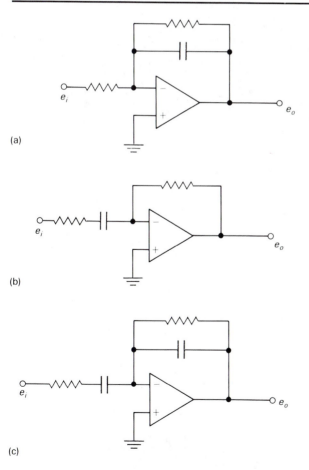

(a)

(b)

(c)

7.23   Differentiators and Integrators

A final op-amp application is in circuits that respond to the rate of change or the time history of an input signal, called *differentiators* and *integrators*, respectively [Figs 7.35(a) and (b)].

In the differentiator, the currents through the resistor and capacitor are equal, and $e_- = e_+ = 0$. Thus

$$C \frac{d}{dt} e_i = -\frac{e_o}{R},$$

or

$$e_o = -RC \frac{d}{dt} e_i. \tag{7.36}$$

In the integrator, the capacitor charges in proportion to the time summation of e_i. Again, the resistor and capacitor currents are equal:

$$\frac{e_i}{R} = C \frac{d}{dt}(-e_o),$$

Figure 7.35   (a) Op-amp differentiator, (b) op-amp integrator

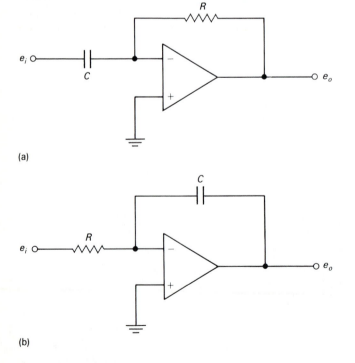

(a)

(b)

or

$$e_o = -\frac{1}{RC}\int e_i \, dt + \text{constant}.\qquad(7.36a)$$

To prevent drift in the capacitor's charge over long time intervals, a large resistor, R_f, may be placed in parallel with it. In that case, the integrator circuit is restricted to signal frequencies $f \gg 2\pi R_f C$.

7.24 Component Coupling Methods

When electrical circuit elements are connected, special attention must often be given to the coupling methods used. In certain cases, transducer-amplifier, amplifier-recorder, or other component combinations are inherently incompatible, making direct coupling impossible or, at best, causing nonoptimal operation. Coupling problems include obtaining proper impedance matching and maintaining circuit requirements such as damping. These problems are usually caused by the desire for maximum energy transfer and optimum fidelity of response.

The importance of impedance matching, however, varies considerably from application to application. For example, the input impedances of most cathode-ray oscilloscopes and electronic voltmeters are relatively high, but satisfactory operation may be obtained from directly connected low-impedance transducers. In this case, voltage is the measured quantity and power transfer is incidental. *In most cases, driving a high-impedance circuit component with a low-impedance source presents fewer problems than does the reverse.*

As a simple example of transfer, consider Fig. 7.36. Shown is a *source* of energy E_S and a *sink* or *load* having impedance Z_L. Z_S is the source impedance. To simplify our example further, let the impedances be simple electrical resistances, R_S and R_L, respectively. Then the voltage across R_L will be

$$E_L = E_S\left(\frac{R_L}{R_L + R_S}\right)$$

Figure 7.36 Simple circuit for demonstrating transfer concepts

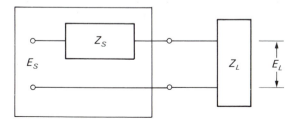

and the power delivered to R_L is

$$P = \frac{E_L^2}{R_L} = \frac{E_S^2}{R_L}\left(\frac{R_L}{R_L + R_S}\right)^2 \tag{7.37}$$

To determine the values of R_L for maximum power transfer, set dP/dR_L to zero and solve. We find that maximum power is transferred if $R_L = R_S$. Although this is not really a proof, in general terms, maximum power is transferred when $Z_L = Z_S$.

In addition to, or instead of, optimizing *power transfer,* proper coupling may be important in providing adequate dynamic response. Three special methods of coupling, depending on the circuit elements are common; they utilize (1) matching transformers (discussed shortly), (2) impedance transforming (see Section 7.17.2), and (3) coupling networks (discussed shortly).

An example of transformer coupling is the use of electronic power amplifiers to drive vibration exciters such as those discussed in Section 17.16.1. The problem is similar to that of connecting a speech amplifier to a loudspeaker. In both cases the output impedance of the driving power source in the amplifier is often higher than the load impedance to which it must be connected. Transformer coupling is generally used as shown in Fig. 7.37(a). Matching requirements may be expressed by the relation

$$\frac{N_S}{N_L} = \frac{Z_S}{Z_L}, \tag{7.38}$$

Figure 7.37 Impedance matching (a) by means of a coupling transformer and (b) by means of a resistance pad

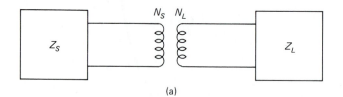

(a)

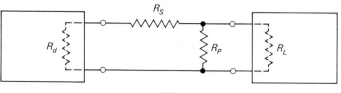

(b)

where

$$Z_S = \text{the source impedance,}$$

$$Z_L = \text{the load impedance,}$$

$$N_S/N_L = \text{the turns ratio of the transformer.}$$

General-purpose devices such as simple voltage amplifiers and oscillators often incorporate a final amplifier stage, called a buffer, to supply a low-impedance output. By reducing the impedance source, one can minimize losses in the connecting lines and the possibility of extraneous signal pickup.

Proper coupling may also be accomplished through use of matching resistance pads. Figure 7.37(b) illustrates one simple form.

If we assume that the driver output and load impedances are resistive, then the matching requirements may be put in simple form as follows: The driving device, which may be a voltage amplifier, *looks* into the resistance network and *sees* the resistance R_s in series with the paralleled combination of R_L and R_p. Hence, for proper matching,

$$R_d = R_s + \frac{R_p R_L}{R_p + R_L}, \qquad (7.39)$$

where

$$R_d = \text{the output impedance of the driver } (\Omega),$$

$$R_L = \text{the load resistance } (\Omega),$$

$$R_p = \text{the paralleling resistance } (\Omega),$$

$$R_s = \text{the series resistance } (\Omega).$$

The driven device *sees* two parallel resistances, made up of R_p and the series-connected resistances R_s and R_d. Hence, for matching,

$$R_L = \frac{R_p(R_s + R_d)}{R_p + (R_s + R_d)}. \qquad (7.39a)$$

Solving for R_s and R_p, we have

$$R_s = [R_d(R_d - R_L)]^{1/2} \qquad (7.40)$$

and

$$R_p = \left[R_L \left(\frac{R_d R_L}{R_d - R_L} \right) \right]^{1/2}. \qquad (7.40a)$$

Now if R_d and R_L are known, values of R_s and R_p may be determined to satisfy the matching requirements by use of Eqs. (7.40) and (7.40a). In using resistive elements, a loss in signal energy is unavoidable. Such losses are often referred to as *insertion losses*. In general, however, by providing proper match, the network will provide optimum gain.

7.25   Summary

Electrical signal conditioning can serve many purposes: to convert sensor outputs to more easily used forms (e.g., resistance change into voltage); to separate small signals from large offsets or noise; to increase signal voltage; to remove unwanted frequency components from the signal; or to enable the signal to drive output devices.

1. For resistance-type transducers, several simple signal-conditioning circuits are useful if the resistance change is relatively large. These include current-sensitive circuits, ballast circuits, and voltage-dividing circuits (Sections 7.5, 7.6, 7.7).

2. For small resistance changes, bridge circuits provide a more sensitive method of detection, in which offsets are eliminated and output can vary linearly with resistance change. When the resistance changes are too large, however, bridge circuits become nonlinear (Sections 7.8, 7.9).

3. AC excitation of signal-conditioning circuits is necessary with inductive and capacitive sensors. AC-excited circuits include reactance bridges (Section 7.10) and resonant circuits (Section 7.11).

4. Decibels (dB) provide a convenient method of quantifying gain or attenuation. The decibel is a logarithmic measure of signal power (Section 7.12).

5. Electronic amplifiers are ubiquitous elements of signal conditioning circuits. Amplifiers serve to increase voltage, to increase power or current, or to change impedance. Today's amplifiers are usually based on the transistor (Sections 7.13–7.16).

6. Operational amplifiers are among the most common elements in instrumentation circuits. Many different op-amp configurations are possible, the choice depending on the characteristics of the specific sensor involved and the type of output response desired (Sections 7.17, 7.18, 7.23).

7. Some signal-conditioning techniques apply primarily to time-varying measurands. These include carrier modulation (Section 7.3), filtering (Sections 7.20–7.22), and differentiating and integrating (7.23).

8. Filters allow unwanted frequency components to be removed from a signal. Filters are characterized by a cutoff frequency (or $-3\,dB$ point) separating the pass and rejection bands and by the steepness of rolloff from the passband to the rejection band. RC filters are the simplest type of passive filter. Common applications of filters include noise removal, dc-offset removal, and carrier demodulation (Sections 7.20–7.22).

9. Many basic signal-conditioning devices are available as integrated circuits. These include op amps, filters, and even bridges (Section 7.16).

10. Shielding and grounding are essential considerations in building and using measurement circuits. Shielding for noise prevention is especially important when precise measurements are to be made (Section 7.19).

11. Component coupling is often designed to maximize power transfer or dynamic response. If voltage detection is of greater importance than power transfer, however, it is often sufficient that a device's input impedance be large compared to the output impedance of the device driving it (Section 7.24).

Suggested Readings

Antoniou, A. *Digital Filters, Analysis and Design*. New York: McGraw-Hill, 1980.

Berlin, H. M. *Operational Amplifiers*. Benton Harbor, Mich.: Heath, 1979.

Booth, S. F. *Precision Measurement and Calibration*, Vol. 1. (Selected NBS technical papers on electricity and electronics). NBS Handbook 77. Washington, D.C.: U.S. Government Printing Office, 1961.

Carlson, A. B., and D. G. Gisser. *Electrical Engineering: Concepts and Applications*. Reading, Mass.: Addison-Wesley, 1981.

Carr, J. J. *Designer's Handbook of Instrumentation and Control Circuits*. San Diego: Academic Press, 1991.

DeMaw, D. (ed.) *The Radio Amateur's Handbook*. Hartford, Conn.: American Radio Relay League (annual editions).

Horowitz, P., and W. Hill. *The Art of Electronics*. 2nd ed. New York: Cambridge University Press, 1989.

Hughes, F. W. *Op-amp Handbook*, 2nd ed. Englewood Cliffs, N.J.: Prentice Hall, 1986.

Jones, B. K. *Electronics for Experimentation and Research*. Englewood Cliffs, N.J.: Prentice-Hall, 1986.

General-Purpose Linear Devices Databook. Santa Clara, California: National Semiconductor Corporation.

Morrison, R. *Grounding and Shielding Techniques in Instrumentation*. 2nd ed. New York: John Wiley, 1977.

Ott, H. W. *Noise Reduction Techniques in Electronic Systems*, 2nd ed. New York: John Wiley, 1988.

Rhodes, J. D. *Theory of Electrical Filters*. New York: John Wiley, 1976.

Smith, R. J. *Circuits, Devices and Systems: A First Course in Electrical Engineering*. 4th ed. New York: John Wiley, 1984.

Taylor, F. J. *Digital Filter Design Handbook*, New York: Marcel Dekker, 1983.

Wait, J. V. et al. *Introduction to Operational Amplifier Theory and Applications*. New York: McGraw-Hill, 1975.

Williams, A. B. *Electronic Filter Design Handbook*. New York: McGraw-Hill, 1981.

Problems

7.1 A force cell uses a resistance element as the sensing element. It is connected in a simple current-sensitive circuit in which the series resistance R_m is 100 Ω, which is one-half the nominal resistance of the force cell. Determine the current for

Figure 7.38

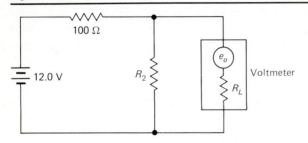

force inputs of (a) 25%, (b) 50%, and (c) 75% of full range if the input voltage is 10 V.

7.2 If the force cell of Problem 7.1 is placed into a ballast circuit $(R_b = R_m)$, determine the output voltage for the conditions of Problem 7.1.

7.3 For the ballast circuit of Problem 7.1, determine the sensitivity, η, for the three percentages of full range.

7.4 Equations (7.5) and (7.6) are derived on the basis of a high-impedance indicator. Analyze the circuit assuming that the indicator resistance R_m is comparable in magnitude to R_t.

7.5 For $E_s = 10$ V and $R_s = 75\,\Omega$, use Eq. (7.37) to plot P versus R_L over the range $0 < R_L < 200\,\Omega$.

7.6 The circuit shown in Figure 7.38 is used to determine the value of the unknown resistance R_2. If the voltmeter resistance, R_L, is 10 MΩ and the voltmeter reads $e_o = 4.65$ V, what is the value of R_2?

7.7 The voltage-dividing potentiometer shown in Fig. 7.5 is modified as shown in Fig. 7.39. Determine the relationship for e_o/e_i as a function of k. Compare the results with Eq. (7.9). What advantages or disadvantages does this circuit have over the general voltage-dividing potentiometer?

7.8 Form a spreadsheet template to solve Eq. (7.15a), permitting each term to be varied by a delta amount. [*Suggestion:* Rewrite the equation, multiplying each term by $(1 + k)$, where k is the delta plus/minus term—for example, $R_i(1 + k_i)$.]

Figure 7.39

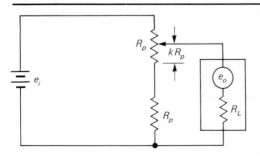

7.9 A simple Wheatstone bridge as shown in Fig. 7.10 is used to determine accurately the value of an unknown resistance R_1 located in leg 1. If upon initial null balance R_3 is 127.5 Ω and if when R_2 and R_4 are interchanged, null balance is achieved when R_3 is 157.9 Ω, what is the value of the unknown resistance R_1?

7.10 Consider the voltage-sensitive bridge shown in Fig. 7.10. If a thermistor whose resistance is governed by Eq. (16.3) is placed in leg 1 of the bridge while $R_2 = R_3 = R_4 = R_0$ determine the bridge output when $T = 400°C$ if $R_0 = 1000\,\Omega$ at $T_0 = 27°C$ and $\beta = 3500$. Plot the bridge output from $T = 27°C$ to $T = 500°C$ and determine the maximum deviation from linearity in this temperature range.

7.11 Referring to Fig. 7.10, show that if initially $R_1 = R_2 = R_3 = R_4 = R$ and if $\Delta R_1 = -\Delta R_2$, the bridge output will be linear. (*Note:* This bridge configuration is very commonly used when strain gages are applied to a beam in bending situations; see Table 12.4).

7.12 Referring to Fig. 7.10, initially let $R_1 = R_2 = R_3 = R_4 = R$. In addition, assume that $\Delta R_4/R = -\Delta R_1/R$. Demonstrate the nonlinearity of the bridge output by plotting e_o/e_i over the range $0 < \Delta R_i/R < 0.1$. (*Suggestion:* Use a computer plotting program, if available.)

7.13 A resistive element of a force cell forms one leg of a Wheatstone bridge. If the no-load resistance is 500 Ω and the sensitivity of the cell is 0.5 Ω/N, what will be the bridge outputs for applied loads of 100, 200, and 350 N if the bridge excitation is 10 V and each arm of the bridge is initially 500 Ω?

7.14 Figure 7.40 shows a differential shunt bridge configuration. One or more of the resistances, R_i, may be resistance-type transducers (thermistor, resistance thermometer, strain gage, etc.), with the remaining resistances fixed. Resistance R_6 is a conventional voltage-dividing potentiometer, usually of the multiturn variety. It may be used either for initial nulling of the bridge output or as a readout means. The variable k is a proportional term varying from 0 to 1 (or 0% to 100%); see

Figure 7.40

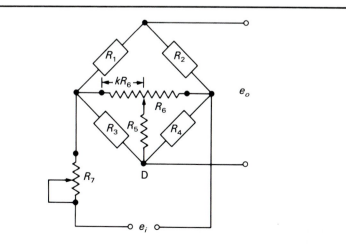

Section 7.9. R_5 is sometimes called a *scaling resistor*. Its value largely determines the range of effectiveness of R_6. R_7 is used to adjust bridge sensitivity. Devise a spreadsheet template to be used for designing a bridge of this type.

7.15 Using the spreadsheet template (or program) created in answer to the preceding problem, determine the null-balance range of ΔR_1 that the bridge can accommodate if

$$R_1 = 1000\ \Omega \text{ (nominal)},$$

$$R_2 = R_3 = R_4 = 1000\ \Omega \text{ (fixed)},$$

$$R_5 = 10{,}000\ \Omega,$$

$$R_6 = 12{,}000\ \Omega,$$

$$R_7 = 0\ \Omega.$$

7.16 Using the template or the program and the nominal resistance values listed in Problem 7.15, investigate changes in measurement range of the bridge as affected by (a) changes in R_5 and (b) changes in R_6. Investigate the linearity of the circuit when used in the null-balance mode, using k as the calibrated readout.

7.17 Referring to Problem 7.14, assume that the tolerances for resistances R_2, R_3, and R_4 are ±5%, what will now be the effective null-balance range, ΔR_1, that can be accommodated?

7.18 Derive the relationships for the Wien bridge circuit shown in Fig. 7.16.

7.19 Derive the relationships for the Maxwell bridge circuit shown in Fig. 7.16.

7.20 Show that an increase of 1 dB corresponds to a power increase of about 26%. Also show that an increase of n dB corresponds to a power increase to approximately $(1.26)^n$.

7.21 Equation (7.28c) may be written as

$$K = \frac{dB_{(corrected)}}{dB_{(indicated)}} = 1 + 10\,\frac{\log(R_r/R_s)}{dB_{(indicated)}}$$

where

$$K = \text{correction factor},$$

$$dB_{(corrected)} = \text{corrected decibels},$$

$$dB_{(indicated)} = \text{indicated decibels},$$

$$R_r = \text{reference impedance},$$

$$R_s = \text{source impedance}.$$

Plot K versus R_r/R_s for $dB_{(indicated)} = 50$, 100, 150, and 200. Make separate families of plots (a) for $R_r/R_s \leq 1$ and (b) for $R_r/R_s \geq 1$.

7.22 Consider the amplifier circuit shown in Fig. 7.41. Determine the amplitude V_o if $V_i = 5\cos(40\pi t)$ if V_i is in megavolts and t is in seconds. What is the amplitude V_o if $V_i = 3.0\sin(4000\pi t)$?

7.23 The circuit shown in Fig. 7.42 is a voltage-to-current converter. Determine the value of R_3 for which i_f in milliamps equals V_i in volts and find V_o.

Figure 7.41   Circuit for Problem 7.22

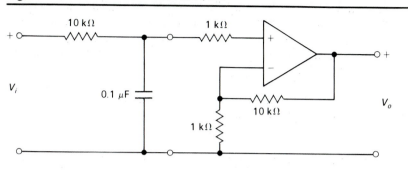

Figure 7.42   Circuit for Problem 7.23

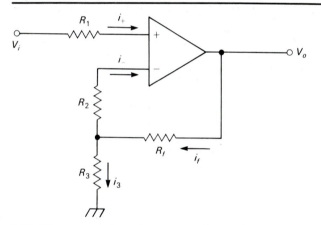

7.24 The current through a semiconductor diode is related to the voltage across it by $i = I_o(\exp(\lambda v) - 1)$ where $I_o = 1 \times 10^{-9}\,A$, $\lambda = 1.17 \times 10^4/T$, and T is the absolute temperature in kelvin. For $T = 300\,K$ and v_i on the order of one volt, show that the output of the circuit in Fig. 7.43 is proportional to the exponential of the input voltage.

Figure 7.43   Circuit for Problems 7.24 and 7.25

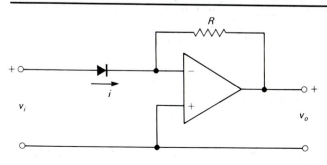

Figure 7.44

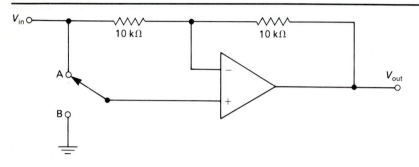

7.25 Show that, if the diode and the resistor in Fig. 7.43 are interchanged, the circuit output is proportional to the logarithm of the input voltage.

7.26 Determine the output of the circuit in Fig. 7.44 when the switch is in position A and when it is in position B.

7.27 In the circuit of Fig. 7.45, a photodiode is reverse-biased with a voltage V_b. In this condition, the photodiode generates a current, i, which is proportional to the intensity, H, of the light which irradiates it: $i = kH$. Determine the relation between the output voltage, V_o, and H.

7.28 For the summing amplifier of Example 7.9, if

$$e_1 = 8.0 \sin(400t) \text{ V},$$

$$e_2 = -2.0 \sin(400t) \text{ V},$$

$$e_3 = 3.0 \cos(400t) \text{ V},$$

and $R_1 = R_2 = R_4 = 5 \text{ k}\Omega$, and $R_3 = 2.5 \text{ k}\Omega$, what is the rms output voltage?

Figure 7.45

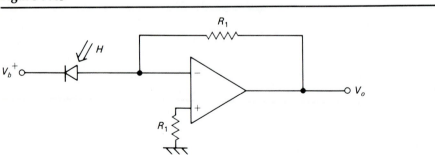

Figure 7.46

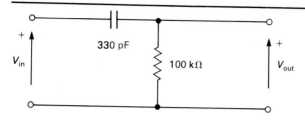

330 pF

V_{in}

100 kΩ

V_{out}

7.29 Consider the filter circuit shown in Fig. 7.46.

a. What type of filter is this? Calculate its cut-off frequency in Hz.

b. The following input signal is applied to the circuit:

$$V_{in} = \{5\sin(2\pi 200t) + 2.5\cos(2\pi 1000t) + +1.5\sin(2\pi 10000t)\}\ mV$$

Determine V_{out}.

7.30 Consider the following data from an RC filter. From experimental testing, these data were obtained for a 1.0 V sine-wave input.

Frequency (Hz)	V_{out} (Volts)
10.0	1.00
20.0	1.00
50.0	0.97
100.0	0.92
200.0	0.71
500.0	0.37
1,000.0	0.21
2,000.0	0.10
5,000.0	0.04
10,000.0	0.02

If the capacitor $C = 0.033\ \mu F$, determine the value of the resistor R.

Figure 7.47

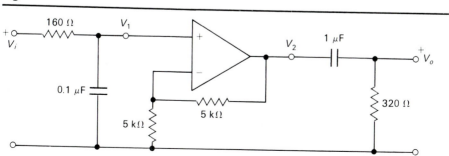

160 Ω

V_1

$+$
V_i

0.1 μF

5 kΩ

5 kΩ

V_2

1 μF

320 Ω

$+$
V_o

7.31 Consider the circuit shown in Fig. 7.47.
    a.   Qualitatively speaking, what does this circuit do to an ac signal?
    b.   Make a Bode plot of the amplitude ratio $|V_1/V_i|$.
    c.   Add a Bode plot of the amplitude ratio $|V_2/V_i|$ to the graph of part (b).
    d.   Add a Bode plot of the amplitude ratio $|V_o/V_i|$ to the graph of parts (b) and (c).

7.32 Referring to Fig. 7.36, show that the maximum power is transferred when $R_L = R_s$.

References

1. Geldmacher, R. C. Ballast circuit design. *SESA Proc.* 12(1): 27, 1954.

2. Meier, J. H. Discussion of Ref. (9), in same source, p. 33.

3. Wheatstone, C. An account of several new instruments and processes for determining the constants of a voltaic circuit. *Phil. Trans. Roy. Soc. (London)* 133: 303, 1843.

4. Wheatstone's bridge. In *Encyclopedia Britannica*, vol. 23. Chicago: William Benton, Publisher, 1957, p. 566.

5. Laws, F. A. *Electrical Measurements.* 2nd ed. New York: McGraw-Hill, 1938, p. 217.

6. Bowes, C. A. Variable resistance sensors work better with constant current excitation. *Instrument Technol.:* March 1967.

7. Sion, N. Bridge networks in transducers. *Instrument Control Systems:* August 1968.

8. Hague, B. *Alternating Current Bridge Methods.* London: Pitman, 1938.

9. Lenkurt Electric Co. dB and Other Logarithmic Units. *The Lenkurt Demodulator* 15: 4, 1966.

10. Keast, D. N. *Measurements in Mechanical Dynamics.* New York: McGraw-Hill, 1967.

11. Kistler Instrument Corporation. *Kistler Piezo-Instrumentation General Catalog.* 1st ed. Amherst, N.Y.: Kistler Instrument Corp., 1989.

12. Morrison, R. *Grounding and Shielding Techniques in Instrumentation.* 2nd ed., New York: John Wiley, 1977.

CHAPTER 8

|||||||||||||||||||||||||||||||||||||||||||||||||||||||||||||||||||||||||||||||||

Digital Techniques in Mechanical Measurements

8.1   Introduction

In this chapter we will discuss some of the basic uses of digital logic and circuitry as they apply to mechanical measurements, but it must be understood at the outset that our purpose is not to attempt an in-depth coverage of digital electronics: The breadth of the subject and space limitations prevent such a treatment. Rather, our intent is to survey the subject sufficiently so that those in fields of engineering other than electrical will gain some appreciation for the advantages, disadvantages, and general workings of solid-state circuitry, especially of the pulsed logic type.

Most measurands originate in analog form. An analog variable or signal is one that varies with time in a smooth and uniform manner without discontinuity. In many cases the amplitude is the basic variable; in others, the frequency or phase might be. A common example of a quantity in analog form is the ordinary 170 V, 60 Hz power-line voltage [Fig. 8.1(a)]. An analog signal, however, need not be simple sinusoidal or periodic in form. The stress-time relationship accompanying a mechanical shock [Fig. 8.1(b)], is considered analog in form. The pressure variations associated with the transmission of the human voice through the air are also analog. In addition, the readout from the ordinary D'Arsonval meter is considered analog because of the possibility of the uninterrupted movement of the pointer over the scale. An analog scale can be compared to the range of brightness between black and white, including all the variations of gray in between. Digital information, on the other hand, would permit only the time variation of the two brightnesses, black or white.

Figure 8.1 Examples of voltage-time relationships for analog-type
signals (a) and (b), and a digital signal (c)

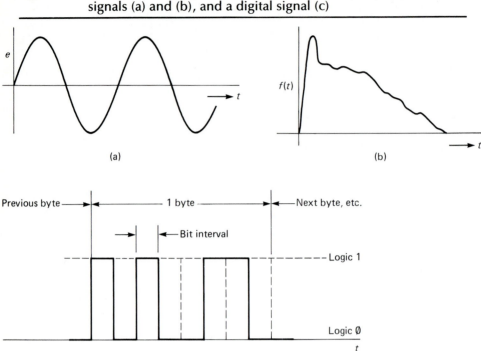

(a)

(b)

(c)

Digital information is transmitted and processed in the form of *bits* [Fig. 8.1(c)], each bit being defined by (1) one or the other of two predefined "logic levels," and (2) the time interval assigned to it, called a *bit interval*. The most common basis for the two logic states is predetermined voltage levels, say 0 and 5 V dc. Current or shifts in carrier frequency are also used. The time rate of the bits is closely controlled, commonly by a crystal-controlled oscillator (or *clock*). The intelligence is then carried by specially coded bit groupings, coded in predetermined sequences; for example, alphanumeric data may be handled by sequences of three, four, or more bits sent in the various possible combinations, each combination or group forming a *word* of information. The term *byte* is applied to an 8-bit word, whereas the term *word* may be applied to *any* unit of digital information. A 16-bit word is two bytes in length. A 4-bit word is sometimes referred to as a *nibble*. (Coding practices are discussed in Section 8.6.)

Figure 8.1(c) shows one possible combination of bits grouped to form one byte. In this case, the sequence of bit values is 1Ø1Ø Ø11Ø. From this we see that a bit need not be a pulse in the sense that it must be a *completed* off/on/off sequence. Indeed a byte of information could well be ØØØØ ØØØØ or 1111 1111, in which case no bit-to-bit changes occur. One bit corresponds to either of two different logic states held constant for one bit interval.

Because most measurement inputs originate in analog form, some type of analog-to-digital (A/D) converter (or ADC) is usually required (Section 8.12). In certain instances it may also be desirable or necessary to use a digital-to-analog (D/A) converter (or DAC) somewhere in the measurement chain. A sophisticated example of digital information handling is *pulse-code modulation* (PCM) of the human voice. The spoken word is converted to digital form for transmission over telephone circuits and then converted back to its original analog form at the receiving end. Some of the motivations for this apparently circular process are discussed in Section 8.2.

8.2 Why Use Digital Methods?

One's first contact with digital instrumentation may occur with devices that provide digital rather than analog readouts. The digital voltmeter (DVM) provides direct numerical display of voltage, whereas the D'Arsonval meter uses the analog pointer and scale. The more advanced DVMs also provide automatic *scaling*—i.e, decimal positioning. Obvious advantages of digital readouts are that interpolation is not required and that the reading is direct and precise, thereby minimizing errors caused by misinterpretation.

A much greater advantage of the digital technique in instrumentation undoubtedly lies in the fact that digitally based devices may be coupled easily with each other and with either a general-purpose or a dedicated computer. A *dedicated* computer is one that is programmed for a specific application; it may be quite simple and inexpensive. The advent of the integrated-circuit central processing unit (CPU) (Section 8.10) has done much to make this result possible.

Computer-based instrumentation makes the recording and printout of data easy, but of equal importance is that many steps necessary in data reduction can be handled automatically. For example, temperatures, flows, etc., acquired from a process may be combined immediately and reduced to provide on-the-spot overall results. This feature does not preclude the automatic recording of the elemental data items for additional study; perhaps to pinpoint the cause of some anomaly. In addition, the outputs may be processed to provide system control. (See Section 8.14.)

Digital instrumentation has additional advantages. Digital signals are inherently noise-resistant. The informational content of the digital signal is *not* amplitude-dependent. Rather, it is dependent on the particular sequence of on/off pulses that apply. Therefore, so long as the sequence is identifiable, the *true and complete* form of the input remains unimpaired. Maintenance of accuracy, lack of distortion induced by signal processing and noise pickup, and greater stability are all enhanced in comparison to analog methods. The nature of the digital signal and circuitry permits the signal to be regenerated or reconstituted from point to point throughout the processing chain. The voltage amplitude of the informational pulses is commonly 5 V dc. Unless noise pulses approach this magnitude—a highly unusual condition—they are ignored. This is of particular value in the central control of a large processing system, such as a refinery or power plant, where signals must be relayed over relatively great distances, perhaps a mile or more. This advantage is even more obvious when radio links are used, as in the ground recording of signals originating from a space vehicle.

Digital circuits imply relatively low operational voltages; 5 and 12 V dc are typical. This contrasts with several hundred volts commonly required by many older analog circuits.

A digital measurement system may hold no particular advantage in cost when a comparison is based on instrumentation alone. However, if savings in a technician's time and the increased reliability and accuracy are considered, digital instrumentation may very well hold a marked cost advantage, particularly as the complexity of the particular problem being solved increases.

8.3   Digitizing Mechanical Inputs

To be digitally processed, analog measurands must be

1. converted to yes/no pulses;
2. coded in a form meaningful to the remainder of the system; and
3. synchronized so as to mesh properly with other inputs or control or command signals.

Meeting these three requirements is collectively referred to as *interfacing*.

When some form of computer is a part of the system, not only must the input be converted to digital pulses, but also the pulses must be converted to the language used by the computer, that is, *binary words*. In addition, of course, the computer is unable to give undivided attention to any one signal source: It will also be receiving inputs from other sources, processing the inputted data, and outputting data and control commands. Input from any one source must wait its turn for attention. In other words, all the inputs and outputs must be synchronized through proper interfacing. Before attempting further coverage of interfacing, however (see Sections 8.11, 8.12, and 8.13), we will discuss some of the fundamentals and a few of the simpler types of digital instrumentation.

Single digital-type instruments whose end purpose is to simply *display* the magnitude of an input in digital form (as opposed to an input to be interfaced into a system) often require only that the input be transduced to a frequency. Conventional transducers may be used to sense the magnitude of the measurand and to convert it into an analogous voltage. The voltage can then be amplified and, by a *voltage-controlled oscillator* (VCO), transduced to a proportional frequency. A frequency-measuring circuit (Section 8.7.3) might then be calibrated to display the magnitude of the input. There are many transducers in the field of mechanical measurements that produce voltage outputs, e.g., strain-gage bridges, thermistor bridges, differential transformers, thermocouples, etc. In addition, mechanical motion, both rotational and translational, may often be quickly, easily, and completely converted to digital voltage pulses by proximity transducers (Sections 6.12 and 6.17), photocells (Section 6.16), optical interrupters (Section 6.16), and so on.

8.4 Fundamental Digital Circuit Elements

8.4.1 Basic Logic Elements

An ordinary single-pole single-throw (SPST) switch [Fig. 8.2(a)] is a digital element in its simplest form. When actuated, it is capable of producing and controlling a *yes/no,* on/off sequence. The ordinary electromechanical relay [Fig. 8.2(b)] is a slightly more-advanced digital device in which an electrical input may be used to change the output condition. More sophisticated switching devices are the triode electronic vacuum tube and the transistor [Fig. 8.2(c)]. When properly biased, a transistor can be made to conduct or not to conduct, depending on the input signal. It is a near-ideal switching device. It can function at relatively low control voltages, is capable of switching at rates of hundreds of MHz, can be made extremely small and rugged, is inexpensive, and does not require a heat-producing filament. Initially, discrete transistors were hardwired into the various circuits. More recently, they have been integrated with other simple elements such as resistors and diodes into special-purpose building blocks—the *integrated-circuit chips,* or ICs. Figure 8.3 illustrates the outward appearance of a typical chip. The one shown has 14 pins, that is, input/output connections. Others may have as many as 40 or more.

A multitude of special-purpose IC chips are available to the electronic design engineer. Their variations and complexity are growing at a tremendous rate, and much of electronic circuitry is rapidly being converted to their use. Only a few simple elements in various combinations are combined to form most chips. Special shorthand symbols are used to depict their various operations; we discuss several in the following paragraphs. Each symbol represents a group of solid-state elements—transistors, diodes, resistors, etc.—combined to perform the indicated functions. In addition, combinations of logic elements are normally assembled in various configurations in a single chip. The so-called MSI (medium-scale integration) chips may incorporate 50 to 100 individual components per chip. The LSI (large-scale integration) chips contain more than 100.

Figure 8.4 illustrates symbols for common logic elements used in various combinations to form many of the IC chips. Elements 8.4(a) through 8.4(f) are also called *gates.* Figure 8.4(g) represents a simple inverter. Also shown are *logic,* or *truth, tables,* which list all the possible combinations of inputs and

Figure 8.2   Digital switching devices: (a) a simple mechanically operated switch, (b) electrically controlled relay, (c) a transistor-type switch. For the latter, when the input is "high" the transistor conducts, thereby effectively shorting the output to ground, hence providing near-zero or "low" output. When the input is low the transistor does not conduct, thereby providing near +5 V dc at the output. Additional arrangements are possible if signal inversion is not desired.

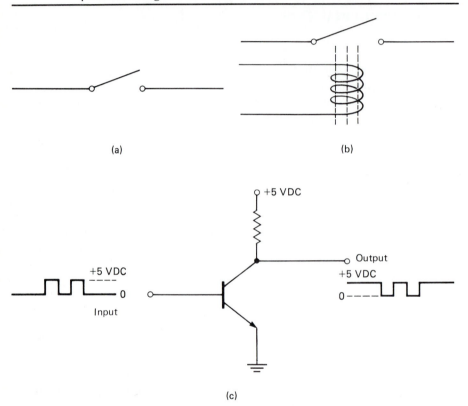

Figure 8.3   Outward appearance of a typical 14-pin IC chip. Others may have as few as six pins or as many as 40. This construction is often referred to as a DIP, or "dual in-line package."

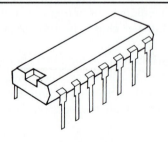

Figure 8.4 Symbols for some common digital logic units. Also shown are the respective truth tables for the units.

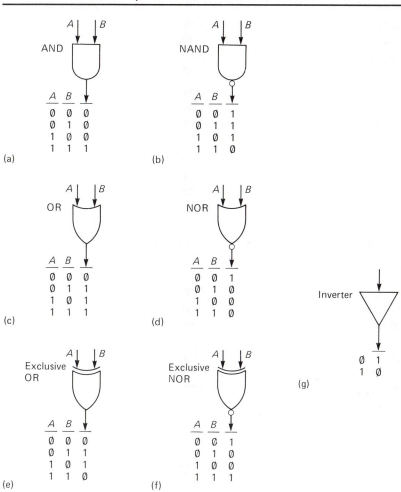

their corresponding outputs. Recall that basic digital operations are based on simple yes/no, 1/Ø, states. For example, the truth table for the AND gate shows that the output is high *only* if the inputs to *both* A and B are high, hence, the *AND rule:*

> For the AND gate, *any* low input will cause a low output; that is, *all* inputs must be high to yield a high output.

In like manner, we can also easily state rules for the other elements. Truth tables are of particular importance to the circuit designer, especially when combinations of circuits increase their complexity.

Figure 8.5   A schematic diagram showing the internal
structure of a NAND gate.

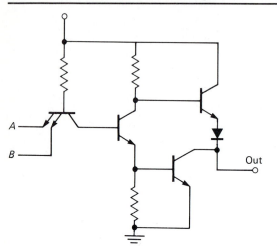

As a matter of interest, the NAND gate actually contains four transistors, three resistors, and a diode, as shown in Fig. 8.5. Suppose a chain of pulses is applied to input *B* of the AND gate. We can see that their passage may be controlled by input *A*. If *A* is high, the chain will be permitted to pass; if low, the chain will be stopped. From this simple example the origin and significance of the term *gate* is clear.

The various gates may be expanded to provide more than two inputs; see, for example, Fig. 8.6. The IC shown is a three-input NAND gate, for which a zero input level at any one of the input ports permits the passage of pulses from any other port; in other words, all inputs must be high in order for the

Figure 8.6   NAND gate with three inputs. NAND gates are available with as many as eight inputs, permitting 256 input combinations.

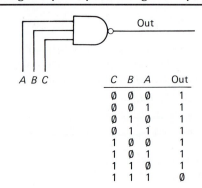

C	B	A	Out
0	0	0	1
0	0	1	1
0	1	0	1
0	1	1	1
1	0	0	1
1	0	1	1
1	1	0	1
1	1	1	0

output to be low. With this arrangement, a combination of several control conditions must be met simultaneously to block passage of a signal.

8.4.2 Some Simple Combinations of Logic Elements

The Flip-Flop

Figure 8.7(a) shows two NAND gates connected to form a very useful circuit. As a simple flip-flop, element A is called the SET gate and element B, the RESET gate. Consideration of the individual truth tables, along with their particular interconnections, shows that the following are the only workable conditions:

	S	R	Q	Q'
Condition I	1	1	1	0
Condition II	1	1	0	1
Condition III	0	1	1	0
Condition IV	1	0	0	1

Now suppose that both S and R are initially at logic 1; then either of Conditions I or II may exist, depending on random or programmed preconditions. If either of two outputs can correspond to a given input, the input is referred to as being *bistable*. In some contexts, the circuit is called a *latch*.

Let us momentarily ground input S—that is, impose logic 0. Condition III, called the SET condition, will result. This is true regardless of whether the circuit is initially in Condition I or II. Return of S to the high state will cause no change in Q or Q': It is *latched*.

Now, if R is momentarily grounded, Condition IV will be instituted, and this state will continue even when R is returned to logic 1. This is called RESET, and we see that the outputs are caused to flip and flop between SET and RESET.

The circuit has various important uses. For example, as a latch it may be used to hold (latch) a count in an electronic events counter, then await a RESET input for initiating the next count. The flip-flop is used as a memory cell, capable of holding one bit of information for later use. It also provides the basis for the *switch debouncer*. When an ordinary electric switch depending on mechanical contacts is closed, numerous contacts are actually made and lost before solid contact is finalized. In a counting circuit, for example, this switch hash cannot be tolerated. By placing a flip-flop or latch in the switch circuit [Fig. 8.7(b)] we cause the latch to respond to that first momentary contact and then to ignore all that follows, until a RESET signal reinitializes the circuit. These are only several of the uses to which the circuit may be applied; and the particular circuit discussed is only one of a number of different circuits referred to as flip-flops.

Figure 8.7   (a) Two NAND gates configured to form a flip-flop, or latch, (b) a switch-debouncer circuit

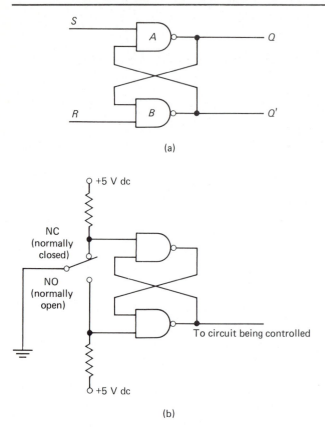

(a)

(b)

8.4.3   IC Families

There are several *families* of integrated chips, each having special characteristics but, in general, all performing essentially the same basic tasks. Common groups are the *resistor-transistor-logic,* or RTL, group; the *diode-transistor-logic,* or DTL, group; the *transistor-transistor-logic,* or TTL, group; and the *complementary metal oxide semiconductor,* or CMOS, group. In each family the various logic units are combined to perform special functions. For example, the TTL family consists of more than 150 different types of chips. Table 8.1 is a partial list. All have more or less the same outward appearance (see Fig. 8.3), but each is designed to perform a different function. Schematics of several of the TTL family are illustrated in Fig. 8.8. The circuits in Fig. 8.8(a), (b), and (c) are simple enough that the functional performance symbols may be shown. The circuitry of Fig. 8.8(d), however, is so complex (because of the number of elements used) that we have made no attempt to

Table 8.1 A partial listing of TTL IC chips

Type Number	Description
7400	Quad 2-input NAND gate
7404	Hex inverter
7408	Quad 2-input AND gate
7414	Hex Schmidt trigger
7416	Hex driver, inverting
7430	8-input NAND gate
7432	Quad 2-input OR gate
7445	BCD to 1-of-10 decoder driver
7447	BCD to 7-segment decoder driver
7474	Dual D-edge-triggered flip-flop
7475	Quad latch
7483	4-bit full adder
7485	4-bit magnitude comparator
7489	64-bit (16 × 4) memory
7490	Decade counter
7492	Base-12 counter
74121	Monostable multivibrator
74150	1-of-16 data selector
74154	1-of-16 data distributor
74181	Arithmetic unit (CPU)

indicate the internal architecture. Application of most of the chips selected for listing in Table 8.1 is covered in later sections of this chapter.

8.4.4   Readout Elements

In some situations we may require only simple, single bulbs or lights for an indication or readout. In other cases we may use combinations of discrete lamps. These may be filament types or, more often, one of the elements discussed below.

Alphanumeric readout elements normally require one of the following: (a) the cold-cathode nixie type, (b) the neon type, (c) the liquid-crystal diode (LCD) type, and (d) the light-emitting diode (LED) type. Types (a) and (b) require relatively high voltages not otherwise needed in solid-state circuitry; hence they have been largely superseded. The liquid crystal has a decided advantage in some cases, as, for example, in a digital watch requiring very low power consumption. A disadvantage is the need for proper external illumination. In instrumentation, the LED is undoubtedly the most common of the four. Figure 8.9 shows an arrangement of a typical LED seven-segment digital

Figure 8.8 Diagram showing four typical IC chips: (a) quad 2-input AND gate, (b) hex inverter (six independent inverter elements), (c) dual 4-input NAND gate, (d) BCD (binary-coded decimal) to 7-segment decoder-driver

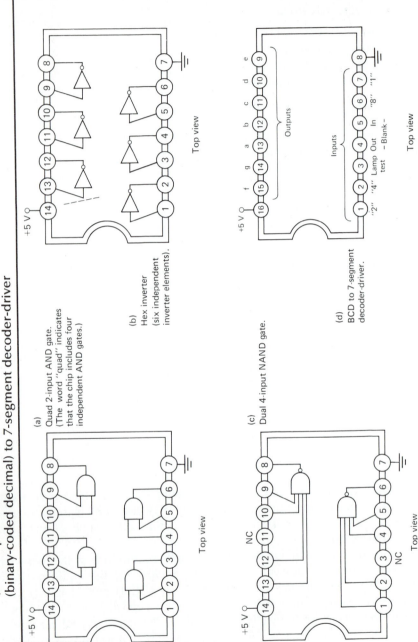

(a)
Quad 2-input AND gate.
(The word "quad" indicates that the chip includes four independent AND gates.)

(b)
Hex inverter
(six independent inverter elements).

(c)
Dual 4-input NAND gate.

(d)
BCD to 7-segment decoder-driver.

Figure 8.9 Typical solid-state numeric display. For example, grounding
pin *a* through a suitable voltage dropping
resistor lights segment *a*.

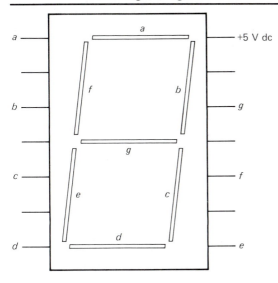

display. Each segment consists of one or more LED elements and with proper
switching supplied by an appropriate IC driver, each of the decimal digits, Ø
through 9, can be formed. Quite commonly, the input signal and a 5 V dc
power source are all that is required for power.

We have presented a smattering of some of the simplest building blocks
used to form the more complex IC circuits. Again we state our purpose for this
chapter as being to give mechanical engineers a somewhat better understanding of the digital devices that are becoming increasingly important in their
measurement systems.

8.5 Number Systems

Whether it is in digital or analog form, a measurement signal conveys a
magnitude. Magnitudes are expressed in numbers and numbers imply some
sort of numbering system or structure. Digital devices with their high/low,
yes/no, on/off sequencing suggest the use of a base 2, or binary, system for
counting. In this section we will present the essentials of number systems other
than decimal that are pertinent to digital operations.

We know that the *position* of each of the digits in a decimal number is important. For example, consider the decimal number 347.25. The 3 is the most significant digit and the 5 is the least significant digit. We know that the number can be expanded to read

$$347.25 = 3 \times 10^2 + 4 \times 10^1 + 7 \times 10^0 + 2 \times 10^{-1} + 5 \times 10^{-2}.$$

It is clear that the various positions of the digits determine the power to which the base ten is raised.

For the binary number, only two different digits are required, a 1 and a $\emptyset$. The digit 1 may correspond to a *high* condition, say $+5\,\mathrm{V}\,\mathrm{dc}$, and the digit $\emptyset$, to a *low* condition, say $0\,\mathrm{V}\,\mathrm{dc}$. We have used these designations in an earlier discussion. In the binary system, as in the decimal system, position has meaning. Consider the binary number 11010.01_2. (We add the subscript 2 to make it clear that we are using the binary system.) The first digit, 1, is the most significant digit and the last digit, 1, the least significant. Each digit is called a *bit* in the sense that it supplies an elemental "bit" of information. Hence, the terms *most significant bit,* or MSB, and *least significant bit,* or LSB, are used. What corresponds to the decimal point in the decimal system is called the *binary point* in the binary system.

As we have observed, in a decimal number each position corresponds to an integral power of 10. By the same token, in a binary number, each position corresponds to an integral power of 2. Whereas the coefficients for the decimal number could be anything between 0 and 9, for the binary system we are limited to $\emptyset$ and 1.

Let's convert the binary number written above to the equivalent decimal number. *Equivalent,* of course, means that the two numbers signify the same true magnitude or quantity (it may be convenient to refer to Table 8.2).

11010.01_2

$$= 1 \times 2^4 + 1 \times 2^3 + 0 \times 2^2 + 1 \times 2^1 + 0 \times 2^0 + 0 \times 2^{-1} + 1 \times 2^{-2}$$

$$= 16 + 8 + 0 + 2 + 0 + 0 + \frac{1}{4}$$

$$= 26.25_{10}.$$

We can see that the positional significance of the 0s and 1s lies in the integral powers to which the base (also called the *radix*) is raised. This is true for both the binary and the decimal systems.

Two other systems are commonly used in digital manipulations: the *octal,* or base 8, system; and the *hexidecimal,* or base 16, system. These two systems have the marked advantage over the decimal system in the ease and convenience of their conversion, by either machine or human, to binary.

For further discussion of number systems, see Appendix C.

Table 8.2 Decimal values of bases raised to various powers

n^*	2^n	4^n	8^n	6^n
⋮	⋮	⋮	⋮	⋮
−3	1/8	1/64	1/512	1/4096
−2	1/4	1/16	1/64	1/256
−1	1/2	1/4	1/8	1/16
0	1	1	1	1
1	2	4	8	16
2	4	16	64	256
3	8	64	512	4,096
4	16	256	4,096	65,536
5	32	1,024	32,768	1,048,576
6	64	4,096	262,144	16,777,216
⋮	⋮	⋮	⋮	⋮

* n is both the *power to which the base is raised* and the *positional weight*.

8.6 Binary Codes

In addition to the binary number, which is, of course, limited in magnitude only by the number of positional bits that may be arbitrarily permitted, there are various binary *codes*. At this point we will consider several of them.

Suppose we limit the number of positions to four, i.e., we provide only four on/off switches or their equivalents. We are then limited to the decimal range of 0 to 15 inclusive. The result is what is known as the four-bit *word*, also referred to as a *four-level code*.

A modification of this code is known as *binary-coded decimal*, or *BCD*. Although BCD is also a 4-bit word, it arbitrarily makes illegal all words greater than 1001_2, or 9_{10}. We see that the BCD code is used because of its convenient relationship to the decimal digits 0 through 9. Table 8.3 lists equivalencies. When the BCD code is used each digit in a decimal number is processed separately as a 4-bit sequence (e.g., the decimal number 875_{10} translates to $1000\,0111\,0101_2$).

These codes are for the transmission and processing of numeric data. Alphanumeric information must also include provision for the letters of the alphabet and perhaps certain other symbols, such as punctuation marks.

One of the simplest binary codes for transmission of general information (as opposed to numeric data only) is the International Morse Code, which uses *pulse-duration modulation*, or PDM (also sometimes called pulse-length or pulse-width modulation). The two different pulse widths, the "dot" and the "dash," are used in various combinations to transmit the alphabet, the decimal digits, and certain other special-purpose telegraphic symbols. International Morse Code is an *uneven-length* code in that various time intervals are

Table 8.3 Four-bit binary codes

Decimal Digit	Four-Bit Binary Equivalent	Binary-Coded Decimal Equivalent
0	0000	0000
1	0001	0001
2	0010	0010
3	0011	0011
4	0100	0100
5	0101	0101
6	0110	0110
7	0111	0111
8	1000	1000
9	1001	1001
10	1010	Illegal
11	1011	Illegal
12	1100	Illegal
13	1101	Illegal
14	1110	Illegal
15	1111	Illegal

required for the transmission of the various characters. This code is not important for mechanical measurements.

The *Baudot,* or common teletypewriter, code is a five-unit, *even-length code.* All characters are formed by a combination of five possible on/off states, each of *uniform* duration. In addition to the five informational intervals, a *start* pulse is required because the idling state is zero logic and the receiving machine would otherwise be unable to distinguish between an unimportant idling bit and an information bit for any character that might begin with a zero condition. Also needed is a stop pulse for synchronization. In the traditional teletype machine the pulse sequences are mechanically produced by cam-actuated switch contacts selected by the keys on the keyboard. At the receiving end an ingenious combination of electrical relay and kinematics selects the proper type-bar for actuation.

A popular computer code is known as the *American Standard Code for Information Interchange (ASCII).* This code is of 7-bit binary form (Table 8.4). Both the upper- and lowercase letters of the English alphabet, plus the decimal digits zero through 9 and certain other control symbols, are included. When writing the binary equivalents of the ASCII code, one generally divides the binary number into two groups of three and four bits each; for example, for the letter *a,* the binary equivalent is written 110 0001. The lefthand binary digit is the MSB (most significant bit) and the righthand, the LSB. In processing, the ASCII number for *a* would appear as 110 0001. As a function of time it would appear as shown in Fig. 8.10.

The International Morse and the Baudot codes are necessarily of a *serial* type, i.e., the various on/off states occur in sequence, one following another in "bucket-brigade" fashion. If they were transferred by an electrical conductor, only a single transmission circuit would be required. In some cases, an alternative to serial transmission (namely, *parallel* transmission) is used. This means that the bits in a single word are transmitted simultaneously. In its

Table 8.4 The American Standard Code for Information Interchange (ASCII)

Binary	Hexadecimal	Character	Binary	Hexadecimal	Character
000 0000	*00*	*NUL*			
·			011 1000	38	8
	Nonprinting control characters		011 1001	39	9
·			011 1010	3A	:
010 0000	20	Space	011 1011	3B	;
010 0001	21	!			
010 0010	22	"	011 1100	3C	<
010 0011	23	#	011 1101	3D	=
			011 1110	3E	>
010 0100	24	$	011 1111	3F	?
010 0101	25	%			
010 0110	26	&	100 0000	40	@
010 0111	27	'	100 0001	41	A
			100 0010	42	B
010 1000	28	(	100 0011	43	C
010 1001	29	)			
010 1010	2A	*	100 0100	44	D
010 1011	2B	+	100 0101	45	E
			100 0110	46	F
010 1100	2C	·	100 0111	47	G
010 1101	2D	−			
010 1110	2E	,	100 1000	48	H
010 1111	2F	/	100 1001	49	I
			100 1010	4A	J
011 0000	30	0	100 1011	4B	K
011 0001	31	1			
011 0010	32	2	100 1100	4C	L
011 0011	33	3	100 1101	4D	M
			100 1110	4E	N
011 0100	34	4	100 1111	4F	O
011 0101	35	5			
011 0110	36	6			
011 0111	37	7			

continued

Table 8.4 *continued*

Binary	Hexadecimal	Character	Binary	Hexadecimal	Character
101 0000	50	P	110 1000	68	h
101 0001	51	Q	110 1001	69	i
101 0010	52	R	110 1010	6A	j
101 0011	53	S	110 1011	6B	k
101 0100	54	T	110 1100	6C	l
101 0101	55	U	110 1101	6D	m
101 0110	56	V	110 1110	6E	n
101 0111	57	W	110 1111	6F	o
101 1000	58	X	111 0000	70	p
101 1001	59	Y	111 0001	71	q
101 1010	5A	Z	111 0010	72	r
101 1011	5B	[	111 0011	73	s
101 1100	5C	\	111 0100	74	t
101 1101	5D	]	111 0101	75	u
101 1110	5E	^	111 0110	76	v
101 1111	5F	—	111 0111	77	w
110 0000	60		111 1000	78	x
110 0001	61	a	111 1001	79	y
110 0010	62	b	111 1010	7A	z
110 0011	63	c	111 1011	7B	{
110 0100	65	d	111 1100	7C	‖
110 0101	65	e	111 1101	7D	}
110 0110	66	f	111 1110	7E	~
110 0111	67	g	111 1111	7F	Rubout

simplest form, this type requires as many transmission circuits as there are code levels, but obviously it saves time. We shall see later that occasionally the one form may be converted into the other. Parallel circuitry is often used within a single instrument or device, whereas serial circuitry is used when long cables are involved, as, for example, for time-sharing via telephone-interfaced computer circuits.

Figure 8.10 The ASCII logic sequence for the letter "a"

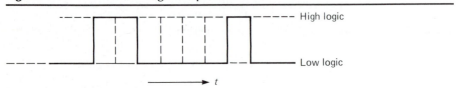

8.6.1 Encoders

Figure 8.11(a) shows a schematic diagram of one type of simple binary displacement encoder. The "card" shown consists of five active tracks plus a reference track, which may or may not be needed. Pickups (not shown) sense the relative displacement. One sensor is used per track. Optical sensors are most commonly used to pick up on/off pulses as the relative motion takes place. The cards may be transparent or opaque for use with transmitted, or reflected, light, respectively. Printed-circuit methods may also be used. Output from the pickups may be fed to lamp-type indicators or to computer circuitry for data processing. Figure 8.11(b) illustrates a circular card offering similar possibilities for rotation.

Figure 8.11 (a) Binary displacement encoder, (b) circular encoder card

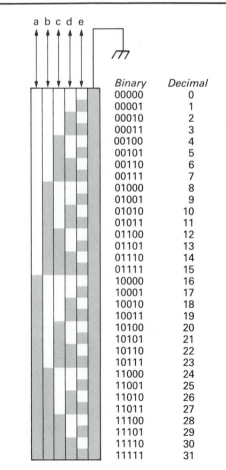

Binary	Decimal
00000	0
00001	1
00010	2
00011	3
00100	4
00101	5
00110	6
00111	7
01000	8
01001	9
01010	10
01011	11
01100	12
01101	13
01110	14
01111	15
10000	16
10001	17
10010	18
10011	19
10100	20
10101	21
10110	22
10111	23
11000	24
11001	25
11010	26
11011	27
11100	28
11101	29
11110	30
11111	31

(a)

continued

Figure 8.11 *continued*

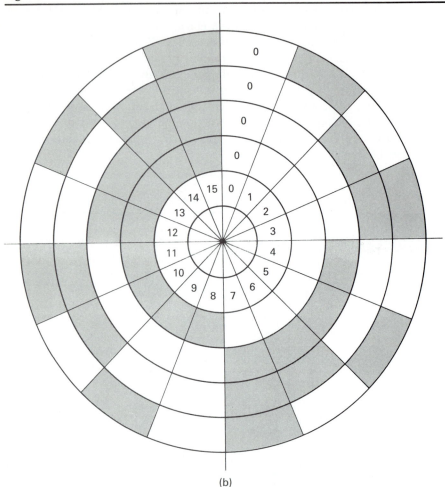

(b)

8.6.2   Bar Code Encoders

Everyone is familiar with at least some of the applications of bar codes. In this case the familiar black/white lines are arranged to codify data of various sorts—addresses, inventory, and virtually any information that can be put in concise binary code.

8.7   Some Simple Digital Circuitry

8.7.1   Events Counter

Figure 8.12 shows the outward simplicity of one form of digital events counter. Each decade consists of a seven-segment LED display, two IC chips from the TTL family, and a few current-limiting resistors. The 7490 chip is called a

Figure 8.12   Schematic circuit for a simple digital events counter

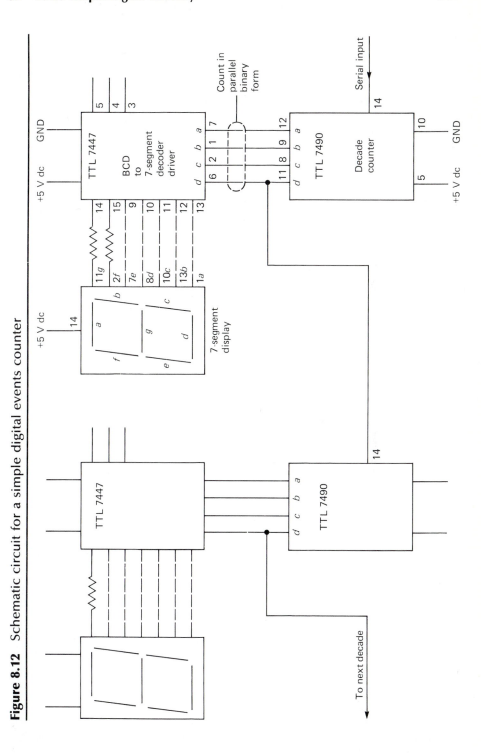

decade counter (see Table 8.1). It accepts input pulses in serial form through pin 14 and sends as output parallel BCD pulses through pins *d, c, b,* and *a,* where *d* corresponds to the most significant bit and *a* to the least significant. In addition, this chip provides an output pulse at the *end* of the ninth input pulse. This pulse is available from pin 11 and can be used as the input to the next higher counting decade. We can see that decades may be cascaded easily. As we will note later, this facility obviously also provides a "divide-by-ten" capability.

The BCD outputs from the 7490 chip are fed to a 7447 chip, called a BCD-to-seven-segment decoder driver (Table 8.1). This chip accepts the parallel BCD input, decodes the input—i.e., converts it to a unit decimal output—and switches on (grounds) the appropriate segments in the seven-segment readout, thereby displaying the required decimal digit. This simple circuit is capable of counting at rates of up to about 100 kHz. In addition to this function, IC 7447 also provides control terminals 3, 4, and 5, which may be used for

1. resetting the readout to zero;

2. blanking unused leading zeros—e.g., making provision so that a five-decade display of the number 25, for instance, would not show as 00025 but simply as 25, with the unnecessary zeros blanked;

3. making provisions for displaying appropriate decimal points.

8.7.2 Gating

Suppose the counter circuit shown in Fig. 8.12 is to be used to count a sample of pulses stemming from a continuing sequence of pulses. Can we not use a simple mechanical switch to *gate* the input line? Probably not directly! When the contacts of the mechanical switch (or relay) are closed, there exists a period of indecision. The contacts touch, break, touch again, etc., many times before a final and complete contact is established. This is the source of *switch hash*, which shows on the screen of a cathode-ray oscilloscope. The counter is unable to distinguish between the "good guys" and the "bad guys" and the speed of the counter is sufficient to count them all. Under such circumstances it is necessary to use *debouncing circuitry* (Section 8.4.2) in the input, which recognizes the first contact, latches, and then ignores the succeeding contacts.

8.7.3 Frequency Meter

Electronic switching or gating would probably be used to switch the counting circuit on and off. Recall the AND gate (see Fig. 8.4a). The input pulses may be connected to input *A* and a control input to *B*. There will be an output only when both are high. This provides an essential part of a digital *frequency meter*. Suppose we introduce an unknown frequency at *A* and control *B* with a square-wave oscillator (Fig. 8.13). When both inputs are high, the counter will

Figure 8.13 Schematic diagram for a simple frequency meter circuit

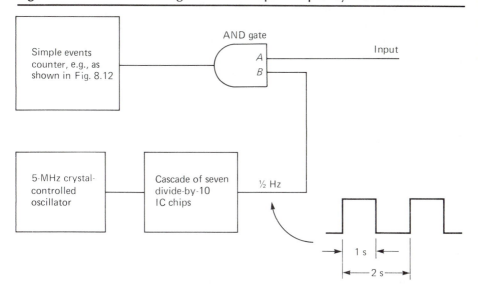

count. When B is low, counting will stop. Obviously the accuracy of the count will be dependent on the accuracy of the gated time interval. The oscillator will probably be crystal-controlled, and crystals oscillate at relatively high rates. For instance, the fundamental oscillator might have a frequency of 5 MHz. The period of $1/10\,\mu s$ would certainly be too short for gating most mechanical inputs. Recall, however, that we already have discussed a divide-by-ten IC, the IC 7490 used in the simple counter. By cascading seven divide-by-ten ICs we can reduce the fundamental period to 2 s, 1 s of which will be high and 1 s low. By using an AND gate as shown in Fig. 8.13, we can sample a second's worth of the input and hence its frequency in hertz. Obviously, with the cascade of divide-by-ten chips we can use panel-controlled switches for different gating times by simply tapping the cascade at other points in the chain.

In the preceding example we have oversimplified the process somewhat. We have not provided for returning to zero between samplings, and we may perhaps obtain controlled single-pulse gating intervals by better means—for example, through use of a monostable oscillator. As we stated earlier in this chapter, our purpose, however, is to give some feel for digital techniques to the mechanical engineer, rather than to cover the multitude of possibilities in depth.

8.7.4   Wave Shaping

Digital logic circuits prefer instant toggling from low to high and back again. Suppose we wish to feed the simple counter we described above with a sine wave or some other cyclic waveform. If the level is sufficient, our counter may work, but then again it might not. We can convert the sine wave to a pulsed signal by using a *Schmidt trigger* (see TTL 7414 in Table 8.1, for instance). It accepts a gradually rising (or falling) input but does not become conducting until its "trigger level" is reached. It then becomes fully conducting (or non-conducting on the down side of the cycle). We are therefore able to reshape the input into a series of nearly true square-wave pulses, much preferred by the common IC. If the input is complex, we can use an attenuator to reduce the peaks in relation to the trigger level. In this case the time characteristic of the waveform would have much to do with our success or lack thereof.

8.7.5   An IC Oscillator or Timer

Figure 8.14 illustrates an interesting IC oscillator—the 555 chip [1]. The single chip contains 23 transistors, 15 resistors, and 2 diodes. The package is about half the size of a common postage stamp, has eight terminals, and costs less than a cup of coffee and a doughnut. When configured with two or three external resistors and a capacitor, it becomes capable of either monostable or astable oscillation. When in astable oscillation it is capable of covering a frequency range from 0.1 Hz to over 100 kHz, with square-wave output. With

Figure 8.14   Example of "special" IC chips: a 555 astable/
monostable or multivibrator oscillator

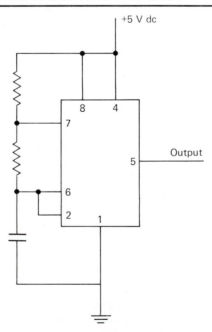

additional outboard circuit elements, it is capable of producing triangular and linear ramp waveforms. Chips such as the 555 are used to provide a time-base reference for a wide range of devices, and they are sometimes referred to as *timers*. The IC oscillator is an example of the wide versatility of a single special-purpose integrated circuit.

8.7.6 Multiplexing and Demultiplexing

Figure 8.15(a) illustrates an IC chip called a *multiplexer,* or a 1-of-16 data selector (see TTL 74150, Table 8.1). In mechanical terms it may be considered a selectable commutator; quite simply, it is similar to a single-pole, 16-position switch [Fig. 8.15(b)]. The particular input (of up to 16) that is permitted to pass is determined by the binary number inserted at the *d, c, b,* and *a* ports, where *d* is the MSB and *a* is the LSB.

The demultiplexer, or 1-of-16 data distributor, is similar but with reversed action (see TTL 74154, Table 8.1). It accepts a digital signal through its one input and then routes it to the particular output selected by the binary value at the control ports.

Why concern ourselves with multiplexers and demultiplexers? In many cases continuous monitoring or recording of a given data source is not necessary: Periodic sampling will suffice. We can see that by sequencing the binary control through 0 to 15, we can consecutively connect 16 different inputs to a given readout/recording/computing system. This approach is economical in many cases.

Multiplexing may also be used, in certain instances, within the circuitry of a single instrument. Recall the events counter we discussed in Section 8.7.1. Each decade required a seven-segment decoder driver plus seven current-limiting resistors. More sophisticated counter circuitry uses a multiplexer-demultiplexer combination arranged so that each readout element is sequentially connected to a single driver-resistor combination. By time sharing in this manner we can reduce the number of circuit elements and realize quite a saving in cost. The sequencing rate is sufficiently high that the readout appears to be continuously illuminated.

Data processing may take place at quite some distance from the data source, as in a large industrial complex such as a refinery or power plant. By using multiplexer-demultiplexer combinations, we may also use single, rather than separate, circuits to connect the two positions (Fig. 8.16). We hasten to add that this may not be quite the case because it would also be necessary to synchronize our binary control circuits at the two locations. A simple solution is to run four additional wires connecting ports *d, c, b,* and *a*. Thus we would have 5 circuits instead of 16 (for the particular combination cited).

There is, however, another possibility through use of more sophisticated ICs. The universal asynchronous receiver/transmitter (UART) contains what amounts to a multiplexer and a demultiplexer—actually a set of two each—on a single chip. It is capable of converting serial to parallel or parallel to serial

Figure 8.15   (a) The TTL 74150 multiplexer or 1 of 16 selector. Each of the 16 possible logic combinations of the *dcba* control port is employed to select a corresponding signal input for connection to the output. (b) Illustration of a nearly equivalent mechanical switching arrangement. A difference, however, lies in the fact that the mechanical device must switch through the sequence in order, whereas the multiplexer need not.

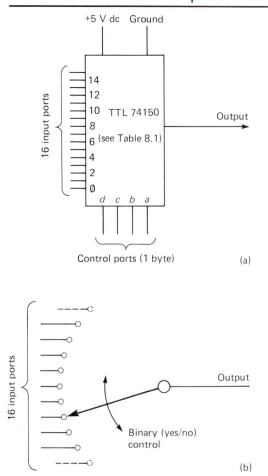

data. The word asynchronous indicates that the operation need not be synchronized. Actually, this is not quite true. What is meant is that the inputs and outputs need not be *perfectly* synchronized: Approximate synchronization, say ±5%, is sufficient. This permits the use of separate control oscillators, or "clocks," at the transmitting and receiving ends, whose frequencies are quite

Figure 8.16 A multiplexer-demultiplexer circuit

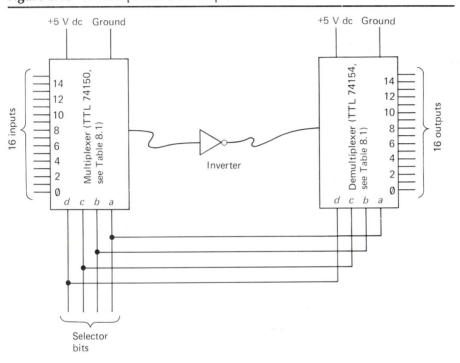

close but not necessarily exactly the same. In this case, directly connected synchronizing circuits between the two locations are not necessary.

8.8 The Digital Computer as a Measurements System Tool

Small computers, either dedicated or general purpose, are a natural adjunct to a measurements system. They may be used simply to monitor and record one or more measurands, or they may serve as an interactive part of a more comprehensive measurements control system.

To connect common measurement circuitry to a computer requires considerably more understanding of the functioning of both systems than it would first appear. Most measurands originate as analog-type signals; however, the computer recognizes only digital quantities. As a result, often the first requirement is to change the language by means of an analog-to-digital converter. In addition, data from circuitry external to the computer must be fed to the computer in a form and order that the computer is prepared to accept. The computer's connections to the outside (its ports) are generally bidirectional. That is, they may either receive or output data, but only in a well-defined and orderly manner. The computer must be *told* through software whether a particular port is to handle incoming or outgoing information. Software assignment of port functions is often called *configuring the ports*. Single lines may be all that are required for handling simple limit-switch/warning-light information, but multiple lines may be necessary for processing analog-converted inputs. In addition, input/output may be handled in either serial or parallel form.

In the next few sections we provide an introduction to the use of computers, specifically microcomputers, for mechanical measurements. Since the subject is so broad we limit our discussion to the basic requirements.

8.9   Data Processors, Computers, Microcomputers: The Computer Hierarchy

To some degree the term computer is misused. In its strictest sense, the word restricts the device to arithmetical manipulations. A more appropriate term is *data processor*. However, because of common usage, we will use the two terms interchangeably.

Consider the following:

Microprocessor: A single chip using digital methods to manipulate numerical data. To be practical the unit requires the services of supporting ICs.

Data processor: A microprocessor-based system of memory, input/output devices, some form of resident operating program, power supply, etc., all tied together by buses, thus forming a system for gathering, sorting, rearranging, and analyzing data, including storage and/or printout.

Logic elements represent the most basic type of computer. The elements may be electromechanical or one or more of the various basic integrated circuits. An overvoltage relay is a simple example. It is set to some maximum voltage that, if exceeded, initiates an action—a switch opening accompanied, perhaps, by a latching. An input is thereby permitted to pass only under a certain set of conditions, i.e., provided a voltage is not too great. An op-amp voltage comparator (Section 7.17.2, Example 7.8) is another simple device whose output is determined on the basis of relative input values. These examples are mentioned simply to describe the least complex data processing.

A *programmable calculator* is intended for number manipulation. It is limited in both the number of steps that may be programmed and the speed with which it operates. In its common form it is constructed to make no more than the most elementary decisions. Sometimes its programs are volatile (they are lost when power is removed); more often, it will possess a nonvolatile memory.

The next step in the hierarchy of data processing is the *microcomputer* or micro data processor. It has the ability to make decisions, to control outputs on the basis of inputs, and to serve, within limits, as a data processor. It is capable of manipulating memories of several thousand words. Its most common form is based on a 16-bit word size, although some use 8 bits and, more recently, some have been introduced with 32-bit capabilities. With proper peripherals, the microcomputer can handle higher languages. The heart of the microcomputer is the LSI (large-scale integration) IC chip, the *microprocessor.*

The microcomputer may be either *general purpose* or *dedicated.* The latter is designed and programmed for a specific application such as controlling a solar heating system or an automobile fuel injection system. Costs may range from as low as $100 for a simple dedicated type to as much as $10,000 (depending on the array of peripheral devices).

The *minicomputer* is one step above the microcomputer. It can handle as many as 32 million 16-bit words and costs from $5,000 up. In large applications such as those required by a chemical processing plant, the minicomputer might very well be served by one or more microcomputers, and, in turn, the minicomputer would feed information to a large-scale computer for processing beyond the minicomputer's capability. The large general-purpose computer, for the most part, has capabilities beyond the direct requirements of mechanical measurement data acquisition and processing and is mentioned here merely to complete our description of the hierarchy. We see, of course, that an initial problem lies in the selection of the proper level of computer required to do a specific job.

We limit the following discussion to the use of microcomputers. Micro-computers are based on microprocessors, which, in turn, must have the support of a family of LSI chips. There are a number of microprocessor families available: the 8086 family of Intel Corporation, the 6800 and 68000 families of Motorola, and others. In general, each system performs similarly; however, in detail, each system uses its own particular set of operational codes. In the following sections, as we refer to specifics, we will use the Motorola 6800 system as our example.

8.10 The Microprocessor

As we stated earlier, the heart of a microcomputer is the LSI IC chip, the *central* processing unit (CPU), or *microprocessor* unit (MPU). Basically, it serves as the control center for directing the flow of digitized information. It is more than a traffic controller because it not only provides the organizational plan for the flow of elemental bits of information but also assigns the pathways, temporary "parking" spaces, and stop/go gating and can perform a limited manipulation of the traffic. It accepts inputs in digital form, either data or command instructions, and routes them to predetermined (programmed) destinations over buses (pathways) to displays, memories, controllable devices, etc. Sources of the data or commands may be external memories, keyboards, transducers, and so on.

There are a wide range of CPUs available, with many levels of complexity—4-bit, 8-bit, 16-bit, 32-bit, and so on. The CPU selected for a given purpose will depend on the application. For some simple low-end dedicated uses a 4-bit CPU may suffice, whereas 16-bit or 32-bit CPUs are selected for general-purpose or high-end microcomputers. To accomplish the intent of this chapter—that is, to provide an introduction to digital manipulation and devices—the 8-bit Motorola 6800 family serves very well. The basic principles involved are common to all levels of CPUs.

Figure 8.17 is a highly simplified schematic diagram of the external connections and some of the internal features of the Motorola 6800. The diagram shows the primary buses into and out of the processor plus some essential internal devices. To be functionally useful, the system requires additional supporting circuitry external to the CPU, including interfacing devices sometimes referred to as buffers, input/output (I/O) facilities, synchronizing clocks, etc. When combined with sufficient peripherals (devices external to the CPU), the combination becomes a microcomputer.

The Motorola 6800 MPU is a single LSI IC chip housed in a dual-in-line package (DIP) similar to, but larger than, the one shown in Fig. 8.3. The dimensions, exclusive of the pins that provide electrical connections, are about $\frac{1}{4} \times \frac{3}{4} \times 2$ in. The power requirement for the MPU alone is 5 V dc at from 0.6 to 1.2 W.

Figure 8.17   Simplified schematic of the Motorola 6800 microprocessor

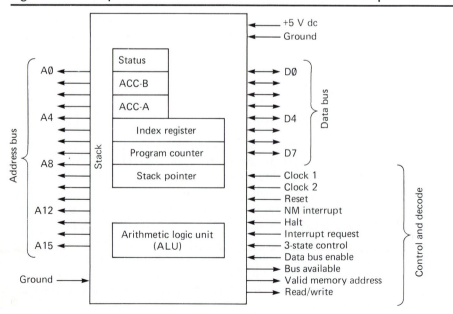

The figure shows the following:

1. An *address bus* consisting of 16 parallel lines for accessing (connecting to) $2^{16} = FFFF_{16} = 65,536_{10}$ different memory locations. The actual number available is, of course, dependent on what may be provided by supporting hardware.

2. A *data bus* consisting of eight bidirectional lines for simultaneously handling 8 bits (one byte) of data. This bus is a two-way street, the direction of flow being controlled by gating (Section 8.7.2). Eight bits provide for $2^8 = FF_{16} = 256_{10}$ combinations.

3. Various *control and decode lines*. Bus synchronization, closely akin to multiplexing-demultiplexing (Section 8.7.6), is controlled through use of two external clock signals. Additional lines are shown, their titles in many cases providing a clue to their uses.

Figure 8.17 also shows some of the internal structure of the 6800 CPU. Various *registers* are used. These may be considered as *temporary* storage bins or momentary "parking" locations for data, instructions, or addresses as they are being shunted from one location to another. Essentially, all data and instructions must pass through at least one of the accumulators, *A* or *B,* each having a one byte (8-bit) handling capability. Primarily for providing manipulation of 2-byte (16-bit) addresses, the *index register* (IR), *program counter* (PC), and *stack pointer* (SP) are added.

As illustrated in Fig. 8.17, the *stack* consists of the registers as shown. Their contents are stored in contiguous addresses, the purpose of the stack pointer being to keep track of where the stack information is stored in the external random access memory (RAM). The *program counter* controls the sequencing of any program steps, including the starting point, and the *index register,* in addition to other functions, provides a channel through which two-byte addresses may be handled in a program. The contents of the registers are always available on command.

The arithmetic logic unit (ALU) has a relatively limited capability of manipulating numbers. It can add one byte to another or determine their difference, or it can perform several logic functions, such as AND, OR, and EOR (Section 8.4.1). To handle data in magnitudes requiring several bytes, the add and subtract functions of the ALU may involve carryovers (for addition) or borrows (for subtraction). It is the responsibility of the status register (SR) (also called the condition code register) to monitor such requirements for possible further program use. *Flags* (primary data bits) in the status register are either set (made equal to 1) or not set (made equal to $\emptyset$), depending on predetermined conditions. If an addition results in a carryover from one byte to the next byte, a bit momentarily indicating that fact, stored in the status register, will be added to the byte of the next higher order. Or, if the operation results in a zero, a zero flag in the SR may be used to trigger the decision to branch (or not to branch) to some other point in a program. These are only two of a number of functions of the status register. It is quite powerful and its full importance cannot be covered in a limited summary such as ours.

8.11 The Microcomputer

As we have stated several times, a microprocessor must be surrounded by a number of servants before it can claim to be a microcomputer. Figure 8.18 is a schematic diagram of the Motorola 6800 system. In addition to the MPU, some combination of the following peripherals is required.

8.11.1 Read-Only Memory (ROM)

Figure 8.18 shows a single ROM; there might be others. ROM contains what is called the *monitor,* or the *executor.* The term executor is particularly apt, because therein are contained the microcomputer's operational orders in the form of various subroutines required for organizing the system and ensuring proper operation. Examples of such subroutines are

1. address building;
2. interrupt sequencing;
3. memory examine and exchange;
4. power-up sequence;
5. code interpretation [e.g., operational instructions, provision for outputting ASCII (Section 8.6) when required];
6. PIA input/output (see Section 8.11.3).

Figure 8.18 The essential parts of a microcomputer

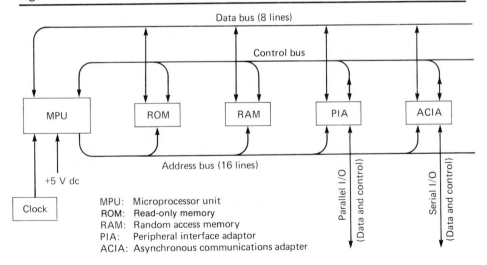

The complete program for a dedicated computer (one assigned a single task only) would also be contained in ROM.

8.11.2 Random Access Memory (RAM)

RAM is the "bank" that is used for temporary deposits and withdrawals of data or information required for the establishment of programs. The program itself may be held in either ROM or RAM. Should the program require modification from time to time, it would require random access, hence RAM. In both RAM and ROM the data or operational instructions are held in the form of single 8-bit bytes.

8.11.3 Peripheral Interface Adapter (PIA)

The PIA provides one form of bridge between the computer and the outside. It handles data in *parallel* fashion in that all 8 bits of each byte of information or data are processed simultaneously. It is obvious, then, that eight separate lines into and out of the PIA are required.

Figure 8.19 shows a simplified schematic diagram of the Motorola 6820 PIA. Its operation is actually rather complex. The device provides two separate sections, each serving a *primary* 8-bit port. The in/out buses, designated A and B, with lines PAØ to PA7 and PBØ to PB7, are shown on the right side of the diagram. *Each* separate line in each port can be made to serve as single input or output lines. Any combination of input/output may be had, as required. For example:

1. lines PAØ, PA1, and PA5 could be selected as inputs for receiving information from outside the computer and the remaining lines used (or not) as output lines; or

2. all A-lines could be made outputs and all B-lines inputs; or

3. any other combination could be used.

Software assignment of the basic line functions is called *configuring*. At appropriate locations in the operating program, the in/out lines must be configured. This leads to two basic and different in/out requirements, as follows:

1. Many situations simply require the handling of limit signals. For example, the input may be simply "yes, a certain event has occurred," or "no, it has not." The two conditions can be signified by a binary 1 for the yes and a binary Ø for the no. In response, through appropriate program manipulation, the output may be simply to trip a relay, light a warning lamp, sound a siren, or shut down the entire plant. In these cases, only *single* input/output lines are required. The ins and outs need not be paired. There can be many inputs, all used to trigger a single output. In an automobile, overheating at any of a number of locations, underpressure in

Figure 8.19   Schematic diagram of a peripheral interface adapter. Not shown are the input/output control terminals or power connections.

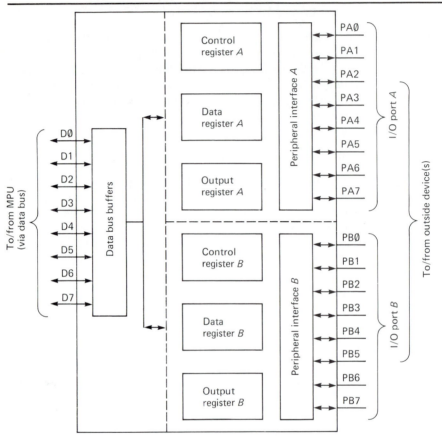

oil systems, a hood unlatch, etc., could all be used to trigger either a warning lamp, an audible sound, or perhaps an ignition shutoff.

2.   On the other hand, the input may stem from an analog quantity from which the application requires the monitoring, sorting, or printout of the input data. In this case a number of lines, the number determined by the desired *resolution,* are required. For example, an analog-to-digital converter (Section 8.12) might provide full 8-bit bytes of information. Then the digital equivalents of the analog input could vary over a range from $00_{16} = 00_{10}$ to a maximum of $FF_{16} = 256_{10}$. This would permit a resolution of 1 in 255_{10} (often rounded to 250_{10}) or 0.4%. We see that in this case the entire primary port, say port A, would be absorbed in handling the single data source. Likewise, an 8-bit *output* would require a second full primary

port, say port *B*. It should be clear that should a lesser resolution be satisfactory, then fewer data lines may be used. For example, four lines would provide a resolution of 1 in 15, or about 7%.

Programming requirements to configure the lines is beyond the scope of this book and must be left to reference material (see especially [2, 3]). Suffice it to say that the data direction register and the control register may be addressed through software, and through proper program manipulation, configuring can be accomplished. In fact, should it be desired, a given line, say PA8, could accept *inputs* for *part* of a program reconfigured by software to provide outputs for an additional part of a program.

Most microcomputers can accommodate multiple PIAs; hence a wide range of flexibility is possible.

8.11.4   Asynchronous Interface Adapter (ASCIA)

The ASCIA is a second type of connection, or bridge, into and out of the computer. Whereas the PIA handles data in parallel form, the ASCIA does so serially. Serial transmission of digital information is generally used between such devices as a computer and keyboards, video displays, printers, and the like. Particularly in the case of printers and keyboards, the time required of each mechanical operation is so much greater than that required by the computer that there is no need for speed of transmission. Serial handling is also used for long-line transmission, where cost of multiple parallel lines becomes prohibitive. Often a UART (Section 8.7.6) is used to convert from the parallel format used by the computer to the serial format for output.

We recommend that the reader consult the Suggested Readings at the end of the chapter for details on ASCIA performance and configuration.

8.12   Analog-to-Digital and Digital-to-Analog Conversion

As we discussed in Section 8.3, some measurands originate in digital form. Most mechanical inputs, however, exist in analog form. Hence, before digital data processing can be accomplished, an analog-to-digital conversion is necessary. In a like manner, if a computer's digital output is used to drive an analog device, a digital-to-analog conversion must be performed.

Analog-to-digital (A/D) and digital-to-analog (D/A) conversions can be executed using a variety of different circuits. A/D and D/A converters are often manufactured as integrated circuits, and their prices range from as little as $10 to as much as several thousand dollars, depending on the number of bits used, the speed of processing, and other features. Here we describe only typical examples of each, leaving further development to the Suggested Readings.

8.12.1   A Digital-to-Analog Converter

A simple D/A converter is based on the summing amplifier (Example 7.9, Section 7.17.2). Referring to Fig. 8.20, we see that the currents summed are controlled by a set of digital switches. Four basic elements are involved:

1. A stable reference voltage (for instance, $E_{ref} = 1$ V dc).

2. A ladder arrangement of summing resistors. For this 8-bit converter, eight ladder resistors are used. The resistance values increase in a sequence of powers of 2 from 2^0R to 2^7R.

3. A series of switches. These are not mechanical switches; rather, they are solid-state gates [e.g., simple AND ICs (Section 8.4.1)]. The eight switches can be activated by TTL inputs, so their operation may be controlled by the respective bits contained in a single byte of data.

4. Op-amp output circuitry. The op-amp output voltage is equal to R_G times the sum of the currents from each ladder branch. The gain controlling resistor, R_G, is selected to *scale* the output voltage, perhaps to a maximum of 10 V dc.

When a switch is closed, a current is delivered to the op amp in proportion to the power of 2 for that circuit's resistor: Switch 0 contributes $i_0 = E_{ref}/128R = 2^{-7}E_{ref}/R$, switch 1 contributes $i_1 = E_{ref}/64R = 2^{-6}E_{ref}/R$, and so on. Hence, switch 0 corresponds to the least significant bit, b_0, switch 7 corresponds to the most significant bit, b_7, etc. By closing selected switches, the output voltage can be made proportional to any particular 8-bit number. For example, if switches 0, 5, and 6 are closed, the output voltage is

Figure 8.20   A simple 8-bit DAC (digital-to-analog converter)

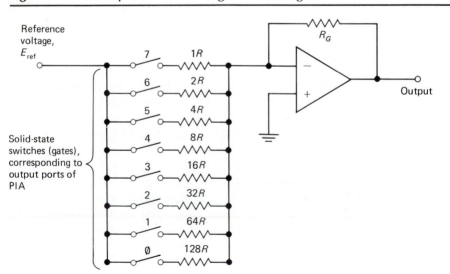

proportional to $0110\,0001_2$ $(= 61_{16} = 97_{10})$. Since R_G sets the constant of proportionality, the output gain can be scaled to provide an appropriate range of analog values.

8.12.2 An Analog-to-Digital Converter

One typical A/D converter is the *parallel encoder* (Fig. 8.21). This circuit uses a set of voltage comparators (Example 7.8, Section 7.17.2) and a series of resistors to simultaneously compare an analog input signal to a set of reference

Figure 8.21 A 3-bit parallel A/D converter

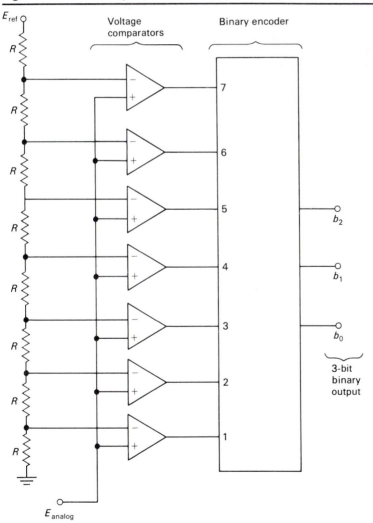

voltages. The basic elements are as follows:

1. A stable reference voltage (such as $E_{ref} = 1$ V dc).

2. A series of equal resistors. These resistors form a voltage-dividing sequence between E_{ref} and ground. For this 3-bit converter, $2^3 = 8$ resistors are used. The voltages at the nodes between resistors increase in increments of $(R/8R) \cdot E_{ref}$, specifically, $(\frac{1}{8})E_{ref}, (\frac{2}{8})E_{ref}, \ldots, (\frac{7}{8})E_{ref}$.

3. A set of voltage comparators. The analog voltage is simultaneously compared to each node's voltage. A comparator's output voltage is high (on) when E_{analog} is above a given reference voltage and low (off) when it is below. Seven $(2^3 - 1)$ comparators are needed for the 3-bit converter.

4. An encoder circuit. The encoder reads the comparator outputs (each a high or low voltage), and produces a 3-bit binary output corresponding to one of the eight possible on/off conditions of inputs 1 through 7. (Note that one condition is to have all seven comparators off.)

For example, if the input signal is between $(\frac{3}{8})E_{ref}$ and $(\frac{4}{8})E_{ref}$, comparators 1–3 read high and 4–7 read low. The input state is three, corresponding to a binary output of 011_2 $(=3_{10})$. If all comparators read high $(E_{analog} > (\frac{7}{8})E_{ref})$, the output is 111_2; and if all are low $(E_{analog} < (\frac{1}{8})E_{ref})$, it is 000_2. The digital output is shown as a function of E_{analog} in Fig. 8.22. Parallel A/D conversion is particularly fast, since all bits are set simultaneously, for which reason it is sometimes called *flash encoding*.

Figure 8.22   Binary output versus analog input voltage for the 3-bit parallel A/D converter

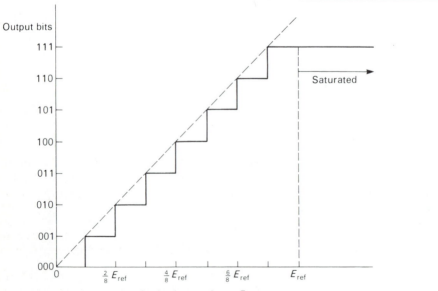

8.12.3 Analog-to-Digital Conversion Considerations

Saturation Error. The most obvious limitation of an A/D converter is that it has definite upper and lower limits of voltage response. Typical full-scale ranges are 0 to 10 V and -10 to $+10$ V. If the input signal exceeds the upper or lower limits of response, the converter saturates and the recorded signal does not vary with the input. This situation can be prevented by appropriate signal conditioning, such as amplitude attenuation or dc offset removal.

Resolution and Quantization Error. As we have seen, an A/D converter responds to discrete changes in the input voltage. For example, the 3-bit parallel encoder's output steps correspond to changes in E_{analog} of $E_{ref}/2^3$ (Fig. 8.22). Thus there is a smallest increment of voltage change that can be resolved by an A/D converter. In general, the voltage resolution per bit, ε_V, depends on the full-scale voltage range and the number of bits of the converter:

$$\varepsilon_V = \frac{\Delta V_{fs}}{2^n}, \tag{8.1}$$

where

$$\Delta V_{fs} = \text{the full-scale voltage range},$$

$$n = \text{the number of bits of the A/D converter}.$$

Typical A/D converters have 8, 12, or 16 bits, corresponding to division of ΔV_{fs} into a total of $2^8 = 256$, $2^{12} = 4096$, and $2^{16} = 65{,}536$ increments. A 16-bit converter with a -10 to $+10$ V range has a voltage resolution of 0.3 mV. The voltage resolution is a known value for a given A/D converter, and it is normally needed when processing the digitized data.

The finite resolution of the A/D converter introduces error in the recorded values, since the actual analog voltage usually lies between the available bit levels. This is called *quantization* error (since the digital data are "quantized"), and it is entirely analogous to the reading error of a digital display (Section 3.2.2). An estimate for the *quantization uncertainty* is $u_q = \varepsilon_V/2$ (95%). Quantization errors may be reduced by using an A/D converter with more bits.

Conversion Errors. A/D converters may also suffer from slight nonlinearity, zero-offset errors, scale errors, or hysteresis. Such errors are a direct by-product of the particular method of input quantization. For example, the conversion illustrated in Fig. 8.22 is, on the average, low by one-half of the least significant bit (the amount by which the solid curve lies beneath the dashed line). Normally, the manufacturer will provide specifications for the potential size of such conversion errors.

Sample Rate. The rate at which an A/D converter records successive values of a time-varying input is called the sample rate. Each A/D converter has a maximum possible sample rate, of which typical values range from about 1000 Hz to more than 100 MHz. Software often allows the user to specify any sample rate up to this maximum value. The influence of sampling rate on the accuracy of the recorded signal is discussed in Sections 4.6.1 and 4.6.2.

Signal Conditioning for A/D Conversion. To make the best use of an A/D converter, conditioning of the analog signal is often required. The most important considerations are prevention of *aliasing* (Section 4.6.2), minimization of quantization errors, and prevention of saturation errors. Aliasing can be prevented by using a low-pass, or *antialiasing,* filter to remove frequencies of $f_{\text{sample}}/2$ or more from the analog signal. Quantization error can be minimized by amplifying the signal to span as much of the full-scale range as possible. However, this approach sometimes conflicts with the need to avoid saturation errors.

For example, a hot-wire anemometer signal (Section 15.10) may include both a 3 V dc component and important ac components of only 5 mV rms. A ±10 V, 16 bit A/D converter is to be used for sampling. The ac components are near the A/D converter's resolution (=0.3 mV), but large amplification of the sum of the ac and dc components will saturate the converter. To resolve the ac components accurately, an offset-and-gain (or "buck-and-gain") amplifier may be used. This amplifier subtracts (or "bucks") a precisely specified dc voltage from the analog signal and amplifies the remaining ac signal by a gain of several hundred. The dc offset voltage is recorded with one A/D converter channel and the amplified ac component is recorded with another. In this way, both components can be accurately digitized.

8.13 Buses

When designed to be compatible, black boxes such as power supplies, voltmeters, counters, amplifiers, computers, and the like may be connected to form a *measurement system.* The system may be tailored for specific measurement or control duties and programmed to provide near-automatic functioning. To assemble such a system requires the use of *buses,* the cabling over which the devices communicate. Basically, the buses must provide for (1) data and/or command transfer, plus (2) whatever communication or synchronization for proper functioning is required. The latter requirement is commonly referred to as *handshaking* between the units.

A given device may be a transmitter (talker), a receiver (listener), a controller, or any combination of the three. In the first case, the device is a source of information; in the second, it is a receptor; and in the third, it outputs control signals. Quite often, data lines are bidirectional; that is, information may pass in either direction over the same line. Therefore, a sequencing or synchronization of the line assignments is mandatory, and that is the handshaking function.

Handshaking is a complex matter, to say the least. Certain handshaking actions are dependent on bit level—i.e., either HIGH or LOW—whereas others are determined by the transmitting *direction,* either HIGH-to-LOW or vice versa. And all takes place at a rate of microseconds.

Various bus configurations have been "standardized"; two of these, the RS-232C and IEEE 488-1975, are briefly considered next.

The RS-232C is used for transfer of *bit-serial* data and specifies a maximum of 25 separate lines, 10 of which are seldom used. It is widely used to transmit digital data over telephone lines, between printers, keyboards, and CRT terminals. In its simplest application, the RS-232C may consist of the following:

1. TRANSMIT DATA (output from A)*
2. RECEIVE DATA (input to A)
3. SIGNAL GROUND
4. CARRIER DETECT (synchronous input to A)
5. DEVICE READY (synchronous output from A)

The IEEE 488-1975 bus is sometimes referred to as an *8-bit parallel, character-serial* bus. This means that it simultaneously transfers the 8 bits of a single character or byte of information, with the bytes (usually ASCII-coded) being transmitted in serial fashion. Obviously, the first requirement is eight separate data lines. In addition, the 488-1975 bus provides eight additional control lines, making a total of 16, as shown in Fig. 8.23. Some hint of the complexity of the bus application may be inferred from the fact that to state the present standard requires a booklet of 80 pages [3].

More and more commercially available devices are provided with bus connections. Although devices made by the same company are compatible, devices from different sources may not be. That buses are "standardized" does not always mean that such devices will work together. Bits are not always assigned to data lines in a consistent manner, control signals may or may not be similar, some systems may be ASCII-coded and others not, etc. A commercial system with a growing following is the Hewlett-Packard "HP-IB" bus configuration, where IB stands for interface bus.

It is clear from this brief discussion that system assembly may follow either of two directions: (1) Compatible commercial units may be used and simply plugged together with little or no thought about detailed bus operations, or (2) systems may be "home-brewed" for the particular application. For the former it is necessary only that the user ensure that the devices are truly compatible. For the latter, the ground-up design and use of an integrated measurements

---

* In general, at least two devices, say A and B, are assumed.

Figure 8.23   Simplified schematic of the IEEE-488-1975 bus. Devices I,
II, III, and so on may be counters, signal
generators, or voltmeters.

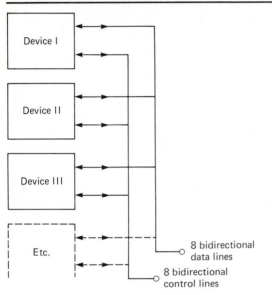

system of any but the most simple form requires specialized knowledge, in
depth, of digital techniques. As a start, the Suggested Readings at the end of
the chapter will help.

8.14   Getting It All Together

The possibilities for putting together an advanced measurements system for a
given project are entirely open-ended. The degree of completeness is usually
limited by resources. In many situations the design and assembly of a
sophisticated system may be more costly in time and money than the project
warrants. On the other hand, when great masses of data are to be collected,
requiring extensive computational time to digest, funds and time expended in
putting together an "automatic" system may very well be cost-conservative.

Figure 8.24 shows the tremendous possibilities for advanced data gathering
and processing. Since it is based on materials discussed in this chapter and in
Chapter 9, the diagram is self-explanatory.

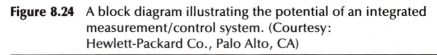

Figure 8.24 A block diagram illustrating the potential of an integrated measurement/control system. (Courtesy: Hewlett-Packard Co., Palo Alto, CA)

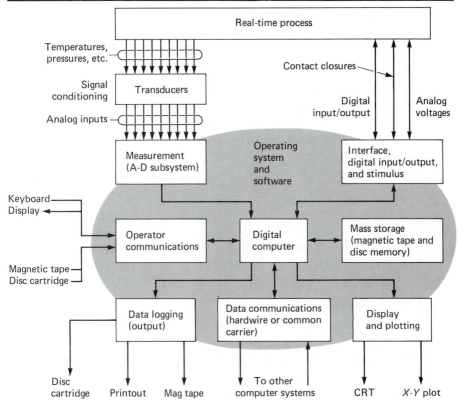

8.15 Summary

In bringing this chapter to a close, we reiterate that it is presumptuous to attempt to summarize digital techniques in so few pages. We hope, however, that the material presented will serve as an introduction to further study. There is no doubt as to the value of the topic as it relates to mechanical measurements, and further developments are continually increasing this importance.

1. A digital signal has only two values—on or off, 0 or 1, yes or no, black or white, low voltage or high. Digital information may be composed of sets of bits, each of which takes one of the two possible values (Section 8.1).

2. Digital instrumentation has the advantage of direct computer interfacing, inherent noise resistance, low system voltages, and direct numerical

display of readings (Section 8.2). Analog signals must be converted to an appropriate digital form to be processed by a digital measuring system (Section 8.3).

3. The basic elements of digital circuits are switches (transistors) and logic gates. These elements are the building blocks for more complex devices, such as flip-flops, innumerable integrated circuits, and digital displays (Section 8.4).

4. Digital information, when expressed as a string of bits, may be interpreted using the binary number system by assigning the values 0 and 1 to the bit levels (Section 8.5). Binary numbers, in turn, allow information to be expressed in digital codes, such as binary-coded decimal (BCD) and the American Standard Code for Information Interchange (ASCII). Digital patterns (strips of black and white) may be used to encode positional information or product labels (Section 8.6).

5. Simple digital circuits include events counters, frequency meters, wave-form-shaping devices, oscillators or timers, and multichannel switches, or multiplexers (Section 8.7).

6. The digital computer is a natural component of measuring systems, both for recording and processing of data. The microcomputer is composed of a microprocessor, permanent and temporary memory (ROM and RAM), and parallel or serial interface adaptors (PIA or ASCIA) (Sections 8.8–8.11).

7. Conversion of signals between analog and digital form is essential when analog transducers or devices are coupled to a digital computer or microprocessor. These operations are achieved using either analog-to-digital (A/D) or digital-to-analog (D/A) converters (Section 8.12).

8. Interconnection of digital devices in a measuring system is achieved using an appropriate communications bus (Section 8.13).

Suggested Readings

Artwick, B. A. *Microcomputer Interfacing.* Englewood Cliffs, N.J.: Prentice Hall, 1980.

Berlin, H. M. *The 555 Timer Applications Sourcebook, with Experiments.* Derby, Connecticut: E&L Instruments, 1976.

Bibbero, R. J. *Microprocessors in Instruments and Control.* New York: John Wiley, 1977.

CMOS Integrated Circuits. Santa Clara, California: National Semiconductor Co., 1979.

Data Acquisition and Conversion Handbook. Mansfield, Massachusetts: Datel-Intersil, Inc., 1979.

Dempsey, J. A. *Experimentation with Digital Electronics.* Reading, Mass.: Addison Wesley, 1977.

Driscoll, F. F. *Introduction to 6800/68000 Microprocessors.* Boston: Breton, 1987.

Ferguson, John. *Microprocessor Systems Engineering.* Reading, Mass.: Addison Wesley, 1987.

General-Purpose Linear Devices Data Book. Santa Clara, California: National Semi-conductor Co., 1989.

Hilburn, J. L., and P. M. Julich. *Microcomputers/Microprocessors.* New York: Prentice-Hall, 1976.

Hoeschele, D. F. *Analog-to-Digital/Digital-to-Analog Conversion Techniques.* New York: John Wiley, 1968.

Horowitz, P., and W. Hill. *The Art of Electronics.* 2nd ed. New York: Cambridge University Press, 1989.

Introduction to Digital Techniques. Benton Harbor, Michigan: Heath, 1975.

Lesea, A., and R. Zaks. *Microprocessor Interfacing Techniques.* Berkeley, California: Sybex, 1977.

Leventhal, L. A. *Microcomputer Experimentation with the Motorola MEK 6800D2.* Englewood Cliffs, N.J.: Prentice Hall, 1981.

Libes, S. *Digital Logic Circuits.* Rochelle Park, N.J.: Hayden, 1975.

Microprocessors. Benton Harbor, Mich.: Heath, 1977.

Motorola Semiconductor Products, Inc. *Microprocessor Applications Manual.* New York: McGraw-Hill, 1975.

Newell, S. B. *Introduction to Microcomputing.* 2nd ed. New York: John Wiley, 1989.

Roberts, R. A., and C. T. Mullis. *Digital Signal Processing.* Reading, Mass.: Addison Wesley, 1987.

Titus, J. A. *Microcomputer-Analog Converter Software and Hardware Interfacing.* Indianapolis: Howard W. Sam, 1978.

Tocci, R. J. *Digital Systems—Principles and Applications.* 4th ed. Englewood Cliffs, N.J.: Prentice Hall, 1988.

Triebel, W. A. *Integrated Digital Electronics.* 2nd ed. Englewood Cliffs, N.J.: Prentice Hall, 1985.

TTL Data Book for Design Engineers. Dallas, Texas: Texas Instruments, Inc., 1979.

Problems

8.1 Write the following binary numbers as base 10 numbers.
- **a.** 1010111
- **b.** 1010
- **c.** 1111
- **d.** 10111011

8.2 Write the following decimal numbers as binary numbers.
- **a.** 16
- **b.** 87
- **c.** 419
- **d.** 40177

Figure 8.25   AND/OR circuit for Problem 8.3

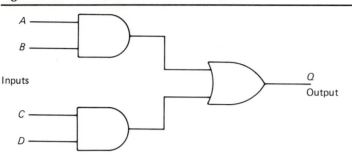

8.3   Use a truth table to show that the output from the circuit in Fig. 8.25 is high except when the two inputs to either one or both of the AND gates are simultaneously high.

8.4   Use a truth table to show that the output from the circuit in Fig. 8.26 will be high only when both inputs of either of the OR gates are simultaneously low.

8.5   Find the percent resolution of each.
    a.   An 8-bit A/D converter
    b.   A 12-bit A/D converter

8.6   The output from a temperature sensor is expected to vary from 2.500 mV to 3.500 mV. If the signal is fed to a 12-bit A/D converter having a ±5.0 V range, estimate the voltage increment represented by LSB. At what gain should the signal be amplified?

8.7   Each step of an 8-bit D/A converter represents 0.10 V. What will the output of the D/A converter be for the following digital inputs?
    a.   01000111
    b.   10101101

8.8   Use a truth table to determine under what set of conditions the output from the three-input NAND gate shown in Fig. 8.27 will be low.

Figure 8.26   OR/NAND circuit for Problem 8.4

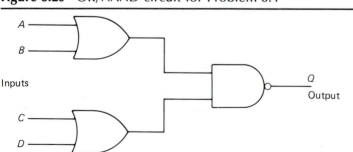

Figure 8.27 Circuit for Problem 8.9

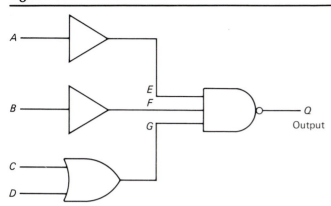

References

1. Berlin, H. M. *The 555 Timer Application Source Book with Experiments.* Derby, Conn.: E&L Instruments, Inc., 1976.

2. Southern, R. *Programming the 6800 Microprocessor.* Ottawa, Ont.: Southcroft, 1977.

3. *IEEE Standard Interface for Programmable Instrumentation.* IEEE Std. 488-1975 (also ANSI MC 1.1-1975). New York: Institute of Electrical and Electronics Engineers.

CHAPTER 9

Readout and Data Processing

9.1 Introduction

Final usefulness of any measuring system depends on its ability to present the measured output in a form that is comprehensible to the human operator or the controlling device. The primary function of the terminating device is to accept the analogous driving signal presented to it and to either provide the information in a form for immediate reading or record it for later interpretation.

For direct human interpretation, except for simple yes-or-no indication, the terminating device presents the readout (1) as a relative displacement, or (2) in digital form. Examples of the first are a pointer moving over a scale, a scale moving past an index, a light beam and scale, and a liquid column and scale. Examples of digital output are an odometer in an automobile speedometer, an electric decade counter, and a rotating drum mechanical counter.

Examples of exceptions to the two forms above are any form of yes-or-no limiting-type indicator, such as the red oil-pressure lights in some automobiles, pilot lamps on equipment, and—an unusual kind—litmus paper, which also provides a crude measure of magnitude in addition to a yes-or-no answer. Perhaps the reader can think of other examples.

As we have stated on numerous occasions in previous chapters, measurement of *dynamic* mechanical quantities practically presupposes use of electrical equipment for stages one and two. In many cases the electronic components used consist of rather elaborate systems within themselves. This is true, for example, of the cathode-ray oscilloscope. Sweep circuitry is involved, providing a time basis for the measurement. In addition, the input is carried through further stages of amplification before final presentation. The primary purpose of the complete system, however, is to present the input analogous signal in a form acceptable for interpretation. Such a self-contained system will therefore be classified as an integral part of the terminating device itself.

For the most part, dynamic mechanical measurement requires some form of voltage-sensitive terminating device. Rapidly changing inputs preclude strictly mechanical, hydraulic-pneumatic, and optical systems, either because of their extremely poor response characteristics or because the output cannot be interpreted. Therefore, the major portion of this chapter will be concerned with electric indicators and recorders.

The most basic readout device is undoubtedly the simple counter of items or events. Mechanically constructed counting devices are quite familiar. Most automobile odometers are of this type, simply counting the turns of a drive shaft through a gear reduction, which scales the readout to miles or kilometers. Modern laboratory-type counters, however, are electronic, such as those discussed in Section 8.7 and in the following section.

9.2   The Electronic Counter

The electronic counter is a multipurpose digital counting device. A basic understanding of how the electronic counter operates suggests uses for this instrument and also makes the user aware of its limitations.

To determine the principles of operation let us first consider the astable oscillator or timer (see Section 8.7.5). The output from the device is a square wave, which may be used as a timing signal [see Fig. 9.1(a)]. A second IC of importance is the J-K flip-flop, which is a variation of the S-R flip-flop discussed in Section 8.4.2. This circuit may be used as a divide-by-two device. The J-K output is triggered by the negative side of the input square wave and the flip-flops may be cascaded with the output from one providing the input for the next. For example, for each *period* of the clock input, the value of the 4-bit binary output, A3, A2, A1, A0, is as shown in Fig. 9.1(b). A3 is the most significant bit (MSB) and A0 is the least significant bit (LSB).

This circuitry is called a binary counter and is the governing circuit for the universal counter. As shown, the counter will count only to 1111_2, or 15_{10}. By cascading *eight* J-K flip-flops, the count may be increased to $1111\,1111_2$, or 255_{10}. See Section 8.7 for further discussion of electronic counters.

The schematic diagram for the electronic counter is shown in Fig. 9.1(c). The binary numbers are converted to decimals and displayed by means of light-emitting diodes (LEDs) or via a liquid-crystal display (LCD).

9.2.1   Event Counter

The event count is the simplest measurement to perform using an electronic counter. If the clock is disconnected and replaced by a device that produces a square pulse (presumably due to a single physical event), the result is an instrument that totalizes events, or an event counter, as shown in Fig. 9.2(a).

Figure 9.1   (a) Frequency-dividing circuit, (b) four-bit binary counter,
(c) schematic diagram for the electronic counter

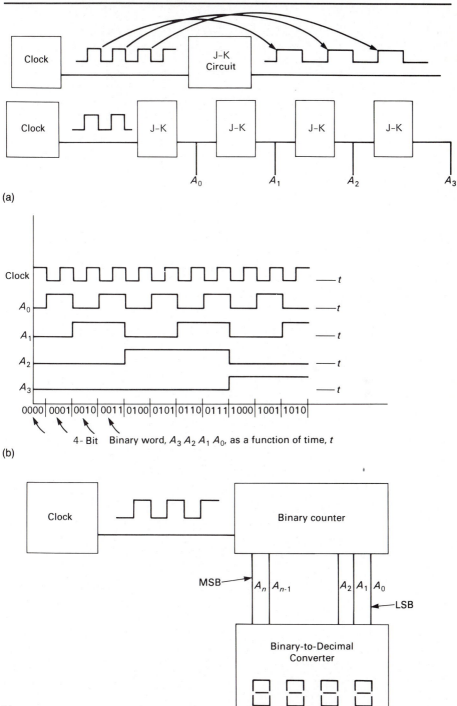

(a)

4- Bit   Binary word, $A_3 A_2 A_1 A_0$, as a function of time, t

(b)

(c)

Figure 9.2   (a) The electronic counter used as an event counter,
                (b) time-interval meter, (c) EPUT meter

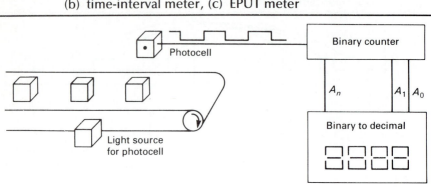

(a)

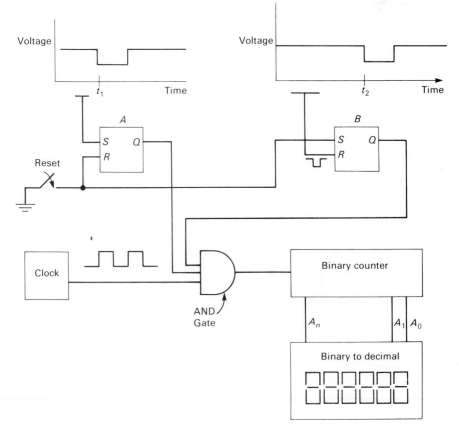

(b)

continued

Figure 9.2 *continued*

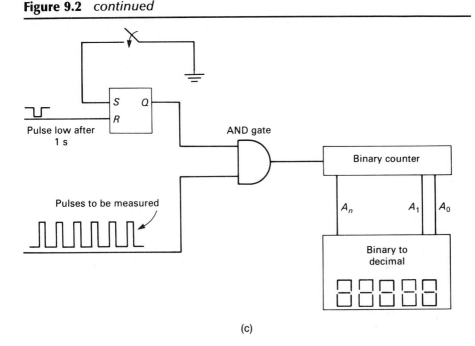

(c)

9.2.2 Time-Interval Meter

Figure 9.2(b) illustrates a binary counter with the addition of an AND gate (see Fig. 8.4) and an electronic circuit called an S-R latch [see Fig. 8.7(a)]. If the counter is armed (by pulsing the reset), $Q = 0$ on latch A and $Q = 1$ on latch B. At instant of time t_1, the voltage to latch A causes Q for latch A to go high. This, in turn, switches the AND gate on, and the clock signal passes through the AND gate, causing the counter to start counting. As soon as the voltage to latch B reaches time t_2, latch B output (Q) goes low, which stops the clock signal from passing through the AND gate. The final count is held in the counter, which represents the time $(t_2 - t_1)$. Obviously, the higher the clock frequency, the greater the accuracy of the measurement.

9.2.3 Events Per Unit Time (EPUT) Meter

By rearranging the basic elements in an electronic counter, an EPUT meter [see Fig. 9.2(c)] may be obtained. In this mode the electronic counter is configured to count pulses in a specified length of time, called the *gate time*. The gate time is the interval between closing the switch on S and receiving the low pulse at R. If the gate time is 1 s, then the number of pulses on the counter is in pulses per second.

9.2.4   Count Error

Counting accuracy may involve time base error, trigger error, and a "±1 count ambiguity." Time base error is concerned with any deviation of the time base oscillator frequency from intended frequency. This may result from a lack of both short- and long-term frequency stability. For most mechanical measurements, this error may be negligible (e.g., after proper warm-up a drift or aging rate of less than 3 parts in 10^7 per month may be attained). Trigger error is concerned with the preciseness with which the gating action is known or controlled. Uncertainties may be reduced by averaging over a longer period.

Finally, a ±1 count error in electronic counting often exists because of the normal lack of synchronization between the gating and the measured pulses (whether from internal or external sources). It results from the possibility that the gate closing (or opening) may occur so as to barely miss the count of a passing cycle, but still, in fact, include (or exclude) the greater part of that particular cycle's period. For frequency measurement this source of error may be minimized by designing the instrument to measure the *period* of a cycle and then *compute* the reciprocal and display frequency. This feature minimizes the effect of the 1 count resolution, particularly at the low frequencies where it might become serious. This method helps to maintain the accuracy of frequency measurement over the entire counter range.

9.3   Analog Electric Meter Indicators

The common analog electric meter used for measuring either current or voltage is based on the *D'Arsonval movement*. It consists of a coil assembly mounted on a pivoted shaft whose rotation is constrained by spiral hairsprings, as shown in Fig. 9.3. The coil assembly is mounted in a magnetic field, as shown. Electric current, the measurand, passes through the coil, and the two interacting magnetic fields result in a torque applied to the pivoted assembly. Rotation occurs until the driving and constraining torques balance. The resulting displacement is calibrated in terms of electric current. The D'Arsonval movement forms the basis for some electric meters and is also the basis of the stylus and light-beam oscillographs (Section 9.9).

Meter-type indicators may be classified as (1) simple current meters (ammeters) or voltage meters (voltmeters); (2) ohmmeters and volt-ohm-milliammeters (VOM or multimeters); and (3) meter systems whose readouts are preceded by some form of amplification. In the past, the latter type used vacuum tube amplifiers; hence they became known as "vacuum-tube volt-meters," or VTVMs. The abbreviation is still heard in spite of the fact that solid-state amplifiers are now used.

The simple D'Arsonval meter movement is often used as the final indicating device. However, moving-iron meters may be used for measuring alternating current. In the more versatile types, such as the volt-ohm-

Figure 9.3   The D'Arsonval meter movement

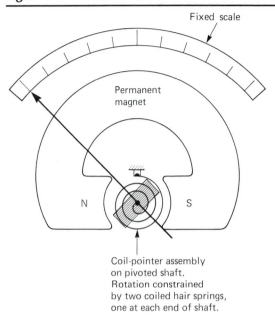

Fixed scale

Permanent
magnet

N   S

Coil-pointer assembly
on pivoted shaft.
Rotation constrained
by two coiled hair springs,
one at each end of shaft.

milliammeter or multimeter, internal shunts or multipliers are provided with switching arrangements for increasing the usefulness of the instrument.

Basically, the D'Arsonval movement is current-sensitive; hence, regardless of the application, whether it be as a current meter or as a voltmeter, current must flow. Naturally, in most applications, the smaller the current flow, the lower will be the *loading* on the circuit being measured. The meter movement itself possesses internal resistance varying from a few ohms for the less sensitive milliammeter to roughly $2000\,\Omega$ for the more sensitive microammeter. Actual meter range, however, is primarily governed by associated range resistors.

Figure 9.4 shows schematically the basic dc voltmeter and dc ammeter circuits. Either multiplier or shunt resistors are used in conjunction with the same basic meter movement. To minimize circuit loading, it is desirable that total *voltmeter* resistance be much greater than the resistance of the circuit under test. For the same reason, the *ammeter* resistance should be as low as possible. In both cases, meter movements providing large deflections for given current flow through the meter are required for high sensitivity.

9.3.1   Voltmeter Sensitivity

Voltmeter resistance is determined primarily by the series multiplier resistance. High multiplier resistance means that the current available to actuate the meter movement is low and that a sensitive basic movement is required.

Figure 9.4   (a) dc voltmeter circuit, (b) dc ammeter circuit

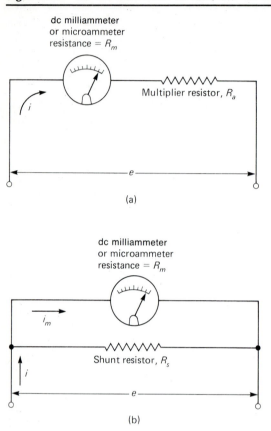

(a)

(b)

Because sensitivity may differ from meter to meter even though the meters may be of the same range, it is insufficient to rate voltmeters simply by stating total resistance. Rating is commonly stated in terms of *ohms per volt. This value may be thought of as the total voltmeter resistance that a given movement must possess in order for the application of* 1 V *to provide full-scale deflection.* This value combines both resistance and movement sensitivity, and the higher the value, the lower will be the loading effect for a given meter indication.

Simple pocket multimeters generally use a meter of 1 mA and 1000 Ω/V rating, whereas more expensive multimeters may use movements with a rating of 50 μA and 20,000 Ω/V.

The value of the series multiplying resistor, R_a, as shown in Fig. 9.4(a), may be determined from the relation

$$R_a = \frac{e}{i} - R_m. \tag{9.1}$$

9.3.2   The Current Meter

Since current meters are connected in series with the test circuit, the voltage drop across the meter must be kept as low as possible. This means that the combination of meter and shunt must have as low a combined resistance as practical. Referring to Fig. 9.4(b), we may write the following relation, based on equal voltage drops across meter and shunt:

$$R_s = \frac{i_m R_m}{i - i_m}. \tag{9.2}$$

9.3.3   AC Meters

Provision for measuring ac voltages is made by using a rectifier in conjunction with a dc meter movement. Meters of this type are usually calibrated to read in terms of the root-mean-square (rms) values (see Section 4.7). The rms value of ac current or voltage is the dc value representing the equivalent, or effective *power,* content of the corresponding ac value. This is often described in terms of heating ability. The rms current (or voltage) is the corresponding dc input that possesses the same heating ability as does the ac input. Note that, in this context, a pure resistive load is assumed.

In general, for direct current applied to a pure resistance,

$$P = I^2 R = \frac{E^2}{R}, \tag{9.3}$$

where: P = power; I and E are dc current and voltage, respectively; and R is the resistive load.

For inputs that vary with time (ac inputs),

$$P = \frac{1}{T} \int_0^T i^2 R \, dt = \frac{1}{2\pi} \int_0^{2\pi} i^2 R \, d(\omega t) \tag{9.4}$$

or

$$P = \frac{1}{T} \int_0^T \frac{e^2}{R} \, dt = \frac{1}{2\pi} \int_0^{2\pi} \frac{e^2}{R} \, d(\omega t). \tag{9.5}$$

where i and e are current and voltage as functions of time and T is the period of the cycle.

If we equate the dc powers expressed by Eq. (9.3) to corresponding powers expressed by Eq. (9.4), for effective current I_{eff} we obtain

$$I_{\text{eff}} = I_{\text{rms}}$$

$$= \sqrt{\frac{1}{T} \int_0^T i^2 \, dt} \tag{9.6}$$

$$= \sqrt{\frac{1}{2\pi} \int_0^{2\pi} i^2 \, d(\omega t)}, \tag{9.6a}$$

or, in terms of voltage,

$$E_{eff} = E_{rms} = \sqrt{\frac{1}{T}\int_0^T e^2\, dt} \tag{9.7}$$

$$= \sqrt{\frac{1}{2\pi}\int_0^{2\pi} e^2\, d(\omega t)}. \tag{9.7a}$$

For the most common case (i.e., sinusoidal variations),

$$i = I_0 \cos\frac{2\pi t}{T} = I_0 \cos \omega t, \tag{9.8}$$

or

$$e = E_0 \cos\frac{2\pi t}{T} = E_0 \cos \omega t. \tag{9.8a}$$

where I_0 and E_0 are current and voltage amplitudes, respectively. Substituting in corresponding Eqs. (9.4) and (9.5) and evaluating, we obtain

$$I_{rms} = \frac{I_0}{\sqrt{2}} \approx 0.707 I_0 \tag{9.9}$$

and

$$E_{rms} = \frac{E_0}{\sqrt{2}} \approx 0.707 E_0. \tag{9.9a}$$

The results indicate that a sinusoidal source delivers about 71% of the power that a dc source of like amplitude delivers. Considering an ordinary line voltage of 120 V ac, we see that the voltage amplitude is

$$E_0 = \frac{120}{0.707} \approx 170\,\text{V}.$$

This is the voltage that must be applied when required insulation is considered. The peak-to-peak magnitude in this case is about 340 V.

The following two cases are examples of waveforms that are not sinusoidal.

Example 9.1
Consider the waveform shown in Fig. 9.5. Inspection suggests that because of symmetry we need only deal with the shape over the range $0 < \omega t < \pi/2$.

Solution. For this range,

$$e = \frac{E_0}{\pi/2}\, \omega t.$$

Figure 9.5 Waveform for Example 9.1

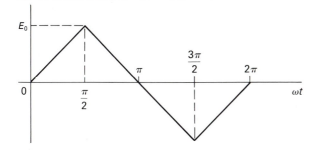

Substituting in Eq. (9.7a) and evaluating, we have

$$E_{rms} = \left[\frac{1}{\pi/2} \int_0^{\pi/2} \left(\frac{E_0}{\pi/2} \right)^2 (\omega t)^2 \, d(\omega t) \right]^{1/2}$$

$$= \frac{E}{\sqrt{3}} \approx 0.577 E_0;$$

for equal amplitudes the waveform shown in Fig. 9.5 is only about 58% as effective as dc of the same amplitude.

Example 9.2

Figure 9.6 shows a pulsed square wave for which we wish to determine the rms voltage.

Figure 9.6 Waveform for Example 9.2

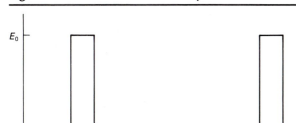

Solution. There are three distinct intervals in each cycle, which will be treated separately.

$$0 < \omega t < \frac{\pi}{2}, \qquad e = 0,$$

$$\frac{\pi}{2} < \omega t < \frac{3\pi}{4}, \qquad e = E_0,$$

$$\frac{3\pi}{4} < \omega t < 2\pi, \qquad e = 0.$$

Applying Eq. (9.7a),

$$E_{\text{rms}} = \left\{ \frac{1}{2\pi} \left[\int_0^{\pi/2} (0)\, d(\omega t) + \int_{\pi/2}^{3\pi/4} E_0^2\, d(\omega t) + \int_{3\pi/4}^{2\pi} (0)\, d(\omega t) \right] \right\}^{1/2}$$

$$= E_0 \left\{ \frac{1}{2\pi} \left(\frac{3\pi}{4} - \frac{\pi}{2} \right) \right\}^{1/2}$$

$$= \frac{E_0}{\sqrt{8}} \approx 0.35 E_0.$$

It should be noted that the integration circuitry of the ac meter yields rms readouts independent of the sine waveform. Based on a voltage amplitude of E_0, the reader may wish to confirm this statement by using the setup suggested in Problem 9.7 and showing that, conforming with theory,

1. For a square waveform, $E_{\text{rms}} = E_0$;
2. For a sine waveform, $E_{\text{rms}} = 0.707 E_0$;
3. For a triangular waveform, $E_{\text{rms}} = 0.577 E_0$.

9.3.4 The Multimeter and Resistance Measurement

A versatile tool around any laboratory is the basic volt-ohm-milliammeter (VOM), which uses switching arrangements for connecting multiplier and shunt resistors and a rectifier into or out of a circuit in order to cover ranges of dc and ac voltages. In addition, the meter is arranged to measure resistances, using an internal source of current. The VOM applies a known voltage to the resistor and senses the resulting current flow. By switching to the ohmmeter function and connecting the leads to the unknown resistance, one can determine from the meter movement the current flowing through the resistor. In ohmmeter mode, the current flow indication is calibrated in terms of resistance, thereby providing a direct means for measurement. Generally, the resistor should be removed from its circuit before it is measured.

9.4   Meters with Electronic Amplification

There are two reasons for amplifying the input to a voltmeter or VOM. The obvious one, of course, is to increase the instrument's sensitivity. Of equal importance, however, is the fact that the input impedance of the meter can be made very much greater, thereby decreasing the effect of the meter load on the tested item.

Although high input resistance is quite desirable in most cases, it is not an unmixed blessing inasmuch as the instrument becomes more susceptible to *noise,* the most troublesome being an extraneous 60-Hz hum radiated from the power lines. Circuitry providing common-mode rejection (Section 7.17) is very helpful in this case.

9.5   Digital-Readout Multimeters

The advent of the digital counter brought about its application to numerous measurement problems. Basically the counter is simply that—a counter of events. In Section 8.7 we showed how a simple counter circuit can be arranged to display a frequency. A simple way to make use of this capability for measuring a voltage is to combine the meter with a voltage-controlled oscillator (VCO). The frequency output of a VCO (e.g., the National Semiconductor LM566) is determined by the magnitude of the applied voltage. It is easy to visualize that through use of the VCO, a frequency counter, and proper scaling circuitry, a digitally reading voltmeter can be devised. Although this is a simple approach, it possesses certain disadvantages and is not commonly used.

Dual-slope integration is a much more common method. The essential building blocks for this method are an op-amp integrator (Fig. 7.35b, Section 7.23), a clock, and a frequency counter, combined with the necessary scaling and control circuitry. The clock is simply a fixed-frequency oscillator, usually crystal-controlled, that supplies timing pulses.

Through use of IC gating the integrator capacitor is charged for a predetermined length of time (referenced to the clock frequency). It is then discharged at a constant current rate, and the clock pulses occurring during the discharge period are counted. This count becomes the measure of the input voltage. With proper scaling, the count is equated to the input magnitude. Other circuitry can be incorporated within the meter, making it a general-purpose multimeter for measuring dc or ac voltages or currents, or resistances.

Advantages of the double-slope circuit over others is that aging of either the clock or the integrator causes little or no error. We can see that charge and discharge of the integrator capacitor are each dependent on capacitance value and time interval and that changes in either will be self-compensating. Should the clock slow down with age, the charging time will be reduced, but the discharge time will be increased in like proportion. A similar effect results from small changes in capacitor value.

Other approaches include

1. single-slope conversion,

2. charge balance,

3. linear ramp conversion, and

4. successive approximation.

These approaches are not discussed here; however, the reader is referred to the Suggested Readings at the end of the chapter for further information.

Many digital multimeters also include automatic polarity indication and self-ranging ability (automatic placement of the decimal point). Many meters of this type are said to display one-half digits, e.g., a "$3\frac{1}{2}$-digit display." This

Figure 9.7   (a) Digital-readout volt-ohm-milliammeter (VOM)
(Courtesy: Hewlett-Packard Co., Palo Alto. CA)

(a)

continued

Figure 9.7   (b) Typical digital multimeter

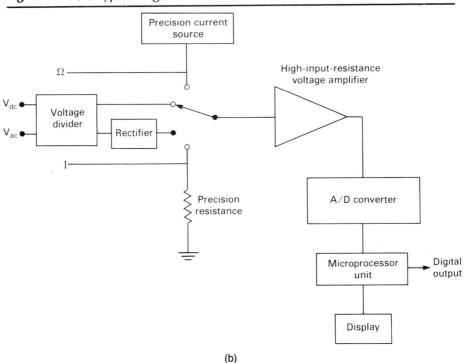

(b)

means the most significant digit can be only a 0 or a 1, excluding all others. A $3\frac{1}{2}$-digit meter, for instance, is not capable of displaying a number greater than 1999.

Figure 9.7(a) illustrates a compact, digital-readout VOM having the following abbreviated specifications:

dc voltmeter ranges to 1200 V dc with accuracies equal to or better than 0.1% of reading + 2 digits

ac voltmeter ranges to 1200 V rms with accuracies equal to or better than 1.5% of reading: 10 digits

dc input resistance = 10 MΩ

dc and ac ammeter ranges to 2.0 A

ohmmeter ranges to 20 MΩ

A/D conversion: dual slope

Figure 9.7(b) illustrates how a typical digital voltmeter operates.

9.6   The Cathode-Ray Oscilloscope (CRO)

Probably the most versatile readout device used for mechanical measurements is the cathode-ray oscilloscope (CRO). This is a voltage-sensitive instrument, much the same as the electronic voltmeter, but with an inertialess (at mechanical frequencies) beam of electrons substituted for the meter pointer and a fluorescent screen replacing the meter scale. Figure 9.8 shows a typical general-purpose CRO.

The heart of the instrument is the cathode-ray tube (CRT), shown schematically in Fig. 9.9. A stream of electrons emitted from the cathode is focused sharply on the fluorescent screen, which glows at the point of impingement, forming a bright spot of light. Deflection plates control the direction of the electron stream and hence the position of the bright spot on the screen. If an electrical potential is applied across the plates, the effect is to bend the pencil of electrons, as shown in Fig. 9.10. With the use of two sets of deflection plates arranged to bend the electron stream both vertically and horizontally, an instantaneous relation between two separate deflection voltages may be obtained.

Figure 9.11 is a block diagram of a typical general-purpose cathode-ray oscilloscope. The nature of the CRO is such that it may appear with many different variations in the form of special controls and input and test terminals. The diagram shown is not for any particular commercial instrument. Certain

Figure 9.8   Model 54502A digitizing oscilloscope
               (Courtesy: Hewlett-Packard Co., Palo Alto, CA)

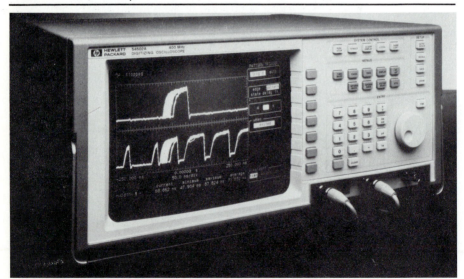

Figure 9.9   Elements of the basic cathode-ray tube (CRT)

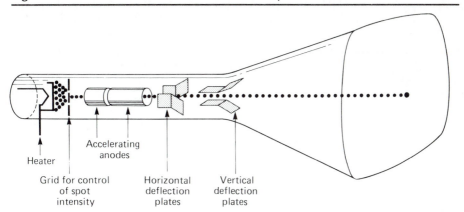

Heater

Accelerating
anodes

Grid for control
of spot
intensity

Horizontal
deflection
plates

Vertical
deflection
plates

oscilloscopes will have features not shown here, and others may not use certain
ones that are shown.

9.6.1   Oscilloscope Amplifiers

The sensitivity of the typical electrostatic cathode-ray *tube* is relatively low,
varying from about 0.010 to 0.15 cm deflection per volt dc, or from about 6 to
100 V/cm of deflection. This means that in order to be widely useful for
measurement work, the CRO should provide means for signal amplification

Figure 9.10   Time varying voltage V_V applied to the vertical plates and
sawtooth waveform voltage V_H applied to the horizontal
plates cause the electron beam to display
patterns on the screen.

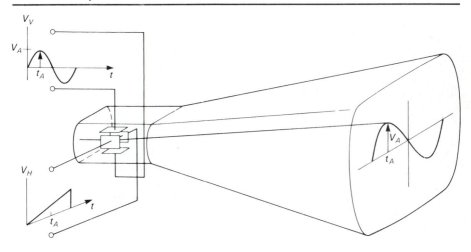

Figure 9.11 Block diagram of a typical general-purpose oscilloscope

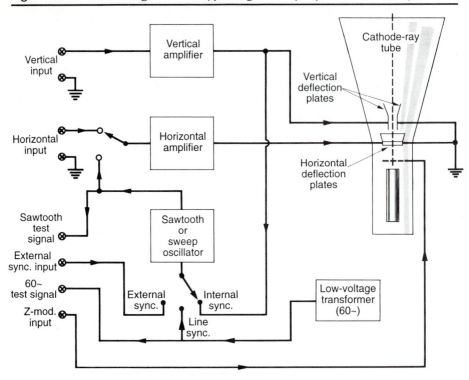

before the signal is applied to the deflection plates. All general-purpose oscilloscopes provide such amplification. Most are equipped for both dc and ac amplification on both the vertical and horizontal plates. During ac amplification, or *ac coupling,* a capacitor is used to block the dc component of the signal, so that only the ac component is amplified.

Some means for varying gain is provided in order to control the amplitude of the trace on the screen. This is often accomplished through use of fixed-gain amplifiers, preceded by variable attenuators.

9.6.2 Sawtooth Oscillator or Time Base Generator

Except for special-purpose applications, the usual cathode-ray oscilloscope is equipped with an integral *sawtooth,* or *sweep,* oscillator. This variable-frequency oscillator produces an output voltage-time relation in the form shown in Fig. 4.9(c). Ideally, the voltage increases uniformly with time until a maximum is reached, at which point it collapses almost instantaneously.

When the output from the sawtooth oscillator is applied to the horizontal deflection plates of the cathode-ray tube, the bright spot of light will traverse the screen face at a uniform velocity. As the voltage reaches a maximum and collapses to zero, the spot is whipped back across the screen to its starting point, from which it repeats the cycle. The length of the path will then be a measure of the period of the oscillator frequency (called sweep frequency) in seconds, and each point along the path will represent a proportional time interval measured from the beginning of the trace. (By convention, increasing time is measured to the right.) In this manner a very useful *time base* is obtained along the x-axis of the tube face.

As a simple example, let us suppose that ordinary 60 Hz line voltage is applied to the y-deflection plates of the tube and the output from a variable-frequency sweep oscillator is applied to the x-deflection plates. (Usually the sawtooth oscillator is within the case of the CRO and a knob is simply set to "Internal Sweep.") With the two voltages applied, the frequency of the sawtooth oscillator would be adjusted by means of the sweep range and sweep vernier controls on the control panel. If the sawtooth frequency is adjusted *exactly* to 60 Hz, then one complete cycle of the vertical input waveform will appear stationary on the screen, as shown in Fig. 9.12(a). If the sweep frequency is slightly greater or slightly less than 60 Hz, then the waveform will appear to creep backward or forward across the screen. *The reciprocal of the time in seconds required for the waveform to creep exactly one complete wavelength on the screen will be the discrepancy in hertz between the*

Figure 9.12 Example of internal triggering: (a) 60 Hz voltage applied to vertical plates and horizontal sweep adjusted to 60 Hz, (b) trace obtained when horizontal sweep is changed to 30 Hz, (c) general internal trigger synchronization

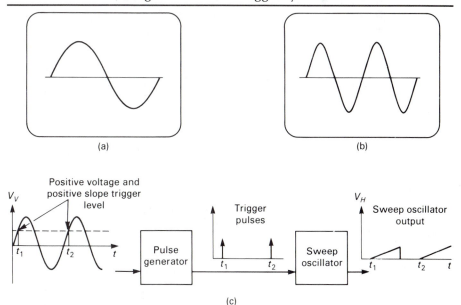

sweep frequency and the input frequency. In certain cases this relationship may be used in making precise measurement of frequency or period.

If the sweep frequency is changed to *exactly* 30 Hz, then two complete cycles of the input signal will appear and remain stationary on the screen, as shown in Fig. 9.12(b).

9.6.3   Synchronization or Triggering

In the example just referred to, one cycle of the 60 Hz waveform will appear stationary on the screen only if the sweep frequency is exactly 60 Hz. Frequencies from all types of electronic generators tend to shift or drift with time. This is caused by a change in component characteristics brought about by temperature changes due to the warm-up of the instrument. Therefore, to hold a pattern on the screen without creep, one must continuously monitor the trace, making adjustments in sweep frequency as required.

When a steady-state signal is applied to the vertical terminals, however, it is possible to lock the sweep oscillator frequency to that of the input frequency, provided the sweep frequency is *first* adjusted to approximately the input frequency or some multiple thereof. This is controlled through use of "Trigger Source" and "Trigger Level" controls.

In our example, we would wish to use the vertical input as our synchronizing signal source, so we would set the trigger source to "Internal" [see Fig. 9.12(c)]. Voltage pulses from the input signal would then be applied to the sweep oscillator and would be used to control the oscillator frequency over a small range. If the frequency is initially adjusted to some integral multiple of the input frequency, the sweep oscillator would then lock in step with the input signal and the trace would be held stationary on the screen. The trigger level voltage would need to be adjusted to lie in the range of the periodic signal in order for trigger pulses to be generated.

Electrical engineers are often concerned with measurements at 60 Hz; hence oscilloscopes are usually equipped to provide a direct synchronizing signal from the power line. This setting on the trigger source selector is often simply marked "Line."

Finally, it is often desirable to trigger a CRO trace from an external source closely associated with the input signal. As an example, suppose some form of electrical pressure pickup is being used for measuring the cylinder pressures in a reciprocating-type air compressor. Although the pressure signal from the pickup may be steady state, making internal triggering a possibility, changing load, erratic valve action, or the like may make this signal an undesirable source for synchronization. An external circuit may be used in a case of this sort. A simple make-and-break contactor could be attached to the compressor shaft, and a voltage pulse could be provided for synchronization through use of a simple dry cell. Such a circuit would be connected between the external trigger input and the ground, and the trigger source would be set at

"External." In this case the horizontal sweeps would take place only when initiated by the external contactor.

This arrangement is also useful when a *single sweep* only is desired. In such a case, the synchronizing contactor or switch may be simply hand-operated, or it may be incorporated in the test cycle. As an example, a photocell circuit could be arranged so that a beam of light intercepted by a projectile or the like would provide the initiating pulse in synchronization with the test signal of interest.

When the driven sweep is initiated as just outlined, through use of an external source of triggering, the sweep occurs once for each triggering pulse. The sweep rate in this case is still controlled by the sweep range and sweep vernier. Of course, the sweep cannot be pulsed at a rate greater than that provided by the sweep control settings. That is, the electron beam must have returned from the previous excursion before it can be triggered again.

9.6.4 Intensity Modulation or Z-Modulation

The fluorescent trace produced by the electron beam may be brightened or darkened by applying a positive or a negative voltage component, respectively, to the grid of the cathode-ray tube. Actually, this is what is done in a television receiver tube to produce the light and dark picture areas. Some oscilloscopes make provision for applying a brightness-modulating voltage from an external source, either through a terminal on the front panel or through a connection on the back of the instrument. This is known as *intensity modulation*, or *Z-modulation*. (Z is used in the sense that along with the x- and y-trace deflections, intensity variation provides the third coordinate.)

If on a normal input trace, say from a pressure pickup, an alternating Z-modulation is superimposed, the trace becomes a dashed line, providing timing calibration as well as the usual y-input information.

9.6.5 External Horizontal Input

It is not always necessary to use the sweep oscillator for the horizontal input. Input terminals are provided for connecting other sources of voltage, thereby permitting a comparison of voltages, frequencies, and phase relations (see Section 10.5).

9.7 Additional CRO Features

9.7.1 Multiple Trace

In many cases it is desirable to make an accurate time comparison between two continuing inputs, and very often CRO multiple-trace capability is the solution to this problem. Although oscilloscopes are available that permit simultaneous writing of more than two traces, the dual-trace type is the most common.

There are two different basic methods for accomplishing double traces: (1) through use of two separate electron "guns" within a single tube envelope, and (2) by high-speed gating (switching) two inputs to the vertical plates of a conventional one-gun cathode-ray tube. In either case, duplicate circuitry (terminals, amplifiers, positioning controls, etc.) is required. The second approach is by far the more common.

When the gating method is used, the oscilloscope design engineer has a choice of either or both of two different schemes. The first, called the *chopped-trace* method, successively switches from input A to input B and back again many times during a single sweep across the CRT screen. A switching rate of 200 kHz is typical. For many sweep rates the gating is so fast that the two traces appear to be continuous. In addition to being dependent on the relative rates, gating to sweep, the illusion of continuity also depends somewhat on the persistence of the phosphor. However, as the sweep rate is increased, depending on the demands of the measurement, the actual discontinuity of the traces may become a problem.

Alternate gating, the second method, alternately displays the entire traces, first for input A, then B, back to A, and so on. Screen persistence will permit simultaneous viewing of near-simultaneous traces. Many dual-trace scopes provide switch selection of either method.

An oscilloscope accessory called an *electronic switch* may also be used to convert a single-trace CRO to a double-trace one. The same chopped-input method as described before is used, the primary difference being that the circuitry is outboard rather than an integral part of the oscilloscope.

9.7.2 Magnification and Delayed Sweep

Amplification may be used to stretch out or magnify the horizontal sweep to a number of times the size of the CRO screen. This means, coupled with adjustment of the horizontal position control, allows us to magnify a portion of the normal sweep for closer inspection. Oscilloscopes with more advanced circuitry (and higher cost) may use what is called *delayed sweep* to accomplish similar results. That is, the operator may select any small portion of the normal display, which may then be shown at a selected higher sweep rate. The effect of the increased rate over the selected portion of the normal sweep is to expand or magnify the portion that has been pinpointed.

9.7.3 Storage Scopes

In many instances measurands are nonrepeating. Examples are the load or strain resulting from an impulsive load, or the sound wave corresponding to the discharge of a gun. Through proper photographic techniques, as described in Section 9.8 and in Chapter 10, a record may be captured on film. These techniques may not be convenient, however, and a series of trials may be required before system adjustments are refined. The answer may be use of a *storage scope,* which is able to "hold" a trace on the CR-tube face for a period of time after it has been written. Eventually the trace fades; however, on some storage scopes the trace may be held for up to an hour.

Digital techniques may also be incorporated in sophisticated oscilloscopes whereby the contents of the entire screen may be stored and held in memory for repetitive writing, for later recall, or for input into a computer system.

Digital Storage Oscilloscopes

The digital storage oscilloscope differs from its analog storage counterpart in that it "digitizes," or converts the analog input waveform into a digital signal that is stored in memory and then converted back into analog form for display on a conventional CRT (Fig. 9.13). The contents of the memory are outputted to a D/A converter and then to the vertical and horizontal (y and x) deflection sections of the CRT circuitry. The data are displayed most frequently in the form of individual dots that collectively make up the CRT trace. The vertical screen position of each dot is given by the binary number stored in each memory location and the horizontal screen position is derived from the binary address of that memory location. The number of dots displayed depends on three factors: the frequency of the input signal with respect to the digitizing rate, the memory size, and the rate at which the memory contents are read out. The greater the frequency of the input signal with respect to the digitizing rate, the fewer the data points captured in the oscilloscope memory in a single pass, and the fewer the dots available in the reconstructed waveform.

Digital storage oscilloscopes and analog oscilloscopes each have distinct advantages, but digital oscilloscopes have created the most recent excitement because of their dramatic improvements in performance. In addition to merely capturing and displaying waveforms, the digital oscilloscopes can perform the

Figure 9.13 Block diagram of a digital storage oscilloscope

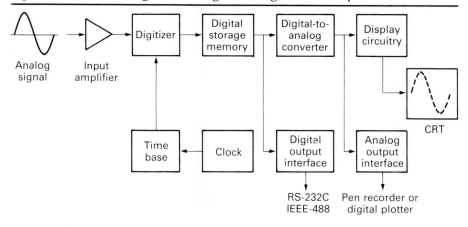

following tasks: indefinite storage of waveform data for comparison, transferring stored data to other digital instruments, and in some cases providing onboard processing of the data. Furthermore, such waveform parameters as maximum, minimum, peak-to-peak, mean, rms, rise time, fall time, waveform frequency, and pulse delay can be computed and made available for presentation in decimal form on the oscilloscope screen. Digital oscilloscopes are also well suited for capturing transient signals. If the oscilloscope is set to the single-sweep mode, data can be automatically captured and stored on the occurrence of the trigger event. As a result, the problem of synchronization inherent in analog oscilloscopes is eliminated.

The analog-to-digital converter of the digital oscilloscope determines some of its most important operating characteristics. The voltage resolution is dictated by the bit resolution of the A/D converter, and the storage speed by the maximum speed of the converter. For example, a 0–10 V converter using 8, 10, or 12 bits is able to resolve to 0.0391 V, 0.0098 V, and 0.0025 V, respectively. The time resolution is selectable, in that the user can define how much memory space is needed for each waveform stored.

The output of digital oscilloscopes is available in other forms than the trace on the CRT. Analog output is provided for driving pen recorders. Digital output formats in RS-232C and IEEE-488 (Section 8.13) are generally available. For more flexible output recording, some oscilloscopes are fitted with either single or twin disk drives so that any captured signals can be immediately stored on floppy disk drives for more complex data analysis.

9.7.4   Single-Ended and Differential Inputs

There are two types of inputs through which a signal can be connected to the oscilloscope: the single-ended input and the differential input.

Single-ended inputs have only one input terminal besides the ground terminal at each amplifier channel. Only voltages relative to ground can be measured with a single-ended input. (Note that the most common input connector to oscilloscopes is the BNC coaxial connector. The external conductor of the BNC connector is the ground terminal of the input.)

A differential input has three terminals (two input terminals besides the ground terminal at each amplifier channel). With a differential input, the voltage between two nongrounded points in a circuit can be measured. The amplifier electronically subtracts the voltage levels applied at the two terminals and displays the difference on the screen. In addition, differential amplifiers are able to reduce unwanted common-mode interference problems. This feature is especially important when it is necessary to measure small signals in the presence of much larger, undesired common-mode signals. The differential input is sometimes available as a plug-in unit on those oscilloscopes that have interchangeable plug-in capabilities.

9.8 CRO Recording Techniques

Direct observation of an oscilloscope trace often provides sufficient information. In other cases, however, particularly when transient conditions are being studied, some form of recording is mandatory. For analog scopes, this normally dictates the use of photographic methods. Various forms of photographic equipment may be used, but the most satisfactory are special-purpose cameras that can be attached directly to the oscilloscope bezel. Several types are available, including those using ordinary photographic film, the Polaroid Land camera, and moving-film cameras.

Only very simple photographic techniques are required in using the first two types. When the trace is from a steady-state source, the sweep may be synchronized to hold the trace stationary on the screen. It is then necessary only to make an appropriate exposure to capture the record.

When a transient input is to be recorded, single sweep, along with "time" or "bulb" shutter setting on the camera, may be used. The camera shutter is opened, the sweep initiated either internally or externally, and the trace recorded.

Often photographic recording techniques are also used with digital storage oscilloscopes. Here the trace can be "frozen" on the screen for easy photographing. The digitized trace may alternatively be transferred to some form of *xy* plotter.

9.9 Oscillographs

The oscillograph is basically an adaptation of the D'Arsonval meter movement (Section 9.3), in which either a writing stylus (Fig. 9.14) or a small mirror (Fig. 9.15) replaces the meter pointer or hand. The stylus writes through direct contact on a moving strip of paper. Either an ink pen or a heated stylus on special paper can be used. The mirror-type oscillograph functions by directing a pencil of light onto photographic paper or film. In both cases the meter movement is commonly referred to as the galvanometer. As the stylus (or light beam) is deflected by the input signal, the paper is moved at a known rate, thereby recording the time function of the input. The complete oscillograph incorporates the galvanometer(s), adjustable-speed paper drive, and power amplifier(s), plus voltage amplifiers and calibration circuits as needed.

Obviously the frictional drag between paper and pen of the stylus requires considerably more driving torque than does a simple meter or a light-beam galvanometer. In any case an important parameter in the design is the magnitude of the magnetic flux from the permanent magnet. This requires a relatively large and heavy magnet. Commercially available stylus-type oscillographs may have as many as 8 channels and provide flat response from dc to about 150 Hz. The light-beam type may have as many as 36 channels, with typical responses and sensitivities as listed in Table 9.1.

Figure 9.14 Essential parts of stylus-type oscillograph

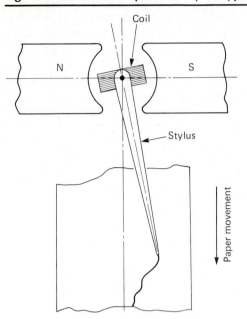

Figure 9.15 Essential parts of a light-beam-type oscillograph

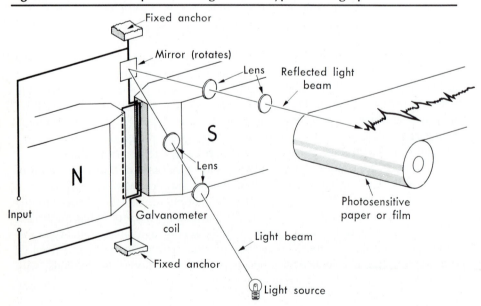

Table 9.1 Typical galvanometer characteristics

Undamped Natural Frequency, Hz	Flat (±5%) Frequency Response, Hz	Sensitivity (with 30-cm Optical Arm)	
		µA/cm	mV/cm
24	0–15	4	0.5
40	0–40	20	0.6
100	0–60	25	1.3
200	0–120	65	4
400	0–240	200	24
600	0–540	330	105
1,000	0–600	5,000	260
5,000	0–3,000	50,000	1,600
8,000	0–4,800	100,000	3,600

9.10   *XY* Plotters

The term *x-y plotter* is very nearly self-explanatory. It refers to an instrument used to produce a Cartesian graph originated by two dc inputs, one plotted along the *x*-axis and the other along the *y*-axis. Of course the great advantage in its use is that the graph is plotted automatically, thereby sidestepping the laborious point-by-point plotting by hand. In addition, families of curves may be plotted easily by varying a third parameter in step fashion from plot to plot. Figure 9.16 shows a typical *xy* plotter. Basic components consist of a platen to which the graph paper is either mechanically attached or held by vacuum or by electrostatic means and one or more servo-driven styluses. In addition, amplification of the input signals is normally required.

Performance variables include input ranges (amplitude and frequency), sensitivity, stylus slewing rate and acceleration limit, resolution, resetability and provision for common mode rejection. Slewing rate in cm/s is the maximum velocity with which the stylus can be driven. This becomes a limiting response characteristic, especially when large amplitudes are to be plotted. Limits on the maximum acceleration of the stylus are more often a factor when low-amplitude, high-frequency inputs are plotted.

Common chart sizes are 22×28 cm ($8\frac{1}{2} \times 11$ in.) and 28×44 cm (11×17 in.). Two-stylus (*xyy*) models are available for the simultaneous plotting of two curves. These models, of course, require three separate drive systems.

9.11   Digital Waveform Recorders

Most analog *xy* recorders are limited by a relatively slow stylus slewing rate and acceleration limit. The digital plotter has generally replaced the older strip-chart recorders because of its improved performance characteristics. The

Figure 9.16   A two-pen *xy* recorder
(Courtesy: Hewlett-Packard Co., Palo Alto, CA)

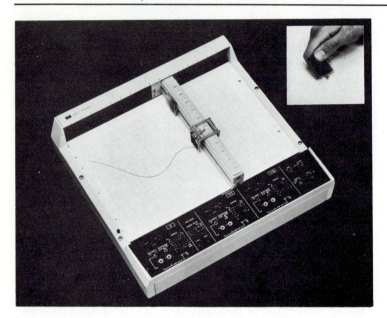

inked pen or hot stylus strip-chart recorder plotted analog time-dependent
input voltages in real time on long rolls of chart paper. (The strip-chart
oscillograph recorder is still used where long recording times are necessary.)
Because of the pen or stylus inertia, the top recording frequency is below
50 Hz in order to maintain the full amplitude swing of the pen or stylus arm.
These features are minimized by the use of a digital plotter. It combines the
features of an *xy* recorder, a waveform recorder or storage oscilloscope, and a
digital voltmeter. It samples analog signals, digitizes them, stores them in
digital memory, and plots them in analog form.

Figure 9.17 illustrates the versatility of an HP 7090A low-frequency
measurement-plotting system, which can be used for recording most signals
from sensors used in mechanical measurements. The figure illustrates just one
of three channels which may be input simultaneously. Each channel has a
buffer, which can capture and store 1000 data-point values. During a buffered
recording the input buffers are filled simultaneously. This buffered data can in
turn be recalled and plotted or viewed on an oscilloscope.

A direct recording mode is also available to produce real-time recordings
as a hard-copy plot. The input signal is converted to digital data at a fixed rate
of 250 samples/s (but not stored in memory), which are used to drive the

Figure 9.17 HP 7090A plotter system

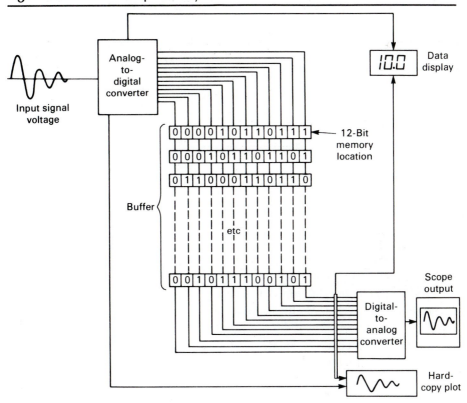

plotter stepper motors. The data of any two channels can be plotted versus time or versus a third channel.

9.12 The Spectrum Analyzer

Spectrum analyzers, which measure the frequency spectrum of a signal, are of two forms: the swept-type and the digital-type. Figure 9.18 shows a simplified block diagram of the workings of a swept-type spectrum analyzer. Pertinent items are as follows:

1. A *sawtooth waveform generator* running at a fixed frequency, but whose voltage output varies linearly in ramp fashion.

2. A *voltage-controlled oscillator* (VCO), whose output frequency, f_{VCO}, sweeps linearly across a given frequency range. Its voltage amplitude is constant.

Figure 9.18 Block diagram of a spectrum analyzer

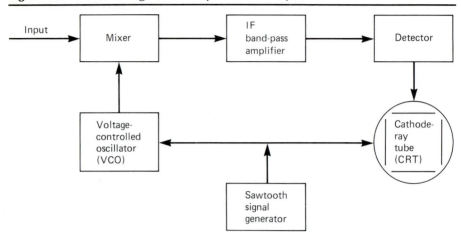

3. A *mixer* that combines (mixes) the input signal with the VCO output. This produces sum-and-difference frequency components. For an input frequency, f_{in}, two side frequencies, $(f_{VCO} - f_{in})$ and $(f_{VCO} + f_{in})$, are generated (see Section 10.8).

4. An *IF* (intermediate-frequency) *band-pass amplifier,* whose passband is designed to accommodate a single (ideally) value of $(f_{VCO} - f_{in})$ to the exclusion of other frequencies.

5. A *detector,* which is basically a voltage rectifier, passing a voltage of one polarity (say, positive).

6. A *cathode-ray tube* (CRT), used for display.

In operation, as the output of the sawtooth generator linearly rises from zero, it drives the CRT electron beam along the x-axis, across the face of the tube. At the same time, the output frequency from the VCO sweeps linearly upward. When the VCO frequency and the frequency component of the input produce a difference frequency, $(f_{VCO} - f_{in})$, matching the passband frequency of the IF amplifier, a signal component passes whose amplitude is proportional to that of the input component. This is then rectified and displayed on the CRT screen as a spike located on the horizontal frequency axis at a point corresponding to the difference frequency. Both axes can be calibrated in terms of the input parameters.

To help us visualize the result, let us consider a two-component input:

$$A(t) = 10 \cos \Omega t + 5 \cos 2\Omega t.$$

As an amplitude-time plot, the function would appear as shown in Fig. 9.19(a). Figure 9.19(b) represents the corresponding amplitude-frequency plot as it would be shown by a spectrum analyzer. For an ideal input and an ideal

Figure 9.19 (a) Time domain plot of $A(t) = 10 \cos \Omega t + 5 \cos 2\Omega t$,
(b) frequency domain plot of $A(t) = 10 \cos \Omega t + 5 \cos 2\Omega t$

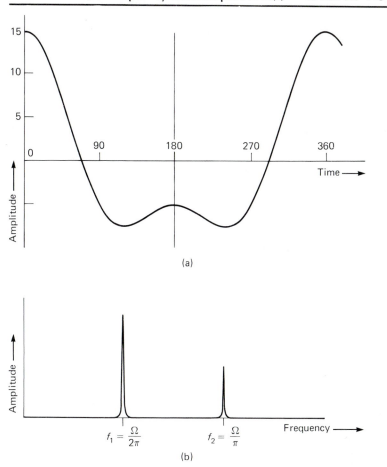

(a)

(b)

response, the two spikes would be indicated by perfect vertical lines. However, as with all instrumentation, there are limitations and as input frequencies are increased, a broadening of the spikes becomes apparent.

Various adjustments are provided on even the simplest analyzers to accommodate ranges of frequency and amplitude. In addition, the IF bandpass analyzer can often be varied to help in isolating frequency components in the input signal. Analyzers are selected on the basis of application: low-frequency analyzers for vibration and sound work, radio-frequency analyzers for RF work, UHF for TV, gigahertz for microwaves, etc.

Material on digital fast fourier transform (FFT) analysis and digital spectrum analyzers is given in Section 4.6 and Section 18.6.

Suggested Readings

Analog and Digital Meters. Benton Harbor, Mich.: Heath, 1979.

Batholomew, D. *Electrical Measurements and Instrumentation*. Boston: Allyn and Bacon, 1963.

Ibrahim, K. F. *Instruments and Automatic Test Equipment: An Introductory Textbook*. New York: John Wiley, 1986.

Sessions, K. W., and W. Fischer. *Understanding Oscilloscopes and Display Waveforms*. New York: John Wiley, 1978.

Van Erk, R. *Oscilloscopes, Functional Operation and Measuring Examples*. New York: McGraw-Hill, 1978.

Wolf, S., and R. F. M. Smith. *Student Reference Manual For Electronic Instrumentation Laboratories*. Englewood Cliffs, N.J.: Prentice Hall, 1990.

Problems

9.1   Verify Eqs. (9.6a) and (9.7a).

9.2   Determine the rms voltage for the waveform shown in Fig. 9.20.

Figure 9.20   Waveform for Problem 9.2

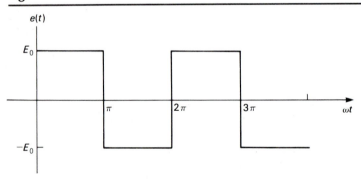

Figure 9.21   Waveform for Problem 9.3

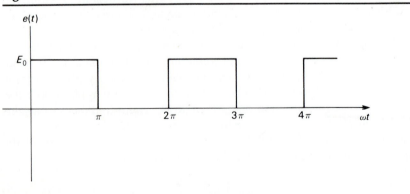

Figure 9.22 Waveform for Problem 9.4

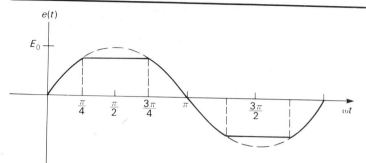

9.3 Determine the rms voltage for the waveform shown in Fig. 9.21.

9.4 Figure 9.22 illustrates a clipped sine wave. Determine the rms voltage for this waveform.

9.5 What is the rms value for the waveform defined by

$$y = \sin \omega t \cos 2\omega t?$$

9.6 What is the rms value for the waveform defined by

$$y = 10 \sin \omega t + 10 \sin 2\omega t.$$

9.7 Assemble the apparatus shown in Fig. 9.23.

    a. Input selected waveforms and record the rms values indicated by the ac voltmeter (ACVM) and E_0 as determined by the CRO.

    b. Use Eq. (9.7a) and calculate the rms value for the selected waveform and compare with the meter readout.

9.8 Assemble the following equipment: a general-purpose oscilloscope (preferably a single-channel basic CRO) and two variable-frequency signal generators, along with an assortment of appropriate leads. Connect one signal source to the vertical

Figure 9.23 Circuit for Problem 9.7 (signal generator has sine, square, and triangular waveform capabilities)

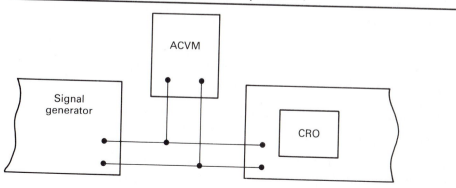

Figure 9.24 Experimental setup for Problem 9.10

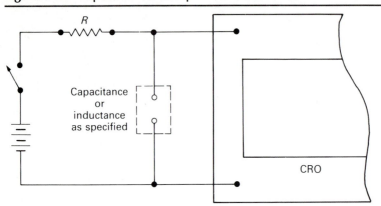

input terminals of the CRO and the other to the horizontal terminals. Proceed to "turn the knobs." Experiment until you are familiar with the purpose and action of each of the controls. (Feel free to make any front-panel adjustments that are available, except for possible screwdriver balance adjustments. In addition, avoid holding an intense, concentrated, fixed spot on the screen. To do so could cause local burning of the phosphor.) Can you reproduce these patterns? If the scope has provision for Z-modulation, experiment with this control.

9.9 The cathode-ray oscilloscope and electronic voltmeters are considered "high-input-impedance" devices. The simple D'Arsonval meter is usually considered to be of "low impedance." What is meant by "high" and "low" in this sense? Discuss the relative merits and disadvantages of each category.

9.10 Figure 9.24 illustrates an experimental setup for determining the time constant for various combinations of R and C (or L). Insert a range of components and compare experimentally determined values for the resulting time constants with theoretical values. (Refer to Section 5.18.)

Problems 9.11 through 9.16 specify experimental exercises to be performed. Although the circuits are prescribed, specific component values are not specified since a variety of values as may be available will usually be quite satisfactory. So-called decade boxes of resistances, capacitances, and inductances are particularly useful because they permit a wide range of values. A decade box is simply an assembly of components, in switch-selectable decade steps of value.

9.11 Insert the circuit shown in Fig. 9.25 in the circuit of Fig. 9.26. Using components of known value, find the LC resonance frequency experimentally. Check against Eq. (7.26).

9.12 Duplicate the experiment given in Problem 9.11; however, use the experimentally determined resonance frequency and a known value of capacitance to determine an unknown inductance.

9.13 Using the circuit in Fig. 9.26, experimentally determine the characteristics of the circuit shown in Fig. 7.27(a).

Figure 9.25 Parallel *LC* circuit to be used in Problems 9.11 and 9.12

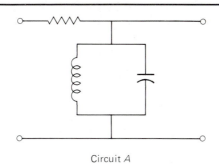

Circuit *A*

9.14 Using the circuit in Fig. 9.26, experimentally determine the characteristics of the circuit shown in Fig. 7.27(b).

9.15 Using the circuit in Fig. 9.26, experimentally determine the characteristics of the circuit shown in Fig. 7.31(a).

9.16 Using the circuit in Fig. 9.26, experimentally determine the characteristics of the selected circuits shown in Fig. 7.32.

9.17 A pressure pickup is used for measuring the pressure-time relationship in the cylinder of an internal combustion engine. The output is amplified and then applied to the vertical plates of an oscilloscope. Describe arrangements for synchronizing the scope trace with the engine speed (a) using internal sweep and (b) using external sweep. If the pickup is applied to the determination of the pressure-time history resulting from detonations of explosive charges, how should the oscilloscope be configured to obtain satisfactory traces? Assume that the charges are "one-shot," but that they may be repeated as desired.

Figure 9.26 Circuit for Problems 9.11 through 9.16 (signal generator has sine, square, and triangular waveform capabilities)

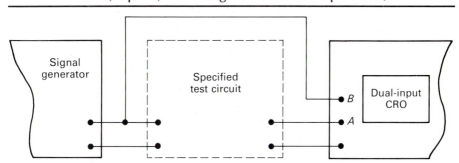

PART II

Applied Mechanical Measurements

CHAPTER 10

Determination of Count, Events per Unit Time, and Time Interval

10.1   Introduction

To be able to count items or events is basic to engineering. Items or events to be counted may be pounds of steam, cycles of displacement, number of lightning flashes, or anything divisible into discrete units. Also, time is often introduced, and the number of items or *events per unit of time* (EPUT) must be measured. The expressions "EPUT" and "frequency" usually have slightly different connotations. Frequency is thought of as being the events per unit of time for phenomena under steady-state oscillations, such as mechanical vibrations or ac voltage or current. EPUT, however, is not dependent on a steady rate, and the term includes the counting of events that take place intermittently or sporadically. An example of this is the counting of any of the various particles radiated from a radioactive source.

Time interval is often desired, and this becomes *period* if it is the duration of a cycle of a periodic event. Or the time interval desired may be that which occurs between events in an erratic phenomenon, or perhaps the duration of a "one-shot" event such as an impulsive pressure or force.

Problems in counting or timing emerge primarily when the events are too rapid to determine by direct observation, or the time intervals are of very short duration, or unusual accuracy is desired. In general, counting and timing-measurement problems may be classified as follows:

1. *Basic counting,* either to determine a total or to indicate the attainment of a predetermined count.

413

2. *Number of events or items per unit of time* (EPUT) independent of rate of occurrence.

3. *Frequency,* or the number of cycles of uniformly recurring events per unit of time.

4. *Time interval* between two predetermined conditions or events.

5. *Phase relation,* or percentage of period between predetermined recurring conditions or events.

10.2   Use of Counters

Several examples of general-purpose counting equipment, including the various forms of mechanical, electrical, and electronic counters, were discussed in Chapter 9. In addition, general laboratory equipment such as oscilloscopes and oscillographs, used in conjunction with frequency standards, may also be used in various EPUT and time-interval measuring systems, limited only by the ingenuity of the user.

The use of simple mechanical counters or electrically energized mechanical counters requires no particular technique, and further discussion should be unnecessary.

10.2.1   Electronic Counters

Electronic counters used as either basic counting devices or EPUT meters require that the counted input be converted to simple voltage pulses, a count being recorded for each pulse. It should be clear that input functions used to trigger the counter need not be analogous to any quantity other than the count; hence even a simple switch may be used, actuated by the function to be counted. In addition, photocells; variable resistance, inductance, or capacitance devices; Geiger tubes; and like may be employed. Simple amplifiers may be used, if necessary, to raise the voltage level to that required by the counter—and, because most electronic counters have a high-impedance input, no particular power requirement is imposed. Signal inputs may include almost any mechanical quantity, such as displacement, velocity, acceleration, strain, pressure, and load, so long as distinct cycles or pulses of the input are provided. The starting or stopping of the counting cycle may be controlled by direct manual-switch operation on the panel or by remote switching. One must not overlook, however, the ±1 count ambiguity referred to in Section 9.2.4.

A variation of the simple electronic counter is the *count-control* instrument. Provision is made for setting a predetermined count, and when the count is reached, the instrument supplies an electrical output that may be used as a control signal. Figure 10.1 shows how such a device could be used to prepare predetermined batches or lots for packaging.

Figure 10.1 Counter arrangement to provide a
control of predetermined count

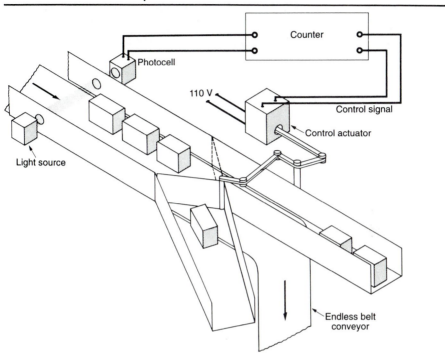

10.2.2 EPUT Meters

EPUT meters combine the simple electronic counter and an internal time base with a means for limiting the counting process to preset time intervals. This permits direct measurement of frequency and is quite useful for accurate determinations of rotational speeds (see Sections 8.6.1 and 10.9). The instrument is not limited, however, to an input varying at a regular rate; intermittent or sporadic events per unit of time may also be counted. Other applications include its use as a readout device for frequency-sensitive pickups such as resonant wire pressure pickups and turbine-type flowmeters (see Sections 14.8.4 and 15.7.1).

10.2.3 Time-Interval Meter

By modifying the arrangement of circuitry of an electronic counter, one can obtain a *time-interval meter*. In this case input pulses start and stop the counting process, and the pulses from an internal oscillator make up the counted information. In this manner the time interval taking place between starting and stopping may be determined, provided the frequency of the internal oscillator is known.

Figure 10.2   Time-interval meter arranged to count the number of hundred-thousandths of a second required for the projectile to traverse a known distance between photocells

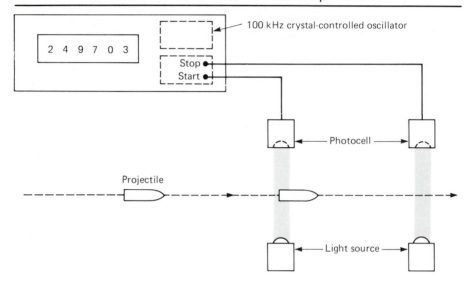

Figure 10.2 illustrates a simple application of the time-interval meter. Photocells are arranged so that the interruption of the beams of light provide pulses—first to start the counting process and second to stop it. The counter records the number of cycles from the oscillator, which has an accurately known stable output. In the example shown, the count would represent the number of hundred-thousandths of a second required for the projectile to traverse the distance between the light beams. Refer to Fig. 9.2(b) for a more detailed picture of the electronic counter's operation in this application.

10.3   The Stroboscope

The term *stroboscope* is derived from two Greek words meaning "whirling" and "to watch." Early stroboscopes used a whirling disk as shown in Fig. 10.3. During the intervals when openings in the disk and the stationary mask coincided, the observer would catch fleeting glimpses of an object behind the disk. If the disk speed was synchronized with the motion of the object, the object could be made to appear to be motionless. In some ways the action is the inverse of the illusion produced by the motion picture projector. Also, if the disk were made to rotate with a period slightly less than, or greater than, the period of the observed object, the object could be made to apparently creep either forward or backward. This made possible direct observation of

Figure 10.3 Essential parts of early disk-type stroboscope

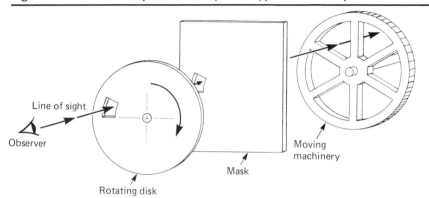

such things as rotating gears, shaft whip, helical spring surge, and the like, while the devices were in operation.

Modern stroboscopes operate on a somewhat different principle. Instead of the whirling disk, a controllable, intense flashing light source is used. Repeated short-duration (10 to 40 μs) light flashes of adjustable frequency are supplied by the light source. The frequency, controlled by an internal oscillator, is varied to correspond to the cyclic motion being studied. The readout is the flashing rate required for synchronization. These devices are often called *strobe lights*.

Two different cautions require mention. The first involves a minor problem concerned with the geometry of the item being studied. Suppose, for example, that the gear illustrated in Fig. 10.3 is the study subject, and suppose that the spokes are used as the target for synchronization. A moment's thought makes it clear that each of the six spokes in this example, will, in succession, occupy a given position. One must use care in making certain that one, and only one, spoke is identified. The usual practice is to place a distinctive mark on one of the spokes and to use that spoke alone in searching for one-on-one synchronization.

The second caution concerns multiple ratios of flashing rate to the object's true cycling rate. As an example, consider the rotation of a crank arm. Suppose that the arm is rotating at 1200 revolutions per minute (rpm) and the flashing rate is 600 cycles per minute (cpm). Also, if the rate were 400 cpm or 300 cpm, the arm would occupy the same position for successive flashes and would appear to be stationary. This is another example of *aliasing* (see Section 4.6.2).

An obvious approach is to "stop" the motion, note the rate, then double the rate and check again. Another approach is to use the following convenient procedure:

1. Determine a flashing rate f_1 that freezes the motion.

Figure 10.4 Photo obtained by "open-shutter" camera technique and
using a Strobotac® set at a flash rate of 20 Hz
(Courtesy: GenRad, Inc., Concord, MA)

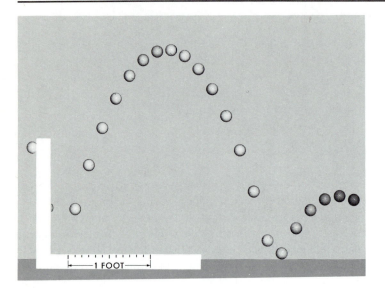

2. Slowly reduce the rate until the motion is frozen once more. Note this
rate, f_2.

Then

$$f_0 = \frac{f_1 f_2}{f_1 - f_2},$$ (10.1)

where

f_0 = actual cycling rate of the object.

It should be noted that it is not always necessary to obtain a one-to-one
synchronization. Note also that the procedure described makes it possible to
extend the upper measurement limit beyond the stroboscope's normal range.
In addition, as shown in Fig. 10.4, stroboscopic lighting can be used to study
nonrepeating action. By using a still camera with the shutter locked in the
open position, stroboscopic lighting can be used to track the position of a
moving object. For example, if the flashing rate is known, the position or
displacement of the object can be determined at various instants of time.
These data can be numerically differentiated to determine the instantaneous
velocity or acceleration.

10.4 Frequency Standards

The cesium "clock" is a basic frequency standard (Section 2.7). Pendulums,
tuning forks, electronic oscillators, and other devices may be used as secondary
standards. *Frequency* is the number of recurrences of a phenomenon or series

of events during a given time interval, and the reciprocal of frequency is *period*. A frequency standard *chops* time into discrete bits that may be used as time standards and, through comparative means, for timing events. The actual source of such a frequency may be mechanical or electrical or, in fact, pneumatic, hydraulic, or thermal. In certain cases mechanical frequency sources are used because of their long-time stability. The mechanical source, such as a pendulum or tuning fork, may be combined with the electrical, the mechanical being used to control the electrical. Or a strictly electronic source may be used. An electromechanical device that has become very common for providing closely fixed frequencies in the radio-frequency range is the piezo-electric crystal. It can be used to maintain very precise frequencies in the range of from 4 kHz to 100 MHz. Such crystals possess the ability to convert mechanical energy into electrical energy or vice versa (see Section 6.14). Materials exhibiting this characteristic include quartz, barium titanate, and various crystalline salts. When a small plate or bar of such materials is mechanically strained, a voltage develops across its faces; conversely, when a voltage is applied to the faces, a mechanical strain results. If the voltage is an alternating voltage, the plate or bar may be made to vibrate, and because of its mechanical mass-elastic characteristics, such a member will have a natural frequency of vibration. This fundamental frequency (including overtones) is often used as the basis for very stable control of electronic oscillators.

10.4.1 Electronic Oscillators

Electronic oscillators are sources of periodic voltage variation of either fixed or variable frequency. The rotating ac generator is a form of nonelectronic oscillator whose primary purpose is to provide a source of power rather than voltage. However, 60-Hz line voltage is often quite useful as a frequency reference. For the purposes of mechanical measurement, it is the voltage output from the oscillator that is of primary value.

In general, electronic oscillators are used in a wide variety of applications: as energy sources for circuitry measurement, as audio sources for electronic musical instruments, as sweep generators for oscilloscopes and TV receivers, as carriers for radio and TV signal propagation, as "clocks" for synchronizing computer actions, and so forth. They can also be employed as frequency references that, by suitable comparative means, may be used for timing and phase measurements.

Electronic oscillators may be classified as follows:

1. Fixed-frequency oscillators
   a. Simple electronic
   b. Tuning-fork-controlled
   c. Crystal-controlled

2. Variable-frequency oscillators
   a. Sine wave
      i. Audio frequency (0 to 20,000 Hz)
      ii. Supersonic (20,000 to 50,000 Hz, roughly)
      iii. Radio frequency (50,000 to 10,000,000,000 Hz or 50 kHz to 10 GHz)
   b. Nonsine wave
      i. Square wave
      ii. Sawtooth wave
      iii. Random noise

10.4.2   Fixed-Frequency Oscillators

Fixed-frequency oscillators are of primary value in mechanical measurements for calibrating and recording standard timing signals.

Precise frequency and time standards are available to all through reception of transmissions from the National Institute of Standards and Technology radio stations. Stations WWV, WWVB, and WWVL are located at Fort Collins, Colorado. Station WWVH is in Hawaii. WWV and WWVH are classified as *high-frequency* (HF) stations, whereas WWVB transmits at 60 kHz and is classified as *low-frequency,* and the frequency of WWVL is 20 kHz, which is called *very-low-frequency* (VLF). Low and very low frequencies have the advantage of providing more stable reception at a distance because of reduced variations in the transmission paths peculiar to those frequencies. Eventually the National Institute of Standards and Technology expects the two lower-frequency stations to have sufficient power to provide a worldwide frequency and time service.*

10.4.3   Variable-Frequency Oscillators (VFO)

Used for mechanical measurements, these normally produce sine-wave outputs covering a frequency range from about 1 Hz to 100 kHz. Although this exceeds the audible range, oscillators of this type are often referred to as *audio oscillators* to distinguish them from higher-frequency *RF oscillators*. A typical audio oscillator has an output of 1 W at a maximum of 25 V rms.

10.4.4   Complex-Wave Oscillators

Outputs from sine-wave oscillators may be shaped to provide a variety of waveforms for special applications. Ramp or sawtooth waveforms are used for sweep generators (Section 9.6.2); square waves may be used for evaluating

---

* In addition to the United States, many other nations also broadcast timing signals of various types, e.g., CHU in Canada (3.330, 7.335, 14.670 MHz). JJY in Japan (2.5, 5.0, 10.0, 15.0 MHz). MSF in the United Kingdom (2.5, 5.0, 10.0 MHz), and VNG in Australia (4.5, 7.5, 12.0 MHz).

signal conditioner responses (Section 5.20) or for providing synchronization and coding in digital computers. IC chips that provide most or all of these functions are available.

10.5   Direct Application of Frequency Standards by Comparative Methods

Probably the simplest and most basic method for measuring frequencies and short time intervals is to make a direct comparison of the unknown with a frequency standard. The problem lies in selecting a usable method for making the comparison.

When multichannel recording equipment is available, the solution is easily obtained. The input or inputs to be measured are simply recorded in terms of time in separate channels. In many cases the speed of the chart-recorder paper is known accurately enough to be used as the time reference, and the necessary time information is obtained automatically. In other cases it may be desirable simultaneously to record the output from a stable oscillator whose frequency is known, and to use this record as the measure of time. Some oscillographs provide a time base as a built-in part of the instrument.

Figure 10.5   Stamping press instrumented to produce synchronized outputs on a strip chart. The signal generator supplies a timing reference.

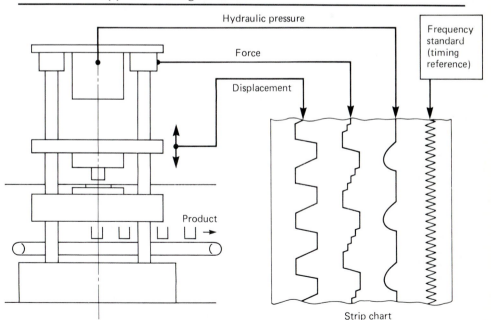

Figure 10.5 shows an arrangement that is simple but important because of its fundamental nature. Various inputs are fed to separate channels on a strip-chart recorder, one of which is derived from a single generator that supplies a timing reference.

A greater challenge is presented to the engineer equipped only with the simple test equipment found in many laboratories, including the basic item, the general-purpose oscilloscope. The following several examples are presented to illustrate methods for accurately determining frequencies, short time intervals, and phase relations. These by no means exhaust the many possibilities, and the examples given will undoubtedly suggest other equally good arrangements or modifications.

10.5.1   Time Calibration by Substitution and Comparison

Equipment. (1) Single-trace cathode-ray oscilloscope, with provision for single-sweep triggering; (2) calibrated frequency standard capable of producing a frequency several times that expected from the unknown; (3) a means of recording the oscilloscope trace.

Method. Known and unknown signals may be introduced in succession to an oscilloscope, and calibration may be made through delayed comparison. Figure 10.6 shows a possible arrangement.

Figure 10.6   Arrangement of equipment for frequency or time interval
          determination through use of successive sweeps

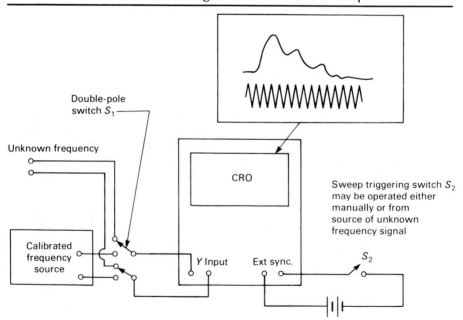

With the camera in place, a record is made, first for the unknown signal and then for the known. If the camera position can be shifted slightly, the timing trace can be displaced on the film so that the two traces are not superimposed. Single sweeps should be used, either through an external synchronization circuit as shown, or perhaps by internal synchronization for the unknown signal and external synchronization for the calibration signal. In many cases the synchronization may be obtained through a switch, S_2, actuated by some movement occurring in the system originating the signal source. For example, a lug attached to the side of a rotating gear could be used to close a spring-loaded microswitch. Exact synchronization could be provided by arranging the switch mounting so that the switch could be moved forward or backward, thereby controlling the relation between sweep and cycle. If sound is involved, use of a microphone for sweep triggering may prove feasible.

If we assume that the CRO sweep settings have remained unchanged between the two exposures, it is a simple matter to determine either an unknown frequency or a time interval by direct comparison.

We remind the reader at this point that we are assuming *minimum equipment and budget.* It is clear that application of a dual-trace scope or a storage scope (Section 9.7) would make the task much easier.

10.5.2   Time Calibration Using an Electronic Switch

Equipment. The same equipment as that used in the preceding example, including an electronic switch (Section 9.7.1). (It is assumed that a dual-trace scope is not available.)

Method. The equipment is assembled as shown in Fig. 10.7. By proper adjustment of the oscilloscope, the electronic switch, and the frequency standard a pattern may be obtained as shown. This may be recorded by approximately the same procedure outlined in the preceding example. In this case, however, only a single photographic exposure will be required. In addition, only external synchronization can be considered. If internal synchronization is attempted, the transients introduced by the electronic switch will cause recurring sweep, resulting in a confused record.

Another limitation in the use of the method lies in the maximum switching rate of the electronic switch. This should be at least *ten times* that of either of the input frequencies. The maximum rate for typical switches is about 500 Hz, which would therefore limit the input frequency to about 50 Hz.

When this method is used, the possibility of changing sweep-rate settings between recording of signal and calibration traces is eliminated.

The unknown frequency or time interval may be determined easily by direct comparison of traces.

Figure 10.7   Application of electronic switch for obtaining comparative frequency or time-interval information

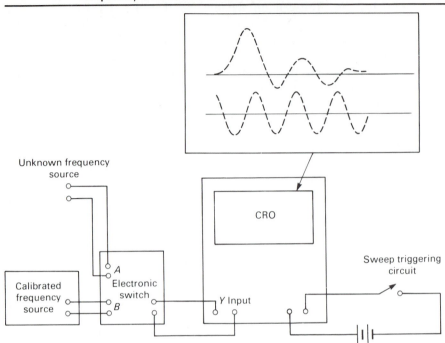

10.5.3   Frequency Determination by Z-Modulation (Primarily for Transient Inputs)

Equipment. (1) Cathode-ray oscilloscope, with provision for Z-modulated input (see Section 9.6.4); (2) calibrated frequency standard, preferably with square-wave output; (3) an oscilloscope camera for recording.

Method. Oscilloscope trace intensity may be increased or decreased by applying a voltage to the grid of the cathode-ray tube. This technique provides a very convenient method for supplying a timing calibration. For example, voltage from a calibrated oscillator may be applied to the Z-modulation terminals, as shown in Fig. 10.8. If the oscillator is set, say, at 10,000 Hz, the CRO intensity and the oscillator voltage output may be adjusted so that blanking occurs at intervals of one every 1/10,000 s. Hence the time required for the trace, or any portion of it, may be determined by counting the markers.

Square-wave oscillators are preferred for this purpose because the sharp voltage changes provide corresponding intensity changes, thus supplying good definition to the blanking. Sweep triggering may be accomplished by any of the previously described methods, using either internal or external synchronization.

Figure 10.8 Arrangement for use of Z-modulation for time interval or frequency determination

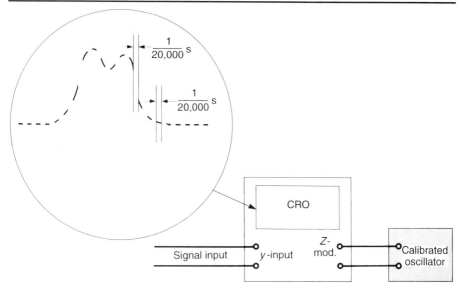

10.6 Use of Lissajous Diagrams for Determination of Frequency and Phase Relations

Equipment. (1) Cathode-ray oscilloscope; (2) calibrated variable-frequency standard.

Procedure. Lissajous (Liss-a-ju) diagrams, first studied by Nathanial Bowditch [1], and their interpretation form a basic approach to determining relative characteristics of two different frequency sources, primarily their frequency and phase relations.

Suppose two 60-Hz sinusoidal voltages from different sources are connected to a cathode-ray oscilloscope, one to the vertical and the other to the horizontal plates. Any of the following several patterns may result.

In-Phase Relations

If the two voltages are in phase, then as the x-voltage increases, so also does the y-voltage. The x-voltage will deflect the beam along the horizontal axis, and the y-voltage will deflect it in the vertical direction. The resulting trace, then, will be a line diagonally placed across the face of the tube, as shown in Fig. 10.9(a). The angle that the line makes with the horizontal will depend on the relative voltage magnitudes and the oscilloscope gain settings.

Figure 10.9 (a) In-phase Lissajous diagram, (b) Lissajous diagram for
sinusoidal inputs ±90° out of phase, (c) Lissajous
diagram for inputs 180° out of phase

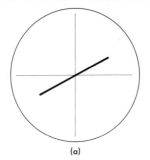

(a)

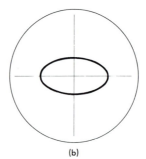

(b)

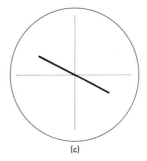

(c)

90° Phase Relations

Suppose the two 60-Hz sinusoidal voltages are 90° out of phase. Then as one
voltage passes through zero, the other will be at a maximum and vice versa.
The resulting trace will be that shown in Fig. 10.9(b). In general, it will be an
ellipse with axes placed horizontal and vertical.

180° Phase Relations

Figure 10.9(c) shows the pattern that results when the two voltages are 180° out of phase.

Other Forms of Lissajous Diagrams

Intermediate forms are ellipses with axes inclined to the horizontal. A study of Fig. 10.10 shows that when the horizontal input is at midsweep, the vertical precedes it by θ degrees, corresponding to a vertical input of y_1.

From the sine-wave plot of the curve we see that

$$\sin \theta = \frac{y_1}{y_2} = \frac{y\text{-intercept}}{y\text{-amplitude}} .$$

Figure 10.10   Lissajous diagram for sinusoidal inputs of the same frequency, but with a phase relation of θ degrees

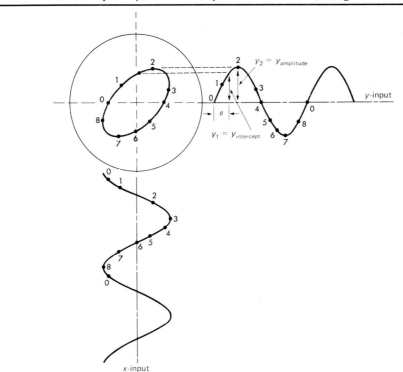

Figure 10.11   Arrangement for measuring the phase shift in an amplifier

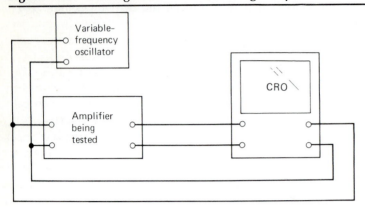

Therefore, by determining the values of y_1 and y_2 from the ellipse, we may determine the phase relation between the two inputs.

An example of the application of this method is the determination of phase shift through an amplifier. A sampling of the amplifier input signal would be applied to the x-input terminals of a CRO, and the amplifier output would be connected to the y-input terminals as shown in Fig. 10.11. By scanning the frequency range for which the amplifier is intended, one could detect any shift in phase relation. Of course, it would be necessary to know that no shift occurs with frequency in the oscilloscope circuitry or in any of the circuitry external to the amplifier.

It should be obvious by this time how Lissajous diagrams may be used to determine frequencies. Suppose an unknown frequency source with voltage output is connected to the y-input terminals of an oscilloscope, and that the output of a variable-frequency oscillator is connected to the x-input terminals. In general, the two frequencies would be different. However, by adjusting the oscillator frequency, we may obtain equal frequency diagrams such as those shown in Figs. 10.9 and 10.10. When some form of ellipse results, proof would be established that the oscillator and unknown frequencies are equal. With one known, so too would be the other.

Fortunately the method is not limited to equal frequencies. Figure 10.12 shows Lissajous diagrams for several other frequency ratios. By studying these figures, we see that a basic relation may be written as follows:

$$\frac{\text{Vertical input frequency}}{\text{Horizontal input frequency}}$$

$$= \frac{\text{number of vertical maxima on Lissajous diagram}}{\text{number of horizontal maxima on Lissajous diagram}}.$$

Figure 10.12 Lissajous displays for sinusoidal inputs
at various frequency ratios

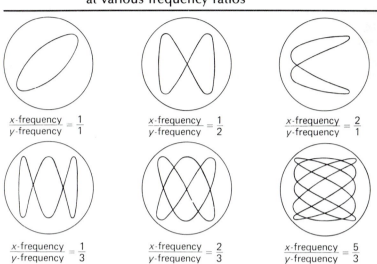

We also see that for the diagram to remain fixed on the screen, either the two input frequencies must each be fixed or they must be changing at proportional rates. In addition, the symmetry of the diagram will depend on the phase relation between the two inputs.

If the frequencies are reasonably fixed, ratios as high as ten to one may be determined without undue difficulty.

10.7 Calibration of Frequency Sources

These methods suggest means whereby a variable-frequency source, such as an oscillator or signal generator, may be calibrated. By use of a fixed-frequency source, such as the 60 Hz line voltage (through a small step-down transformer), any variable-frequency source may be calibrated for a number of points. Using the ten-to-one relation mentioned above, 60 Hz line voltage may be used to spot-calibrate from 6 to 600 Hz.

In Section 10.4 we discussed various sources of "standard" frequencies. IC divide-by-X solid-state chips, as discussed in Section 8.7.3, are useful for providing a range of frequencies from a single source. National Institute of Standards and Technology radio stations are also available, with proper receiving equipment. Of course, they provide the basic standard for the nation. The stations alternately broadcast signals at 440 and 600 Hz, which provide calibration points between 44 and 6000 Hz. If two or more variable-frequency sources are available, they may be used to extrapolate calibration points to higher frequencies.

Figure 10.13   A method for determining fan speed employing a
photoelectric sensor and a frequency standard

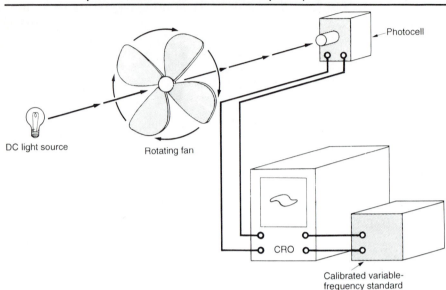

It should also be pointed out that pure sine-wave inputs are not always necessary. Figure 10.13 shows a simple arrangement for determining the speed of a fan. In this case the resulting "one-to-one" Lissajous diagram approximates a distorted parallelogram rather than an ellipse.

Various other standardizing sources suggest themselves. In Section 8.7.3 we discussed a crystal-controlled oscillator followed by IC dividers for the purpose of gating a counter. Such a system may be used to form a frequency standard whose accuracy would correspond to that of the crystal.

10.8   The Heterodyne Method of Frequency Measurement

Suppose we have two sources of pure audio tones, the two tones having nearly the same frequency. When mixed, a third or beat note (see Fig. 4.7) is produced. The frequency of the beat is a function of the *difference* in the two original notes. If the frequency of one of the sources is known and is adjusted to produce *zero beat*, then the frequency of the other source is also known by comparison. This procedure for determining frequency is called the *heterodyne* method: The two signals are *heterodyned* (see Section 4.4.1). Piano tuners make use of this method when adjusting a piano string to zero beat with a tuning fork.

The method is particularly useful for determining frequencies well above the audio range. Radio frequencies are often measured by this method. In this case the standardizing signal originates from a carefully calibrated, variable-frequency oscillator. For radiated signals an ordinary radio receiver covering the desired frequency range may serve as a mixer. The generator frequency is adjusted until the difference between the known and unknown frequencies falls within the audio range, thereby producing the well-known amplitude-modulated squeal so familiar when two radio stations interfere. The generator is then fine-tuned to produce zero beat. True zero beat may be determined by ear within 20 or 30 Hz. The uncertainty, of course, would also include the uncertainty inherent in the standard. By using an oscilloscope or an analog-type electronic voltmeter, the resolution may be reduced to near zero. Provision is made in most signal generators of this type for spot calibration with one or more of the National Institute of Standards and Technology radio signals (Section 10.4.1). One caution should be noted: The signal generator may very well produce many harmonics, any one of which would be suitable as a calibrating source, *provided* one knows which harmonic is being used. This problem is generally minimal because the experimenter usually has a fairly good idea of the approximate value of the unknown frequency.

The heterodyne principle is very important in a wide variety of frequency measurements. One especially important application is in ultrasound and laser velocity measurements where the determination of the Doppler-shift frequencies is needed (Section 15.11).

10.9 Measurement of Angular Motion

Probably the most common example of direct counting and EPUT determination is the measurement of angular motion. Many different devices have been used for this purpose, most of which fall under one of the headings in the following classification:

1. Mechanical
    a. Direct counters
    b. Centrifugal speed indicators
2. Electrical
    a. Generators (ac and dc)
    b. Reluctance-type proximity pickups
    c. Hall-effect sensors
3. Optical
    a. Stroboscope
    b. Photocells
    c. Optical shaft encoders.

Mechanical counters may be of the *direct-counting* digital type or may be counters with a gear reducer. In the latter case, angular motion available at a shaft end is reduced by a worm and gear, and the output is indicated by rotating scales. In both examples, rpm is measured by simply counting the revolutions for a length of time as measured with a stopwatch, and calculating the turns per minute from the resulting data.

Modifications incorporate the timing mechanism in the counter. The timer is used to actuate an internal clutch that controls the time interval during which the count is made.

Centrifugal rpm indicators use the familiar *flyball-governor* principles, which balance centrifugal force against a mechanical spring. An appropriate mechanism transmits the resulting displacement to a pointer, which indicates the speed on a calibrated scale. The term *tachometer* is often applied to an instrument of this sort, or to any *direct-indicating* angular-speed measuring device.

Electrical tachometers generally make use of a small permanent magnet-type dc or ac generator connected to a simple voltmeter. The dc generator requires some form of commutation, which presents the problem of brush maintenance. On the other hand, the ac generator requires an instrument rectifier if a simple dc meter is to be used for indication. Of course, the advantage of the electrical kind over the mechanical is that the former provides continuous indication that may be displayed or recorded remotely.

A variable-reluctance pickup (discussed in Section 6.12) or a Hall-effect device such as the sensor illustrated in Figure 6.23 permit the measurement of angular speed or position by noncontacting means. If the pickup is placed near the teeth of a rotating gear, for example, extremely accurate speed measurements may be made by either an electronic counter, a frequency-sensitive indicator (frequency meter) or, if the speeds are constant, Lissajous techniques (Section 10.6).

Photocells or optical interruptors (Figure 6.20), may also be used to provide voltage pulses originating by the interruption of a light beam from rotation or movement of a machine member. These pulses may be treated in a manner similar to those from a reluctance or Hall-effect sensor. Optical shaft encoders, based on circular binary patterns [Fig. 8.11(b)], may be similarly adapted to angular speed measurement. Finally, any of the various strobo-scopic methods discussed in Section 10.3 may also be used for speed measurements.

Suggested Readings

ASME PTC 19.13–1961, *Measurement of Rotary Speed.* New York: ASME, 1961.

Techniques for Digitizing Rotary and Linear Motion. Wilmington, Mass.: Dynamics Research Corp., 1976.

Wolf, S., and R. F. M. Smith. *Student Reference Manual for Electronic Instrumentation Laboratories.* Englewood Cliffs, N.J.: Prentice Hall, 1990.

Problems

10.1 Prove the validity of Eq. (10.1).

10.2 Using an electronic counter, monitor the local power-line frequency. Use a step-down transformer to avoid the danger of the line voltage. What variations are noted over the period of the test? Compare results using (a) the setting for frequency readout and (b) the setting for period readout.

10.3 Use a flashing-light-type stroboscope and determine the time-speed relationship for a shop-bench-type grinder as it *decelerates*. A suggested technique is as follows: Set the stroboscope flashing rate to a predetermined frequency; then turn off the grinder power, simultaneously starting a timer (e.g., a stopwatch). Determine the time required for the first synchronization between wheel speed and flashing rate. Repeat for a range of flashing rates until sufficient data are obtained to plot a deceleration versus time curve. This approach is a very simple one that requires a minimum of test equipment. Why would this approach not be viable to determine the acceleration-time plot for the typical bench grinder? Other more sophisticated schemes may be devised; propose other approaches to measuring the decelerating speed-time characteristics.

10.4 Using a radio receiver and a CRO, compare the local power-line frequency with the 600 Hz audio tone transmitted by WWV the radio station of the National Institute for Standards and Technology. *Be sure to use an isolation transformer between the power line and the CRO.*

10.5 Use the power-line frequency to calibrate a signal generator over the range 10 to 600 Hz. *Be sure to use a step-down transformer to isolate the line and to obtain a reasonably low voltage (say, no more than 6 V ac).*

10.6 Suggest a means of using an oscilloscope for measuring the actuating time of a simple double-throw electric relay. [*Hint:* Use the relay actuation voltage to trigger the sweep and a circuit using the relay contacts for the timing-limit switches. Would it be practical to use a debouncing element in the circuit? (See Section 8.4.2.)]

10.7 Show that an ellipse is generated on a CRO display when the y-input is $A_0\sin(\omega t - \phi)$ and the x-input is $A_i\sin \omega t$.

10.8 Make a test arrangement similar to that shown in Fig. 10.13 (or with modifications as desired) and determine the time-speed relationship for a common office fan as it accelerates from rest to full speed. Observe the CRO screen, determining the time for the Lissajous diagram to transit the one-to-one ratio. Adapt the procedural suggestions given in Problem 10.2 to this setup. Use curve-fitting methods given in Sections 3.14 and 3.15 to find a relationship, $\omega = f(t)$. Check to see whether, by chance, the relationship approximates first-order characteristics. Run a similar set of tests to determine the speed-time curve for deceleration.

10.9 Devise an experimental method for determining the *accelerating* speed-time characteristics of a shop-bench-type tool grinder (see Problem 10.2).

10.10 Reference [2] describes a simple method for determining the fundamental resonance frequency of a turbine blade. In essence, the method consists of striking the blade with a soft mallet and using a microphone to pick up the sound.

The microphone output is fed to a CRO and its frequency is compared with that of a signal generator using the Lissajous technique. Any harmonics of significance would also be displayed. Locate a complex elastic member and use the method to study its characteristics of frequency of vibration.

10.11 Using the basic approach and the "complex elastic member" suggested in the final sentence of Problem 10.10, make a harmonic analysis of the waveform captured from the CRO screen. Use of a storage scope would be particularly useful.

References

1. Curves, Special. In *Encyclopedia Britannica,* vol. 6. Chicago: William Benton, Publisher, 1957, p. 892.

2. Rosard, D. D. *Natural Frequencies of Twisted Cantilever Beams.* ASME Paper 52-A-15, 1952.

CHAPTER 11

Displacement and
Dimensional
Measurement

11.1 Introduction

The determination of linear displacement is one of the most fundamental of all measurements. The displacement may determine the extent of a physical part, or it may establish the extent of a movement. It is characterized by the determination of a component of space. In *unit* form it may be a measure of either strain (Chapter 12) or angular displacement.

Probably to a greater extent than any other quantity, displacement lends itself to the simplest process of measurement: *direct comparison*. Certainly the most common form of displacement measurement is by direct comparison with a secondary standard. Measurements to least counts of the order of 0.5 mm (about 0.02 in.) may be accomplished without undue difficulty with use of nothing more than a steel rule for a standard. For greater resolutions or sensitivities, measuring systems of varying degrees of complexity are required. Measurements to least counts on the order of 0.005 mm (about 0.0002 in.) may be achieved with precision mechanical mechanisms, which typically use a system of gears to amplify a small displacement. For example, a micrometer converts the linear displacement of its shaft to a much larger rotation along the scale on the shaft barrel. Higher resolution devices (to a few microinches) may provide direct comparison to a set of calibrated gage blocks; very high resolution may be obtained with an optical interferometer, which enables displacement measurements to a fraction of the wavelength of light (about 0.2 microinch).

Table 11.1 Classification of Displacement Measuring Devices

Low-Resolution Devices (to 1/100 in.) (0.25 mm)

1. Steel rule used directly or with assistance of
   a. Calipers
   b. Dividers
   c. Surface gage
2. Thickness gages

Medium-Resolution Devices (to 1/10,000 in.) (2.5×10^{-3} mm)

1. Micrometers (in various forms, such as ordinary, inside, depth, screw thread, etc.) used directly or with assistance of accessories such as
   a. Telescoping gages
   b. Expandable ball gages
2. Vernier instruments (various forms, such as outside, inside, depth, height, etc.)
3. Specific-purpose gages (variously named, such as plug, ring, snap, taper, etc.)
4. Dial indicators
5. Measuring microscopes

High-Resolution Devices (to a Few Microinches) (2.5×10^{-5} mm)

Gage blocks used directly or with assistance of some form of comparator, such as
a. Mechanical comparators
b. Electronic comparators
c. Pneumatic comparators
d. Optical flats and monochromatic light sources

Super-Resolution Devices

Various forms of interferometers used with special light sources.

Figure 11.1 Typical dimensioning specifications for
                an internal diameter (English units)

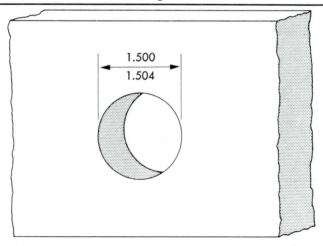

For purposes of discussion, we classify various measuring devices in Table 11.1. With a few exceptions, these systems are used for measuring fixed physical dimensions. Before we discuss any of them in detail, let us consider the following measurement problem.

11.2   A Problem in Dimensional Measurement

Suppose a hole is to be bored to the dimensions shown in Fig. 11.1, and that the part is to be produced in quantity. Such a dimension would probably be checked with some form of plug gage, illustrated in Fig. 11.2. One end of the plug gage is the *go* end, and the other the *no-go* end. If the *go* end of the gage

Figure 11.2   A go/no-go plug-type gage

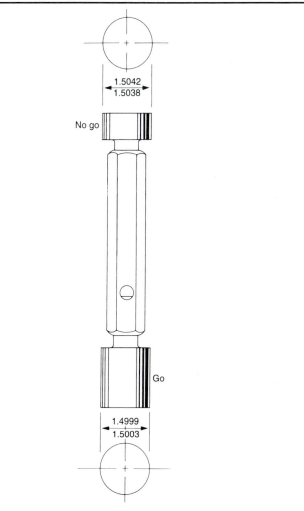

fits the hole, we know that the hole has been bored large enough; if the *no-go* end cannot be inserted, the hole has not been bored too large.

Now, the plug gage itself would have to be manufactured, and no doubt drawings of it would be made. A rule of thumb is to dimension the plug gage with tolerances on the order of 10% of the tolerance of the part to be measured. If this rule is followed, the ends of the plug gage may be dimensioned as shown in Fig. 11.2.

It will be noted that the gage tolerance of 0.0004 in. (10% of the part tolerance, 0.004 in.) is applied symmetrically to the *no-go* end, corresponding to the upper limiting dimension of 1.504 in. On the other hand, the gage tolerance as applied to the *go* end penalizes the machinist somewhat because, in effect, an extra ten-thousandth of an inch is taken away from what is the machinist's. This is often done to increase the life of the gage by letting the gage wear toward the specified limit. Ideally, the *go* end will be inserted every time the gage is used and hence will wear, whereas the *no-go* end will never be inserted and will therefore experience no wear.

Provision has now been made for satisfactorily gaging the bored hole, provided the gages themselves are accurately made. How will we know if the gages are within tolerance? We can find out only by measuring them. This leads directly to the *gage block* listed under "High-resolution devices" in Table 11.1. A gage-block set is the basic "company" standard for any small (0.01 to 10 in.) dimension.

By use of one of the comparison methods (to be described in a later section), the plug gage would be checked dimensionally. But, of course, we must not overlook the fact that to be useful, the gage blocks themselves must be measured, and so on ad infinitum, at least back to the basic length standard (Section 2.5).

An example may be used to illustrate the extreme importance of measurement standards. Suppose that the 1.500-in. hole described above is in a part to be used by an automobile manufacturer. Very probably the gage would be made by some other company, one that specializes in making gages. Both the gage maker and the automobile manufacturer would undoubtedly "standardize" their measurements by using gage blocks. It is clear that unless the different gage-block sets are accurately derived from the same basic standard, the dimension specified by the automobile manufacturer will not be reproduced by the gage maker.

11.3   Gage Blocks

Gage-block sets are industry's dimensional standards. They are the *known* quantities used for calibration of dimensional measuring devices, for setting special-purpose gages, and for direct use with accessories as gaging devices. They are simply small blocks of steel having parallel faces and dimensions accurate within the tolerances specified by their class. Blocks are normally available in the following classes.

Figure 11.3   A set of 81 gage blocks
              (Courtesy: The DoAll Co., Des Plaines, IL)

For the English system of units:

Grade of Block		Tolerance, μin. (microinches)
Class B	"Working" blocks	±8*
Class A	"Reference" blocks	±4
Class AA	"Master" blocks	±2 for all blocks up to 1 in. and ±2 μin./in. for larger blocks

For the SI system:

Grade of block	Tolerance, μm (micrometers)†
0	±0.10 to ±0.25
1	±0.15 to ±0.40
2	±0.25 to ±0.70
3	±0.50 to ±1.30

Gage blocks are supplied in sets, with those sets having the largest number of blocks being the most versatile. Figure 11.3 shows a set made up of 81

---

* For very small displacement or tolerances, the mechanical engineer will commonly use the μm (the micrometer) or the microinch (1×10^{-6} in.). Physicists commonly use the angstrom (1×10^{-7} mm). Equivalents are $1\,\mu$m = 39.37 μin., and 1 Å = 0.003937 in. *Unit displacement* (e.g., strain and certain coefficients) is unitless. In this case parts per million, or ppm, is conveniently employed.

† A range of tolerances is listed. Precise values depend on the size of the block.

blocks (plus 2 wear blocks) having dimensions as follows:

9 blocks with 0.0001-in. increments from 0.1001 to 0.1009 inclusive

49 blocks with 0.001-in. increments from 0.101 to 0.149 inclusive

19 blocks with 0.050-in. increments from 0.050 to 0.950 inclusive

4 blocks with 1-in. increments from 1 to 4 inclusive

2 tungsten-steel wear blocks, each 0.050 in. thick

Blocks are made of steel that has been given a stabilizing heat treatment to minimize dimensional change with age. This consists of alternate heating and cooling until the metal is substantially without "built-in" strain. They are hardened to about 65 Rockwell C.

Distribution of sizes within a set is carefully worked out beforehand, and for the set in Fig. 11.3 accurate combinations are possible in steps of one ten-thousandth in over 120,000 dimensional variations.

11.4   Assembling Gage Block Stacks

Blocks may be assembled by *wringing* two or more together to make up a given dimension. Suppose that a dimensional standard of 3.7183 in. is desired. The procedure for arriving at a suitable combination might be determined by successive subtraction, as indicated immediately below:

	Blocks Used
Desired dimension = 3.7183	
Ten-thousandths place = 0.1003	0.1003
Remainder = 3.618	
Thousandths place = 0.108	0.108
Remainder = 3.51	
Hundredths place = 0.11	0.11
Remainder = 3.4	
Two wear blocks = 0.1 (0.05 each)	0.1
Remainder = 3.3	
Tenths place = 0.3	0.3
Remainder = 3.0	
Units place = 3.0	3.0
Remainder = 0.0 Check ⋯	3.7183

Blocks are not stacked by simply resting them one on top of another. They must be wrung together in such a way as to eliminate all but the thinnest oil film between them. This oil film, incidentally, is an integral part of the block itself; it cannot be completely eliminated, since it was present even at manufacture. The thickness of the oil film is always of the order of 0.2 μin. [1].

Figure 11.4 Two methods for using gage blocks for direct
comparison of the length dimension

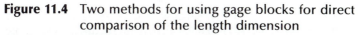
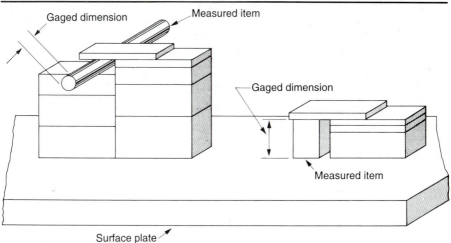

Properly wrung blocks markedly resist separation because the adhesion between the surfaces is about 30 times that due to atmospheric pressure. Unless the assembled blocks exhibit this characteristic, they have not been properly combined. The resulting assembly of blocks may be used for *direct* comparison in various ways. Two simple ways are shown in Fig. 11.4.

11.5 Surface Plates

When blocks are used as shown in Fig. 11.4, some accurate reference plane is required. Such a flat surface, known as a *surface plate,* must be made with an accuracy comparable to that of the blocks themselves. In years past, carefully aged cast-iron plates with adequate ribbing on the reverse side were used. Such plates were prepared in sets of three, carefully ground and lapped together. When combinations of two are successively worked together, the three surfaces gradually approach the only possible surface common to all three—the true flat.

Machine-lapped and polished granite surface plates have largely replaced the hand-produced cast-iron type. Granite has several advantages. First, it is probably more nearly free from built-in residual stresses than any other material because it has had the advantage of a long period of time for relaxing. Hence, there is less tendency for it to warp when the plates are prepared. Second, should a tool or work piece be accidentally dropped on its surface, residual stresses are not induced, as they are in metals, causing warpage; the granite simply powders somewhat at the point of impact. Third, granite does not corrode.

Optical flats, the uses of which are discussed in Section 11.11, are very similar to surface plates except that they are usually made of fused quartz and are generally much smaller in size.

11.6 Temperature Problems

Temperature differences or changes are major problems in accurate dimensional gaging. The coefficient of expansion of gage-block steels is about 11.2 ppm/°C (6.4 ppm/°F). Hence even a shift of one degree in temperature would cause dimensional changes of the same order or magnitude as the gage tolerances. The standard gaging temperature has been established as 20°C (68°F).

Several solutions to the temperature problem are possible. First, the most obvious solution is to use air-conditioned gaging rooms, with temperature maintained at 68°F. This procedure generally is followed when the volume of work warrants it; however, it is not a complete solution, for mere handling of the blocks causes thermal changes requiring up to 20 min to correct. For this reason, use of insulating gloves and tweezers is recommended. In addition, care must be exercised to minimize radiated heat from light bulbs, etc. [2].

A constant-temperature bath of kerosene or some other noncorrosive liquid may be used to bring the blocks and work to the same temperature. They may be removed from the bath for comparison; or, in extreme cases, measurement may be made with the items submerged.

On the other hand, if temperature control is not feasible, corrections may be used, based on existing conditions. A moment's thought will indicate that *if the gage blocks and work piece are of like materials, there will be no temperature error so long as the two parts are at the same temperature.* In by far the greatest number of applications, steel parts are gaged with steel gage blocks, and although there will probably be a slight difference in coefficients of expansion and the gage and parts may be at slightly different temperatures, appreciable compensation exists and the problem is not always as great as suggested in the preceding several paragraphs.

If both the part being gaged and the blocks are at temperature T_r (room temperature), corrections may be made by application of the following:

$$L = L_b[1 - (\Delta\alpha)(\Delta T)(10^{-6})], \tag{11.1}$$

where

$\Delta\alpha = (\alpha_p - \alpha_b),$

$\Delta T = (T_r - \text{standard reference temperature}),$

$L =$ the true length of the dimension being gaged (at reference temperature),

$L_b =$ the nominal length of gage blocks determined by summation of dimensions etched thereon,

α_p = the temperature coefficient of expansion of the part being gaged (ppm/°),

α_b = the temperature coefficient of expansion of the gage-block material (ppm/°),

T_r = the ambient temperature.

In using these relations, it is very necessary that proper signs be applied to $\Delta\alpha$ and ΔT.

Example 11.1
Let L_b = 10 cm, α_p = 13 ppm/°C, α_b = 11.2 ppm/°C, and T_r = 24°C. Find L.

Solution. Substituting, we have

$$L = 10[1 - (1.8 \times 10^{-6})(4)]$$

$$= 9.999928 \text{ cm}.$$

Example 11.2
Let L_b = 9.7153 in., α_p = 5.9 ppm/°F, α_b = 6.4 ppm/°F, and T_r = 62°F. Find L.

Solution. Substituting, we have

$$L = 9.7153[1 - (-0.5 \times 10^{-6})(-6)]$$

$$= 9.715271 \text{ in.}$$

11.7 Use of Gage Blocks with Special Accessories

Gage blocks are sometimes used with special accessories, including clamping devices for holding the blocks. When so used, height gages, snap gages, dividers, pin gages, and the like may be assembled, using the basic gage blocks for establishing the essential dimensions. Use of devices of this type eliminates the necessity for transferring the dimension from the gage-block stack to the measuring device.

11.8 Use of Comparators

One of the primary applications of gage blocks is that of calibrating a device called a *comparator*. As the name suggests, a comparator is used to compare known and unknown dimensions. One form of mechanical comparator is

Figure 11.5   Comparator employed to measure the difference between
                    a known and an unknown linear dimension

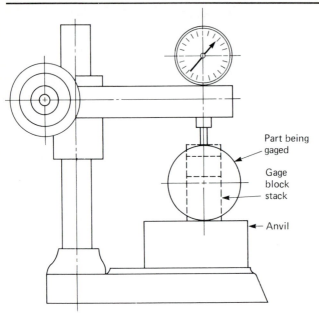

Part being
gaged

Gage
block
stack

Anvil

shown in Fig. 11.5. Some form of displacement sensor is used (the figure shows
an ordinary dial indicator) to indicate any dimensional differences as described
in the next paragraph. A variety of different sensing devices are found on
commercial comparators, including strain-gage types, purely mechanical devices, variable inductance, etc. The resolution and accuracy of the sensing
device would add to the gage tolerance in establishing the minimum uncertainty. We see that to use a sensor with the ability to sense discrepancies less
than block tolerances would be quite unnecessary.

As an example of a comparator's use, suppose that the diameter of a plug
gage is required. The nominal dimension may first be determined by use of an
ordinary micrometer. Gage blocks would be stacked to the indicated rough
dimension and placed on the comparator anvil, and the indicator would be
adjusted on its support post until a zero reading is obtained. The gage blocks
would then be removed and the part to be measured substituted. A change in
the indicator reading would show the difference between the unknown
dimension and the height of the stack of blocks and would thereby establish
the value of the dimension in question. Inasmuch as most gage-block sets
contain series having very small changes in base dimensions (e.g., in steps of
ten-thousandths of an inch for the English system), the gage set may be used
quite conveniently for calibrating the comparator.

Pneumatic Comparators

Pneumatic gaging is based on a double-orifice arrangement such as that illustrated in Fig. 11.6. Intermediate pressure, P_i, is dependent on the source pressure, P_s, and the pressure drops across the orifices O_1 and O_2. The effective size of orifice O_2 may be varied by a change in distance d. As d is changed, pressure P_i will change, and this change can be used as a measure of dimension d. Figure 11.7 shows schematically how this arrangement may be used as a comparator.

Using a double orifice as shown in Fig. 11.7, Graneek [3] determined the following empirical equation for $P_s = 15\,\text{lbf/in.}^2$ absolute (or psia) and an orifice diameter $O_1 = 0.033\,\text{in.}$:

$$\left(\frac{A_2}{A_1}\right)^2 = \frac{P_s}{P_i} - \frac{P_i}{P_s}, \tag{11.2}$$

in which A_1 and A_2 are the areas of orifices O_1 and O_2, respectively. For values of P_i/P_s between 0.4 and 0.9, the relation corresponds very nearly to the straight line

$$\frac{P_i}{P_s} = 1.10 - 0.50\frac{A_2}{A_1}.$$

Hence, for fixed values of P_s and A_1, P_i may be expected to vary linearly with the value of A_2. In addition, the device may be made quite sensitive, capable of measuring displacements of a few microinches.

Figure 11.6 Schematic diagram showing the principle of operation of the pneumatic comparator

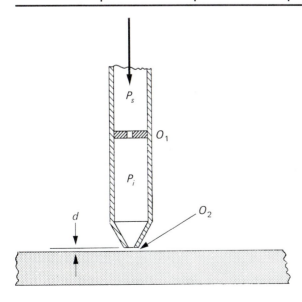

Figure 11.7 Circuit diagram for a pneumatic comparator

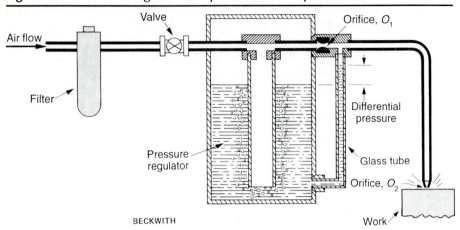

Pneumatic gaging is widely used for production work, where reliability and ruggedness are a requirement. Gage blocks are used as standards for calibration. (References [4] and [5] contain additional material on pneumatic gaging methods.)

11.9 Optical Methods

Optical methods applied to linear measurement may be divided into two general areas as follows: (1) very accurate measurement of small dimension or displacement by methods using light-wave interference or image magnification, and (2) measurement of large dimensions by use of alignment telescopes with accessories and projection systems. Although such a classification can only be quite general, a dimension of about three feet or one meter may be suggested as a rough dividing line, with the realization that there will be overlap in many cases.

11.10 Monochromatic Light

The method of interferometry, used for accurate measurement of small linear dimension, is described in subsequent sections. A required tool is a source of monochromatic (one color or wavelength) light.

There are various sources of monochromatic light. Optical filters may be used singly or in combination to isolate narrow bands of approximately single-wavelength light. Or a prism may be employed to "break down" white light into its components and, in conjunction with a slit, to isolate a desired wavelength; however, both these methods are quite inefficient. Most practical sources rely on the electrical excitation of atoms of certain elements that radiate light at discrete wavelengths. The excitation system may be either a lamp or a laser.

Possible elements for lamps include mercury, mercury 198, cadmium, krypton, krypton 86, thallium, sodium, helium, and neon [4]. Means are provided for vaporizing the element, if not already gaseous, and to produce a visible light through the application of electric potential, often at a high voltage and/or frequency.

A common industrial standard is the helium lamp. This standard is obtained by means not unlike those used in the familiar neon signs. A tube is charged with helium and connected to a high-voltage source, which causes it to glow. The resulting light has a narrow range of wavelengths and is intense enough for practical use. The wavelength of this source is $0.589\,\mu$m ($23.2\,\mu$in.). Table 11.2 lists the approximate wavelengths for the various primary colors, and the wavelengths of several specific sources are given in Table 11.3.

Laser light sources are of various forms, among which gas lasers are probably the type most often used for dimensional measurements. Common examples include the red helium-neon (HeNe) laser and the blue-green argon-ion laser. Gas lasers have several major advantages over lamps as sources of monochromatic light: (1) They can be extremely monochromatic, containing only a very narrow wavelength band; (2) their light is highly collimated, forming a narrow and directional beam; (3) they are usually polarized; and (4) their light is highly coherent.

The property of coherence is particularly valuable for interference measurements. A light beam is *coherent* when the light wave remains in phase with itself over the length of the beam. Thus, one speaks of the *coherence length* of a light beam as the distance over which the beam stays in phase with itself; beyond this length, the beam shows some phase shift. Interference measurements, which relate the phase difference of two coherent beams to a difference in their path lengths, are possible only for path differences less than the coherence length. Because lasers have much greater coherence lengths than lamps, they vastly increase the distances which can be measured by

Table 11.2 Approximate wavelengths of light of the various primary colors

Color	Range of Wavelengths	
	Micrometers	*Microinches*
Violet	0.399 to 0.424	15.7 to 16.7
Blue	0.424 to 0.490	16.7 to 19.3
Green	0.490 to 0.574	19.3 to 22.6
Yellow	0.574 to 0.599	22.6 to 23.6
Orange	0.599 to 0.645	23.6 to 25.4
Red	0.645 to 0.699	25.4 to 27.5

Table 11.3 Wavelengths from specific sources

Source	Wavelengths		Fringe Interval	
	μin.	μm	μin./fringe	μm/fringe
Lamps				
Mercury Isotope 198	21.5	0.546	10.75	0.273
Helium	23.2	0.589	11.6	0.295
Sodium	23.56	0.598	11.78	0.299
Krypton 86	23.85	0.606	11.92	0.303
Cadmium Red	25.38	0.644	12.69	0.322
Lasers				
Argon Ion	18.03	0.4579	9.01	0.2290
	19.21	0.4880	9.606	0.2440
	20.26	0.5145	10.13	0.2573
Krypton Ion	16.26	0.4131	8.13	0.2066
	25.48	0.6471	12.74	0.3236
	26.63	0.6764	13.32	0.3382
Helium-Neon	24.91	0.6328	12.46	0.3164

interference. For example, a monochromatic lamp has a coherence length no more than several tens of centimeters and is useful for interference measurements of similar distances at most. A frequency-stabilized HeNe laser, on the other hand, can have a coherence length of more than a kilometer; in principle, laser interference allows distances of hundreds of meters to be measured to a fraction of the wavelength of light.

11.11 Optical Flats

An important accessory in the application of monochromatic light to measurement problems is the optical flat (see Fig. 11.8), which is made of a material such as strain-free glass or fused quartz. As the name indicates, at least one of the surfaces is lapped and polished to a close flatness tolerance. Either square or circular flats are available in sizes ranging from 1 to 16 in. (or larger on special order) across a side or diameter. Thicknesses range from $\frac{1}{4}$ in. (6.35 mm) for the small flats to $2\frac{3}{4}$ in. (70 mm) for the large flats. Manufacturing tolerances are as follows: commercial, 8 μin. (0.203 μm); working, 4 μin. (0.101 μm); master, 2 μin. (0.051 μm); reference, 1 μin. (0.025 μm).

Figure 11.8   A set of optical flats
(Courtesy: The Van Keuren Co., Watertown, MA)

11.12   Applications of Monochromatic Light and Optical Flats

An optical flat and a monochromatic light source may be used to compare gage-block dimensions with unknown dimensions; that is, they may be used together as a form of dimensional comparator. They may also be used to measure variation from flatness or to determine the contour of an almost flat surface. For these applications the principles involved are as follows:

When light waves are applied to measurement problems, principles of interferometry are used. In general, light waves from a single source may be caused to add or subtract, increasing or decreasing the light intensity, depending on the phase relation. The arrangement shown in Fig. 11.9, making use of an optical flat and a reflective surface, illustrates the basic principles.

Two requirements must be met: (1) *an air gap (a wedge) of varying thickness must exist between the two surfaces,* and (2) *the work surface must be reflective.*

As shown in Fig. 11.9, the light is reflected from both the working face of the flat and the work surface of the part being inspected. At the particular points where multiples of half wavelengths occur, we can see dark interference bands or fringes. Figure 11.10 shows a typical pattern. We see that a fringe represents a locus of separation between work and flat of a definite integral number of half wavelengths of the light used. *Adjacent fringes may be interpreted, therefore, as representing contours of elevation differing by one-half wavelength.* We shall call this distance the *fringe interval.*

Figure 11.9   Sketch illustrating the basic light-interference principle

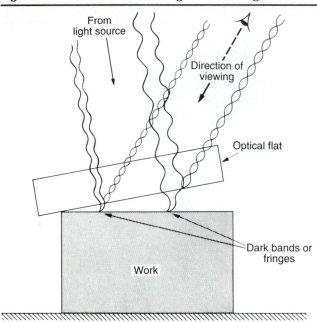

Figure 11.10   Photograph showing the interference pattern for a badly
worn comparator anvil
(Courtesy: The Van Keuren Co., Watertown, MA)

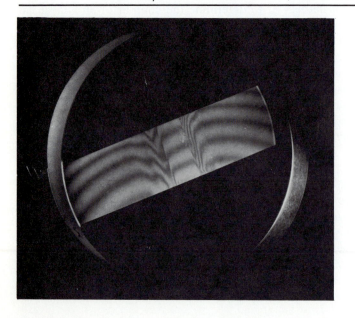

Point or Line of Contact

Another item of information that is generally desirable is the point of contact between the flat and the work surface. There will be at least one primary support point for the flat. Sometimes, instead of making contact at a point, the flat will make contact with the work along a line. This can be determined by gently rocking the flat on the surface. Actually, of course, this motion must be very slight in order to maintain an observable pattern, and only a very light pressure need be applied. The general rule is that as the flat is rocked, the point of contact is determined as that spot or line in the pattern that does not shift. Actually in some situations the point will move slightly, but in general it does not stray far from its starting point.

Figure 11.11(a) illustrates the pattern that would be observed if the work surface were spherically convex. The flat rests on the central high spot, and a fringe pattern of concentric circles results. It will be remembered that each adjacent fringe represents a change in elevation, or *fringe interval*, of one-half wavelength of the light that is used. For a helium-lamp source, this value would be 0.295 μm (11.6 μin.).

If one edge of the optical flat is pressed gently, it will rock on the high spot and the fringe pattern will shift into a form like that shown in Fig. 11.11(b). The high spot, indicated by the center of the concentric circles, will shift

Figure 11.11   (a) Appearance of interference fringes that occur when an optical flat is placed on a convex spherical surface, (b) shift of fringe pattern caused by rocking an optical flat.

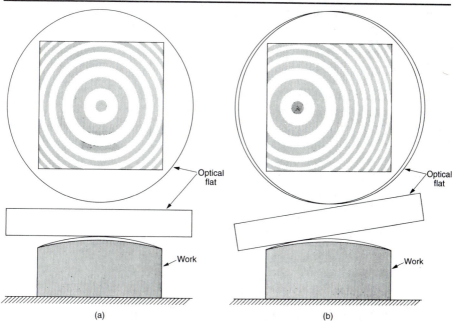

(a)   (b)

slightly. However, it remains as the primary center of the pattern. In addition, it should be noted *that many of the outer fringes run out at the edge of the work.*

On the other hand, suppose that the surface is spherically concave and that the flat is resting on a line extending all the way around the edge, or at least on several points around the edge. If the edge of the optical flat is gently pressed down, it will be observed that although the central point may shift slightly, the edge points (or line) remain stationary and do not move as pressure is varied. From this simple test it can only be concluded that the surface is concave. Having determined the points of contact, we may now map the general contour.

We may put this on a more formal basis as follows: Let us define the term *fringe order* as the number we would assign a given fringe if we counted them in sequence, starting with a contact point as zero order. We may write the relation

$$\Delta d = (\text{fringe interval}) \times N$$
$$= \frac{\lambda}{2} N, \tag{11.3}$$

in which

$\Delta d =$ the difference in elevation between contact and the point in question (the contact point will always be the highest point; hence this difference in elevation will always be negative, or away from the face of the optical flat),

$N =$ the fringe order at the point in question,

$\lambda =$ the wavelength of the light source used.

11.13   Use of Optical Flats and Monochromatic Light for Dimensional Comparison

Suppose that the diameter of a plug gage (Section 11.2) is to be determined to an uncertainty of several microinches or so. Gage blocks and some form of comparator could be used or, in place of the comparator, an optical flat and a monochromatic light source would serve. If the latter method were used, the procedure would be as follows:

1. The gage diameter is first approximated by use of a micrometer. An estimate to the nearest half-thousandth of an inch (about 0.01 mm) could be expected.

2. Gage blocks are then stacked to the "miked" dimension. At this point it would also be a good idea to check the micrometer against the blocks. If a significant discrepancy is found, the stack of blocks can be re-formed.

3. The stack of blocks, along with the plug gage, is then arranged as shown in Fig. 11.12, with the flat resting on top of the combination.

4. If the dimensions of the stack and the plug are nearly the same, an observable fringe pattern will appear over the surface of the gage block.

Figure 11.12   Arrangement for measurement using gage blocks,
optical flat, and monochromatic light source

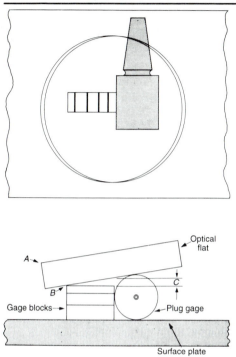

Sometimes considerable patience is required to obtain a good pattern, and experience is certainly helpful. Probably the most important single factor in obtaining good results is cleanliness, because a small bit of lint or a foreign particle will be very large in comparison with the sensitivity of the interference fringes. Ideally, a pattern such as that shown in Fig. 11.12 should finally be obtained.

5.   Although Fig. 11.12 indicates that the plug dimension is greater than the stack dimension, this fact must be proved by determining the point of contact between flat and gage block. If the flat were gently pressed downward at point *A*, the fringes would either crowd more closely together (if the plug dimension is the larger) or spread apart (if the stack dimension is the larger). A moment's thought will confirm this if it is remembered that the difference in elevation between adjacent fringes is a fixed quantity. If the situation illustrated exists, then the fringes will crowd more closely together, and contact at *B* between block and flat will be indicated.

6.   Now that the point of contact between flat and gage blocks has been determined, the difference in elevation between flat and blocks, *C*, may be

determined by counting the number of fringes, say to the nearest $\frac{1}{5}$ fringe. This number, multiplied by the half wavelength of the light used, gives the height of the flat above the block at C.

7. Simple application of similar triangles can now be used to determine the difference between the height of the stack of gage blocks and the plug-gage diameter.

11.14   The Interferometer

Measurement of length to, say, 0.5μin., or about 10^{-5} mm, is not commonly required in mechanical development work, but by this time the reader is undoubtedly aware that some accurate and reliable means must be available to establish the absolute length of gage blocks. In other words, how are gage blocks calibrated? We have already discussed methods whereby they may be compared with other gage blocks, but somehow a comparison must be made with a more fundamental standard, such as the wavelength of a specific light source, as discussed in Section 2.5. Interferometers make possible this very basic measurement.

An optical interferometer of great historical significance was devised and used by Albert A. Michelson (1852--1931). Until 1960 the length standard was the meter defined by two finely scribed lines on the platinum-iridium prototype meter bar. Gages of this sort are called *line standards,* whereas the ordinary gage block is an *end standard.* Michelson's primary objective was to determine the wavelengths of light derived from certain sources by comparison to the meter bar. Of course, the tables are now turned. Light now provides the standard, and the problem is to use it to measure lengths, such as gage-block dimensions.

The fundamental elements of a Michelson-type interferometer are shown in Fig. 11.13. A laser beam passes through a beam splitter, which directs 50% of its light to each of two mirrors or cube-corner reflectors (shown). The light beams reflect back to the beam splitter, and a portion of each again passes through to an aperture and photodetector. At the detector, the two beams will interfere constructively (bright) or destructively (dark), depending on the number of wavelengths by which their paths differ.

One reflector can be traversed along the length of an unknown dimension. If it moves a distance δ, the path of its light beam increases by 2δ. The number of successive dark fringes that occur at the photodetector during this motion is equal to the number of wavelengths, N, in the path change:

$$2\delta = N\lambda$$

By counting the passing fringes, N is obtained and the distance δ is measured. With care, changes of only 1/100 of a fringe can be resolved.

Many variations on this basic arrangement have been implemented. For example, Michelson reduced the problem of counting large numbers of fringes by devising stepped gages, each representing a known number of fringes,

Figure 11.13 Michelson-type interferometer

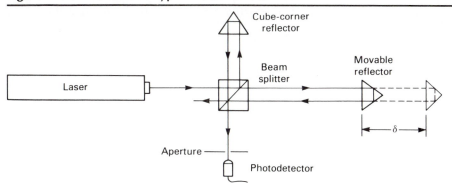

which he termed *etalons*. His procedure was analogous to using a yard- or meter-stick to step off longer distances.

Gage-block interferometers are available from manufacturers of optical apparatus. Although differing in design details, all use essentially the same operating principles. For more information on the construction of gage-block interferometers, see Refs. [4], [5], and [6].

11.15 Measuring Microscopes

Figure 11.14 shows a section through a general-purpose low-power microscope. Basically the instrument consists of an objective cell containing the objective lens, an ocular cell containing the eye and field lenses, and a reticle mounting arrangement, all assembled in optical and body tubes. The ocular cell is adjustable in the optical tube, thereby allowing the eyepiece to be focused sharply on the reticle. The complete optical tube is adjustable in the body tube, by means of a rack and pinion, for focusing the microscope on the work.

Aside from the necessary optical excellence required in the lens system, the heart of the measuring microscope lies in the reticle arrangement. The reticle itself may involve almost any type of plane outline, including scales, grids, and lines. Figure 11.15 illustrates several common forms. In use, the images of the reticle and the work are superimposed, making direct comparison possible. If a scale such as Fig. 11.15(a) is used and if the relation between scale and work is known, the dimension may be determined by direct comparison.

Microscopes used for mechanical measurement are of relatively low power, usually less than $100 \times$ and often about $40 \times$. They may be classified as follows: (1) fixed-scale, (2) filar, (3) traveling, (4) traveling-stage, and (5) draw-tube. The first two, fixed-scale and filar, are intended for measurement of

Figure 11.14   Section through a simple low-power microscope
(Courtesy: The Gaertner Scientific Corp., Chicago, IL)

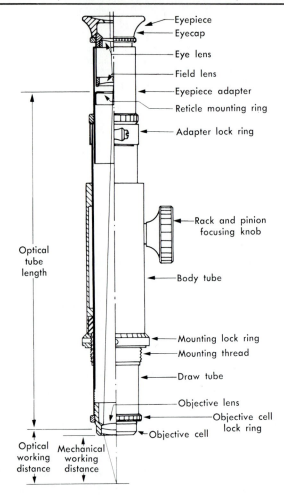

- Eyepiece
- Eyecap
- Eye lens
- Field lens
- Eyepiece adapter
- Reticle mounting ring
- Adapter lock ring
- Rack and pinion focusing knob
- Body tube
- Mounting lock ring
- Mounting thread
- Draw tube
- Objective lens
- Objective cell lock ring
- Objective cell

Optical tube length

Optical working distance   Mechanical working distance

Figure 11.15   Examples of measuring microscope reticles
(Courtesy: The Gaertner Scientific Corp., Chicago, IL)

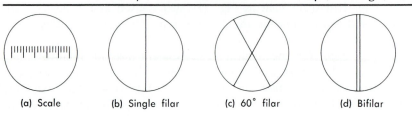

(a) Scale   (b) Single filar   (c) 60° filar   (d) Bifilar

relatively small dimensional magnitudes, from 0.050 to 0.200 in. (1 to 5 mm) in most cases.

11.15.1 Fixed-Scale Microscopes

The fixed-scale measuring microscope uses reticles of the type shown in Fig. 11.15(a). After proper focusing has been accomplished, the scale is simply compared with the work dimension, and the number of scale units is thereby determined. The scale units, of course, must be translated into full-scale dimensions; that is, the instrument must be calibrated. This is accomplished by focusing on a calibration scale, which is generally made of glass with an etched scale. Typical calibration scales are: 100 divisions with each division 0.1 mm long, and 100 divisions with each division 0.004 in. long. Comparison of the calibration and reticle scales provides a positive calibration. Some microscopes are precalibrated and expected to maintain their calibration indefinitely. However, if the objective is changed or tampered with, recalibration will be necessary.

11.15.2 Filar Microscopes

Filar microscopes make use of moving reticles. Actually in most cases a single or double hairline is moved by a fine-pitch screw thread, with the micrometer drum normally divided into 100 parts for subdividing the turns, as shown in Fig. 11.16. A total range of about 0.25 in. (6 mm) is common. The kind using the double hairline, called a bifilar type, is more common. In use, the double hairline is aligned with one extreme of the dimension, then moved to the other extreme, with the movement indicated by the micrometer drum. In general, the bifilar type is more easily used than the single-hairline type. A comparison of the views obtained by both the filar and the bifilar microscopes is shown in Fig. 11.15.

One of the problems in using a filar-measuring microscope is to keep track of the number of turns of the micrometer screw. Two methods for accomplishing this are used. In the first case, a simple counter is attached to the microscope barrel, thus providing direct indication of the number of turns. The more common method is to use a built-in notched bar, or *comb,* in the field of view, as shown in Fig. 11.17. Each notch on the comb corresponds to one complete turn of the micrometer wheel. Further minor and major divisions corresponding to 5 and 10 turns of the wheel are also provided. When the comb is used, a mental scale is applied. As an example, referring to Fig. 11.17, assume that for the initial position the micrometer wheel reads 85. The user might mentally designate the major divisions, reading to the right, as 0, 1000, 2000, 3000, etc. The initial reading would therefore be 085. The hairlines are then moved to the other extreme of the dimension being measured. Suppose now that the micrometer scale reads 27. Reference to the mental scale applied to the comb supplies the hundreds and thousands places, and the reading

Figure 11.16 Filar-type measuring microscope (Courtesy: The
Gaertner Scientific Corp., Chicago, IL)

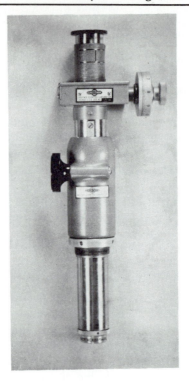

should therefore be 3127. The dimension then is 3127 less 85, or 3042 micrometer divisions. From previous calibration, each micrometer division has been determined to be, say, 0.000032 in. Therefore the actual dimension is 3042 × 0.000032, or 0.09734 in.

An example of a special application of the filar microscope is described in Section 17.10.

11.15.3 Traveling and Traveling-Stage Microscopes

A traveling microscope is moved relative to the work by means of a fine-pitch lead screw, and the movement is measured in a manner similar to that used for the ordinary micrometer. In this case the microscope is used merely to provide a magnified index. The traveling-stage type is similar, except that the work is moved relative to the microscope. In both cases the microscope simply serves as an index, and the micrometer arrangement is the measuring means. About 4 or 5 in. is the usual limit of movement, with a least count of 0.0001 in.

Figure 11.17 View through eyepiece of a filar-type microscope showing reference comb for indicating turns of the microscope drum (Courtesy: The Gaertner Scientific Corp., Chicago, IL)

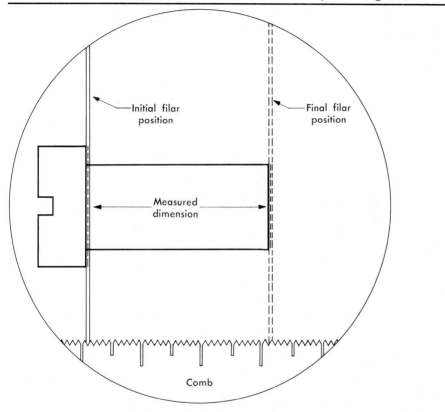

Initial filar
position

Final filar
position

Measured
dimension

Comb

An instrument called a *toolmaker's* microscope is an elaborate version of the moving-stage microscope. Special illuminators are used, along with a protractor-type eyepiece.

11.15.4 The Draw-Tube Microscope

The draw-tube microscope uses a scale on the side of the optical tube to give a measure of the focusing position. The microscope is used to determine displacements in a direction along the optical axis. For example, the height of a step could be measured. The instrument would be focused on the first level, or elevation, and a reading made; then it would be moved to the second elevation and a second reading made. The difference in readings would be the height of the step.

Vernier scales are normally used, along with microscopes having very shallow depth of focus. A typical range of measurement is $1\frac{1}{2}$ in., with a least count of 0.005 in.

11.15.5   Focusing

Proper focusing of any measuring microscope is essential. First, the eyepiece is carefully adjusted on the reticle without regard to the work image. This adjustment is accomplished by sliding the ocular relative to the optical tube, up or down, until maximum sharpness is achieved. Next, the complete optical system is adjusted by means of the rack and pinion until the work is in sharp focus.

Positive check on proper focus may be obtained by checking for parallax. *When the eye is moved slightly from side to side, the relative positions of the reticle and work images should remain unchanged.* If the reticle image appears to move with respect to the work when this check is made, the focusing has not been done properly.

11.16   Optical Tooling and Long-Path Interferometry

Precision alignment of parts of relatively large dimension is often of great importance. Manufacturers of large aircraft are confronted with problems concerning the assembly of such components as the wings, fuselage, engine mounts, tail surfaces, etc., each made by different companies. Space vehicle assembly presents similar problems, as does the assembly of large machine tools, turbogenerators, and the like. Tolerances of a few thousandths of an inch or less in a number of feet are often specified. To produce dimensions of this precision, special gaging procedures are required; hence, optical methods are widely used.

Special equipment available for optical tooling includes the following: (1) alignment telescopes, (2) collimators, (3) autocollimators, and (4) accessories. A simple alignment telescope is very similar to the familiar surveyor's transit; in fact, surveyor's transits are often used. Basically, the instrument consists of a medium- to high-power telescope with a cross-hair reticle (other special reticles may be used) at the focal point of the eyepiece. The telescopes may be used in the same manner as the surveyor's transit for establishing datum lines and levels. They are particularly useful, however, when used in conjunction with a *collimator*.

A collimator is simply a source of a bundle of *parallel* light rays. Essential parts of the device are a lens tube, a light source, and a lens system for projecting the bundle of rays. Also included are reticles whose images are projected by the collimator. Figure 11.18 shows a schematic representation, with the relative positions of collimator and telescope indicated. Reticle R_2 is at the focal point of the collimator lens system, whereas reticle R_3 is in the collimated light beam. One important feature of the setup is that when reticle R_2 is in place, the observed image at the telescope is a function of *angular alignment only*, independent of lateral or transverse positioning. Another important feature is that when reticle R_3 is observed, its image is dependent only on *lateral* position and is independent of angular alignment. It is therefore possible to first establish correct angular relation between the collimator and the telescope and then to determine the magnitude of any lateral misalignment. Magnitudes are independent of distance of separation and are read from

Figure 11.18 Sketch showing arrangement of alignment telescope and collimator

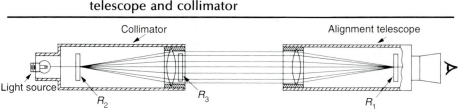

scales inscribed on the reticles (Fig. 11.19). Through the use of this type of optical means, widely separated reference points may be established whose relative locations are accurately known. Such points may then be used for establishing shorter local dimensions by means of the more conventional methods.

Another form of instrument is the *autocollimator*. Basically, this device is a combined telescope and collimator that (a) projects a bundle of parallel light rays and (b) uses the same lens system for viewing a reflected image. An important accessory is some form of mirror that is used as a target for reflecting the light beam. Figure 11.20 shows schematically how it may be used. The autocollimator-target combination provides an optical reference line to which important dimensions may be referred. Intermediate targets may be

Figure 11.19 Examples of reticle designs

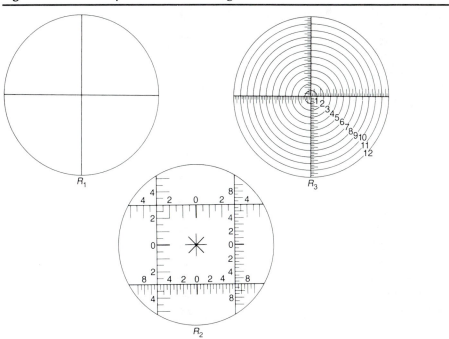

Figure 11.20   Sketch illustrating the use of the autocollimator

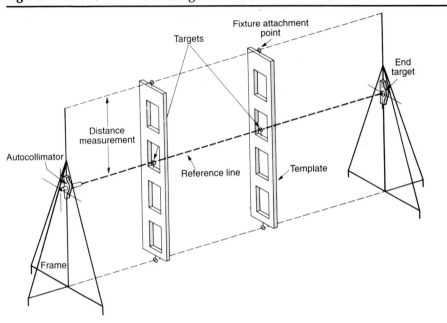

set up to provide reference points from which dimensional measurements may be made. Instruments of this type also provide for making direct horizontal and vertical measurements of about $\pm\frac{1}{8}$ in. with an accuracy of 0.001 in.

Many accessories are available for these instruments, including special mounts, reticles, optical squares, tooling bars and carriages, targets, etc. A device of great usefulness is the *cube-corner,* a trihedral prism that may be substituted for a mirror to reflect light back toward its origin. The reflected beam emerges parallel to the direction of the incident beam, regardless of alignment (see Fig. 11.13).

The laser interferometer represents an important advance in optical tooling. As discussed in Section 11.10, the high coherence length of a gas laser makes it possible to measure distances two or three orders of magnitude greater than are possible with a monochromatic-lamp interferometer. The improved coherence length is closely related to the very narrow linewidth of lasers as compared to lamps. For example, a Hg-198 source has a wavelength spread of approximately 0.0005 nm about a center wavelength of 546.1 nm ($\Delta\lambda/\lambda \approx 10^{-6}$). Because of this lack of preciseness, the fringes become more poorly defined as the path difference increases. In fact, the coherence length is somewhat less than $L_c \approx \lambda^2/\Delta\lambda = 60$ cm. In contrast, a moderately priced frequency-stabilized helium-neon laser has a linewidth of about 5×10^{-6} nm about a center wavelength of 632.8 nm ($\Delta\lambda/\lambda \approx 7.5 \times 10^{-9}$), for a coherence length of $L_c \approx 85$ m. The actual coherence length or linewidth of a particular light source will depend greatly on the efforts made to minimize the spread.

A commercially available laser interferometer system* makes use of the heterodyning (Section 10.8) of two laser beams, originating from a single source, but of slightly different frequencies. The specifications list a range of up to 260 ft (80 m) with an uncertainty of 5 parts in 10^7. Electronic counting is used and to minimize the problem of air turbulence, a smoothing or averaging mode is provided.

Reflected pulses of laser light are also used for distance measurement. In this *pulse-echo* technique, a short pulse of light (a few nanoseconds in duration) is directed at the object whose distance is to be found. The time Δt elapsing before the reflected pulse returns is measured, and the distance is calculated as $L = c\,\Delta t/2$ for c the speed of light. This approach has, for example, been used to measure the distance between the earth and the moon (to ± 15 cm) [7], and it is currently being developed as an automotive "radar" for collision prevention [8]. The pulse-echo technique is also used with ultrasound, for example, to make inexpensive handheld distance meters.

11.17   Whole-Field Displacement Measurement

Most often mechanical displacements are measured as relative movements between discrete points: Point A is displaced by some amount in relation to some reference or datum point. On occasion the relative movements of an array of points may be desired, including whole-field map, which provides the movements of all points within its bounds. Very commonly the end purpose is related to some form of experimental stress analysis. Section 12.18 provides a discussion of photoelasticity and moiré, which are whole-field techniques.

Laser holography is the basis of another whole-field displacement measurement. A three-dimensional image of the initial, undisplaced surface of an object is stored on a holographic plate. The surface is then displaced (perhaps by application of a straining load) and the image of the displaced surface is allowed to interfere with the stored original image. The resulting interference pattern is essentially a map of the displacement of the surface. More information on *holographic interferometry* can be found in the Suggested Readings for this chapter.

11.18   Displacement Transducers

In Chapter 6, we listed a number of devices that are basically displacement-sensitive. These include

1.   Resistance potentiometers (Section 6.6);

---

* Hewlett-Packard 5527/5528A Laser Measurement System.

2. Resistance strain gages (Section 6.7 and the subject of Chapter 12);
3. Variable-inductance devices (Section 6.10);
4. Differential transformers (Section 6.11 and the subject of further discussion in the next section);
5. Capacitive transducers (Section 6.13);
6. Piezoelectric transducers (Section 6.14); and
7. Hall-effect transducers (Section 6.17).

Most of the other transducers listed in Chapter 6 can be configured to sense displacement. For example, the variable-reluctance transducer is basically velocity-sensitive. Combined with an integrating circuit, the output can be made displacement-related, etc.

Usually variable-inductance, capacitance, piezoelectric, and strain-sensitive transducers are suitable only for small displacements (a few microinches to perhaps $\frac{1}{4}$ in.). The differential transformer may be used over intermediate ranges, say a few microinches to several inches. Although wire-wound resistance potentiometers are not as sensitive to small displacements as most of the others, conductive-film potentiometers are sensitive to a few microinches, and there is practically no limit on the maximum displacement for which potentiometers may be used [9, 10]. With the exception of the piezoelectric type, all may be used for both static and dynamic displacements.

Another important class of displacement transducers are the *binary optical encoders* described in Section 8.6.1. Encoders detecting either linear or angular displacement are available. Rotary encoders have resolutions from 20 to 100,000 pulses per rotation. Linear encoders have accuracies reaching a few tens of microinches.

11.19 The Differential Transformer

Because of its singular importance and because it is fundamentally a mechanical displacement transducer, detailed discussion of the differential transformer was reserved for this chapter.

The device, often referred to as a linear-variable differential transformer, or LVDT, provides an ac voltage output proportional to the relative displacement of the transformer core and the windings. Figure 11.21 illustrates the simplicity of its construction. It is a mutual-inductance device using three coils and a core, as shown.

The center coil is energized from an external ac power source, and the two end coils, connected together in phase opposition, are used as pickup coils. Output amplitude and phase depend on the relative coupling between the two pickup coils and the power coil. Relative coupling is, in turn, dependent on the position of the core. Theoretically, there should be a core position for which the voltage induced in each of the pickup coils will be of the same magnitude,

Figure 11.21   The differential transformer: (a) schematic arrangement,
(b) section through a typical transformer

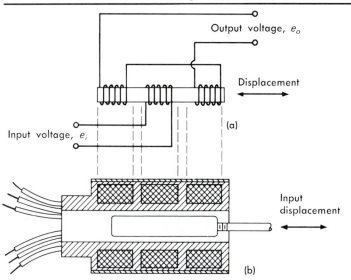

Output voltage, e_o

Displacement

Input voltage, e_i

(a)

Input
displacement

(b)

and the resulting output should be zero. As we will see later, this condition is
difficult to attain perfectly.

Typical differential transformer characteristics are illustrated in Fig. 11.22,
which shows output versus core movement. Within limits, on either side of the
null position, core displacement results in proportional output. In general, the

Figure 11.22   Typical differential transformer performance characteristics

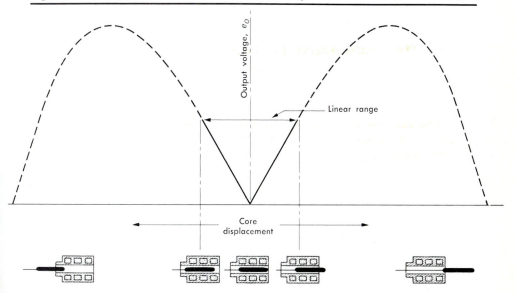

Output voltage, e_O

Linear range

Core
displacement

Table 11.4 Typical variable differential transformer specifications

Linear Range, Inches	Transformer Size (OD × Length), Inches	Core Size (Diameter × Length), Inches	Sensitivity, mV/0.001 in./V Input into High-Impedance Load Excitation Frequency, Hz				
			60	400	2,000	5,000	10,000
±0.005	$\frac{3}{8} \times \frac{9}{16}$	0.10 × 0.20	0.40		1.9		
±0.050	$\frac{7}{8} \times 1\frac{1}{8}$	0.25 × $\frac{7}{8}$	0.70	3.00	3.7	3.7	3.75
±0.020	$\frac{1}{2} \times \frac{5}{8}$	0.10 × $\frac{1}{4}$	0.85		3.5		
±0.200	$\frac{7}{8} \times 2\frac{1}{2}$	0.25 × $1\frac{7}{8}$	1.4	2.5	2.5	2.3	2.3
±0.400	$\frac{7}{8} \times 4\frac{3}{8}$	0.25 × $3\frac{1}{8}$	0.8	1.0	1.0	0.5	0.5
±1.0	$\frac{7}{8} \times 6\frac{5}{8}$	0.25 × $4\frac{1}{4}$	0.1	0.3	0.4	0.4	0.3
±5.0	$\frac{7}{8} \times 18$	0.25 × 6	0.05	0.15	0.15	0.15	0.15

linear range is primarily dependent on the length of the secondary coils. Although the output voltage magnitudes are ideally the same for equal core displacements on either side of null balance, the phase relation existing between power source and output changes 180° through null. It is therefore possible, through phase determination or the use of phase-sensitive circuitry (discussed later), to distinguish between outputs resulting from displacements on either side of null.

Table 11.4 lists typical differential transformer specifications.

11.19.1   Input Power

Input voltage is limited by the current-carrying ability of the primary coil. In most applications, LVDT sensitivities are great enough so that very conservative ratings can be applied. Many commonly used commercial transformers are made to operate on 60 Hz at 6.3 V. Most of the 60-Hz differential transformers draw less than a watt of excitation power. Generally higher frequencies provide increased sensitivities. However, in order to maintain linearity, design differences, primarily core length, may be required for different frequencies; and, in general, a given LVDT is designed for a specific input frequency.

Exciting frequency, sometimes referred to as *carrier* frequency, limits the dynamic response of a transformer. The desired information is superimposed on the exciting frequency, and a minimum ratio of 10 to 1 between carrier and signal frequencies is usually considered to be the limit. For ratios less than 10 to 1, signal definition tends to become lost, and therefore the selection of an operating frequency is important.

Transformer sensitivity is usually stated in terms of *millivolts output per volt input per* 0.001-*in. core displacement.* It is directly proportional to exciting voltage and, as indicated above, also increases with frequency. Of course, the output also depends on LVDT design, and in general the sensitivity will increase with increased number of turns on the coils. There is a limit, however, determined by the solenoid effect on the core. In many applications this effect must be minimized; hence design of the general-purpose LVDT is the result of compromise [11].

Solenoid or axial force exerted by the core is zero when the core is centered and increases linearly with displacement. Increasing the excitation frequency reduces this force. Typically, an LVDT having a linear range of ±0.03 in. exerts an axial force of about 1.80×10^{-4} lbf at 60 Hz and about 1.80×10^{-5} lbf at 1000 Hz, for a driving voltage of 7 V rms.

When utmost sensitivity is required, attainment of a sharp null balance may be difficult without the addition of external components. First of all, in addition to a reactive balance, resistive balance may also be required. This balance can be accomplished through use of a paralleled potentiometer, inserted as shown in Fig. 11.23(a), whose total resistance is high enough to minimize output loading: $20,000\,\Omega$ or higher may be used. In addition to resistive balance, small reactive unbalances may remain at null. They may be

Figure 11.23   (a) Arrangement for improving the sharpness of
null balance, (b) demodulated output

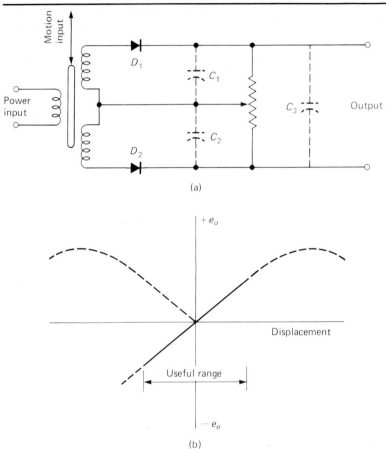

(a)

(b)

caused by unavoidable differences in physical characteristics of the two pickup
coils or from external sources present in a particular installation. These
unbalances can usually be nulled through use of small capacitances whose
values and locations are determined by trial. One or more of the capacitors
shown dotted in Fig. 11.23(a) may be required. The figure also shows two
diodes, D_1 and D_2. These may or may not be added, as desired, to effectively
demodulate the output signal, providing plus and minus voltages on either side
of null, as shown in Fig. 11.23(b).

Integrated-circuit techniques make it possible to incorporate a sine-wave
oscillator, a demodulator, and amplifier circuits within the LVDT housing. In
this form, the LVDT is powered by a regulated dc power supply (± 15 V,
typically) and provides a dc output voltage linearly proportional to the core
position (typically 0.1 to 1 V/mm displacement). The onboard circuitry

incorporates phase discrimination, so that the sign of the output voltage changes when the null position is passed.

The LVDT offers several distinct advantages over many competitive transducers. First, serving as a primary detector-transducer, it converts mechanical displacement into a proportional electric voltage. As we have found, this is a fundamental conversion. In contrast, the electrical strain gage requires the assistance of some form of elastic member. In addition, the LVDT cannot be overloaded mechanically, since the core is completely separable from the remainder of the device. It is also relatively insensitive to high or low temperatures or to temperature changes, and it provides comparatively high output, often usable without intermediate amplification. It is reusable and of reasonable cost.

Probably its greatest disadvantages lie in the area of dynamic measurement. Its core is of appreciable mass, particularly compared with the mass of the bonded strain gage.

11.20 Surface Roughness [4]

Surface finish may be measured by many different methods, using several different units of measurement. Following is a list of some of the basic methods that have been used or suggested.

1. *Visual comparison* with a *standard* surface. This method is based on appearance, which involves more than the surface roughness.

2. The *tracer method,* which uses a stylus that is dragged across the surface. This method is the most common for obtaining quantitative results.

3. The *plastic-replica method,* wherein a soft, transparent, plastic film is pressed onto the surface, then stripped off. Light is then passed through the replica and measured. Refraction caused by the roughened surface reduces the transparency, and the intensity of transmitted light is used as the measure.

4. *Reflection of light* from the surface measured by a photocell.

5. *Magnified inspection,* using a binocular microscope or an electron microscope.

6. *Adsorption* of gas or liquid, wherein the magnitude of adsorption is used as the surface roughness criterion. Radioactive materials have been used for providing a method of quantitative measurement.

7. *Parallel-plane clearance.* Leakage of low-viscosity liquid or gas between the subject surface and a reference flat is used as the measure of roughness.

8. The *electrolytic method,* which assumes that the electrical capacitance is a function of the actual surface area, the rough surface providing a greater capacitance than a smooth surface.

9. *The scanning-tunneling microscope.* This device detects quantum-mechanical (tunneling) effects on a tiny stylus passing over a surface to determine the surface structure on an atomic scale [12].

Suppose Fig. 11.24 represents a sample contour of a machined surface. The values listed thereon may be thought of as actual deviations of the surface from the reference plane x-x, which is located such that the sectional areas above and below the line are equal. These values are also listed in Table 11.5 as absolute values, and their average is calculated to be 12.89 μin. In addition, the root-mean-square (rms) average is calculated as 14.58 μin. We also see that the peak-to-peak height is 27 + 22, or 49 μin. Each of the values—the peak-to-peak height, the arithmetical average deviation, or the root-mean-square average—may be used as measures of roughness.

The American National Standards Institute specifies the arithmetical average deviation, defined by the equation

$$Y = \frac{1}{l} \int_0^l |y|\, dx \tag{11.4}$$

as the standard unit for surface roughness [13]. In this equation,

Y = the arithmetical average deviation,

y = the ordinate of the curve profile from the centerline,

l = the length over which the average is taken.

The term centerline, as used in defining the distance y, corresponds to the x-x line in Fig. 11.24. We see therefore that the value 12.89 μin. in our example is a practical evaluation of Eq. (11.4).

Figure 11.24 Assumed contour of a finished metal surface

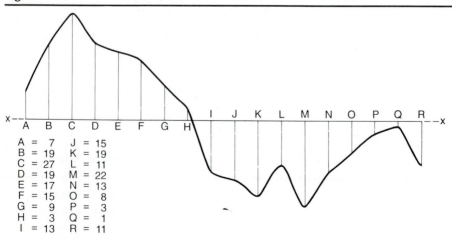

A = 7   J = 15
B = 19   K = 19
C = 27   L = 11
D = 19   M = 22
E = 17   N = 13
F = 15   O = 8
G = 9   P = 3
H = 3   Q = 1
I = 13   R = 11

Table 11.5   Calculation of mean absolute height
             and root-mean-square average

Position	Absolute Elevation from x-x, μin.	Square of Elevation
A	7	49
B	19	361
C	27	729
D	19	361
E	17	289
F	15	225
G	9	81
H	3	9
I	13	169
J	15	225
K	19	361
L	11	121
M	22	484
N	13	169
O	8	64
P	3	9
Q	1	1
R	11	121
	Total = 232	Total = 3828

$$\text{Average absolute height} = \frac{232}{18} = 12.89$$

$$\text{Root-mean-square average} = \sqrt{\frac{3828}{18}} = 14.58$$

Many roughness measuring systems using the tracer method yield the rms average. On a given surface, this value will be approximately 11% greater than the arithmetical average deviation [13]. The difference between the two values, however, is less than the normal variations from one piece to another and is commonly ignored. An idea of the relative value for practical surface finishes may be obtained from Table 11.6.

Although the tracer method for measuring surface roughness is used more than any of the others, it does present several important problems. First, in order for the scriber to follow the contour of the surface, it should have as sharp a point as possible. Some form of conical point with a spherical end is most common. A point radius of 0.0005 in. is often used. Therefore the stylus will not always follow the true contour. If the surface irregularities are primarily what might be referred to as *wavy* or *rolling* hills and valleys of appreciable vertical radius, the stylus may indeed follow the actual contour. If

Table 11.6   Relative values of surface finish

Common Name for Finish	Roughness (rms), μin.	Average Peak-to-Peak Height, μin.	Usual Tolerance Specified for Finished Part, in.
Mirror	4	15	0.0002
Polished	8	28	0.0005
Ground	16	56	0.001
Smooth	32	118	0.002
Fine	63	220	0.003
Semifine	125	455	0.004
Medium	250	875	0.007
Semirough	500	1750	0.013
Rough	1000	3500	0.025

the surface is rugged, on the other hand, the stylus will not extend fully into the valleys.

Second, the stylus will probably actually round off the peaks as it is dragged over the surface. This problem increases as the radius of the tip is decreased. It would seem, therefore, that there are two conflicting requirements with regard to stylus-tip radius. The marring of the surface will in part be a function of the material constants, which of course should have no bearing on the measure of surface roughness.

A third problem lies in the fact that the stylus can never inspect more than a very small percentage of the overall surface.

In spite of these problems, the tracer method is undoubtedly the most commonly used. Various forms of secondary transducers have been employed, including piezoelectric elements (Section 6.14), variable-inductance (Section 6.10), and variable-reluctance (Section 6.12). In each case the stylus motion is transferred to the transducer element, which converts it to an analogous electric signal.

Readout devices may be any of the ordinary voltage indicators or recorders, such as a simple meter, an oscillograph, or a CRO. The basic indicated value depends on the particular combination of transducing element and indicator used. Piezoelectric variable-inductance and electronic transducers are basically displacement-sensitive, whereas the variable-reluctance type provides an output proportional to velocity. Outputs from the latter, however, may be integrated (Section 7.23), thereby providing displacement information.

An ordinary electronic voltmeter provides a convenient and inexpensive indicator. Outputs from both the piezoelectric and electronic pickups are of sufficient level to provide adequate drive without additional amplification. In addition, if the meter has reasonable frequency response characteristics, the speed with which the displacement-type pickups are moved over the work need not be carefully controlled.

Of course, the meter, because of its inherent characteristics, indicates rms amplitudes. When some form of oscillograph or oscilloscope is used, peak-to-peak values may be indicated or also recorded. Because of the lower cost, increased portability, and simple operating technique, combinations of displacement-type transducers and meter indicators are the most popular.

Suggested Readings

ASME, B89.1.9M-1984. *Precision Inch Gage Blocks for Length Measurement.* New York: 1984.

ASME, PTC 19.14-1958. *Linear Measurements.* New York, 1958.

American Society for Tool Engineers, *Handbook of Industrial Metrology.* Englewood Cliffs, N.J.: Prentice Hall, 1967.

Anthony, D. M. *Engineering Metrology.* Elmsford, N.Y.: Pergamon Press, 1987.

Booth, S. F. *Precision Measurement and Calibration,* vol. 3 (Selected NBS technical papers on optics, metrology and radiation), NBS Handbook 77. Washington, D.C.: U.S. Government Printing Office, 1961.

Busch, T. *Fundamentals of Dimensional Metrology. Laboratory Experiments.* Albany, N.Y.: Delmar, 1965.

Fullmer, I. H. *Dimensional Metrology* (subject classified with abstracts). NBS Misc. Publ. 265. Washington, D.C.: U.S. Government Printing Office, 1966.

Scarr, A. J. T. *Metrology and Precision Engineering.* London: McGraw-Hill, 1967.

Slocum, A. H. *Precision Machine Design.* Englewood Cliffs, N.J.: Prentice Hall, 1992.

Smith, W. J. *Modern Optical Engineering.* New York: McGraw-Hill, 1966.

Vest, C. M. *Holographic Interferometry.* New York: John Wiley, 1979.

Problems

11.1 Using the dimensioning "rules" given in Section 11.2 for go/no-go gages, sketch and dimension a plug gage for gaging a hole that carries the dimension 1.125/1.132 in.

11.2 Assuming that the dimensions specified for the plug gage in Fig. 11.2 correspond to a temperature of 68°F, calculate the limiting dimensions corresponding to (a) 90°F and (b) 40°F. Use data for chrome-vanadium steel listed in Table 6.3.

11.3 Determine the temperature change (°F) that will cause a change in the length of a 1-in. gage block equal to the block's nominal tolerance for (a) working blocks, (b) reference blocks, and (c) master blocks.

11.4 Using the set of blocks listed in Section 11.3, specify a combination to provide a dimension of 2.7816 in.: (a) including wear blocks, and (b) without wear blocks.

11.5 Apply the following data to Figure 11.12: (a) height of gage-block stack (the sum of stamped values) = 3.147 in.; (b) $7\frac{1}{2}$ fringes from a helium light source are read; (c) nominal width and thickness of the blocks are 1.25 and $\frac{1}{2}$ in., respectively; and (d) the entire assembly is at 68°F. Applying the procedure described in the text, calculate the diameter of the cylindrical plug gage.

11.6 Refer to figure 11.12. It is the usual practice to simplify the true geometry of the situation by assuming that the optical flat contacts the plug gage on the vertical centerline of the circular section, which, of course, is not precisely the case. Using the data given in Problem 11.5, make sufficient calculations to convince yourself of the practicality of the assumption.

11.7 Equation (11.1) is written assuming that both the blocks and the gaged dimension are at the same nonstandard temperature. Modify the relation to care for a situation in which the blocks and the gaged part are at different temperatures, neither of which is the standard temperature.

11.8 Referring to Problem 11.5, calculate the dimension at standard temperature, 20°C (68°F), if all parts are measured at 35°C (95°F), $7\frac{1}{2}$ fringes are read, and (a) the measured part is of type 302 stainless steel; (b) the measured part is of Invar. (See Table 6.3 for necessary data.)

11.9 Referring to Problem 11.5 determine how many fringes would appear if the entire assembly were raised to a temperature of 35°C (95°F) and (a) the blocks and the gaged part have identical coefficients of expansion; (b) the gaged part is of aluminum; and (c) the gaged part is of type 302 stainless steel. (See Table 6.3 for data.)

11.10 Referring to Figure 11.25, if $h_B - h_A = 0.0004$ in., how many fringes would be produced over block A? Over block B? How does the block width w affect the fringe pattern?

11.11 Figure 11.26 illustrates an arrangement of gage blocks, optical flat, and surface plate for measuring the dimension of a hex bar across flats. If the true dimensions are shown, what number of interference fringes would be seen if a helium light source were used?

Figure 11.25   Arrangement for measuring the difference in lengths between two gage blocks (see Problem 11.10)

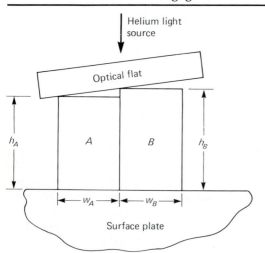

Figure 11.26 Arrangement for measuring the "across-flats" dimension of a hexagonal section (see Problem 11.11)

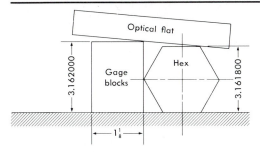

11.12–11.19 Various combinations of values for the parameters defined in Fig. 11.27 are given in Table 11.7. For each case determine the missing numbers.

11.20 A Michelson-type interferometer is shown in Fig. 11.28. The laser beam passes through a beam splitter, travels to each of two mirrors, and is reflected back to a photodetector. Mirror A is movable; as its position changes, the interference at

Figure 11.27 Gage block and optical flat arrangement to be used for Problems 11.12–11.19

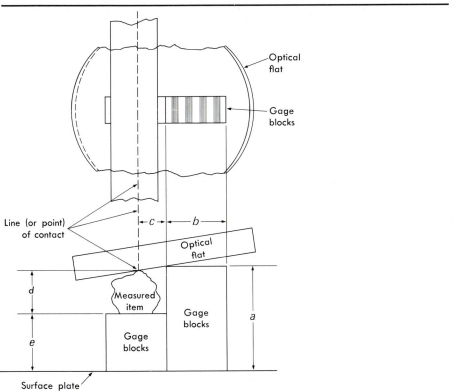

Table 11.7   Data for Problems 11.12–11.19

Problem Number	Geometry of Measured Item and Material	a	b	c	d	e	Is a > (d + e)?	Light Source	Fringes	Temp., °F
				Inches						
11.12	Rectangular: Carbon steel	1.7839	0.875	1.250	?	0.0	No	Helium	$6\frac{1}{2}$	68
11.13	Cylindrical: Carbon steel	1.7619	1.250	d/2	?	1.4387	Yes	Helium	8	68
11.14	Cylindrical: Aluminum	2.7814	1.000	d/2	2.7805	0	...	Krypton 86	?	68
11.15	Cylindrical: Aluminum	3.7892	1.250	d/2	?	0	No	Hg 198	$15\frac{1}{2}$	90
11.16	Spherical: Brass	1.3470	0.875	d/2	?	1.2750	Yes	Cad. Red	12	75
11.17	Rectangular: Stainless steel	6.3400	1.125	1.500	6.3400*	0	?	Helium	?	98
11.18	Rectangular: Carbon steel	1.7839	0.875	1.250	?	0.0	Yes	Helium	$6\frac{1}{2}$	68
11.19	Cylindrical: Carbon steel	1.7619	1.250	0.400	?	1.4387	Yes	Helium	8	68

Note: Use the following values for thermal coefficients of linear expansion, α, μin./in. · °F: gage block steel, 6.4; carbon steel, 6.7; stainless steel, 9.4; aluminum, 13.0; brass, 10.0.

* At 68°F.

Figure 11.28 (a) Michelson interferometer, (b) AC-coupled CRO trace when mirror A moves at constant speed, (c) DC-coupled CRO trace when mirror A, initially at rest, moves a distance δ and stops

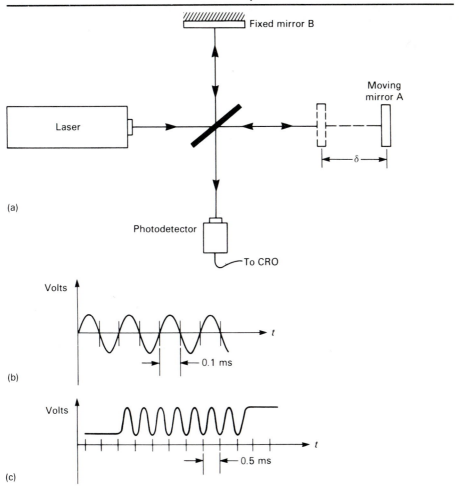

the photodetector changes. This interference produces either light or dark fringes, corresponding to high and low voltages at the photodetector output.

a. If the laser wavelength is $0.5145 \, \mu m$, how many dark fringes will occur as mirror A is moved a distance of $\delta = 100 \, \mu m$?

b. When the mirror is moved at a constant speed, the photodetector voltage seen on an ac-coupled CRO tracer is shown in Fig. 11.28(b). At what speed is the mirror moving?

c. Mirror A, initially at rest, moves a short distance and stops. The dc-coupled CRO output voltage trace is shown in Fig. 11.28(c). How far did mirror A move?

References

1. Peters, C. G., and W. B. Emerson. Interference methods for producing and calibrating end standards. *NBS J. Res.* 44: 427, April 1950.

2. Metrology of gage blocks. *NBS Circular* 581: 67, April 1, 1957. Washington, DC: U.S. Government Printing Office.

3. Graneek, M. A pneumatic comparator of high sensitivity. *The Engineer* 172: 414, 1951.

4. American Society of Tool and Manufacturing Engineers. *Handbook of Industrial Metrology.* Englewood Cliffs, N.J.: Prentice Hall, 1967.

5. Scarr, A. J. T. *Metrology and Precision Engineering.* New York: McGraw-Hill, 1967.

6. *Precision Measurement and Calibration,* Vol. III. National Bureau of Standards Handbook 77, Washington, DC: U.S. Government Printing Office, 1966.

7. Jenkins, F. A., and H. E. White. *Fundamentals of Optics.* 4th ed., pp. 655–56.

8. Yanagisawa, T., K. Yamamoto, and Y. Kubota. Development of a laser radar system for automobiles. *SAE Technical Paper* No. 920745, February 1992.

9. Kneen, W. A review of electric displacement gages used in railroad car testing. *ISA Proc.* 6: 74, 1951.

10. Todd, C. D. *The Potentiometer Handbook.* New York: McGraw-Hill, 1975.

11. Boggis, A. G. Design of differential transformer displacement gauges. *SESA Proc.* 9(2): 171, 1952.

12. Hansma, P. K., and J. Tersoff. Scanning tunneling microscopy. *J. Applied Physics* 61: R1, 1987.

13. *American National Standard Surface Texture,* ANSI B46. 1-1962. New York: ANSI.

CHAPTER 12

Strain and Stress: Measurement and Analysis

12.1 Introduction

All machine or structural members deform to some extent when subjected to external loads or forces. The deformations result in relative displacements that may be normalized as percentage displacement, or strain. For simple axial loading (Fig. 12.1),

$$\varepsilon_a = \frac{dL}{L} \approx \frac{L_2 - L_1}{L_1} = \frac{\Delta L}{L_1}, \tag{12.1}$$

in which

ε_a = axial strain,

L_1 = linear dimension or gage length,

L_2 = final strained linear dimension.

More correctly, the term *unit strain* should be used for the preceding quantity and is generally intended when the word strain is used alone. Throughout the following discussion, when the word strain is used, we mean the quantity defined by Eq. (12.1). If the net change in a dimension is required, the term *total strain* will be used.

Because the quantity strain, as applied to most engineering materials, is a very small number, it is commonly multiplied by one million; the resulting number is then called *microstrain*,* or parts per million (ppm).

---

* Considerable use of the term *microstrain* will be made throughout this chapter and elsewhere in the book. For convenience, the abbreviation μ-strain will often be used.

Figure 12.1   Defining relations for axial and lateral strain

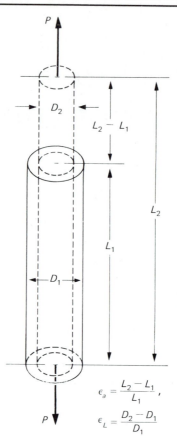

$$\epsilon_a = \frac{L_2 - L_1}{L_1},$$

$$\epsilon_L = \frac{D_2 - D_1}{D_1}$$

The stress-strain relation for a uniaxial condition, such as exists in a simple tension test specimen or at the outer fiber of a beam in bending, is expressed by

$$E = \frac{\sigma_a}{\varepsilon_a},$$ (12.2)

where

E = Young's modulus,

σ_a = uniaxial stress,

ε_a = the strain in the direction of the stress.

This relation is linear; i.e., E is a constant for most materials so long as the stress is kept below the proportional limit.

When a member is subjected to simple uniaxial stress in the elastic range (Fig. 12.1), lateral strain results in accordance with the following relation:

$$v = \frac{-\varepsilon_L}{\varepsilon_a} \tag{12.2a}$$

where

$$v = \text{Poisson's ratio,}$$

$$\varepsilon_L = \text{lateral strain.}$$

A more general condition commonly exists on the *free surface* of a stressed member. Let us consider an element subject to orthogonal stresses σ_x and σ_y, as shown in Fig. 12.2. Suppose that the stresses σ_x and σ_y are applied one at a time. If σ_x is applied first, there will be a strain in the x-direction equal to σ_x/E. At the same time, because of Poisson's ratio there will be a strain in the y-direction equal to $-v\sigma_x/E$.

Now suppose that the stress in the y-direction, σ_y, is applied. This stress will result in a y-strain of σ_y/E and an x-strain equal to $-v\sigma_y/E$. The net strains are then expressed by the relations

$$\varepsilon_x = \frac{\sigma_x - v\sigma_y}{E} \quad \text{and} \quad \varepsilon_y = \frac{\sigma_y - v\sigma_x}{E}. \tag{12.3}$$

Figure 12.2 An element taken from a biaxially stressed
condition with normal stresses known

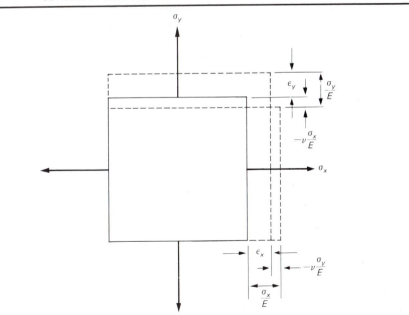

If these relations are solved simultaneously for σ_x and σ_y, we obtain the equations

$$\sigma_x = \frac{E(\varepsilon_x + v\varepsilon_y)}{1 - v^2} \quad \text{and} \quad \sigma_y = \frac{E(\varepsilon_y + v\varepsilon_x)}{1 - v^2}. \tag{12.4}$$

When a stress σ_z exists, acting in the third orthogonal direction, the more general three-dimensional relations are

$$\varepsilon_x = \frac{1}{E}[\sigma_x - v(\sigma_y + \sigma_z)],$$

$$\varepsilon_y = \frac{1}{E}[\sigma_y - v(\sigma_z + \sigma_x)], \tag{12.5}$$

$$\varepsilon_z = \frac{1}{E}[\sigma_z - v(\sigma_x + \sigma_y)].$$

12.2 Strain Measurement

Strain may be measured either directly or indirectly. Modern strain gages are inherently sensitive to *strain*; that is, the unit output is directly proportional to the unit dimensional change (strain). However, until about 1930 the common experimental procedure consisted of measuring the displacement ΔL over some initial gage length L and then calculating the resulting average strain using Eq. (12.1). An apparatus called an *extensometer* was used. This device generally incorporated either a mechanical or optical lever system and sensed displacements over gage lengths ranging from about 50 mm to as great as 25 cm (about 10 in.). The Huggenberger and the Tuckerman extensometers are representative of the more advanced mechanical and optical types, respectively. Reference [1] provides a good summary of extensometer practices.

Electrical-type strain gages are devices that use simple resistive, capacitive [2, 3], inductive [4], or photoelectric principles. The resistive types are by far the most common and are discussed in considerable detail in the following sections. They have advantages, primarily of size and mass, over the other types of electrical gages. On the other hand, strain-sensitive gaging elements used in calibrated devices for measuring other mechanical quantities are often of the inductive type, whereas the capacitive kind is used more for special-purpose applications. Inductive and capacitive gages are generally more rugged than resistive ones and better able to maintain calibration over a long period of time. Inductive gages are sometimes used for permanent installations, such as on rolling-mill frames for monitoring roll loads. Torque meters often use strain gages in one form or another, including inductive [5] and capacitive [2].

Other strain measuring techniques include optical methods, such as photoelasticity (Section 12.18), the Moiré technique (Section 12.18), and holographic interferometry (Section 11.17).

12.3   The Electrical Resistance Strain Gage

In 1856 Lord Kelvin demonstrated that the resistances of copper wire and iron wire change when the wires are subject to mechanical strain. He used a Wheatstone bridge circuit with a galvanometer as the indicator [6]. Probably the first wire resistance strain gage was that made by Carlson in 1931 [7]. It was of the unbounded type: Pillars were mounted, separated by the gage length, with wires stretched between them. (An application of this type is shown in Fig. 17.7.) What was probably the first bonded strain gage was used by Bloach [8]. It consisted of a carbon film resistance element applied directly to the surface of the strained member.

In 1938 Edward Simmons made use of a bonded wire gage in a study of stress-strain relations under tension impact [9]. His basic idea is covered in U.S. Patent No. 2,292,549. At about the same time, Ruge of M.I.T. conceived the idea of making a preassembly by mounting wire between thin pieces of paper. Figure 12.3 shows the general construction.

During the 1950s advances in materials and fabricating methods produced the foil-type gage, soon replacing the wire gage. The common form consists of a metal foil element on a thin epoxy support and is manufactured using printed-circuit techniques. An important advantage of this type is that almost unlimited plane configurations are possible; a few examples are shown in Fig. 12.4.

Figure 12.3   Construction of bonded-wire-type strain gage

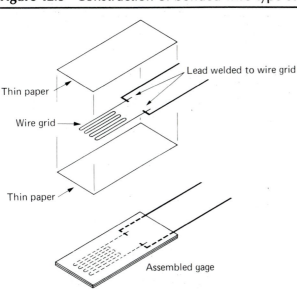

Thin paper

Lead welded to wire grid

Wire grid

Thin paper

Assembled gage

Figure 12.4 Typical foil gages illustrating the following types: (a) single element, (b) two-element rosette, (c) three-element rosette, (d) one example of many different special-purpose gages. The latter is for use on pressurized diaphragms.

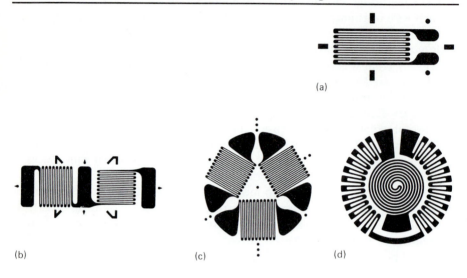

(a)

(b)   (c)   (d)

12.4 The Metallic Resistance Strain Gage

The theory of operation of the metallic resistance strain gage is relatively simple. When a length of wire (or foil) is mechanically stretched, a *longer* length of *smaller* sectioned conductor results; hence the electrical resistance changes [10]. If the length of resistance element is intimately attached to a strained member in such a way that the element will also be strained, then the measured change in resistance can be calibrated in terms of strain.

A general relation between the electrical and mechanical properties may be derived as follows: Assume an initial conductor length L, having a cross-sectional area CD^2. (In general the section need not be circular; hence D will be a sectional dimension and C will be a proportionality constant. If the section is square, $C = 1$; if it is circular, $C = \pi/4$, etc.) If the conductor is strained axially in tension, thereby causing an increase in length, the lateral dimension should reduce as a function of Poisson's ratio.

We will start with the relation [Eq. (6.2)]

$$R = \frac{\rho L}{A} = \frac{\rho L}{CD^2}. \tag{12.6}$$

If the conductor is strained we may assume that each of the quantities in Eq.

(12.6) except for C may change. Differentiating, we have

$$dR = \frac{CD^2(L\,d\rho + \rho\,dL) - 2C\rho\,LD\,dD}{(CD^2)^2}$$

$$= \frac{1}{CD^2}\left((L\,d\rho + \rho\,dL) - 2\rho L\frac{dD}{D}\right). \tag{12.7}$$

Dividing Eq. (12.7) by Eq. (12.6) yields

$$\frac{dR}{R} = \frac{dL}{L} - 2\frac{dD}{D} + \frac{d\rho}{\rho}, \tag{12.8}$$

which may be written

$$\frac{dR/R}{dL/L} = 1 - 2\frac{dD/D}{dL/L} + \frac{d\rho/\rho}{dL/L}. \tag{12.8a}$$

Now

$$\frac{dL}{L} = \varepsilon_a = \text{axial strain},$$

$$\frac{dD}{D} = \varepsilon_L = \text{lateral strain},$$

and

$$v = \text{Poisson's ratio} = -\frac{dD/D}{dL/L}.$$

Making these substitutions gives us the basic relation for what is known as the *gage factor*, for which we shall use the symbol F:

$$F = \frac{dR/R}{dL/L} = \frac{dR/R}{\varepsilon_a} = 1 + 2v + \frac{d\rho/\rho}{dL/L}. \tag{12.9}$$

This relation is basic for the resistance-type strain gage.

Assuming for the moment that resistivity should remain constant with strain, then according to Eq. (12.9) the gage factor should be a function of Poisson's ratio alone, and in the elastic range should not vary much from $1 + (2)(0.3) = 1.6$. Table 12.1 lists typical values for various materials. Obviously, more than Poisson's ratio must be involved, and if resistivity is the only other variable, apparently its effect is not consistent for all materials. Note the value of the gage factor for nickel. The negative value indicates that a stretched element with increased length and decreased diameter (assuming elastic conditions) actually exhibits a reduced resistance.

In spite of our incomplete knowledge of the physical mechanism involved, the factor F for metallic gages is essentially a *constant* in the usual range of required strains, and its value, determined experimentally, is reasonably consistent for a given material.

Table 12.1 Representative properties of various grid materials

Grid Material	Composition	Approx. Gage Factor, F	Approximate Resistivity		Approximate Temperature Coefficient of Resistance, ppm/°C	Maximum Operating Temp., °C (approx.)
			Microhm · cm	Ohms per mil · foot		
Nichrome V*	80% Ni; 20% Cr	2.0	108	650	400	1100
Constantan*; Copel*;	45% Ni; 55% Cu	2.0	49	290	11	480
Advance*						—
Isoelastic*	36% Ni; 8% Cr; 0.5% Mo; Fe remainder	3.5	112	680	470	—
Karma*	74% Ni; 20% Cr; 3% Al; 3% Fe	2.4	130	800	18	815
Manganin*	4% Ni; 12% Mn; 84% Cu	0.47	48	260	11	—
Platinum-Iridium	95% Pt; 5% Ir	5.1	24	137	1250	1100
Monel*	67% Ni; 33% Cu	1.9	42	240	2000	—
Nickel		−12†	7.8	45	6000	—
Platinum		4.8	10	60	3000	—

* Trade names.

† Varies widely with cold work.

By rewriting Eq. (12.9) and replacing the differential by an incremental resistance change, we obtain the following equation:

$$\varepsilon = \frac{1}{F}\frac{\Delta R}{R}. \tag{12.10}$$

In practical application, values of F and R are supplied by the gage manufacturer, and the user determines ΔR corresponding to the input situation being measured. This procedure is the fundamental one for using resistance strain gages.

12.5 Selection and Installation Factors for Bonded Metallic Strain Gages

Performance of bonded metallic strain gages is governed by five gage parameters: (a) grid material and configuration; (b) backing material; (c) bonding material and method; (d) gage protection; and (e) associated electrical circuitry.

Desirable properties of grid material include: (a) high gage factor, F; (b) high resistivity, ρ; (c) low temperature sensitivity; (d) high electrical stability; (e) high yield strength; (f) high endurance limit; (g) good workability; (h) good solderability or weldability; (i) low hysteresis; (j) low thermal emf when joined to other materials; and (k) good corrosion resistance.

Temperature sensitivity is one of the most worrisome factors in the use of resistance strain gages. In many applications, compensation is provided in the electrical circuitry; however, this technique does not always eliminate the problem. Two factors are involved: (1) the differential expansion existing between the grid support and the grid proper, resulting in a strain that the gage is unable to distinguish from load strain, and (2) the change in resistivity ρ with temperature change.

Thermal emf superimposed on gage output obviously must be avoided if dc circuitry is used. For ac circuitry this factor would be of little importance. Corrosion at a junction between grid and lead could conceivably result in a miniature rectifier, which would be more serious in an ac than in a dc circuit.

Table 12.1 lists several possible grid materials and some of the properties influencing their use for strain gages. Commercial gages are usually of constantan or isoelastic. The former provides a relatively low temperature coefficient along with reasonable gage factors. Isoelastic gages are some 40 times more sensitive to temperature than are constantan gages. However, they have appreciably higher output, along with generally good characteristics otherwise. They are therefore made available primarily for dynamic applications where the short time of strain variation minimizes the temperature problem.

The gage factor listed to nickel is of particular interest, not only because of its relatively high value, but also because of its negative sign. It should be noted, however, that the value of F for nickel varies over a relatively wide range, depending on how it is processed. Cold working has a rather marked effect on the strain- and temperature-related characteristics of nickel and its alloys, and this feature is used advantageously to produce special temperature self-compensating gages (see Section 12.10.2).

Common backing materials include phenolic-impregnated paper, epoxy-type plastic films, and epoxy-impregnated fiberglass. Most foil gages intended for a moderate range of temperatures ($-75°C$ to $100°C$) use an epoxy film backing. Table 12.2 lists commonly recommended temperature ranges.

No particular difficulty should be experienced in mounting strain gages if the manufacturer's recommended techniques are carefully followed. However, we may make one observation that is universally applicable. *Cleanliness* is an absolute requirement if consistently satisfactory results are to be expected. The mounting area must be cleaned of all corrosion, paint, etc., and bare base material must be exposed. All traces of greasy film must be removed. Several of the gage suppliers offer kits of cleaning materials along with instructions for their use. These materials are very satisfactory.

Most gage installations are not complete until provision is made to protect the gage from ambient conditions. The latter may include mechanical abuse, moisture, oil, dust and dirt, and the like. Once again, gage suppliers provide recommended materials for this purpose, including petroleum waxes, silicone resins, epoxy preparations, and rubberized brushing compounds. A variety of materials is necessary because of the many types of protection required—from such things as hot oil, immersion in water, liquefied gases, etc. An extreme requirement for gage protection is found in the case of gages mounted on the

Table 12.2   General recommendations for strain-gage backing
materials and adhesives

Grid and Backing Materials	Recommended Adhesive	Permissible Temperature Range, °C
Foil on epoxy	Cyanoacrylate	−75 to 95
Foil on phenol-impregnated fiberglass	Phenolic	−240 to 200
Strippable foil or wire	Ceramic	−240 to 400 (to 1000 for short-time dynamic tests)
Free filament wire	Ceramic	−240 to 650 (to 1100 for short-time dynamic tests)

exterior of a ship or submarine hull for the purpose of sea trials [11]. Special methods of protection are used, including vulcanized rubber boots over the gages.

Gages may have one or more elements (see Fig. 12.4). When used for stress analysis, the single-element gage is applied to the uniaxial stress condition; the two-element rosette is applied to the biaxial condition when either the principal axes or the axes of interest are known, and the three-element rosette is applied when a biaxial stress condition is completely unknown (see Section 12.15.2 for a more complete discussion).

12.6 Circuitry for the Metallic Strain Gage

When the sensitivity of a metallic resistance gage is considered, its versatility and reliability are truly amazing. The basic relation as expressed by Eq. (12.10) is

$$\varepsilon = \frac{1}{F}\frac{\Delta R_g}{R_g}. \tag{12.11}$$

Typical gage constants are

$$F = 2.0, \qquad R_g = 120\ \Omega.$$

Strains of 1 ppm (1 microstrain) are detectable with commercial equipment; hence, the corresponding resistance change that must be measured in the gage will be

$$\Delta R_g = FR_g\varepsilon = (2)(120)(0.000001) = 0.00024\ \Omega,$$

which amounts to a resistance change of 0.0002%. Obviously, to measure changes as small as this, instrumentation more sensitive than the ordinary ohmmeter will be required.

Three circuit arrangements are used for this purpose: the simple voltage-dividing potentiometer or ballast circuit (Section 7.6), the Wheatstone bridge (Section 7.9), and the constant-current circuit. Some form of bridge arrangement is the most widely used.

12.7 The Strain-Gage Ballast Circuit

Figure 12.5 illustrates a simple strain-gage ballast arrangement. Using Eq. (7.4) and substituting R_g for kR_t, we may write

$$e_o = e_i\frac{R_g}{R_b + R_g}$$

and

$$de_o = \frac{e_i R_b\,dR_g}{(R_b + R_g)^2} = \frac{e_i R_b R_g}{(R_b + R_g)^2}\frac{dR_g}{R_g}.$$

From Eq. (12.11),

$$de_o = \frac{e_i R_b R_g}{(R_b + R_g)^2} F\varepsilon,$$ (12.12)

where

e_i = the exciting voltage,

e_o = the voltage output,

R_b = the ballast resistance (Ω),

R_g = the strain-gage resistance (Ω),

F = the gage factor,

ε = strain.

Some of the limitations inherent in this circuit may be demonstrated by the following example. Let

$$R_b = R_g = 120\ \Omega.$$

A resistance of 120 Ω is common in a strain gage, and it will be recalled that in Section 7.6 we showed that equal ballast and transducer resistances provide maximum sensitivity. Also, let

$$e_i = 8\ \text{V}$$

and let

$$F = 2.0,$$

which is a common value. Then

$$e_o = 8\left(\frac{120}{120 + 120}\right) = 4\ \text{V}, \quad \text{and}$$

$$de_o = \frac{8 \times 120 \times 120 \times 2 \times \varepsilon}{(120 + 120)^2} = 4\varepsilon.$$

If our indicator is to provide an indication for strain of, say, 1 μ-strain, it must sense a 4-μV variation in 4 V, or 0.00010%. This severe requirement

Figure 12.5   Ballast circuit for use with strain gages

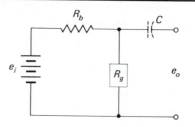

practically eliminates the ballast circuit for *static* strain work. We may use it, however, in certain cases for dynamic strain measurement when any static strain component may be ignored. If a capacitor is inserted into an output lead, the dc exciting voltage is blocked and only the variable component is allowed to pass (Fig. 12.5). Temperature compensation is not provided; however, when only transient strains are of interest, this type of compensation is often of no importance.

12.8   The Strain-Gage Bridge Circuit

A resistance-bridge arrangement is particularly convenient for use with strain gages because it may be easily adjusted to a null for zero strain, and it provides means for effectively reducing or eliminating the temperature effects previously discussed (Section 12.5). Figure 12.6 shows a minimum bridge arrangement, where arm 1 consists of the strain-sensitive gage mounted on the test item. Arm 2 is formed by a similar gage mounted on a piece of unstrained material as nearly like the test material as possible and placed near the test location so that the temperature will be the same. Arms 3 and 4 may simply be fixed resistors selected for good stability, plus portions of slide-wire resistance, D, required for balancing the bridge.

If we assume a voltage-sensitive deflection bridge with all initial resistances nominally equal, using Eq. (7.17) we have

$$\frac{\Delta e_o}{e_i} = \frac{\Delta R_1/R}{4 + 2(\Delta R_1/R)}.$$

Figure 12.6   Simple resistance-bridge arrangement for strain measurement

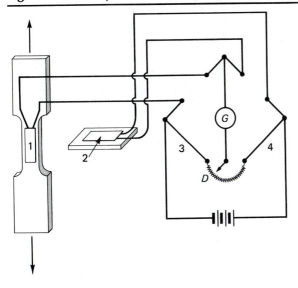

In addition,

$$\varepsilon = \frac{1}{F}\frac{\Delta R_1}{R} \quad \text{or} \quad \Delta R = FR\varepsilon.$$

Then

$$\Delta e_o = \frac{e_i F\varepsilon}{4 + 2F\varepsilon}.$$

For $e_i = 8$ V and $F = 2$,

$$\Delta e_o = \frac{8 \times 2 \times \varepsilon}{4 + (2)(2)\varepsilon}.$$

If we neglect the second term in the denominator, which is normally negligible, then

$$\Delta e_o = e_o = 4\varepsilon \text{ V},$$

or for $\varepsilon = 1\,\mu$-strain, $e_o = 4\,\mu$V.

We see that under similar conditions the output increment for the bridge and ballast arrangements is the same. The tremendous advantage that the bridge possesses, however, is that the incremental output is not superimposed on a large fixed-voltage component. Another important advantage, which is discussed in Section 12.10, is that temperature compensation is easily attained through the use of a bridge circuit incorporating a "dummy," or compensating, gage.

12.8.1   Bridges with Two and Four Arms Sensitive to Strain

In many cases bridge configuration permits the use of more than one arm for measurement. This is particularly true if a known relation exists between two strains, notably the case of bending. For a beam section symmetrical about the neutral axis, we know that the tensile and compressive strains are equal except for sign. In this case, both gages 1 and 2 may be used for strain measurement. This is done by mounting gage 1 on the tensile side of the beam and mounting gage 2 on the compressive side, as shown in Fig. 12.7 (see also case F in Table 12.4). The resistance changes will be alike but of opposite sign, and a doubled bridge output will be realized.

This may be carried further, and all four arms of the bridge made strain-sensitive, thereby *quadrupling* the output that would be obtained if only a single gage were used. In this case gages 1 and 4 would be mounted to record like strain (say tension) and 2 and 3 to record the opposite type (case G in Table 12.4).

Bridge circuits of these kinds may be used either as null-balance bridges or as deflection bridges (Section 7.9). In the former the slide-wire movement becomes the indicated measure of strain. This bridge is most valuable for strain-indicating devices used for static measurement. Most dynamic strain-

Figure 12.7 Bridge arrangement with two gages sensitive to strain

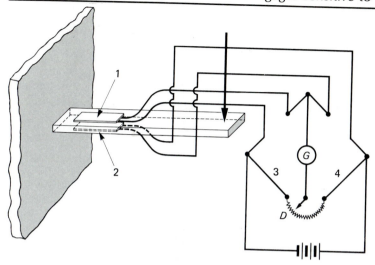

measuring systems, however, use a voltage- or current-sensitive deflection bridge. After initial balance is accomplished, the output, amplified as necessary, is used to deflect an indicator, such as a cathode-ray oscilloscope beam or input to a digital recorder. In addition, the constant-current bridge may offer certain advantages (Section 7.9.3).

12.8.2 The Bridge Constant

At this point we introduce the term *bridge constant,* which we shall define by the following equation:

$$k = \frac{A}{B},$$

(**12.13**)

where

k = the bridge constant,

A = the actual bridge output,

B = the output from the bridge if only a single gage, sensing maximum strain, were effective.

In the example illustrated in Fig. 12.7, the bridge constant would be 2. This is true because the bridge provides an output double of that which would be obtained if only gage 1 were strain-sensitive. If all four gages were used, quadrupling the output, the bridge constant would be 4. In certain other cases (Section 12.16), gages may be mounted sensitive to lateral strains that are functions of Poisson's ratio. In such cases bridge constants of 1.3 and 2.6 (for Poisson's ratio = 0.3) are common.

12.8.3   Lead-Wire Error

When it is necessary to use unusually long leads between a strain gage and other instrumentation, *lead-wire error* may be introduced. The reader is referred to Sections 7.9.5 and 16.4.2, where solutions to this problem are discussed.

12.9   The Simple Constant-Current Strain-Gage Circuit

Measurement of dynamic strains may be accomplished by the simple circuit shown in Fig. 12.8. It is assumed that the power source is a true constant-current supply and that the indicator (CRO is shown) possesses near-infinite input impedance compared to the gage resistance. As the gage resistance changes as a result of strain, the voltage across the gage, hence the input to the CRO, will be

$$e_i = i_i R \tag{12.14}$$

and

$$\Delta e_i = i_i \Delta R. \tag{12.14a}$$

Dividing Eq. (12.14a) by Eq. (12.14), we have

$$\frac{\Delta e_i}{e_i} = \frac{\Delta R}{R}.$$

Inserting $\Delta R/R$ in Eq. (12.10) gives us

$$\varepsilon = \frac{1}{F} \frac{\Delta e_i}{e_i}. \tag{12.14b}$$

The CRO should be set in the ac mode to cancel the direct dc component, and, of course, the CRO amplification capability must be sufficient to provide an adequate readout.

Figure 12.8 Single-gage constant-current circuit

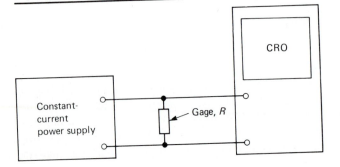

12.10   Temperature Compensation

As already implied, resistive-type strain gages are normally quite sensitive to temperature. Both the differential expansion between the grid and the tested material and the temperature coefficient of the resistivity of the grid material contribute to the problem. It has been shown (Table 12.1) that the temperature effect may be large enough to require careful consideration. Temperature effects may be handled by (1) cancellation or compensation or (2) evaluation as a part of the data reduction problem.

Compensation may be provided (1) through use of adjacent-arm balancing or compensating gage or gages or (2) by means of self-compensation.

12.10.1   The Adjacent-Arm Compensating Gage

Consider bridge configurations such as those shown in Figs. 12.6 and 12.7. Initial electrical balance is obtained when

$$\frac{R_1}{R_2} = \frac{R_3}{R_4}.$$

If the gages in arms 1 and 2 are *alike* and *mounted on similar materials* and if both gages experience the same resistance shift, ΔR_t, caused by temperature change, then

$$\frac{R_1 + \Delta R_t}{R_2 + \Delta R_t} = \frac{R_3}{R_4}.$$

We see that the bridge remains in balance and the output is unaffected by the change in temperature. When the compensating gage is used merely to complete the bridge and to balance out the temperature component, it is often referred to as the "dummy" gage.

12.10.2   Self-Temperature Compensation

In certain cases it may be difficult or impossible to obtain temperature compensation by means of an adjacent-arm compensating or dummy gage. For example, temperature gradients in the test part may be sufficiently great to make it impossible to hold any two gages at similar temperatures. Or, in certain instances, it may be desirable to use the ballast rather than the bridge circuit, thereby eliminating the possibility of adjacent-arm compensation. Situations of this sort make *self-compensation* highly desirable.

The two general types of self-compensated gages available are the *selected-melt* gage and the *dual-element* gage. The former is based on the discovery that through proper manipulation of alloy and processing, particularly through cold working, some control over the temperature sensitivity of the grid material may be exercised. Through this approach grid materials may be prepared that show very low apparent strain versus temperature change over certain

Figure 12.9   Approximate range of apparent strain versus temperature for a typical selected-melt gage mounted on the appropriate material (e.g., steel at about 11 ppm/°C)

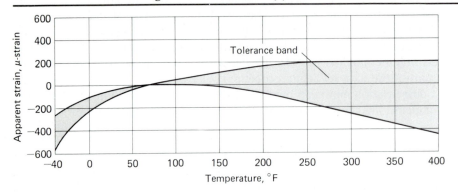

temperature ranges when the gage is mounted on a particular test material. Figure 12.9 shows typical characteristics of selected-melt gages compensated for use with a material having a coefficient of expansion of 6 ppm/°F, which corresponds to the coefficient of expansion of most carbon steels. In this case, practical compensation is accomplished over a temperature range of approximately 50°F–250°F. Other gages may be compensated for different thermal expansions and temperature ranges. These curves give some idea of the degree of control that may be obtained through manipulation of the grid material.

The second approach to self-compensation makes use of two grid elements connected in series in one gage assembly. The two elements have different temperature characteristics and are selected so that the net temperature-induced strain is minimized when the gage is mounted on the specified test material. In general, the performance of this type of gage is similar to that of the selected-melt gage shown in Fig. 12.9.

Neither the selected-melt nor the dual-element gage has a distinctive outward appearance. One company uses color-coded backings to assist in identifying gages of different specifications.

12.11   Calibration

Ideally, calibration of any measuring system consists of introducing an accurately known sample of the variable that is to be measured and then observing the system's response. This ideal cannot often be realized in bonded resistance strain-gage work because of the nature of the transducer. Normally, the gage is bonded to a test item for the simple reason that the strains (or stresses) are unknown. Once bonded, the gage can hardly be transferred to a *known* strain situation for calibration. Of course, this is not necessarily the case if the gage or gages are used as secondary transducers applied to an appropriate elastic member for the purpose of measuring force, pressure, torque, etc. In cases of this sort, it may be perfectly feasible to introduce known inputs and carry out satisfactory calibrations. When the gage is used for the purpose of experimentally determining strains, however, some other approach to the calibration problem is required.

Resistance strain gages are manufactured under carefully controlled conditions, and the gage factor for each lot of gages is provided by the manufacturer within an indicated tolerance of about $\pm0.2\%$. Knowing the gage factor and gage resistance makes possible a simple method for calibrating any resistance strain-gage system. The method consists of determining the system's response to the introduction of a known small resistance change at the gage and of calculating an equivalent strain therefrom. The resistance change is introduced by shunting a relatively high-value precision resistance across the gage, as shown in Fig. 12.10. When switch S is closed, the resistance of bridge arm 1 is changed by a small amount, as determined by the following calculations.

Let

$$R_g = \text{the gage resistance,}$$

$$R_s = \text{the shunt resistance.}$$

Then the resistance of arm 1 before the switch is closed equals R_g, and the resistance of arm 1 after the switch is closed equals $(R_g R_s)/(R_g + R_s)$, as determined for parallel resistances. Therefore, the change in resistance is

$$\Delta R = \frac{R_g R_s}{(R_g + R_s)} - R_g = -\frac{R_g^2}{R_g + R_s}.$$

Figure 12.10 Bridge employing a shunt resistance for calibration

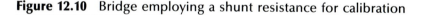

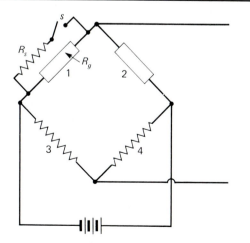

Now to determine the equivalent strain, we may use the relation given by Eq. (12.11):

$$\varepsilon = +\frac{1}{F}\frac{\Delta R_g}{R_g}.$$

By substituting ΔR for ΔR_g, the equivalent strain is found to be

$$\varepsilon_e = -\frac{1}{F}\left(\frac{R_g}{R_g + R_s}\right). \qquad\qquad (12.15)$$

Example 12.1

Suppose that

$$R_g = 120\ \Omega,$$

$$F = 2.1,$$

$$R_s = 100\ k\Omega\ (\text{i.e., } 100{,}000\ \Omega).$$

What equivalent strain will be indicated when the shunt resistance is connected across the gage?

Solution. From Eq. (12.15),

$$\varepsilon_e = -\frac{1}{2.1}\left[\frac{120}{100{,}000 + 120}\right] = -0.00057$$

$$= -570\ \mu\text{-strain}$$

Dynamic calibration is sometimes provided by replacing the manual calibration switch with an electrically driven switch, often referred to as a *chopper*, which makes and breaks the contact 60 or 100 times per second. When displayed on a CRO screen or recorded, the trace obtained is found to be a square wave. The *step* in the trace represents the equivalent strain calculated from Eq. (12.15).

There are other methods of electrical calibration. One system replaces the strain-gage bridge with a substitute load, initially adjusted to equal the bridge load [12]. A series resistance is then used for calibration. Another method injects an accurately known voltage into the bridge network.

12.12   Commercially Available Strain-Measuring Systems

Commercially available systems intended for use with metallic-type gages fall within four general categories:

1.   The basic strain indicator, useful for static, single-channel readings

2. The single-channel system either external to or an integral part of a cathode-ray oscilloscope.

3. Oscillographic systems incorporating either a stylus-and-paper or light-beam and photographic paper readout.

4. Data acquisition systems (e.g., see Fig. 8.24), whereby the strain data may be:

    a. displayed (digitally and/or by a video terminal),

    b. recorded (magnetic tape or hard-copy printout),

    c. fed back into the system for control purposes.

The wide range of availability and divergence of such systems makes it impractical to attempt any but a superficial coverage in this text. The better source of state-of-the-art details is brochures and technical "aids" provided by many of the commercial suppliers.

12.12.1   The Basic Indicator

Typically, the basic indicator consists of a manually or self-balancing Wheatstone bridge with meter-type or digital readout, an amplifier, and adjustments to accommodate a range of gage factors. Provision is also common for handling bridges with a single active-gage, two-gage, and four-gage configurations. For fewer than four gages, the bridge loop is completed within the instrument. The measurement process consists of zeroing the bridge under initial conditions, then, after applying test conditions, rebalancing the bridge. The difference between initial and final readings provides the strain increment. Such instruments are generally precalibrated to provide direct strain readout, often in digital form.

12.12.2   Bridge and Amplifier for Use
with Cathode-Ray Oscilloscope

Figure 12.11 shows a simplified circuit diagram for providing single-channel strain input to a CRO. Basically the circuit consists of a dc bridge with provision for resistance balance and shunt calibration. Output from the amplifier provides both a meter readout and an input for driving an oscilloscope. The meter may be used for initial balancing and also for static strain readout, whereas dynamic inputs may be displayed on the oscilloscope screen.

12.12.3   Multichannel Recording Systems

Either stylus-type or digital recorders are often used for strain-measurement systems. These are commonly general-purpose systems adaptable to various measurement problems through use of plug-in preamplifiers—the preamplifiers configured for the particular purpose—e.g., strain measurement.

Figure 12.11   Schematic circuit diagram for a basic strain-measuring system. Provision is made for meter and/or CRO readouts.

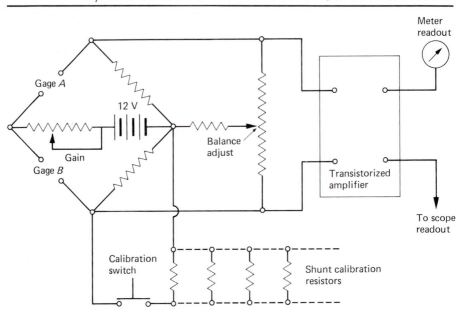

Figure 12.12(a) is a block diagram of such a system. As shown, the bridge is ac excited and, after voltage amplification, the signal is demodulated and fed to a power amplifier needed to drive the recorder. Systems such as this are often multichannel, requiring separate bridge-amplifier-recorder circuits for each channel, all channels sharing a common power supply, chart drive, and enclosure. Figure 12.12(b) illustrates a typical readout.

12.13   Strain-Gage Switching

Mechanical development problems often require the use of many gages mounted throughout the test item, and simultaneous or nearly simultaneous readings are often necessary. Of course, if the data must be recorded at precisely the same instant, it will be necessary to provide separate channels for each gage involved. However, frequently steady-state conditions may be maintained or the test cycle repeated, and readings may be made in succession until all the data have been recorded. In other cases, the budget may prohibit duplication of the required instrumentation for simultaneous multiple readings or recordings, and it becomes desirable to switch from gage to gage, taking data in sequence.

Two basic switching arrangements are possible when resistance bridges are used. They are *intrabridge* switching and *interbridge* switching. Figure 12.13 illustrates the first arrangement, in which various gages are switched into and

Figure 12.12 (a) Block diagram of a single-channel recording system with carrier amplifier. Multichannel recording requires duplication of all of the elements shown above except the recorder. (b) Typical record obtained from a multichannel system.

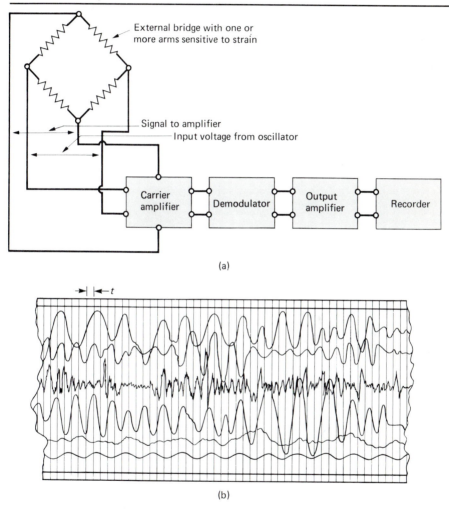

(a)

(b)

out of a single bridge circuit. Figure 12.14 illustrates the second, in which switch connections are entirely outside the bridges.

When metallic gages and intrabridge switching are simultaneously used, variations in switch resistance can be a very annoying problem. As was shown in Section 12.8, the resistance change resulting from strain of the metallic gage is very small. It is quite possible that the change in switch resistance from one "switching" to the next may be of the same order of magnitude as the quantity

Figure 12.13   Intrabridge switching, where gages are
switched into and out of a bridge

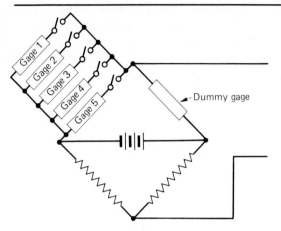

Dummy gage

Figure 12.14   Interbridge switching, where complete bridges
are switched into or out of a circuit

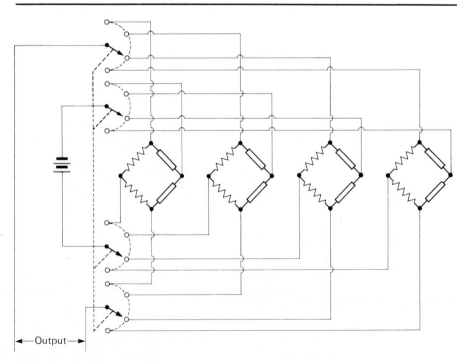

Output

of interest. Unless extreme care and the highest quality of switch components are used, this method of switching may prove entirely impractical.

For really trouble-free strain-gage switching, the method illustrated in Fig. 12.14 is recommended. Here all switch contacts are placed completely outside the system. Although the same variations in switch resistance occur as do in intrabridge switching, in this case they do not alter bridge balance. Their effect is to alter bridge sensitivity slightly, but this is not normally measurable. The disadvantage in the latter method lies in the more complicated installation. This disadvantage must be weighed against the possibly questionable data yielded by the simpler setup.

12.14   Use of Strain Gages on Rotating Shafts

Strain-gage information may be conducted from rotating shafts in at least three different ways: (1) by direct connection, (2) by telemetering, and (3) by use of slip rings. When a shaft rotates slowly enough and when only a sampling of data is required, direct connections may be made between the gages and the remainder of the measuring system. Sufficient lead length is provided, and the cable is permitted to wrap itself onto the shaft. In fact, the available time may be doubled with a given length of cable if it is first wrapped on the shaft so that the shaft rotation causes it to unwrap and then to wrap up again in the opposite direction. If the machine cannot be stopped quickly enough as the end of the cable is approached, a fast or automatic disconnecting arrangement may be provided. This actually need be no more than soldered connections that can be quickly peeled off. Shielded cable should be used to minimize reactive effects resulting from the coil of cable on the shaft. This technique is somewhat limited, of course, but should not be overlooked, because it is quite workable at slow speeds and avoids many of the problems inherent in the other methods.

A second method is that of actually transmitting the strain-gage information through the use of a radio-frequency transmitter mounted on the shaft and picking up the signal by means of a receiver placed nearby. This method has been used successfully [13], and the procedure and equipment will undoubtedly be perfected for more general use. There is commercially available transistorized FM equipment of relatively small size in which the frequency change of the transmitter RF is a function of the strain. Such a system is quite practical when the added cost can be justified.

Undoubtedly the most common method for obtaining strain-gage information from rotating shafts is through the use of slip rings. Slip-ring problems are similar to switching problems, as discussed in the preceding section, except that additional variables make the problem more difficult. Such factors as ring and brush wear and changing contact temperatures make it imperative that the full bridge be used at the test point and that the slip rings be introduced externally to the bridge, as shown in Fig. 12.15.

Figure 12.15   Slip rings external to the bridge

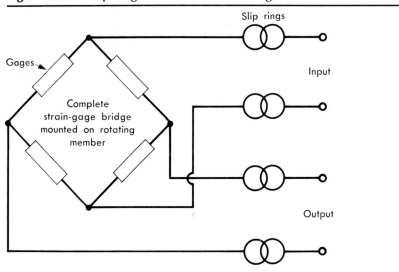

Commercial slip-ring assemblies are available whose performances are quite satisfactory. Their use, however, presents a problem that is often difficult to solve. The assembly is normally self-contained, consisting of brush supports and a shaft with rings mounted between two bearings. The construction requires that the rings be used at a free end of a shaft, which more often than not is separated from the test point by some form of bearing. This arrangement presents the problem of getting the leads from the gage located on one side of a bearing to the slip rings located on the opposite side. It is necessary to feed the leads through the shaft in some manner, which is not always convenient. Where this presents no particular problem, the commercially available slip-ring assemblies are practical and also probably the most inexpensive solution to the problem.

12.15   Stress-Strain Relationships

As previously stated, strain gages are generally used for one of two reasons: to determine stress conditions through strain measurements or to act as secondary transducers calibrated in terms of such quantities as force, pressure, displacement, and the like. In either case, intelligent use of strain gages demands a good grasp of stress-strain relationships. Knowledge of the *plane*, rather than of the general three-dimensional case, is usually sufficient for strain-gage work because it is only in the very unusual situation that a strain gage is mounted anywhere except on the *unloaded surface* of a stressed member. For a review of the plane stress problem, the reader is directed to Appendix E.

12.15.1   The Simple Uniaxial Stress Situation

In bending, or in a tension or compression member, the unloaded outer fiber is subject to a uniaxial stress. However, this condition results in a triaxial strain condition, because we know that there will be lateral strain in addition to the strain in the direction of stress. Because of the simplicity of the ordinary tensile (or compressive) situation and its prevalence (see Fig. 12.1), the fundamental *stress-strain relationship* is based on it. Young's modulus is defined by the relation expressed by Eq. (12.2), and Poisson's ratio is defined by Eq. (12.2a). It is important to realize that both these definitions are made on the basis of the simple one-direction stress system.

For situations of this sort, calculation of stress from strain measurements is quite simple. The stress is determined merely by multiplying the strain, measured in the axial direction in microstrains, by the modulus of elasticity for the test material.

Example 12.2
Suppose the tensile member in Fig. 12.6 is of aluminum having a modulus of elasticity equal to 6.9×10^{10} Pa $(10 \times 10^6 \text{ lbf/in.}^2)$ and the strain measured by the gage is 326 μ-strain. What axial stress exists at the gage?

Solution.

$$\sigma_a = E\varepsilon_a = (6.9 \times 10^{10}) \times (325 \times 10^{-6})$$
$$= 22.4 \times 10^6 \text{ Pa } (3250 \text{ lbf/in.}^2).$$

Example 12.3
Strain gages are mounted on a beam as shown in Fig. 12.7. The beam is of steel having an estimated modulus of elasticity of 20.3×10^{10} Pa $(29.5 \times 10^6 \text{ lbf/in.}^2)$. If the total readout from the two gages is 390 μ-strain, what stress exists at the longitudinal center of the gage? Note that the bridge constant is 2.

Solution.

$$\sigma_b = E\varepsilon_b = (20.3 \times 10^{10}) \times (390 \times 10^{-6}/2)$$
$$= 3958 \times 10^4 \text{ Pa } (5700 \text{ lbf/in.}^2).$$

Figure 12.16   Element located on the shell of a cylindrical pressure vessel

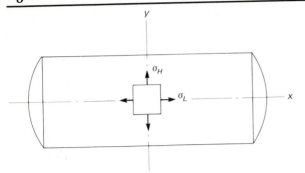

12.15.2   The Biaxial Stress Situation

Often gages are used at locations subject to stresses in more than one direction. If the test point is on a free surface, as is usually the case, the condition is termed *biaxial*. A good example of this condition exists on the outer surface, or shell, of a cylindrical pressure vessel. In this case, we know that there are *hoop* stresses, acting circumferentially, tending to open up a longitudinal seam. There are also longitudinal stresses tending to blow the heads off. The situation may be represented as shown in Fig. 12.16.

The stress-strain condition on the outer surface corresponds to that shown in Fig. 12.2. The two stresses σ_L and σ_H are principal stresses (no shear in the longitudinal and hoop directions), and the corresponding stresses may be calculated using Eqs. (12.4), if we know (or can estimate) Young's modulus and Poisson's ratio.

Example 12.4
Suppose we wish to determine, by strain measurement, the stress in the circumferential or hoop direction on the outer surface of a cylindrical pressure vessel. The modulus of elasticity of the material is 10.3×10^{10} Pa, and Poisson's ratio is 0.28. By strain measurement the hoop and longitudinal strains (in microstrain) are determined to be

$$\varepsilon_H = 425 \quad \text{and} \quad \varepsilon_L = 115.$$

Solution. Using Eqs. (12.4), we have

$$\sigma_H = \frac{E(\varepsilon_H + \nu\varepsilon_L)}{1 - \nu^2} = \frac{10.3 \times 10^{10}(425 + 0.28 \times 115) \times 10^{-6}}{1 - 0.28^2}$$

$$= 5.11 \times 10^7 \text{ Pa } (7.42 \times 10^3 \text{ lbf/in.}^2).$$

Although we may not be directly interested, we have the necessary information to determine the longitudinal stress also, as follows:

$$\sigma_L = \frac{E(\varepsilon_L + v\varepsilon_H)}{1 - v^2} = \frac{10.3 \times 10^{10}(115 + 0.28 \times 425) \times 10^{-6}}{1 - 0.28^2}$$

$$= 2.61 \times 10^7 \text{ Pa } (3.8 \times 10^3 \text{ lbf/in.}^2).$$

It may be noted that the 2-to-1 stress ratio traditionally expected for the thin-wall cylindrical pressure vessel does not yield a like ratio of strains. The strain ratio is more nearly 4 to 1.

Use of Eqs. (12.4) permits us to determine the stresses in two orthogonal directions. However, this information gives the *complete* stress-strain picture only when the two right-angled directions coincide with the *principal directions* (see Appendix E). If we do not know the principal directions, our readings would only by chance yield the maximum stress. In general, if a plane stress condition is completely unknown, at least three strain measurements must be made, and it becomes necessary to use some form of three-element *rosette* (see Fig. 12.4). From the strain data secured in the three directions, we obtain the complete stress-strain picture. Stress-strain relations for rosette gages are given in Table 12.3.

Although only three strain measurements are necessary to define a stress situation completely, the *T*-delta rosette, which includes a fourth gage element, is sometimes used to advantage for the following reasons:

1. The fourth gage may be used as a check on the results obtained from the other three elements.

2. If the principal directions are approximately known, gage *d* may be aligned with the estimated direction. Then, if the readings from gages *b* and *c* are of about the same magnitude, it is known that the estimate is reasonably correct, and the principal stresses may be calculated directly from Eqs. (12.4), greatly simplifying the arithmetic. If the estimate of direction turns out to be incorrect, complete data are still available for use in the equations from Table 12.3.

3. If the four readings are used in the T-delta equations in Table 12.3, an averaging effect results in better accuracy than if only three readings are used.

In spite of the advantages of the T-delta rosette, the rectangular one is probably the most popular, with the equiangular (delta) kind receiving second greatest use.

Table 12.3 Stress-strain relations for rosette gages*†

Type of Rosette:	Rectangular	Equiangular (Delta)	T-Delta
Principal strains, $\varepsilon_1, \varepsilon_2$	$\dfrac{1}{2}\left[\varepsilon_a + \varepsilon_c\right.$ $\left.\pm \sqrt{2(\varepsilon_a - \varepsilon_b)^2 + 2(\varepsilon_b - \varepsilon_c)^2}\right]$	$\dfrac{1}{3}\left[\varepsilon_a + \varepsilon_b + \varepsilon_c\right.$ $\left.\pm \sqrt{2(\varepsilon_a - \varepsilon_b)^2 + 2(\varepsilon_b - \varepsilon_c)^2 + 2(\varepsilon_c - \varepsilon_a)^2}\right]$	$\dfrac{1}{2}\left[\varepsilon_a + \varepsilon_d\right.$ $\left.\pm \sqrt{(\varepsilon_a - \varepsilon_d)^2 + \frac{4}{3}(\varepsilon_b - \varepsilon_c)^2}\right]$
Principal stresses, σ_1, σ_2	$\dfrac{E}{2}\left[\dfrac{\varepsilon_a + \varepsilon_c}{1 - \nu} \pm \dfrac{1}{1 + \nu}\right.$ $\left.\sqrt{2(\varepsilon_a - \varepsilon_b)^2 + 2(\varepsilon_b - \varepsilon_c)^2}\right]$	$\dfrac{E}{3}\left[\dfrac{\varepsilon_a + \varepsilon_b + \varepsilon_c}{1 - \nu} \pm \dfrac{1}{1 + \nu}\right.$ $\left.\sqrt{2(\varepsilon_a - \varepsilon_b)^2 + 2(\varepsilon_b - \varepsilon_c)^2 + 2(\varepsilon_c - \varepsilon_a)^2}\right]$	$\dfrac{E}{2}\left[\dfrac{\varepsilon_a + \varepsilon_d}{1 - \nu} \pm \dfrac{1}{1 + \nu}\right.$ $\left.\sqrt{(\varepsilon_a - \varepsilon_d)^2 + \frac{4}{3}(\varepsilon_b - \varepsilon_c)^2}\right]$
Maximum shear, τ_{max}	$\dfrac{E}{2(1 + \nu)}$ $\sqrt{2(\varepsilon_a - \varepsilon_b)^2 + 2(\varepsilon_b - \varepsilon_c)^2}$	$\dfrac{E}{3(1 + \nu)}$ $\sqrt{2(\varepsilon_a - \varepsilon_b)^2 + 2(\varepsilon_b - \varepsilon_c)^2 + 2(\varepsilon_c - \varepsilon_a)^2}$	$\dfrac{E}{2(1 + \nu)}$ $\sqrt{(\varepsilon_a - \varepsilon_d)^2 + \frac{4}{3}(\varepsilon_b - \varepsilon_c)^2}$
$\tan 2\theta$	$\dfrac{2\varepsilon_b - \varepsilon_a - \varepsilon_c}{\varepsilon_a - \varepsilon_c}$	$\dfrac{\sqrt{3}\,(\varepsilon_c - \varepsilon_b)}{(2\varepsilon_a - \varepsilon_b - \varepsilon_c)}$	$\dfrac{2}{\sqrt{3}}\dfrac{(\varepsilon_c - \varepsilon_b)}{(\varepsilon_a - \varepsilon_d)}$
$0 < \theta < +90°$	$\varepsilon_b > \dfrac{\varepsilon_a + \varepsilon_c}{2}$	$\varepsilon_c > \varepsilon_b$	$\varepsilon_c > \varepsilon_b$

* References: [1, 14, 15].

† Note: θ = the angle of reference, measured positive in the counterclockwise direction from the a-axis of the rosette to the axis of the algebraically larger stress.

Example 12.5

Figure 12.17 illustrates a rectangular rosette used to determine the stress situation near a pressure vessel nozzle. For thin-walled vessels, the assumption that principal directions correspond to the hoop and longitudinal directions is valid for the shell areas removed from discontinuities. Near an opening, however, the stress condition is completely unknown, and a rosette with at least three elements must be used.

Let us assume that the rosette provides the following data (in microstrain):

$$\varepsilon_a = 72, \qquad \varepsilon_b = 120, \qquad \varepsilon_c = 248.$$

In addition, we shall say that

$$v = 0.3, \qquad E = 20.7 \times 10^{10}\,\text{Pa}.$$

Solution. A study of the equation forms in Table 12.3 shows that for each case, the principal strain, the principal stress, and maximum shear relations involve similar radical terms. Therefore, in evaluating rosette data, it is convenient to calculate the value of the radical as the first step. It will also be noted that the second term in the principal stress relations is equal to the shear stress; thus arithmetical manipulations may be kept to a minimum if the shear stress is calculated before the principal stresses are determined. Hence,

$$\sqrt{2(\varepsilon_a - \varepsilon_b)^2 + 2(\varepsilon_b - \varepsilon_c)^2} = \sqrt{2(72 - 120)^2 + 2(120 - 248)^2}$$

$$= 193\,\mu\text{-strain}$$

and

$$\varepsilon_1 = \frac{1}{2}[72 + 248 + 193] = 256\,\mu\text{-strain},$$

$$\varepsilon_2 = \frac{1}{2}[72 + 248 - 193] = 63\,\mu\text{-strain},$$

Figure 12.17   Rosette installation near a pressure vessel nozzle

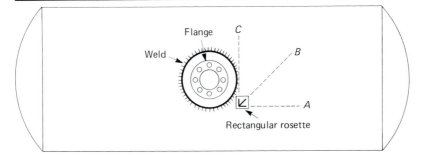

Flange   C

Weld

B

A

Rectangular rosette

$$\tau_{max} = \frac{20.7 \times 10^{10}}{2(1 + 0.3)}(193) \times 10^{-6} = 1537 \times 10^4 \, \text{Pa} \quad (2230 \, \text{lbf/in.}^2),$$

$$\sigma_1 = \frac{20.7 \times 10^{10}}{2} \times \frac{72 + 248}{0.7} \times 10^{-6} + 1537 \times 10^4$$

$$= (4731 + 1537) \times 10^4 = 6268 \times 10^4 \, \text{Pa} \quad (9091 \, \text{lbf/in.}^2),$$

$$\sigma_2 = (4731 - 1537) \times 10^4 = 3194 \times 10^4 \, \text{Pa} \quad (4632 \, \text{lbf/in.}^2).$$

To determine the principal planes, we have

$$\tan 2\theta = \frac{(2\varepsilon_b - \varepsilon_a - \varepsilon_c)}{(\varepsilon_a - \varepsilon_c)}$$

$$= \frac{(2 \times 120) - 72 - 250}{(72 - 150)} = 0.46,$$

$$2\theta = 25° \quad \text{or} \quad 205°,$$

Figure 12.18 Stress conditions determined from data obtained by the rosette shown in Fig. 12.17

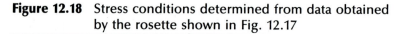

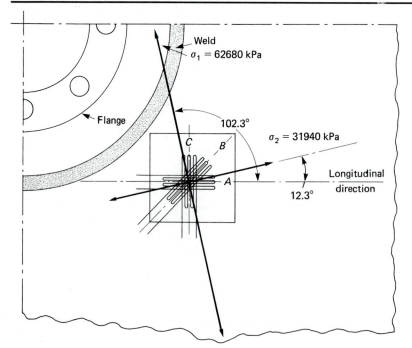

or
$$\theta = 12.5° \quad \text{or} \quad 102.5°,$$

measured counterclockwise from the axis of element A. We must test for the proper quadrant as follows (see the last line in Table 12.3):

$$\frac{\varepsilon_a + \varepsilon_c}{2} = \frac{72 + 248}{2} = 16 \times 10^1,$$

which is greater than ε_b. Therefore, the axis of maximum principal stress does *not* fall between $0°$ and $90°$. Hence, $\theta = 102°$. Figure 12.18 illustrates this condition.

12.16 Gage Orientation and Interpretation of Results

In a given situation it is often possible to place gages in several different arrangements to obtain the desired data. Often there is a best way, however, and in certain instances unwanted strain components may be canceled by proper gage orientation. For example, it is often desirable to eliminate unintentional bending when only direct axial loading is of primary interest; or, perhaps, only the bending component in a shaft is desired, to the exclusion of torsional strains.

The following discussion should be helpful in determining the proper positioning of gages and interpretation of the results. We will assume a *standard* bridge arrangement as shown in Table 12.4, and the gages will be numbered in the following examples according to this standard. When fewer than four gages are used, it is assumed that the bridge configuration is completed with fixed resistors insensitive to strain.

Recall the relationship given in Eq. (7.15a), Section 7.9.1, which evaluates bridge output e_o for a given input e_i, namely,

$$e_o = e_i \frac{R_1 R_4 - R_2 R_3}{(R_1 + R_2)(R_3 + R_4)}. \tag{12.16}$$

If we assume the resistance of each bridge arm to be variable, then

$$de_o = \frac{\partial e_o}{\partial R_1} dR_1 + \frac{\partial e_o}{\partial R_2} dR_2 + \frac{\partial e_o}{\partial R_3} dR_3 + \frac{\partial e_o}{\partial R_4} dR_4. \tag{12.17}$$

By using Eq. (12.16) we can evaluate the various partial derivatives and write

$$\frac{de_o}{e_i} = \frac{R_2 \, dR_1}{(R_1 + R_2)^2} - \frac{R_1 \, dR_2}{(R_1 + R_2)^2} - \frac{R_4 \, dR_3}{(R_3 + R_4)^2} + \frac{R_3 \, dR_4}{(R_3 + R_4)^2}, \tag{12.17a}$$

where dR_1, dR_2, dR_3, and dR_4 are the various resistance changes in each of the bridge arms.

Table 12.4 Strain-gage orientation

Standard Bridge Configuration

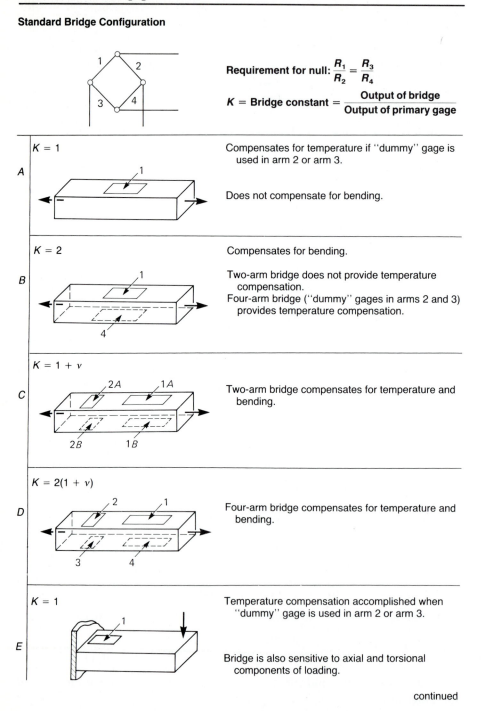

Requirement for null: $\dfrac{R_1}{R_2} = \dfrac{R_3}{R_4}$

K = Bridge constant = $\dfrac{\text{Output of bridge}}{\text{Output of primary gage}}$

A $K = 1$

Compensates for temperature if "dummy" gage is used in arm 2 or arm 3.

Does not compensate for bending.

B $K = 2$

Compensates for bending.

Two-arm bridge does not provide temperature compensation.
Four-arm bridge ("dummy" gages in arms 2 and 3) provides temperature compensation.

C $K = 1 + \nu$

Two-arm bridge compensates for temperature and bending.

D $K = 2(1 + \nu)$

Four-arm bridge compensates for temperature and bending.

E $K = 1$

Temperature compensation accomplished when "dummy" gage is used in arm 2 or arm 3.

Bridge is also sensitive to axial and torsional components of loading.

continued

Table 12.4 *continued*

Standard Bridge Configuration

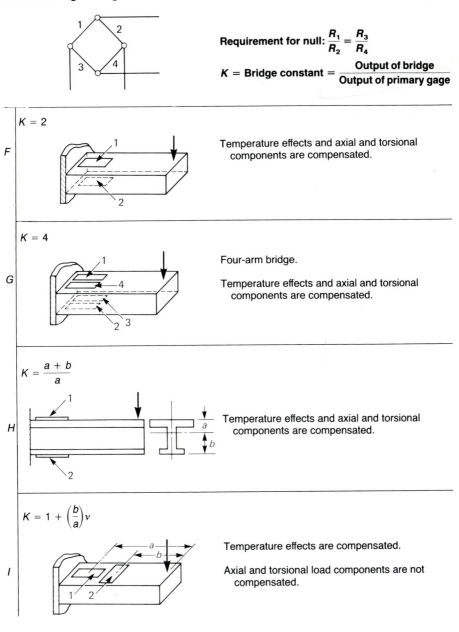

Requirement for null: $\dfrac{R_1}{R_2} = \dfrac{R_3}{R_4}$

K = Bridge constant = $\dfrac{\text{Output of bridge}}{\text{Output of primary gage}}$

F $\quad K = 2$

Temperature effects and axial and torsional components are compensated.

G $\quad K = 4$

Four-arm bridge.

Temperature effects and axial and torsional components are compensated.

H $\quad K = \dfrac{a + b}{a}$

Temperature effects and axial and torsional components are compensated.

I $\quad K = 1 + \left(\dfrac{b}{a}\right)v$

Temperature effects are compensated.

Axial and torsional load components are not compensated.

continued

Table 12.4 *continued*

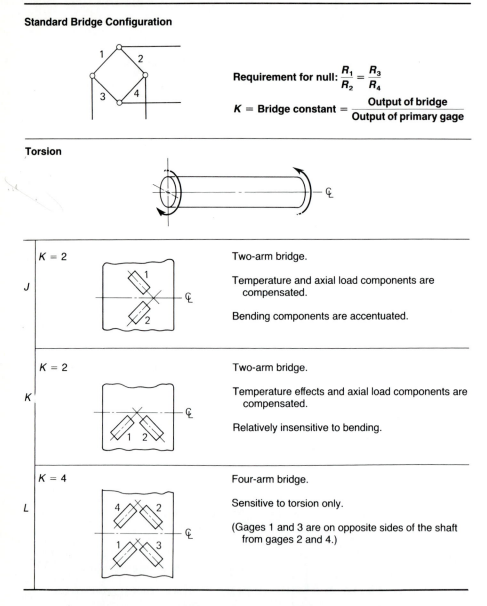

Standard Bridge Configuration

Requirement for null: $\dfrac{R_1}{R_2} = \dfrac{R_3}{R_4}$

K = Bridge constant = $\dfrac{\text{Output of bridge}}{\text{Output of primary gage}}$

Torsion

J	$K = 2$	Two-arm bridge. Temperature and axial load components are compensated. Bending components are accentuated.
K	$K = 2$	Two-arm bridge. Temperature effects and axial load components are compensated. Relatively insensitive to bending.
L	$K = 4$	Four-arm bridge. Sensitive to torsion only. (Gages 1 and 3 are on opposite sides of the shaft from gages 2 and 4.)

Ordinarily the gages used to make up a bridge will be from the same lot and

$$R_1 = R_2 = R_3 = R_4 = R.$$

Each gage may experience a different resistance change; hence we must retain the subscripts on the dR's; however, we can drop them from the R's. Doing so yields

$$\frac{de_o}{e_i} = \frac{dR_1 - dR_2 - dR_3 + dR_4}{4R}. \tag{12.17b}$$

From Eq. (12.10), we have

$$\frac{dR_n}{R_n} = F\varepsilon_n, \tag{12.18}$$

where

$$F = \text{the gage factor,}$$

$$\varepsilon_n = \text{the strain sensed by gage } n.$$

Combining Eqs. (12.18) and (12.17b) gives us

$$\frac{de_o}{e_i} = \frac{F}{4}[\varepsilon_1 - \varepsilon_2 - \varepsilon_3 + \varepsilon_4], \tag{12.19}$$

where the ε's are the strains sensed by the respective gages.

Equation (12.19) aids in the proper interpretation of the strain results obtained from the standard four-arm bridge in addition to assisting the stress analyst in the proper placement and orientation of gages for experimental measurements.

For example, when only one active gage is used, Eq. (12.19) reduces to

$$\frac{de_o}{e_i} = \frac{F\varepsilon}{4}.$$

In further discussions we will assume that the term *bridge constant* abides by the definition given in Section 12.8.2.

Example 12.6
The simplest application uses a single measuring gage with an external compensating gage as shown in Fig. 12.6 (also Case A, Table 12.4). This arrangement is primarily sensitive to axial strain; however, it will also sense any unintentional bending strain. The compensating gage, mounted on a sample of unstrained material identical to the test material, is located so that its temperature and that of the specimen will be the same. In this case,

$$\varepsilon_1 = \varepsilon_a + \varepsilon_b + \varepsilon_T,$$

$$\varepsilon_2 = \varepsilon_T,$$

$$\varepsilon_3 = 0 \quad \text{(a fixed resistor),}$$

$$\varepsilon_4 = 0 \quad \text{(a fixed resistor),}$$

where

ε_a = the strain caused by axial loading,

ε_b = the strain caused by any bending component,

ε_T = the strain caused by temperature changes.

Solution. Substituting in Eq. (12.19) gives us

$$\frac{de_o}{e_i} = \frac{F}{4}[\varepsilon_a + \varepsilon_b].$$

If the bending strain is negligible,

$$\frac{de_o}{e_i} = \frac{F\varepsilon_a}{4}$$

and the bridge constant is unity. Note that any strains caused by temperature effects cancel.

Example 12.7

The arrangement shown in Case G, Table 12.4, uses gages in each of the four bridge arms. Gages 1 and 4 experience positive-bending strain components and gages 2 and 3 sense negative-bending components. All gages would sense the same strains derived from axial load and/or temperature should these be present. In addition, should the member be subjected to an axial torque, gages 1 and 2 would sense like strains from this source, as would gages 3 and 4. All gages would sense strain components of like magnitude from any torque acting about the longitudinal axis of the member; however, strains sensed by gages 1 and 2 would be of opposite sign to those sensed by 3 and 4.

Solution. Substitution of all these effects into Eq. (12.19) yields

$$\frac{de_o}{e_i} = \left(\frac{F}{4}\right)(4\varepsilon_b) = F\varepsilon_b.$$

We see that only bending strains will be sensed and that the bridge constant is 4.

12.16.1   Gages Connected in Series

Figure 12.19 shows a load cell element using six gages, three connected in series in each of the bridge arms 1 and 2. At first glance it might be thought that the three gages in series would provide an output three times as great as

Figure 12.19 Load cell employing three series-connected axial gages and three series-connected Poisson-ratio gages

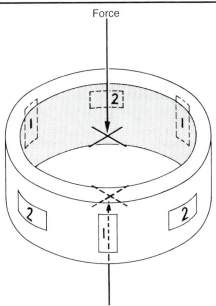

that from a single gage under like conditions. Such is not the case, for it will be recalled that it is the percentage change in resistance, or dR/R, that counts, not dR alone. It is true that the resistance change for one arm, in this case, is three times what it would be for a single gage, but so also is the total resistance three times as great. Therefore, the only advantage gained is that of *averaging* to eliminate incorrect readings resulting from eccentric loading. The remaining two arms (not shown in the figure) may be made up of either inactive strain gages or fixed resistors. The bridge constant is $1 + v$.

12.17 Special Problems

12.17.1 Cross-Sensitivity

Strain gages are arranged with most of the strain-sensitive filament aligned with the sensitive axis of the gage. However, unavoidably, a part of the grid is aligned transversely. The transverse portion of the grid senses the strain in that direction and its effect is superimposed on the longitudinal output. This is known as *cross-sensitivity*. The error is small, seldom exceeding 2 or 3% and the overall accuracy of many applications does not warrant accounting for it. For more detailed consideration the reader is referred to [15], [16], and [17].

12.17.2   Plastic Strains and the Post-Yield Gage

The average commercial strain gage will behave elastically to strain magnitudes as high as 3%. This represents a surprising performance when it is realized that the corresponding uniaxial elastic stress in steel would be almost 1,000,000 lbf/in.^{2} (if elastic conditions in the steel were maintained). It is not very great, however, when viewed by the engineer seeking strain information beyond the yield point. When mild steel is the strained material, strains as great as 15% may occur immediately following attainment of the elastic limit, before the stress again begins to climb above the yield stress. Hence, the usable strain range of the common resistance gage is quickly exceeded.

Gages known as *post-yield* gages have been developed, extending the usable range to approximately 10% to 20%. Grid material in very ductile condition is used, which is literally caused to flow with the strain in the test material. The primary problem, of course, in developing an "elastic-plastic" grid is to obtain a gage factor that is the same under both conditions. Data reduction presents special problems, and for coverage of this aspect see Refs. [18] and [19].

12.17.3   Fatigue Applications of Resistance Strain Gages

Strain gages are subject to fatigue failure in the same manner as are other engineering structures. The same factors are involved in determining their fatigue endurance. In general, the vulnerable point is the discontinuity formed at the juncture of the grid proper and the lead wire to which the user makes

Figure 12.20   Relationship of endurance limit to strain level for gages of various materials and constructions (data from various sources including manufacturer's literature)

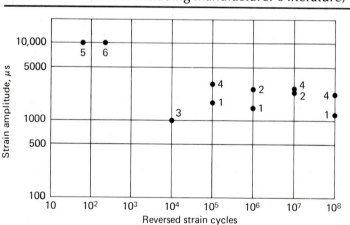

connection. Of course, as with any fatigue problem, strain level is the most important factor in determining life.

Isoelastic grid material performs better under fatigue conditions than does constantan; the carrier material is also an important factor. Figure 12.20 illustrates the effects of most of the factors just discussed.

12.17.4   Cryogenic Temperature Applications

Extreme cryogenic temperatures often cause relatively unpredictable performances of resistance strain gages. Adhesives and backings become glass-hard and quite brittle. Whereas the mechanical properties of certain grid materials are drastically curtailed, those of others remain only slightly affected. Large changes in resistivities may be encountered, and the effective values are dependent to a great degree on trace elements and previous mechanical working of the materials. Much work is being conducted in this area and the state of the art is rapidly changing. Even if all the temperature-related properties were known, however, there would still remain the difficult problem of either controlling the temperature or measuring it.

Telinde reports on a comprehensive evaluation of strain-gage use at temperatures as low as −452°F [20]. His work favors Karma as a grid material, supported on fiberglass-reinforced epoxy.

12.17.5   High-Temperature Applications

Maximum temperatures for short-period use of paper, epoxy, and glass-filled phenolic-base gages with appropriate cements are about 180°F, 250°F, and 600°F, respectively. Primary limiting factors are decomposition of cement and carrier materials. At these temperatures grid materials present no particular problems. For applications at higher temperatures (to 1800°F) some form of ceramic-base insulation must be used. The grid may be of the strippable support, free-element type with the bonding as described below, or the gage may be of the "weldable" type.

Use of the free-element-type gage involves "constructing" the gage on the spot. Either brushable or flame-sprayed ceramic bonding materials are used. Application of the former consists of laying down an insulating coating upon which the free-element grid is secured with more cement. The process demands considerable skill and carefully controlled baking or curing-temperature cycling.

Flame spraying involves the use of a plasma-type oxyacetylene gun [21]. Molten particles of ceramic are propelled onto the test surface and used as both the cementing and insulating material for bonding the grid element to the test item. In both cases, leads must be attached by spot-welding to provide the necessary high-temperature properties to the connections. Lead-wire temperature-resistance variations may also present problems. It is obvious that considerable technique must be developed to use either of these types satisfactorily.

A weldable strain gage consists of a resistance element surrounded by a ceramic-type insulation and encapsulated within a metal sheath. The gage is applied by spot-welding the edges of the assembly to the test member [22].

A novel laser-based extensometer usable at 3500°F or higher and having an overall accuracy of ±0.0002 in. over a 0.3-in. gage length is described in [23].

12.17.6   Creep

Creep in the bond between gage and test surface is a factor sometimes ignored in strain-gage work. This problem is approximately diametrically opposite to the fatigue problem in that it is of importance only in static strain testing, primarily of the long-duration variety. For example, residual stresses are occasionally determined by measuring the dimensional relaxation as stressed material is removed. In this case, the strain is applied to the gage once and once only. The loading cycle cannot be repeated. Under these circumstances, gage creep will result in direct errors equal to the magnitude of the creep. If the load can be slowly cycled, the creep will appear as a hysteresis loop in the results. This effect is a function of several things but is primarily determined by the strain level and the cement used for bonding.

12.17.7   Residual Stress Determination

Occasionally it is necessary to determine the residual stresses existing in a structure or machine element. These stresses are generally developed during mechanical forming processes, such as casting or heat treatment. These stresses can be determined by using the strain-measuring techniques previously described, although they generally destroy the structure being analyzed.

Consider the pressure vessel of Example 12.5. If it is desired to estimate the residual stresses near a pressure-vessel nozzle due to welding, the rectangular rosette may be applied to the unpressurized vessel as shown. After the various strain-gage lead wires are attached to strain readout equipment, the region of pressure vessel containing the rosette is removed (cut away) from the rest of the material, and the resulting change in strains from the gages is recorded. Using these data (note the change in sign), the residual stresses existing in the unpressurized vessel at this location may be estimated. (See Problem 12.30 for an example of this process.)

Most strain-gage manufacturers provide a special strain rosette, whereby the strain-gage elements are arranged in such a fashion that a single hole may be drilled, relieving the stresses in the region and thus eliminating the need to completely cut away the material.

12.18 Whole-Field Methods

12.18.1 Photoelasticity

Photoelasticity is an experimental technique based on the fact that when certain materials transparent to light are strained, they become optically double-refracting (polarizing*) and:

1. The two orthogonal polarizing planes at a given point coincide with the principal stress planes;

2. The two rays of polarized light corresponding to the two planes emerge from the material out of phase in proportion to the maximum shear stress at the point. Since the maximum shear stress in a two-dimensional stress case is equal to the difference in principal stresses, this difference is also known.

Through use of a polariscope (Fig. 12.21), the two emerging rays are combined, resulting in two sets of interference fringes: one called the *isochromatics* and the other the *isoclinics*. Figure 12.22 illustrates a typical pattern of isochromatics. Basically, the technique is two-dimensional; however, various methods have been developed whereby three-dimensional conditions may be investigated.

The experimental procedure is as follows:

1. A two-dimensional model of the prototype is prepared from a suitable material, usually an epoxy, and is loaded in a manner simulating the prototype's loading. Note that the stress distribution is not material-dependent so long as elastic conditions prevail.

2. The loaded model is placed between polarizing filters, *A* and *B* in Fig. 12.21. Near-monochromatic light (commonly a greenish color) is passed through the model and the resulting fringes are photographed.

Figure 12.21 Photoelastic polariscope

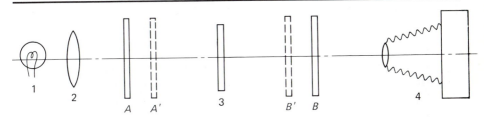

---

* Distinction must be made between a *polarizer* and a *polarizing filter*. An optical polarizer breaks a random ray of light into two orthogonal components and permits *both* to pass. A polarizing filter breaks the ray into two orthogonal components, suppresses one, but permits the other to pass. The familiar Polaroid® sunglasses use polarizing filters. On the other hand, many materials, particularly when stressed, are polarizers: Glass is one.

Figure 12.22 Isochromatic pattern for diametrically loaded ring

When a simple plane-polariscope is used, two superimposed sets of fringes are formed:

1. *Isochromatics,* which are loci of constant principal stress difference, and
2. *Isoclinics,* which are loci of points where the principal stress directions are aligned with the optical polarizing axes. We see that by simply rotating the filters *A* and *B,* isoclinics for various directions may be obtained.

When what are called quarter-wave plates, *A'* and *B'* in Fig. 12.21, are inserted, a circularly polarized polariscope is formed. Under these conditions the isoclinics are suppressed and only the isochromatics remain. In this manner the two sets of fringes may be separated. Figure 12.22 was obtained using a circular polariscope.

To determine stress magnitudes it is necessary to calibrate the stress sensitivity of the material used for the model. The calibration is accomplished by using a circular polariscope and applying the above procedure to a specimen such as a simple tension model, a specimen for which the applied stress may be easily and reliably calculated. The corresponding fringe orders, beginning with fringe 0 (zero stress), are then compared with the calculated stresses, as the load is slowly applied. Actually, the principal stress difference is the criterion; however, for a simple tension member the transverse stress is zero.

The photoelastic method may be used quite simply to determine free-boundary stresses for two-dimensional models of any shape, as follows:

1. Stress differences along isochromatics are easily determined from the calibration constant and the fringe order. (Fringe order is easily determined by simply noting the order of their appearances: Fringe N remains fringe N regardless of where it may migrate to as the load is changed. There are also other means for confirming the order of a given fringe.

2. On any *free boundary* the two principal planes (two-dimensional case) are normal to and tangent to the boundary, and the stress on the tangent plane is zero. Hence, knowing the difference (from the fringe orders) and the fact that the one stress is zero reveals the magnitude of the remaining stress.

Various methods may be used to separate the principal stresses at interior locations; however, the details are beyond the space limitations of this book. The reader is referred to the Suggested Readings at the end of the chapter for additional information on the subject.

12.18.2   The Moiré Technique

We have all noted the wavy fringe patterns that are produced by the overlapping of two layers of something like window screening. The term *moiré* (a French word for watery) is commonly applied to such a pattern. There is a fabric called silk moiré that is formed by permanently pressing lines nearly, but not exactly, aligned with the threads of the fabric. A "modern" version of moiré occurs when the television image of a newscaster's suit involves a pattern of lines roughly approximating the alignment and pitch of the television receiver's raster. In each case a pattern of fringes is produced, and relative movement causes them to shimmer like the reflections from the surface of water.

Moiré patterns properly produced and analyzed may be used for evaluating whole-field *displacements*. The displacements may then be converted to strains and stresses. To accomplish this, a *master* or *reference* pattern, called a *grating,* is required, usually in the form of a photographic film or glass plate. Various patterns may be used; however, gratings consisting of closely packed straight, parallel lines are most common. The line widths and the widths of the intervening spaces are made equal.

Moiré analysis is a two-dimensional technique in much the same sense that the photoelastic method is two-dimensional. The idea on which the method is based is quite simple, but its implementation requires skill and practice. To obtain required sensitivity, one must most often use a low-modulus material, commonly a polyurethane. In addition, sensitivity is a function of the fineness of the grating, i.e., the number of lines per unit length. As would be expected, the greater the number of lines per unit length, the greater the sensitivity of the method and the greater the skill required to produce satisfactory results.

The procedure is as follows:

1. A two-dimensional model of the prototype is prepared using a low-modulus material. If the material is not transparent to light, a reflective surface is required.

2. A working grating is photographically deposited on the surface of the model using the reference grating as the "negative."

3. The master or reference grating is then placed in intimate contact with the model, thereby forming a reference for determining relative displacements. The model is strained by loading, resulting in a pattern of moiré fringes. This pattern can then be photographed and the print can be used for further analysis. Although it is not entirely necessary, we will assume perfect initial alignment between reference and working gratings. Should we take a right section through the two gratings under zero-load conditions we might see something like what is shown in Fig. 12.23(a). With lines in perfect alignment we see no fringes.

4. After straining, we find the condition shown in Fig. 12.23(b). At some point where a *line* on the working grating exactly coincides with a *space* on the reference grating, passage of light is effectively blocked. At other points there will be partial or no blockage. In this manner a pattern of dark and light fringes is formed. As in the case of the photoelastic method,

Figure 12.23   Diagram showing how a moiré pattern is generated

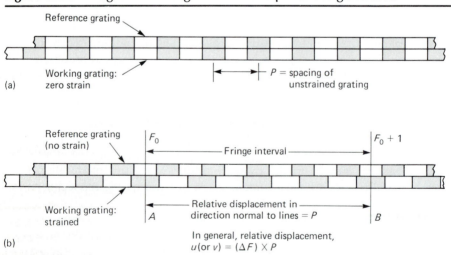

fringe orders, F, may be assigned in integral numerical order, beginning with an arbitrarily chosen zero fringe, F_0.

5. From Fig. 12.23(b) we see that *relative displacements of adjacent fringe sites, in the direction normal to the grating lines, are equal to the pitch of the reference grating.* In other words, if the reference grating lines run in the *y-y* direction, the relative *x*-displacements along fringe sites equal $(F_m - F_n) \times P$, where F_m and F_n are fringe orders and P is the pitch of the reference grating. For total displacements, both *x*- and *y*-components must be measured and their vector sums determined. Two separate test runs are required. If whole-field *displacements* are the objective, the testing would be completed at this point.

6. Should whole-field strains be desired, further analysis would be required as follows. Figure 12.24(a) shows a random section of fringe pattern obtained with *y-y*-oriented gratings. An origin O may be arbitrarily selected. Fringe intersections along the *x*- and *y*-axes are identified. The u_x and u_y displacements corresponding to these points are plotted in Figs. 12.24(b) and (c) versus their respective *x*- and *y*-coordinate positions. The slopes yield the information shown in the diagrams. In like manner, similar *y*-displacements are obtained using *x-x*-oriented gratings, as shown in Fig. 12.25(a), (b), and (c).

Figure 12.24 Graphical procedure for determining *x*-strains

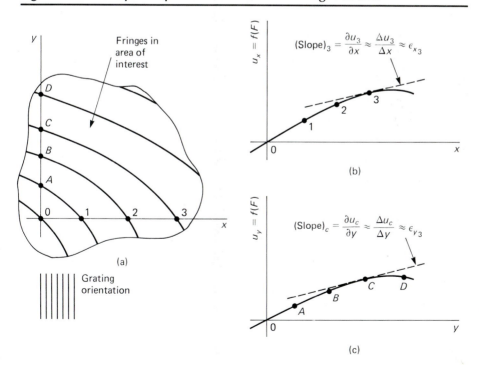

(a)

Grating orientation

(b)

$$(\text{Slope})_3 = \frac{\partial u_3}{\partial x} \approx \frac{\Delta u_3}{\Delta x} \approx \epsilon_{x_3}$$

(c)

$$(\text{Slope})_c = \frac{\partial u_c}{\partial y} \approx \frac{\Delta u_c}{\Delta y} \approx \epsilon_{y_3}$$

Figure 12.25   Graphical procedure for determining y-strains

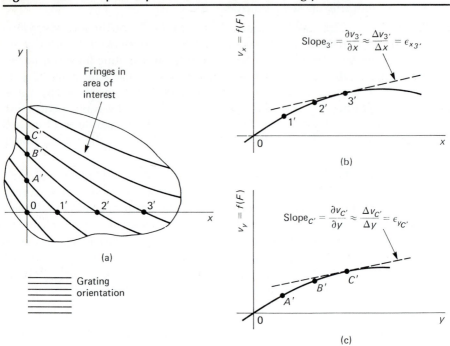

(a)

Fringes in area of interest

Grating orientation

(b)

$$\text{Slope}_{3'} = \frac{\partial v_{3'}}{\partial x} \approx \frac{\Delta v_{3'}}{\Delta x} = \epsilon_{x3'}$$

(c)

$$\text{Slope}_{C'} = \frac{\partial v_{C'}}{\partial y} \approx \frac{\Delta v_{C'}}{\Delta y} = \epsilon_{yC'}$$

7. As shown in Figs 12.24(b) and (c) and 12.26(b) and (c) the following strains may be determined:

$$\varepsilon_x = \frac{\Delta u}{\Delta x},$$

$$\varepsilon_y = \frac{\Delta v}{\Delta y},$$

$$\gamma_{xy} = \frac{\Delta u}{\Delta y} + \frac{\Delta v}{\Delta x}.$$

From these values, stresses may be determined using the following relations:

$$\sigma_{1,2} = \frac{E}{2}\frac{\varepsilon_x + \varepsilon_y}{1 - v} \pm \frac{\sqrt{(\varepsilon_x - \varepsilon_y)^2 + (\gamma_{xy})^2}}{1 + v},$$

$$\tan 2\phi = \frac{\gamma_{xy}}{\varepsilon_x - \varepsilon_y}.$$

12.19   Final Remarks

In addition to being the key to experimental stress analysis, strain can be made an analog for essentially any of the various mechanical-type inputs of interest to the engineer: force, torque, displacement, pressure, temperature, motion, etc. For this reason strain gages are very widely and successfully used as secondary transducers in measuring systems of all types. Their response characteristics are excellent, and they are reliable, relatively linear, and inexpensive. It is important, therefore, that the engineer concerned with experimental work be well versed in the techniques of their use and applications.

Suggested Readings

Characteristics and Applications of Resistance Strain Gages, NBS Circular 528. Washington, D.C.: U.S. Government Printing Office, 1954.

Dally, J. W., and W. F. Riley. *Experimental Stress Analysis.* New York: McGraw-Hill, 1965.

Dove, R. C., and P. H. Adams. *Experimental Stress Analysis and Motion Measurement.* Columbus, Ohio: Charles E. Merrill, 1964.

Durelli, A. J. *Applied Stress Analysis.* Englewood Cliffs, N.J.: Prentice Hall, 1967.

Durelli, A. J., and V. J. Parks. *Moiré Analysis of Strain.* Englewood Cliffs, N.J.: Prentice Hall, 1970.

Frocht, M. M. *Photoelasticity,* vols. 1 and 2. New York: John Wiley, 1941, 1948.

Hendry, A. W. *Elements of Experimental Stress Analysis.* London: The Macmillan Co., 1964.

Hetenyi, M. (ed.). *Handbook of Experimental Stress Analysis.* New York: John Wiley, 1950.

Holister, G. S. *Experimental Stress Analysis.* Cambridge, England: Cambridge University Press, 1967.

Hurst, R. C. et al. (eds). *Strain Measurement at High Temperatures.* New York: Elsevier Science Publishers, 1986.

Lee, G. H. *An Introduction to Experimental Stress Analysis.* New York: John Wiley, 1950.

Perry, C. C., and H. R. Lissner. *The Strain Gage Primer,* 2nd ed. New York: McGraw-Hill, 1962.

Principles of Stresscoat, Chicago, Ill. Magnaflux Corp., 1967.

Problems

12.1   A simple tension member with a diameter of 0.505 in. is subjected to an axial force of 7215 lbf. Strains of 1640 and −485 μ-strain are measured in the axial and transverse directions, respectively. Assuming elastic conditions, determine the

Figure 12.26   Strain-gage/beam configuration described in Problem 12.7

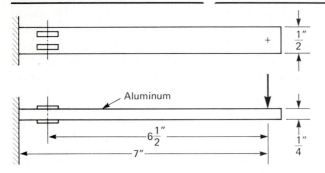

values of Young's modulus and Poisson's ratio for the material. (*Note:* The diameter that is specified is commonly considered a "standard" for circularly sectioned metal specimens. Do you know why?)

12.2 A single strain gage is mounted on a tensile member, as shown in Fig. 12.6. If the readout is 425 μ-strain, what is the axial stress (a) if the member is of steel, and (b) if the member is of aluminum? (See Appendix D for values of E.)

12.3 A strain gage is centered along the length of a simply supported beam carrying a centrally positioned, concentrated load. The beam is four gage lengths long. What correction factor should be applied to care for the strain gradients if the purpose of the measurement is to determine the maximum strain?

12.4 Referring to the previous problem, what correction should be applied if the beam carries a uniformly distributed load?

12.5 Referring to Problem 12.3, what correction factor should be applied if, instead of being simply supported, the beam has built-in ends?

12.6 A resistance-type strain gage having a factor of 2.00 ± 0.05 and a resistance of 121 ± 2 Ω is used in conjunction with an indicator having an uncertainty of ±2%. What maximum uncertainty may be introduced by these tolerances? What probable uncertainty?

12.7 The sensing element of a weighing scale is described in Problem 6.11. Four strain gages are located as shown in Fig. 12.26. The gages are connected in a full bridge (see Case G, Table 12.4). Their nominal resistances are 300 Ω and their gage factors are 3.5. If the bridge is powered with a regulated 5.6 V dc source, what will be the voltage output corresponding to the maximum design load of 300/18 = 16.67 lbf?

12.8 Two strain gages are mounted on a cantilever beam as shown in Fig. 12.7. If the *total* strain readout is 620 μ-strain, what are the outer fiber stresses (a) if the member is of steel, and (b) if the member is of aluminum?

12.9 Assume a system configured as shown in Fig. 12.11, using a conventional oscilloscope for readout.

    a.   Make a list of variables that you feel will have a measurable effect on the overall uncertainty of the system. Indicate those that you would expect to change with input magnitude and those that will be relatively constant; see

Figure 12.27 Detail of arrangement described in Problem 12.11

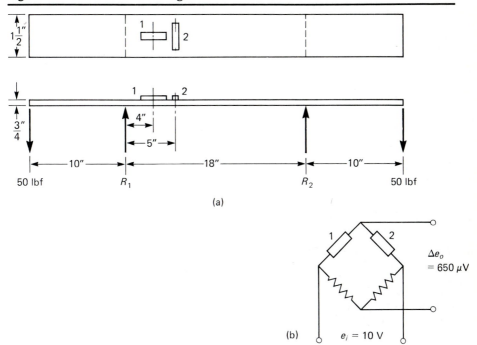

(a)

(b) $e_i = 10 \text{ V}$

$\Delta e_o = 650 \, \mu\text{V}$

Eq. (12.11). Include the uncertainty due to limits of resolution of the readout method and note that some form of system calibration must be used, with its attendant uncertainty.

b. Assign what you believe to be reasonable uncertainties to each factor in your list and determine the overall uncertainty in the final readout. Finally, divide the uncertainties into two categories; those having a major effect on the overall uncertainty and those of minor importance.

12.10 A plastic specimen is subjected to a biaxial stress condition for which $\sigma_x = 1380$ and $\sigma_y = 605 \text{ lbf/in.}^2$. Measured strains (in microstrain) are $\varepsilon_x = 1780$ and $\varepsilon_y = 139$. Calculate Poisson's ratio and Young's modulus.

12.11 Two identical strain gages are mounted on a constant-moment beam as shown in Fig. 12.27. They are connected into a Wheatstone bridge as shown in Fig. 12.27(b). With no load on the beam the bridge is nulled with all arms having equal resistances. When the loads are applied, a bridge output of 650 μV is measured. Determine the gage factor for the gages on the basis of the following additional data: $E = 29.7 \times 10^6 \text{ lbf/in.}^2$, Poisson's ratio = 0.3, and the gage nominal resistance is 120 Ω.

12.12 A two-element strain rosette is mounted on a simple tensile specimen of steel. One gage is aligned in an axial direction and the other in a transverse direction. The gages are connected in adjacent arms of the bridge. If the total bridge readout (based on single-gage calibration is 900 μ-strain, what is the axial stress in pascal? $E = 20 \times 10^{10} \text{ Pa } (29 \times 10^6 \text{ lbf/in.}^2)$ and $v = 0.29$.

Table 12.5 Data for Problem 12.13

ε_a	ε_b	ε_c	E	v
−320	210	680	10×10^6 lbf/in.^{2}	0.29
1585	470	0	29×10^6 lbf/in.^{2}	0.3
1250	−820	425	15×10^6 lbf/in.^{2}	0.28
−1020	985	−420	30×10^6 lbf/in.^{2}	0.3
2220	0	0	30×10^6 lbf/in.^{2}	0.29
0	850	−990	7.5×10^{10} Pa	0.28
−1010	−125	1440	20×10^{10} Pa	0.3
−210	−510	−212	10×10^{10} Pa	0.28
ε	ε	ε	E	v
ε	ε	$-\varepsilon$	E	v
ε	$-\varepsilon$	$-\varepsilon$	E	v

12.13 Each line in Table 12.5 represents a set of data corresponding to a given plane stress condition. The first three columns are strains (in microstrain) obtained using a three-element *rectangular* rosette. The final two items are material properties. For a selected set of data, determine the following:

a.   The principal strains
b.   The principal stresses
c.   The maximum shear stress
d.   Principal directions referred to the axis of gage *a*
Also, sketch the following:
e.   Mohr's circles for stress
f.   Mohr's circles for strain
g.   An element similar to that shown in Fig. 12.18
Use units corresponding to those given for *E*.

12.14 Repeat Problem 12.13(a) through (g), assuming an equiangular rosette.

12.15 If a rectangular-type rosette happens to be aligned such that elements *a* and *c* coincide with the principal directions, then measured values of ε_a and ε_c will be ε_1 and ε_2 (or vice versa). Show that under these circumstances the strain sensed by *b* will be $(\varepsilon_1 + \varepsilon_2)/2$.

12.16 Devise a spreadsheet template and/or a computer program to evaluate the rectangular strain rosette relationships in Table 12.3.

12.17 Devise a spreadsheet template and/or a computer program to evaluate the T-delta strain rosette relationships in Table 12.3.

12.18 Devise a spreadsheet template and/or a computer program to evaluate the equiangular strain rosette relationships in Table 12.3.

12.19 Values of v and E must be used in equations for converting strains to stresses. For steel, $v = 0.3$ and $E = 30 \times 10^6$ lbf/in.^{2} are often assumed. In fact, however, for steel v may vary over the approximate range 0.27 to 0.32 and E may vary over a range of about 28×10^6 to 32×10^6 lbf/in.^{2}. Using the rectangular rosette strain values given in Example 12.5 in Section 12.15.2, analyze the effects of variations in assumed values v and E on the calculated principal strains, stresses,

Figure 12.28 Arrangement of strain gages described in Problem 12.22

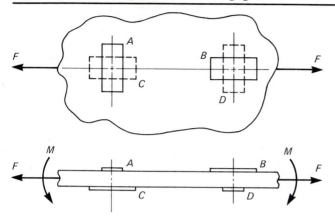

and directions. It is suggested that the spreadsheet template (or program) written for Problem 12.16 be used to minimize the drudgery of number crunching.

12.20 Show how strain gages may be mounted on a simple beam to sense temperature change while being insensitive to variations in beam loading.

12.21 Two strain gages are mounted on a steel shaft ($E = 20 \times 10^{10}$ Pa and Poisson's ratio $= 0.29$), as shown in Case J, Table 12.4. The gage resistance is 119 Ω and $F = 1.23$. When a 250,000-Ω resistor is shunted across gage 1, a 3.4-cm upward shift is recorded on the face of the CRO. When the shaft is torqued, a 5.7-cm shift is measured. For these conditions and assuming bending and axial loading may be neglected:

a. Calculate the maximum torsional stress.

b. What are the three principal stresses on the shaft surface?

c. Plot Mohr's circles for stress.

d. Should a bending moment and/or axial load be present, how would the results be affected?

12.22 Strain gages A, B, C, and D are mounted on a plate subjected to a simple bending moment M and an axial load F, as shown in Fig. 12.28. How should the gages be inserted into a standard bridge (see Table 12.4) in order to accomplish the following:

a. To sense bending only and, under this requirement, provide maximum bridge output?

b. To sense axial stress only (eliminating bending stress)?

In each case, what will be the bridge constant, and will adjacent-arm temperature compensation be accomplished?

12.23 To determine the power transmitted by a 10 cm (3.94 in.) shaft, four strain gages are mounted as shown in Case L, Table 12.4. They are connected as a four-arm bridge and the output is fed to a recording oscillograph. Gage resistances are 118 Ω with a gage factor of 2.1. A 210,000 Ω calibration resistor may be shunted across one of the gages. Figure 12.29(a) and (b) shows the calibration and strain records, respectively. The chart speed is 100 mm/s. The shaft is of steel with

Figure 12.29 Oscillographic readout for conditions of Problem 12.23

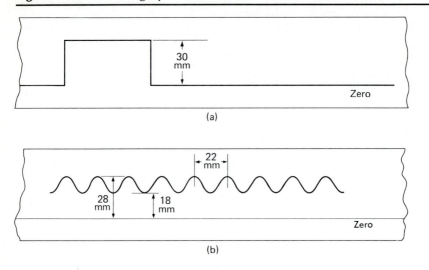

(a)

(b)

Figure 12.30 Configuration of strain gages described in Problem
12.24. Values of ε are in microstrain.

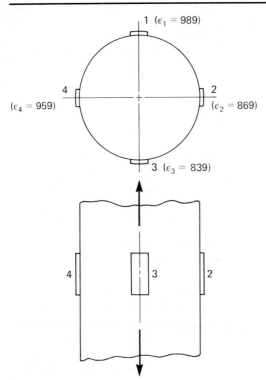

$E = 20 \times 10^{10}$ Pa and Poisson's ratio $= 0.3$. Determine the extreme and mean values of transmitted power in watts.

12.24 Four axially aligned, identical strain gages are equally spaced around a $1\frac{1}{4}$ in. (31.75 mm) diameter bar, as shown in Fig. 12.30. The basic load on the bar is tensile; however, because of a small load eccentricity a bending moment also exists. If the strain readings shown on the sketch are determined for the individual gages, what axial load and bending moment must exist? Also determine the position of the neutral axis of bending.

12.25 Four gages are mounted on a thin-wall cylindrical pressure vessel. Two of the gages are aligned circumferentially (these are gages 1 and 4 in the standard bridge, Table 12.4), and the remaining gages 2 and 3 are aligned in the axial direction. (Note that this is not necessarily an optimal configuration.) If the bridge output is 27.8 units when a 300,000 Ω resistor is shunted across gage 1, and an output from the bridge of 47 units is recorded when the vessel is pressurized, what is the circumferential stress? Use $F = 3.5$, $R_g = 180\,\Omega$, $E = 7 \times 10^{10}$ Pa, and Poisson's ratio $= 0.3$. Assume that the conventional 2-to-1, circumferential-to-longitudinal stress ratio applies. (See Example E.2, Appendix E.)

12.26 Strain readouts from a rectangular strain rosette are $\varepsilon_a = 620$, $\varepsilon_b = -200$, and $\varepsilon_c = 410\,\mu$-strain. Assume that under the same conditions an equiangular rosette is mounted and that its a element is aligned with the direction of the a element of the original rectangular rosette. What readouts should be expected from the delta gage? Assume the same gage factor and resistances for both rosettes. (*Hint:* See Appendix E for Mohr's circles for strain.)

12.27 A strain gage having a resistance $R_g = 120\,\Omega$ and a gage factor $F = 2.0$ is used in an optimum ballast circuit. What is the maximum error over a range of $0 < \varepsilon < 1500\,\mu$-strain relative to a "best" straight line referenced to $\varepsilon = 0$?

12.28 Analyze the effect of lead wire length and wire gage on the sensitivity of the following strain gage circuits:
a. Ballast circuit
b. Circuit shown in Fig. 12.6
c. Circuit shown in Fig. 12.7
d. A four-arm bridge such as shown in Fig. 12.28
The following data may be useful if a quantitative analysis is being made.

Wire Size A.W.G.*	Ohms per 1000 ft at 25°C
12	1.62
15	3.25
18	6.51
20	10.35
24	26.17

* American Wire Gage

12.29 Analyze the uncertainty inherent in shunt calibration of strain-gage circuits.

12.30 A mechanical engineering student wishes to determine the internal pressure existing in a diet soda can. She proceeds by carefully mounting a single-element

Figure 12.31 Instrumented soda can

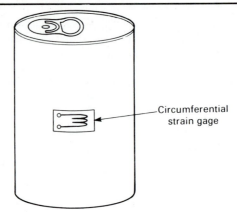

Circumferential
strain gage

strain gage aligned in circumferential direction on the center of the soda can, as shown in Fig. 12.31. After wiring the gage properly to a commercial strain indicator, she "pops" the flip-top lid, which relieves the internal pressure. She notes that the strain indicator reads $-400\,\mu$-strain. If the can body is made of aluminum with a thickness of 0.010 in. and a diameter of 2.25 in., what was the original internal pressure of the sealed can?

12.31 Another student also performed the experiment described in Problem 12.30. Unfortunately, he did not have access to the commercial strain indicator, and instead he had to construct his own Wheatstone bridge circuit. His strain gage had an initial resistance of $120\,\Omega$ and a gage factor of 2.05. He used the single gage as one leg of the bridge, which he powered with a 6 V battery. The bridge output was fed to an amplifier (gain = 1000), and the amp's output was read by a DVM. The student balanced the bridge circuit before he opened the can. After the can was opened, the DVM indicated a voltage of -1.57 V. What was the measured strain for his can?

References

1. Hetenyi, M. *Handbook of Experimental Stress Analysis*. New York: John Wiley, 1950.

2. Brookes-Smith, C. H. W., and J. A. Colls. Measurement of pressure, movement, acceleration and other mechanical quantities by electrostatic systems. *J. Sci. Inst. (London)* 14: 361, 1939.

3. Carter, B. C., J. F. Shannon, and J. R. Forshaw. Measurement of displacement and strain by capacity methods. *Proc. Inst. Mech. Eng.* 152: 215, 1945.

4. Langer, B. F. Design and application of a magnetic strain gage. *SESA Proc.* 1(2): 82, 1943.

5. Langer, B. F. Measurement of torque transmitted by rotating shafts. *J. Appl. Mech.* 67(3): A.39, March 1945.

6. Thompson, K. On the electro-dynamic qualities of metals. *Phil. Trans. Roy. Soc. (London)* 146: 649–751, 1856.

7. Eaton, E. C. Resistance strain gage measures stresses in concrete. *Eng. News Rec.* 107: 615–616, Oct. 1931.

8. Bloach, A. New methods for measuring mechanical stresses at higher frequencies. *Nature* 136: 223–224, Aug. 19, 1935.

9. Clark, D. S., and G. Datwyler. Stress-strain relations under tension impact loading. *Proc. ASM* 38: 98–111, 1938.

10. Krammer, E. W., and T. E. Pardue. Electric resistance changes of fine wires during elastic and plastic strains. *SESA Proc.* 7(1): 7, 1949.

11. Mills, D., III. Strain gage waterproofing methods and installation of gages on propeller strut of USS Saratoga. *SESA Proc.* 16(1): 137, 1958.

12. Frank, E. Series versus shunt bridge calibration. *Instr. Automation* 31: 648, 1958.

13. Campbell, W. R., and R. F. Suit, Jr. A transistorized AM-FM radio-link torque telemeter for large rotating shafts. *SESA Proc.* 14(2): 55, 1957.

14. Baumberger, R., and F. Hines. Practical reduction formulas for use on bonded wire strain gages in two-dimensional stress fields. *SESA Proc.* 2(1): 133, 1944.

15. Perry, C. C., and H. R. Lissner. *The Strain Gage Primer.* 2nd ed. New York: McGraw-Hill, 1962, p. 157.

16. Meier, J. H. On the transverse sensitivity of foil gages. *Exp. Mech.* 1: July 1961.

17. Wu, C. T. Transverse sensitivity of bonded strain gages. *Exp. Mech.* 2: 338, Nov. 1962.

18. Pian, T. H. H. Reduction of strain rosettes in the plastic range. *J. Aerospace Sci.* 26: 842, December 1959.

19. Ades, C. S. Reduction of strain rosettes in the plastic range. *Exp. Mech.* 2: 345, November 1962.

20. Telinde, J. C. *Investigation of Strain Gages at Cryogenic Temperatures.* Douglas Paper No. 3835. Huntington Beach. CA: Douglas Missile and Space Systems Division, 1966.

21. Leszynski, S. W. The development of flame sprayed sensors. *ISA J.* 9: 35, July 1962.

22. Rastogi, V., K. D. Ives, and W. A. Crawford. High-temperature strain gages for use in sodium environments. *Exp. Mech.* 7: 525, December 1967.

23. Karnie, A. J., and E. E. Day. A laser extensometer for measuring strain at incandescent temperatures. *Exp. Mech.* 7: 485, November 1967.

CHAPTER 13

Measurement of Force and Torque

13.1 Introduction

Mass, time, and displacement are fundamental measurement dimensions. *Mass is the measure of quantity of matter. Force* is a derived unit and *weight* is a force having distinctive characteristics.*

Mass is one of the fundamental parameters determining the gravitational attraction (force) exerted between two bodies. Newton's law of universal gravitation is expressed by the relation

$$F = \frac{Cm_1m_2}{r^2},$$ (13.1)

or

$$C = \frac{Fr^2}{m_1m_2}.$$ (13.1a)

The units of C are $\text{N} \cdot \text{m}^2/\text{kg}^2$ or $\text{lbf} \cdot \text{ft}^2/\text{lbm}^2$, where

m_1 and m_2 = the masses of bodies 1 and 2, respectively,

r = the distance separating them,

F = the mutual gravitational force exerted, one on the other,

C = the gravitational constant.

Henry Cavendish (1731–1810), an English scientist, used a sensitive torsional balance (see Fig. 13.1) to determine the value of C. In SI units $C = 6.67 \times 10^{-11}\,\text{N} \cdot \text{m}^2/\text{kg}^2$.

---

* At this point it may be well for the reader to review Section 2.10, "Conversion Between Systems of Units."

Figure 13.1   Balance used by Cavendish to measure gravitational constant

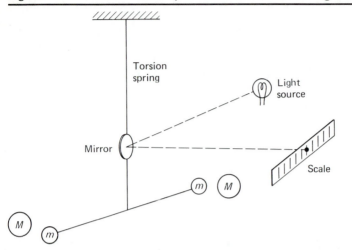

When one of the attracting masses is the earth and the second is that of some object on the surface of the earth, the resulting force of mutual attraction is called *weight*. Mass and weight are related through Newton's laws of motion. The gist of his first law is contained in the statement: *If the resultant of all forces applied to a particle is other than zero, the motion of the particle will be changed.*

His second law may be stated as follows: *The acceleration of a particle is directly proportional to and in the same direction as the resultant applied force.* This may be expressed as

$$\frac{F_1}{a_1} = \frac{F_2}{a_2} = \frac{m}{g_c} = \frac{w}{g}. \tag{13.2}$$

To help establish the correctness of Eq. (13.2) let us look at the units (refer to Table 2.6). For the SI system of units we have

$$\frac{F}{a}\left(\frac{N \cdot s^2}{m}\right) = \frac{m}{g_c}\left(\frac{kg \cdot N \cdot s^2}{1\ kg \cdot m}\right).$$

Using the English Engineering system we have

$$\frac{F}{a}\left(\frac{lbf \cdot s^2}{ft}\right) = \frac{m}{g_c}\left(\frac{lbm \cdot lbf \cdot s^2}{32.2\ lbm \cdot ft}\right).$$

A most convenient force to apply is the earth's gravitational attraction for the body or particle, which is the weight. If this is the *only* force, then the resulting acceleration is that of the falling body *in vacuo* at the particular location. Both the weight and the gravitational attraction will vary from

Figure 13.2   (a) Definition of moment, (b) definition of torque or couple

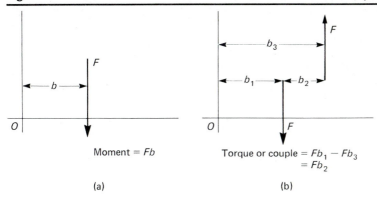

Moment $= Fb$

Torque or couple $= Fb_1 - Fb_3$
$= Fb_2$

(a)   (b)

location to location. Their ratio, however, remains constant and is proportional to the mass, m, as expressed in Eq. (13.2).

As we can see from Eq. (13.2), neither the ratio m/g_c nor w/g is required
for application of Newton's second law. Any ratio F/a, where F is an applied
force and a is the resulting acceleration, is just as valid for establishing the
necessary value. The ratio w/g is particularly convenient, however, because
over the surface of the earth, g is reasonably constant and indeed is often
considered a constant in many engineering calculations. As a result, measurement of weight suffices for determining the ratio w/g. The "standard" value of
g is 32.1739 ft/s^2, or 9.80665 m/s^2. Rounded values of 32.2 ft/s^2 or 9.81 m/s^2
are commonly used.

Force, in addition to its effect along its line of action, may exert a turning
effort relative to any axis other than those intersecting the line of action. Such
a turning effect is variously called *torque, moment,* or *couple,* depending on
the manner in which it is produced. The term *moment* is applied to conditions
such as those illustrated in Fig. 13.2(a), whereas the terms *torque* and *couple*
are applied to conditions involving counterbalancing forces, such as those
shown in Fig. 13.2(b).

Mass Standards

As stated previously (Section 2.6), the fundamental unit of mass is the
kilogram, equal to the mass of the International Prototype Kilogram located
at Sèvres, France. A gram is defined as a mass equal to one-thousandth of
the mass of the International Prototype Kilogram. The commonly used
avoirdupois* pound is 0.435 592 37 kilogram, as agreed to in 1959 (Section
2.6). Various classifications and tolerances for laboratory standards are
recommended by the National Institute of Standards and Technology [1, 2].

---

* From the French, meaning "goods of weight."

13.2   Measuring Methods

As in other areas of measurement, there are two basic approaches to the problem of force and weight measurement: (1) direct comparison, and (2) indirect comparison through use of calibrated transducers. Directly comparative methods use some form of beam balance with a null-balance technique. If the beam neither amplifies nor attenuates, the comparison is *direct*. The simple analytical balance is of this type. Often, however, as in the case of a platform scale, the force is attenuated through a system of levers so that a smaller weight may be used to *balance* the unknown, with the variable in this case being the magnitude of attenuation. This method requires calibration of the system.

Question: When an equal-arm balance scale is used, are forces or are masses being compared? (Problem 13.2 expands on this query.)

13.3   Mechanical Weighing Systems

Mechanical weighing systems originated in Egypt, and were probably used as early as 5000 B.C. [3]. The earliest devices were of the cord and *equal-arm* type, traditionally used to symbolize justice. *Unequal-arm* balances were apparently first used in the form shown in Fig. 13.3(a). This device, called a Danish steelyard, was described by Aristotle (384–322 B.C.) in his *Mechanics*. Balance is accomplished by moving the beam through the loop of cord, which acts as the fulcrum point, until balance is obtained. A later unequal-arm balance, the Roman steelyard, which employed fixed pivot points and movable balance weights, is still in use today; see Fig. 13.3(b).

13.3.1   The Analytical Balance

Probably the simplest weight- or force-measuring system is the ordinary equal-arm beam balance (Fig. 13.4). Basically this device operates on the principle of *moment comparison*. The moment produced by the unknown weight or force is compared with that produced by a known value. When null balance is obtained, the two weights are equal, provided the two arm lengths are identical. A check on arm equivalence may easily be made by simply interchanging the two weights. If balance was initially achieved and if it is maintained after exchanging the weights, it can only be concluded that the weights are equal, as are the arm lengths. This method for checking the true null of a system is known as the method of *symmetry*.

A common example of the equal-arm balance is the analytical scale used principally in chemistry and physics. Devices of this type have been constructed with capacities as high as 400 lbm, having sensitivities of 0.0002 lbm [4]. In smaller sizes the analytical balance may be constructed to have sensitivities

Figure 13.3 (a) Danish steelyard, (b) Roman steelyard

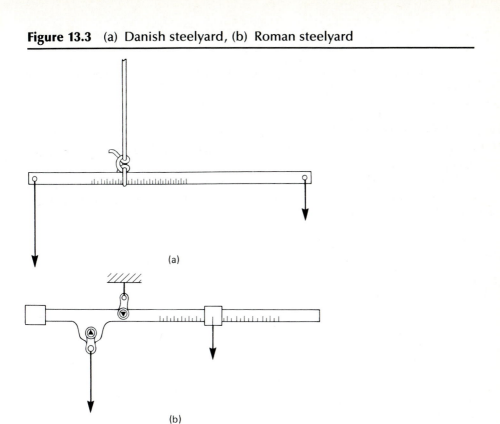

(a)

(b)

Figure 13.4 Requirement for equilibrium of an analytical balance

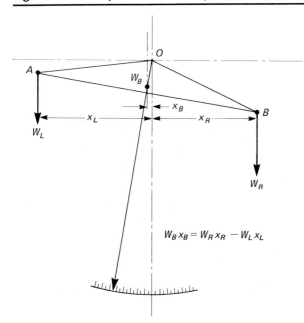

$$W_B x_B = W_R x_R - W_L x_L$$

of 0.001 mg. Some of the factors governing operation of this type of balance were discussed in Section 5.10.

13.3.2   Multiple-Lever Systems

When large weights are to be measured, neither the equal-arm nor the simple unequal-arm balance is adequate. In such cases, multiple-lever systems, shown schematically in Fig. 13.5, are often used. With such systems, large weights W may be measured in terms of much smaller weights W_p and W_s. Weight W_p is called the *poise weight* and W_s the *pan weight*. An adjustable counterpoise is used to obtain an initial zero balance.

We will assume for the moment that W_p is at the zero beam graduation, that the counterpoise is adjusted for initial balance, and that W_1 and W_2 may be

Figure 13.5   Multiple-lever system for weighing

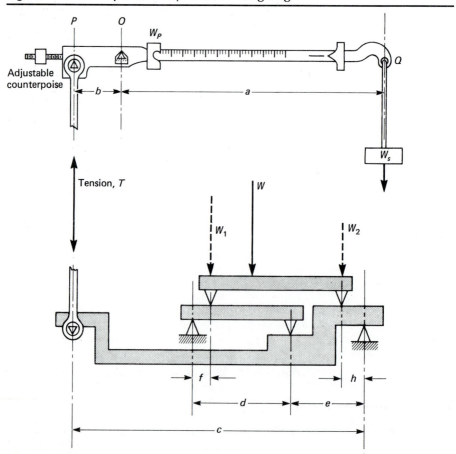

substituted for W. With W on the scale platform and balanced by a pan weight W_s, we may write the relations

$$T \times b = W_s \times a \tag{13.3}$$

and

$$T \times c = W_1 \frac{f}{d} e + W_2 h. \tag{13.4}$$

Now if we proportion the linkage such that

$$\frac{h}{e} = \frac{f}{d},$$

then

$$T \times c = h(W_1 + W_2) = hW. \tag{13.4a}$$

From this we see that W may be placed anywhere on the platform and that its position relative to the platform knife edges is immaterial.

Solving for T in Eqs. (13.3) and (13.4a) and equating, we obtain

$$\frac{W_s a}{b} = \frac{Wh}{c}$$

or

$$W = \frac{ac}{bh} W_s = RW_s. \tag{13.5}$$

The constant

$$R = \frac{ac}{bh}$$

is the scale *multiplication ratio*.

Now if the beam is divided with a scale of u lb/in., then a poise movement of v in. should produce the same result as a weight W_p placed on the pan at the end of the beam. Hence,

$$W_p v = uva \quad \text{or} \quad u = \frac{W_p}{a}.$$

This relation determines the required scale divisions on the beam for any poise weight W_p.

Dynamic response of a scale of this sort is a function of the natural frequency and damping. The natural frequency will be a function of the moving masses, multiplication ratio, and restoring forces. The latter are determined by the relative vertical placement of the pivot points, primarily those of the balance beam O, P, and Q. If O is below a line drawn from P to Q, then the beam will be unstable, and balance will be unattainable. Pivot O is normally above line PQ, and as the distance above the line is increased, the natural frequency and sensitivity are both reduced.

13.3.3   The Pendulum Force-Measuring Mechanism

Another type of moment comparison device used for measurement of force and weight is shown in Fig. 13.6. This is often referred to as a *pendulum scale*. Basically, the pendulum mechanism is a force-measuring device of the multiple-lever type, with the fixed-length levers replaced by ribbon- or tape-connected sectors. The input, either a direct force or a force proportional to weight and transmitted from a suitable platform, is applied to the load rod. As the load is applied, the sectors rotate about points *A*, as shown, moving the counter-weights outward. This movement increases the counterweight effective moment until the load and balance moments are equalized. Motion of the equalizer bar is converted to indicator movement by a rack and pinion, the sector outlines being proportioned to provide a linear dial scale. This device may be applied to many different force-measuring systems, including dynamometers (Section 13.9).

Figure 13.6   Essentials of a pendulum scale

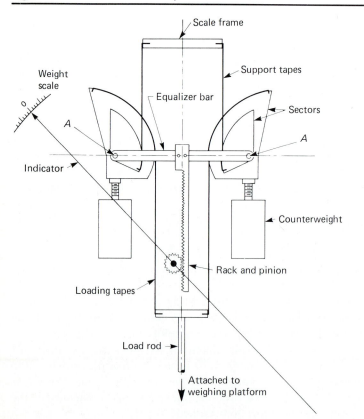

13.4   Elastic Transducers

Many force-transducing systems make use of some mechanical elastic member or combination of members. Application of load to the member results in an analogous deflection, usually linear. The deflection is then observed directly and used as a measure of force or load, or a secondary transducer is used to convert the displacement into another form of output, often electrical.

Most force-resisting elastic members adhere to the relation

$$K = \frac{F}{y},\tag{13.6}$$

in which

F = the applied load,

y = the resulting deflection at the location of F,

K = the deflection constant.

To determine the value of the deflection constant of an element, it is necessary to write only the deflection equation, and if the deflection is a linear function of the load, K may be found. Table 13.1 lists representative relations indicating the general form.

Design detail of the detector-transducer element is largely a function of capacity, required sensitivity, and the nature of any secondary transducer, and depends on whether the input is static or dynamic. Although it is impossible to discuss all situations, there are several general factors we may consider.

It is normally desirable that the detector-transducer be as sensitive as possible; i.e., maximum output per unit input should be obtained. An elastic member would be required that deflects considerably under load, indicating as low a value of K as possible. There are usually conflicting factors, however, with the final design being a compromise. For example, if we were to measure rolling-mill loads by placing cells between the screwdown and bearing blocks, our application could scarcely tolerate a *springy* load cell—that is, one that deflected considerably under load. It would be necessary to construct a stiff cell at the expense of elastic sensitivity and then attempt to make up for the loss by using as sensitive a secondary transducer as possible.

Another factor involving sensitivity is response time, or time required to come to equilibrium. This is a function of both damping and natural frequency (see Section 5.11). Fast response corresponds to high natural frequency, and thus a stiff elastic member is needed.

Stress, also, may be a limiting factor in any loaded member. It is especially important that the stresses remain below the elastic limit, not only in gross section but also at every isolated point. In this respect residual stresses are often of significance. Although load stresses may be well below the elastic limit for the material, it is possible that when they are added to *locked-in* stresses,

Table 13.1

	Elastic Element	Deflection Equation	Deflection Constant K_1
A	F = Load L = Length A = Cross-sectional area y = Deflection at load E = Young's modulus	$y = \dfrac{FL}{AE}$	$K = \dfrac{AE}{L}$
B	F = Load L = Length E = Young's modulus I = Moment of inertia	$y = \dfrac{1}{48}\dfrac{FL^3}{EI}$	$K = \dfrac{48EI}{L^3}$
C	F = Load L = Length E = Young's modulus I = Moment of inertia	$y = \dfrac{1}{3}\dfrac{FL^3}{EI}$	$K = \dfrac{3EI}{L^3}$
D	F = Load D_m = Mean coil diameter N = Number of coils E_s = Shear modulus D_w = Wire diameter	$y = \dfrac{8FD_m^3 N}{E_s D_w^4}$	$K = \dfrac{E_s D_w^4}{8D_m^3 N}$

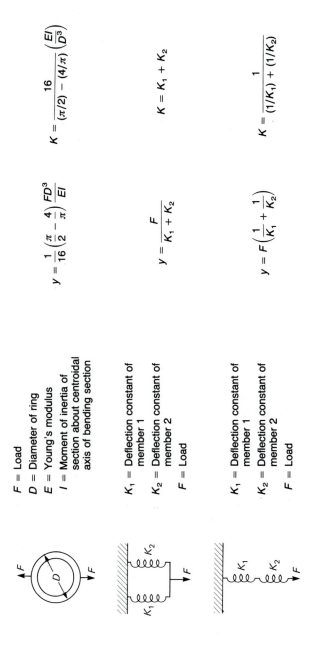

E

F = Load
D = Diameter of ring
E = Young's modulus
I = Moment of inertia of section about centroidal axis of bending section

$$y = \frac{1}{16}\left(\frac{\pi}{2} - \frac{4}{\pi}\right)\frac{FD^3}{EI}$$

$$K = \frac{16}{(\pi/2) - (4/\pi)}\left(\frac{EI}{D^3}\right)$$

F

K_1 = Deflection constant of member 1
K_2 = Deflection constant of member 2
F = Load

$$y = \frac{F}{K_1 + K_2}$$

$$K = K_1 + K_2$$

G

K_1 = Deflection constant of member 1
K_2 = Deflection constant of member 2
F = Load

$$y = F\left(\frac{1}{K_1} + \frac{1}{K_2}\right)$$

$$K = \frac{1}{(1/K_1) + (1/K_2)}$$

the total may be too great. Even though such a situation occurs only at a single isolated point, hysteresis and nonlinearity will result.

Manufacturing tolerances are yet another factor of importance in the design and application of elastic load elements. Tolerances were discussed in some detail in Section 6.18.

13.4.1   Calibration Adjustment

Various calibration adjustments may be made to account for variation in characteristics of elastic load members. Sometimes a simple check at the time of assembly and the selection of one of several standard scale graduations may suffice. For the coil spring tolerance example (Section 6.18.1) we determined the deflection constant uncertainty from dimensional tolerances to be 9%. At the time of assembly a quick single calibration check and a choice of two faceplates could cut the uncertainty from this source in half. Four plates would reduce it to ±2.5%. This scheme is often used not only for load-measuring devices, but for all varieties of inexpensive instruments employing a scale. It does not provide for calibration adjustment in use, however.

When coil springs are used, means are sometimes provided to adjust the number of effective coils through use of an end connection that may be screwed into or out of the spring, thereby changing the number of *active* coils and hence the stiffness of the spring. In other cases, the springs may be purposely overdesigned with regard to stress, and the number of coils specified so that in no case may the tolerances add up to give a spring that is too flexible. Then at the time of assembly the springs are buffed on a wheel to obtain the required deflection constant.

Figure 13.7   Approximate calibration method employing
                  paralleling vernier members

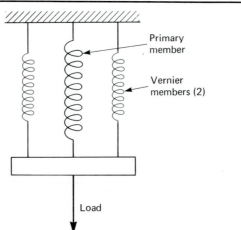

If a secondary transducer is used, we may be able to provide for calibration by making adjustments in its characteristics. As an example, we could use a voltage-dividing potentiometer to sense the load deflection of the spring just discussed. We might employ this procedure to provide remote indication or recording. A circuit arrangement could be used in which an adjustable series resistor would be employed to provide calibration for the complete system.

Figure 13.7 illustrates a calibration adjustment scheme that minimizes the need for holding unduly close tolerances. Here the total load is shared between a primary member (a coil spring in this case) and one or more *vernier* members (the small springs in the figure). The design may call on the vernier members to carry 10% or less of the total load. At the time of assembly the verniers are selected from a range of stiffnesses so as to make the uncertainty of the assembly less than some specified value, say ±0.2%. Schemes of this sort are adaptable to a wide range of elastic devices.

13.4.2 The Proving Ring

This device has long been the *standard* for calibrating materials testing machines and is, in general, the means whereby accurate measurement of large static loads may be obtained. Figure 13.8 shows the construction of a

Figure 13.8 Compression-type proving ring with vibrating reed

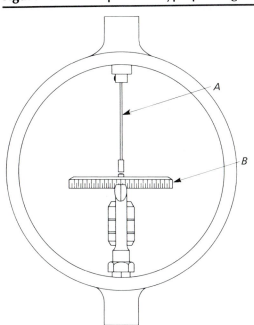

compression-type ring. Capacities generally fall in the range of from 300 to 300,000 lbf (1334 N to 1.334 MN) [5].

Here, again, deflection is used as the measure of applied load, with the deflection measured by means of a precision micrometer. Repeatable micrometer settings are obtained with the aid of a vibrating reed. In use, the reed A is plucked (electrically driven reeds are also available), and the micrometer spindle B is advanced until contact is indicated by the marked damping of the vibration. Although different operators may obtain somewhat different individual readings, consistent differences in readings still will be obtained provided both zero and loaded readings are made by the same person. With 40 to 64 micrometer threads per inch, readings may be made to one or two hundred-thousandths of an inch [5].

The equation given in Table 13.1 for circular rings is derived with the assumption that the radial thickness of the ring is small compared with the radius. Most proving rings are made with a section of appreciable radial thickness. However, Timoshenko [6] shows that use of the thin-ring rather than the thick-ring relations introduces errors of only about 4% for a ratio of section thickness to radius of $\frac{1}{2}$. Increased stiffness on the order of 25% is introduced by the effects of integral bosses [5]. It is, therefore, apparent that use of the simpler thin-ring equation is normally justified.

Stresses may be calculated from the bending moments M determined by the relation [6]

$$M = \frac{PR}{2}\left(\cos\phi - \frac{2}{\pi}\right). \tag{13.7}$$

Symbols correspond to those shown in Fig. 13.9.

Figure 13.9 Ring loaded diametrically in compression

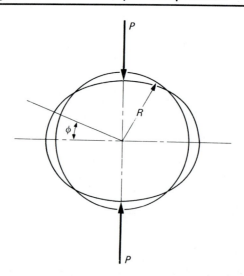

13.5   Strain-Gage Load Cells

Instead of using total deflection as a measure of load, the strain-gage load cell measures load in terms of unit strain. Resistance gages are very suitable for this purpose (see Chapter 12). One of the many possible forms of elastic member is selected, and the gages are mounted to provide maximum output. If the loads to be measured are large, the direct tensile-compressive member may be used. If the loads are small, strain amplification provided by bending may be used to advantage.

Figure 13.10 illustrates the arrangement for a tensile-compressive cell using all four gages sensitive to strain and providing temperature compensation for the gages. The bridge constant (Section 12.8.2) in this case will be $2(1 + v)$, where v is Poisson's ratio for the material. Compression cells of this sort have been used with a capacity of 3 million pounds [7]. Simple beam arrangements may also be used, as illustrated in Table 12.5.

Figure 13.11 illustrates proving-ring strain-gage load cells. In Fig. 13.11(a) the bridge output is a function of the bending strains only, the axial components being cancelled in the bridge arrangement. By mounting the gages as shown in Fig. 13.11(b), somewhat greater sensitivity may be obtained

Figure 13.10   Tension-compression resistance strain-gage load cell

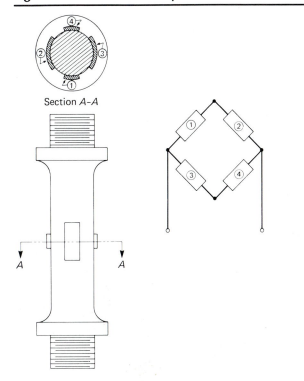

Figure 13.11   Two arrangements of circular-shaped load cells employing
resistance strain gages as secondary transducers

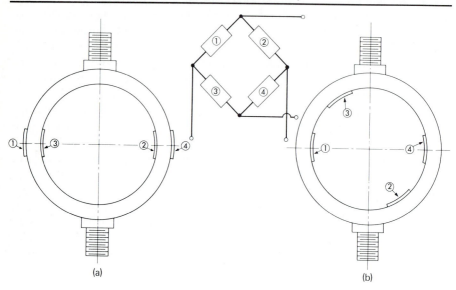

(a)   (b)

because the output includes both the bending and axial components sensed by
gages 1 and 4.

Temperature Sensitivity

The sensitivity of elastic load-cell elements is affected by temperature
variation. This change is caused by two factors; variation in Young's modulus
and altered dimensions. Variation in Young's modulus is the more important
of the two effects, amounting to roughly $2\frac{1}{2}\%$ per 100°F. On the other hand,
the increase in cross-sectional area of a tension member of steel will amount to
only about 0.15% per 100°F change.

Obviously, when accuracies of $\pm\frac{1}{2}\%$ are desired, as provided by certain
commercial cells, a means of compensation, particularly for variation in
Young's modulus, must be supplied. When resistance strain gages are used as
secondary transducers, this compensation is accomplished electrically by
causing the bridge's electrical sensitivity to change in the opposite direction to
the modulus effect. As temperature increases, the deflection constant for the
elastic element decreases; it becomes more *springy* and therefore deflects a
greater amount for a given load. This increased sensitivity is offset by reducing
the sensitivity of the strain gage bridge through use of a thermally sensitive
compensating resistance element, R_s, as shown in Fig. 13.12.

As discussed in Section 7.9.6, the introduction of a resistance in an input-
lead reduces the electrical sensitivity of an equal-arm bridge by the factor

Figure 13.12 Schematic diagram of a strain-gage bridge
with a compensating resistor

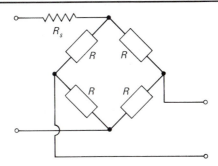

expressed as

$$n = \frac{1}{1 + (R_s/R)}.$$

Requirements for compensation may be analyzed through use of the relation for the initially balanced equal-arm bridge, Eq. (7.17). If we assume

$$2\frac{\Delta R}{R} \ll 4,$$

Eq. (7.17) may be reduced to Eq. (7.18):

$$\frac{\Delta e_o}{e_i} = \frac{k}{4}\frac{\Delta R}{R}.$$

This is true, particularly for a *strain-gage bridge* for which $\Delta R/R$ is always small. A bridge constant, k, is included to account for use of more than one active gage. If all four gages are equally active, $k = 4$. For the arrangement shown in Fig. 13.10, $k = 2(1 + v)$, where v is Poisson's ratio. If we account for the compensating resistor, the equation will then read

$$\frac{\Delta e_o}{e_i} = \frac{k}{4}\frac{\Delta R}{R}\left[\frac{1}{1 + (R_s/R)}\right]. \tag{13.8}$$

Rewriting Eq. (12.10), we have

$$\varepsilon = \left(\frac{1}{F}\right)\left(\frac{\Delta R}{R}\right),$$

and from the definition of Young's modulus, E [Eq. (12.2)],

$$P = EA\varepsilon,$$

we may solve for sensitivity:

$$\frac{\Delta e_o}{P} = \left(\frac{e_i}{4}\right)\left(\frac{FRk}{A}\right)\left[\frac{1}{E(R + R_s)}\right]. \tag{13.9}$$

If it is assumed that the gages are arranged for compensation of resistance variation with temperature and that the gage factors F remain unchanged with temperature, and, further, that any change in the cross-sectional area of the elastic member may be neglected, then complete compensation will be accomplished if the quantity $E(R + R_s)$ remains constant with temperature.

Using Eqs. (6.15) and (6.23), we may write

$$E(R + R_s) = E(1 + c\,\Delta T)[R + R_s(1 + b\,\Delta T)], \qquad \textbf{(13.10)}$$

from which we find

$$\frac{R_s}{R} = -\frac{c}{b + c}. \qquad \textbf{(13.11)}$$

This equation indicates that temperature compensation may possibly be accomplished through proper balancing of the temperature coefficients of Young's modulus, c, and electrical resistivity, b. Because c is usually negative (see Table 6.3), and because the resistances cannot be negative, it follows that

$$b > -c.$$

In addition, we may write [see Eq. (6.2)]

$$R_s = \rho\frac{L}{A} = -R\left(\frac{c}{b + c}\right), \qquad \textbf{(13.12)}$$

from which

$$L = -\frac{RA}{\rho}\left(\frac{c}{b + c}\right). \qquad \textbf{(13.12a)}$$

From these relations, specific requirements for compensation may be derived. After a resistance material, generally in the form of wire, is selected, the required length may be determined through use of Eq. (13.12a).

Although a single resistor would serve, commercial cells normally use two *modulus resistors*, as shown in Fig. 13.13. This technique ensures proper connections regardless of instrumentation and also permits electrical calibra-

Figure 13.13   Strain-gage bridge with two compensating resistors

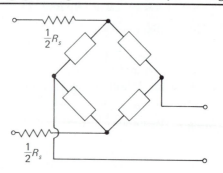

Figure 13.14 Schematic diagram of a strain-gage bridge showing how calibration may be accomplished

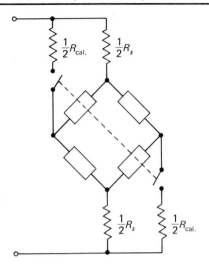

tion of the gages by shunt resistances as described in Section 12.11. It is necessary, however, to use two calibration resistors, as shown in Fig. 13.14. If each resistor is considered as one-half the total calibration resistance, then the relation given, Eq. (12.15), will remain legitimate.

13.6 Piezoelectric Load Cells

Load cells that employ piezoelectric secondary transducers are particularly useful for measuring dynamic loading, especially of an impact or abruptly applied nature. The transducer produces an electrostatic charge (Section 6.14), which is generally conditioned through use of a charge amplifier (Section 7.18.2). Transducer outputs are in terms of coulombs per unit input, with 10 to 20 pC/lbf being typical. Desirable qualities include wide ranges of working load in a given unit, excellent frequency response, great stiffness, high resolution, and relatively small size. An important limitation is that piezoelectric devices are inherently of a dynamic, rather than a static, nature. Long-term static output stability is not generally practical.

Multiaxis cells are available. When a quartz master crystal is sliced to produce transducer elements, the selection of slicing planes yields elements with different properties. Slices may be taken to produce elements selectively sensitive to tension-compression, shear, or bending (see Section 6.14). By taking advantage of these characteristics, load cells may be designed that provide various combinations of orthogonal load and/or torque outputs.

13.7   Ballistic Weighing

Figure 13.15 represents a very simplified example of what is termed *ballistic weighing*. Such systems are particularly adaptable to certain production applications [8].

Theoretically, if a mass is suddenly applied to a resisting member having a linear load-deflection characteristic, the dynamic deflection will be exactly twice the final static deflection. This is true so long as damping is absent. This fact may be used as the basis for a weighing system. The basic equation for a system of this type is

$$\frac{m}{g_c}\frac{d^2y}{dt^2} + ky = \frac{mg}{g_c},\qquad\qquad(13.13)$$

Figure 13.15   A ballistic weighing system

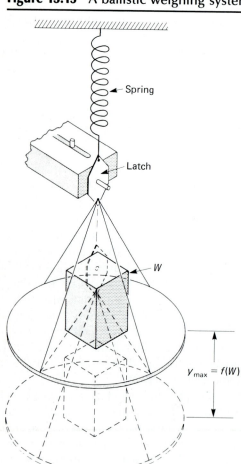

in which
$$m = \text{mass,}$$
$$k = \text{the deflection constant,}$$
$$g_c = \text{the dimensional constant,}$$
$$g = \text{local acceleration due to gravity,}$$
$$y = \text{deflection,}$$
$$t = \text{time.}$$

A solution is

$$y = \frac{mg}{g_c k}(1 - \cos \omega_n t) \qquad \textbf{(13.13a)}$$

for which the maximum value is

$$y_o = \frac{2mg}{g_c k} = 2y_{\text{static}} \qquad \textbf{(13.14)}$$

when $t = \pi/\omega_n$ for $\omega_n = $ the undamped natural frequency. The period of oscillation will be

$$T = 2\pi\sqrt{\frac{m}{g_c k}}. \qquad \textbf{(13.15)}$$

In operation, the platform is locked (Fig. 13.15) with the spring unstretched; then the weight to be measured is put in place, the system is unlocked, and the maximum excursion is measured. If damping is minimized, the maximum displacement will be linearly proportional to the weight and can be used to measure the weight. Of course, the system is useful only for mass measurement and cannot be used to measure force.

13.8 Hydraulic and Pneumatic Systems

If a force is applied to one side of a piston or diaphragm, and a pressure, either hydraulic or pneumatic, is applied to the other side, some particular value of pressure will be necessary to exactly balance the force. Hydraulic and pneumatic load cells are based on this principle.

For hydraulic systems, conventional piston and cylinder arrangements may be used. However, the friction between piston and cylinder wall and required packings and seals is unpredictable, and thus good accuracy is difficult to obtain.* Use of a *floating* piston with a diaphragm-type seal practically eliminates this variable.

---

* An exception to this statement applies to the so-called dead-weight tester (Fig. 14.8). The prescribed procedure is to make certain that the weights, hence piston, is rotating in the cylinder as readings are taken. This practice essentially eliminates seal friction.

Figure 13.16   Section through a hydraulic load cell

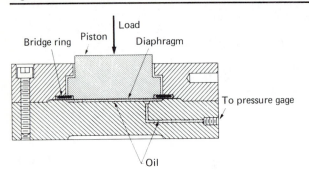

Figure 13.16 shows a hydraulic cell in section. This cell is similar to the type used in some materials-testing machines. The piston does not actually contact a cylinder wall in the normal sense, but a thin elastic diaphragm, or bridge ring, of steel is used as the positive seal, which allows small piston movement. Mechanical stops prevent the seal from being overstrained.

When force acts on the piston, the resulting oil pressure is transmitted to some form of pressure-sensing system such as the simple Bourdon gage. If the system is completely filled with fluid, very small transfer or flow will be required. Piston movement may be less than 0.002 in. at full capacity. In this respect, at least, the system will have good dynamic response; however, overall response will be determined very largely by the response of the pressure-sensing element.

Very high capacities and accuracies are possible with cells of this type. Capacities to 5,000,000 lbf (22.2 MN) and accuracies of the order of $\pm\frac{1}{2}\%$ of reading or $\pm\frac{1}{10}\%$ of capacity, whichever is greater, have been attained. Since hydraulic cells are somewhat sensitive to temperature change, provision should be made for adjusting the zero setting. Temperature changes during the measuring process cause errors of about $\frac{1}{4}\%$ per 10°F change.

Pneumatic load cells are quite similar to hydraulic cells in that the applied load is balanced by a pressure acting over a resisting area, with the pressure becoming a measure of the applied load. However, in addition to using air rather than liquid as the pressurized medium, these cells differ from the hydraulic ones in several other important respects.

Pneumatic load cells commonly use diaphragms of a flexible material rather than pistons, and they are designed to regulate the balancing pressure automatically. A typical arrangement is shown in Fig. 13.17. Air pressure is supplied to one side of the diaphragm and allowed to escape through a position-controlling *bleed* valve. The pressure under the diaphragm, therefore, is controlled both by source pressure and bleed valve position. The diaphragm seeks the position that will result in just the proper air pressure to support the load, assuming that the supply pressure is great enough so that its value multiplied by the effective area will at least support the load.

We see that as the load changes magnitude, the measuring diaphragm must

Figure 13.17   Section through a pneumatic load cell

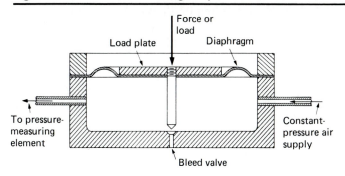

change its position slightly. Unless care is used in the design, a nonlinearity may result, the cause of which may be made clear by referring to Fig. 13.18(a). As the diaphragm moves, the portion between the load plate and the fixed housing will alter position as shown. If it is assumed that the diaphragm is of a perfectly flexible material, incapable of transmitting any but tensile forces,

Figure 13.18   (a) A section through a diaphragm showing how a change in effective area may take place. (b) When sufficient "roll" is provided, the effective area remains constant.

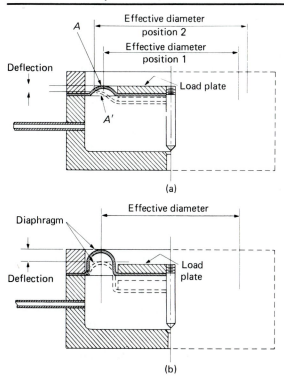

then the division of vertical load components transferred to housing and load plate will occur at points A or A', depending on diaphragm position. We see then that the effective area will change, depending on the geometry of this portion of the diaphragm. If a complete semicircular roll is provided, as shown in Fig. 13.18(b), this effect will be minimized.

Since simple pneumatic cells may tend to be dynamically unstable, most commercial types provide some form of viscous damper to minimize this tendency. Also, additional chambers and diaphragms may be added to provide for *tare* adjustment.

Single-unit capacities to 80,000 lbf (356 kN) may be obtained, and by use of parallel units practically any total load or force may be measured. Errors as small as 0.1% of full scale may be expected.

13.9   Torque Measurement

Torque measurement is often associated with determination of mechanical power, either power required to operate a machine or power developed by the machine. In this connection, torque-measuring devices are commonly referred to as *dynamometers*. When so applied, both torque and angular speed must be determined. Another important reason for measuring torque is to obtain load information necessary for stress or deflection analysis.

There are three basic types of torque-measuring apparatus—namely, absorption, driving, and transmission dynamometers. *Absorption dynamometers* dissipate mechanical energy as torque is measured; hence they are particularly useful for measuring power or torque developed by power sources such as engines or electric motors. *Driving dynamometers,* as their name indicates, both measure torque or power and also supply energy to operate the tested devices. They are, therefore, useful in determining performance characteristics of such things as pumps and compressors. *Transmission dynamometers* may be thought of as passive devices placed at an appropriate location within a machine or between machines, simply for the purpose of sensing the torque at that location. They neither add to nor subtract from the transmitted energy or power, and are sometimes referred to as *torque meters*.

13.9.1   Mechanical and Hydraulic Dynamometers

Probably the simplest type of absorption dynamometer is the familiar *prony brake,* which is strictly a mechanical device depending on dry friction for converting the mechanical energy into heat. There are may different forms, two of which are shown in Fig. 13.19.

Another form of dynamometer operating on similar principles is the *water brake,* which uses fluid friction rather than dry friction for dissipating the input energy. Figure 13.20 shows this type of dynamometer in its simplest form. Capacity is a function of two factors, speed and water level. Power absorption

Figure 13.19 Two forms of the prony brake

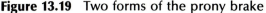

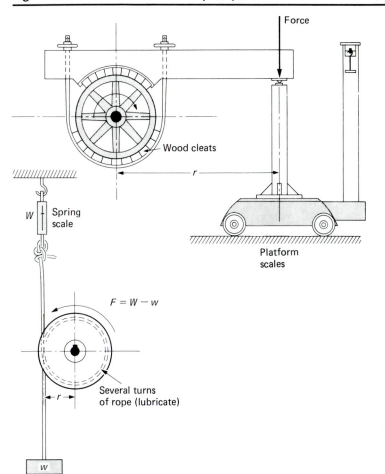

is approximately a function of the *cube* of the speed, and the absorption at a given speed may be controlled by adjustment of the water level in the housing. This type of dynamometer may be made in considerably larger capacities than the simple prony brake because the heat generated may be easily removed by circulating the water into and out of the casing. Trunnion bearings support the dynamometer housing, allowing it freedom to rotate except for restraint imposed by a reaction arm.

In each of the foregoing devices the power-absorbing element tends to rotate with the input shaft of the driving machine. In the case of the prony brake, the absorbing element is the complete brake assembly, whereas for the water brake it is the housing. In each case such rotation is constrained by a force-measuring device such as some form of scales or load cell, placed at the

Figure 13.20 Section through a typical water brake

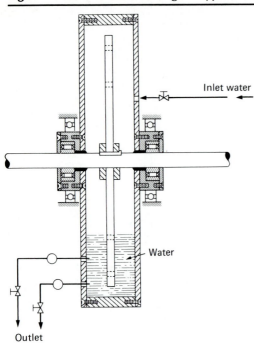

Inlet water

Water

Outlet

end of a reaction arm of radius r. By measuring the force at the known radius, the torque T may be computed by the simple relation

$$T = Fr. \tag{13.16}$$

If the angular speed of the driver is known, power may be determined from the relation

$$P = 2\pi(T)\,(\text{rev/s}), \tag{13.17}$$

where

T = torque,

F = the force measured at radius r,

P = power,

rev/s = revolutions per second.

At this point it may be wise to carefully consider the units to be used in the above relationships. We may rewrite Eq. (13.17) as follows:

$$P = F(2\pi r \cdot \text{rev/s}) = \text{force} \times \text{distance/time}$$

$$= \text{work/unit time}.$$

Using the SI system of units, work is measured in joules (J), where 1 J is equal to 1 N multiplied by 1 m, or

$$J = N \cdot m.$$

Mechanical power then becomes $N \cdot m/s = J/s =$ watts (W). Checking the units in Eq. (13.17) yields watts. Using the English system of units we find power as determined from Eq. (13.17) to yield units of $lbf \cdot ft/s$. The English system often goes an additional step by assigning the term *horsepower* (hp) to $550 \, lbf \cdot ft/s$, or

$$hp = \frac{2\pi(T)(rev/s)}{550}. \tag{13.18}$$

Conversion from watts to horsepower may be made using the relation

$$W = 7.457 \times 10^2 \times hp. \tag{13.19}$$

Example 13.1
Calculate the power if $F = 120 \, N$ (or 26.98 lbf), $r = 75 \, cm$ (or 2.46 ft), and rev/s $= 20$.

Solution. Using SI units we have

$$P = 2\pi \cdot 120 \cdot 0.75 \cdot 20 = 11,310 \, W.$$

Using English units we have

$$P = 2\pi \cdot 26.98 \cdot 2.46 \cdot 20 = 8340 \, lbf \cdot ft/s$$

$$= 15.16 \, hp.$$

A check on equivalence yields

$$\frac{11,310}{15.16} = 746 \, W/hp.$$

13.9.2 Electric Dynamometers

Almost any form of rotating electric machine can be used as a driving dynamometer, or as an absorption dynamometer, or as both. As expected, those designed especially for the purpose are most convenient to use. Four possibilities are: (1) eddy-current dynamometers, (2) cradled dc dynamometers, (3) dc motors and generators, (4) ac motors and generators.

Eddy-current dynamometers are strictly of the absorption type. They are incapable of driving a test machine such as a pump or compressor; hence they are only useful for measuring the power from a source such as an internal combustion engine or electric motor.

The eddy-current dynamometer is based on the following principles. When a conducting material moves through a magnetic flux field, voltage is generated, which causes current to flow. If the conductor is a wire forming a part of a complete circuit, current will be caused to flow through that circuit, and with some form of commutating device a form of ac or dc generator may be the result. If the conductor is simply an isolated piece of material, such as a short bar of metal, and not a part of a complete circuit as generally recognized, voltages will still be induced. However, only local currents may flow in practically short-circuit paths within the bar itself. These currents, called eddy currents, become dissipated in the form of heat.

An eddy-current dynamometer consists of a metal disk or wheel that is rotated in the flux of a magnetic field. The field is produced by field elements or coils excited by an external source and attached to the dynamometer housing, which is mounted in trunnion bearings. As the disk turns, eddy currents are generated, and the reaction with the magnetic field tends to rotate the complete housing in the trunnion bearings. Torque is measured in the same manner as for the water brake, and Eqs. (13.16), (13.17), and (13.18) are applicable. Load is controlled by adjusting the field current. As with the water brake, the mechanical energy is converted to heat energy, presenting the problem of satisfactory dissipation. Most eddy-current dynamometers must use water cooling. Particular advantages of this type are the comparatively *small size* for a given capacity and characteristics permitting *good control at low rotating speeds*.

Undoubtedly the most versatile of all types is the *cradled dc dynamometer*, shown in Fig. 13.21. This type of machine is usable both as an absorption and as a driving dynamometer in capacities to 5000 hp (3730 kW). Basically the device is a dc motor generator with suitable controls to permit operation in either mode. When used as an absorption dynamometer, it performs as a dc generator and the input mechanical energy is converted to electrical energy, which is dissipated in resistance racks. This latter feature is important, for unlike the eddy-current dynamometer, the heat is dissipated outside of the machine. Cradling in trunnion bearings permits the determination of reaction torque and the direct application of Eqs. (13.16), (13.17), and (13.18).

Figure 13.21   The general-purpose electric dynamometer

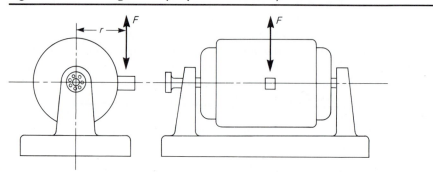

Provision is made for measuring torque in either direction, depending on the direction of rotation and mode of operation. As a driving dynamometer, the device is used as a dc motor, which presents a problem in certain instances of obtaining an adequate source of dc power for this purpose. Use of either an ac motor-driven dc generator set or a rectified source is required. *Ease of control* and *good performance at low speeds* are features of this type of machine.

Ordinary *electric motors* or *generators* may be adapted for use in dynamometry. This is more feasible when dc rather than ac machinery is used. Cradling the motor or generator may be used for either driving or absorbing applications, respectively. By measuring torque reaction and speed, power may be computed. This, of course, requires special effort in designing and fabricating a minimum-friction arrangement. Adjustment of driving speed or absorption load could be provided through control of field current. Load-cell mounting may be used.

Knowledge of motor or generator characteristics versus speed presents another approach. If a dc generator is used as an *absorption dynamometer,* then

$$\text{Power (absorbed)} = \frac{(e)(i)}{\text{Efficiency}}, \tag{13.20}$$

where

e = the output voltage, in volts,

i = the output current, in amperes,

Efficiency = the efficiency of the generator.

In like manner, Eq. (13.20) holds if a dc motor is used as a *driving* dynamometer, except that e and i are *input* voltage and current, respectively. Both e and i may be measured separately, or a wattmeter may be used and the electric power measured directly.

In many applications, only approximate results may be required, in which case *typical* motor or generator efficiencies supplied by the manufacturer should suffice. For more accurate results, some form of dynamometer would be required to determine the efficiencies for the particular machine to be used. The use of ac motors or generators, while feasible, is considerably more difficult and will not be discussed here. In any case, application of *general-purpose* electrical rotating machinery to dynamometry must be considered special and will not yield as satisfactory results as equipment particularly designed for the purpose.

13.10   Transmission Dynamometers

As mentioned earlier, transmission dynamometers may be thought of as passive devices neither appreciably adding to nor subtracting from the energy involved in the test system. Various devices have been used for this purpose, including gear-train arrangements and belt or chain devices.

Any gear box producing a speed change is subjected to a reaction torque equal to the difference between the input and output torques. When the reaction torque of a cradled gear box is measured, a function of either input or output torque may be obtained.

Belt or chain arrangements, in which reaction is a function of the difference between the tight and loose tensions, may also be used. Torque at either main pulley is also a function of the difference between the tight and loose tensions; hence the measured reaction may be calibrated in terms of torque, from which, with speed information, power may be determined. Mechanical losses introduced by arrangements of these types, combined with general awkwardness and cost, make them rather unsatisfactory except for an occasional special application.

More common forms of transmission dynamometers are based on calibrated measurement of unit or total strains in elastic load-carrying members. A popular dynamometer of the elastic type uses bonded strain gages applied to a section of torque-transmitting shaft [9, 10], as shown in Table 12.4. Such a dynamometer, often referred to as a *torque meter,* is used as a coupling between driving and driven machines or between any two portions of a machine. A complete four-arm bridge is used, incorporating modulus gages to minimize temperature sensitivity (Section 13.5). Electrical connections are made through slip rings, with means provided to lift the brushes when they are not in use, thereby minimizing wear. Any of the common strain-gage

Figure 13.22 Transmission dynamometer that employs beams
                and strain gages for sensing torque

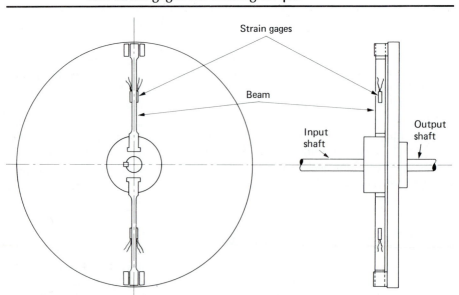

indicators or recorders are usable to interpret the output. Dynamometers of this type are commercially available in capacities of 100 to 30,000 in. · lbf (12 to 3500 N · m). Accuracies to $\frac{1}{4}\%$ are claimed.

In most cases resistance strain-gage transducers are most sensitive when bending strains can be used. Figure 13.22 suggests methods whereby torsion may be converted to bending for measurement.

Slip rings are subject to wear and may present annoying maintenance problems when permanent installations are required. For this reason many attempts have been made to devise electrical torque meters that do not require direct electrical connection to the moving shaft. Inductive [11, 12] and capacitive [13] transducers have been used to accomplish this (see Fig. 6.13).

In addition to temperature sensitivity resulting from variation in elastic constants, further variation may be caused in the inductive type by change in magnetic constants with temperature. This may be compensated for by resistors in a manner similar to that used for strain-gage load cells (Section 13.5).

These types are relatively expensive, and cannot be considered general-purpose instruments. However, in permanent installations they provide the advantage of long service without maintenance problems.

Suggested Readings

ASME PTC 19.7-1961. *Measurement of Shaft Horsepower.* New York, 1961.

ASME PTC 19.5.1-1964. *Weighing Scales.* New York, 1964.

Specifications, Tolerances and Other Technical Requirements for Weighing and Measuring Devices. Washington, D.C.: U.S. Government Printing Office, 1955.

Problems

13.1 There are various sources of data on local gravity accelerations, such as geological surveys, university physics departments, research organizations, and oil, gas or mining companies. Research the values of gravity acceleration for your particular locality.

13.2 Consider a simple balance-beam-type scale (Fig. 13 4). Does the scale compare "weights" or does it compare "masses"? Is the scale sensitive to local gravity? Is it as functional on a mountain top as it is at sea level? Would the scale perform its function in gravity-free space?

13.3 A mass of volume V and unit density d is weighed on a sensitive scale. Write a short summary of the problem that may be presented by air buoyancy as it affects measurement accuracy. Include consideration of the type of scale that is used.

13.4 What will 1 kg of water weigh (a) in Ft. Egbert, Alaska? (b) In Key West, Florida? (c) On the moon? (See Appendix D for data.)

13.5 What will 1 lbm of water weigh in each of the locations listed in Problem 13.4?

13.6 Very often spring scales of the type shown in Fig. 6.24 carry divisions marked in kilograms. Is this practice fundamentally correct? Basically, what does such a device measure when a mass is suspended from it? What is the relationship between mass and weight?

13.7 Assume that a spring of the type referred to in Problem 13.6 is properly calibrated to measure force in newtons. If, on the surface of the moon (gravitational acceleration = 1.67 m/s), a reading of 50 N is obtained when an item is suspended from the scale, what weight would be indicated if the measurement were made under standard conditions on the surface of the earth? What would be the mass of the item in kilograms?

13.8 Assign tolerances to the values given (or determined) in Problems 6.11 and 12.7 and calculate an overall uncertainty to apply to the readout from the beam. (*Note:* Any "electronics" used to evaluate the strain-gage output will also contain uncertainties. Make an estimate for this and include it in the final calculation.) A spreadsheet solution is suggested.

13.9 Referring to Fig. 13.5, show that the scale reading is independent of the location of W on the platform.

13.10 Prepare a spreadsheet template for designing a cantilever-beam-type load cell (see Case C, Table 13.1). Assume a beam with a rectangular cross section.

13.11 Using the template prepared for Problem 13.10, determine the deflection constant for a steel beam 6 in. long, $\frac{3}{8}$ in. wide, and $\frac{1}{8}$ in. thick. Investigate the effect of tolerances on each dimension and on the modulus of elasticity. Assign tolerances on each variable with the aim to control the value of the deflection constant to ±3%.

13.12 Referring to Fig. 13.10, show that small transverse and/or angular misalignments of the load relative to the centerline of the cell will not affect the readout.

13.13 A proving-ring-type force transducer is a very reliable device for checking the calibration of material-testing machines. An equation for estimating the deflection constant of the elemental ring, loaded in compression, is given in Table 13.1. If D = 10 in. (25.4 cm) ± 0.010 in. (0.25 mm), t = the radical thickness of the section = 0.6 in. (15.24 mm) ± 0.005 in. (0.127 mm), w = the axial width of the section = 2 in. (5.08 cm) ± 0.015 in. (0.381 mm), and $E = 30 \times 10^6$ lbf/in.^{2} $(20.68 \times 10^{10} \text{ N/m}^2)$ ± 0.5×10^6 lbf/in.^{2} $(0.34 \times 10^{10} \text{ N/m}^2)$, calculate the value of K and its uncertainty, using English units.

13.14 Solve Problem 13.13 using SI units.

13.15 Figure 13.7 shows a scheme for adjusting the calibration of an elastic force measuring system. Prepare a spreadsheet template for the purpose of designing a two-element coil spring arrangement of the type illustrated.

13.16 Using the template devised for Problem 13.15, design a two-element coil spring system to meet the following specifications:

$$K = 100 \text{ lbf/in.} \pm 0.2\%.$$

Note that the solution will consist of a primary spring, along with several selectable vernier springs. Each vernier spring should be designed to adjust for a range of primary spring tolerances. The smaller the number of verniers required, the better.

13.17 Review Problems 6.11 and 12.7. Using data from these two problems and from Tables 6.3 and 6.4, select a resistance material and determine the value for a series resistor to provide compensation for temperature-derived variations in Young's modulus for the beam.

13.18 Determine the bridge constant for the arrangement shown in Fig. 13.11(b).

13.19 A torque meter is incorporated in the coupling between an electric motor and a dc generator. If the effect of a small 60-Hz torque component is to be limited to no more than 3% of the readout, what are the limiting natural frequencies for the two mass system? Damping is negligible. (Ref.: Section 5.16.2).

References

1. Lashof, T. W., and L. B. Macurdy. Precision laboratory standards of mass and laboratory weights. *NBS Circular 547* (1954).

2. *Precision Measurement and Calibration, Optics, Metrology and Radiation,* Handbook 77, Vol. III. National Bureau of Standards, pp. 588 and 615, 1961.

3. Weighing machines, *Encyclopedia Britannica,* Encyclopedia Britannica. Inc. Chicago, Ill.: William Benton Publisher, 23: 483, 1957.

4. *Instruments,* 25: 1300, Sept. 1952.

5. Wilson, B. L., D. R. Tate, and G. Borkowski. Proving rings for calibrating testing machines. *NBS Circular C454*. Washington, D.C.: U.S. Government Printing Office, 1946.

6. Timoshenko, S. *Strength of Materials,* Part II, 2nd ed. New York: Van Nostrand, 1941, p. 88.

7. High capacity load calibrating devices. *NBS Tech. News Bull.* 37: Sept. 1953.

8. Bell, R. E., and J. A. Fertle. Electronic weighing on the production line. *Electronics,* 28(6): p. 152.

9. Ruge, A. C. The bonded wire torquemeter. *SESA Trans.* 1(2): 68, 1943.

10. Rebeske, J. J., Jr. *Investigation of a NACA High-Speed Strain-Gage Torquemeter,* NACA Tech. Note 2003. Jan. 1950.

11. Langer, B. F. Measurement of torque transmitted by rotating shafts. *J. Appl. Mech.* 67: A.39, March 1945.

12. Langer, B. F., and K. L. Wommack. The magnetic-coupled torquemeter. *SESA Trans.* 2(2): 11, 1944.

13. Hetenyi, M. *Handbook of Experimental Stress Analysis.* New York: John Wiley, 1950, Chaps. 6 and 7.

CHAPTER 14

Measurement of Pressure

14.1 Introduction

Pressure is the normal force exerted by a medium, usually a fluid, on a unit area. In engineering, pressure is most often expressed in pascal ($Pa = 1 N/m^2$) or pounds-force per square inch ($lbf/in.^2$, or psi). Typically, pressure is detected as a differential quantity, that is, as the difference between an unknown pressure and a known reference pressure. Atmospheric pressure is the most common reference, and the resulting pressure difference, known as *gage pressure*, is of obvious importance in determining net loads on pressure vessel and pipe walls. In other cases, the reference pressure is taken to be zero (a complete absence of pressure), and the pressure measured is called *absolute*. In the English system of units, gage and absolute pressure are distinguished by writing psig and psia, respectively. Figure 14.1 illustrates these relationships.

Pressure is often measured in terms of the hydrostatic force per unit area at the base of a column of liquid, usually mercury or water. For example, standard atmospheric pressure (101,325 Pa or 14.696 psia [1]) is approximately equal to the pressure exerted at the bottom of a mercury column 760 mm (or 29.92 in.) in height.* Therefore, one often finds standard atmospheric pressure specified as 760 mm Hg or 29.92 in. Hg, even though the fundamental unit of pressure is neither millimeters nor inches. Pressure measurement using liquid columns is called *manometry* (see Section 14.5).

An absolute pressure less than atmospheric pressure is often referred to as a *vacuum*. Vacuum is occasionally measured in terms of a negative gage pressure (so that -7 psig $= 7$ psi vacuum). When the vacuum is nearly complete, however, small variations in atmospheric pressure can produce large errors in the measured gage pressure. Hence, absolute pressure is always used

---

* The pressure is equal to $\rho(g \neq gc)h$, for ρ the liquid density, g the gravitational body force, and h the column height. Since liquid density varies with temperature, the precise height for 1 atm pressure depends on both local gravity and ambient temperature.

Figure 14.1   Relations between absolute, gage, and barometric pressures

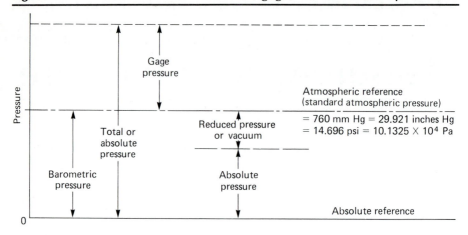

to describe a high vacuum. The low absolute pressures of a high vacuum are commonly evaluated in units of torr (1 torr = 1 mm Hg) or micrometers of mercury (μm Hg).

High pressures are often expressed in units of atmospheres (1 atm = 1.01325×10^5 Pa), bar (1 bar = 10^5 Pa), megapascal (1 MPa = 10^6 Pa), or even megabar (1 Mbar = 10^6 bar). Selected units of pressure measurement are summarized in Table 14.1.

14.2   Static and Dynamic Pressures

When a fluid is at rest, a small pressure sensor in it will read the same *static pressure* at a given position in the fluid no matter how it is oriented. In other words, at any particular point in the fluid, the small surface experiences the same pressure whether it faces upward or downward or left or right.

Table 14.1   Relation of various units of pressure to the Pascal
(H$_2$O at 60°F, Hg at 0°C)

1 microbar = 0.1 Pa	1 in. H$_2$O = 248.8 Pa
1 μm Hg = 0.1333 Pa	1 kPa = 1000 Pa
1 N/m^2 = 1 Pa	1 ft H$_2$O = 2986 Pa
1 mm H$_2$O = 9.795 Pa	1 in. Hg = 3386 Pa
1 mbar = 100 Pa	1 psi = 6895 Pa
1 mm Hg = 133.3 Pa	1 bar = 10^5 Pa
1 torr = 133.3 Pa	1 atm = 101325 Pa

Gravitational force can produce a vertical pressure gradient, causing a higher pressure at lower levels in the fluid, but at any particular level the pressure on the small surface remains independent of its orientation.

When the fluid is in motion, a surface placed in it may experience not only the static pressure, but also a *dynamic pressure*. For example, if the surface is perpendicular to the direction of flow, the fluid must come to rest at the surface. This stagnation of the flow results in the conversion of kinetic energy into an additional pressure on the surface, much like the pressure you experience when standing in the wind. On the other hand, if the surface is parallel to the flow direction, the fluid is not stagnated and flows across the surface without creating any additional pressure. Thus, a pressure transducer's reading in a moving fluid will depend on its orientation.

In Fig. 14.2, two small tubes each sample the pressure in an air duct. Pressure tap *B* senses only the static pressure in the duct. Tube *A,* on the other hand, is aligned so that the flow impacts against its opening, and it senses a *total* or *stagnation pressure.* The static pressure is identical to the pressure one would sense if moving along with the airstream. The stagnation pressure can be defined as that which would be obtained if the stream were brought to rest isentropically. The difference between the stagnation and static pressures results from the motion of the fluid and is called the *velocity pressure* or *dynamic pressure*:

$$\text{Dynamic pressure} = \text{stagnation pressure} - \text{static pressure}$$

As discussed in Section 15.9, this pressure difference can even be exploited for measurement of the fluid's velocity.

We see, therefore, that to obtain and interpret pressure measurements properly, flow conditions must be taken into account. Conversely, to interpret flow measurements properly, the pressure conditions must be considered.

Figure 14.2 Impact-pressure and static-pressure tubes

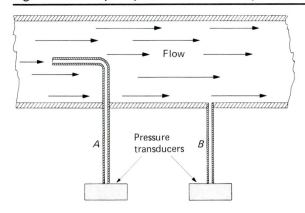

Sound Pressure. Sound waves propagate in an elastic medium as longitudinal pressure variations (along the path of propagation), with pressure fluctuating above and below the static pressure. The instantaneous difference between the pressure at any point and the average pressure there is called the *sound pressure*. Because sound pressures are normally relatively small, they are often expressed in units of microbar ($1\,\mu\text{bar} = 10^{-1}\,\text{Pa}$). Measurement of sound pressure is accomplished with microphones and related apparatus, as discussed in Chapter 18.

14.3   Pressure-Measuring Systems

Pressure-measuring systems probably vary over a greater range of complexity than any other type of measuring system. On the one hand, the ordinary manometer (Fig. 14.3) is one of the most elementary measuring devices imaginable. It is simple, inexpensive, and relatively free from error, and yet it may be arranged to almost any degree of sensitivity. Its major disadvantages lie in its pressure ranges and in its poor dynamic response. It is not very practical for measuring pressures greater than, say, 100 psig, and it is incapable of following any but slowly changing pressures. Another familiar pressure-measuring device, the common Bourdon-tube gage (Fig. 6.1), is quite useful over a wide pressure range, but only for static or slowly changing pressures.

Figure 14.3   Simple U-tube manometer

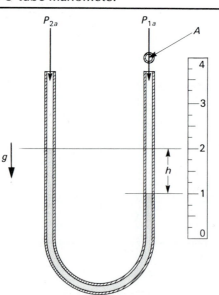

In general, it can be said that when the pressure is *dynamic,* some form of pressure-measuring *system* utilizing electromechanical transducer methods is required. A major portion of this chapter is devoted to discussing applications of devices of this kind.

In accounting for the dynamic response of a pressure-measuring system, the instrumentation and the application must be considered as a whole. The response is not determined by the isolated physical properties of the instrument components alone, but must include the mass elastic damping effects of the pressurized media and connecting passageways.

As an example, a diaphragm-type pickup may be used for measuring the pressure at a specific point on an aircraft skin. In such an application, it may be undesirable to place the diaphragm flush with the aircraft surface. Possibly the size of the diaphragm is too great in comparison with the pressure gradients existing; or perhaps flush mounting would disturb the surface to too great a degree; or it may be necessary to mount the pickup internally to protect it from large temperature variations. In such cases, the pressure would be conducted to the sensing element of the pickup through a passageway, and a small space or cavity would exist over the diaphragm. The passageway and cavity become, in essence, an integral part of the transducer, and the mass elastic damping properties contribute to the determination of the overall response of the system. It is obvious that it would be insufficient to know only the transducer characteristics.

Ideally, a pressure pickup should be insensitive to temperature change and acceleration; friction should be minimized, and any that is unavoidable should have a predictable form. Damping should remain constant for all operating conditions. These items are discussed in more detail later in the chapter.

14.4   Pressure-Measuring Transducers

Often pressure is measured by transducing its effect to a deflection through use of a pressurized area and either a gravitational or elastic restraining element. A comprehensive classification of basic pressure-measuring methods is difficult to make. However, the following should suffice for our purposes.

I. Gravitational types
  A. Liquid columns
  B. Pistons or loose diaphragms, and weights
II. Direct-acting elastic types
  A. Unsymmetrically loaded tubes
  B. Symmetrically loaded tubes
  C. Elastic diaphragms
  D. Bellows
  E. Bulk compression
III. Indirect-acting elastic type, a piston with elastic restraining member.

14.5   Gravitational-Type Transducers

The simple well-type manometer (Fig. 14.4) is one of the most elementary forms of pressure-measuring device. A force-equilibrium expression for the net liquid column is

$$(P_{1a}A - P_{2a}A) = Ah\rho\left(\frac{g}{g_c}\right) \tag{14.1}$$

or

$$(P_{1a} - P_{2a}) = P_d = h\rho\left(\frac{g}{g_c}\right), \tag{14.1a}$$

where

P_{1a} and P_{2a} = the applied absolute pressures,

P_d = the difference or differential pressure,

ρ = the density of the fluid, mass/volume,

h = the net column height, or "head,"

g = the gravitational body force.

In practice, pressure P_{2a} is commonly atmospheric and

$$(P_{1a} - P_{\text{atm}}) = P_{1g} = h\rho - \left(\frac{g}{g_c}\right), \tag{14.2}$$

Figure 14.4   Well-type manometer

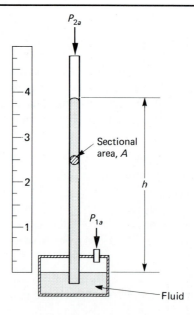

where
$$P_{1g} = \text{the gage pressure at point 1.}$$

Perhaps it would be wise at this point to make sure we understand the units to be used. In simplified form the preceding equations may be written as

$$P_d = h\rho\left(\frac{g}{g_c}\right) \qquad \text{(14.2a)}$$

Substituting units in the right-hand side of the equation, we have, for the SI system,

$$(\text{m})(\text{kg/m}^3)(\text{m/s}^2)(\text{N} \cdot \text{s}^2/\text{kg} \cdot \text{m}) = \text{N/m}^2 = \text{Pa}.$$

Using the English system of units, we have

$$(\text{ft})(\text{lbm/ft}^3)(\text{ft/s}^2)(\text{lbf} \cdot \text{s}^2/\text{lbm} \cdot \text{ft}) = \text{lbf/ft}^2.$$

Example 14.1

Calculate the pressure at the base of a column of water 1 m (3.281 ft) in height if the local gravity acceleration is 9.75 m/s² (31.99 ft/s²) and the temperature is 20°C (68°F).

Solution. From Table D.1 (see Appendix D), we find that the density of water at 20°C = 998.2 kg/m³ (62.316 lbm/ft³). Using SI units, we have

$$P_{SI} = (1)(998.2)(9.75/1) = 9732 \text{ Pa} \qquad (\text{or N/m}^2);$$

using English units, we have

$$P_{Eng} = (3.281)(62.316)(31.99/32.17) = 203.3 \text{ lbf/ft}^2 = 1.412 \text{ psi}.$$

Because the fluid density is involved, accurate work will require consideration of temperature variation: The manometer possesses a certain amount of temperature sensitivity.

When the applied absolute pressure P_{2a} is made to be zero, and P_{1a} is atmospheric, we obtain the ordinary barometer. In this case the fluid is generally mercury.

Figure 14.5 illustrates the function of the simple U-tube manometer. Pressures are applied to both legs of the U, and the manometer fluid is displaced until force equilibrium is attained. Pressures P_{1a} and P_{2a} are transmitted to the manometer legs through some fluid of density ρ_t, while the manometer fluid has some greater density ρ_m. In general we see that

$$P_{1a} - P_{2a} = h(\rho_m - \rho_t)\left(\frac{g}{g_c}\right). \qquad \text{(14.3)}$$

Figure 14.5   Dual-fluid U-tube manometer

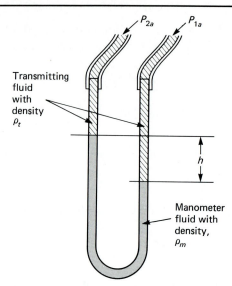

In many cases the density difference between the two fluids is great enough that the lesser density may be ignored—when air is the transmitted fluid and water is the measuring fluid, for instance. In that case, Eq. (14.3) reverts to Eq. (14.1a).

Example 14.2
Suppose the manometer fluids in Fig. 14.5 are water and mercury. This situation might occur when a manometer is used to measure the differential pressure across a venturi meter (see Section 15.3) through which water is flowing. We will consider both systems of units used in this book along with the following pertinent data:

$h = 10$ in. or $\frac{5}{6}$ ft (0.254 m),

Density of water = 62.38 lbm/ft³ (999.2 kg/m³),

Specific gravities of H_2O and Hg = 1 and 13.6, respectively,

Standard gravity will be used (32.174 ft/s² and 9.80665 m/s²).

Determine the differential pressure.

Solution. In the English system of units,

$$P_{1a} - P_{2a} = \left(\frac{5}{6}\right)(13.6 - 1)(62.38)\left(\frac{32.17}{32.17}\right)$$

$$= 655 \text{ lbf/ft}^2 = 4.55 \text{ psi}$$

For the SI system of units, we have

$$P_{1a} - P_{2a} = (0.254)(12.6)(999.2)\left(\frac{9.80665}{1}\right)$$

$$= 31{,}360 \text{ Pa.}$$

It is left for the reader to show that the two answers represent the same physical quantity and that the unit balance is proper in each case.

In general, a U-tube manometer will have a greater pressure range when a more dense measuring fluid is used and a greater sensitivity (change in height per unit change in pressure) when a less dense fluid is used.

Greater sensitivity may also be obtained through a displacement amplification scheme, two of which are shown in Figs. 14.6 and 14.7. For the single inclined leg (Fig. 14.6),

$$P_{1a} = \frac{\rho(L \sin \theta)g}{g_c} + P_{2a}. \tag{14.4}$$

In the case of the two-fluid type manometer (Fig. 14.7),

$$\text{Sensitivity} = \frac{h}{\Delta P} = \frac{1}{[(d/D)^2(\rho_2 + \rho_1) + (\rho_2 - \rho_1)](g/g_c)}. \tag{14.5}$$

When the reservoir diameters are large and the fluid densities are similar, the sensitivity can be substantial (a *micromanometer*). In comparison to the simple U-tube manometer, the deflection amplification equals

$$M = \left[\frac{\rho}{(d/D)^2(\rho_2 + \rho_1) + (\rho_2 - \rho_1)}\right] \tag{14.5a}$$

where $\rho = $ the density of the fluid in the simple manometer and $\rho_1 < \rho_2$.

Figure 14.6 Inclined-type manometer

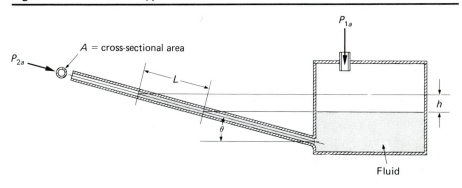

P_{1a}

P_{2a}

$A = $ cross-sectional area

L

θ

h

Fluid

Figure 14.7   Two-fluid manometer with reservoirs

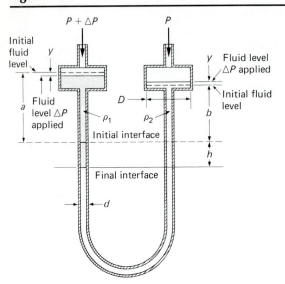

Figure 14.8   Dead-weight-type tester

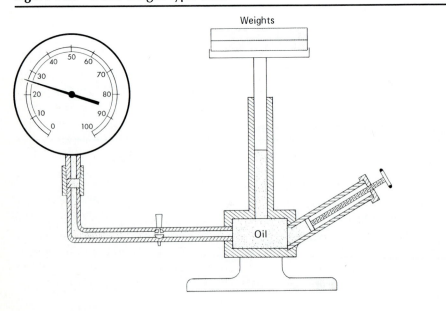

Figure 14.9 Inverted-bell pressure-measuring device

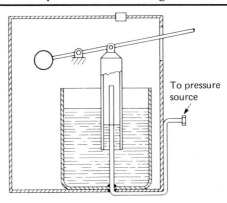

To pressure source

Figure 14.8 illustrates the familiar dead-weight tester that is commonly used as a source of static pressure for calibration purposes but is basically a pressure-producing and pressure-measuring device. When the applied weights and piston area are known, the resulting pressure may be readily calculated.

Figure 14.9 illustrates the principle of operation of the inverted-bell pressure-measuring system. In this case, the force exerted by the pressure against the inner top of the bell is balanced against the net weight of the bell. The net weight depends on the depth of immersion and, as the pressure varies, the bell rises or falls according to pressure magnitude. The primary application of this device is for actuating industrial pressure recorders and controllers. Of course, all gravitational-type pressure transducers are sensitive to the local value of gravity acceleration.

14.6 Elastic-Type Transducers

Elastic transducers operate on the principle that the deflection or deformation accompanying a balance of pressure and elastic forces may be used as a measure of pressure. A familiar example is the ordinary Bourdon tube (see Fig. 6.1). A tube, normally of oval section, is initially coiled into a circular arc of radius R, as shown in Fig. 14.10. The included angle of the arc is usually less than 360°; however, in some cases, when increased sensitivity is desired, the tube may be formed into a helix of several turns.

As a pressure is applied to the tube, the oval section tends to round out, becoming more circular in section. The inner and outer arc lengths will remain approximately equal to their original lengths, and hence the only recourse is for the tube to uncoil. In the simple pressure gage, the movement of the end of the tube is communicated through linkage and gearing to a pointer whose movement over a scale becomes a measure of pressure. Rigorous treatment of the mechanics of Bourdon-tube action is complex, and only approximate analyses have been made [2].

Figure 14.10   Basic Bourdon tube

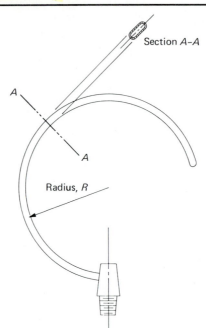

Section A-A

A

A

Radius, R

14.7   Elastic Diaphragms

Many dynamic pressure-measuring devices use an elastic diaphragm as the primary pressure transducer. Such diaphragms may be either flat or corrugated; the flat type [Fig. 14.11(a)] is often used in conjunction with electrical secondary transducers whose sensitivity enables detection of very small diaphragm deflections, whereas the corrugated type [Fig. 14.11(b)] is particularly useful when larger deflections are required.

Diaphragm displacement may be transmitted by mechanical means to some form of indicator, perhaps a pointer and scale as is used in the familiar aneroid barometer. For engineering measurements, particularly when dynamic results are required, diaphragm motion is more often sensed by some form of electrical secondary transducer, whose principle of operation may be resistive, capacitive, inductive, piezoelectric or piezoresistive, as discussed in the following section. The output from the secondary transducer is then processed by appropriate intermediate devices and fed to an indicator, recorder, or controller.

Diaphragm design for pressure transducers generally involves all the following requirements to some degree:

1.   Dimensions and total load must be compatible with physical properties of the material used.

Figure 14.11 (a) Flat diaphragm, (b) corrugated diaphragm

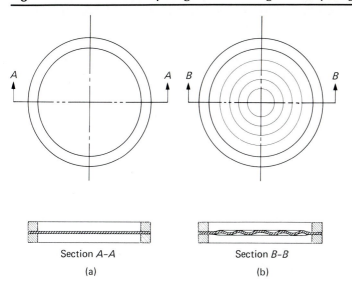

Section *A–A*

(a)

Section *B–B*

(b)

2. Flexibility must be such as to provide the sensitivity required by the secondary transducer.

3. Volume of displacement should be minimized to provide reasonable dynamic response.

4. Natural frequency of the diaphragm should be sufficiently high to provide satisfactory frequency response.

5. Output should be linear.

14.7.1 Flat Metal Diaphragms

Deflection of flat metal diaphragms is limited either by stress requirements or by deviation from linearity. It has been found that as a general rule the maximum deflection that can be tolerated maintaining a linear pressure-displacement relation is about 30% of the diaphragm thickness [3].

In certain cases secondary transducers require physical connection with the diaphragm at its center. This is generally true when mechanical linkages are used and is also necessary for certain types of electrical secondary transducers. In addition, auxiliary spring force is sometimes introduced to increase the diaphragm deflection constant. These requirements make necessary some form of boss or reinforcement at the center of the diaphragm face, which reduces a diaphragm flexibility and complicates theoretical design analysis.

When a central connection is made, a concentrated force F will normally be applied. In general, therefore, the diaphragm may be simultaneously subjected to two deflection forces, the distributed pressure load and a central concentrated force. Design relationships for the fixed-edge, pressurized diaphragm may be found in [4]; for diaphragms with central bosses, in [5].

Calculations for diaphragm dimensions should not be relied on as representing more than a rough guide for design purposes. There are several factors that cannot be accurately predicted. Among these are: (1) the rigidity of the outer supporting ring and inner boss, which is seldom as complete as assumed; and (2) the material physical properties, which are seldom accurately known. In addition, an undesirable characteristic of simple flat diaphragms that is often encountered is a nonlinearity referred to as *oil canning*. The term is derived from the action of the bottom of a simple oil can when it is pressed. A slight unintentional dimpling in the assembly of a flat-diaphragm pressure pickup is difficult to eliminate unless special precautions are taken. In addition, oil canning may be aggravated by differential expansions due to changing ambient conditions. It is desirable, therefore, to construct a pressure cell from materials having the same coefficient of expansion. Even this, however, may not always solve the problem because temperature gradients within the instrument itself may result in a different expansion. One solution to this problem is obtained by using a stretched or *radially* preloaded diaphragm [6]. Theoretical solutions for the radially preloaded diaphragm are considerably more involved than those for the simple flat type. Another solution to the oil-canning problem is to use a small external spring load to *bias* the diaphragm; however, this practice adds mass and thereby sacrifices dynamic response. In all cases, care must be exercised to minimize undesirable temperature effects.

14.7.2   Corrugated Diaphragms

Corrugated diaphragms are normally used in larger diameters than the flat types. Corrugations permit increased linear deflections and reduced stresses. Since the larger size and deflection reduce the dynamic response of the corrugated diaphragms as compared with the flat type, they are more commonly used in static applications.

Adding convolutions to a diaphragm increases the complexity of the theoretical design approach. Grover and Bell [7] have used brittle coatings as a means for evaluating approximate theoretical solutions for stresses.

Two corrugated diaphragms are often joined at their edges to provide what is referred to as a *pressure capsule*. This is the type commonly used in aneroid barometers.

Metal bellows are sometimes used as pressure-sensing elements. Bellows are generally useful for pressure ranges from about $\frac{1}{2}$ psi to 150 psi full scale. Hysteresis and zero shift are somewhat greater problems with this type of element than with most of the others.

14.8 Secondary Transducers Used with Diaphragms

Most electromechanical transducer principles have been applied to diaphragm pressure pickups. The following examples are only representative of many possible variations.

14.8.1 Use of Resistance Strain Gages with Flat Diaphragms

An obvious approach is simply to apply strain gages directly to a diaphragm surface and calibrate the measured strain in terms of pressure. One drawback of this method that is often encountered is the small physical area available for mounting the gages; for this reason, gages with short gage lengths must be used.

Special spiral grids have been used [3, 8]. Grids are mounted in the central area of the diaphragm, with the elements in tension (see Fig. 12.4).

Wenk [3] has found that a satisfactory method for mounting strain gages is the one illustrated in Fig. 14.12. When pressure is applied to the side opposite the gages, the central gage is subject to tension while the outer gage senses compression. The two gages may be used in adjacent bridge arms, thereby adding their individual outputs and simultaneously providing temperature compensation.

Figure 14.12 Location of strain gages on flat diaphragm

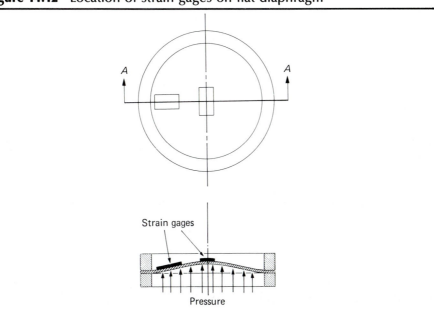

Figure 14.13   Differential pressure cell with inductance-type
                   secondary transducer

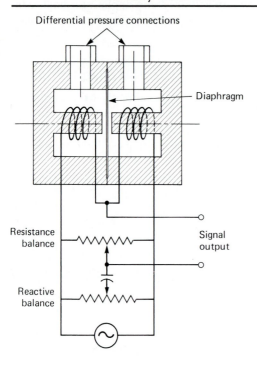

14.8.2   Inductive Types

Variable inductance has also been successfully used as a form of secondary
transducer used with a diaphragm [6]. Figure 14.13 illustrates one arrangement
of this sort. Flexing of the diaphragm due to applied pressure causes it to move
toward one pole piece and away from the other, thereby altering the relative
inductances. An inductive bridge circuit may be used, as shown. Standard
laboratory equipment, such as an oscilloscope or electronic voltmeter, as well
as recorders, may be used to display the gage output. Available ranges are
from 0–1.0 psi to 0–100 psi.

14.8.3   Piezoelectric Pressure Cells

Pressure cells using piezoelectric-type secondary transducers (Section 6.14)
have the advantage of very high sensitivities coupled with high natural
frequencies. These desirable qualities permit wide ranges of working pressures
and excellent frequency response. Typical maximum pressures are as high as

100,000 psi with resolutions of less than 1 psi. Outputs are in terms of coulombs per unit input and may be on the order of 0.2 pC/psi. A typical resonance frequency is 150,000 Hz. Inasmuch as the output impedance is inherently very high, some form of impedance transformation, such as a charge amp (Section 7.18.2), is required in proximity to the transducer.

14.8.4 Other Types of Secondary Transducers

Flexing diaphragms have been used to alter capacitance as a means of producing an electrical output (see Fig. 6.14). This method is not so common as those previously discussed, primarily because of low sensitivity and the problems accompanying the requirement for relatively high carrier frequencies.

Semiconductor chips may be etched to produce a pressure-sensitive diaphragm. The diaphragm itself can be instrumented with diffused piezoresistive strain gages (see Fig. 6.16). Typical ranges for these sensors are 0–1 psi and 0–100 psi.

Another successful method uses an electromechanical resonant system consisting of a fine wire under tensile load vibrating at its natural frequency. One end of the wire is connected to the center of a pressure-sensing diaphragm, which varies the wire tension, depending on the applied pressure. Small permanent magnets provide a magnetic field in which the wire vibrates, causing an ac potential to be developed in the wire. After amplification, a portion of this voltage is fed back to energize driving coils that maintain the vibration. Output *frequency* is the measure of pressure.

14.9 Strain-Gage Pressure Cells

Any form of closed container will be strained when pressurized. Sensing the resulting strain with an appropriate secondary transducer, such as a resistance strain gage, will provide a measure of the applied pressure. The term *pressure cell* has gradually become applied to this type of pressure-sensing device, and various forms of elastic *containers* or cells have been devised.

For low pressures, a pinched tube may be used (Fig. 14.14). This arrangement supplies a bending action as the tube tends to round out. Gages may be placed diametrically opposite on the flattened faces, as shown, with two unstressed temperature-compensating gages mounted elsewhere. This arrangement completes the electrical bridge. Cells of this general design are commercially available.

Probably the simplest form of strain-gage pressure transducer is a cylindrical tube such as that shown in Fig. 14.15. In this application two active gages mounted in the hoop direction may be used for pressure sensing, along with two temperature-compensating gages mounted in an unstrained location. Temperature-compensating gages are shown mounted on a separate disk

Figure 14.14   Flattened-tube pressure cell that employs resistance
strain gages as secondary transducers

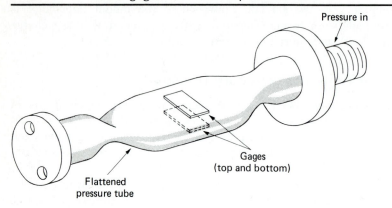

fastened to the end of the cell. Design relationships may be found in most
mechanical design texts.

The sensitivity of a pair of circumferentially mounted strain gages (Fig.
14.15) with gage factor F is expressed by the relationship*

$$\frac{\Delta R}{P_i} = \frac{2FRd^2}{E}\left[\frac{2 - v}{D^2 - d^2}\right],$$ **(14.6)**

where

ΔR = the strain-gage resistance change,

R = the nominal gage resistance,

P_i = the internal pressure,

d = the inside diameter of the cylinder,

D = the outside diameter of the cylinder,

E = Young's modulus,

v = Poisson's ratio.

The bridge constant, 2, appears because two circumferential gages are
assumed. If a single strain-sensitive gage is to be used, the sensitivity will be
one-half that given by Eq. (14.6). Of course, these relations are true only if
elastic conditions are maintained and if the gages are located so as to be
unaffected by end restraints.

Improved frequency response may be obtained for a cell of this type by
minimizing the internal volume. This may be accomplished by use of a solid

---

* Equation (14.6) is based on the Lamé equations for heavy-wall pressure cylinders. See Problem
14.27.

Figure 14.15 Cylindrical-type pressure cell

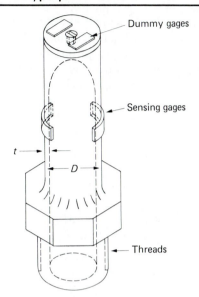

"filler" such as a plug, which will reduce the flow into and out of the cell with pressure variation.

Figure 14.16 shows the electrical circuitry used for a transducer of this type. Gage M is a modulus gage, discussed in Section 13.5, used to compensate for variation in Young's modulus with temperature. The calibration and output resistors are adjusted to provide predetermined bridge resistance and calibration.

Figure 14.16 Strain-gage circuitry for pressure cells employing a modulus gage

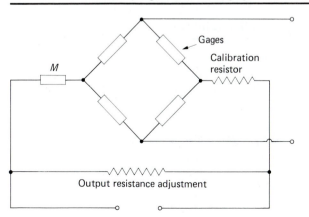

14.10   Measurement of High Pressures

The high-pressure range has been defined as beginning at about 700 atm and extending upward to the limit of present techniques [9], which is on the order of 10^7 atm. Conventional pressure-measuring devices, such as strain-gage pressure cells and Bourdon-tube gages, may be used at pressures as high as 3500 to 7000 atm. Bourdon tubes for such pressures are nearly round in section and have a high ratio of wall thickness to diameter. They are, therefore, quite stiff, and the deflection per turn is small. For this reason, high-pressure Bourdon tubes are often made of a number of turns.

Electrical Resistance Pressure Gages

Very high pressures may be measured by electrical resistance gages, which make use of the resistance change brought about by direct application of pressure to the electrical conductor itself. The sensing element consists of a loosely wound coil of relatively fine wire. When pressure is applied, the bulk-compression effect results in an electrical resistance change that may be calibrated in terms of the applied pressure.

Figure 14.17 shows a bulk modulus gage in section. The sensing element does not actually contact the process medium but is separated therefrom by a kerosene-filled bellows. One end of the sensing coil is connected to a central terminal, as shown, while the other end is grounded, thereby completing the necessary electrical circuit.

Although Eq. (12.8) was written with a somewhat different application in mind, it also applies to the situation being discussed. Let us rewrite this relation:

$$\frac{dR}{R} = \frac{dL}{L} - 2\frac{dD}{D} + \frac{d\rho}{\rho}, \tag{14.7}$$

Figure 14.17   Section through a bulk-modulus pressure gage

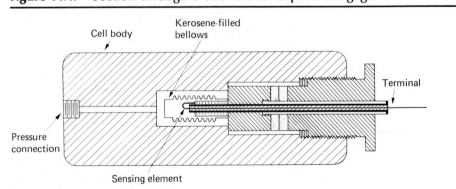

Cell body

Kerosene-filled bellows

Terminal

Pressure connection

Sensing element

in which

$$R = \text{the electrical resistance,}$$
$$L = \text{the length of the conductor,}$$
$$D = \text{a sectional dimension,}$$
$$\rho = \text{the resistivity.}$$

The wire will be subject to a biaxial stress condition because the ends, in providing electrical continuity, will generally not be subject to pressure. Using relations of the form expressed by Eqs. (12.3) and assuming that $\sigma_x = \sigma_y = -P$ and $\sigma_z = 0$, we may write

$$\varepsilon_x = \varepsilon_y = \frac{dD}{D} = -\frac{P}{E}(1 - v) \tag{14.8}$$

and

$$\varepsilon_z = \frac{dL}{L} = \frac{2vP}{E}. \tag{14.8a}$$

Combining the above relations gives us

$$\frac{dR}{R} = \frac{2P}{E} + \frac{d\rho}{\rho} \tag{14.9}$$

or

$$\frac{dR/R}{P} = \frac{2}{E} + \frac{d\rho/\rho}{P}. \tag{14.10}$$

Two metals are commonly used for resistance gages: manganin and an alloy of gold and 2.1% chromium. Both methods provide linear outputs with the following sensitivities: 1.692×10^{-7} and $0.673 \times 10^{-7}\ \Omega/\Omega \cdot \text{psi}$ for manganin and the gold alloy, respectively. Although the former possesses the greater pressure sensitivity, final selection must also be based on temperature sensitivity. Whereas manganin exhibits a resistance change of about 0.2% for the temperature range of 70°F–180°F, the corresponding change for the gold alloy is on the order of 0.01% [10]. Because of the difference, the gold alloy is generally preferred. The lower output is compensated for by greater electrical amplification.

14.11 Measurement of Low Pressures

Atmospheric pressure serves as a convenient reference datum, and, in general, pressures below atmospheric may be called low pressures or vacuums. We know, of course, that a *positive* magnitude of absolute pressure exists at all times, even in a vacuum. It is impossible to reach an absolute pressure of zero.

A common unit of low pressure is the micrometer, which is one-millionth of a meter (0.001 mm) of mercury column. *Very low* pressure may be defined as any pressure below 1 mm of mercury, and an *ultralow* pressure as less than a nanometer ($10^{-3}\,\mu$m). The torr is also used (Section 14.1).

Two basic methods are used for measuring low pressure: (1) *direct* measurement resulting in a displacement caused by the action of force, and (2) *indirect* or *inferential* methods wherein pressure is determined through the measurement of certain other pressure-controlled properties, including volume and thermal conductivity. Devices included in the first category are spiral Bourdon tubes, flat and corrugated diaphragms, capsules, and various forms of manometers. Since these have been discussed in the preceding pages, they need not be discussed further here except to say that their use is generally limited to a lowest pressure value of about 1 mm of Hg. For measurement of pressures below this value, one of the inferential methods is normally dictated.

14.11.1   The McLeod Gage

Operation of the McLeod gage is based on Boyle's fundamental relation

$$P_1 = \frac{P_2 V_2}{V_1}, \qquad (14.11)$$

where P_1 and P_2 are pressures at initial and final conditions, respectively, and V_1 and V_2 are volumes at corresponding conditions. By compressing a known volume of the low pressure gas to a higher pressure and measuring the resulting volume and pressure, one can calculate the initial pressure.

Figure 14.18 illustrates the basic construction and operation of the McLeod gage. Measurement is made as follows. The unknown pressure source is connected at point A, and the mercury level is adjusted to fill the volume represented by the darker shading. Under these conditions the unknown pressure fills the bulb B and capillary C. Mercury is then forced out of the reservoir D, up into the bulb and reference column E. When the mercury level reaches the cutoff point F, a *known* volume of gas is trapped in the bulb and capillary. The mercury level is then further raised until it reaches a zero reference point in E. Under these conditions the volume remaining in the capillary is read directly from the scale, and the difference in heights of the two columns is the measure of the trapped pressure. The initial pressure may then be calculated by use of Boyle's law.

Pressure of gases containing vapors cannot normally be measured with a McLeod gage, for the reason that the compression will cause condensation. By use of instruments of different ranges, a total pressure range of from about 0.01 μm to 50 mm of mercury may be measured with this type of gage.

14.11.2   Thermal Conductivity Gages

The temperature of a given wire through which an electric current is flowing will depend on three factors: the magnitude of the current, the resistivity, and the rate at which the heat is dissipated. The latter will be largely dependent on

Figure 14.18 McLeod vacuum gage

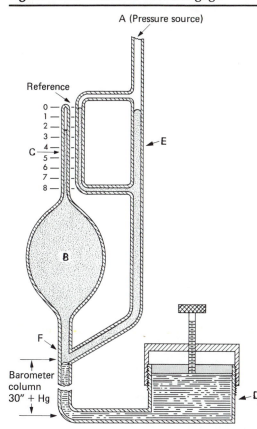

the conductivity of the surrounding media. As the density of a given medium is reduced, its conductivity will also reduce and the wire will become hotter for a given current flow.

This is the basis for two different forms of gages for measurement of low pressures. Both use a heated filament but differ in the means for measuring the temperature of the wire. A single platinum filament enclosed in a chamber is used by the *Pirani gage*. As the surrounding pressure changes, the filament temperture, and hence its resistance, also changes. The resistance change is measured by use of a resistance bridge that is calibrated in terms of pressure, as shown in Fig. 14.19. A compensating cell is used to minimize variations caused by ambient temperature changes.

A second gage also depending on thermal conductivity is of the thermocouple type. In this case the filament temperatures are measured directly by means of thermocouples welded directly to them. Filaments and thermocouples are arranged in two chambers, as shown schematically in Fig. 14.20.

Figure 14.19   The Pirani-type thermal conductivity gage

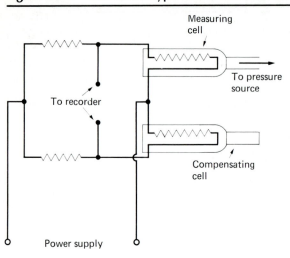

Figure 14.20   Thermocouple-type conductivity gage

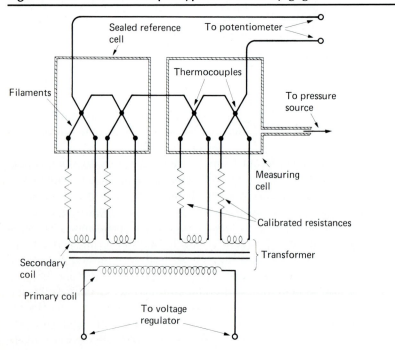

When conditions in both the measuring and reference chambers are the same, no thermocouple current will flow. When the pressure in the measuring chamber is altered, changed conductivity will cause a change in temperature, which will then be indicated by a thermocouple current.

In both cases the gages must be calibrated for a definite pressurized medium, for the conductivity is also dependent on this factor. Gages of these types are useful in the range of 1 to 1000 μm Hg.

14.11.3   Ionization Gages

For measurement of extremely low pressures, an ionization gage, which is usable to pressures down to $0.000001\,\mu$m (one-billionth of a millimeter of mercury), is used. The maximum pressure for which an ionization gage may be used is about 1 μm Hg. An ionization cell for pressure measurement is very similar to the ordinary triode electronic tube. It possesses a heated filament, a positively biased grid, and a negatively biased plate in an envelope evacuated by the pressure to be measured. The grid draws electrons from the heated filament, and collision between them and gas molecules causes ionization of the molecules. The positively charged molecules are then attracted to the plate of the tube, causing a current flow in the external circuit, which is a function of the gas pressure.

Disadvantages of the heated-filament ionization gage are that (1) excessive pressure (above 1 or 2 μm Hg) will cause rapid deterioration of the filament and a short life; and (2) the electron bombardment is a function of filament temperature, and therefore careful control of filament current is required. Another form of ionization gage minimizes these disadvantages by substituting a radioactive source of alpha particles for the heated filament.

14.12   Dynamic Characteristics of Pressure-Measuring Systems

Basic pressure-measuring transducers are driven, damped, spring mass systems whose isolated dynamic characteristics are theoretically similar to the generalized systems discussed in Chapter 5. In application, however, the actual dynamic characteristics of the complete pressure-measuring system are usually controlled more by factors extraneous to the basic pickup than by the pickup characteristics alone. In other words, overall dynamic performance is determined less by the transducer than by the manner in which it is inserted into the complete system.

When the pickup is used to measure a dynamic air or gas pressure, system damping will be determined to a considerable extent by factors external to the pickup. The extraneous pneumatic circuitry will have frequency characteristics of its own, affecting system response. When liquid pressures are measured, the effective sprung mass of the system will necessarily include some portion of the liquid mass. In addition, the elasticity of any conducting tubing will act to change the overall spring constant. Connecting tubing and unavoidable cavities in the pneumatic or hydraulic circuitry introduce losses and phase lags, causing differences between measured and applied pressures. Much theoretical work has been done in an attempt to evaluate these effects [11–15]. Each application, however, must be weighed on its own individual merits; for this reason only a general summary of some of the factors involved is practical in this discussion.

14.12.1   Gas-Filled Systems

As outlined before, the response of a pressure-measuring system involves more than the pickup characteristics alone. The complete system, including the method of conducting the pressure variation to the pickup, must be accounted for. In many applications it is necessary to transmit the pressure through some form of passageway or connecting tube. Figure 14.21 illustrates typical cases.

If the pressurized medium is a gas, such as air, acoustical resonances may occur in the same manner in which the air in an organ pipe resonates. If sympathetic driving frequencies are present, nodes and antinodes will occur, as shown in the figure. A node, characterized by a point of zero air motion, will occur at the blocked end (assuming that the displacement of the pressure-sensing element, such as a diaphragm, is negligible). Maximum pressure variation takes place at this point. Maximum oscillatory motion will occur at the antinodes, and the distance between adjacent nodes and antinodes equals one-fourth the wavelength of the resonating frequency. Theoretical resonant

Figure 14.21   (a) Gas-filled pressure measuring system, (b) gas-filled pressure-measuring system with cavity

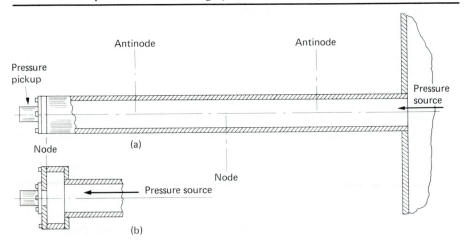

frequencies may be determined from the relation

$$f = \frac{C}{4L}(2n - 1),$$ (14.12)

in which

f = the resonance frequencies (including both fundamental and harmonics), in hertz,

C = the velocity of sound in the pressurized medium,

L = the length of the connecting tube,

n = any positive integer. (It will be noted from the equation that only odd harmonics occur.)

In many cases a cavity is required at the pickup end to adapt the instrument to the tubing, as shown in Fig. 14.21(b). If we assume that the medium is a gas, and that the elasticity of the containing system, including the pickup device, is relatively stiff compared with that of the gas, we have what is known as a Helmholtz resonator. The column of gas with its mass and elasticity form a spring–mass system having an acoustical resonance whose fundamental frequency may be expressed by the relation [16]

$$f = \frac{C}{2\pi}\sqrt{\frac{a}{V(L + \frac{1}{2}\sqrt{\pi a})}},$$ (14.13)

where

a = the cross-sectional area of the connecting tube,

V = the net internal volume of the cavity, excluding the volume of the tube.

By proper configuration and proportioning, connecting systems of this type may be used for acoustical filtering [17]. In certain applications, quite sensitive or fragile sensing devices are required to measure small differential pressures. At the same time, high-energy pressure cycles may be present at frequencies above the range of interest and of sufficient intensity to cause pickup failure. The situation is often present in aircraft testing.

14.12.2 Liquid-Filled Systems

When a pressure-measuring system is filled with liquid rather than a gas, a considerably different situation is presented. The liquid becomes a major part of the total sprung mass, thereby becoming a significant factor in determining the natural frequency of the system.

If a single degree of freedom is assumed,

$$f_n = \frac{1}{2\pi}\sqrt{\frac{k_s g_c}{m}},$$ (14.14)

in which

$$f_n = \text{the natural frequency, in hertz,}$$

$$m = \text{the equivalent moving mass} = m_1 + m_2,$$

$$m_1 = \text{the mass of moving transducer elements,}$$

$$m_2 = \text{the equivalent mass of the liquid column,}$$

$$k_s = \frac{k_t k_1}{k_t + k_1} = \text{the overall system stiffness,}$$

$$k_t = \text{the transducer stiffness,}$$

$$k_1 = \text{the transmitting medium stiffness.}$$

By simplified analysis, White [18] has determined the following approximate relation for the effective mass of the liquid column:

$$m_2 = \frac{4}{3} \rho a L \left(\frac{A}{a}\right)^2, \tag{14.15}$$

in which

$$\rho = \text{the fluid density,}$$

$$a = \text{the sectional area of the tube,}$$

$$L = \text{the length of the tube,}$$

$$A = \text{the effective area of the transducer-sensing element.}$$

It will be noted that A is the *effective* area, which is not necessarily equal to the actual diaphragm or bellows area but may be defined by the relation

$$A = \frac{\Delta V}{\Delta y}, \tag{14.16}$$

where

$$\Delta V = \text{the volume change accompanying sensing-element}$$
$$\text{deflection,}$$

$$\Delta y = \text{the significant displacement of the sensing element.}$$

By substitution, we have

$$f_n = \frac{1}{2\pi} \sqrt{\frac{k_s g_c}{m_1 + \frac{4}{3} \rho a L (A/a)^2}}. \tag{14.17}$$

In many cases the equivalent mass of the liquid, m_2, is of considerably greater magnitude than m_1, and the latter may be ignored without introducing an appreciable discrepancy. By so doing and substituting, we get

$$a = \pi D^2/4,$$

$$D = \text{tubing I.D.},$$

$$f_n = \frac{D}{8A} \sqrt{\frac{3k_s g_c}{\pi \rho L}}. \tag{14.18}$$

As mentioned before, pressure pickups involve spring-restrained masses in the same manner as do galvanometers and seismic-type accelerometers, and therefore good frequency resonance is obtainable only in a frequency range well below the natural frequency of the measuring system itself. For this reason it is desirable that the pressure-measuring system have as high a natural frequency as is consistent with required sensitivity and installation requirements. Inspection of Eq. (14.18) indicates that the diameter of the connecting tube should be as large as practical and that its length should be minimized.

In addition, it has been shown that optimum performance for systems of this general type requires damping in rather definite amounts. White [18] gives the following relation for the damping ratio ξ of a system of the sort being discussed:

$$\xi = \frac{4\pi L v (A/a)^2}{\sqrt{k_s m/g_c}} \tag{14.19}$$

$$= \frac{4\pi L v (A/a)^2}{\sqrt{(k_s/g_c)[m_1 + \frac{4}{3}\rho a L (A/a)^2]}}, \tag{14.20}$$

where v = the viscosity of the fluid. If we ignore m_1 and insert $a = \pi D^2/4$, we may write the equation as

$$\xi = \frac{16 v A}{D^2} \sqrt{\frac{3 L g_c}{\pi k_s \rho}}. \tag{14.21}$$

14.13   Calibration Methods

14.13.1   Methods for Static Pressures

Static calibration of pressure gages presents no particular problems unless the upper pressure limits are unusually high. The familiar dead-weight tester (Fig. 14.8) may be used to accurately supply reference pressures with which transducer outputs may be compared. Testers of this type are useful to pressures as high as 10,000 psi, and by use of special designs, this limit may be extended to 100,000 psi.

Although static calibration is desirable, pickups used for dynamic measurement should also receive some form of dynamic calibration. Dynamic calibration problems consist of (1) obtaining a satisfactory source of pressure, either periodic or pulsed, and (2) reliably determining the true pressure-time relation produced by such a source. These two problems will be discussed in the next few paragraphs.

To be a standard, the precise pressure-time relation must be known. Some sources of dynamic pressure are as follows:

I. Steady-state periodic sources
  A. Piston and chamber
  B. Cam-controlled jet
  C. Acoustic resonator
  D. Siren disk
II. Transient sources
  A. Quick-release valve
  B. Burst diaphragm
  C. Closed bomb
  D. Shock tube

14.13.2   Steady-State Methods

One source of steady-state periodic calibration pressure is simply an ordinary piston and cylinder arrangement, shown schematically in Fig. 14.22 [19]. If the piston stroke is fixed, pressure amplitude may be varied by adjusting the cylinder volume. Although such a system is normally special-purpose, existing

Figure 14.22   Schematic diagram of a piston and cylinder steady-state pressure source

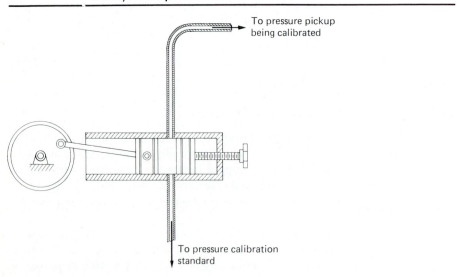

equipment, such as a CFR* engine, may be adapted for this use. Amplitude and frequency ranges will depend on the mechanical design; however, peak pressures of 1000 psi and frequencies to 100 Hz may be obtained.

A method very similar to this is used for microphone calibration. In this case required pressure amplitudes are quite small, and instead of the piston being driven with a mechanical linkage, an electromagnetic system is used [20]. Piston excursion may be determined by the technique described in Section 17.10. This method suggests the possibility of using vibration test shakers (Section 17.16) as a source of piston motion [21].

A variation of the piston source is to drive a diaphragm, bellows, or Bourdon tube. The latter has been successfully used by flexing it by means of an eccentric attached to the end of the tube through a link or connecting rod [22].

Figure 14.23 illustrates another method for obtaining a steady-state periodic pressure. A source of this type has been used to 3000 Hz with amplitudes to 1 psi [6]. A variation of this method uses a motor-driven siren-type disk having a series of holes drilled in it so as alternately to vent a pressure source to atmosphere and then shut it off [23].

Another successful system is to use a variable-speed motor to drive a pressure transmitter by means of a circular cam [24]. The transmitter is essentially an adjustable servo valve that controls the output from a constant-pressure source.

Steady-state sinusoidal pressure generators consisting of an acoustically driven resonant system (see Section 14.12.1) have been used. Hylkema and

Figure 14.23 Jet and cam steady-state pressure source

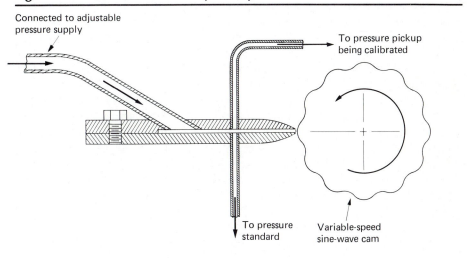

Connected to adjustable
pressure supply

To pressure pickup
being calibrated

To pressure
standard

Variable-speed
sine-wave cam

---

* Cooperative Fuels Research.

Bowersox [14] obtained pressure fluctuations of the order of 0.5 psi rms, using a 40-in.-long pipe energized with a 35 W loudspeaker type of driving unit.* Usable frequencies to about 2000 Hz in integral multiples of the fundamental were obtained.

All the methods suggested here simply supply sources of pressure variation, but in themselves do not provide means for determining magnitudes or time characteristics. They are particularly useful, however, for comparing pickups having unknown characteristics with those of proven performance.

14.13.3   Transient Methods

Steady-state periodic sources used to determine dynamic characteristics of pressure transducers are limited by amplitude and frequency that can be produced. High amplitudes and steady-state frequencies are difficult to obtain simultaneously. For this reason it is necessary to resort to some form of step function in order to determine high-frequency response of pressure transducers in the higher amplitude ranges.

Various methods are used to produce the necessary pulse. One of the simplest is to use a fast-acting valve between a source of hydraulic pressure and the pickup. Rise times, from 0 to 90% of full pressure, of 10 ms are reported [25].

Pressure steps may also be obtained through use of bursting diaphragms. Two chambers are separated by a thin plastic diaphragm or plate whose failure is mechanically induced by a plunger or knife. It has been found that a pressure drop, rather than a rise, produces a more nearly ideal step function. A drop time of about 0.25 ms has been obtained [14].

Still another source of stepped-pressure function is the closed bomb, in which a pressure generator such as a dynamite cap is exploded. Peak pressure is controlled by net internal volume, and pressure steps as high as 700 psi in 0.3 ms have been obtained [14].

Undoubtedly the *shock tube* provides the nearest thing to a transient pressure "standard." Construction of a shock tube is quite simple: it consists of a long tube, closed at both ends, separated into two chambers by a diaphragm, as shown in Fig. 14.24. A pressure differential is built up across the diaphragm, and the diaphragm is burst, either directly by the pressure differential or initiated by means of an externally controlled probe, or *dagger*. Rupturing of the diaphragm causes a pressure discontinuity, or *shock wave*, to travel into the region of the lower pressure and a rarefaction wave to travel through the chamber of initially higher pressure. The reduced pressure wave is reflected from the end of the chamber and follows the stepped pressure down the tube at a velocity that is higher because it is added to the velocity already possessed by the gas particles from the pressure step. Figure 14.25 illustrates the sequence of events immediately following the bursting of the diaphragm.

---

* See Problem 18.11.

Figure 14.24 Basic shock tube

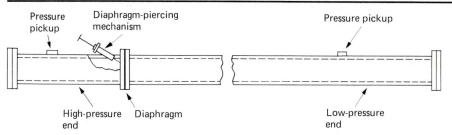

A relationship between pressures and shock-wave velocity may be expressed as follows [26]:

$$\frac{P_1}{P_0} = 1 + \frac{2k}{k+1}(M_0^2 - 1), \qquad (14.22)$$

Figure 14.25 Pressure sequence in a shock tube before and immediately after diaphragm is ruptured. Abscissa represents longitudinal axis of tube.

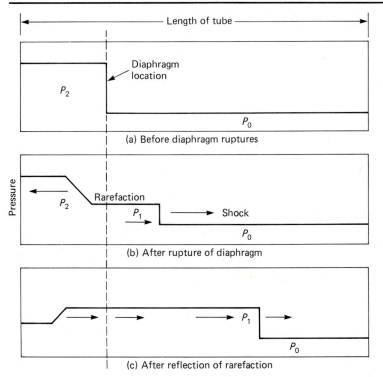

in which

P_1 = the intermediate transient pressure,

P_0 = the lower initial pressure,

k = the ratio of specific heats,

M_0 = the Mach number corresponding to the lower initial conditions.

We see, then, that if the gas properties are known, measurement of the propagation velocity will be sufficient to determine the magnitude of the pressure pulse. Propagation velocity may be determined from information supplied by accurately positioned pressure pickups in the wall of the tube. By this means, a known transient pressure pulse may be applied to a pressure transducer or to a complete pressure-measuring system simply by mounting the pickup in the wall of the shock tube. The response characteristics, as determined in this manner, may then be used to calculate the general response of the device or system over a spectrum of frequencies [27]. The methods for doing this, however, are beyond the scope of this book, and the large amount of computation required is particularly adaptable to a digital computer.

14.14   Final Remarks

An attempt has been made in the preceding pages to introduce the reader to some of the problems attending accurate experimental determination of pressure. We realize that many approaches to the problem have been omitted and that in certain respects the coverage has been brief and somewhat superficial. Such is a penalty that must be paid in assembling a book of this nature. For more detailed discussions, the reader is referred to the Suggested Readings for this chapter.

Suggested Readings

ASME MC88.1-1972 (R1987). *Guide for Dynamic Calibration of Pressure Transducers.* New York, 1972.

ASME 19.2-1987. Instruments and Apparatus: Part 2. *Pressure Measurement.*

Ametek. *Pressure Gage Handbook.* New York: Marcel Dekker, 1985.

Benedict, R. P. *Fundamentals of Temperature, Pressure and Flow Measurements,* 2nd ed. New York: John Wiley, 1977.

Bridgeman, P. W. *The Physics of High Pressure.* New York: Macmillan, 1931.

Peggs, G. N. (ed.). *High-pressure Measurement Techniques.* New York: Elsevier Science Publishers, 1983.

Rombacher, W. G. *Survey of Micromanometers.* NBS Monograph 114. Washington, D.C.: U.S. Government Printing Office, 1970.

Schweppe, L. C., et al. *Methods for the Dynamic Calibration of Pressure Transducers.* NBS Monograph 67. Washington, D.C.: U.S. Government Printing Office, 1963.

Spain, I. L. and J. Paauwe. *High Pressure Technology,* vol 1 and vol 2. New York: Marcel Dekker, 1977.

Problems

14.1 Standard atmospheric pressure is 1.01325×10^2 kPa. What are the equivalents in (a) newtons per square meter, (b) pounds-force per square foot, (c) meters of water (head), (d) inches of oil, with 0.89 specific gravity, (e) millibars, (f) micrometers, and (g) torr?

14.2 Determine the factors for converting pressure in pascals to "head" in (a) meters of water; (b) centimeters of mercury.

14.3 The following are some commonly encountered pressures (approximate). Convert each to the SI units, Pa or kPa: (a) automobile tire pressure of 32 psig; (b) household water pressure of 120 psia; (c) regulation football pressure of 13 psig.

14.4 First in SI units and then in English units:

    a.   Write expressions relating the height of a fluid column in terms of a reduced gage pressure (a vacuum).

    b.   Under standard conditions of atmosphere and gravity, what is the maximum height to which water may be raised by suction alone?

    c.   Under similar conditions, to what height may a column of mercury be raised?

    d.   On the surface of the moon, what is the height to which water could be raised by suction alone? (See Appendix D for data.)

14.5 The common mercury barometer may be formed by sealing the upper end of a tube (e.g., Fig. 14.4), inverting it and filling it with mercury, then righting the tube into a mercury-filled reservoir. A vacuum is formed over the column and the height of the column is governed primarily by the pressure (air pressure) applied at the base. Under these conditions is a true zero absolute pressure (a complete vacuum) formed over the column? Investigate the vapor pressure of mercury and determine the degree of error introduced if it is ignored.

14.6 Rewrite Eq. (14.3) in terms of specific gravities.

14.7 Write a few sentences explaining the mechanics of "suction."

14.8 Write a few sentences explaining the operation of a syphon.

14.9 The U-tube-type manometer (Fig. 14.5) uses mercury and water as the manometer and transmitting fluids, respectively. What value of h should be expected at 20°C, if the applied differential pressure is 80 kPa (11.6 psig)? (See Appendix D for data.) Use SI units.

14.10 Solve Problem 14.9 using English units.

14.11 A dual-fluid U-tube manometer, Fig. 14.5, located at Fort Egbert, Alaska, displays a pressure, $\Delta h = 5.400$ in. fluid displacement. Under identical conditions, except for location, determine what pressure would be indicated (a) at Key West, Florida, and (b) on the moon. (Refer to Appendix D for data.)

Figure 14.26   Manometer arrangement referred to in Problem 14.12

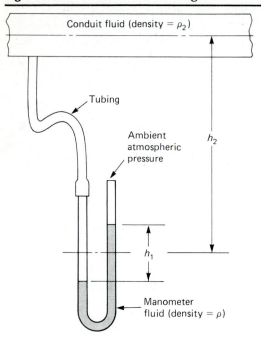

14.12 Figure 14.26 illustrates a manometer installation. Write an expression for determining the static pressure in the conduit in terms of h_1, h_2, and the other pertinent parameters.

14.13 For the conditions shown in Fig. 14.26, if the manometer fluid is Hg and the conduit fluid is H_2O (both at 20°C), $h_1 = 18.4$ cm (7.24 in.), and $h_2 = 0.7$ m (2.30 ft), what pressure exists in the conduit? Solve using SI units.

14.14 Solve Problem 14.13 using English units.

14.15 Express the ratio of sensitivities of an inclined manometer (Fig. 14.6) to that of a simple manometer in terms of the angle θ. For an inclined manometer six times more sensitive than a simple manometer, what should be the angle of incline?

14.16 Verify Eqs. (14.5) and (14.5a).

14.17 The manometer shown in Fig. 14.7 uses water and carbon tetrachloride as the two fluids (see Appendix D for data). If the area ratio is 0.01, what magnification will result as compared to the simple manometer using (a) water as the fluid and (b) carbon tetrachloride as the fluid?

14.18 Derive an expression for the two-fluid manometer in Fig. 14.7, substituting reservoirs of diameters D_1 and D_2 for the like-sized reservoirs shown in the figure.

14.19 A two-fluid manometer as shown in Fig. 14.7 uses a combination of kerosene (specific gravity, 0.80) and alcohol-diluted water (specific gravity, 0.83). Also, $d = \frac{1}{4}$ in. (6.35 mm) and $D = 2$ in. (50.8 mm). What amplification ratio is obtained with this arrangement as compared to a simple water manometer? What error would be introduced if the ratio of diameters was ignored?

14.20 Confirm Eq. (14.4).

14.21 Note that Eq. (14.4) is based on a moving datum—namely, the liquid level in the reservoir. Derive an equation for the differential pressure based on the movement of the liquid in the inclined column only. (Note that a practical solution would be to make provision for adjusting the reservoir level to an index or "zero" line.)

14.22 Figure 14.15 shows a cylindrical pressure cell using two sensing strain gages. If the cylinder may be assumed to be "thin wall," then

$$\sigma_H = \frac{Pd}{2t}$$

and

$$\sigma_L = \frac{Pd}{4t}.$$

(See Appendix E, Figure E.10, for symbol meanings.)
For this case, show that

$$\frac{\Delta R}{P} = \frac{FRd}{Et}\left[1 - \frac{v}{2}\right]$$

where

$$F = \text{gage factor,}$$

$$R = \text{gage resistance,}$$

$$E = \text{Young's modulus,}$$

$$v = \text{Poisson's ratio.}$$

14.23 For a circular diaphragm of the type and loading shown in Fig. 14.11(a), the maximum normal stress occurs in the radial direction at the outer boundary, expressed as follows [28]:

$$\sigma_r = \frac{3}{4}\left(\frac{a}{t}\right)^2 P.$$

Greatest linear deflection occurs at the center and is equal to

$$Y_{max} = \left(\frac{3}{16}\right)\left(\frac{Pa^4}{Et^3}\right)(1 - v^2),$$

where

$$P = \text{pressure,}$$

$$a = \text{radius,}$$

$$E = \text{Young's modulus,}$$

$$v = \text{Poisson's ratio.}$$

For $a = \frac{1}{4}$ in. (6.35 mm), $E = 30 \times 10^6$ psi (20.68×10^7 kPa), and $v = 0.3$, and for a design stress of 9×10^4 psi (6.2×10^5 kPa), what maximum deflection may be expected if $P = 300$ psi (2.07×10^3 kPa)?

14.24 Solve Problem 14.23 using SI units and reconcile the two answers.

14.25 A thin-wall, cylindrically sectioned tube of nominal diameter D and wall thickness t is subjected to a pressure P. Circumferential (hoop) and longitudinal stresses may be determined from the relations $\sigma_H = PD/2t$ and $\sigma_L = PD/4t$ (see Example E.2, Appendix E). Using Eqs. (12.3) show that the corresponding circumferential and longitudinal strains are

$$\varepsilon_H = \left(\frac{PD}{2Et}\right)\left(1 - \frac{1}{2}v\right) \quad \text{and} \quad \varepsilon_L = \left(\frac{PD}{2Et}\right)\left(\frac{1}{2} - v\right).$$

14.26 A pressure transducer is constructed from a steel tube having a nominal diameter of 15 mm (0.59 in.) and a wall thickness of 2 mm (0.0787 in.).

    a.   If the design stress is limited to 2.75×10^8 Pa (39,885 psi), what maximum pressure may be applied to the transducer?

    b.   For the maximum pressure calculated in part (a), determine the circumferential and longitudinal strains that should be expected. Use $E = 20 \times 10^{10}$ Pa (29×10^6 psi) and $v = 0.3$.

14.27 The simple stress relations given in Problem 14.25 assume a uniform stress distribution through the wall of the cylinder. For so-called "heavy-wall" cylinders, this simplifying assumption leads to error. The following, more complex relations, often referred to as the Lamé equations, must be used.

$$\sigma_H = \frac{P(D^2 + d^2)}{D^2 - d^2} \quad \text{on the inner surface,}$$

$$\sigma_H = \frac{2Pd^2}{D^2 - d^2} \quad \text{on the outer surface,}$$

$$\sigma_L = \frac{Pd^2}{D^2 - d^2},$$

$$\sigma_r = -P \quad \text{on the inner surface.}$$

All are principal stresses (see Appendix E).

    a.   For a design stress of 2.75×10^8 Pa (39,885 psi), $d = 2$ cm (0.787 in.), and $D = 5$ cm (1.968 in.), what is the maximum pressure that may be applied?

    b.   If the maximum pressure is applied, what circumferential and longitudinal strains should be expected on the outer surface? Use 0.3 for Poisson's ratio and 20×10^{10} Pa for Young's modulus.

14.28 Solve Problem 14.27 using English units.

14.29 Derive Eq. (14.6). Note that this equation is based on the Lamé equations for heavy-wall pressure vessels. See Problem 14.27.

14.30 Confirm Eqs. (14.9) and (14.10).

14.31 The speed of sound in a gas, C, may be expressed by the relation [1]

$$C = [kRTg_c]^{1/2},$$

where

$$k = \text{the ratio of specific heats } (= 1.4 \text{ for air}),$$

$$R = \text{the gas constant } (= 287 \text{ J/kg} \cdot \text{K for air}),$$

T = absolute temperature,

g_c = the dimensional constant.

Using the above equation and Eq. (14.12), determine the change in frequency in percent, corresponding to a temperature change from 10°C to 40°C.

14.32 A pressure-measuring system involves a $\frac{1}{4}$-in.-diameter tube, 24 in. long, connecting a pressure source to a transducer. At the transducer end there is a cylindrical cavity $\frac{1}{2}$ in. in diameter and $\frac{1}{2}$ in. long. Proper performance requires that the frequency of applied pressures be such as to avoid resonance. Calculate the resonance frequency of the system.

14.33 Assume that the 24-in. connecting tube used in Problem 14.32 is reduced to zero length. What will be the resonance frequency? [Use Eq. (14.13), letting $L = 0$.]

14.34 A Helmholtz resonator consists of a spherical cavity to which a circularly sectioned tube is attached. It may be considered as approximating the tube and cavity of Problem 14.32. The resonance frequency of a Helmholtz resonator may be estimated by the relation

$$f = \frac{C}{2\pi} \sqrt{\frac{a}{VL}}$$

(see Ref. [2] for Chapter 18). The symbols have the same meaning as in Eq. (14.13). Use this equation to estimate the resonance frequency of the system described in Problem 14.32.

References

1. Anderson, J. D., Jr. *Introduction to Flight*. New York: McGraw-Hill, 1978.

2. Wolfe, A. An elementary theory of the bourdon gage. *J. Appl. Mech.* 68: A.207, Sept. 1946.

3. Wenk, E. Jr. A diaphragm-type gage for measuring low pressures in fluids. *SESA Proc.* 8(2): 90, 1951.

4. Roark, R. J. *Formulas for Stress and Strain*. 4th ed. New York: McGraw-Hill, 1965.

5. Stedman, C. K. The characteristics of flat annular diaphragms. *Instrument Notes* 31. Los Angeles: Statham Laboratories, January 1957.

6. Patterson, J. L. A miniature electrical pressure gage utilizing a stretched flat diaphragm, *NACA Tech. Note 2659:* April 1952.

7. Grover, H. J., and J. C. Bell. Some evaluations of stresses in aneroid capsules. *SESA Proc.* 5(2): 125, 1958.

8. Werner, F. D. The design of diaphragms for pressure gages which use the bonded wire resistance strain gage. *SESA Proc.* 11(1): 137, 1935.

9. Howe, W. H. What's available for high pressure measurement and control. *Control Eng.* 2: 53, April 1955.

10. Howe, W. H. The present status of high pressure measurement. *ISA J.* 2: 77, 109, March, April 1955.

11. Wildhack, W. A. Pressure drop in tubing in aircraft instrument installations. *NACA Tech. Note 593:* 1937.

12. Iberall, A. S. Attenuation of oscillatory pressures in instrument lines. *NBS J. Res.* 45: 85, July 1950.

13. Moise, J. C. Pneumatic transmission lines. *ISA Proc.* 8: 152, 1953.

14. Hylkema, C. G., and R. B. Bowersox. Experimental and mathematical techniques for determining the dynamic response of pressure gages. *ISA Proc.* 8: 115, 1953, and *ISA J.* 1: 27, February 1954.

15. Stedman, C. K. Alternating flow of fluids in tubes. *Instrument Notes* 30. Los Angeles: Statham Laboratories, January 1956.

16. Lord Rayleigh. *The Theory of Sound,* vol. II. 2nd ed. New York: Dover Publications, 1945, p. 188.

17. Mylius, R. D., and R. J. Reid. Acoustical filters protect pressure transducers. *Control Eng.* 4: 115, January 1957.

18. White, G. Liquid filled pressure gage systems. *Instrument Notes* 7. Los Angeles: Statham Laboratories, January–February 1949.

19. Taback, I. The response of pressure measuring systems to oscillating pressure. *NACA Tech. Note 1819:* February 1949.

20. Badmaieff, A. Techniques of microphone calibration. *Audio Eng.* 38: Dec. 1954.

21. Reid, R. Use standard functions to test pneumatic systems. *Control Eng.* 5: 117, January 1958.

22. Baird, R. C., R. L. Solnich, and J. R. Amiss. Calibrator for dynamic pressure transducers. *Inst. Automation* 27: 1074, July 1954.

23. Meyer, R. D. Dynamic pressure transmitter calibrator. *Rev. Sci. Inst.* 17: 199, May 1946.

24. Eckman, D. P., and J. C. Moise. A pneumatic sine-wave generator for process control study. *ISA Proc.* 7: 13, 1952.

25. Davis, W. R. Measuring high pressure transients. *Auto. Control* 4: 24, January 1956.

26. National Bureau of Standards Monograph 67, *Methods for the Dynamic Calibration of Pressure Transducers,* Washington, D.C.: U.S. Department of Commerce, 1963.

27. Bowersox, R. Calibration of high frequency response pressure transducers. *ISA J.* 5: Nov. 1958.

28. Roark, R. J. *Formulas for Stress and Strain.* New York: McGraw-Hill, 1965, p. 217.

CHAPTER 15

‖‖‖

Measurement of Fluid Flow

15.1 Introduction

Accurate measurement of flow presents many and varied problems. The flowing medium may be a liquid, a gas, a granular solid, or any combination thereof. The flow may be laminar or turbulent, steady-state or transient. In addition, there are several very different *basic* approaches to the problem of flow measurement. This section, therefore, will present only an outline of some of the more important aspects of the general topic.

The most direct way to measure flow rate is to capture and record the volume or mass that flows during a fixed time interval (a *primary measurement* of flow rate). More often, some other quantity, such as a pressure difference or mechanical response, is used to infer the flow rate (a *secondary measurement*). Another distinction is that between flowmeters and velocity sensors. *Flowmeters* determine volume or mass flow rates (e.g., gallons per minute) through tubes and ducts, whereas *velocity-sensing probes* measure fluid speed (e.g., meters per second) at a point in the flow. Although velocity-sensing probes can be used as building blocks for flowmeters, the converse is rarely true. In addition, *flow-visualization* techniques are sometimes employed to obtain an image of the overall flow field. A categorization of flow-measurement methods is as follows:

1. Primary or quantity methods
   a. Weight or volume tanks, burettes, etc.
   b. Positive-displacement meters
2. Flowmeters
   a. Obstruction meters
      i. Venturi meters

        ii. Flow nozzles
        iii. Orifices
    b. Variable-area meters
    c. Turbine and propeller meters
    d. Magnetic flowmeters (liquids only)
    e. Vortex-shedding meters
3. Velocity probes
    a. Pressure probes
        i. Total pressure and Pitot-static tubes
        ii. Direction-sensing probes
    b. Hot-wire and hot-film anemometers
    c. Scattering techniques
        i. Laser-Doppler anemometer (LDA)
        ii. Ultrasonic anemometer (primarily liquid)
4. Flow-visualization techniques
    a. Smoke trails and smoke wire (gases)
    b. Dye injection, chemical precipitates (liquids)
    c. Hydrogen bubble (liquids)
    d. Laser-induced fluorescence
    e. Refractive-index change: interferometry, schlieren, shadowgraph

The above outline does not exhaust the list of flow-measuring methods, but it does attempt to include those of primary interest to the mechanical engineer. Obstruction meters are probably those most often used in industrial practice. Application of some of the methods listed is so obvious that only passing note will be made of them. This is particularly true of *quantity* methods. Weight tanks are especially useful for steady-state calibration of liquid flow-meters, and no particular problems are connected with their use.

Displacement meters have many forms and variations. Common examples are the water and gas meters used by suppliers to establish charges for services. Basically, displacement meters are hydraulic or pneumatic motors whose cycles of motion are recorded by some form of counter. Only such energy from the stream is absorbed as is necessary to overcome the friction in the device, and this is manifested by a pressure drop between inlet and outlet. Most of the configurations used for motors have been applied to metering. These include reciprocating and oscillating pistons, vane arrangements, including the nutating (or nodding) disk, helical screw devices, etc.

15.2 Flow Characteristics

When fluids move through uniform conduits at very low velocities, the motions of individual particles are generally along lines paralleling the conduit walls. Actual particle velocity is greatest at the center and zero at the wall, with the velocity distribution as shown in Fig. 15.1(a). Such a flow is called *laminar*.

Figure 15.1   Velocity distribution for: (a) laminar flow in a pipe or
tube and, (b) turbulent flow in a pipe or tube

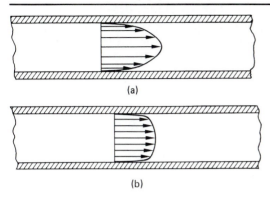

(a)

(b)

As the flow rate is increased, a point is reached where the particle motion
becomes more random and complex. Although this change in the nature of
flow may appear to occur at a definite velocity, careful observation will show
that the change can be somewhat gradual and may occur over a relatively
narrow range of velocities. The *approximate* velocity at which the change
occurs is called the *critical velocity,* and the flow at higher rates is referred to
as *turbulent.* The corresponding time-average velocity distribution across a
circular tube is shown in Fig. 15.1(b).

It has been found that the critical velocity is a function of several factors
that may be put in a dimensionless form called the Reynolds number, Re_D,* as
follows:

$$Re_D = \frac{\rho D V}{\mu},\qquad\qquad (15.1)$$

where

D = a sectional dimension of the fluid stream (normally the diameter
      if the conduit is a pipe of circular section),†

ρ = the density of the fluid,

V = the fluid velocity,

μ = the absolute viscosity of the fluid.

See Example 2.1 for examples of commonly used units.

---

* The units for absolute viscosity are $N \cdot s/m^2$ (or $lbf \cdot s/ft^2$), depending on the system of units
used. We see that although the form of Eq. (15.1) as written is most common, inclusion of g_c is
required to obtain a proper unit balance in the English Engineering system.

† The subscript D is used to indicate nominal pipe diameter. When Reynolds number is based, for
example, on the throat diameter of a venturi or an orifice, a lowercase d is commonly used, e.g.,
Re_d.

It has been shown by many investigators [1] that below the critical-velocity range, friction loss in pipes is a function of Re_D only, whereas for turbulent flow, the Reynolds number and the surface roughness determine the losses. The critical Reynolds number for pipes is usually between 2100 and 4000.

The volume flow rate, Q, through a pipe or duct is just the integral of the velocity distribution, $V(x,y)$, over the cross-sectional area A:

$$Q = \int_A V(x,y)\, dA \tag{15.2}$$

For engineering purposes, an average velocity, V, over the cross section is often used:

$$V = \frac{Q}{A} = \frac{1}{A} \int_A V(x,y)\, dA \tag{15.2a}$$

Flowmeters measure Q and/or V, while velocity probes measure $V(x,y)$. However, the output of a velocity probe can be integrated to obtain Q or V.

Bernoulli's equation for the flow of incompressible fluids between points 1 and 2 (Fig. 15.2) may be written in terms of local or average velocity* as

$$\frac{P_1 - P_2}{\rho} = \frac{V_2^2 - V_1^2}{2g_c} + \frac{(Z_2 - Z_1)g}{g_c}, \tag{15.3}$$

in which

P = absolute pressure,	lbf/ft²	N/m² (or Pa)
ρ = density,	lbm/ft³	kg/m³
V = linear velocity,	ft/s	m/s
Z = elevation,	ft	m
g = acceleration due to gravity,	32.17 ft/s²	9.807 m/s²
g_c = dimensional constant.	32.17 lbm · ft/lbf · s²	1 kg · m/N · s²

Figure 15.2   Section through a restriction in a pipe or tube

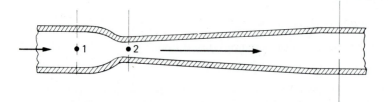

---

* Bernoulli's equation applies to lossless flow along a streamline, expressed in terms of local velocity. It may be applied approximately in terms of the average velocity in a duct, so long as the losses are negligible.

As written here, the relationship assumes that there is no mechanical work done on or by the fluid and that there is no heat transferred to or from the fluid as it passes between points 1 and 2, This equation provides the basis for evaluating the operation of flow-measuring devices generally classified as *obstruction meters* and of velocity sensors classified as *pressure probes*.

15.3 Obstruction Meters

Figure 15.3 shows three common forms of obstruction meters: the venturi, the flow nozzle, and the orifice. In each case the basic meter acts as an obstacle placed in the path of the flowing fluid, causing localized changes in velocity.

Figure 15.3   (a) A venturi, (b) a flow nozzle, (c) an orifice flowmeter

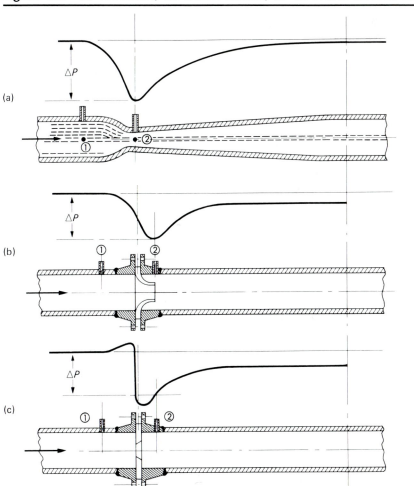

Concurrently with velocity change, there will be pressure change, as illustrated in the figure. At points of maximum restriction, hence maximum velocity, minimum pressures are found. A certain portion of this pressure drop becomes irrecoverable owing to dissipation of kinetic energy; therefore, the output pressure will always be less than the input pressure. This is indicated in the figure, which shows the venturi, with its guided reexpansion, to be the most efficient. Losses of about 30%–40% of the differential pressure occur through the orifice meter.

15.3.1   Obstruction Meters for Incompressible Flow

For *incompressible fluids,*

$$\rho_1 = \rho_2 = \rho \quad \text{and} \quad Q = A_1 V_1 = A_2 V_2,$$

where

$$Q = \text{volume/unit time},$$

$$A = \text{area}.$$

If we let $Z_1 = Z_2$ and substitute $V_1 = (A_2/A_1)V_2$ in Eq. (15.3), we obtain

$$P_1 - P_2 = \frac{V_2^2 \rho}{2g_c} \left[1 - \left(\frac{A_2}{A_1} \right)^2 \right] \tag{15.4}$$

and

$$Q_{\text{ideal}} = A_2 V_2 = \left[\frac{A_2}{\sqrt{1 - (A_2/A_1)^2}} \right] \sqrt{\frac{2g_c(P_1 - P_2)}{\rho}}. \tag{15.5}$$

For a given meter, A_1 and A_2 are established values, and it is often convenient to calculate

$$E = \frac{1}{\sqrt{1 - (A_2/A_1)^2}}. \tag{15.5a}$$

For circular sections, the area $= \pi(\text{diameter})^2/4$; hence

$$E = \frac{1}{\sqrt{1 - \beta^4}} \tag{15.5b}$$

where

$$\beta = \frac{d}{D}$$

and

$$D = \text{the larger diameter},$$

$$d = \text{the smaller diameter}.$$

We call E the *velocity of approach factor.*

Two additional factors used in obstruction meter calculations are the *discharge coefficient, C,* and the *flow coefficient, K.* These are defined as follows:

$$C = \frac{Q_{\text{actual}}}{Q_{\text{ideal}}} \tag{15.5c}$$

and

$$K = CE = \frac{C}{\sqrt{1 - \beta^4}}. \tag{15.5d}$$

The discharge coefficient C is the factor that accounts for losses through the meter, and the flow coefficient K is used as a matter of convenience, combining the loss factor with the meter constants.

Therefore, we may write

$$Q_{\text{actual}} = KA_2 \sqrt{\frac{2g_c}{\rho}} \sqrt{P_1 - P_2}. \tag{15.5e}$$

15.3.2   Venturi Characteristics

Venturi proportions are not standardized; however, the dimensional ranges shown in Fig. 15.4 include most cases. They are high-efficiency devices with discharge coefficients falling within a narrow range, depending on the finish of

Figure 15.4   Recommended proportions of Herschel-type venturi tubes (Source: ASME, Fluid Meters, 6th ed., 1971)

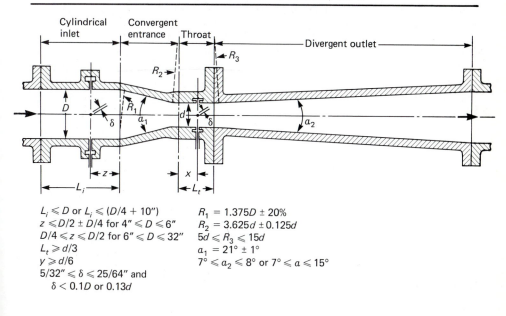

$L_i \leqslant D$ or $L_i \leqslant (D/4 + 10'')$
$z \leqslant D/2 \pm D/4$ for $4'' \leqslant D \leqslant 6''$
$D/4 \leqslant z \leqslant D/2$ for $6'' \leqslant D \leqslant 32''$
$L_t \geqslant d/3$
$y \geqslant d/6$
$5/32'' \leqslant \delta \leqslant 25/64''$ and
  $\delta < 0.1D$ or $0.13d$

$R_1 = 1.375D \pm 20\%$
$R_2 = 3.625d \pm 0.125d$
$5d \leqslant R_3 \leqslant 15d$
$a_1 = 21° \pm 1°$
$7° \leqslant a_2 \leqslant 8°$ or $7° \leqslant a \leqslant 15°$

Figure 15.5 Dimensional relations for ASME long-radius flow nozzles
(Source: ASME, Fluid Meters, 6th ed., 1971)

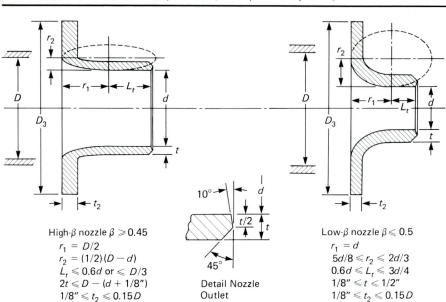

High-β nozzle $\beta \geqslant 0.45$
$r_1 = D/2$
$r_2 = (1/2)(D - d)$
$L_t \leqslant 0.6d$ or $\leqslant D/3$
$2t \leqslant D - (d + 1/8")$
$1/8" \leqslant t_2 \leqslant 0.15D$

Detail Nozzle
Outlet

Low-β nozzle $\beta \leqslant 0.5$
$r_1 = d$
$5d/8 \leqslant r_2 \leqslant 2d/3$
$0.6d \leqslant L_t \leqslant 3d/4$
$1/8" \leqslant t \leqslant 1/2"$
$1/8" \leqslant t_2 \leqslant 0.15D$

the entrance cone. For the venturi [2],

$$0.95 < C < 0.98.$$

15.3.3 Flow-Nozzle Characteristics

Figure 15.5 illustrates examples of two "standard" types of flow nozzles. The approach curve must be proportioned to prevent separation between the flow and the wall, and the parallel section is used to ensure that the flow fills the throat. The usual range of discharge coefficients is shown in Fig. 15.6. In addition, the following empirical equation is listed in Ref. [2] for evaluating the discharge coefficient:

$$C = 0.99622 + 0.00059D - \frac{(6.36 + 0.13D - 0.24\beta^2)}{\mathrm{Re}_d}. \tag{15.6}$$

15.3.4 Orifice Characteristics

The primary variables in the use of flat-plate orifices are the ratio of orifice to pipe diameter, tap locations, and characteristics of orifice sections. Various configurations of bevel and rounded edges are used in seeking particular performance characteristics, especially constant coefficients at low Reynolds numbers. Figure 15.7 illustrates typical orifice installations. Three tap locations

Figure 15.6   Range of discharge coefficients for long-radius flow nozzles

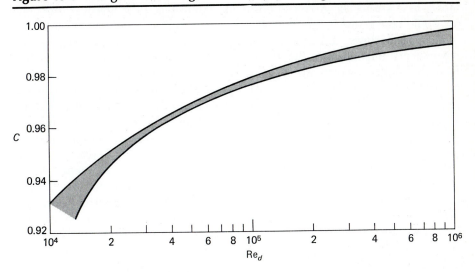

Figure 15.7   Locations of pressure taps for use with concentric, thin-plate,
square-edge orifices (Source: ASME,
Fluid Meters, 6th ed., 1971)

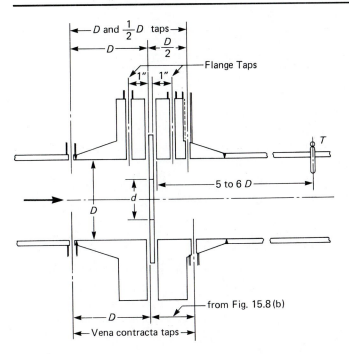

are indicated: (1) flange taps, (2) "1D" and "½D" taps, and (3) vena contracta taps. These are all shown in composite fashion in Fig. 15.7; however, only one set would be used for a given installation.

As fluid flows through an orifice, the necessary transverse velocity components imparted to the fluid as it approaches the obstruction carry through to the downstream side. As a result, the minimum stream section occurs not in the plane of the orifice, but somewhat downstream, as shown in Fig. 15.8(a). The term *vena contracta* is applied to the location and conditions of this minimum stream dimension. This is also the location of minimum pressure; hence it explains the interest in the vena contracta tap location. A guide for the location of the vena contracta is given in Fig. 15.8(b).

Figure 15.9 shows ranges for typical orifice flow coefficients versus Reynolds number, based on the diameter d. The dashed lines are loci of average values for the diameter ratios indicated. For example, the values for $\beta = 0.5$ range on either side of the $\beta = 0.5$ plot, depending on the tap locations and particular characteristics. It is important to note that this figure is

Figure 15.8   (a) Diagram illustrating vena contracta location for an orifice,
(b) guide for locating vena contracta as
measured from orifice face

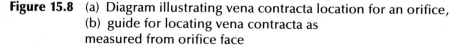
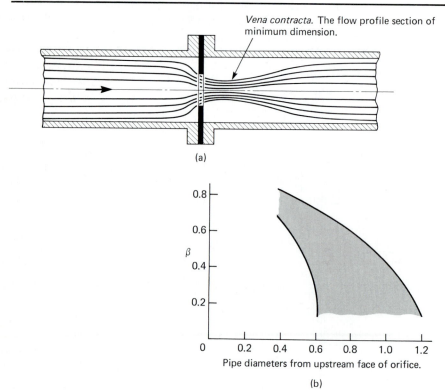

Figure 15.9 Range of flow coefficients for flat-plate orifices

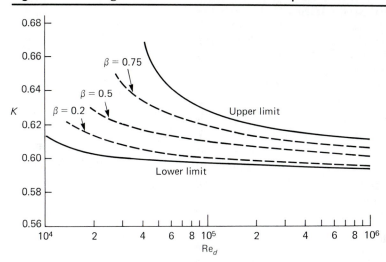

not intended for precise flow coefficient predictions. Rather, it displays the *ranges* within which the values may be expected to fall. The figure may be used for estimates; however, for more precise values the reader is directed to the Suggested Readings at the end of the chapter. In any case, however, the experimenter should be aware that accurate work *requires* careful calibration of each installation (See also Sections 3.11.2 and 15.12).

An orifice plate is vulnerable to damage caused by pressure surges, entrained debris, and the like. An estimate of the maximum stress due to differential pressure may be found from the following. The relationship is adapted from a rather complex equation [3] and assumes Poisson's ratio = 0.3.

$$\sigma_{\max} = \frac{FD^2 \, \Delta P}{t^2} \tag{15.7}$$

where

$\sigma_{\max}$ = maximum normal stress (radial direction at the clamped edge),

t = plate thickness,

ΔP = differential pressure across the plate,

F = a factor, the value of which may be estimated from the following:

β	0.2	0.3	0.4	0.5	0.6	0.7	0.8
F	0.18	0.17	0.15	0.12	0.09	0.06	0.04

15.3.5   Relative Merits of the Venturi, Flow Nozzle, and Orifice

High accuracy, good pressure recovery, and resistance to abrasion are the primary advantages of the venturi. They are offset, however, by considerably greater cost and space requirements than with the orifice and nozzle. The orifice is inexpensive, and may often be installed between existing pipe flanges. However, its pressure recovery is poor, and it is especially susceptible to inaccuracies resulting from wear and abrasion. It may also be damaged by pressure transients because of its lower physical strength. The flow nozzle possesses the advantages of the venturi, except that it has lower pressure recovery, and it has the added advantage of shorter physical length. It is expensive compared with the orifice and is relatively difficult to install properly.

Example 15.1

A venturi designed according to the specifications of Fig. 15.4 is placed in an 8-in.-diameter (20.32-cm) line passing 500 gal (1.893 m³) of water per minute. If the throat diameter is 4 in. (10.16 cm), what differential pressure may be expected across the pressure taps? The water temperature is 70°F (21.1°C).

Solve the problem (a) using the English engineering system of units and (b) using the SI system; and finally, (c) reconcile the two answers. Use the tables in Appendix D for the properties of water.

Solution.

a.
$$A_1 = 0.349 \text{ ft}^2,$$

$$A_2 = 0.087 \text{ ft}^2,$$

$$Q = 500 \text{ gal/min} = 1.114 \text{ ft}^3/\text{s}.$$

For water at 70°F,

$$\rho = 62.3 \text{ lbm/ft}^3,$$

$$\beta = \frac{4}{8} = 0.5,$$

$$E = \frac{1}{\sqrt{1 - \beta^4}} = 1.033.$$

From Section 15.3.2, estimate $C = 0.97$:

$$K = CE = 0.97 \times 1.033 = 1.002.$$

Rewriting Eq. (15.5e), solving for $P_1 - P_2$, we have

$$P_1 - P_2 = \left(\frac{Q}{KA_2}\right)^2\left(\frac{\rho}{2g_c}\right)$$

$$= \left[\frac{1.114}{1.002 \times 0.087}\right]^2\left[\frac{62.3}{2 \times 32.17}\right]$$

$$= 158.13 \text{ lbf/ft}^2 = 1.098 \text{ lbf/in.}^2$$

b. $\qquad A_1 = 0.0324 \text{ m}^2,$

$\qquad\qquad A_2 = 0.0081 \text{ m}^2,$

$\qquad\qquad Q = 1.893 \text{ m}^3/\text{min} = 0.0316 \text{ m}^3/\text{s}.$

For water at 21.1°C,

$$\rho = 997.9 \text{ kg/m}^3,$$

$$K = 1.002,$$

as before. Substituting, we have

$$P_1 - P_2 = [(0.0316/(1.002 \times 0.0081)]^2[997.9/(2 \times 1)]$$

$$= 7563.6 \text{ Pa}.$$

c. From Appendix A, we find that 1 psi = 6894.8 Pa; thus converting the answer for part (b), we obtain

$$P_1 - P_2 = \frac{7563.6}{6894.8} = 1.097 \text{ psi}.$$

(See Problem 15.11.)

15.4 Obstruction Meters for Compressible Fluids

When compressible fluids flow through obstruction meters of the types discussed in Section 15.3, the density does not remain constant during the process; that is, $\rho_1 \neq \rho_2$. The usual practice is to base the energy relation, Eq. (15.3), on the density at condition 1 (Fig. 15.2) and to introduce an *expansion factor, Y,* as follows:

$$W = KA_2Y\sqrt{2g_c\rho_1(P_1 - P_2)}, \qquad\qquad (15.8)$$

where W = the mass flow rate.

The expansion factor, Y, may be determined theoretically for gases flowing through nozzles and venturis and experimentally for gases in orifice meters. For nozzles and venturis values may be calculated from the following

Figure 15.10   (a) Expansion factors for venturis and nozzles ($k = 1.4$),
                   (b) expansion factors for square-edged concentric orifices
                   (Source: ASME, Fluid Meters, 6th ed., 1971)

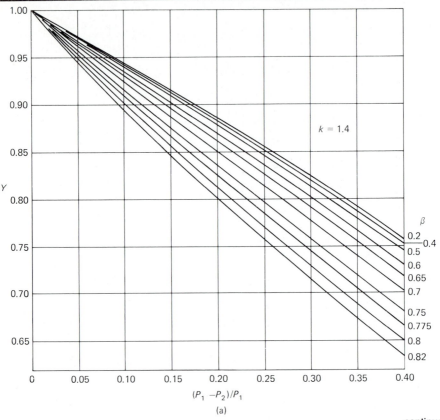

(a)

continued

relation [2]:

$$Y = \left[\left(\frac{P_2}{P_1}\right)^{2/k}\left(\frac{k}{k-1}\right)\left(\frac{1-(P_2/P_1)^{(k-1)/k}}{1-(P_2/P_1)}\right)\left(\frac{1-\beta^4}{1-\beta^4(P_2/P_1)^{2/k}}\right)\right]^{1/2}, \quad \textbf{(15.8a)}$$

in which

$$k = \frac{\text{specific heat at constant pressure}}{\text{specific heat at constant volume}}.$$

For square-edged orifices, an empirical relation has been developed that is expressed as follows [2]:

$$Y = 1 - [0.41 + 0.35\beta^4]\left[\frac{P_1 - P_2}{kP_1}\right]. \qquad \textbf{(15.8b)}$$

Figure 15.10   *continued*

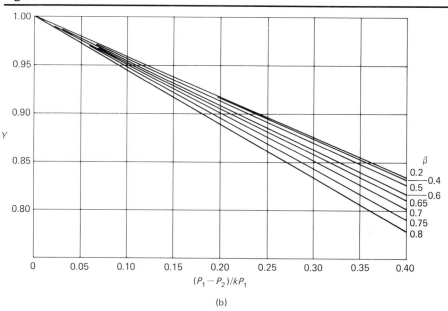

(b)

Expansion factors, Y, for venturis and nozzles with $k = 1.4$ are shown plotted against pressure ratio in Fig. 15.10(a). Similar values for square-edged orifices are given in Fig. 15.10(b).

Example 15.2
A sharp-edged orifice is used to measure flow of 30°C air through a 0.25 m diameter circular-sectioned duct. Estimate the flow rate if the differential pressure between vena contracta taps is 20 cm of 0.92 specific gravity oil. The upstream pressure P_1 is 4×10^5 Pa(abs) and $\beta = 0.6$.

Solution.

$$A_1 = 0.0491 \text{ m}^2, \qquad A_2 = 0.0177 \text{ m}^2$$

From Fig. 15.9 we may estimate $C \approx 0.6$. (*Note:* At this point we are unable to determine Re$_d$.)

$$E = \frac{1}{\sqrt{1 - \beta^4}} = 1.0719$$

$$K = CE = 0.643$$

From Table D.2 in Appendix D, we have

$$\rho = \rho_{atmos}\left(\frac{P}{P_{atmos}}\right) = 1.14[(4 \times 10^5)/(1.01325 \times 10^5)]\,kg/m^3$$

$$= 4.5\,kg/m^3,$$

$$g_c = 1\,kg \cdot m/N \cdot s^2.$$

Using Eq. (14.1a), we know that

$$P_1 - P_2 = h\rho_0\left(\frac{g}{g_c}\right),$$

where ρ_0 is the density of manometer oil. From Table D.1 in Appendix D, we find that the density of water at 30°C is 995.7 kg/m^3 and

$$P_1 - P_2 = \left(\frac{20}{100}\right)(0.92 \times 995.7)\left(\frac{9.807}{1}\right) = 1796.7\,Pa.$$

To determine Y from Fig. 15.10(b) with $k = 1.4$, we have

$$\frac{P_1 - P_2}{kP_1} = \frac{1796.7}{1.4 \times 4 \times 10^5} = 0.0032.$$

Entering Fig. 15.10(b) gives us $Y \approx 1$, and using Eq. (15.8), we have

$$W = 0.643 \times 0.0177 \times 1 \times \sqrt{2 \times 1 \times 4.5 \times 1796.7}$$

$$= 1.45\,kg/s.$$

The value of Re$_d$ may now be calculated and the accuracy of our estimate of C determined.

$$\text{Volume flow rate} = \frac{W}{density} = \frac{1.45}{4.5} = 0.32\,m^3/s,$$

$$\text{Velocity} = \frac{\text{volume flow rate}}{area} = \frac{0.32}{0.0177} = 18.2\,m/s,$$

$$\text{Reynolds number} = \frac{\rho DV}{\mu g_c}$$

$$= (4.5)(0.25 \times 0.6)(18.2)/(1.85 \times 10^{-5})$$

$$= 6.64 \times 10^5.$$

Referring to Fig. 15.9, we have $C \approx 0.61$. This compares reasonably with our assumed value of 0.6, and greater refinement is not warranted.

Choked Flow

One particular issue necessitates caution when considering isentropic flow of compressible fluids. As flow rates are increased, the flow velocity through a constriction eventually reaches the sonic velocity for the existing conditions

(Mach number = 1.0). The condition of Mach number equal to unity at a constriction such as the throat of a nozzle is called "choked-flow" [4]. When this occurs, variations in pressure downstream of the constriction or throat, as shown in Fig. 15.3, no longer influence the mass flow rate. One might interpret this condition by stating that once a Mach number of unity is reached at the throat, the effects of exit pressure variations can no longer be propagated back upstream.

The *critical pressure ratio* for the choked-flow condition is expressed in terms of the static upstream and throat pressures as

$$\frac{P_2}{P_1} = \left(\frac{1+k}{2}\right)^{k/(1-k)} \tag{15.9}$$

This ratio can be interpreted as being the limiting ratio for achieving sonic flow in a nozzle; that is, so long as the pressure ratio P_2/P_1 is greater than the value given in Eq. (15.9), the flow may be predicted by Eq. (15.8). However, when the pressure ratio given by Eq. (15.9) is reached, the flow is choked and the mass flow rate cannot be increased by lowering the pressure ratio further. For air, $k = 1.4$ and this ratio becomes 0.528.

15.5   Predictability of Obstruction Meter Performance

The foregoing discussion of obstruction meters is based to a great extent on practical information assembled and published by the American Society of Mechanical Engineers (see Suggested Readings at the end of this chapter). Tables and charts of coefficients for various and precisely prescribed metering methods have evolved through the years from accumulated experiences of many different people and interested commercial and research organizations. This material probably forms as useful and reliable a guide to flow measurements as is available, and when diligently adhered to, yields satisfactory and practical results. However, a "limit of accuracy" must be applied to the values of any generally expressed discharge or flow coefficients or expansion factors.

Table 15.1 Tolerance ranges for theoretical flow measurement coefficients and factors

Type of Element	Tolerances in Percent Coefficient or Factor	
	For Discharge or Flow Coefficients	For Expansion Factors
Venturi tubes	$\pm\frac{3}{4}$ to $1\frac{1}{2}$	± 0.2 to ± 1.2
Flow nozzles, long radius	± 1 to ± 2	± 0.2 to ± 1.2
Square-edge concentric orifices	± 1 to $\pm 5^*$	$\pm 1\frac{1}{2}$

* Larger tolerances correspond to smaller sizes and lower values of Re_d.

The tolerance ranges in Table 15.1 have been abstracted from Ref. [2] and apply to coefficient and factor data presented therein. From this tabulation we see that the coefficients may be quite accurately predicted. Of course, it must be remembered that these figures do not include errors of observation, errors of subsequent signal conditioning apparatus, etc. For more accurate work, calibration of individual meters along with associated instrumentation is required.

15.6   The Variable-Area Meter

A major disadvantage of the common forms of obstruction meters (the venturi, the orifice, and the nozzle) is that the pressure drop varies as the square of the flow rate [Eq. (15.5e)]. This means that if these meters are to be used over a wide range of flow rates, pressure-measuring equipment of very wide range will be required. In general, if the pressure range is accommodated, accuracy at low flow rates will be poor: the small pressure readings in that range will be limited by pressure transducer resolution. One solution would be to use two (or more) pressure-measuring systems: one for low flow rates and another for high rates.

Figure 15.11   Variable-area flowmeter

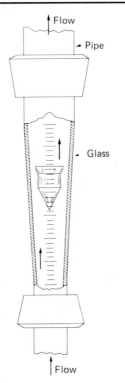

A device whose indication is essentially *linear* with flow rate is shown in section in Fig. 15.11. This instrument is a variable-area meter, commonly called a *rotameter*. Two parts are essential, the float and the tapered tube in which the float is free to move. The term *float* is somewhat a misnomer in that it must be heavier than the liquid it displaces. As flow takes place upward through the tube, four forces act on the float: a downward gravity force, an upward buoyant force, and pressure and viscous drag forces.

For a given rate of flow, the float assumes a position in the tube where the forces acting on it are in equilibrium. Through careful design, the effects of changing viscosity or density may be minimized, leaving only the pressure force as a variable. The latter is dependent on flow rate and the annular area between the float and the tube. Hence its position will be determined by the flow rate alone. A basic equation for the rotameter has been developed [5] in the following form:

$$Q = A_w C \left[\frac{2g v_f (\rho_f - \rho_w)}{A_f \rho_w} \right]^{1/2}, \tag{15.10}$$

where

Q = the volumetric rate of flow,

v_f = the volume of the float,

g = acceleration due to gravity,

ρ_f = the float density,

ρ_w = the liquid density,

A_f = the area of the float,

C = the discharge coefficient,

A_w = the area of the annular orifice

$\quad = \left(\frac{\pi}{4} \right) [(D + by)^2 - d^2],$

D = the diameter of the tube when the float is at the zero position,

b = the change in tube diameter per unit change in height,

d = the maximum diameter of the float,

y = the height of the float above zero position.

Normally, the values of D, b, and d will be selected to produce an essentially linear variation of A_w with y. Thus, the flow rate is a *linear* function of the reading, in contrast to the obstruction meters. Because the rotameter's response is linear, its resolution is the same at both high and low flow rates.

Certain disadvantages of the rotameter as compared with the other, constant-area obstruction meters are: the meter must be installed in a vertical position; the float may not be visible when opaque fluids are used; it cannot be used with liquids carrying large percentages of solids in suspension; for high pressures or temperatures, it is expensive. Advantages include the following: There is a uniform flow scale over the range of the instrument, with the pressure loss fixed at all flow rates; the capacity may be changed with relative ease by changing float and/or tube; many corrosive fluids may be handled without complication; the condition of flow is readily visible. Rotameters are typically accurate to between 1 and 10% of full scale.

15.7   Additional Flowmeters

The preceding discussion covers the more common types of flowmeters, but many additional types are in use. In this article, we consider a few of the more important ones. Note that each of the following devices produces an output that is linear in flow rate.

15.7.1   Turbine-Type Meters

The familiar anemometer used by weather stations to measure wind velocity is a simple form of free-steam turbine meter. Somewhat similar rotating-wheel flowmeters are used by civil engineers to measure water flow in rivers and streams [6]. Both the cup-type rotors and the propeller types are used for this purpose. In each case the number of turns of the wheel per unit time is counted and used as a measure of the flow rate.

Figure 15.12   Turbine-type flowmeter

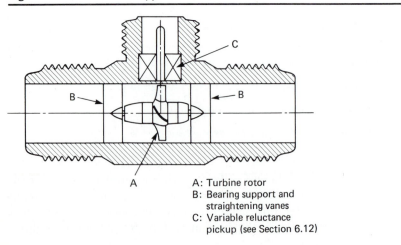

A: Turbine rotor
B: Bearing support and
   straightening vanes
C: Variable reluctance
   pickup (see Section 6.12)

Figure 15.12 illustrates a modern adaptation of these methods to the measure of flow in tubes and pipes. Rotor motion, proportional to flow rate, is sensed by a reluctance-type pickup coil. A permanent magnet is encased in the rotor body, and each time a rotor blade passes the pole of the coil, change in permeability of the magnetic circuit produces a voltage pulse at the output terminal. The pulse rate may then be indicated by a frequency meter, displayed on a CRO screen, or counted by some form of EPUT meter. Frequency converters are also available that convert flowmeter pulses to a proportional dc output, permitting use of simple meters for indication. Accuracies within $\pm\frac{1}{2}\%$ are claimed for these devices within a specific flow range. Available sizes cover the range from $\frac{1}{8}$ to 8 in. Transient response is good; time constants for stepped pressure pulses are on the order of 2 to 12 ms [7].

In addition to bearing maintenance, a major problem inherent in this type of meter is reduced accuracy at low flow rates. Maximum to minimum capacities vary from about 8 to 1 for small meters to about 40 to 1 for the large sizes.

15.7.2 Magnetic Flowmeters

Magnetic flowmeters are based on Faraday's law of induced voltage for a conductor moving through a magnetic field, expressed by the relation [8]

$$e = BDV \times 10^{-8}, \tag{15.11}$$

where

$e = $ the induced voltage, in volts,

$B = $ the magnetic flux density, in gauss,

$D = $ the length of the conductor, in cm,

$V = $ the velocity of the conductor, in cm/s.

Basic flowmeter arrangement is as shown in Fig. 15.13. The flowing medium is passed through a pipe, a short section of which is subjected to a transverse magnetic flux. The fluid itself acts as the conductor having dimension D equal to pipe diameter and velocity V roughly equal to the average fluid velocity. Fluid motion relative to the field causes a voltage to be induced proportional to the fluid velocity. This emf is detected by electrodes placed in the conduit walls. Either an alternating or direct magnetic flux may be used. However, if amplification of the output is required, the advantage lies with the alternating field.

Two basic types have been developed. In the first, the fluid need be only slightly electrically conductive, and the conduit must be of glass or some similar nonconducting material. The electrodes are placed flush with the inner conduit surfaces making direct contact with the flowing fluid. Output voltage is quite low, and an alternating magnetic field is used for amplification and to

Figure 15.13   (a) Schematic arrangement of a magnetic flowmeter,
                   (b) section showing electrodes and magnetic field

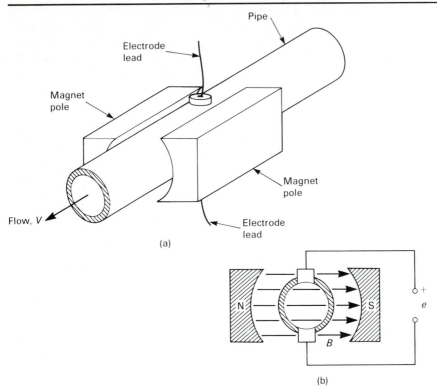

(a)

(b)

eliminate polarization problems. Special circuitry is required to separate the
no-flow output from the signal caused by flow, but most commercial units
incorporate it [9].

A second form of magnetic flowmeter is primarily intended for use with
highly conductive fluids such as liquid metals. This meter operates on the same
basic principle but may use electrically conducting materials for the conduit.
Stainless steel is commonly used. A permanent magnet supplies the necessary
flux, and the electrodes may be simply attached to diametrically opposite
points on the *outside* of the pipe. This arrangement provides for easy
installation at any time and at any point along the pipe. The output of this type
is sufficient to drive ordinary commercial indicators or recorders, and zero
output for nonflow conditions is an added advantage [10].

Commercial magnetic flowmeters have rated accuracies of 0.5% to 1%.
They are particularly useful for corrosive liquids and slurries.

15.7.3   Pulse-Producing Methods

The advent of digital electronics has increased the need for flow-measuring devices that provide pulsed outputs. The turbine meter (Fig. 15.12) produces pulses proportional to flow rate; however, its continuously moving rotor subjects it to mechanical maintenance problems. Several meters that circumvent this disadvantage have been developed.

Vortex shedding meters are based on the fact that when a bluff body is placed in a stream, vortices are alternately formed, first to one side of the obstruction and then to the other (Fig. 15.14). The frequency of formation, f, is a function of flow rate. For incompressible fluids [11],

$$f = \left(\frac{\text{St}}{D}\right)V, \qquad (15.12)$$

where

St = a calibration constant called the Strouhal number,

V = the flow velocity,

D = the dimension of the obstruction transverse to the flow direction.

For compressible fluids the relation is more complicated, because St is a function of the Reynolds number.

Various schemes are used to sense the frequency of vortex formation. The obstructing body may be mounted on an elastic support and the support oscillation sensed by one of a number of means. Heated thermistors downstream, with one to each side of the obstruction, and the flexing of diaphragms have been used; fluctuation in resistance is the output parameter. A patented sensing method makes use of an ultrasonic beam that is amplitude modulated by pulses. In another scheme, a differential piezoelectric pressure transducer is implanted in the rear of the bluff body. Commercially available vortex shedding meters are accurate to about 1% of flow rate.

Figure 15.14   Vortex shedding caused by bluff body in flow stream

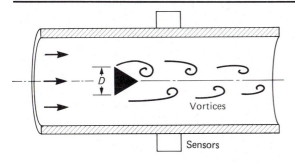

Yet another approach is the Coanda effect, which is the tendency for a fluid jet to remain in contact with a guiding surface, once contact is established. The effect can be used to produce a hydraulic or pneumatic flip-flop analogous to the electronic counterpart. If designed for self-oscillation, the action indicates a measure of flow rate.

15.8   Measurements of Fluid Velocities

Flow rate is generally proportional to some flow velocity; hence, by measuring the velocity, a measure of flow rate is obtained. Often velocity per se is desired, particularly velocity *relative* to a fluid. An example of the latter is an aircraft moving through the air. The following sections deal with measurement of absolute and relative velocities of fluids. The instruments considered generally measure local velocity (rather than an average velocity), which we shall refer to as V for simplicity of notation.

15.9   Pressure Probes

Point measurement of pressure is accomplished by the use of tubes joining the location in question with some form of pressure transducer. A *probe* or sampling device is intended insofar as possible to obtain a reliable and interpretable indication of the pressure at the signal source. Therein lies a difficulty, however, for the mere presence of the probe will alter, to some extent, the quantity being measured.

A common reason for desiring point-pressure information is to determine flow conditions. Flowing media may be gaseous or liquid in a symmetrical conduit or pipe or in a more complex situation such as a jet engine or compressor.

There are many different types of pressure probes, with the selection depending on the information required, space available, pressure gradients, and constancy of flow magnitude and direction. Basically, pressure probes measure either of two different pressures or some combination thereof (Fig. 15.15). In Section 14.2 we briefly discussed *static* and *total* pressures and indicated that the difference is a result of the flow velocity; that is,

$$P_t = P_s + P_v. \tag{15.13}$$

15.9.1   Incompressible Fluids

Referring to Eq. (15.3) for incompressible fluids, we may write

$$P_t = P_s + \frac{\rho V^2}{2g_c},$$

Figure 15.15   Total and static pressure probes

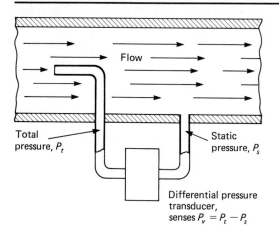

in which

P_t = the total pressure (often called *stagnation* pressure),

P_s = the static pressure,

P_v = the velocity pressure,

ρ = the fluid density (static),

V = the velocity,

g_c = the conversion constant.

Solving for velocity, we obtain

$$V = \sqrt{\frac{2g_c(P_t - P_s)}{\rho_s}} = \sqrt{\frac{2g_c(\Delta P)}{\rho_s}}. \tag{15.14}$$

From this we see that velocity may be determined simply by measuring the difference between the total and static pressures.

15.9.2   Compressible Fluids

For flow of a compressible fluid an isentropic compression occurs at the probe tip as the pressure changes from P_s to P_t. Under these conditions [2, 12],

$$V = \sqrt{2\left(\frac{k}{k-1}\right)\left(\frac{P_s}{\rho_s}\right)\left[\left(\frac{P_t}{P_s}\right)^{(k-1)/k} - 1\right]g_c}$$

$$= \sqrt{2\left(\frac{k}{k-1}\right)\left(\frac{P_s}{\rho_s}\right)\left[\left(1 + \frac{\Delta P}{P_s}\right)^{(k-1)/k} - 1\right]g_c}, \tag{15.14a}$$

where k = the ratio of specific heats = c_p/c_v and $\Delta P = P_t - P_s$.

When velocity is used to measure flow rate, consideration must be given to the velocity distribution across the channel or conduit. A mean may be found by traversing the area to determine the velocity profile, from which the average may be calculated, or a multiplication constant may be determined by calibration (see Problem 15.33).

15.9.3   Total-Pressure Probes

Obtaining a measure of total or impact pressure is usually somewhat easier than getting good measure of static pressure, except in cases such as an open jet, when a barometer reading may be used for the static component. The simple Pitot tube (named for Henri Pitot) shown at A in Fig. 14.2 is usually adequate for determining impact pressure. More often, however, the Pitot tube is combined with static openings, constructed as shown in Fig. 15.16. This is known as a *Pitot-static tube,* or sometimes as a Prandtl-Pitot tube. For

Figure 15.16   A Pitot-static tube

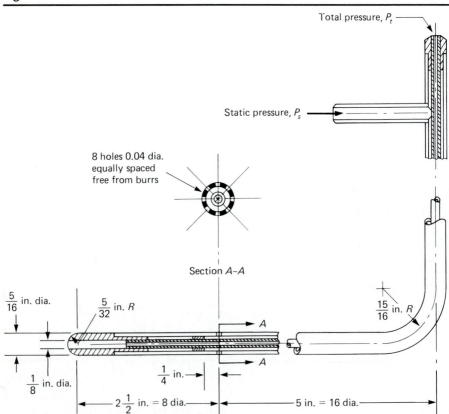

steady-flow conditions, a simple differential manometer, often of the inclined type, suffices for pressure measurement, and $P_t - P_s$ is determined directly. When variable conditions exist, some form of pressure transducer, such as one of the diaphragm types, may be used. Of course, care must be exercised in providing adequate response, particularly in the connecting tubing (see Section 14.12).

A major problem in the use of an ordinary Pitot-static tube is to obtain proper alignment of the tube with flow direction. The angle formed between the probe axis and the flow streamline at the pressure opening is called the *yaw angle*. This angle should be zero, but in many situations it may not be constant: The flow may be fixed neither in magnitude nor in direction. In such cases, yaw sensitivity is very important. The Pitot-static tube is particularly sensitive to yaw, as shown in Fig. 15.17. Although sensitivity is influenced by orientation of both impact and static openings, the latter probably has the greater effect.

The Kiel tube, designed to measure total or impact pressure only (there are no static openings), is shown in Fig. 15.18. It consists of an impact tube surrounded by what is essentially a venturi. The curve demonstrates the

Figure 15.17   Yaw sensitivity of a standard Pitot-static tube (Courtesy: The Airflo Instrument Company, Glastonbury, CT)

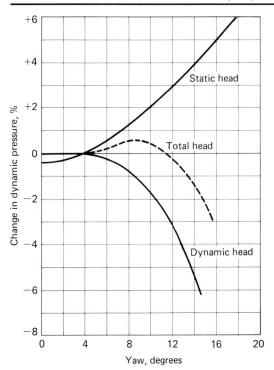

Figure 15.18 Kiel-type total-pressure tube and plot of yaw sensitivity
(Courtesy: The Airflo Instrument Company, Glastonbury, CT)

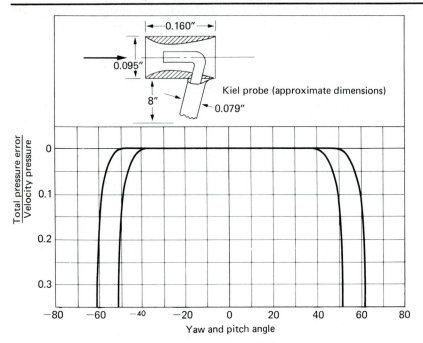

striking insensitivity of this type to variations in yaw. Modifications of the Kiel tube make use of a cylindrical duct, beveled at each end, rather than the streamlined venturi. This appears to have little effect on the performance and makes the construction much less expensive.

15.9.4 Static-Pressure Probes

Static-pressure probes have been used in many different forms [13]. Ideally, the simple opening with axis normal to flow direction should be satisfactory. However, slight burrs or yaw introduce appreciable errors. As mentioned previously, in many situations yaw angle may be continually changing. For these reasons, special static-pressure probes may be used. Figure 15.19 shows several probes of this type and corresponding yaw sensitivities.

As mentioned earlier, the mere presence of the probe in a pressure-flow situation alters the parameters to be measured. Probes interact with other probes, with their own supports, and with duct or conduit walls. Such interaction is primarily a function of geometry and relative dimensional proportions; it is also a function of Mach number. Much work has been conducted in this area, and the interested reader is referred to Refs. [14] and [15] for a review of some of these efforts.

Figure 15.19   Angle characteristics of certain static-pressure-sensing elements (Courtesy: Instrument Society of America, Research Triangle Park, NC)

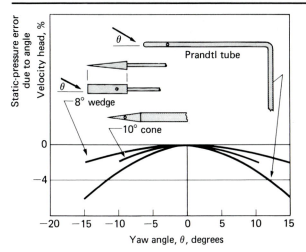

Figure 15.20   Special direction-sensing elements and their yaw characteristics (Courtesy: Instrument Society of America, Research Triangle Park, NC)

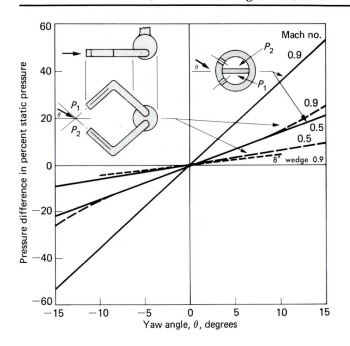

15.9.5 Direction-Sensing Probes

Figure 15.20 illustrates two forms of direction-sensing or yaw-angle probes. Each of these probes uses two impact tubes. In each case the probe is placed transverse to flow and is rotatable around its axis. The angular position of the probe is then adjusted until the pressures sensed by the openings are equal. When this is the case, the flow direction will correspond to the bisector of the angle between the openings. Probes are also available with a third opening midway between the other two. The additional hole, when properly aligned, senses maximum impact pressure.

Example 15.3

A Pitot-static tube is used to determine the velocity of air at the center of a pipe. Static pressure is 18 psia (124,106 Pa), the air temperature is 80°F (26.7°C), and a differential pressure of 3.8 in. of water (9.65 cm) is measured. What is the air velocity? Perform the calculations using (a) the English system of units and (b) the SI system.

Solution. (a) Using the English engineering system of units, we find from Table D.2 that $\rho_{80} = 0.0735 \, \text{lbm/ft}^3$ at a pressure of 14.7 psia. At 18 psia,

$$\rho_{80} = (0.0735)(18/14.7) = 0.090 \, \text{lbm/ft}^3,$$

$$\Delta P = (3.8 \times 144)/(12 \times 2.31) = 19.74 \, \text{lbf/ft}^2,$$

$$P_s = 18 \times 144 = 2592 \, \text{lbf/ft}^2,$$

$$P_t = P_s + \Delta P = 2592 + 19.74 = 2612 \, \text{lbf/ft}^2,$$

$$k = 1.4.$$

Substituting in Eq. (15.14a) gives us

$$V = \sqrt{2(1.4/0.4)(2592/0.09)[(2612/2592)^{(0.4/1.4)} - 1] \times 32.17}$$

$$= 119 \, \text{ft/s (or 36.4 m/s)}.$$

(b) Using the SI system of units, we find from Table D.2 that $\rho_{26.7} = 1.153 \, \text{kg/m}^3$ at standard atmospheric pressure. At 124 kPa,

$$\rho_{26.7} = (124/101.35) \times 1.153 = 1.41 \, \text{kg/m}^3,$$

$$1 \, \text{cm H}_2\text{O} = 98.06 \, \text{Pa} \; (\textit{Note:} \text{ This is true at 4°C. More accurate}$$
$$\text{calculations would require a correction.})$$

$$\Delta P = 9.65 \times 98.06 = 946.3 \, \text{Pa},$$

$$P_t = P_s + \Delta P = 124,106 + 946.3 = 125,052 \, \text{Pa}.$$

Using Eq. (15.14a), we have

$$V = \sqrt{2(1.4/0.4)(124,106/1.41)[(125,052/124,106)^{(0.4/1.4)} - 1] \times 1}$$
$$= 36.58 \, \text{m/s}.$$

Close scrutiny of the arithmetical manipulations required in this example clearly shows that calculation errors of considerable size may easily result from the fact that P_t and P_s are quite commonly of very nearly the same magnitudes: Calculation of the ratio must be quite precise. A much simpler approach would be to use Eq. (15.14) along with an appropriate multiplying coefficient, which commonly falls between 0.96 and 1. If we substitute the values for part (b) of the above example directly into Eq. (15.14) we obtain

$$V = \sqrt{2 \times 1 \times (946.3)/1.41} = 36.6 \, \text{m/s}.$$

For our particular example we see that essentially the same answer is obtained. (Problem 15.31 bears on this matter.) Of course, one should always be cautious when using shortcuts of this sort. It has been suggested that the practical use of Eq. (15.14) for compressible fluids and without correction be limited to velocities for which the applicable Mach number is less than 0.3 [16].

15.10 Thermal Anemometry

A heated object in a moving stream loses heat at a rate that increases with the fluid velocity. If the object is electrically heated at a known power, it will reach a temperature determined by the rate of cooling. Thus, its temperature will be a measure of the velocity. Conversely, the heating power may be controlled by a feedback system to hold the temperature constant. In that case, heating power is a measure of velocity. These relations are the basis for *thermal anemometry*.

The most well-known thermal velocity probes are the *hot-wire* and *hot-film anemometers*. The hot-wire anemometer consists of a fine wire supported by two larger-diameter prongs; an electric current heats the wire to a temperature well above the fluid temperature (Fig. 15.21). Typically, the wire is 4 to 10 μm in diameter, is 1 mm in length, and is made of platinum or tungsten. These fine wires are extremely fragile, so hot-wire probes are used only in clean gas flows. In liquids, or in rugged gas-flow applications, the hot-film probe is used instead. Here, a quartz fiber is suspended between the prongs, and a platinum film coated onto the fiber surface provides the electrically heated element. The fiber, with a diameter of 25 to 150 μm, has much greater mechanical strength than fine wire.

Hot-wire and hot-film probes are most often operated at constant temperature using a feedback-controlled bridge (Fig. 15.22). The probe forms one leg of a voltage-sensitive deflection bridge (Section 7.9). Current flowing

Figure 15.21   Two forms of hot-wire anemometer probes:
(a) wire mounted normal to probe axis,
(b) wire mounted parallel to probe axis

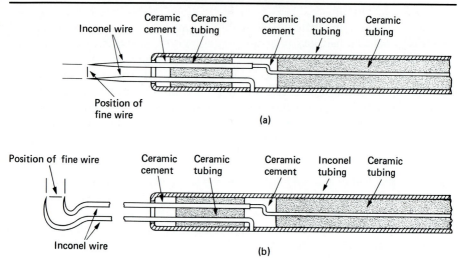

(a)

(b)

Figure 15.22   Constant-temperature-anemometer bridge circuit

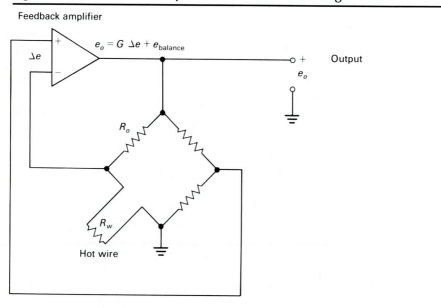

through the bridge provides heating power to the wire. The resistance of the wire is a function of temperature (Section 16.4.1), and any increase in flow velocity tends to lower the wire's temperature, reducing its resistance and causing bridge imbalance. The voltage imbalance drives a feedback amplifier, which increases the voltage and current supplied to the bridge; the added current increases the heating power, thus raising the wire temperature and resistance and restoring bridge balance. The voltage supplied to the bridge also serves as the circuit output. High-quality bridges are available commercially; alternatively, acceptable bridges can be built at minimal cost [17].

The relation between flow speed and bridge output is obtained by equating the electrical power to the heat loss. The heating power is

$$\text{Electrical power} = \frac{e_w^2}{R_w} = \left(\frac{R_w}{R_o + R_w} e_o\right)^2 \frac{1}{R_w}$$

$$= e_o^2\left(\frac{R_w}{(R_o + R_w)^2}\right)$$

where

$$e_w = \text{the voltage across the wire,}$$

$$R_w = \text{the wire resistance,}$$

$$R_o = \text{the upper leg's resistance,}$$

$$e_o = \text{the output voltage.}$$

The heat loss is largely by convection to the fluid: Radiation is negligible, and conduction to the supporting prongs is accounted for by calibration. This loss takes the following form:

$$\text{Heat loss} = A_w h(T_w - T_a)$$

where

$$A_w = \text{the wire surface area,}$$

$$T_w = \text{the wire temperature,}$$

$$T_a = \text{the ambient temperature,}$$

$$h = \text{convective heat transfer coefficient.}$$

The heat transfer coefficient for small-diameter wires is given by

$$h = A + B\sqrt{\rho V},$$

where

$$A \text{ and } B = \text{constants that depend on ambient temperature,}$$

$$\rho = \text{density of fluid,}$$

$$V = \text{velocity of fluid approaching the wire.}$$

Setting electric power equal to heat loss and solving, we obtain:

$$e_o^2 = \left(\frac{(R_o + R_w)^2}{R_w}(T_w - T_a)A_w\right)(A + B\sqrt{\rho V}).\qquad(15.15)$$

Since the bridge holds the wire temperature and resistance constant, the resistances and temperatures may be lumped into single constants, C and D:

$$e_o^2 = C + D\sqrt{\rho V}\qquad(15.15a)$$

This result is usually called *King's law* [18].

Hot wires and hot films must be calibrated before use. Typically, a side-by-side comparison to a secondary standard, such as a Pitot tube, is made at different flow speeds. The results are used to fit values of C and D.

Significant complications arise if the ambient temperature varies. Small changes in T_a may be corrected for by placing a fine-wire resistance thermometer (a "cold wire") adjacent to the hot wire and allowing the temperature difference in Eq. (15.15) to vary. Larger temperature changes necessitate further modification of King's law [19].

The primary value of the hot-wire anemometer lies in its high frequency response and excellent spatial resolution. The frequency response of a hot wire can easily reach 10 kHz; if the loss of spatial resolution [20] is unimportant, hot wires can be applied at frequencies even several times higher.* On the other hand, owing to their relatively high cost and inherent fragility, hot wires are usually justifiable only when their response characteristics are essential to the measurement at hand.

Thermal flowmeters are also available. In one form, a section of a pipe wall is electrically heated; the resulting increase in fluid temperature, downstream, is proportional to the mass-flow rate in the pipe. Automotive fuel-injection systems use a somewhat different thermal flowmeter as part of the microprocessor control system. A hot-film sensor is located in the intake manifold; this sensor is of relatively rugged construction, lowering its frequency response in favor of improved reliability. The hot-film signal identifies ρV; when the latter is multiplied by the manifold cross section, the total mass-flow rate is obtained. A temperature sensor is incorporated to compensate for changes in environmental conditions. The measured mass-flow rate enables the microprocessor to set the fuel injectors for the correct fuel mixture.

15.11   Scattering Measurements

Light or sound waves undergo a Doppler shift when they are scattered off of particles in a moving fluid. The Doppler frequency shift in the scattered waves is proportional to the speed of the scattering particle. Thus, by measuring the

---

*These issues, and many other aspects of thermal anemometry, have been examined in great detail in hundreds of research papers [21].

frequency difference between the scattered and unscattered waves, the particle or flow speed can be found. The wave sources most often used in fluid velocimetry are laser light and ultrasound.

The Doppler shift is responsible for the familiar change in pitch as a moving source of sound passes, such as that heard in a siren or car horn. The amount of Doppler shift in a wave scattered from a moving particle depends on the particle's direction relative to the incident wave, as well as the observer's position (Fig. 15.23). A calculation shows that the frequency shift observed is [22]:

$$\Delta f = \left(\frac{2V}{\lambda}\right) \cos\beta \sin\left(\frac{\alpha}{2}\right), \tag{15.16}$$

where

$\quad \Delta f$ = the Doppler frequency shift,

$\quad V$ = the particle velocity,

$\quad \lambda$ = the wavelength of the original wave before scattering,

$\quad \beta$ = the angle between the velocity vector and the bisector of the angle SPQ,

$\quad \alpha$ = the angle between the observer and the axis of the incoming wave.

The important aspect of this shift is its proportionality to the particle velocity.

Figure 15.23 When an incident wave of frequency f is scattered from a particle moving at speed V, the observer sees a scattered wave of frequency $f + \Delta f$, where Δf is the Doppler frequency shift (Eq. 15.16).

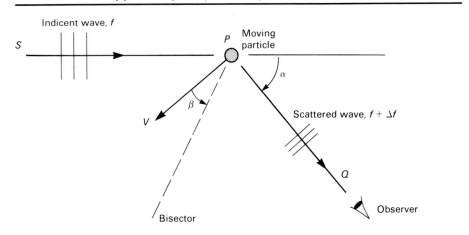

If we can measure the shift, we can find the particle speed; if the particle moves with the flow, this speed should equal the fluid speed.

Because laser light and ultrasonic waves have relatively high frequencies, the Doppler shift is only a small fraction of the original wave's frequency. For example, the fractional frequency change in scattered laser light may be only $1:10^8$ (see Example 4.3). The Doppler frequency is usually resolved by *heterodyning* the scattered wave with an unshifted reference wave to produce a measurable beat frequency (see Section 4.4.1).

15.11.1   Laser-Doppler Anemometry

The early *laser-Doppler anemometer* (LDA) used separate scattering and reference beams to create an optical heterodyne at a photodetector [Fig. 15.24(a)]. The light scattered by particles in the flow interfered with the light from the reference beam to produce beats at a frequency of one-half the Doppler shift. Unfortunately, these systems were fairly difficult to align because the intensity of the reference beam must nearly equal that of the scattered light in order to achieve an acceptable heterodyne [compare Fig. 4.7(a) to Fig. 4.7(b)]. Consequently, reference beam systems have largely given way to the *differential Doppler* approach shown in Fig. 15.24(b).

The differential doppler system splits the laser into two equal intensity beams, which are focused into an intersection point. A particle passing through the intersection scatters light from *both* beams, and this light is collected by a photomultiplier tube (PMT). The Doppler shift for each beam is different, by virtue of their different angles, but the intensities of the two scattered waves are now identical. The resulting beat frequency at the detector is just the difference in the Doppler shifts of the two beams. Figure 15.25 shows a commercial LDA transmitter/receiver arrangement.

The signal of the differential LDA can be interpreted in terms of the light and dark *interference fringes* produced at the beam crossover point (see Fig. 15.25). The distance between these fringes is

$$\delta = \frac{\lambda}{2 \sin(\theta/2)},\qquad(15.17)$$

where

δ = the fringe spacing,

λ = the laser wavelength,

θ = the angle between the two beams.

A small particle crossing the fringe pattern produces a burst of scattered light whose intensity varies as the particle crosses each fringe (Fig. 15.25). The frequency of this *Doppler burst* is just the particle velocity divided by the

Figure 15.24 Laser-Doppler optical systems: (a) reference-beam arrangement, (b) differential-Doppler arrangement

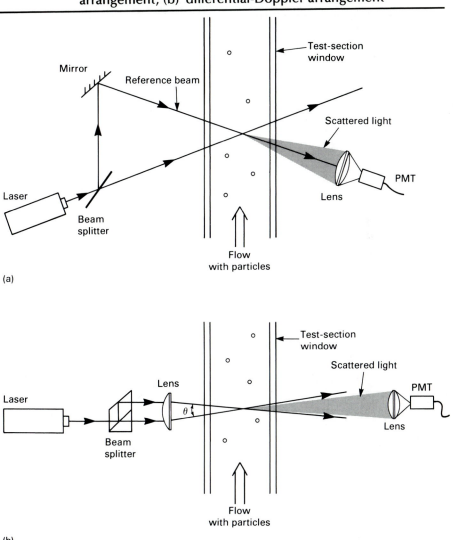

(a)

(b)

fringe spacing:

$$f_D = \frac{V_x}{\delta} = \left(\frac{2V_x}{\lambda}\right) \sin\left(\frac{\theta}{2}\right), \qquad (15.18)$$

where

f_D = the Doppler-shift frequency,

V_x = the particle velocity in the direction normal to the fringes.

Figure 15.25   LDA transmitter and receiver packages (Courtesy of
David Carr, Aerometrics Inc., Sunnyvale, CA)

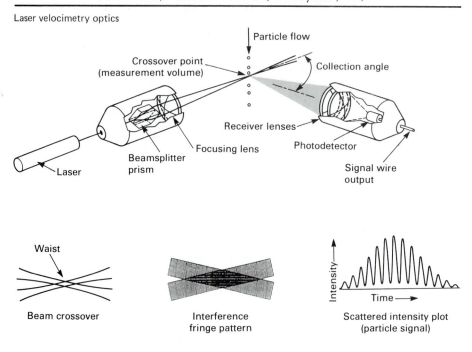

Note that the burst frequency depends only on the velocity component *normal*
to the plane of the fringes. Also note that the Doppler frequency is now
independent of the position of the photomultiplier tube.

The PMT signal is processed electronically to find the frequency, f_D, and
thus the particle velocity, V_x. One electronic scheme identifies the Doppler
frequency by high-pass-filtering the burst signal and then counting the number
of zero crossings during a time interval (a *counter* processor). Another system
calculates the Fourier transform of the signal (Section 4.6) to identify the
Doppler frequency (a *frequency-domain processor*). Because particles in the
flow cross the beams' intersection point at random time intervals, the final
velocity data consist of a sequence of individual velocity measurements that
must usually be analyzed statistically.

The beam intersection volume can be quite small, depending on the
focusing optics. Typical beam *probe volumes* are elliptical with a major axis of
0.1 to 1 mm, and have fringe spacings measured in micrometers. The scattering
particles are usually seeded into the flow. For liquids, natural impurities may
provide acceptable seed particles; if not, adding small polystyrene spheres or
even a little milk will work. In gases, an aerosol of nonvoliti e oil can be used.
For good signal quality, the scattering particles should generally be of diameter

smaller than the fringe spacing; seed particle diameters of about $1\,\mu$m are common for gas flows.

The actual intensity of the scattered light depends upon the angle from which the scattering particle is viewed as well as the ratio of the laser wavelength to particle diameter and the particle's index of refraction [23]. The strongest signals are obtained when the scattering particle's diameter is several times the wavelength and when the particle is viewed from the forward direction using a small collection angle (see Fig. 15.25). However, the collection angle for scattered light may be increased up to 180° if space limitations prevent the use of a forward receiver.

Finally, it should be noted that ordinary LDA cannot distinguish the flow direction: Positive and negative values of V_x will produce the same Doppler shift. This difficulty is overcome by applying a known *frequency shift* to one or both laser beams to create an offset frequency that is increased or decreased by the Doppler-shift. The measured frequency no longer passes through zero when the flow direction changes; instead, it becomes larger or smaller than the offset frequency. With frequency shift, LDA becomes one of very few velocimeters capable of measuring flow reversals.

Commercial LDA systems are expensive (usually from \$40,000 to \$250,000), and their use is justified only when local and nonintrusive measurements are absolutely imperative. If intrusion can be tolerated and if flow reversal or flowborne particulates are not an issue, thermal anemometers are a more economical alternative.

15.11.2   Ultrasonic Anemometry

Ultrasonic waves in the range of hundreds of kilohertz to several megahertz have been applied to Doppler flow measurements in liquids. The ultrasonic transmitter and receiver may be piezoelectric and are often designed to be clamped to the outside of a pipe. Ultrasonic Doppler techniques have been widely adapted to flow metering, and portable models are available at moderate cost.

Like LDA, ultrasonic Doppler requires that particles be present in the flow. Since ultrasonic flowmeters are often used in industrial settings where deliberate seeding may be inconvenient, they have found greatest application to the measurement of slurries and dirty liquids which already include particulates. Ultrasonic Doppler flowmeters are accurate to between 1 and 5% of the flow rate.

Ultrasound is also used for time-of-travel measurements of mean flow velocity [24]. A pair of piezoelectric or magnetostrictive transducers are placed on the outside of a conduit a few inches apart. One serves as a 100-kHz source and the other serves as the pickup. As the sound wave travels from the source to the receiver, its ordinary velocity in the stationary fluid, c, will be either increased or decreased due to the Doppler effect resulting from the liquid velocity V. For example, if the sound wave crosses the pipe at an angle θ

relative to the flow direction, then the effective velocity of the wave is $c \pm V \cos \theta$, depending on whether the wave moves upstream or downstream. Since a wave travels more slowly in the upstream direction than in the downstream direction, the flow velocity can be determined from the difference in travel time or the relative phase shift between upstream and downstream waves. To obtain both upstream and downstream waves, the function of the two ultrasonic transducers is reversed ten times per second.

Additional information on ultrasonic flow measurement may be found in the Suggested Readings for this chapter.

15.12   Calibration of Flow-Measuring Devices

Facilities for producing standardized flows are required for flowmeter calibration. Fluid at known rates of flow must be passed through the meter and the rate compared with the meter readout. When the basic flow input is determined through measurement of time and either linear dimensions (volumetric flow) or weight (mass flow), the procedure may be called *primary calibration.* After receiving a primary calibration, a meter may then be used as a *secondary standard* for standardizing other meters through *comparative calibration* (See Section 3.11.2).

Primary calibration is usually carried out at a constant flow rate with the procedure consisting of an integration or summing of the total flow for a predetermined period of time. Volumetric displacement of a liquid may be measured in terms of the liquid level in a carefully measured tank or container. For a gas, at moderate rates, volume may be determined through use of an inverted bell-type *gasometer,* or "meter prover." Primary calibration in terms of mass is commonly accomplished by means of the familiar *weigh tank,* in which the liquid is collected and weighed. Although the latter method is normally used only for liquids, with proper facilities it may also be used for calibration with gases [25].

Figure 15.26 illustrates a method obviating the requirement for direct weight measurement. A standpipe of known capacity (diameter) is used as a collector. We see that the pressure or head at the base is the analog of the mass as it is collected. Calibrations within 0.1% for meters handling 50 to 600,000 lbm/h are claimed [26].

Primary calibration is usually considered only at relatively low flow rates; however, on occasion, the cost of large-scale facilities can be justified by the importance of the application. An unusual system adapted to liquid hydrogen flow calibrations, at rates to 7000 gallons per minute, is described in Ref. [27]. Two 50,000 gallon dewars and an interconnecting flow loop were used and volume rates were determined by means of liquid-level gages. Calibration errors of no more than 4% were claimed for this unusual application.

Secondary calibration may be either *direct* or *indirect.* Direct secondary calibration is accomplished by simply placing a secondary standard in series

Figure 15.26 Standpipe employed for flow-meter calibration

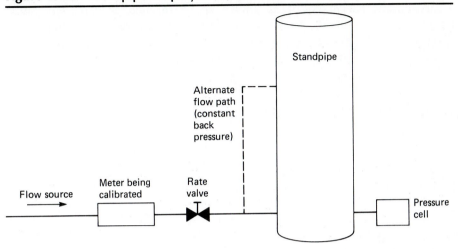

with the meter to be calibrated and comparing their respective readouts over the desired range of flow rates. Turbine-type meters are particularly useful as secondary standards for "field" calibration of orifice or venturi meters [28]. It is clear that this procedure requires careful consideration of meter installations, minimizing interactions or other forms of disturbances such as might be caused by nearby line obstructions, e.g., elbows or tees.

Indirect calibration is based on the equivalencies of two different meters. The requirement for similarity is met through maintenance of equal Reynolds numbers, or

$$\frac{\rho_1 D_1 V_1}{\mu_1} = \frac{\rho_2 D_2 V_2}{\mu_2},$$

where the subscripts 1 and 2 refer to the "standard" and the meter to be calibrated, respectively. The practical significance lies in the fact that, provided that similarity is maintained, discharge coefficients of the two meters will be directly comparable.

We see then that for geometrically similar meters it is possible to predict the performance of one meter on the basis of the experimental performance of another. Meters that are small in physical size may be used to determine the discharge coefficient of large meters. Indeed, coefficients for one fluid may be determined through test runs with another fluid, provided that similarity is maintained through Reynolds numbers. However, when a liquid is used to calibrate a meter intended for a gas, corrections for density and expansion must also be made.

Velocity-sensing probes are often calibrated by direct comparison to another velocity sensor, as, for example, a hot-wire may be calibrated by comparison to a Pitot tube. Alternatively, a velocity sensor may be mounted in a mechanism that moves it at known speed through a fluid. The latter approach is adaptable to dynamic calibration (Section 5.20). For example, hot wires are sometimes oscillated at known frequency in an otherwise steady flow by mounting them in a Scotch-yoke mechanism (Section 4.3) [29].

15.13   Flow Visualization

Many flows are so complex that the designer may have difficulty predicting their form. An experimental visualization of the flow field then becomes an integral part of the design process. For example, an understanding of the flow past an automobile is essential to determining drag and improving fuel efficiency, particularly for the flow toward the rear of the car and in its wake. Normally, a scale model of the proposed body design is tested in a wind tunnel. Smoke trails introduced upstream of the car are used to make the flow visible as it passes the model. The designer can then identify regions of separated flow and adjust the body contour to reduce the drag.

The advantage, of course, is that flow visualization can illustrate an entire flow field, whereas velocity probes yield information at only a single point. Regions of separation, recirculation, and pressure loss may be eliminated without detailed (and more expensive) calculations or velocity measurements.

Techniques of flow visualization are widely varied. A number introduce some visible material, such as particles or a dye, into the flow. In other cases, density variations (and thus refractive index changes) in the fluid itself may be rendered visible. The following lists a few of the common techniques. Further discussion may be found in the Suggested Readings for this chapter.

1. *Smoke wire* visualization: A thin steel wire ($\sim$0.1 mm) is coated with oil and placed in an air flow. An electric pulse resistively heats the wire, causing the oil to form smoke. The thin line of smoke is carried with the flow, showing the fluid pathlines.

2. *Hydrogen bubble* visualization: A very fine wire, often platinum of 25 to 100 μm diameter, is placed in a water flow. A second, flat electrode is placed nearby in the flow. When dc power of about 100 V is applied to the wire, a current passes through the water, causing electrolysis at the wire surface. Tiny hydrogen bubbles are created. These bubbles are too small to experience much buoyancy, and they instead follow the water as a visible marker [30].

3. *Particulate tracer* visualization: Reflective or colored particles make the flow pattern visible. In liquids, particles of near-liquid density are preferred; diameters of 25 to 200 μm are typical. Examples are polystyrene spheres, hollow glass spheres, fish scales, and aluminum or magnesium flakes. In gases, very-small-diameter particles must be used, since larger particles tend to settle out of the flow. Smokes of hydrocarbon oils or

titanium dioxide (which are 0.01 to 0.5 μm in diameter) have been used, as have oil droplets, various hollow spheres, and even helium-filled soap bubbles [31].

4. *Dye injection*: A colored dye is bled into a liquid flow through a small hole or holes in the surface of a test object. The dye track shows the path taken by the liquid as it passes the object.

5. *Chemical indicators*: A chemical change is produced, often electrolytically, to cause a solution to change color or to cause formation of a fine colloidal precipitate. For example, a hydrogen-bubble electrical arrangement can cause a thymol-blue solution to change from orange to blue.

6. *Laser-induced fluorescence*: Fluorescence is the tendency of some molecules to absorb light of one color (or frequency) and reemit light of a different color (a lower frequency). Fluorescent dyes are added to water, as in regular dye injection, but a now thin sheet of laser light is used to excite the dye in a specific plane of the flow. The resulting fluorescence provides visualization of the flow in that plane alone. Typical dyes for use with a blue-green argon-ion laser include rhodamine (fluoresces dark red or yellow) and fluoresceine (fluoresces green). Laser-induced fluorescence can be adapted to gas flows as well.

7. *Refractive-index-change* visualizations: The refractive index changes with fluid density (or temperature). Variations in refractive index will deflect or phase-shift light passing through a fluid; with an appropriate optical arrangement, these effects can be made visible. For example, the abrupt density change at a shock wave deflects light and can be made to appear as a thin shadow in a photograph (a *shadowgraph*). Smaller density changes in compressible flows or temperature gradients in buoyancy-driven flows may also be visualized (and quantitatively measured) using refractive-index methods such as the schlieren technique and holographic interferometry.

8. *Digital image processing* is a useful adjunct to many of the preceding methods. For example, a time series of digitized images can be processed to trace the time history of particle motions, yielding a whole-field velocity measurement.

Suggested Readings

ASME PTC 19.5, Application, Part II of Fluid Meters. New York: 1972.

ASME Fluid Meters, Their Theory and Applications. 6th ed. ASME, 1971.

Benedict, R. P. *Fundamentals of Temperature, Pressure and Flow Measurements.* 2nd ed. New York: John Wiley, 1977.

Cheremisinoff, N. P., and P. N. Cheremisinoff (eds.). *Flow Measurement for Engineers and Scientists.* New York: Marcel Dekker, 1988.

Cheremisinoff, N. P. *Applied Fluid Flow Measurement: Fundamentals and Technology.* New York: Marcel Dekker, 1981.

Cusick, C. F. (ed.). *Flow Meter Engineering Handbook.* 5th ed. Ft. Washington, Penn.: Honeywell, Inc., Process Control Div., 1977.

Drain, L. E. *The Laser Doppler Technique.* New York: John Wiley, 1980.

Goldstein, R. J. *Fluid Mechanics Measurements.* Washington, D.C.: Hemisphere, 1983.

Harada, M. (ed.). *Fluid Control and Measurement,* vol 1 and vol 2. Elmsford, N.Y.: Pergamon Press, 1986.

Katys, G. P. *Continuous Measurement of Unsteady Flow.* New York: Macmillan, 1964.

Lynnworth, L. C. *Ultrasonic Measurements for Process Control: Theory, Techniques, and Applications.* San Diego: Academic Press, 1989.

Merzkirch, W. *Flow Visualization.* 2nd ed. Orlando, Fla.: Academic Press, 1982.

Miller, R. W. *Flow Measurement Engineering Handbook.* 2nd ed. New York: McGraw-Hill, 1989.

Perry, A. E. *Hot-wire Anemometry.* New York: Oxford University Press, 1982.

Spink, L. K. *Principles and Practice of Flow Meter Engineering.* 9th ed. Foxboro, Massachusetts: The Foxboro Co., 1978.

Yang, W.-J. (ed.) *Handbook of Flow Visualization.* New York: Hemisphere Publishing Co., 1989.

Problems

15.1 Water at 30°C (86°F) flows through a 10-cm (3.94-in.) pipe at an average velocity of 6 m/s (19.68 ft/s). Calculate the value of Re_D, using SI units. Check for proper unit balance. Obtain required data from Appendix D.

15.2 Solve Problem 15.1 using English units and check for unit balance. (Note: The same numerical result should be obtained in both cases.)

15.3 If Problem 15.1 had specified 4-in., schedule 140 pipe, what would be the resulting Re_D? (Check any general engineering handbook for the significance of pipe "schedule" numbers.)

15.4 Check Eq. (15.3) for a balance of units using (a) the SI system of units and (b) the English system.

15.5 An 8-in. ID (20.32-cm) pipe is connected by means of a reducer to a 6-in. ID pipe. Kerosene (see Appendix D for data) flows through the system at 2000 gal/min (7.57 m³/min.). What pressure differential in head of Hg should exist between the larger and the smaller pipes? What differential pressure should be found if the fluid is changed to water at the same flow rate?

15.6 Solve Problem 15.5 using SI units and reconcile the answers.

15.7 A section of horizontally oriented pipe gradually tapers from a diameter of 16 cm (6.3 in.) to 8 cm (3.15 in.) over a length of 3 m (9.84 ft). Oil having a specific gravity of 0.85 flows at 0.05 m³/s (1.77 ft³/s). Assuming no energy loss, what pressure differential should exist across the tapered section? Check unit balance.

15.8 Solve Problem 15.7 using English units.

15.9 If the conduit described in Problem 15.7 is oriented into a vertical position with the larger diameter at the lowest point, what differential pressure across the section should be found? Solve (a) using SI units and (b) using English units. (c) Check the two answers for equivalency. (d) Would there be a difference if the smaller diameter is at the lowest position?

15.10 Show that Eq. (15.5e) may be written as follows

$$Q = KA_2\sqrt{2gh} \, ,$$

where h = the differential pressure across the meter, measured in the "head" of the flowing fluid. Check the unit balance.

15.11 When an obstruction meter is placed in a vertical run of pipe (as opposed to a horizontal run), what precautions must be made in measuring the differential pressure?

15.12 Prepare a spreadsheet template for solving venturi differential-pressure problems.

15.13 Use the spreadsheet template prepared in answer to Problem 15.12 to check the calculations in the venturi example in Section 15.3.5.

15.14 A venturi meter is placed in a horizontal run of $2\frac{1}{2}$-inch IPS (Iron Pipe Size) pipe [ID = 2.47 in. (62.74 mm)] for the purpose of metering heating oil as it is pumped into a storage tank. The throat diameter of the venturi is $1\frac{3}{8}$ in. (34.93 mm). If the differential pressure is held to the equivalent of 22 in. (55.9 cm) of water for one-half hour, how many gallons (liters) of oil should have been pumped? The oil temperature is 60°F (15.6°C) and its specific gravity is 0.86.

15.15 Solve Problem 15.14 using SI units.

15.16 Water at 15°C and 650 kPa flows through a 15 × 10 cm (15-cm pipe and 10-cm throat) venturi. A differential pressure of 25 kPa is measured. Calculate the flow rate (a) in kg/min and (b) in m³/h.

15.17 A venturi with a 40-cm (15.75-in.) diameter throat is used to meter 15°C (59°F) air in a 60-cm (23.62-in.) duct. If the differential pressure measured across vena contracta taps is 84 mm (3.31 in.) of water and the upstream pressure (absolute) is 125 kPa (18.13 psi), what is the flow rate in kg/s? In m³/s?

15.18 Solve Problem 15.17 using English units.

15.19 Prepare a spreadsheet template to be used for solving flat-plate orifice problems.

15.20 A sharp-edged concentric orifice is used to measure the flow of kerosene in a 6-cm (2.36-in.) diameter line. If $\beta = 0.4$, what differential pressure may be expected for a flow rate of 0.9 m³/min (31.78 ft³/min) when the temperature of the fluid is 10°C (50°F)?

15.21 Solve Problem 15.20 using English units.

15.22 A sharp-edged concentric orifice plate is used to measure the flow of 60°F water in a 4-in.-diameter pipe. Prepare a plot of differential pressure readout (inches of H₂O) vs. β over a range $0.3 > \beta > 0.95$. (*Note:* A spreadsheet solution is recommended.)

15.23 A sharp-edged concentric orifice is to be used in a 4-in. (ID) pipe carrying water whose temperature may vary between 40°F and 120°F. The range of flow rate is from 150 to 600 gal/min. It is realized that the differential pressure across the orifice will, among other things, depend on the value of β. To minimize losses, it is desirable to use as large an orifice diameter as is feasible.

Investigate and specify the type of differential pressure sensor to be used. Considering sensitivity limits, determine the largest practical value of β that will satisfy your specifications. Write a statement explaining the selections you have made. It is suggested that a spreadsheet template be used.

15.24 A sharp-edged circular orifice plate has a nominal diameter of 2.75 in. If the orifice is calibrated at 60°F, plot the percent error introduced caused by temperature change over a range of 35°F to 120°F for: (a) a steel plate, and (b) for a brass plate. See Table 6.3 for coefficients.

15.25 If the steel plate used for a $\frac{3}{16}$-in.-thick concentric orifice plate in a 5-in.-diameter pipe has a yield strength of 45,000 lbf/in.^{2} and $\beta = 0.5$, estimate the differential pressure above which distortion may be expected.

15.26 Direct secondary calibration is used to determine the flow coefficient for an orifice to meter the flow of nitrogen. For an orifice inlet pressure of 25 psia (172 kPa), the flow rate determined by the primary meter is 9 lbm/min (19.8 kg/m) at 68°F (20°C). If the differential pressure across the orifice being calibrated is 3.1 in. (7.87 cm) of water, the conduit diameter is 4 in. (10.16 cm), and $\beta = 0.5$, what is the value of $(K \times Y)$? (Use $\rho = 0.0726$ lbm/ft^3 at 68°F and standard pressure.) Solve using English units.

15.27 Solve Problem 15.26 using SI units.

15.28 Using Eq. (15.9), plot P_2/P_1 versus k over a range of $k = 1$ to 1.4.

15.29 A Pitot-static tube is used to measure the velocity of 20°C (68°F) water flowing in an open channel. If a differential pressure of 6 cm (2.36 in.) of H_2O is measured, what is the corresponding flow velocity? Check the result by using English units and comparing your answer to the SI result.

15.30 Using the data given in Problem 15.29, what would be the result if the temperature of the flowing water were 50°C, the water in the manometer were at 5°C, and the measurements are made at (a) Key West, Florida, and (b) Ft. Egbert, Alaska? (*Note:* See Appendix D for data and overlook "significant figures," by carrying the results to the degree required to show a difference.)

15.31 A Pitot-static tube is used to measure the velocity of an aircraft. If the air temperature and pressure are 5°C (41°F) and 90 kPa (13.2 psia), respectively, what is the aircraft velocity in km/h if the differential pressure is 450 mm (17.5 in.) of water? (a) Solve using Eq. (15.14), then (b) using Eq. (15.14a).

15.32 Solve Problem 15.31 using English units.

15.33 The velocity profile for turbulent flow in a smooth pipe is sometimes given as [32]

$$\frac{V(r)}{V_{center}} = \left[1 - \left(\frac{2r}{D}\right)\right]^{1/n},$$

where D = the pipe diameter, r = the radial coordinate from the center of the pipe, and n ranges in value from about 6 to 10, depending on Reynolds number. For $n = 8$, determine the value of r at which a Pitot tube should be placed to provide the velocity V_{ave}, such that $Q = AV_{ave}$.

15.34 A one-fifth size, geometrically similar model is made of the orifice described in the example in Section 15.4. If water at 70°F is used to experimentally determine the flow coefficient, to what velocity should the flow be adjusted for dynamic similarity?

Figure 15.27   Angular orientation of a hot-wire probe

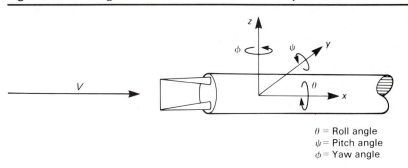

θ = Roll angle
ψ = Pitch angle
ϕ = Yaw angle

15.35 Like a Pitot-static tube, a hot-wire probe may also suffer from angular misalignment errors. Figure 15.27 defines three possible angles that the probe may have relative to its desired alignment. Describe separately what error is incurred when each angle is nonzero. Give qualitative answers and assume that the angles remain less than 45°.

References

1. Streeter, V. L., and E. B. Wylie. *Fluid Mechanics*. New York: McGraw-Hill, 1975.

2. ASME PTC 19.5, Application—Pt. II of Fluid Meters: Interim Supplement on Instruments and Apparatus, 1972, p. 232.

3. Roark, R. J. *Formulas for Stress and Strain*. 4th ed. New York: McGraw-Hill, 1965, p. 221, Case 17.

4. Li, W. H., and S. H. Lam. *Principles of Fluid Mechanics*. Reading, Mass.: Addison-Wesley, 1964, p. 318.

5. Schoenborn, E. M., and A. P. Colburn. The flow mechanism and performance of the rotameter. *Trans. AIChE* 35(3): 359, 1939.

6. Nagler, F. A. Use of current meters for precise measurement of flow. *ASME Trans.* 57: 59, 1935.

7. Grey, J. Transient response of the turbine flowmeter. *Jet Propulsion* 26: Feb. 1956.

8. Knowlton, A. E. *Standard Handbook for Electrical Engineers*. 8th ed. New York: McGraw-Hill, 1949, pp. 36–40.

9. James, W. G. An A-C induction flow meter, *ISA Proc.* 6: 5, 1951.

10. Gray, W. C., and E. R. Astley. Liquid metal magnetic flowmeters. *ISA J.* 1: 15, June 1954.

11. Kiverson, G. Promising newcomers for tough flow measurements. *Mach. Des.*, January 8, 1976.

12. Binder, R. C. *Advanced Fluid Dynamics and Fluid Machinery*. Englewood Cliffs, N.J.: Prentice Hall, 1951.

13. Gracey, W. Measurement of static pressure on aircraft. *NACA Tech. Note 4184:* November 1957.

14. Krause, L. N., and C. C. Gettelman. Effect of interaction among probes, supports, duct walls and jet boundaries on pressure measurements in ducts and jets. *ISA Proc.* 7: 138, 1952.

15. Gettleman, C. C., and L. N. Krause. Considerations entering into the selection of probes for pressure measurement in jet engines. *ISA Proc.* 7: 138, 1952.

16. Anderson, J. D., Jr. *Introduction to Flight.* New York: McGraw-Hill, 1978, p. 103.

17. Itsweire, E. C., and K. N. Helland. A high-performance low-cost constant-temperature hot-wire anemometer. *J. Phys. E: Sci. Instrum.* 16: 549–553, 1983.

18. King, L. V. On the convection of heat from small cylinders in a stream of fluid, with applications to hot-wire anemometry. *Phil. Trans. Roy. Soc. (London)* 214, 14, Ser. A: 373–432, 1914.

19. Lienhard V, J. H. *The decay of turbulence in thermally stratified flow.* Doctoral dissertation, University of California, San Diego, 1988, Chapter 3.

20. Wyngaard, J. C. Measurement of small-scale turbulence structure with hot wires. *J. Phys. E: Sci. Instrum.* 1: 1105–8, 1968.

21. Freymuth, P. *A Bibliography of Thermal Anemometry.* St. Paul, Minn.: TSI, Inc., 1982. Also, *1983 Addendum to "A Bibliography of Thermal Anemometry."* St. Paul, Minn.: TSI, Inc., 1983.

22. Drain, L. E. *The Laser Doppler Technique.* New York: John Wiley, 1980, Chapter 3.

23. Van de Hulst, H. C. *Light Scattering by Small Particles.* New York: Dover Publications, 1981.

24. *NBS Tech. News Bull.* 37: March 1953.

25. Collins, W. T., and T. W. Selby. A gravimetric flow standard. *Flow Measurement Symposium.* New York: ASME, p. 290, 1966.

26. Jarret, F. H. Standpipes simplify flowmeter calibration. *Control Eng.* 1: 37, December 1954.

27. Liebenberg, D. H., R. W. Stokes, and F. J. Edeskuty. The calibration of flowmeters with liquid hydrogen in the region between 1000 and 7000 GPM. *Flow Measurement Symposium.* New York: ASME, p. 155, 1966.

28. Bowen, R. P. Designing portability into a flow standard. *ISA J.* 18: 40, May 1961.

29. Perry, A. E. *Hot-wire Anemometry.* New York: Oxford University Press, 1982, pp. 107–108.

30. Geller, E. W. An electrochemical method of visualizing the boundary layer. *J. Aero. Sci.* 22: 869–870, 1955.

31. Merzkirch, W. *Flow Visualization.* 2nd ed. Orlando, Fla.: Academic Press, 1982, p. 46.

32. Hansen, A. G. *Fluid Mechanics.* New York: John Wiley, 1967, p. 422.

CHAPTER 16

Temperature
Measurements

16.1   Introduction

Temperature change is usually measured by observing the change in another temperature-dependent property. Unlike the direct comparison of other fundamental physical quantities to a calibrated standard (as mass is measured by comparison to the International Prototype Kilogram), direct comparison of an unknown temperature to a reference temperature is relatively difficult. In formal thermodynamics, this comparison is executed by connecting a Carnot engine between two systems at different temperature. In practical thermometry such direct comparison is abandoned, and temperature is instead gauged by its effect on quantities such as volume, pressure, electrical resistance, or radiated energy.

Section 2.8 describes the international practical temperature scale (ITS-90). This scale *assigns* values of temperature to a few highly reproducible states of matter, such as certain freezing-point and triple-point conditions. These defined reference temperatures then provide calibration points for various special thermometers, which are used to interpolate between the reference points. The interpolating thermometers undergo a change in some other physical property, such as pressure or electrical resistance, as their temperature changes, and the resulting value of that property is used to infer the corresponding temperature. In a sense, temperature itself is never directly sampled in practical thermometry.

In this chapter, we survey a selection of common temperature-sensing techniques. These thermometers are based on changes in a broad range of physical properties, among which are the following:

1. Changes in physical dimensions
   a. Liquid-in-glass thermometers
   b. Bimetallic elements

2. Changes in gas pressure or vapor pressure
    a. Constant-volume gas thermometers
    b. Pressure thermometers (gas, vapor, and liquid-filled)
3. Changes in electrical properties
    a. Resistance thermometers (RTD, PRT)
    b. Thermistors
    c. Thermocouples
    d. Semiconductor-junction sensors
        i. Diodes
        ii. Integrated circuits
4. Changes in emitted thermal radiation
    a. Thermal and photon sensors
    b. Total-radiation pyrometers
    c. Optical and two-color pyrometers
    d. Infrared pyrometers
5. Changes in chemical phase
    a. Fusible indicators
    b. Liquid crystals
    c. Temperature-reference (fixed-point) cells

Of these methods, electrical sensors are perhaps the most broadly used, particularly when automatic or remote recording is desired and when temperature sensors are incorporated into control systems. Bimetallic elements are used in various low-accuracy, low-cost applications. Radiant sensors are used for noncontact temperature sensing, either in high-temperature applications like combustors or for infrared sensing at lower temperatures; since radiant sensors are optical in nature, they are also adaptable to whole-field temperature measurement (thermal imaging). The familiar liquid-in-glass thermometer continues to appear in both laboratory and household situations, primarily because of its ease of use and low cost. Changes in chemical phase are somewhat less often applied in engineering work.

Table 16.1 outlines approximate ranges and uncertainties of various temperature-measuring devices. The values listed in the table are only approximate, and many untabulated factors will cause deviations from the values listed. Among those factors are the precise form of electrical signal-conditioning employed, the influence of manufacturer's and/or laboratory calibration techniques, and dedicated efforts to extend the operating range of a particular sensor.

16.2   Use of Bimaterials

16.2.1   Liquid-in-Glass Thermometers

The ordinary thermometer is an example of the liquid-in-glass type. Essential elements consist of a relatively large bulb at the lower end, a capillary tube with scale, and a liquid filling both the bulb and a portion of the capillary. In

Table 16.1 Characteristics of various temperature-measuring elements and devices (data from various sources)

Type	Useful Range*	Limits of Uncertainty*	Comments
Liquid in Glass			
Mercury filled	−35 to 600°F −37 to 320°C	0.5°F 0.3°C	Low cost. Remote reading not practical.
Pressurized mercury	−35 to 1200°F −37 to 650°C	"	Lower limit of mercury-filled thermometers determined by freezing point of mercury.
Alcohol	−100 to 200°F −75 to 129°C	1°F 0.6°C	Upper limit determined by boiling point.
Bimetal	−80 to 800°F −65 to 430°C	1 to 20°F $\frac{1}{2}$ to 12°C	Rugged. Inexpensive.
Pressure systems			
Gas (laboratory)	−450 to 212°F −270 to 100°C	0.005 to 0.5°F 0.002 to 0.2°C	Very accurate. Quite fragile. Not easily used. Used as an interpolating standard for ITS-90 (see Section 2.8).
Gas (industrial)	−450 to 1400°F −270 to 760°C	0.5 to 2% "	Bourdon pressure gage used for readout. Rugged, with wide range.
Liquid (except mercury)	−125 to 700°F −90 to 370°C	2°F 1°C	Relative elevations of read-out and sensing bulb are critical. Smallest bulb. Up to 10 ft (3 m) capillary.
Liquid (mercury)	−35 to 1200°F −37 to 630°C	$\frac{1}{2}$ to 2% "	Relative elevations of read-out and sensing bulb are critical. Smallest bulb. Up to 10 ft (3 m) capillary.
Vapor pressure	−100 to 650°F	$\frac{1}{2}$ to 2%	Fast response. Nonlinear. Lowest cost.
Thermocouples			
General	−420 to 4400°F −250 to 2400°C	1°F 0.6°C	Extreme ranges—all types.
Type B (Pt, 30% Rh(+) vs. Pt, 6% Rh(−))	1600 to 3100°F	$\pm\frac{1}{2}$%	Not for reducing atmosphere or vacuum. Generates high emf per degree.
Type E (Chromel†(+) vs. Constantan†(−))	−300 to 1600°F −184 to 870°C	$\pm\frac{1}{2}$%	Highest output of common thermocouples.
Type J (Fe(+) vs. Alumel†(−))	32 to 1400°F	2 to 10°F‡ 1 to 6°C	For reducing or neutral atmosphere. Popular and inexpensive.
Type K (Chromel(+) vs. Alumel†(−))	32 to 2300°F 0 to 1260°C	$\pm\frac{3}{4}$%‡	For oxidizing or neutral atmosphere. Attacked by sulfur. Most linear of all thermocouples.
Type R(Pt(+) vs. Pt, 13% Rh(−))	32 to 2700°F 0 to 1480°C	$\frac{1}{2}$%‡	Requires protection in all atmospheres. Higher output than Type S. Linearity poor below 1000°F (540°C).
Type S(Pt(+) vs. Pt, 10% Rh(−))	32 to 2700°F 0 to 1480°C	$\frac{1}{2}$%‡	Requires protection in all atmospheres. Under proper conditions yields highest precision.

continued

Table 16.1 *continued*

Type	Useful Range*	Limits of Uncertainty*	Comments
Type T(Cu(+) vs. Constantan†(−))	−420 to 650°F −250 to 340°C	1°F 0.6°C	May be used in either oxidizing or reducing atmospheres. Good stability.
W, 5% Rh(+) vs. W, 26% Rh(−)	−450 to 4200°F −270 to 2310°C	—	No standards. Reducing or neutral atmospheres. Highest temperature limit of all thermocouples.
Resistance			
Platinum	−435 to 1800°F −260 to 980°C	0.04 to 0.4°F‡ 0.02 to 0.2°C	High repeatability. Linear. Used as an interpolating device for ITS-90 (see Section 2.8). Sensor can be used as far as 5000 ft (1500 m) from readout.
Nickel	−300 to 600°F −180 to 320°C	—	High repeatability. Nonlinear. Produces greater resistance change per degree than does Pt. Sensor can be as far as 5000 ft (1500 m) from readout.
Thermistor (Metal Oxide)	−150 to 600°F −100 to 315°C	$\frac{1}{2}$°F 0.2°C	Negative temperature coefficient. Highly non-linear. Less stable than metal types.
Thermistor (Doped Germanium)	−459 to −280°F −273 to −173°C	0.05°F 0.03°C	High repeatability. Nonlinear. Negative temperature coefficient. Cryogenic sensor.
Thermistor (Carbon-Glass)	−458 to 125°F −272 to 50°C	0.1°F 0.05°C	High repeatability. Nonlinear. Negative temperature coefficient. Cryogenic sensor.
Semiconductor Junction			
Diode (Silicon, GaAlAs)	−457 to 125°F −272 to 50°C	0.1°F 0.05°C	Nonlinear. High accuracy requires calibration. Cryogenic sensor.
Linear Integrated Circuit	−60 to 300°F −50 to 150°C	1°F 0.5°C	Inexpensive. Linear. Easily integrated into electronics. Limited temperature range.
Pyrometers			
Optical	1400 to 6300°F	$\frac{1}{2}$ to 2%	Used only for high temperatures. Requires manual manipulation by operator.
Total Radiation	0 to 7000°F −15 to 3870°C	$\frac{1}{2}$ to 2%	Can measure "spot" or average temperatures.
Infrared	0 to 6000°F −15 to 3300°C	$\frac{1}{2}$ to 2%	Portable. Self-contained.

* Approximate values. Actual values depend on many factors such as environment, physical size of sensor, purity of materials, etc. Types such as thermocouples and resistance thermometers require additional signal-conditioning apparatus. Values given are for sensors only.

† Trade names.

‡ In higher ranges.

addition, a smaller bulb is generally incorporated at the upper end to serve as a safety reservoir when the intended temperature range is exceeded.

As the temperature is raised, the greater expansion of the liquid compared with that of the glass causes it to rise in the capillary or stem of the thermometer, and the height of rise is used as a measure of the temperature. The volume enclosed in the stem above the liquid may either contain a vacuum or be filled with air or another gas. For the higher temperature ranges, an inert gas at a carefully controlled initial pressure is introduced in this volume, thereby raising the boiling point of the liquid and increasing the total useful range. In addition, it is claimed that such pressure minimizes the potential for column separation.

Several desirable properties for the liquid used in a glass thermometer are as follows:

1. The temperature-dimensional relationship should be linear, permitting a linear instrument scale.

2. The liquid should have as large a coefficient of expansion as possible. For this reason, alcohol is better than mercury. Its larger expansion makes possible larger capillary bores, and hence provides easier reading.

3. The liquid should accommodate a reasonable temperature range without change of phase. Mercury is limited at the low-temperature end by its freezing point, $-37.97°F$ (or $-38.87°C$), and spirits are limited at the high-temperature end by their boiling points.

4. The liquid should be clearly visible when drawn into a fine thread. Mercury is obviously acceptable in this regard, whereas alcohol is usable only if dye is added.

5. Preferably, the liquid should not adhere to the capillary walls. When rapid temperature drops occur, any film remaining on the wall of the tube will cause a reading that is too low. In this respect, mercury is better than alcohol.

Within its temperature range, mercury is undoubtedly the best liquid for liquid-in-glass thermometers and is generally used in the higher-grade instruments. Alcohol is usually satisfactory. Other liquids are also used, primarily for the purpose of extending the useful ranges to lower temperatures.

16.2.2 Calibration and Stem Correction

High-grade liquid-in-glass thermometers are made with the scale etched directly on the thermometer stem, thereby making it mechanically impossible to shift the scale relative to the stem. The care with which the scale is laid out depends on the intended accuracy of the instrument (and to a large extent governs its cost). The process of establishing *bench marks* from which a scale is determined is known as *pointing,* and two or more *marks* or *points* are required. In spite of contrary intentions, a particular thermometer will exhibit some degree of nonlinearity. This may be caused by nonlinear temperature-dimension characteristics of liquid or glass or by the nonuniformity of the bore of the column. In the simplest case, two points may be established, such as the freezing and boiling points of water, and equal divisions used to interpolate (and extrapolate) the complete scale. For a more accurate scale, additional points—sometimes as many as five—are used. Calibration points for this purpose are obtained through use of known phase-equilibrium temperatures, as discussed in Section 16.12.

Greatest sensitivity to temperature is at the bulb, where the largest volume of liquid is contained; however, all portions of a glass thermometer are temperature-sensitive. With temperature variation, the stem and upper bulb (if present) will also change dimensions, thereby altering the available liquid space and hence the thermometer reading. For this reason, if maximum accuracy is to be attained, it is necessary to prescribe how a glass thermometer is to be subjected to the temperature. Greatest control is obtained when the complete thermometer is entirely immersed in a uniform temperature medium. Often this is not possible, especially when the medium is liquid. A common practice, therefore, is to calibrate the thermometer for a given partial immersion, with the proper depth of immersion indicated by a scribed line around the stem. Thermometer accuracy is then prescribed for this condition only. This technique does not ensure absolute uniformity because the upper portion of the stem is still subject to some variation in ambient conditions.

When the immersion employed is different from that used for calibration, an *estimate* of the correct reading may be obtained from the following relation (for mercury-in-glass thermometers only):

$$T = T_1 + kT'(T_1 - T_2),\qquad\qquad (16.1)$$

where

T = the correct temperature, in degrees,

T_1 = the actual temperature reading, in degrees,

k = the differential expansion coefficient between liquid and glass (for mercury thermometers, commonly used values are 0.00009 for the Fahrenheit scale and 0.00016 for the Celsius scale),

T_2 = the ambient temperature surrounding the emergent stem (this may be determined by attaching a second thermometer to the stem of the main thermometer), in degrees,

T' = degrees of mercury thread emergence to be corrected.

The value T' is determined as follows: For a *total immersion thermometer*, T' should be the actual length of the thread of mercury that is emerging, measured in scale degrees. For the *partial immersion thermometer*, T' should be the number of scale degrees between the scribed calibration immersion line and the actual point of emergence. When the thermometer is too deeply immersed, the value of T' will be *negative*.

Another factor influencing liquid-in-glass thermometer calibration is a variation in the applied pressure, particularly in pressure applied to the bulb. The resulting elastic deformation causes displacement of the column and hence an incorrect reading. Normal variation in atmospheric pressure is not usually of importance, except for the most precise work. However, if the thermometer is subjected to system pressures of higher values, considerable error may be introduced.

16.2.3 Bimetal Temperature-Sensing Elements

When two metal strips having different coefficients of expansion are brazed together, a change in temperature will cause a free deflection of the assembly [1]. Such bimetal strips form the basis for control devices such as the common home heating system thermostat. They are also used to some extent for temperature measurement. In the latter case, the sensing strip is commonly wrapped into a helical form, similar to a simple helical spring. As the temperature changes, the free end of the helix rotates (the diameter of the helix either increasing or decreasing due to the differential action). The rotational motion is directly indicated by the movement of a pointer over a circular scale.

Thermometers with bimetallic temperature-sensitive elements are often used because of their ruggedness, their ease of reading, their low cost, and the convenience of their particular form. (See Problem 16.4.)

16.3 Pressure Thermometers

Figure 16.1 illustrates a simple *constant-volume* gas thermometer. Gas, usually hydrogen or helium, is contained in bulb A. A mercury column, B, is adjusted so that reference point C is maintained. In this manner, a constant volume of gas is held in the bulb and adjoining capillary. Mercury column h is a measure of the gas pressure and can be calibrated in terms of temperature.

In this form the apparatus is fragile, difficult to use, and restricted to the laboratory. It does, however, illustrate the working principle of a group of practical instruments called *pressure thermometers*.

Figure 16.2 shows the essentials of the practical *pressure thermometer*. The necessary parts are bulb A, tube B, pressure-sensing gage C, and some sort of filling medium. Pressure thermometers are called *liquid-filled, gas-filled,* or *vapor-filled,* depending on whether the filling medium is completely liquid, completely gaseous, or a combination of a liquid and its vapor. A primary advantage of these thermometers is that they can provide sufficient force output to permit the direct driving of recording and controlling devices. The

Figure 16.1   Sketch illustrating the essentials of a
                  constant-volume gas thermometer

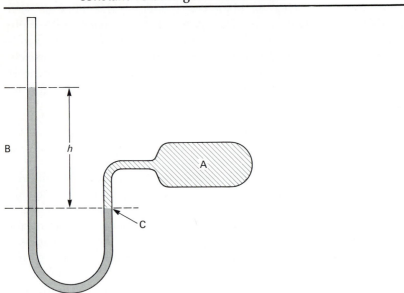

Figure 16.2   Schematic diagram showing the operation of
                  a practical pressure thermometer

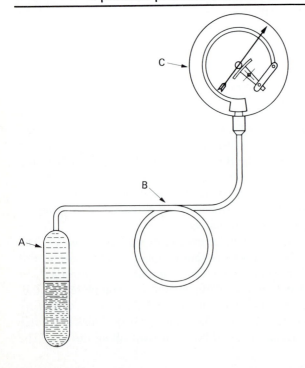

pressure-type temperature-sensing system is usually less costly than other systems. Tubes as long as 200 ft may be used successfully.

Expansion (or contraction) of bulb A and the contained fluid or gas, caused by temperature change, alters the volume and pressure in the system. In the case of the liquid-filled system, the sensing device C acts primarily as a differential volume indicator, with the volume increment serving as an analog of temperature. For the gas- or vapor-filled systems, the sensing device serves primarily as a pressure indicator, with the pressure providing the measure of temperature. In both cases, of course, both pressure and volume change.

Ideally the tube or capillary should serve simply as a connecting link between the bulb and the indicator. When liquid- or gas-filled systems are used, the tube and its filling are also temperature-sensitive, and any difference from calibration conditions along the tube introduces output error. This error is reduced by increasing the ratio of bulb volume to tube volume. Unfortunately, increasing bulb size reduces the time response of a system, which may introduce problems of another nature. On the other hand, reducing tube size, within reason, does not degrade response particularly because, in any case, flow rate is negligible. Another source of error that should not be overlooked is any pressure gradient resulting from difference in elevation of bulb and indicator not accounted for by calibration.

Temperature along the tube is not a factor for vapor-pressure systems, however, so long as a free liquid surface exists in the bulb. In this case, Dalton's law for vapors applies, which states that if both phases (liquid and vapor) are present, only one pressure is possible for a given temperature. This is an important advantage of the vapor-pressure system. In many cases, though, the tube in this type of system will be filled with liquid, and hence the system is susceptible to error caused by elevation difference.

16.4 Thermoresistive Elements

We have already seen (Section 6.18.2) that the electrical resistance of most materials varies with temperature. In Sections 12.10 and 13.5 we found this to supply a troublesome extraneous input to the output of strain gages. It can only follow that this relation, which proves so worrisome when unwanted, should be the basis for a good method of temperature measurement.

Traditionally, resistance elements sensitive to temperature are made of metals generally considered to be good conductors of electricity. Examples are nickel, copper, platinum, and silver. A temperature-measuring device using an element of this type is commonly referred to as a *resistance thermometer,* or a *resistance temperature detector,* abbreviated *RTD.* Of more recent origin are elements made from semiconducting materials having large—and usually negative—resistance coefficients. Such materials are usually some combination of metallic oxides of cobalt, manganese, and nickel. These devices are called *thermistors.*

One important difference between these two kinds of material is that, whereas the resistance change in the RTD is small and positive (increasing temperature causes increased resistance), that of the thermistor is relatively large and usually negative. In addition, the RTD type provides nearly a linear temperature-resistance relation, whereas that of the thermistor is nonlinear. Still another important difference lies in the temperature ranges over which each may be used. The practical operating range for the thermistor lies between approximately −100°C to 275°C (−150°F to 500°F). The range for the resistance thermometer is much greater, being from about −260°C to 1000°C (−435°F to 1800°F). Finally, the metal resistance elements are more time-stable than the semiconductor oxides; hence they provide better repro-ducibility with lower hysteresis.

16.4.1   Resistance Thermometers (RTDs)

Evidence of the importance and reliability of the resistance thermometer may be had by recalling that the International Temperature Scale of 1990 specifies a platinum resistance thermometer as the interpolation standard over the range from −259.35°C to 961.78°C (−484.52°F to 1763.20°F) (see Section 2.8).

Certain properties are desirable in material used for resistance thermom-eter elements. The material should have a resistivity permitting fabrication in convenient sizes without excessive bulk, which would degrade time response. In addition, its thermal coefficient of resistivity should be high and as constant as possible, thereby providing an approximately linear output of reasonable magnitude. The material should be corrosion-resistant and should not undergo phase changes in the temperature range of interest. Finally, it should be available in a condition providing reproducible and consistent results. In regard to this last requirement, it has been found that to produce precision resistance thermometers, great care must be exercised in minimizing residual strains, requiring careful heat treatment subsequent to forming.

As is generally the case in such matters, no material is universally acceptable for resistance-thermometer elements. Undoubtedly, platinum, nickel, and copper are the materials most commonly used, although others such as tungsten, silver, and iron have also been employed. The specific choice normally depends upon which compromises may be accepted.

The temperature-resistance relation of an RTD must be determined experimentally. For most metals, the result can be accurately represented as

$$R(T) = R_0[1 + A(T - T_0) + B(T - T_0)^2] \qquad (16.2)$$

where

$R(T)$ = the resistance at temperature T,

R_0 = the resistance at a reference temperature T_0,

A and B = temperature coefficients of resistance depending on material.

Figure 16.3   Section illustrating the construction of a simple RTD

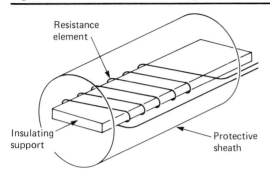

Resistance element

Insulating support

Protective sheath

Over a limited temperature interval (perhaps 50°C for platinum), a linear approximation to the resistance variation may be quite acceptable,

$$R(T) = R_0[1 + A(T - T_0)], \tag{16.2a}$$

but for the highest accuracy, a high-order polynomial fit is required [2].

The resistance element is most often a metal wire wrapped around an electrically insulating support of glass, ceramic, or mica. The latter may have a variety of configurations, ranging from a simple flat strip, as shown in Fig. 16.3, to intricate "bird-cage" arrangements [3]. The mounted element is then provided with a protective enclosure. When permanent installations are made and when additional protection from corrosion or mechanical abuse is required, a *well* or *socket* may be used, such as shown in Fig. 16.4.

More recently, thin films of metal-glass slurry have been used as resistance elements. These films are deposited onto a ceramic substrate and laser trimmed. Film RTDs are less expensive than the wire RTDs and have a larger resistance for a given size; however, they are also somewhat less stable [4]. Resistance elements similar in construction to foil strain gages are available as well. The resistance grid is deposited onto a supporting film, such as Kapton, which may then be cemented to a surface. These sensors are generally designed to have low strain sensitivity and high temperature sensitivity.

Table 16.2 describes characteristics of several typical commercially available resistance thermometers.

16.4.2   Instrumentation for Resistance Thermometry

Some form of electrical bridge is normally used to measure the resistance change in the RTD. However, particular attention must be given to the manner in which the thermometer is connected into the bridge. Leads of some length appropriate to the situation are required, and any resistance change therein due to any cause, including temperature, may be credited to the thermometer element. It is desirable, therefore, that the lead resistance be

Figure 16.4 Installation assembly for an industrial-type
                 resistance thermometer

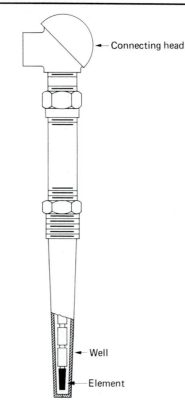

Connecting head

Well

Element

kept as low as possible relative to the element resistance. In addition, some
modification may be used, providing lead compensation.

Figure 16.5(a), (b), and (c) illustrate three different bridge arrangements
used to minimize lead error. Inspection of the diagrams indicates that arms AD
and DC each contain the same lead lengths. Therefore, if the leads have
identical properties to begin with and are subject to like ambient conditions,
the effects they introduce will cancel. In each case the battery and voltmeter
may be interchanged without affecting balance. When the Siemens arrangement is used, however, no current will be carried by the center lead at balance,
as shown. This may be considered an advantage. The Callender arrangement is
quite useful when thermometers are used in both arms AD and DC to provide
an output proportional to temperature differential between the two thermometers. The four-lead arrangement is used in the same way as the one with
three leads. Provision is made, however, for using any combination of three,
thereby permitting checking for unequal lead resistance. By averaging readings, more accurate results are possible. Some form of this arrangement is used
where highest accuracies are desired.

Table 16.2 Typical properties of resistance-thermometer elements

Type of Element	Case Material	Temperature Range °C (°F)	Resistance, Ω	Temperature coefficient, A, Ω/(Ω°C) (approx.)	Limits of* Error, °C	Response,[†] s
Platinum (Laboratory)	Pyrex Glass	−190 to 540 (−310 to 1000)	25 at 0°C	0.0039	±0.01	
Platinum (Industrial)	Stainless Steel	−200 to 125 (−325 to 260)[‡] −18 to 540 (0 to 1000)[§]	25 at 0°C 25 at 0°C	0.0039	±1 ±2	10 to 30 10 to 30
Platinum (Film)	Ceramic Coating	−50 to 600 (−60 to 1100)	1000 at 0°C	0.0039	±0.25	~1
Rhodium-Iron	Alumina and Glass	−272 to 200 (−458 to 390)	27 at 0°C	0.0037	±0.04	
Copper	Brass	−75 to 120 (−100 to 250)	10 at 25°C	0.0038	±0.5	20 to 60
Nickel	Brass	0 to 120 (32 to 250)	100 at 20°C	0.0067	±0.3	20 to 60

* Typical values.

† Time required to detect 90% of any temperature change in water moving at 30 cm/s. The lower value is for the thermometer case only, whereas the higher value is for the thermometer in a protective well. Actual values vary with sensor packaging and flow conditions.

‡ Low range.

§ High range.

Figure 16.5 Four methods for compensating for lead resistance

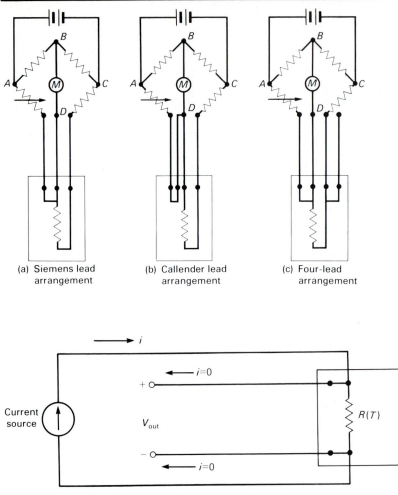

(a) Siemens lead
   arrangement

(b) Callender lead
   arrangement

(c) Four-lead
   arrangement

(d) Four-wire constant-current circuit

The general practice is to use the bridge in the null-balance form, but the deflection bridge may also be used (see Section 7.9). In general, the null-balance arrangement is limited to measurement of static or slowly changing temperatures, whereas the deflection bridge is used for more rapidly changing inputs. Dynamic changes are most conveniently recorded rather than simply indicated, and for this purpose either the self-balancing or the deflection types may be used, depending on time rate of temperature change.

When a resistance bridge is used for measurement, current will necessarily flow through each bridge arm. An error may, therefore, be introduced, caused by i^2R heating. For resistance thermometers such an error will be of opposite sign to that caused by conduction and radiation from the element (Section 16.10.1), and in general it will be small because the gross effects in individual arms will be largely balanced by similar effects in the other arms. An estimate of the overall error resulting from ohmic heating may be had by making readings at different current values and extrapolating to zero current.

Figure 16.5(d) shows a four-wire constant-current circuit for RTD resistance measurement. In this case, the current source holds i constant, irrespective of changes in either lead or sensor resistance. The output voltage, V_{out}, is read with a high-input-impedance meter, so that no current is drawn through the output leads and no voltage drop occurs along them. Thus, the output voltage is a linear function of sensor resistance, $V_{out} = iR(T)$, and it is independent of the lead resistance. However, because this circuit is essentially a ballast-type circuit, it lacks a bridge circuit's sensitivity to small resistance changes (Section 7.8). Also, ohmic heating effects are still present.

16.4.3   Thermistors

The thermistor is a thermally sensitive variable resistor made of a ceramic-like semiconducting material. Unlike metals, thermistors respond negatively to temperature: As the temperature rises, the thermistor resistance decreases (also see Section 6.15.1). Figure 16.6 shows typical temperature-resistance relations.

Thermistors are often composed of oxides of manganese, nickel, and cobalt in formulations having resistivities of 100 to 450,000 $\Omega \cdot$ cm. In cryogenic applications, doped germanium and carbon-impregnated glass are used. Thermistors are available in various forms, such as shown in Fig. 16.7. Table 16.3 lists the properties of certain commercially available thermistors.

The temperature-resistance function for a thermistor is given by the relationship [5]

$$R = R_0 \exp\left[\beta\left(\frac{1}{T} - \frac{1}{T_0}\right)\right] \tag{16.3}$$

in which

$R =$ the resistance at any temperature T, in K,

$R_0 =$ the resistance at reference temperature T_0, in K,

$\beta =$ a constant.

The constant β depends on the thermistor formulation or grade.

When a thermistor is used in an electrical circuit, current normally flows through it, and ohmic heating is generated by its resistance. The temperature of the element is then raised, the amount depending on the rate with which the heat is dissipated. For given ambient conditions, a temperature equilibrium will occur at which a definite resistance value will exist. Through proper

Figure 16.6   Typical thermistor temperature-resistance relations

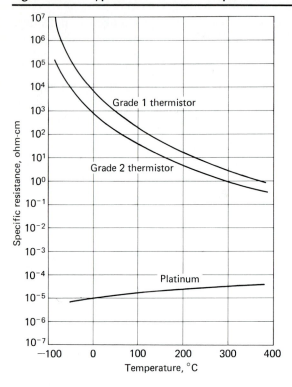

application of thermistor and electrical circuit characteristics, the devices may be used for temperature measurement or control. In addition, they are quite useful for compensating electrical circuitry for changing ambient temperature—largely because of the *negative* temperature characteristics of the thermistor in contrast to *positive* characteristics possessed by most electrical components. Also, time-delay actions over large ranges are possible through proper balancing of electrical and heat-transfer conditions. Figure 16.8 illustrates typical thermistor self-heating response characteristics. Of course, the environment (the heat transfer condition) is a major factor in an actual application. Thermistors can be quite small (a few millimeters in diameter), so their response to changes in ambient temperature may potentially be very rapid.

The inherently high sensitivity possessed by thermistors permits use of very simple electrical circuitry for measurement of temperature. Ordinary ohm-meters may be used within the limits of accuracy of the meter itself. More often one of the various forms of resistance bridge is used (Section 7.9), either in the null-balance form or as a deflection bridge. Simple ballast circuits (Section 7.6) are also usable. In some cases, special *linearizing* circuits are used to obtain an output voltage that varies linearly with temperature.

Figure 16.7   Various thermistor forms commercially available

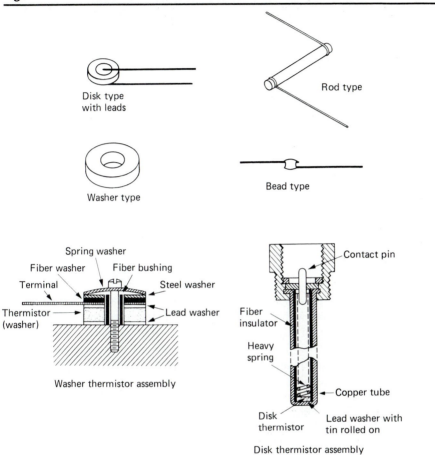

Table 16.3   Typical thermistor specifications

Type (see Fig. 16.7)	Resistance			Approximate Maximum Continuous Temperature, °C
	At 0°C	At 25°C	At 50°C	
Bead	165 kΩ	60 kΩ	25 kΩ	—
Glass-coated bead	8.8 kΩ	3.1 kΩ	1.3 kΩ	300
Washer	28.3 Ω	10 Ω	4.1 Ω	150
Washer	3270 Ω	1000 Ω	360 Ω	150
Rod	103 kΩ	31.5 kΩ	11.3 kΩ	150
Rod	327 kΩ	100 kΩ	36 kΩ	150
Disk	283 Ω	100 Ω	40.7 Ω	125

Figure 16.8 Typical current-time relations for thermistors

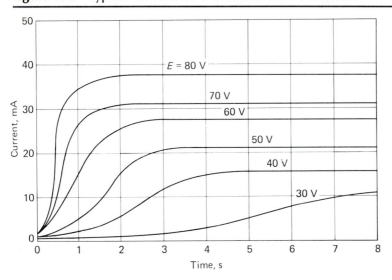

Through use of the thermistor's temperature-resistance characteristics alone, or in conjunction with controlled heat transfer, thermistors have been used for measurement of many quantities, including pressure, liquid level, power, and others. They are also used for temperature control, timing (through use of their delay characteristics in combination with relays), overload protectors, warning devices, etc.

By changing the composition, thermistors can be made to have large *positive* temperature coefficients. Sometimes these *PTC sensors* are made from semiconductor oxides having barium titanate as the main component; in other instances they are made with heavily doped silicon. PTC sensors show an enormous increase of resistance with increasing temperature; this resistance change can be tailored to occur abruptly at a given temperature, which makes PTC thermistors useful as temperature-controlled switching elements. In conjunction with ohmic self-heating, they are also applied as current-limiting devices.

16.5 Thermocouples

In 1821, T. J. Seebeck discovered that an electromotive force exists across a junction formed of two unlike metals [6]. Later it was shown [7, 8] that the potential actually comes from two different sources: that resulting solely from *contact* of the two dissimilar metals at the *junction temperature* and that due to *temperature gradients* along the conductors in the circuit. These two effects are named the Peltier and Thomson effects after their respective discoverers. In most cases the Thomson emf is quite small relative to the Peltier emf, and

Figure 16.9 Elementary thermocouple circuit

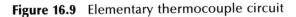

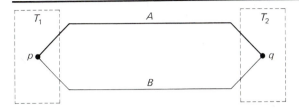

with proper selection of materials may be disregarded. These effects form the basis for a very important temperature-measuring element, the *thermocouple*, often abbreviated TC.

If a circuit is formed including a thermocouple, as shown in Fig. 16.9, a minimum of two conductors will be necessary, unavoidably resulting in two junctions, p and q. If we disregard the Thomson effect, the net emf will be the result of the difference between the two Peltier emf's occurring at the two junctions. If the temperatures T_1 and T_2 are equal, the two emf's will be equal but opposed, and no current will flow. However, if the temperatures are different, the emf's will not balance, and a current *will* flow. The net emf is a function of the two materials used to form the circuit and the *temperatures* of the two junctions. The actual relations, however, are empirical, and the temperature-emf data must be based on experiment. An important fact is that the results are reproducible and therefore provide a reliable method for measuring temperature.

Note particularly that *two* junctions are *always* required. In general, one senses the desired or unknown temperature; this one we shall call the *hot* or *measuring* junction. The second will usually be maintained at a known fixed temperature; this one we shall refer to as the *cold* or *reference* junction.

16.5.1 Application Laws for Thermocouples

In addition to the Seebeck effect, there are certain laws by which thermo-electric circuits abide, as follows:

Law of intermediate metals [9]. Insertion of an intermediate metal into a thermocouple circuit will not affect the net emf, provided the two junctions introduced by the third metal are at identical temperatures.

Applications of this law are shown in Fig. 16.10. As shown in part (a) of the figure, if the third metal C is introduced and if the new junctions r and s are both held at temperature T_3, the net potential for the circuit will remain unchanged. This feature permits insertion of a measuring device or circuit without upsetting the temperature function of the thermocouple circuit. In Fig. 16.10(b) the third metal may be introduced at either a *measuring* or *reference* junction, so long as couples p_1 and p_2 are maintained at the same temperature

Figure 16.10 Diagrams illustrating the law for intermediate metals

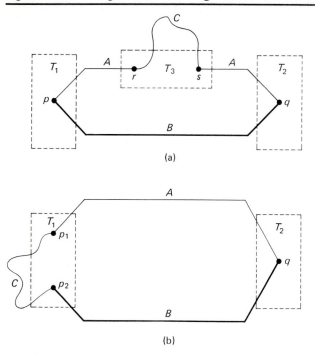

(a)

(b)

T_1. This makes possible the use of joining materials, such as soft or hard solder, in fabricating the thermocouples. In addition, the thermocouple may be actually embedded directly into the surface or interior of either a conductor or nonconductor without altering the thermocouple's usefulness.

> **Law of intermediate temperatures [9].** If a simple thermocouple circuit develops an emf e_1 when its junctions are at temperatures T_1 and T_2 and an emf e_2 when its junctions are at temperatures T_2 and T_3, it will develop an emf $e_1 + e_2$ when its junctions are at temperatures T_1 and T_3.

This makes possible direct correction for secondary junctions whose temperatures may be known but are not directly controllable. It also makes possible the use of thermocouple tables based on a "standard" reference temperature (say 0°C) although neither junction may actually be at the "standard" temperature.

16.5.2 Thermocouple Materials and Installation

Theoretically, any two unlike conducting materials could be used to form a thermocouple. Actually, of course, certain materials and combinations are better than others, and some have practically become standard for given

Figure 16.11 Common forms of thermocouple construction (See
Section 16.5.5 for significance of point *j*)

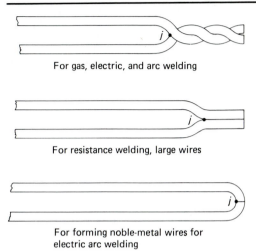

For gas, electric, and arc welding

For resistance welding, large wires

For forming noble-metal wires for
electric arc welding

Figure 16.12 Methods for insulating thermocouple leads

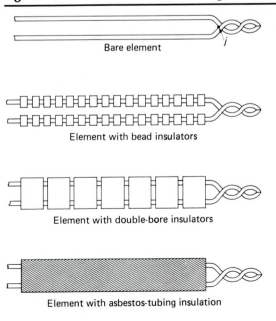

Bare element

Element with bead insulators

Element with double-bore insulators

Element with asbestos-tubing insulation

temperature ranges. Materials and combinations are listed in Table 16.1. Indicated letter designations are ANSI standards.

Wire size is of importance. Usually the higher the temperature to be measured, the heavier should be the wire. As the size increases, however, the time response of the couple to temperature change increases. Therefore, some compromise between response and life may be required.

Thermocouples may be prepared by twisting the two wires together and brazing, or preferably welding, as shown in Fig. 16.11. Low-temperature couples are often used bare; however, for higher temperatures, some form of protection is generally required. Figure 16.12 illustrates common methods for separating the wires, and Fig. 16.4 shows a section through a typical protective tube.

16.5.3 Measurement of Thermal EMF

The actual magnitude of electrical potential developed by thermocouples is quite small when judged in terms of many standards. Table 16.4 provides some idea of the range of values to be expected. Expanded tables have been

Table 16.4 Values of EMF in absolute millivolts for selected metal combinations based on reference junction temperature at 32°F (0°C)*

	Thermocouple Type				
Temperature °F (°C)	Cu vs. Constantan (T)	Chromel vs. Constantan (E)	Iron vs. Constantan (J)	Chromel vs. Constantan (K)	Platinum vs. Platinum, 10% Rhodium (S)
−300 (−184.4)	−5.341	−8.404	−7.519	−5.632	
−200 (−128.9)	−4.149	−6.471	−5.760	−4.381	
−100 (−73.7)	−2.581	−3.976	−3.492	−2.699	
0 (−17.8)	−0.674	−1.026	−0.885	−0.692	−0.092
100 (37.8)	1.518	2.281	1.942	1.520	0.221
200 (93.3)	3.967	5.869	4.906	3.819	0.597
300 (148.9)	6.647	9.708	7.947	6.092	1.020
400 (204.4)	9.523	13.748	11.023	8.314	1.478
500 (260.0)	12.572	17.942	14.108	10.560	1.962
700 (371.1)	19.095	26.637	20.253	15.178	2.985
1000 (537.8)		40.056	29.515	22.251	4.609
1500 (815.6)		62.240		33.913	7.514
2000 (1093.3)				44.856	10.675
2500 (1371.1)				54.845	14.018
3000 (1648.9)					17.347

* See note, Table 16.7.

Table 16.5 Values of thermal EMF in millivolts vs. temperature in degrees celsius for type T thermocouples [Cu (+) vs. constantan (−)] and a reference temperature of 0°C*

°C	0	5	10	15	20
−200	−5.603	−5.522	−5.439	−5.351	−5.261
−175	−5.167	−5.069	−4.969	−4.865	−4.758
−150	−4.648	−4.535	−4.419	−4.299	−4.177
−125	−4.051	−3.923	−3.791	−3.656	−3.519
−100	−3.378	−3.235	−3.089	−2.939	−2.788
−75	−2.633	−2.475	−2.315	−2.152	−1.987
−50	−1.819	−1.648	−1.475	−1.299	−1.121
−25	−0.94	−0.757	−0.571	−0.383	−0.193
0	0	0.195	0.391	0.589	0.789
25	0.992	1.196	1.403	1.611	1.822
50	2.035	2.250	2.467	2.687	2.908
75	3.131	3.357	3.584	3.813	4.044
100	4.277	4.512	4.749	4.987	5.227
125	5.469	5.712	5.957	6.204	6.452
150	6.702	6.954	7.207	7.462	7.718
175	7.975	8.235	8.495	8.757	9.021
200	9.286	9.553	9.820	10.090	10.360
225	10.632	10.905	11.180	11.456	11.733
250	12.011	12.291	12.572	12.854	13.137
275	13.421	13.707	13.993	14.261	14.570
300	14.860	15.151	15.443	15.736	16.030
325	16.325	16.621	16.919	17.217	17.516
350	17.816	18.118	18.420	18.723	19.027
375	19.332	19.638	19.945	20.252	20.560

* See note, Table 16.7.

developed by the National Institute of Standards and Technology [10]. These are adopted as ANSI and ASTM standards. Table 16.5 is adapted from this source and lists emf values for the type T (copper-constantan) thermocouples. The referenced source contains extended tables for all commonly used thermocouples, typically listing values to six significant places at 1°C increments.

Computer processing of data makes it desirable to incorporate the thermocouple data into computer memory in some manner. To do this the power series given in Tables 16.6 and 16.7 may be used [10]. Values derived from Table 16.6 are referred to as being "exact" and those in Table 16.7 are exact within ±0.2°C or better. Additional functional relationships are included in reference [10].

Table 16.6 Power expansion of $E = f(C)$*

TC Type	Applicable Range	$E = f(C)$
(B), Pt. 30% Rh(+) vs. Pt, 6% Rh(−)	0 to 1820°C	$E = (-2.4674601620 \times 10^{-1} \times C$ $- 1.4307123430 \times 10^{-6} \times C^3$ $- 3.1757800720 \times 10^{-12} \times C^5$ $- 9.0928148159 \times 10^{-19} \times C^7$ $+ 5.9102111169 \times 10^{-3} \times C^2$ $+ 2.1509149750 \times 10^{-9} \times C^4$ $+ 2.4010367459 \times 10^{-15} \times C^6$ $+ 1.32990505137 \times 10^{-22} \times C^8) \times 10^{-3}$
(E), Chromel (+) vs. Constantan (−)	0 to 1000°C	$E = (+5.8695857799 \times 10 \times C$ $+ 5.7220358202 \times 10^{-5} \times C^3$ $+ 1.5425922111 \times 10^{-9} \times C^5$ $+ 2.3389721459 \times 10^{-15} \times C^7$ $+ 2.5561127497 \times 10^{-22} \times C^9) \times 10^{-3}$ $+ 4.3110945462 \times 10^{-2} \times C^2$ $- 5.4020668085 \times 10^{-7} \times C^4$ $- 2.4850089136 \times 10^{-12} \times C^6$ $- 1.1946296815 \times 10^{-18} \times C^8$
(J), Fe (+) vs. Constantan (−)	−210 to 760°C	$E = (5.0372753027 \times 10 \times C$ $- 8.5669750464 \times 10^{-5} \times C^3$ $- 1.7022405966 \times 10^{-10} \times C^5$ $- 9.6391844859 \times 10^{-17} \times C^7) \times 10^{-3}$ $+ 3.0425491284 \times 10^{-2} \times C^2$ $+ 1.3348825735 \times 10^{-7} \times C^4$ $+ 1.9416091001 \times 10^{-13} \times C^6$

(K), Chromel (+) vs. Alumel (−)   0 to 1372°C

$$E = (-1.8533063273 \times 10$$
$$+ 1.6645154356 \times 10^{-2} \times C^2$$
$$+ 2.2835785557 \times 10^{-7} \times C^4$$
$$+ 2.9932909136 \times 10^{-13} \times C^6$$
$$+ 2.2239974336 \times 10^{-20} \times C^8$$
$$+ 3.8918344612 \times 10 \times C$$
$$- 7.8702374448 \times 10^{-5} \times C^3$$
$$- 3.5700231258 \times 10^{-10} \times C^5$$
$$- 1.2849848798 \times 10^{-16} \times C^7$$
$$+ 125 \exp[-\tfrac{1}{2}((C - 127)/65)^2]) \times 10^{-3}$$

(R), Pt(−) vs. Pt, 13% Rh(+)   630 to 1064°C

$$E = (-2.6418007025 \times 10^2$$
$$+ 2.9892293723 \times 10^{-3} \times C^2$$
$$+ 8.0468680747 \times 10^0 \times C$$
$$- 2.6876058617 \times 10^{-7} \times C^3) \times 10^{-3}$$

1064 to 1665°C

$$E = (+1.4901702702 \times 10^3$$
$$+ 8.0823631189 \times 10^{-3} \times C^2$$
$$+ 2.8639867552 \times 10^0 \times C$$
$$- 1.9338477638 \times 10^{-6} \times C^3) \times 10^{-3}$$

(S), Pt, 10% Rh(+) vs. Pt(−)   630 to 1064°C

$$E = (-2.9824481615 \times 10^2$$
$$+ 1.6453909942 \times 10^{-3} \times C^2) \times 10^{-3}$$
$$+ 8.2375528221 \times 10^0 \times C) \times 10^{-3}$$

1064 to 1665°C

$$E = (+1.2766292175 \times 10^3$$
$$+ 6.3824648666 \times 10^{-3} \times C^2$$
$$+ 3.4970908041 \times 10^0 \times C$$
$$- 1.5722424599 \times 10^{-6} \times C^3) \times 10^{-3}$$

(T), Cu(+) vs. Constantan (−)   0 to 400°C

$$E = (+3.8740773840 \times 10 \times C$$
$$+ 2.0714183645 \times 10^{-4} \times C^3$$
$$+ 1.1031900550 \times 10^{-8} \times C^5$$
$$+ 4.5653337165 \times 10^{-14} \times C^7$$
$$+ 3.3190198092 \times 10^{-2} \times C^2$$
$$- 2.1945834823 \times 10^{-6} \times C^4$$
$$- 3.0927581898 \times 10^{-11} \times C^6$$
$$- 2.7616878040 \times 10^{-17} \times C^8) \times 10^{-3}$$

* E = thermocouple emf is millivolts referred to 0°C, and C = temperature in °C. Also, see note, Table 16.7.

Table 16.7 Power expansion of $C = g(E)$*

TC Type	Applicable Range	$C = g(E)$
E	0 to 400°C	$C = (+1.7022525 \times 10 \times E - 2.2097240 \times 10^{-1} \times E^2 + 5.4809314 \times 10^{-3} \times E^3 - 5.7669892 \times 10^{-5} \times E^4)$
	400 to 1000°C	$C = (+2.9347907 \times 10 + 1.3385134 \times 10 \times E - 2.6669218 \times 10^{-2} \times E^2 + 2.3388779 \times 10^{-4} \times E^3)$
J	0 to 400°C	$C = (+1.9750953 \times 10 \times E - 1.8542600 \times 10^{-1} \times E^2 + 8.3683958 \times 10^{-3} \times E^3 - 1.3280568 \times 10^{-4} \times E^4)$
	400 to 760°C	$C = (9.2808351 \times 10 + 5.4463817 \times E + 6.5254537 \times 10^{-1} \times E^2 - 1.3987013 \times 10^{-2} \times E^3 + 9.9364476 \times 10^{-5} \times E^4)$
K	400 to 1000°C	$C = (-2.4707112 \times 10 + 2.9465633 \times 10 \times E - 3.1332620 \times 10^{-1} \times E^2 + 6.5075717 \times 10^{-3} \times E^3 - 3.9663834 \times 10^{-5} \times E^4)$
T	0 to 400°C	$C = (+2.5661297 \times 10 \times E - 6.1954869 \times 10^{-1} \times E^2 + 2.2181644 \times 10^{-2} \times E^3 - 3.5500900 \times 10^{-4} \times E^4)$

*C = temperature in °C, and E = thermocouple emf in millivolts. Referred to 0°C.

Note: The values in Tables 16.4–16.7 are based on the International Practical Temperature Scale of 1968 (ITPS-68; see Section 2.8). These relations will be superseded by the International Temperature Scale of 1990 (ITS-90) when the National Institute of Standards and Technology releases Monograph 175 [12] in mid-1993. For temperatures below 600°C, the new values will differ by less than 0.1°C. For temperatures between 600°C and 2000°C, the difference is less than 0.75°C.

At a given temperature, type E thermocouples have the highest output voltage among common types, but this voltage is still a value measured in millivolts. The sensitivity of thermocouples is also relatively low. For example, type E voltage (in Table 16.4) increases from 2.281 to 5.869 mV as temperature increases from 100°F to 200°F; the average change per degree Fahrenheit is only about 36 μV. Because of these factors, thermocouples require accurate and sensitive voltage measurement and, in practice, cannot be used reliably for temperature changes of less than about 0.1°C.

Traditionally, thermocouple output has been measured through use of a voltage-balancing potentiometer (see [11]). Today, a high-quality digital voltmeter is sufficient. Moreover, solid-state or integrated-circuit devices have largely replaced the cumbersome manual methods in most applications.

Solid-state temperature measurement instrumentation includes digital readout "thermometers" and recorders and controllers. TC thermometers are available for most thermocouple types and provide direct digital readout. Typical specifications are

Range	-100 to 1000°C
Resolution	0.1 to 1°C
Response time	Less than 2 s
Input impedance	100 MΩ
Selectable scale (either °F or °C)	

For discussion of basic thermocouple behavior, we shall ignore digital devices and focus instead on voltage-measuring approaches.

Figure 16.13 Schematic diagram illustrating use of potentiometer terminals as a reference junction

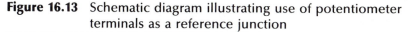

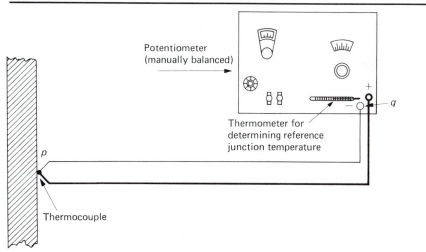

Figure 16.13 shows a simple temperature-measuring system using a thermocouple as the sensing element and an old-style potentiometer for indication. In this illustration, the thermoelectric circuit consists of a measuring junction, p, and a somewhat less obvious reference junction, q, at the potentiometer. Comparison with Fig. 16.10(b) indicates that the instrument box may be considered an intermediate conductor in the same sense as C in the figure. If we assume the two instrument binding posts to be at identical temperature, the cold junction will then be formed by the ends of the two thermocouple leads as they attach to the posts. If a reference temperature is determined by use of a good liquid-in-glass thermometer placed near the binding posts, application of the law of intermediate metals and use of the tables referred to 0°C permits determination of the hot-junction temperature.

Example 16.1

Let us assume an arrangement as shown in Fig. 16.13, using a type T (copper-constantan) thermocouple, a reference temperature of 20°C, determined as just described, and a potentiometer reading of 2.877 mV. The object is to find T_m, the temperature sensed by the measuring couple.

Solution. Because our readout is referenced to 20°C and the TC tables are referenced to 0°C, we must use the law of intermediate temperatures to correct our emf value, as follows:

$$E_{x_0} = E_{x_{20}} + E_{20_0},$$

where E_{x_0} and $E_{x_{20}}$ are emf's corresponding to the unknown temperature referred to 0 and 20°C, respectively, and E_{20_0} is the emf corresponding to 20°C referred to 0°.

Using Table 16.5 we find $E_{20_0} = 0.789 \text{ mV}$; hence,

$$E_{x_0} = 2.887 + 0.789 = 3.666 \text{ mV}.$$

Inspection of the tabulated values yields $T_m = 86 + \text{ °C}$.

Although we would not normally use the relationships in Tables 16.6 and 16.7 for longhand calculations, we will illustrate the use of the polynomials to check a value obtained above. What emf do we obtain from the equation, for 20°C, using a type T thermocouple?

$$E_{20} = 774.8155 + 13.2761 + 1.6571 - 0.3511 + \cdots \text{ mV}$$
$$= 0.78943 \text{ V}.$$

16.5.4 Extension Wires

Thermocouple wire is relatively expensive compared to most common materials, such as ordinary copper. It is therefore often desirable to minimize the use of the more costly materials by employing leads. Arrangements of this are shown in Fig. 16.14. In these cases the measuring junction is shown at p and the reference junction(s) at q. Comparison with Fig. 16.10(a) indicates the similarity. Of course, a requirement for accuracy is that q_1 and q_2 be maintained at the same temperature, and further, that the temperature be accurately known.

An iron-constantan couple is indicated in Fig. 16.14(a), although any of the common materials could be used. Should a copper-constantan couple be used, with copper extension leads, the arrangement would become that shown in Fig. 16.10(b). As before, temperature T_r must be accurately known. This corresponds to the simplified version shown in Fig. 16.10(a).

As indicated above, reference temperature T_r should be known. This may not always be strictly true. In certain commercial equipment, particularly of the recording type, electrical or electromechanical compensation is sometimes

Figure 16.14 Diagrams showing use of extension leads

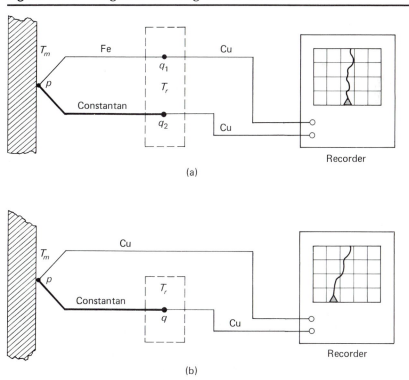

(a)

(b)

Figure 16.15 Systems with fixed reference temperature (ice bath)

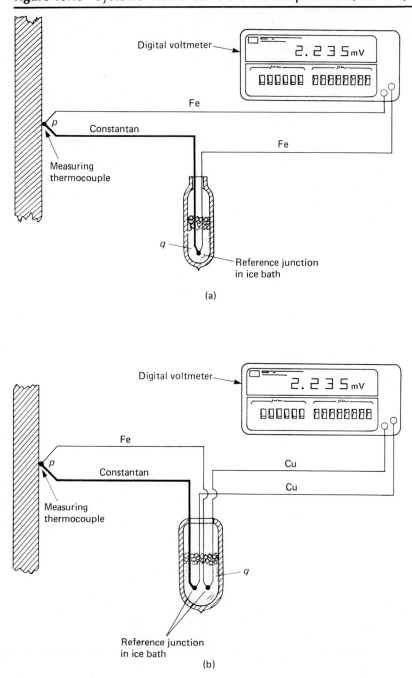

(a)

(b)

built in. When this is done, T_r may not actually be indicated, but its effect is nevertheless recognized and compensated.

Special formulated extension wires are available for each type of TC, minimizing the effects of small temperature variations at intermediate junctions. These wires are less expensive than primary wire for an entire run, but more expensive than copper wire. Use of the special wires is particularly advisable in critical installations.

Laboratory methods for using thermocouples often employ reference junctions at accurately controlled temperatures. The common arrangements make use of ice baths, such as shown in Fig. 16.15. These systems correspond to simplified circuits shown in Figs. 16.10(a) and (b), respectively. (One circuit uses extension leads, while the other does not). For the most accurate work, distilled water with ice made therefrom is advisable to eliminate shifts in the freezing point caused by contaminants. Some form of dewar flask is convenient to use to reduce melting rate.

Although the ice bath supplies an easily obtained, accurately controlled reference temperature, any other controlled source could be employed using the same procedures. For example, a stable reference voltage may be used in place of the reference thermocouple. Such *electronic icepoints* are elements of solid-state TC thermometers.

16.5.5 Effective Junction

In certain instances it may be highly desirable to know, as precisely as possible, the location of the effective thermocouple, junction—that is, where, within the dimensional extent of the couple, the indicated temperature occurs. This requirement becomes of greater importance as both the temperature gradient and the size of the couple are increased. In general terms, the *effective location* is at the point of junction symmetry nearest the leads (points j in Figs. 16.11 and 16.12). Baker, Ryder, and Baker [13] define the area $D \times d$ (Fig. 16.16) as the region of uncertainty in a "bead-type" couple, when there is a temperature gradient through the junction.

Figure 16.16 Region of uncertainty for a bead-type thermocouple

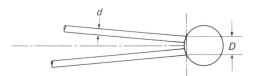

16.5.6   Thermopiles and Thermocouples Connected in Parallel

Thermocouples may be connected electrically in series or parallel, as seen in Fig. 16.17. When connected in series, the combination is generally called a *thermopile*, whereas parallel-connected couples have no particular name.

The total output from *n* thermocouples connected to form a thermopile [Fig. 16.17(a)] will be equal to the sums of the individual emf's, and if the thermocouples are identical, the total output will equal *n* times the output of a single couple. The purpose of using a thermopile rather than a single thermocouple is, of course, to obtain a more sensitive element.

When the couples are combined in the form of a thermopile, it is usually desirable to cluster them together as closely as possible in order to measure the temperature at an approximate point source. It is obvious, however, that when thermocouples are combined in series, the law for intermediate metals, as illustrated in Fig. 16.10(b), cannot be applied to combinations of thermo-couples, for the individual thermocouple emf's would be shorted. Care must therefore be used to ensure that the individual couples are electrically insulated one from the other.

Figure 16.17   (a) Series-connected thermocouples forming a thermopile, (b) parallel-connected thermocouples

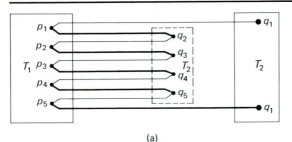

(a)

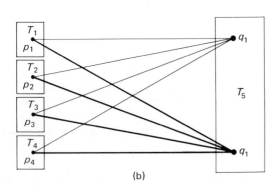

(b)

Parallel connection of thermocouples provides an averaging, which in certain cases may be advantageous. This form of combination is *not* usually referred to as a thermopile.

16.6 Semiconductor-Junction Temperature Sensors

The junction between differently doped regions of a semiconductor has a voltage-current curve that depends strongly on temperature (see Section 6.15). This dependence has been harnessed in two types of temperature sensors: diode sensors and monolithic integrated-circuit sensors [Fig. 16.18(a)]. Like many semiconductor sensors, these devices have maximum operating temperatures of 100 to 150°C. Both types can be small, having dimensions of a few millimeters.

Semiconductor diode sensors, when properly calibrated, are the more accurate. The diode is powered with a fixed forward current of about 10 μA, and the resulting forward voltage is measured with a four-wire constant-current circuit [the diode replaces the resistor in Fig. 16.5(d)]. The diode's forward voltage is a decreasing function of temperature, known from the calibration [Fig. 16.18(b)]. Diode sensors can be accurate to about 50 mK for temperatures between 1.4 K and 300 K [14]. Typical diodes are either silicon or gallium-aluminum-arsenide, and they are often applied in cryogenic temperature measurements. Precision diodes are relatively expensive.

Monolithic integrated circuit devices use silicon transistors to generate an output current proportional to absolute temperature. A modest voltage (4 to 30 V) is applied to the sensor and the current through the circuit is monitored with an ammeter [Fig. 16.18(c)]. One such sensor is the Analog Devices AD590, which produces a current in microamperes numerically equal to the absolute temperature in kelvin (e.g., 298 μA at 298 K or 25°C). Because the device is a current source, its susceptibility to voltage noise and lead-wire errors is minimal. IC temperature sensors are quite inexpensive (a few dollars). They are applied as sensors for control circuits, as temperature-compensation elements in precision electronics, and even as electronic ice points for thermocouple circuits. Accuracy is about 0.5°C.

16.7 The Linear-Quartz Thermometer

The relationship between temperature and the resonating frequency of a quartz crystal has long been recognized. In general, the relationship is nonlinear, and for many applications very considerable effort has been expended in attempts to minimize the frequency drift caused by temperature variation. Hammond [15] discovered a new crystal orientation called the "LC" or "linear cut," which provides a temperature-frequency relationship of 1000 Hz/°C with a deviation from the best straight line of less than 0.05% over

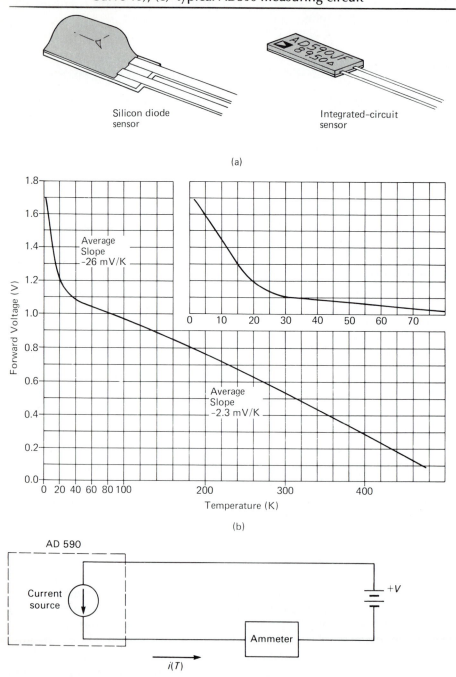

Figure 16.18 (a) Semiconductor junction sensors (Courtesy of Lake Shore Cryotronics, Inc., and Analog Devices, Inc.), (b) diode forward voltage versus temperature (Lake Shore Standard Curve 10), (c) typical AD590 measuring circuit

Silicon diode sensor

Integrated–circuit sensor

(a)

Average Slope −26 mV/K

Average Slope −2.3 mV/K

Forward Voltage (V)

Temperature (K)

(b)

AD 590

Current source

Ammeter

+V

i(T)

(c)

a range of $-40°C$ to $230°C$ ($-40°F$ to $446°F$). This linearity may be compared with a value of 0.55% for the platinum-resistance thermometer.

Nominal resonator frequency is 28 MHz, and the sensor output is compared to a reference frequency of 28.208 MHz supplied by a reference oscillator. The frequency difference is detected, converted to pulses, and passed to an electronic counter, which provides a digital display of the temperature magnitude. Various probes are available, all with time constants of 1 s. Resolution is dependent on repetitive readout rate, with a value of $0.0001°C$ attainable in 10 s. Readouts as fast as four per second may be obtained. Absolute accuracy is rated at $\pm0.040°C$. Remote sensing to 3000 m is possible.

16.8 Pyrometry

The term *pyrometry* is derived from the Greek words *pyros*, meaning "fire" and *metron*, meaning "to measure." Literally, the term means general temperature measurement. However, in engineering usage, the word normally (not always) refers to the measurement of temperatures in the range extending upward from about $500°C$ ($\approx1000°F$). Although certain thermocouples and resistance-type thermometers can be used above $500°C$, pyrometry generally implies thermal-radiation measurement of temperature.

Electromagnetic radiation extends over a wide range of wavelengths (or frequencies), as illustrated in Fig. 16.19. Pyrometry is based on sampling the

Figure 16.19 The electromagnetic radiation spectrum

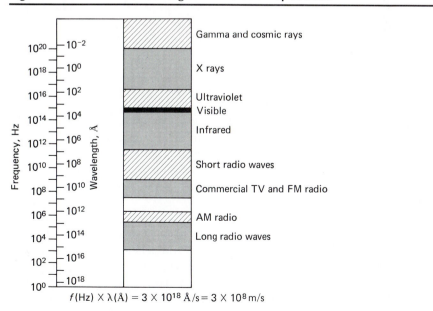

$f(\text{Hz}) \times \lambda(\text{Å}) = 3 \times 10^{18} \text{ Å/s} = 3 \times 10^8 \text{ m/s}$

energies in certain bandwidths of this spectrum. At any given wavelength, a body radiates energy of an intensity that depends on the body's temperature. By evaluting the emitted energy at known wavelengths, the temperature of the body can be found.

Pyrometers are essentially photodetectors designed specifically for temperature measurement. Like photodetectors (Section 6.16), pyrometers are of two general types: *thermal detectors* and *photon detectors*. Thermal detectors are based on the temperature rise produced when the energy radiated from a body is focused onto a target, heating it. The target temperature may be sensed with a thermopile, a thermistor or RTD, or a pyroelectric element (Section 6.14). Photon detectors use semiconductors of either the photoconductive or photodiode type. In those devices, the sensor responds directly to the intensity of radiated light by a corresponding change in its resistance or in its junction current or voltage.

Pyrometers may also be classified by the set of wavelengths measured. A *total-radiation pyrometer* absorbs energy at all wavelengths or, at least, over a broad range of wavelengths (such as all visible wavelengths). An *optical* (or *brightness*) *pyrometer* measures energy at only one specific wavelength; a variant of this approach, the *two-color pyrometer,* compares the energy at two specific wavelengths. Perhaps the most commonly used type is the *infrared pyrometer,* which, rather obviously, determines temperature from measurements over a range of infrared wavelengths.

Because bodies near room temperature radiate most of their energy at infrared wavelengths, the infrared pyrometer has overturned the traditional perception of pyrometry as a strictly high-temperature technique. In fact, the Greek meaning of pyrometry, mentioned before, is no longer so inappropriate. However, pyrometers retain the distinguishing feature of finding an object's temperature without having direct contact with it.

16.8.1 Pyrometry Theory

All bodies above absolute zero temperature radiate energy. Not only do they radiate or emit energy, but they also receive and absorb it from other sources. We all know that when a piece of steel is heated to about 550°C it begins to glow; i.e., we become conscious of visible light being *radiated* from its surface. As the temperature is raised, the light becomes brighter or more intense. In addition, there is a change in color; it changes from a dull red, through orange to yellow, and finally approaches an almost white light at the melting temperature (1430°C to 1540°C).

We know, therefore, that through the range of temperatures from approximately 550°C to 1540°C, energy in the form of *visible light* is radiated from the steel. We can also sense that at temperatures below 550°C and almost down to room temperature, the piece of steel is still radiating energy or *heat* in the form of *infrared radiation,* for if the mass is large enough we can feel it even though we may not be touching it. We know, then, that energy is radiated through certain temperature ranges because our senses provide the necessary information. Although our senses are not as good at lower temperatures, on occasion one can actually "feel" the presence of cold walls in a room because heat is being radiated from one's body *to* the walls. Energy transmission of this sort does not require an intervening medium for conveyance; in fact, intervening substances actually interfere with transmission.

The energy of which we are speaking is transmitted as electromagnetic waves or photons traveling at the speed of light. All substances emit and absorb radiant energy at a rate depending on the absolute temperature and physical properties of the substance. Radiation striking the surface of a material is partially absorbed, partially reflected, and partially transmitted. These portions are measured in terms of *absorptivity* (α), *reflectivity* (ρ), and *transmissivity* (τ), where

$$\alpha + \rho + \tau = 1. \tag{16.4}$$

For an ideal reflector, a condition approached by a highly polished surface, $\rho \to 1$. Many gases represent substances of high transmissivity, for which $\tau \to 1$. A small opening into a large cavity approaches an ideal absorber, or *blackbody*, for which $\alpha \to 1$.

A body in radiative equilibrium with its surroundings emits as much energy as it absorbs. It follows, therefore, that a good absorber is also a good radiator, and it may be concluded that the *ideal radiator* is one for which the value of α is equal to unity. When we refer to emitted radiation as distinguished from absorption, the term *emissivity* (ε) is used rather than absorptivity (α). However, the two are directly related by *Kirchhoff's law*,

$$\varepsilon = \alpha.$$

Table 16.8 lists values of emissivities for certain materials.

Table 16.8 Total emissivity for certain surfaces [16]

Surface	Temperature, °C	Emissivity
Polished silver	225–625	0.0198–0.0324
Platinum filament	25–1225	0.036–0.192
Polished nickel	23	0.045
Aluminum foil	100	0.087
Concrete	21	0.63
Roofing paper	20	0.91
Plaster	10–88	0.91
Rough red brick	21	0.93
Asbestos paper	38–371	0.93–0.945
Smooth glass	22	0.937
Water	0–100	0.95–0.963
Blackbody	—	≈1.00

According to the Stefan-Boltzmann law, the net rate of exchange of energy between two ideal radiators A and B that view only each other is

$$q = \sigma(T_A^4 - T_B^4). \tag{16.5}$$

In the case when a nonideal object A radiates to a perfectly absorbing object B (as well as to other relatively cool objects), the expression must be modified:

$$q = \varepsilon_A F_{BA} \sigma(T_A^4 - T_B^4), \tag{16.6}$$

in which

$\quad q$ = net radiant heat transfer or heat flux to B from A, in W/m^2,

$\quad \varepsilon_A$ = the emissivity of object A,

$\quad F_{BA}$ = configurational factor to allow for relative position and geometry of bodies,

T_A and T_B = absolute temperatures of objects A and B, in K,

$\quad \sigma$ = the Stefan-Boltzmann constant, $5.6697 \times 10^{-8}\,W/m^2K^4$.

This expression forms the basis of thermal-detector total-radiation pyrometry. The detecting element, B, receives heat flux q from the measured object A. A prior calibration has established the temperature, T_B, to which the detecting element rises when exposed to a given flux. Hence, the unknown temperature T_A can be inferred from the detector temperature.

A number of complicating factors must be considered in practice. The emissivity of the body viewed may be unknown. Gases and dust between the two objects may scatter and absorb radiation. Heat radiated from surrounding bodies may be reflected from the unknown object into the pyrometer, particularly if the unknown body has a low emissivity. The geometric relation of the unknown body and the detector may vary. Each of these effects will alter the heat flux predicted by Eq. (16.6). *Calibration* of the pyrometer is thus essential to accurate measurement; and the user must always maintain an awareness of the potential complications.

As we mentioned earlier, the *color* changes with increasing temperature. Change in color, of course, corresponds to change in wavelength and the wavelength of *maximum* radiation decreases with increase in temperature. A decrease in wavelength shifts the color from the reds toward the yellows. Steel at 540°C has a deep red color. At 815°C the color is a bright red, and at 1200°C the color appears white. The corresponding radiant energy *maximums* occur at wavelengths of 3.5, 2.6, and 1.9 μm, respectively.

If we should heat an ideal radiator to various temperatures and determine the relative intensities at each wavelength, we would obtain characteristic energy-distribution curves such as those shown in Fig. 16.20. Not only is the radiation intensity of the higher-temperature body increased, but the wavelength of maximum emission is also shifted toward shorter waves (from red

Figure 16.20 Graphical representation illustrating the basis for Wien's displacement law

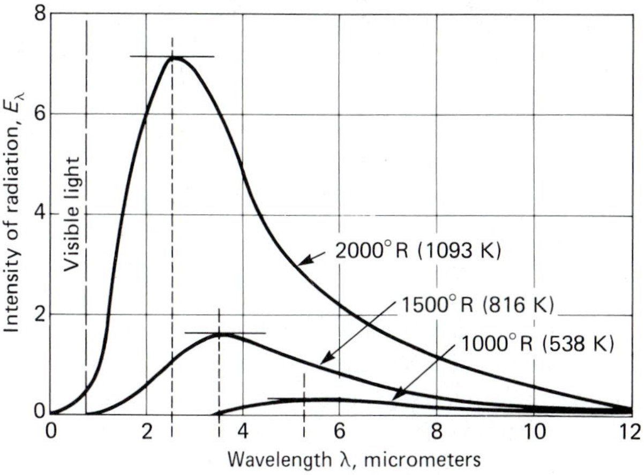

toward blue). The intensity distribution for an ideal radiator (blackbody) may be expressed as follows [17]:

$$E_\lambda = \frac{C_1}{\lambda^5(e^{C_2/\lambda T} - 1)} \tag{16.7}$$

in which

$\quad E_\lambda$ = the energy emitted by wavelength λ, in W/m^2 μm,

$\quad T$ = the absolute temperature, in K,

$\quad \lambda$ = the wavelength, in μm,

$\quad C_1$ = 374.15 MW μm^4/m^2,

$\quad C_2$ = 14388 μm K.

For a nonideal body, the intensity distribution must be multiplied by the emissivity, ε. The wavelength of peak intensity is given by the *Wien displacement law*:

$$\lambda_{max} T = 2898 \ \mu\text{m K} \tag{16.7a}$$

These relations are the basis for optical and two-color pyrometers. Filters are used to eliminate all but the wavelengths of interest, whose intensities are then measured. The body's temperature is calculated from the measured intensity. As in the total-radiation approach, potential sources of interference must be considered, and a calibration must usually be performed.

16.8.2   Total-Radiation Pyrometry

Figure 16.21 shows, in simplified form, the method of operation of a thermal-detector total-radiation pyrometer. Essential parts of the device consist of some light-directing means, shown here as baffles but which are more often lenses, and an approximate blackbody receiver with means for sensing temperature. Although the sensing element may be any of the types discussed earlier in this chapter, it is generally some form of thermopile or pyroelectric sensor; occasionally, a thermistor or gas-pressure thermometer is used. A balance is quickly established between the energy absorbed by the receiver and that dissipated by conduction through leads and emission to surroundings. The receiver equilibrium temperature then becomes the measure of source temperature, with the scale established by calibration.

Figure 16.22 shows a sectional view of a commercially available pyrometer. Although total-radiation pyrometry is primarily used for temperatures above 550°C, the pyrometer shown is selected to illustrate an instrument sensitive to very low-level radiation (50°C–375°C). The arrangement, however, is typical of general radiation pyrometry practice. A lens-and-mirror system is used to focus the radiant energy on a thermopile. Thermocouple reference temperature is supplied by maintaining the assembly at constant temperature through use of a heater controlled by a resistance thermometer. In many cases compensation is obtained through use of temperature-compensating resistors in the electrical circuit.

Particular attention must be given to the optical system of a radiation pyrometer, and appropriate optical glasses must be selected to pass the

Figure 16.21 A simplified form of total-radiation pyrometer

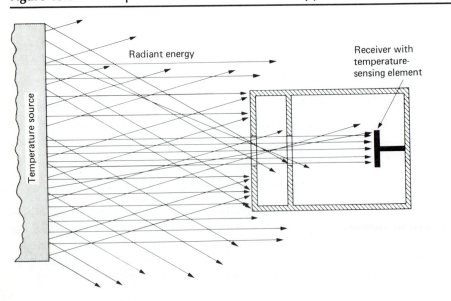

Figure 16.22 Section through a commercially available low-temperature, total radiation pyrometer (Courtesy: Honeywell, Inc., Process Control Division, Ft. Washington, PA)

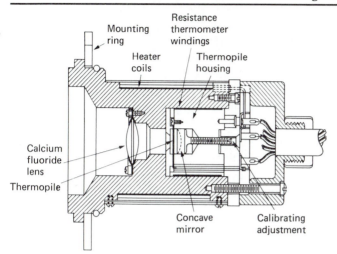

necessary range of wavelengths. Pyrex glass may be used for the range of 0.3 to 2.7 μm, fused silica for 0.3 to 3.8 μm, and calcium fluoride for 0.3 to 10 μm. Although Pyrex glass may be used for high-temperature measurement, it is practically opaque to low-temperature radiation, say below 550°C. By choosing other lens materials or introducing appropriate optical filters, a thermal-detector device like this one may alternatively be made to sense only infrared wavelengths (2 to 14 μm, for instance).

Radiation pyrometers are used ideally in applications where the sources approach blackbody conditions—that is, where the source has an emissivity ε approaching unity. In general, however, the radiated energy is a good measure of temperature only if the applicable value of ε is accounted for. This may be made clear by inspection of Eq. (16.6). Although pyrometers calibrated for blackbody conditions are available, in general they must be calibrated for the particular application. Calibration consists of comparing the pyrometer read-out with that of some standardized device, such as a thermocouple. Often single-point calibration suffices. Devices for adjusting total-radiation pyrometer calibration include the following:

1. Movable aperture in front of the thermopile.

2. Variable thermopile aperture area or pyrometer lens or window area.

3. Movable metal plug screwed into the thermopile housing adjacent to the hot junction.

4. Movable concave mirror reflecting varying amounts of energy back to the thermopile of a lens-type pyrometer (see Fig. 16.22).

5. Variable shunt resistor in the electrical circuit.

Although radiation pyrometers may theoretically be used at any reasonable distance from a temperature source, there are practical limitations that should be mentioned. First, the size of target will largely determine the degree of temperature averaging, and in general, the greater the distance from the source, the greater the averaging. Second, the nature of the intervening atmosphere will have a decided effect on the pyrometer indication. If smoke or dust is present or if certain gases or solids, even though they may appear to be transparent, are in the path, considerable energy absorption may occur. This problem will be particularly troublesome if such absorbents are not constant but vary with time. For these reasons, minimum practical distance is advisable, along with careful selection of pyrometer sighting methods.

There are three common arrangements used to obtain a sample of radiated energy for the thermopile to sense:

1. A lens system that has the power to concentrate the sampled energy over a smaller area, but is subject to aberrations and must be kept carefully cleaned.

2. An open-end tube (illustrated schematically in Fig. 16.21). In effect, this is a selective "baffle" arranged to transmit energy from a selected area only.

3. A closed-end sighting tube, usually of ceramic, that may be inserted into a furnace or immersed in a liquid bath.

It should be clear that the primary purpose of these sighting methods and devices is to obtain a true sample from the area or point of interest, uninfluenced by any surrounding conditions.

16.8.3   Optical Pyrometry

Optical pyrometers measure radiant intensity at only one or two specific wavelengths, which are isolated by use of appropriate filters. The intensity is found either by visual comparison to a calibrated source or by using the output of a calibrated thermal or photon detector. The temperature dependence of the intensity distribution, Eq. (16.7), provides the necessary relation between measured intensity and temperature.

The *disappearing filament pyrometer* is an example of the visual comparison type (Fig. 16.23). The intensity of an electrically heated filament is varied to match the source intensity at a particular wavelength. In use, the pyrometer is sighted at the unknown temperature source at a distance such that the objective lens focuses the source in the plane of the lamp filament. The eyepiece is then adjusted so that the filament and the source appear superimposed and in focus to the observer. In general, the filament will appear either hotter than or colder than the unknown source, as shown in Fig. 16.24. When the battery current is adjusted, the filament (or any prescribed portion such as the tip) may be made to disappear, as indicated in Fig. 16.24(c). The current indicated by the milliammeter to obtain this condition may then be used as the temperature readout. Other readout arrangements use the rheostat

Figure 16.23 Schematic diagram of an optical pyrometer

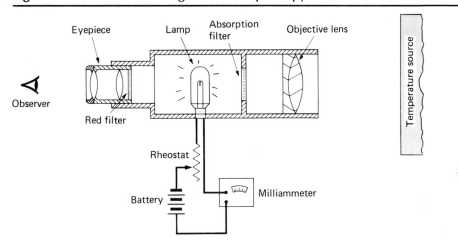

setting, in which case battery standardization or some form of potentiometric null-balancing system is required.

A red filter is generally used to obtain approximately monochromatic conditions, and an absorption filter is used so that the filament may be operated at reduced intensity, thereby prolonging its life.

The source emissivity, ε, is just as important for optical pyrometers as for total-radiation pyrometers. At a given wavelength, imperfect emissivity reduces the source intensity [Eq. (16.7)] by a factor of ε, to $\varepsilon(\lambda)E_\lambda$, where in general ε may depend on wavelength. *Two-color pyrometry* is an optical technique that minimizes the influence of the emissivity. Specifically, the two-color pyrometer measures source intensity at two wavelengths, λ_1 and λ_2. If the emissivity is independent of wavelength or if the wavelengths are nearly

Figure 16.24 Appearance of filament when (a) filament temperature is too high, (b) filament temperature is too low, and (c) filament temperature is correct.

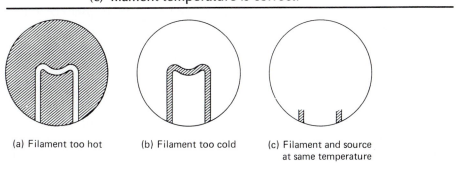

equal, then the ratio of measured intensities depends only on temperature:

$$\frac{\varepsilon(\lambda_1)E_{\lambda_1}}{\varepsilon(\lambda_2)E_{\lambda_2}} \approx \frac{E_{\lambda_1}}{E_{\lambda_2}} = \left(\frac{\lambda_2}{\lambda_1}\right)^5 \frac{(e^{C_2/\lambda_2 T} - 1)}{(e^{C_2/\lambda_1 T} - 1)}$$

Color-ratio pyrometry is the defining temperature standard for ITS-90 in the range above 1234.93 K (Section 2.8); it is usually executed with a high-precision optical pyrometer [2].

16.8.4 Infrared Pyrometry and IR Thermography

The infrared region begins at a wavelength of about $0.75\,\mu m$—where the visible region ends—and extends upward to wavelengths of about $1000\,\mu m$. *Infrared pyrometry* is simply an adaptation of either total-radiation or optical pyrometry to infrared wavelengths, usually by filtering the radiation received from a particular source. The benefit of infrared sensing is found in Wien's displacement law (Fig. 16.20), which shows that the peak radiant intensity of low-temperature bodies occurs at infrared wavelengths. For example, a body at 25°C (298 K) radiates at a peak wavelength of $9.7\,\mu m$. Thus, infrared detection is essential to radiant measurements of near-room-temperature objects.

Common infrared detectors sense some portion of the interval between 2 and $14\,\mu m$. This interval corresponds to peak radiant temperatures between about 200 and 1400 K. However, this interval is also dictated by the need to avoid infrared absorption by air between the source and the detector: atmospheric absorption is at a minimum in several bands between 2 and $5\,\mu m$ and in the interval from 8 to $14\,\mu m$. Most commercial pyrometers are centered on one of these transmitting bandwidths. Specialized devices, such as satellite-based far-infrared detectors, may respond to wavelengths up to $100\,\mu m$; however, detectors for such long wavelengths generally must be operated at cryogenic temperatures.

Pyroelectric materials are particularly useful as broadband IR (infrared) sensors. These materials have an intrinsic electrostatic polarization that decreases with an increase in temperature. Pyroelectric elements, then, act as thermal detectors that respond to a change in temperature by developing a charge, in much the same way that piezoelectric materials respond to strain. Pyroelectric materials include ferroelectric crystals, such as triglisine sulfate (TGS) and lithium tantalate ($LiTaO_3$), and some organic polymer films, such as polyvinylidene fluoride (PVDF) [18].

Like piezoelectric charge, pyroelectric charge dissipates in time. To obtain measurements of steady temperatures, a rotating "chopper" may be used to periodically interrupt the radiation, producing a square-wave signal whose peak amplitude is related to the target temperature [19]. Electrical signal conditioning requirements are otherwise similar to those of piezoelectric devices (Section 7.18.2). Apart from their use in temperature measurement, pyroelectric sensors are commonly used in motion sensors and burglar alarms.

Focusing optics for infrared applications are complicated by the poor infrared transmissivity of ordinary glasses. Infrared lenses and windows are sometimes made from germanium, which has a broadband IR transmissivity. Mirrors are usually of the first-surface type, having the reflective coating on top rather than beneath a layer of glass [19].

One important aspect of infrared pyrometry is its application to whole-field temperature measurement, called *thermal imaging* or *infrared thermography*. IR thermography is used in medical imaging, in testing buildings for heat leakage, in satellite surveys, and in measuring temperature distributions in electronic equipment. Many devices operate by scanning the image with a single, cooled photon detector. More recent designs take advantage of the CCD (charge-coupled diode) array technology developed for silicon-diode digital video cameras. Here, a two-dimensional array of detectors is positioned behind a camera lens to record the thermal image of the field viewed. At present, arrays of thermally detecting pyroelectric sensors are usually employed. Surprisingly, photon detectors (such as silicon diodes) are not currently used in IR arrays; this is because the candidate materials either lack broadband sensitivity above 1μm (like silicon) or are difficult to fabricate as detector arrays. Future developments in fabrication technology (particularly for photoconductive HgCdTe) may change this state of affairs.

16.9   Other Methods of Temperature Indication

One method of temperature measurement given in the introduction to this chapter has not been referred to in the intervening pages. It is the application of changes chemical state or phase. Several devices based on this principle should be mentioned.

Seger cones have long been used in the ceramic industry as a means of checking temperatures. These devices are simply small cones made of an oxide and glass. When a predetermined temperature is reached, the tip of the cone softens and curls over, thereby providing the indication that the temperature has been reached. Seger cones are made in a standard series covering a range from 600°C to 2000°C.

Somewhat similar temperature-level indicators are available in the forms of crayonlike sticks, lacquer, and pill-like pellets. Each may be calibrated at temperature intervals through a range of about 50°C to 1100°C. The crayon or lacquer is stroked or brushed on the part whose temperature is to be indicated. After the lacquer dries, it and the crayon marks appear dull and chalky. When the calibration temperature is reached, the marks become liquid and shiny. The pellets are used in a similar manner, except that they simply melt and assume a shiny liquidlike appearance as the stated temperature is reached. By using crayons, lacquer, or pellets covering various temperatures within a range, the maximum temperature attained during a test may be rather closely determined.

Liquid crystals are perhaps the most colorful of temperature indicators. The liquid crystal is a meso-phase state of certain organic compounds that shares properties of both liquids and crystals. As temperature increases past a threshold, liquid crystals successively scatter reds, yellows, greens, blues, and violets until an upper threshold temperature is reached. By changing the crystal composition, the entire color change can made occur over an interval of as much as 50°C or as little as 1°C. Liquid crystals are useful from roughly −20°C to 100°C; they may resolve temperature changes as small as 0.1°C. Liquid crystals are usually coated in a thin film over a blackened plastic sheet that makes the scattered light more visible; the sheet is then affixed to the surface whose temperature distribution is to be observed.

16.10   Special Problems

The number of special problems associated with temperature measurement is unlimited. However, several are significant enough to warrant special note. These will be discussed in the next several pages.

16.10.1   Errors Resulting from Conduction and Radiation

In considering this issue, let us think in terms of using a thermocouple to measure the gas temperature in a furnace, bearing in mind, however, that the principles discussed apply to other temperature probes and to many other situations.

Basically, any temperature element senses temperature because heat is transferred between the surroundings and the element until some kind of equilibrium condition is reached. When a bare thermocouple is inserted through the wall of a furnace (assume it to be gas- or coal-fired), heat is transferred to it from the immersing gases by convection. Heat also reaches the element through radiation from the furnace walls and from incandescent solids such as a fuel bed or those carried along by the swirling gases. Finally, heat will flow from the element through any connecting leads by conduction. The temperature indicated by the probe therefore will be a function of all these environmental factors, and consideration must be given to their effects in order to intelligently interpret or control the results.

First of all, in the more common case, the major heat flow will occur directly by forced convection between the gases and the probe. This may be expressed by the relation [20]

$$Q = h_c A(T_g - T_i), \tag{16.8}$$

in which

Q = the heat transferred, in W,

h_c = the coefficient of heat transfer, in W/m² K,

A = the surface area of the probe, in m²,

T_g = the gas temperature, in K,

T_t = the probe temperature, in K.

Although the transfer coefficient, h_c, is a function of a number of things, including viscosity, density, and specific heat of the gas, of particular importance in the application under discussion is the fact that it also is a function of a power of the velocity of gas over the probe.

Radiation Effects

Radiation between the probe and any source or sink of different temperature is a function of the difference in the fourth powers of the (absolute) temperatures; see Eq. (16.6). It is generally true, therefore, that radiation becomes an increasingly important source of temperature error as the temperatures and their differences increase. Increased temperature differences generally result from temperature extremes, either high or low, and in either of these cases particular attention must be given to radiation effects.

As discussed in Section 16.8.1, radiant-heat transfer is also a function of the emissivities of the members involved. For this reason, a bright, shiny probe is less affected by thermal radiation than is one tarnished or covered with soot.

Radiation error may be largely eliminated through proper use of thermal shielding. This consists of placing barriers to thermal radiation around the probe, which prevent the probe from "seeing" the radiant source or sink, as the case may be. For low-temperature work, such shields may simply be made of sheet metal appropriately formed to provide the necessary protection. At higher temperatures, metal or ceramic sleeves or tubes may be used. In applications where gas temperatures are desired, however, care must be exercised in placing radiation shields so as not to cause stagnation of flow around the probe. As pointed out earlier, desirable convection transfer is a function of gas velocity.

Consideration of these factors led quite naturally to the development of an aspirated high-temperature probe known as the *high-velocity thermocouple* (HVT) [21]. Figure 16.25 illustrates an aspirated probe with several types of tips. Gas is induced through the end, over the temperature-sensing element, and either is exhausted to the exterior or, if it will not alter process or

Figure 16.25 (a) Section through an aspirated high-velocity thermocouple, HVT, (b) various tips used on high-velocity thermocouples (Courtesy: Babcock & Wilcox, Barberton, OH)

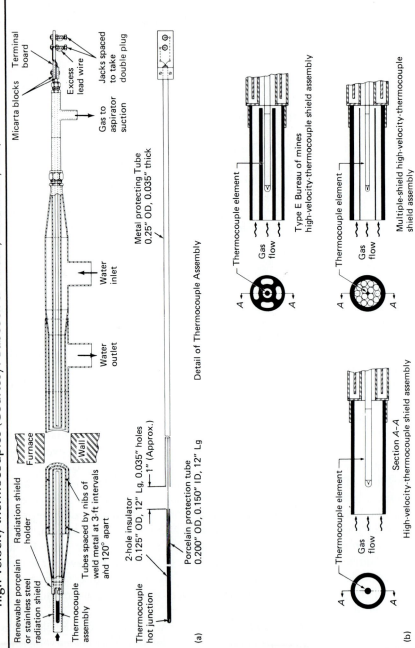

measuring functioning, may be returned to the source. A renewable shield provides radiation protection for the element, and through use of aspiration, convective transfer is enhanced. Gas mass-flow over the element should be not less than 15,000 lbm/h/ft^2 for maximum effectiveness [21].

When a single shield is used, as shown in Fig. 16.25(b), the shield temperature is largely controlled by convective transfer from the aspirated gas through it. Its exterior, however, is subject to thermal-radiation effects, and thus its equilibrium temperature, and hence that of the sensing element, will still be somewhat influenced by radiation. Maximum shielding may be obtained through use of multiple shields, as shown in the lower two sections of Fig. 16.25(b). Thermocouples using multiple shielding are known as *multiple high-velocity thermocouples* (MHVT) [21]. The effectiveness of both the HVT and MHVT relative to a bare thermocouple is graphically illustrated in Fig. 16.26.

Our discussion here of radiation effects has been centered largely on high-temperature application of thermocouples. However, once again it should be made clear that the principles involved apply to *any* temperature-measuring system or situation to one extent or another. Radiation may introduce errors at low temperatures as well as at high ones and will present similar problems to all types of sensing elements. When the fluids are liquid rather than gaseous, the problem is essentially eliminated because most liquids, and to some degree water vapor in air, act as effective thermal-radiation filters.

Errors Caused by Conduction

All temperature-measuring elements of the probe type must have mechanical support, and, in general, some connection must be made to external indicating apparatus. Such connections provide conduction paths through which heat may be transferred to or away from the sensing element. Such transfer of heat will result in a discrepancy between the indicated temperature and that desired— namely, the temperature that would exist were the instrument not present. Factors influencing such errors may be itemized as follows:

1. Conductivity of lead or support material

2. Lead and element sizes

3. Properties of surrounding media

4. Flow conditions over the probe

5. Presence of lead insulation or protective well

6. Configuration of the immersed leads

7. Temperature magnitudes and the form of temperature gradient along the leads or support

8. Depth of immersion of the probe

Figure 16.26   Graphical representation of the effectiveness of the high-velocity thermocouple (Courtesy: Babcock & Wilcox, Barberton, OH)

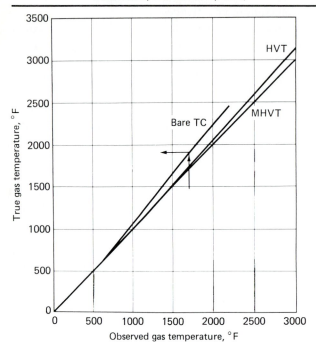

Johnson, Weinstein, and Osterle [22] have found that, except for extreme conditions, variation in the last two factors may generally be ignored. In addition, they show that increasing the ratio of element to lead size reduces the error, or more specifically, a large value of ηL minimizes error ($\eta^2 = 2h/kd$, where h = the convection coefficient, k = the conductivity of the lead wire, L = a length indicating gradient intensity, and d = the diameter of the lead wire). It is also shown that the lead error may be reduced to zero by using a reversed lead configuration of proper proportions, as illustrated in Fig. 16.27, wherein the leads are bent back and brought downstream.

16.10.2   Measurement of Temperature in Rapidly Moving Gas

When a temperature probe is placed in a stream of gas, the flow will be partially stopped by the presence of the probe. The lost kinetic energy will be converted to heat, which will have some bearing on the indicated temperature. Two "ideal" states may be defined for such a condition. A *true* state would be that observed by instruments moving with the stream, and a *stagnation* state would be that obtained if the gas were brought to rest and its kinetic energy

Figure 16.27 Reversed lead configuration, which may be
used to reduce lead error

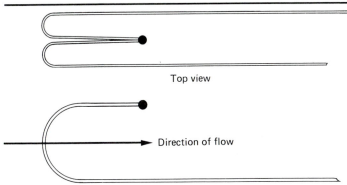

Top view

Direction of flow

completely converted to heat, resulting in a temperature rise. A fixed probe
inserted into the moving stream will indicate conditions lying between the two
states. For exhaust gases from internal combustion engines, we find that
temperature differences between the two states may be as great as 200°C [23].

An expression relating stagnation and true temperature for a moving gas,
assuming adiabatic conditions, may be written as follows [24]:

$$T_t - T_s = \frac{V^2}{2g_c J C_p}. \tag{16.9}$$

This relation may also be written

$$\frac{T_s}{T_t} = 1 + \tfrac{1}{2}(k - 1)M^2, \tag{16.9a}$$

in which

T_s = the stagnation or total temperature, in K (or °R),

T_t = the true or static temperature, in K (or °R),

V = the velocity of flow, in m/s (or ft/s),

g_c = the dimensional constant, $1 \text{ kg} \cdot \text{m/N} \cdot \text{s}^2$ (or $32.2 \text{ lbm} \cdot \text{ft/lbf} \cdot \text{s}^2$),

J = the mechanical equivalent of heat, $1 \text{ N} \cdot \text{m} = 1 \text{ J}$
  (or $778 \text{ ft} \cdot \text{lbf/Btu}$),

C_p = the mean specific heat at constant pressure, in J/kg · K
  (or Btu/lbm · °R),

k = the ratio of specific heats,

M = the Mach number.

A measure of the effectiveness of a probe in bringing about kinetic energy conversion may be expressed by the relation

$$r = \frac{T_i - T_t}{T_s - T_t},$$

(16.10)

where

T_i = the temperature indicated by the probe,

r = a term called the *recovery factor*, which is proportional to the energy conversion.

If $r = 1$, the probe would measure the stagnation temperature, and if $r = 0$, it would measure the true temperature. Experiment has shown that for a given instrument, the recovery factor is essentially a constant and is a function of the probe configuration. It changes little with composition, temperature, pressure, or velocity of the flowing gas [24].

Combining Eqs. (16.9) and (16.10), we obtain

$$T_t = T_i - \frac{rV^2}{2g_c C_p J},$$

(16.11)

or

$$T_s = T_i + \frac{(1 - r)V^2}{2g_c C_p J}.$$

(16.11a)

The recovery factor, r, for a given probe may be determined experimentally [23]. However, this approach does not generally provide sufficient information to determine either the true or the stagnation temperature. Inspection of Eqs. (16.11) and (16.11a) indicates that in addition to knowing the indicated temperature T_i and the recovery factor r, we must know the stream velocity and certain properties of the fluid. When these values are known, the relations yield the desired temperatures directly. In many cases, however, it is particularly difficult to determine the flow velocity, and further theoretical consideration of the situation is required.

It has been shown [24] that for sonic velocities ($M = 1$),

$$T_s = \phi T_i,$$

(16.12)

in which

$$\phi = \frac{k + 1}{2 + r(k - 1)}.$$

(16.13)

One solution to the temperature measurement of high-velocity gases has been to make the measurement at Mach 1, through use of an instrument called a *sonic-flow pyrometer.* Such a device is shown in Fig. 16.28. The basic instrument comprises a temperature-sensing element (thermocouple) located at the throat of a nozzle. Gas whose temperature is to be measured is aspirated (or pressurized by the process) through the nozzle to produce critical or sonic

Figure 16.28 Schematic diagram of a sonic-flow pyrometer (Courtesy: of the National Institute of Standards and Technology)

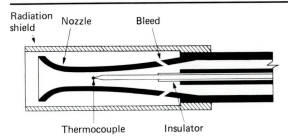

Radiation shield   Nozzle   Bleed

Thermocouple   Insulator

velocity at the nozzle throat. Under these conditions, Eqs. (16.12) and (16.13) apply, and in this manner determination of flow velocity need not be made. It is still necessary to know the ratio of specific heats, but these can usually be determined or estimated with sufficient accuracy. (It may be observed that the dependence of ϕ on k reduces as r is increased.)

16.10.3 Temperature Element Response

An ideal temperature transducer would faithfully respond to fluctuating inputs regardless of the time rate of temperature change; however, the ideal is not realized in practice. A time lag exists between cause and effect, and the system seldom, if ever, actually indicates true temperature input. Figure 16.29 illustrates quite graphically the magnitude of errors that may result from poor response.

The time lag that exists is determined by the particular heat transfer circumstances that apply, and the complexity of the situation depends to a large extent on the relative importance of the convective, conductive, and radiative components. If we assume that radiation and conduction are minimized by design and application, we may equate the heat accepted by the probe per unit time to the rate of heat transfer by convection [25]:

$$Wc\left(\frac{dT_p}{dt}\right) = hA(T_g - T_p) \tag{16.14}$$

or

$$\tau\left(\frac{dT_p}{dt}\right) + T_p = T_g, \tag{16.14a}$$

in which

T_p = the temperature of the probe, in K (or °R),

T_g = the temperature of the surrounding gases, in K (or °R),

c = the specific heat of the probe, in J/kg · K (or Btu/lbm · °R),

Figure 16.29 Temperature-time record made from two thermocouples of different size and location during the starting cycle of a large jet engine (Courtesy: Instrument Society of America)

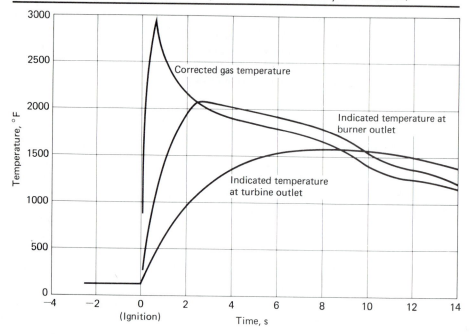

W = the mass of the probe, in kg (or lbm),

t = the time, in s,

h = the convective heat-transfer coefficient, in W/m²K
    (or Btu/s · ft² · °R),

A = the surface area of the probe exposed to gases, in m² (or ft²),

$\tau = Wc/hA$ = the time constant, in s.

We may write Eq. (16.14a) as follows:

$$\int_0^t dt = \tau \int_{T_{p_0}}^T \frac{dT_p}{T_g - T_p}. \qquad (16.14b)$$

Solving gives us

$$T_p = T_g - (T_g - T_{p_0})e^{-t/\tau}.$$

If we let

$$\Delta T_p = T_p - T_{p_0},$$

then

$$\Delta T_p = (T_g - T_{po})(1 - e^{-t/\tau}). \qquad (16.15)$$

This relation corresponds to suddenly exposing the probe at temperature T_{po} to a gas temperature T_g. This would be approximated if the probe were quickly inserted through the wall of a furnace or immersed in a liquid bath.

In response to a steady sinusoidal variation in temperature of angular frequency ω, indicated temperature will oscillate with reduced amplitude and will lag in phase and time [25] (see also the example in Section 5.15.2).

The value of τ will be recognized as the *time constant* or *characteristic* time [26] for the probe, or the time in seconds required for 63.2% of the maximum possible change $(T_g - T_p)$ (see Section 5.15.1). Obviously, τ should be as small as possible, and inspection shows, as should be expected, that this condition corresponds to low mass, low heat capacity, high transfer coefficient, and large area. Probes with low time constant provide fast response, and vice versa.

Even under idealized conditions (convective transfer only), as assumed, the time constant for a given probe is not determined by the probe alone. The convective heat transfer coefficient is also dependent on the character of the gas flow. For this reason, a given probe may show different time constants when subjected to different conditions.

In general, two parameters, total temperature (Section 16.10.2) and mass velocity, are sufficient to describe the flow. Moffat [26] gives the following empirical equation for evaluating the time constant for bare-wire thermocouples:

$$\tau = \frac{3500 \rho c d^{1.25}}{T} G^{-15.8/\sqrt{T}}, \qquad (16.16)$$

where

d = the wire diameter, in in.,

G = the mass velocity, in lbm/s · ft^2,

T = the total temperature, in °R,

ρ = the *average* density for the two wires, in lbm/ft^3,

c = the *average* specific heat for the two wires, in Btu/lbm · °F.

A comparison between time constants calculated by Eq. (16.16) and determined from test data is shown in Fig. 16.30.

Although practical probe response characteristics may, in many cases, be closely approximated by the application of Eq. (16.15), in many other cases more complicated situations exist. Other elements in addition to the actual temperature-sensing element may be involved, resulting in the multiple-time-constant problem.

The case of the common thermometer in a well or a thermocouple or resistance thermometer in a protective sheath (Fig. 16.4) may be better

Figure 16.30   A comparison of time constants calculated by Eq. (16.16)
to those determined from test data (Courtesy:
Instrument Society of America)

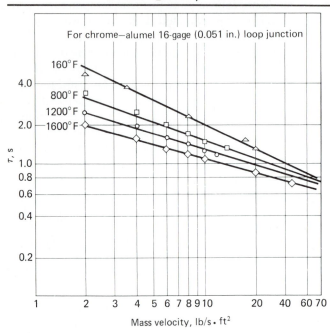

approximated by a two-time-constant model. Both probe and jacket will have
characteristic time constants. Let us analyze this situation as follows. We will
assume that a probe-jacket assembly (Fig. 16.31) at temperature T_1 is suddenly
inserted into a medium at temperature T_2. In the manner of Eq. (16.14), we
may write two relationships, as follows:

$$W_j c_j \frac{dT_j}{dt} = h_j A_j (T_2 - T_j) - h_p A_p (T_j - T_p) \tag{16.17}$$

and

$$W_p c_p \frac{dT_p}{dt} = h_p A_p (T_j - T_p), \tag{16.17a}$$

where subscripts j and p refer to the protective jacket and the probe,
respectively.

The relationships may be rewritten as

$$\tau_j \frac{dT_j}{dt} = T_2 - T_j - \frac{h_p A_p}{h_j A_j} (T_j - T_p) \tag{16.18}$$

and

$$\tau_p \frac{dT_p}{dt} = T_j - T_p. \tag{16.18a}$$

Figure 16.31 Temperature probe in jacket subjected to a step change in temperature

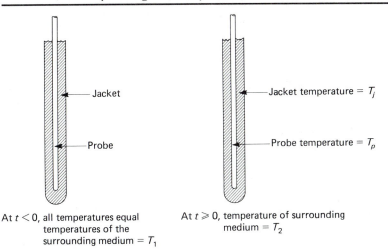

At $t < 0$, all temperatures equal temperatures of the surrounding medium $= T_1$

At $t \geqslant 0$, temperature of surrounding medium $= T_2$

Simplification may be obtained if we assume that the last term in Eq. (16.18) may be neglected. This assumption will be legitimate if

$$A_p \ll A_j \quad \text{and/or} \quad T_j - T_p \ll T_2 - T_j.$$

If this assumption is made, Eqs. (16.18) and (16.18a) may be combined, yielding

$$\tau_j \tau_p \frac{d^2 T_p}{dt^2} + (\tau_j + \tau_p) \frac{dT_p}{dt} + T_p = T_2. \tag{16.19}$$

A solution to this relationship is

$$\frac{T_2 - T_p}{T_2 - T_1} = \frac{\Delta T}{\Delta T_{max}} = \left(\frac{\zeta}{\zeta - 1}\right) e^{-t/\zeta \tau_p} - \left(\frac{1}{\zeta - 1}\right) e^{-t/\tau_p}, \tag{16.20}$$

where

$\Delta T =$ the momentary difference between the indicated and actual temperatures,

$\Delta T_{max} =$ the difference between the temperature of the medium and the probe temperature at $t = 0$,

$$\zeta = \frac{\tau_j}{\tau_p}.$$

Characteristics for various values of ζ are shown in Fig. 16.32. It is seen that for $\zeta = 0$, Eq. (16.20) reverts to Eq. (16.15). In addition, as the time constant for the well is increased, the overall lag is increased, as one would suspect it should be.

Figure 16.32 Two-time constant problem: Plot of $\Delta T/\Delta T_{max}$ versus t/τ_p for various ratios of $\zeta = \tau_j/\tau_p$

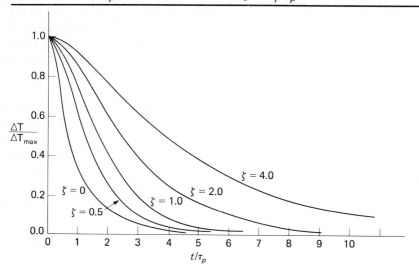

Figure 16.33 Curves illustrating compensating action of a simple RC network (Courtesy: Instrument Society of America)

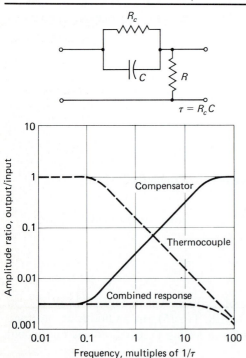

Still more refined methods are sometimes applied, using multiple time constants and using dead times [27, 28]. For example, careful consideration of the simple mercury-in-glass thermometer indicates that the glass envelope, in addition to functioning as a necessary part of the differential-expansion pair, also acts as a thermal shield for the mercury.

16.10.4   Electrical Compensation

Lag in electrical temperature-sensing elements (thermocouples and resistance thermometers) may be compensated approximately by use of appropriate electrical networks. The technique involves selecting a type of filter (Section 7.21) whose electrical-time characteristics complement those of the sensing element [29, 30, 31]. Figure 16.33 illustrates a simple form of such a compensator. In the example illustrated, thermocouple response drops off with increased input frequency (as shown in terms of multiples of time-constant reciprocals). By proper choice of resistors and capacitance, satisfactory combined response may be extended approximately 100 times.

16.11   Measurement of Heat Flux

Heat flux is the *rate* of heat flow per unit area. The common units are W/m^2 or $Btu/h \cdot ft^2$. We can write an expression for heat flux as follows:

$$q = -k\frac{dT}{dx}, \tag{16.21}$$

where

q = heat flux, in W/m^2,

k = the thermal conductivity of the material, in W/mK,

T = temperature, in K,

x = material dimension in direction of flow, in m.

Knowledge of heat flux rather than temperature is of particular value in designing systems to *avoid* excessive temperatures. Examples might be at locations on supersonic aircraft, gas turbine blades, combustor walls in rocket motors, etc.

Heat flux gages are of several forms [32, 33, 34], of which three have particular importance: the slug type (Fig. 16.34), the foil or membrane type (Fig. 16.35), which is also known as the Gardon gage [35], and the thin-film layered type (Figure 16.36).

As shown in Fig. 16.34, the essentials of the slug-type meter include a concentrated mass or slug that is thermally insulated from its surroundings and a temperature sensor, commonly a thermocouple. As heat flows in, the thermal isolation of the slug produces a temperature differential between the

Figure 16.34   Section through a slug-type heat flux sensor

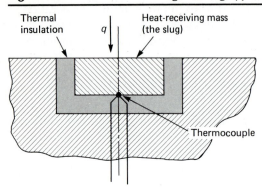

slug and its surroundings. The governing relation is

$$q = \frac{Mc}{A}\frac{dT}{dt} + U\,\Delta T, \qquad\qquad (16.22)$$

where

    M = the mass of the slug, in kg,

    c = the specific heat of the slug, in J/kg K,

    A = the surface area of the slug, in m^2,

    T = the slug temperature, in K,

    t = the time, in s,

    U = the coefficient of loss to surroundings, in W/m^2 K,

  ΔT = the temperature difference between the slug and the surroundings.

Figure 16.35   Section through a foil- or membrane-type
                   heat flux sensor (Gardon gage)

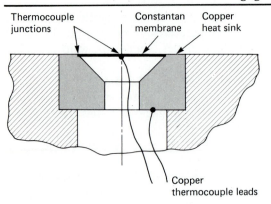

Figure 16.36   Thin-film layered heat-flux gage (vertical scale exaggerated)

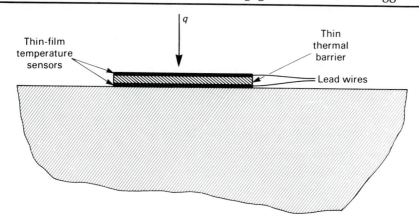

Slug temperature is measured by the sensor and, through calibration, is the analogue of flux. This gage is also called a *thermal capacitance calorimeter.*

This type of flux meter has two primary disadvantages: (1) the assumption is that temperature throughout the slug is uniform at all times, but for high fluxes this will not be true, and (2) the meter is clearly not usable for steady-state conditions.

Construction of the Gardon gage [35] or *asymptotic calorimeter,* is shown in Fig. 16.35. It consists of an embedded copper heat sink, a thin membrane of constantan, and an integral thermocouple. The nature of the construction provides two copper-constantan thermocouple junctions, one at the center of the membrane and the other at the interface between the membrane and the heat sink. Thermocouple output, therefore, is a function of the temperature differential between the center and the periphery of the membrane. This, in turn, is a function of the rate of heat flow from the membrane into the sink. The governing relationship is

$$q = 2\left(\frac{Sk}{R^2}\right)\Delta T$$

$$= Ce, \tag{16.23}$$

where

S = the membrane thickness,

k = the thermal conductivity of the membrane material,

R = the membrane radius,

ΔT = the temperature difference between the center and the edge of the membrane,

C = a calibration constant,

e = the millivolt output of the TC.

The layered gage is shown in Fig. 16.36. Here temperature sensors are attached to the upper and lower surfaces of a thin, thermally resisting layer. The heat flux is obtained directly from the measured temperature difference by approximating Eq. (16.21):

$$q = -k\frac{dT}{dx} \approx k\frac{\Delta T}{\delta} \tag{16.24}$$

where

k = the thermal conductivity of the resisting film,

ΔT = the temperature difference between the upper and lower surface,

δ = the thickness of the resisting film.

Layered gages became practical as high-frequency response sensors only with the advent of microfabrication technology, which made it possible to deposit thin-film temperature sensors (0.1 to 0.5 μm thick) onto relatively thin (1 to 75 μm) thermal barriers. In some applications, the sensors are sputtered onto each side of a Kapton film [36]; in other designs, the sensors and the thermal barrier are sequentially sputtered onto a supporting ceramic substrate [37]. The thin-film sensors are usually either RTDs or thermocouples. Because the temperature difference across the thin barrier is very small, a differential thermopile may be used to obtain ΔT [37, 38]. Between 20 and 100 thermocouple junctions are connected in series, with successive junctions located on opposite sides of the film.

Calibration of heat flux meters involves radiant, conductive, and convective factors and is not simple [39]. One form of standard is patterned after the slug-type meter and uses a gold or single-crystal copper slug. Emissivities must be carefully controlled and short exposure to a heat source is used to avoid large temperature rise. Water and blackbody standards are also used. Application details are beyond the scope of this book.

16.12   Calibration of Temperature-Measuring Devices

As stated in Section 1.7, for the results to be meaningful, measuring procedure and apparatus must be provable. This statement is true for all areas of measurement, but for some reason the impression seems prevalent that it is less true for temperature-measuring systems than for others. For example, it is generally thought that the only limitation in the use of thermocouple tables is in satisfying the requirement for metal combination indicated in the table heading. Mercury-in-glass scale divisions and resistance-thermometer characteristics are commonly accepted without question. And it is assumed that once proved, the calibrations will hold indefinitely.

Of course, we know that these ideas are incorrect. Thermocouple output is very dependent on purity of elementary metals and consistency and homogeneity of alloys. Alloys of supposedly like characteristics but manufactured by different companies may have temperature-emf relations sufficiently at variance to require different tables. In addition, aging with use will alter thermocouple outputs. Resistance-thermometer stability is very dependent on the degree of freedom from residual strains in the element, and comparative results from like elements require very careful use and control of the metallurgy of the materials.

Methods used to calibrate temperature-measuring systems fall into two general classifications: (1) comparison with the primary standards or the fixed temperature points as specified by the International Temperature Scale of 1990, and (2) comparison with reliably calibrated secondary standards. (See Section 2.8.)

Basically, the primary temperature standard consists of the fixed physical conditions (melting points, triple points, etc.) discussed in Section 2.8. Intermediate points are established by specified interpolation procedures. Therefore, for primary calibration, the problems are those of technique attending reproduction of these fundamental points and reproduction of interpolation methods.

In addition to the primary fixed points established by ITS-90, numerous secondary fixed points have been tabulated (see, for example, [40]). Some examples include the sublimation point of carbon dioxide or dry ice ($-78.5°C$), the boiling point of water ($100°C$), and the boiling point of napthalene ($218°C$). For many fixed points, commercial "standards" are available. These standards consist of a sealed container enclosing the reference materials. Glass is used for the container material for the lower temperatures and graphite is used for the higher temperatures. Integral heating coils are employed. To use the standard, the element to be calibrated is placed in a well extending into the center of the container. The heater is then turned on, and the temperature carried above the melting point of the reference substance and held until melting is completed. It is then permitted to cool, and when the freezing point is reached, the temperature stabilizes and remains constant at the specified value as long as liquid and solid are both present (several minutes). It is claimed that accuracies of approximately $0.1°C$ may be easily attained and that $0.01°C$ may be attained if care is exercised.

Suggested Readings

ASME 19.3-1974 (R1986) Instruments and Apparatus: Part 2. *Temperature Measurement.* New York: 1986.

ASTM *Standards on Thermocouples.* 2nd ed. Philadelphia: ASTM.

Benedict, R. P. *Fundamentals of Temperature, Pressure and Flow Measurements.* 2nd ed. New York: Wiley, 1977.

Booth, S. F. (ed.). *Precision Measurement and Calibration,* vol. 2 (selected NBS technical papers on heat and mechanics), NBS Handbook 77. Washington, D.C.: U.S. Government Printing Office, 1961.

Burns, G. W. and M. G. Scroger. *Temperature-Electromotive Force Reference Functions and Tables for Letter-Designated Thermocouple Types Based on the ITS-90.* NIST Monograph 175. Washington, D.C.: U.S. Department of Commerce, National Institute of Standards and Technology, 1992. (Supercedes NBS Monograph 125).

Ginnings, D. C. *Precision Measurement and Calibration* (selected NBS papers on heat), NBS Special Publication 300, vol. 6. Washington, D.C.: U.S. Government Printing Office, 1979.

Kinzie, P. A. *Thermocouple Temperature Measurement.* New York: Wiley, 1973.

Kutz, M. *Temperature Control.* New York: Wiley, 1968.

Lienhard, J. H. *A Heat Transfer Textbook,* 2nd ed. Englewood Cliffs, N.J.: Prentice Hall, 1987.

Mangum, B. W., and G. T. Furukawa. *Guidelines for Realizing the International Temperature Scale of 1990 (ITS-90).* NIST Technical Note 1265. Washington, D.C.: U.S. Department of Commerce, National Institute of Standards and Technology, 1990.

McGee, T. D. *Principles and Methods of Temperature Measurement.* New York: John Wiley, 1988.

Preston-Thomas, H. The International Temperature Scale of 1990 (ITS-90). *Metrologia* 27: 3–10, 1990 (with corrections in *Metrologia* 27: 107, 1990).

Swindells, J. F. *Precision Measurement and Calibration* (selected NBS papers on temperature), NBS Special Publication 300, vol. 2, Washington, D.C.: U.S. Government Printing Office, 1968.

Problems

16.1 At what temperature readings do the Celsius and Fahrenheit scales coincide?

16.2 The temperature indicated by a "total immersion" mercury-in-glass thermometer is 70°C (158°F). Actual immersion is to the 5°C (41°F) mark. What correction should be applied to account for the partial immersion? Assume ambient temperature is 20°C (68°F).

16.3 The uncertainty of a thermometer is stated to be ±1% of "full scale." If the thermometer range is −20°C to 120°C, plot the uncertainty based "on reading" over the thermometer's range.

16.4 The following relation [41] may be used to determine the radius of curvature, r, of a bimetal strip that is initially flat at temperature T_0:

$$r = \frac{t\{3(1 + m)^2 + (1 + mn)[m^2 + (1/mn)]\}}{6(\alpha_2 - \alpha_1)(T - T_0)(1 + m)^2},$$

where

t = the combined thickness of the two strips,

m = the ratio of thicknesses of low- to high-expansion components.

n = the ratio of Young's modulus values of low- to high-expansion components,

α_1 and α_2 = coefficients of linear expansion, with $\alpha_1 < \alpha_2$,

T = the temperature, in °C or °F, depending on the units for α_1 and α_2.

a. Devise a spreadsheet template (SST) to solve for r, using the preceding equation.

b. If 12-cm-long by 1-mm-thick strips of phos-bronze and Invar are brazed together to form a bimetal temperature sensor, determine the deflection of the free end per degree change in temperature. Recall that for a beam in bending, $1/r = d^2y/dx^2$. See Table 6.3 for material properties. (Let $T_0 = 20°C$ and $T = 100°C$.)

16.5 Search the literature (see the Suggested Readings at the end of the chapter) for the range of values for the constants A and B in Eq. (16.2) for commonly used resistance thermometer materials.

16.6 The element of a resistance thermometer is constructed of a 50-cm (19.7-in.) length of 0.03-mm (0.0012-in.) nickel wire. What will be the nominal resistance of the element? (See Table 12.1 for resistivity.) If we assume that the temperature coefficient of resistivity is constant over the common range of ambient temperatures, what will be the change in resistance of the element per degree C? Per degree F?

16.7 If platinum is substituted for nickel in Problem 16.6, what are the calculated values?

16.8 The circuit shown in Fig. 16.5(a) is used with a platinum resistance thermometer having a resistance of 1200 Ω at 200°C. Also, $R_{AB} = R_{BC} = 8000$ Ω. $R_{DC} = 6800$ Ω. Using the data for platinum listed in Problem 3.40, plot the bridge output voltage (assume a high-impedance readout device) versus temperature over a range of 0°C to 500°C. Refer to Sections 7.9 through 7.9.3 for bridge circuit relationships.

16.9 Investigate the techniques used by the various automobile manufacturers for measuring and displaying engine block temperatures. What accuracies do you think are obtained by the various systems?

16.10 Devise a simple thermistor calibration facility consisting of a variable-temperature environment, an accurate resistance-measuring means that avoids ohmic heating of the element, and a reliable temperature-measuring system to be used as the "standard." Calibrate several thermistors and evaluate their degree of adherence to Eqs. (16.3). (Avoid the problems implied in Problem 16.14.)

16.11 Prepare spreadsheet templates for Eq. (16.3) to solve:
a. for R when T, T_0, R_0, and β are given;
b. for β when T, T_0, R, and R_0 are given.

16.12 The following are data for the calibration of a thermistor.

Temperature T, °F	Resistance R, kΩ
78	3.16
76	3.23
72.5	3.89
68	4.24
65	4.47
61	4.76
58	5.31
54	5.77
50.5	6.37
47.5	6.80

For each line of data, calculate the value of β, using Eq. (16.3), and T_0 and R_0 corresponding to the values of 68°F (20°C). Some spread in the results will be found; however, use the average of the calculated values as the magnitude of β. (Use of the spreadsheets prepared in answer to the previous problem is recommended.)

16.13 Write Eq. (16.3) using the value of β found in answer to Problem 16.12, and plot the result over the range of data. Spot-check several points.

16.14 A small insulated box is constructed for the purpose of obtaining temperature calibration data for thermistors. Provision is made for mounting a thermistor within the box and bringing suitable leads out for connection to a commercial Wheatstone bridge. The bulb of a standardized mercury-in-glass thermometer is inserted into the box for the purpose of determining reference temperatures. A small heating element (a miniature soldering iron tip) is used as a heat source.

After the heater is turned on, thermistor resistances and thermometer readings are periodically made as the temperature rises from ambient to a maximum. The heater is then turned off and further data are taken as the temperature falls.

It is quickly noted, however, that there is a very considerable discrepancy in the "heating" resistance-temperature relationship compared with the corresponding "cooling" data. Why should this have been expected? Criticize the design of the arrangement described above when used for the stated purpose. How would you make a *simple* laboratory setup for obtaining reasonably accurate calibration data for a thermistor over a temperature range of, say, 80°F to 400°F?

16.15 An ice-bath reference junction is used with a copper-constantan thermocouple. For four different conditions, millivolt outputs are read as follows: −4.334, 0.00, 8.133, and 11.130. What are the respective junction temperatures (a) in degrees C and (b) in degrees F?

16.16 Copper-constantan thermocouples are used for measuring the temperatures at various points in an air conditioning unit. A reference junction temperature of 22.8°C is recorded. If the following emf outputs are supplied by the various couples, determine the corresponding temperatures: −1.623, −1.088, −0.169, and 3.250 mV.

16.17 Using the equation for E versus C for the type T thermocouple given in Table 16.6, spot-check several points in Table 16.5 to satisfy yourself that the tabulated and calculated values agree.

16.18 Select a thermocouple type (other than type T) and write a computer program to generate an emf versus temperature table, using the appropriate equation from Table 16.6.

16.19 The temperature difference between two points on a heat exchanger is desired. The measuring and reference junctions of a copper-constantan thermocouple are embedded within the inlet and outlet tubes A and B, respectively, and an emf of 0.381 mV is read. Why does this reading provide insufficient data to determine the differential temperature accurately? What additional information must be obtained before the answer may be found?

16.20 Devise a spreadsheet template and/or write a computer program to evaluate the thermocouple outputs for the various combinations listed in Tables 16.6 and 16.7. Spot-check results using values in Table 16.4.

16.21 Use appropriate equations from Table 16.6 and calculate the emfs that are expected for the situations listed below, assuming a circuit such as that shown in Fig. 16.13.

TC Type	Temperature at the Reference Junction, °C	Temperature at the Measuring Junction, °C
B	0	1500
E	20	750
J	0	−170
K	80	1150
S	700	1580

16.22 Use appropriate equations from Table 16.7 and calculate the temperature at the measuring junction for each of the following situations. Assume a circuit such as that shown in Fig. 16.13.

TC Type	Temperature at the Reference Junction, °C	EMF, mV
K	400	29.74
K	700	−8.14
E	56	60.63
E	90	−4.36
J	15	16.30
J	280	−11.78

16.23 Referring to Eq. (16.9a), plot the ratio of static to total temperatures versus Mach number over the range of $M = 0$ to 3 and for $k = 1.3$, 1.4, and 1.5.

16.24 Referring to Eqs. (16.12) and (16.13), plot the ratio of indicated to static temperatures versus the recovery factor over the range of $r = 0$ to 1, for $k = 1.3$, 1.4, and 1.5.

16.25 If you did not work Problems 5.3 and 5.14, you should do so now.

16.26 The following temperature-time data were recorded:

Time, s	Temperature, °C
0	20
4	83
8	123
12	152
20	182
30	194
40	201
50	203

   a.   Plot the data points.
   b.   From the plot, determine a time constant for the system.
   c.   Write a response equation assuming first-order process.
   d.   Calculate sufficient points to plot the theoretical curve.
   e.   Decide, on the basis of the plot, whether or not the process may be considered a single time-constant first-order type.

16.27 A "two time-constant" temperature transducer has time constants in the ratio $\zeta = 4/1$, where $\tau_p = 1.5$ s. If the transducer, initially at a temperature of 80°C, is suddenly immersed in a 500°C environment, what will be the temperature indicated after 3 s?

16.28 If the transducer in Problem 16.27 is initially at 500°C and is suddenly immersed in an 80°C environment, what temperature will be indicated after 3 s?

16.29 The performance of a temperature-measuring system approximates that dictated by two time-constant theory: $\tau_p = 10$ s and $\tau_j = 25$ s. If the system is subjected to the temperature input shown in Fig. 16.37, we see that the probe will not have sufficient time to produce a readout approximating T_{max}. If, however, at the end of the 18-s pulse, the system indicates 135°C, what must be the value of T_{max}?

Figure 16.37 Temperature-time relationship for Problem 16.29

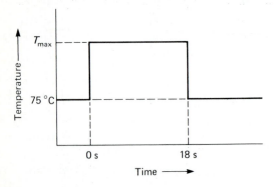

Figure 16.38 Temperature-time relationship for Problem 16.30

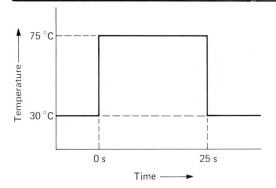

16.30 The behavior of a temperature-measuring system approximates two time-constant theory with $\tau_p = 6\,\text{s}$ and $\tau_j = 14\,\text{s}$. If the system experiences a perturbation as shown in Fig. 16.38, what will be the indicated temperature at $t = 15\,\text{s}$? At $t = 25\,\text{s}$? At 25 s the temperatures of the probe and the jacket will not be the same; however, on the assumption that both are at probe temperature, estimate the indicated temperature at 60 s. Will the calculated value be too high or too low? State your reasoning in answering the last question.

16.31 We wish to have both a continuous record and an instantaneous readout of energy flow rate from heated water passing through a pipe. Temperature, pressure, and rate of flow each vary over a range of values.
   a.   Analyze the problem and prepare a block diagram of the various measurements and functional problems that must be solved.
   b.   Insofar as you can, detail the steps to a solution.
   (*Note:* Figure 8.24 may help in providing a starting point.)

16.32 A thin-film layered heat-flux gage consists of a Kapton film with a single thin-foil thermocouple junction coated on either side (Fig. 16.36). The film has a thickness of 25 μm ($\pm 1\ \mu$m at 20:1 odds) and a conductivity of 0.20 W/m · K (± 0.01 W/m · K at 20:1 odds).
   a.   If the thermocouples operate in a range where their output is 40 μV/°C, and if the thermocouple voltage can be measured to $\pm 1\ \mu$V (20:1), what is the smallest heat flux that can be measured to an uncertainty of no more than 10% (20:1)?
   b.   Describe how the heat-flux gage could be modified to measure a heat flux one tenth as large as that found in part (**a**) with the same percentage uncertainty.
   c.   Thermocouple junctions may have systematic error of up to 2°C. What should be done to ensure accurate readings of the heat flux with this gage?

References

1. Eskin, S. G., and J. R. Fritze. Thermostatic bimetals. *ASME Trans.*, 62: 7, July 1910.

2. Mangum, B. W., and G. T. Furukawa. *Guidelines for Realizing the*

International Temperature Scale of 1990 (ITS-90). NIST Technical Note 1265. Washington, D.C.: U.S. Department of Commerce, National Institute of Standards and Technology, 1990.

3. Swindells, J. F. (ed.) *Precision Measurement and Calibration, Temperature.* NBS Special Publication 300, Washington, D.C.: U.S. Government Printing Office, 2: 164, 1968.

4. *The Temperature Handbook.* Stamford, Conn.: Omega Engineering, Inc., 1989.

5. Dowell, K. P. Thermistors as components open product design horizons. *Elec. Mfg.* 42: 2, August 1948.

6. Seebeck, T. J. *Evidence of the Thermal Current of the Combination Bi-Cu by Its Action on Magnetic Needle.* Berlin: Abt. d. Königl, Akad. d. Wiss. 1822–23, p. 265.

7. Peltier, M. Investigation of the heat developed by electric currents in homogeneous materials and at the junction of two different conductors. *Ann. Chim. Phys.* 56: 371, 1834.

8. Thomson, W. Theory of thermoelectricity in crystals. *Trans. Edinburgh Soc.* 21: 153, 1847. Also in *Math. Phys. Papers* 1: 232, 266, 1882.

9. Dike, P. H. *Thermoelectric Thermometry.* Philadelphia: Leeds and Northrup Company, 1954.

10. Powell, R. L. et al. *Thermocouple Reference Tables Based on the IPTS-68.* National Bureau of Standards Monograph 125. Washington, D.C.: U.S. Government Printing Office, 1974.

11. Beckwith, T. G., and R. D. Marangoni. *Mechanical Measurements.* 4th ed., Section 7.8. Reading, Mass.: Addison-Wesley, 1990.

12. Burns, G. W., and M. G. Scroger. *Temperature-Electromotive Force Reference Functions and Tables for Letter-Designated Thermocouple Types Based on the ITS-90.* NIST Monograph 175. Washington, D.C.: U.S. Department of Commerce, National Institute of Standards and Technology, 1992 (to appear).

13. Baker, H. D., E. A. Ryder, and N. H. Baker. *Temperature Measurement in Engineering,* Vol. 1. New York: John Wiley, 1953, p. 49.

14. *Lake Shore Product Catalog.* Westerville, Ohio: Lake Shore Cryotronics, Inc., 1991.

15. Hammond, D. L., and A. Benjaminson. Linear quartz thermometer. *Instruments and Control Systems,* 38(10): 115, 1965.

16. Lee, J. F., and F. W. Sears. *Thermodynamics.* Reading, Mass.: Addison-Wesley, 1955, p. 292.

17. Edwards, D. K. *Radiation Heat Transfer Notes.* New York: Hemisphere Publishing Corporation, 1981.

18. Keyes, R. J. (ed.) *Optical and Infrared Detectors.* 2nd ed. New York: Springer-Verlag, 1980.

19. *Handbook of Infrared Radiation Measurement.* Stamford, Conn.: Barnes Engineering Company, 1983.

20. Lee, J. F., and F. W. Sears. *Thermodynamics.* Reading, Mass.: Addison-Wesley, 1955, p. 281.

21. *Steam, Its Generation and Use.* 38th ed. New York: The Babcock and Wilcox Company, 1975.

22. Johnson, N. R., A. S. Weinstein, and F. Osterle. *The Influence of Gradient Temperature Fields on Thermocouple Measurements.* ASME Paper No. 57-HT-18, Aug. 1957.

23. Hottel, H. C., and A. Kalitinsky. Temperature measurement in high-velocity air streams. *J. Appl. Mech.,* 67: A25, March 1945.

24. Lalos, G. T., A sonic-flow pyrometer for measuring gas temperatures. *NBS J. Res.* 47(3): 179, Sept. 1951.

25. Seadron, M. D., and I. Warshawsky. Experimental determination of time constants and Nusselt numbers for bare-wire thermocouples in high-velocity air streams and analytic approximation of conduction and radiation errors. *NACA Tech. Note 2599:* January 1952.

26. Moffat, R. J. How to specify thermocouple response. *ISA J.,* 4: 219, June 1957.

27. Lefkowitz, L. Methods of dynamic analysis. *ISA J.* 203: June 1955.

28. Louis, J. R., and W. E. Hartman. *The Determination and Compensation of Temperature Sensor Transfer Functions.* ASME Paper 64-WA/AUT-13, 1964.

29. Shepard, C. E., and I. Warshawsky. Electrical techniques for compensation of thermal time lag of thermocouples and resistance thermometer elements. *NACA Tech. Note 2703:* May 1952.

30. Shepard, C. E., and I. Warshawsky. Electrical techniques for time lag compensation of thermocouples used in jet engine gas temperature measurements. *ISA J.* 1: 119, November 1953.

31. Hopkins, K. H., J. C. LaRue, and G. E. Samuelson. Effect of variable time constants on compensated thermocouple measurements. *NATO Advanced Study Institute on Instrumentation for Combustion and Flow in Engines.* Dordrecht: Kluwer, 1988.

32. Harnbaker, D. R., and D. L. Rall. Heat flux measurements: A practical guide. *Instru. Technol.* 51: February 1968.

33. Baines, D. J. Selecting unsteady heat flux sensors. *Instr. Control Systems* 80: May 1972.

34. Thompson, W. P. "Heat Transfer Gages." In R. J. Emrich (ed.), *Fluid Dynamics,* Vol. 18B of *Methods of Experimental Physics.* New York: Academic Press, 1981.

35. Gardon, R. An instrument for the direct measurement of intense thermal radiation. *Rev. Sci. Instr.:* May 1953.

36. Epstein, A. H., G. R. Guenette, R. J. G. Norton, and Cao Yuzhang. High-frequency response heat-flux gauge. *Rev. Sci. Instr.* 57(4): 639–649, April 1986.

37. Hager, J. M., S. Simmons, D. Smith, S. Onishi, L. W. Langley, and T. E. Diller. *Experimental Performance of a Heat Flux Microsensor.* New York: ASME Paper No. 90-GT-256, 1990.

38. Ortolano, D. J., and F. F. Hines. A simplified approach to heat flow measurement. *Advances in Instrumentation* 38(2), 1449–1456. Research Triangle Park, N.C.: Instrument Society of America, 1983.

39. Stempel, F. C. Basic heat flow calibration. *Instr. Control Systems* 42(5): 105, 1969.

40. Crovini, L., R. E. Bedford, and A. Moser, Extended list of secondary reference points. *Metrologia* 13: 197–206, 1977.

41. Eskin, S. G., and J. R. Fritze. Thermostatic bimetals. *ASME Trans.* 62(5): 433, 1940.

CHAPTER 17

Measurement of Motion

17.1 Introduction

Mechanical motion may be defined in terms of various parameters as listed in Table 17.1. One or more of the values may be constant with time, periodically varying, or changing in a complex manner. Measurement of static displacement was discussed in detail in Chapter 11. Very broadly, if the displacement-time variation is of a generally continuous form with some degree of repetitive nature, it is thought of as being a *vibration*. On the other hand, if the action is of a single-event form, a transient, with the motion generally decaying or damping out before further dynamic action takes place, then it may be referred to as *shock*. Obviously, shock action may be repetitive and in any case the displacement-time relationships will normally contain vibratory characteristics. To be so termed, however, shock must in general possess the property of being discontinuous. Additionally, steep wavefronts are often associated with shock action, although this is not a necessary characteristic.

In any event, both mechanical shock and mechanical vibration involve the parameters of frequency, amplitude, and waveform. Basic measurement normally consists of applying the necessary instrumentation to obtain a time-based record of displacement, velocity, or acceleration. Subsequent analysis can then provide such additional information as the frequencies and amplitudes of harmonic components and derivable displacement-time relationships not directly measured.

In many respects instrumentation used for vibration measurements are directly applicable to shock measurement. On the other hand, testing procedures and methods are quite different. The fundamental aspects of acceleration velocity and displacement measurements can be determined through examination of the most basic device for measuring these quantities, the seismic transducer.

Table 17.1 Motion parameters

Motion Parameter	Defining Relationships	
	For Linear Motion	*For Angular Motion*
Displacement	$s = f(t)$	$\theta = g(t)$
Velocity	$v = \dfrac{ds}{dt}$	$\omega = \dfrac{d\theta}{dt}$
Acceleration	$a = \dfrac{dv}{dt} = \dfrac{d^2 s}{dt^2}$	$\alpha = \dfrac{d\omega}{dt} = \dfrac{d^2\theta}{dt^2}$
Jerk	$\dfrac{da}{dt}$	$\dfrac{d\alpha}{dt}$

17.2 Vibrometers and Accelerometers

Current nomenclature applies the term *vibration pickup* or *vibrometer* to detector-transducers yielding an output, usually a voltage, that is proportional to either displacement or velocity. Whether displacement or velocity is sensed is determined primarily by the secondary transducing element. For example, if a differential transformer (Sections 6.11 and 11.19) or a voltage-dividing potentiometer (Sections 6.6 and 7.7) is used, the output will be proportional to a displacement. On the other hand, if a variable-reluctance element (Section 6.12) is used, the output will be a function of velocity.

The term *accelerometer* is applied to those pickups whose outputs are functions of acceleration. There is a basic difference in design and application between vibration pickups and accelerometers.

17.3 Elementary Vibrometers and Vibration Detectors

In spite of the tremendous advances made in vibration-measuring instrumentation, one of the most sensitive vibration detectors is the human touch. Tests conducted by a company specializing in balancing machines determined that the average person can detect, by means of his or her fingertips, sinusoidal vibrations having amplitudes as low as 12 μin. [1]. When the vibrating member was tightly gripped, the average minimum detectable amplitude was only slightly greater than 1 μin. In both cases, by fingertip touch and by gripping, greatest sensitivity occurred at a frequency of about 300 Hz.

When amplitudes of motion are greater than, say, $\frac{1}{32}$ in. or about 1 mm, a simple and useful tool is the *vibrating wedge,* shown in Fig. 17.1(a). This is simply a wedge of paper or other thin material of contrasting tone, often black, attached to the surface of the vibrating member. The axis of symmetry of the

Figure 17.1 Vibrating-wedge amplitude indicator: (a) stationary wedge, (b) extreme positions of wedge

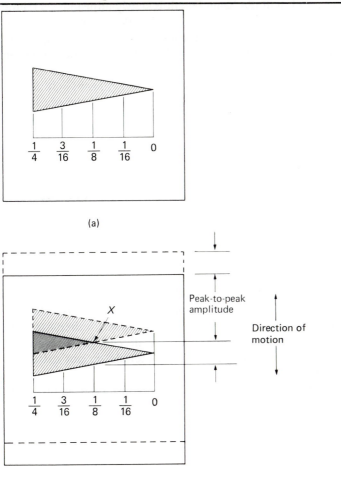

(a)

(b)

wedge is placed at right angles to the motion. As the member vibrates, the wedge successively assumes two extreme positions, as shown in Fig. 17.1(b). The resulting double image is quite well defined, with the center portion remaining the color of the wedge and the remainder of the images a compromise between dark and light. By observing the location of the point where the images overlap, marked X, one can obtain a measure of the amplitude. At this point the width of the wedge is equal to the double amplitude of the motion. This device does not yield any information as to the waveform of the motion. (See Fig. 17.10 for use of a microscope for amplitude measurement.) A simple apparatus for measuring frequency involves a small

variable will be the forcing frequency, Ω. Let us see, then, how the function behaves by plotting the ratio S_{r_0}/S_{s_0} versus Ω/ω_n. This can be done for various damping ratios, thereby obtaining a family of curves. Figure 17.4 is the result. Inspection of the curves shows that for values of Ω/ω_n considerably greater than 1.0, the amplitude ratio is indeed near unity, which is as desired. It may also be observed that the value of the damping ratio is not important for high values of Ω/ω_n. However, in the region near a frequency ratio of 1.0, the amplitude ratio varies considerably and is quite dependent on damping. Below $\Omega/\omega_n = 1.0$, the ratios of amplitude break widely from unity. It may also be observed by inspection of Fig. 17.4 that for certain damping ratios, the amplitude ratio does not stray very far from unity, *even in the vicinity of resonance*.

We may conclude from our inspection that *damping on the order of 65% to 70% of critical is desirable* if the instrument is to be used in the frequency region just above resonance. We also see that, in any case, damping of a general-purpose instrument is a compromise, and inherent errors resulting from the principle of operation will be present. To these would be added errors that may be introduced by the secondary transducer and the second- and third-stage instrumentation.

As an example, let us check the discrepancy for the following conditions:

ξ = the damping ratio = 0.68,
S_{s_0} = 0.015 in.,
f_n = the natural undamped frequency of the instrument = 4.75 Hz,
f_e = the exciting frequency = 7 Hz.

Figure 17.4   Response of a seismic instrument to harmonic displacement

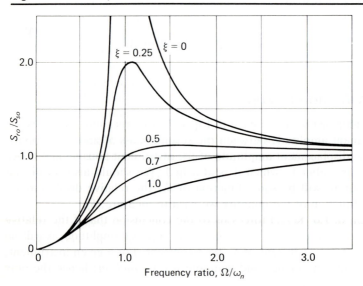

Then

$$\frac{f_e}{f_n} = \frac{7}{4.75} = 1.474, \qquad \left(\frac{f_e}{f_n}\right)^2 = 2.17.$$

Using Eq. (17.7), we get

$$S_{r_0} = \frac{2.17 \times 0.015}{\sqrt{[1 - (2.17)]^2 + (2 \times 0.68 \times 1.472)^2}}$$

$$= 0.01404,$$

$$\text{Inherent error} = \left(\frac{0.01404}{0.015} - 1\right) 100 = -6.38\%.$$

17.6.2 Phase Shift in the Seismic Vibrometer

Let us now turn our attention to the phase relation between relative amplitude and support amplitude. Naturally it would be very desirable to have a zero phase relation for all frequencies. A plot of Eq. (17.6) is shown in Fig. 17.5. This indicates that for *zero damping* the seismic mass moves exactly in phase with the support (but not with the same amplitude), so long as the forcing frequency is below resonance. Above resonance the mass motion is completely out of phase (180°) with the support motion. At resonance there is a sudden

Figure 17.5 Relations between phase angle, frequency ratio, and damping for a seismic instrument

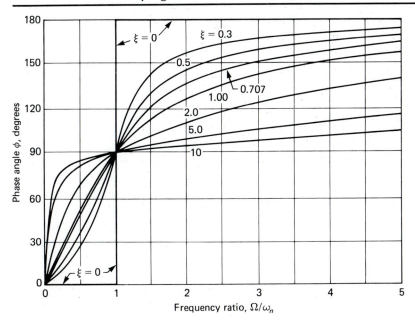

shift in phase. For other damping values, a similar shift takes place, except there is a gradual change with frequency ratio.

A simple experiment verifies this phase-shift relation. A crude support-excited seismic mass can be constructed by tying together five or six rubber bands in series to form a long, soft *spring* and attaching a mass of, say, $\frac{1}{2}$ lb to one end while holding the other end in the hand. When the hand is moved up and down, a relative motion is obtained and the natural frequency of the system can easily be found, it being the frequency that provides greatest amplitudes with least effort. Now try moving the hand up and down at a frequency considerably below the natural frequency. It will be observed that the mass moves up and down at very nearly the same time as the hand. The motion of the seismic mass is approximately *in phase* with the motion of the supporting member, the hand. Now move the hand up and down at a frequency considerably above the natural frequency. It will be observed that the mass now moves downward as the hand moves up, and the weight moves up as the hand moves down. The motions are *out of phase.*

Our observations indicate that from the standpoint of phase shift, a "best" solution would be to design an instrument with zero damping (if that were possible). However, the amplitude relation near resonance would then be in serious error.

Perhaps any amplitude and phase-shift effects near resonance, such as we have been discussing, could be accounted for by a calibration! It would seem at first glance that such a possibility would be good, and indeed it would be feasible if *single-frequency* harmonic motions were always encountered. In fact, if simple sinusoidal motion were always to be measured, phase shift would not be of consequence. We would not care particularly whether the peak relative motion coincided exactly with the peak support motion so long as the waveform and measured amplitudes were correct. The difficulty arises when the input is a complex waveform, made up of the fundamental and many other harmonics, with each harmonic simultaneously experiencing a different phase shift.

As we saw in Section 5.4, if certain harmonic terms in a complex waveform shift relative to the remaining terms, the shape of the resulting wave is distorted, and an incorrect output results. On the other hand, bodily shifts of all harmonics without *relative* changes preserve the true shape, and in most applications no problem results.

We find that there are three possible ways in which distortion from phase shift may be minimized. First, if there is no lag for any of the terms, there will be no distortion. Second, if all components lag by 180°, their relative values remain unchanged. And finally, if the shifts are in proportion to the harmonic orders (i.e., there is a linear shift with frequency), correct relative relations will be retained.

Zero shift requires no further comment, other than to suggest that it rarely, if ever, exists. When a 180° shift takes place, all sine and cosine terms will simply have their signs reversed, and their relative magnitudes will remain unaffected.

A phase shift linear with frequency would be of the type in which the first harmonic lagged by, say, ϕ degrees, the second by 2ϕ, the third by 3ϕ, and so on. Let us consider this situation by means of the following relation:

$$f(t) = A_0 + A_1 \cos \omega t + A_2 \cos 2\omega t + A_3 \cos 3\omega t + \cdots \qquad (17.8)$$

Linear phase shifts would alter this equation to read

$$f(t) = A_0 + A_1 \cos (\omega t - \phi) + A_2 \cos (2\omega t - 2\phi)$$
$$+ A_3 \cos (3\omega t - 3\phi) + \cdots$$
$$= A_0 + A_1 \cos \beta + A_2 \cos 2\beta + A_3 \cos 3\beta + \cdots \qquad (17.8a)$$

where $\beta = \omega t - \phi$. We see, then, that the whole relation is retarded uniformly and that each term retains the same relative harmonic relationship with the other terms. Therefore, there will be no phase distortion. As we shall see, the vibration pickup approximates the second situation (i.e., 180° phase shift), whereas the accelerometer is of the linear phase-shift type.

Figure 17.5 shows that in the frequency region above resonance, used by a seismic-type displacement or velocity pickup, phase shift approaches 180° as the frequency ratio is increased. The swiftness with which it does so, however, is determined by the damping. For zero damping, the change is immediate as the exciting frequency passes through the instrument's resonant frequency. At higher damping rates, the approach to 180° shift is considerably reduced. We see, therefore, that damping requirements for good amplitude and phase response in this frequency area *are in conflict,* and some degree of compromise is required. In general, however, amplitude response is more of a problem than phase response, and commercial instruments are often designed with 60% to 70% of critical damping, although in some cases the damping is kept to a minimum. In any case, the greater the frequency ratio above unity, the more accurately will the relative motion to which the vibrometer responds represent the desired motion.

17.6.3 General Rule for Vibrometers

We may say, therefore, that in order for a vibration pickup of the seismic mass type to yield satisfactory motion information, use of the instrument must be restricted to input forcing frequencies above its own undamped natural frequency. Hence the lower the instrument's undamped natural frequency, the greater its range. In addition, in the frequency region immediately above resonance, compromised amplitude and phase response must be accepted. Of course, the displacement range that can be accommodated is limited by the design of the particular instrument. In general, the vibrometers of larger physical size permit measurement of larger displacement amplitudes. However, as size is increased, so too is the loading on the signal source.

17.7   The Seismic Accelerometer

We now turn our attention to a similar type of seismic instrument: the accelerometer. Basically, the construction of the accelerometer is the same as that of the vibrometer (Fig. 17.3), except that its design parameters are adjusted so that its output is proportional to the applied acceleration.

Let us rewrite Eq. (17.7) as follows:

$$S_{r_0} = \frac{S_{s_0}\Omega^2}{\omega_n^2\sqrt{[1 - (\Omega/\omega_n)^2]^2 + [2\xi\Omega/\omega_n]^2}} \tag{17.9}$$

or

$$S_{r_0} = \frac{a_{s_0}}{\omega_n^2\sqrt{[1 - (\Omega/\omega_n)^2]^2 + [2\xi\Omega/\omega_n]^2}}, \tag{17.10}$$

in which a_{s_0} is the acceleration amplitude of the supporting member.

Inspection of Eq. (17.10) makes the problem of properly designing and using an accelerometer clear. In order that the relative displacement between the supporting member and the seismic mass may be used as a measure of the support acceleration, the radical in the equation should be constant. The term ω_n^2 in the denominator is fixed for a given instrument and does not change with application. Hence, if the radical is a constant, the relative displacement will be directly proportional to the acceleration. Let

$$K = \frac{1}{\sqrt{[1 - (\Omega/\omega_n)^2]^2 + [2\xi\Omega/\omega_n]^2}}. \tag{17.11}$$

Figure 17.6   Response of a seismic instrument to sinusoidal acceleration

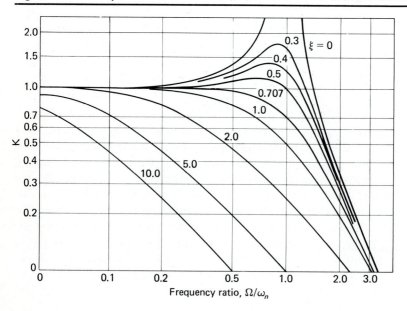

By plotting K versus Ω/ω_n for various damping ratios, we obtain Fig. 17.6. Inspection of the plot indicates that the only possibility of maintaining a reasonably constant amplitude ratio as the forcing frequency changes is over a range of frequency ratio between 0.0 and about 0.40 and for a damping ratio of around 0.7. The extent of the usable range depends on the magnitude of error that may be tolerated.

17.7.1 Phase Lag in the Accelerometer

Referring again to Fig. 17.5 and to the limited accelerometer operating range just indicated—that is, $\Omega/\omega_n = 0$ to about 0.4 and $\xi = 0.70$—we see that the phase changes very nearly linearly with frequency. This relationship is fortunate, for as we have seen, it results in good phase response.

17.7.2 General Rule for Accelerometers

We may now say that in order for a seismic instrument to provide satisfactory acceleration data, it must be used at forcing frequencies *below* approximately 40% of its own undamped natural frequency and the instrument damping should be on the order of 70% of critical damping.

It may be observed that both vibration pickups and accelerometers may use about the same damping: however, the range of usefulness of the two instruments lies on opposite sides of their undamped natural frequencies. The vibration pickup is made to a low undamped natural frequency, which means that it uses a "soft" sprung mass. On the other hand, the accelerometer must be used well below its own undamped natural frequency; therefore, it uses a "stiff" sprung mass. This makes the accelerometer an inherently less sensitive but more rugged instrument than the vibration pickup.

Figure 17.3 may be used to represent either a vibrometer or an accelerometer. As developed in Sections 17.6 and 17.7, the basic readout for a seismic instrument is the relative motion between the mass and the supporting structure. To sense this motion we require a relative-motion secondary transducer. (The mass-spring combination forms the primary transducer.) Although a voltage-dividing potentiometer is shown in the figure, at one time or another essentially all the appropriate transducing principles discussed in this book have been used for this purpose. A list includes variable-reluctance and variable-inductance devices, both bonded and unbonded strain gages, piezoresistive and piezoelectric sensors, variable-capacitance transducers, and some quite uncommon devices not mentioned in the preceding discussion.

Most of the devices listed above are displacement sensors. Variable reluctance is an exception. In this case the output is a function of velocity: the *rate* at which the magnetic lines of flux are cut. Sensitivity of this type is therefore in volts per unit velocity rather than volts per unit displacement. Variable-reluctance transducers have been quite successfully used as vibrometers. There are two basic designs: (1) a permanent magnet forms a part of the seismic mass, which moves relative to pickup coils anchored to the case, or (2) the magnet is fixed to the housing and the coil forms a part of the seismic mass. Neither has a marked advantage, although it is obvious that for the moving coil, the electrical circuit becomes more critical.

17.8   Practical Accelerometers

One form of accelerometer, shown in Fig. 17.7, makes use of an *unbonded* strain-gage bridge. The seismic mass, A, is constrained to single-degree motion by small flexure springs (not shown). The unbonded strain-gage element, B, behaves in the same manner as the bonded type discussed in Chapter 12. Damping is accomplished by use of a silicon fluid surrounding the moving mass, and a small diaphragm is used to provide expansion room required for temperature changes.

An advantage enjoyed by this type of accelerometer is the ease with which the secondary transducer, the strain-sensitive elements, may be calibrated in the field by paralleling calibration resistors (see Section 12.11). Initial calibration of the complete accelerometer is performed by the manufacturer. Instruments of this kind are available covering an acceleration range from 0.5 g to 200 g.

Variable-differential transformers and voltage-dividing potentiometers are also used as secondary transducing elements in accelerometers. When these

Figure 17.7   Internal construction of the Statham Instruments Model A-6
                 accelerometer (Courtesy: Statham Transducer
                 Division of Schlumberger Industries)

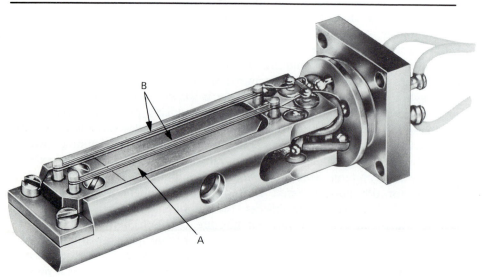

Figure 17.8 Typical piezoelectric-type accelerometer designs (Courtesy: Endevco Corp., San Juan Capistrano, CA)

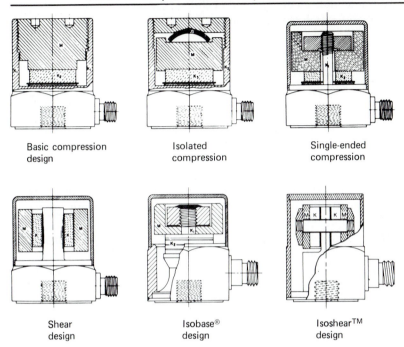

Basic compression design	Isolated compression	Single-ended compression

Shear design	Isobase® design	Isoshear™ design

devices are employed, proper damping is often obtained by means of viscous fluids or through use of eddy-current damping provided by a permanent magnet incorporated into the design.

Undoubtedly the most popular type of accelerometer makes use of a piezoelectric element in some form as shown in Fig. 17.8. Polycrystalline ceramics including barium titanate, lead zirconate, lead titanate, and lead metaniobate are among the piezoelectric materials that have been used [3]. Various design arrangements, as shown in the figure, are also used; the type depends on the characteristics desired, such as frequency range and sensitivity.

Important advantages enjoyed by the piezoelectric type are high sensitivity, extreme compactness, and ruggedness. Although the damping ratio is relatively low (0.002 to 0.25), the useful linear frequency ranges that may be attained are still large because of the high undamped natural frequencies inherent in the design (up to 100,000 Hz).

The output impedance of a piezoelectric device is quite high and presents certain problems associated with proper matching, noise, and connecting-cable motion and length. Either an impedance-transforming amplifier (Section 7.17) or a charge amplifier (Section 7.18.2) is normally required for proper signal conditioning. Each device has both advantages and disadvantages. Greatest effectiveness is attained when the instrumentation is located near the accelerometer. In fact, modern IC technology has made it possible, in some instances, to incorporate the amplifier circuitry within the accelerometer housing. Proper selection of instrumentation will, therefore, depend on application.

17.9   Calibration

To be useful as amplitude-measuring instruments, both vibration pickups and accelerometers must be calibrated by determining the units of output signal (usually voltage) per unit of input (displacement, velocity, or acceleration). For the accelerometer, volts per g could be determined. Also, the calibration should indicate how such "constants" vary over the useful frequency range.

There are two basic approaches to the calibration of seismic-type transducers: (1) by absolute methods (based directly on the physical concepts of mass, length, and time), and (2) by comparative techniques.

The latter approach uses a "standard" against which the subject transducer is compared. It is clear that the standard must have highly reliable characteristics whose own calibration is not questioned. "Identical" motions are then imposed on subject and standard and the two outputs compared. Although this method would appear to be quite simple and is undoubtedly the one most commonly used, there are many pitfalls to be avoided. Error-free results depend on a number of factors [4, 5].

1.   The impressed motions must *indeed* be identical.

2.   Readout apparatus associated with the standard should preferably be and remain a part of the standard, and the *entire* system should have traceable calibration.

3.   Associated readout apparatus in both circuits must have identical responses.

4.   The standard must have long-term reliability.

None of these requirements is easily achieved.

The motion source used for this purpose is generally some form of exciter system as described in Section 17.16.

The following two sections discuss some of the more fundamental methods applied to calibration of seismic-type transducers.

17.10   Calibration of Vibrometers

Vibration pickups are often calibrated by subjecting them to steady-state harmonic motion of known amplitude and frequency. The output of the pickup is then a sinusoidal voltage that is measured either by a reliable voltmeter or a cathode-ray oscilloscope. The primary problem, of course, is in obtaining a harmonic motion of *known* amplitude and frequency.

Electromechanical exciters are commonly used [6]. Devices of this sort are described in detail in Section 17.16. Exciters of this type are capable of producing usable amplitudes at frequencies to several thousand cycles per second.

The only really positive method for determining actual instrument amplitude in a calibration test of this sort is to measure the excursion directly by means of some form of displacement-measuring device. Measuring microscopes (Section 11.15) are very useful for this purpose [7]. Either the filar type or the graduated reticule type, having a magnification of about 40–100 times, may be used. It is necessary that the microscope be mounted on a rigid support so that the pickup will be credited with no more motion than it is actually experiencing. Figure 17.9 shows schematically the general arrangement that may be used for this method of calibration.

A convenient target to observe is a small patch of #320 grit emery cloth cemented to the exciter table or directly on the pickup. A pinpoint of light is then directed on the emery-cloth patch. The light reflected from the emery cloth appears through the microscope as a myriad of small light sources reflected from the rough sides of the individual crystals, as shown in Fig. 17.10(a). As the exciter table moves, the individual points of light each become bright lines [Fig. 17.10(b)] having lengths equal to the double amplitude of the motion. The lengths of these lines may easily be measured

Figure 17.9 Schematic diagram of arrangement for calibrating seismic instruments by use of steady-state harmonic motion

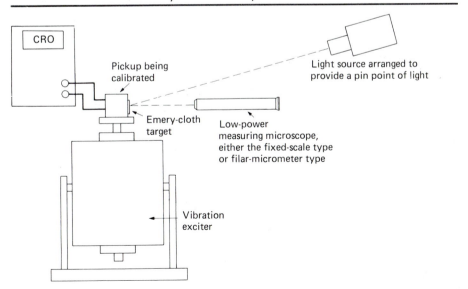

Figure 17.10   Views of emery-cloth target observed through a microscope:
(a) view when exciter table is stationary,
(b) view when table is vibrated

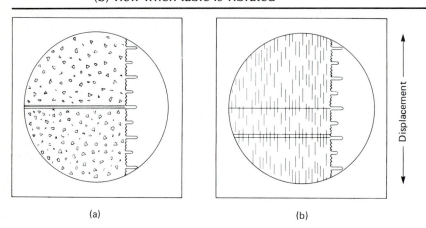

(a)   (b)

through the microscope. (The light source and measuring microscope may also
be replaced by a laser interferometer discussed in Section 11.14).

One of the requirements of this method is that the center of gravity of the
pickup and any mounting fixture must be placed directly on the force axis of
the exciter. Otherwise lateral motion may also occur, which must be avoided.
Lateral motion will not go unnoticed, however, because if it exists, Lissajous
traces (Section 10.6) will be described by the light points and the condition will
immediately become obvious.

The following example illustrates the procedure and data for calibrating a
velocity pickup. A pickup was shaken sinusoidally by a small electromechanical shaker, and the amplitude was measured through use of a filar microscope
by the procedure outlined immediately above. Data were obtained as follows:

$$f = \text{frequency} = 120 \text{ Hz,}$$

$$A_0 = \text{amplitude} = 0.0030 \text{ in. } (0.0060 \text{ in. peak to peak),}$$

$$e = \text{rms voltage measured by VTVM} = 0.150 \text{ V.}$$

Calculations:

$$e_0 = \text{voltage amplitude} = 0.150 \times 1.414 = 0.212 \text{ V,}$$

$$V_0 = \text{velocity amplitude} = 2\pi \times 120 \times 0.0030 = 2.26 \text{ in./s,}$$

$$\text{Sensitivity} = \frac{e_0}{V_0} = \frac{0.212}{2.26} = 0.0938 \text{ V/in./s.}$$

(The manufacturer's nominal rating for the sensitivity of this model was
0.0945 V/in./s.)

17.11 Calibration of Accelerometers

Accelerometer calibration methods may be classified as follows:

1. Static
   a. Plus or minus 1 g turnover method
   b. Centrifuge method
2. Steady-state periodic
   a. Rotation in a gravitational field
   b. Using a sinusoidal shaker or exciter
3. Pulsed
   a. One-g step, using free fall
   b. Multiple spring-mass device
   c. High-g methods

17.11.1 Static Calibration

Plus or Minus 1 g Turnover Method

Low-range accelerometers may be given a $2g$ step calibration by simply rotating the sensitive axis from one vertical position 180° through to the other vertical position, i.e., by simply turning the accelerometer upside down. This method is positive but is, of course, limited in the magnitude of acceleration that may be applied. A simple fixture is described in Ref. [8]. Of course, for precise calibration, the value of local gravity acceleration must be used.

Centrifuge Method

Practically unlimited values of static acceleration may be determined by a centrifuge or rotating table. The normal component of acceleration toward the center of rotation is expressed by the relation

$$a_n = r(2\pi f)^2, \tag{17.12}$$

where

a_n = the acceleration of the seismic mass,

r = the radius of rotation measured from the center of the table to the center of gravity of the seismic mass,

f = table rotation speed, rev/s.

It is assumed here that the axis of rotation is vertical. One of the disadvantages of this method, though not serious, is that of making electrical connections to the instrument.

17.11.2 Steady-State Periodic Calibration

Rotation in a Gravitational Field

This technique is simply a variation of the centrifuge method, in which the turntable is rotated about a horizontal axis [9]. To the average static component as determined by Eq. (17.12), a sinusoidal $1\,g$ gravitational component is superimposed.

Using a Sinusoidal Vibration Exciter

A very satisfactory procedure for obtaining a steady-state periodic calibration is that described in Section 17.10 for calibrating a vibration pickup. The primary difference lies in the fact that the input for the accelerometer is the harmonic acceleration,

$$a = -S_{s_0}\Omega^2 \cos \Omega t \tag{17.13}$$

and

$$a_0 = -S_{s_0}\Omega^2. \tag{17.14}$$

Small portable acceleration calibration units are commercially available. These are single-frequency fixed-amplitude vibration exciters driven by common $9.0\,V$ transistor batteries. The accelerometer is mounted to the exciter surface; when the unit is turned on, it produces a dynamic acceleration amplitude of $1\,g$. The size of such a unit is slightly larger than a size D battery.

17.11.3 Pulsed Calibration

The Free-Fall Method

A $1\,g$ stepped acceleration may be obtained by suspending an accelerometer with something like a string. When the support is suddenly cut, the accelerometer is subjected to an acceleration change of $1\,g$.

High-g Methods

Calibration of accelerometers in the high-g range (up to $40{,}000\,g$ or higher) presents special problems that can be discussed only briefly at this point. Calibration methods are generally based on velocity measurements [10] and use of the following relation:

$$V_2 - V_1 = \int_{t_1}^{t_2} a\,dt. \tag{17.15}$$

The integration covers the time duration of velocity change.

Various arrangements are used for obtaining the necessary acceleration pulse, including ballistic pendulums, drop testers, air guns, and inclined troughs [3, 10, 11]. The problem consists of obtaining the accelerometer calibration factor,

$$K = \frac{e}{a}, \tag{17.15a}$$

Figure 17.11 A simple form of ballistic pendulum
for calibrating accelerometers

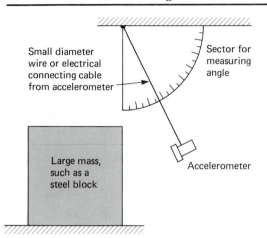

Small diameter
wire or electrical
connecting cable
from accelerometer

Sector for
measuring
angle

Large mass,
such as a
steel block

Accelerometer

in which

  K = the calibration factor, in volts per unit acceleration,

  e = the accelerometer (or accelerometer system) output, in volts,

  a = the acceleration.

  To obtain this value the accelerometer is excited by an impact pulse through some means such as a ballistic pendulum (Fig. 17.11; also see Figs. 17.22 and 17.23), and a record is made of the resulting accelerometer output (Fig. 17.12). Substituting the value of a from Eq. (17.15a) into Eq. (17.15)

Figure 17.12 Typical acceleration-time curve obtained by impact method

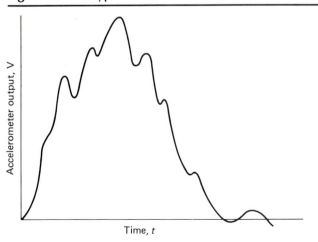

Accelerometer output, V

Time, t

yields

$$K = \frac{1}{(V_2 - V_1) \int_{t_1}^{t_2} e \, dt}.$$ (17.16)

It will be observed that the integral represents the area under the output curve, which can be obtained numerically. By experimentally determining the velocity change resulting from the impact $(V_2 - V_1)$, we can calculate the value of the calibration factor K. The method presumes initially that the relation expressed by Eq. (17.15a) is linear. It should also be clear that this method should be used only when high-g calibrations are required, beyond the practical ranges of the simpler methods described previously.

17.12   Determination of Natural Frequency and Damping Ratio in a Seismic Instrument

On occasion it may be desirable to determine the dynamic characteristics, including damping ratio, for an *existing* vibration pickup or an accelerometer. This will be necessary, for example, when a special-purpose instrument is constructed by the user, or perhaps to check an instrument before a particularly important testing job, or when instrument damage is suspected.

Two quantities must be determined: the damping ratio and the undamped natural frequency. It should be clear at this point that by undamped *natural frequency*, ω_n, we mean the frequency of oscillation that would occur if damping were zero; we do not mean the *damped natural frequency* ω_{nd} that would result if the seismic mass were released from an initially displaced condition. Undamped natural frequency cannot actually be directly measured, because we cannot completely eliminate damping.

Theoretically it is possible to determine the damping ratio and undamped natural frequency of a seismic instrument by subjecting it to a step input and measuring the readout "overshoot" on the first cycle [12]. Practical limitations associated with this method make its application difficult.

A very workable and accurate solution to the problem may be had by determining the instrument's response over a range of driving frequencies [13]. This may be done by sinusoidally exciting the instrument through a frequency range including the instrument's estimated undamped natural frequency. The output in terms of relative amplitude is obtained (Fig. 17.13).

The frequency-amplitude data thus obtained are then compared with theoretical, or "master," curves. If the instrument is a vibration pickup, the master curves would be those shown in Fig. 17.4. If it is an accelerometer, those in Fig. 17.6 would be used. The comparison is most easily made by plotting the measured relative amplitudes as ordinates and the corresponding actual frequencies as abscissas on transparent paper having the same logarith-

Figure 17.13 Experimental response curve obtained from tested instrument

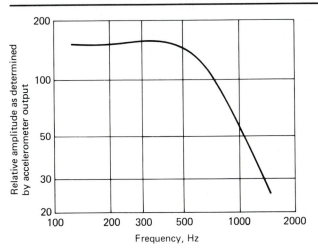

Figure 17.14 Experimental response curve (solid) superimposed on family of theoretical curves (dashed). This shows that the experimental instrument has a natural frequency of 600 Hz and a damping ratio of 0.6.

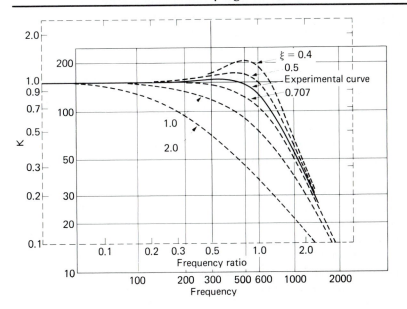

mic scales as the master curves. The experimental curve thus plotted is then superimposed on the theoretical curves and adjusted to take its proper place in the family, as determined by its shape. The damping ratio is determined by interpolation, and the experimental frequency corresponding to the dimensionless frequency ratio of 1.0 is the undamped natural frequency. Figure 17.14 illustrates how this would be done for an accelerometer. The theoretical curves are shown as dashed lines, and the experimental curve, with corresponding scales, is shown as a solid line.

17.13 Response of the Seismic Instrument to Transients

Our discussion of seismic instruments to this point has been largely in terms of simple harmonic motion. How will these instruments respond to complex waveforms and transients? As we saw in Section 4.4 and developed further in Section 5.16, complex waveforms can be analyzed as a series of simple sinusoidal components in appropriate amplitude and phase relationships. It would seem then that a seismic instrument capable of responding faithfully to a range of individual harmonic inputs should also respond faithfully to complex inputs made up of frequency components within that range. Of course if such an assumption is legitimate, the inherent nature of the accelerometer places it initially in a much more restricted area of operation than the vibration pickup. Whereas the accelerometer is limited to a frequency range up to roughly 40% of its own natural frequency, the only frequency restriction on the vibration pickup is that it be operated above its own natural frequency. In comparing the relative merits of the vibrometer and the accelerometer, we must not forget, however, that in terms of phase response the advantage is definitely on the side of the accelerometer (Sections 17.6.2 and 17.7.1). This is an important advantage.

By constructing the vibration pickup with a lightly sprung seismic mass, we can satisfactorily cover almost any frequency range. For the accelerometer, frequency components above the approximate 40% value may be unavoidable. In general, however, higher frequency inputs are attenuated. Figure 17.15 shows the theoretical accelerometer response to square-wave pulses [12, 14, 15]. Results are shown for different damping ratios and undamped natural frequencies.

First of all, these curves confirm our previous conclusion that a damping ratio of about 0.7 is near optimum. They also show that insofar as mass response is concerned, use of an instrument with as high an undamped natural frequency as possible is desirable. This is not surprising, because our previous investigation has pointed to this conclusion. A high undamped natural frequency, however, requires a stiff suspension. The result is that an extra burden is placed on the secondary transducer because of the small relative

Figure 17.15   Response to a square pulse of acceleration of an accelerometer (a), whose natural period is 1.014 times the duration of the pulse, (b) whose natural period is 0.334 times the duration of the pulse, (c) whose natural period is 0.203 times the duration of the pulse. (1) For zero damping. (2) For damping ratio = 0.4. (3) For damping ratio = 0.7. (4) For damping ratio = 1.0 (Courtesy: National Institute of Standards and Technology)

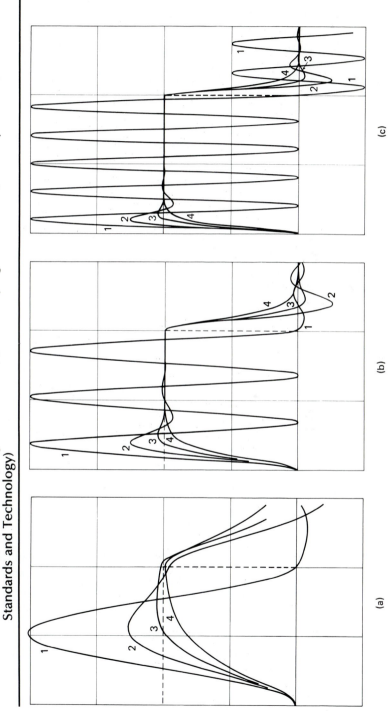

(a)   (b)   (c)

motions between mass and instrument housing. Hence, what is gained in response may be immediately lost in resolution.

The situation emphasizes even more the fact that accelerometer selection must be based on compromise. It means that the accelerometer cannot be selected entirely on its own merits, but that the whole system must be considered and then the accelerometer selected with the highest undamped natural frequency consistent with satisfactory overall response.

17.14   Measurement of Velocity by the Laser Velocity Transducer

Figure 11.13 shows a simplified form of laser displacement sensor. If the movable reflector is attached to a vibrating surface $[\delta = \delta(t)]$, the back-reflected beam is combined with the initially split beam, causing a number of successive dark fringes to be seen by the photodetector. The number of fringes per unit of time represents the surface velocity. The velocity sensed by this transducer is the velocity component of the movable reflector along the direction of the laser beam.

In practice the movable reflector can be a retroreflective tape, which can easily be attached to most surfaces. The operating, or standoff, distance of this device is usually 1.0 m or less. Since this is a noncontacting-type velocity sensor, it can be used for the velocity measurement of structures where the application of seismic-type transducers would greatly alter the structure mass.

Typical uses of this transducer are as follows:

1. Velocity survey of a hot surface such as a combustion engine manifold
2. Velocity survey of a vibrating membrane
3. Orbit analysis of rotating shafts in rotating machinery
4. Measurement of velocities of machine elements where attachment of seismic transducers is impossible.

17.15   Vibration and Shock Testing

Vibration and shock-test systems are particularly important in relation to numerous R&D contracts. Many specifications require that equipment perform satisfactorily at definite levels of steady-state or transient dynamic conditions. Such testing requires the use of special test facilities, often unique for the test at hand but involving principles common to all.

Numerous items for civilian consumption require dynamic testing as part of their development. All types of vibration-isolating methods require testing to determine their effectiveness. Certain material fatigue testing uses vibration test methods. Specific examples of items subjected to dynamic tests include many automobile parts, such as car radios, clocks, headlamps, radiators, ignition components, and larger parts like fenders and body panels. Also, many aircraft components and other items for use by the armed services must meet definite vibration and shock specifications. Missile components are subjected to extremely severe dynamic conditions of both mechanical and acoustic origin.

It might be assumed that dynamic testing should exactly simulate field conditions. However, this situation is not always necessary or even desirable. First of all, field conditions themselves are often nonrepetitive; situations at one time are not duplicated at another time. Conditions and requirements today differ from those of yesterday. Hence, to define a set of *normal* operating conditions is often difficult if not impossible. Dynamic testing, on the other hand, may be used to pinpoint particular areas of weakness under accurately controlled and measurable conditions. For example, such factors as accurately determined resonance frequencies, destructive amplitude-frequency combinations, etc., may be uncovered in the development stage of a design. With such information the design engineer then may judge whether corrective measures are required, or perhaps determine that such conditions lie outside operating ranges and are therefore unimportant. Another factor making dynamic testing attractive is that accelerated testing is possible. Field testing, in many cases, would require inordinate lengths of time.

Our discussion of dynamic testing is divided into two parts: vibration testing and shock testing.

17.16 Vibrational Exciter Systems

In order to submit a test item to a specified vibration, a source of motion is required. Devices used for supplying vibrational excitation are usually referred to simply as *shakers* or *exciters*. In most cases, simple harmonic motion is provided, but systems supplying complex waveforms are also available.

There are various forms of shakers, the variation depending on the source of driving force. In general, the primary source of motion may be electromagnetic, mechanical, or hydraulic-pneumatic or, in certain cases, acoustical. Each is subject to inherent limitations, which usually dictate the choice.

17.16.1 Electromagnetic Systems

A section through a small electromagnetic exciter is shown in Fig. 17.16. This consists of a field coil, which supplies a fixed magnetic flux across the air gap h, and a driver coil supplied from a variable-frequency source. Permanent magnets are also sometimes used for the fixed field. Support of the driving coil is by means of flexure springs, which permit the coil to reciprocate when driven by the force interaction between the two magnetic fields. We see that the electromagnetic driving head is very similar to the field and voice coil arrangement in the ordinary radio loudspeaker.

Figure 17.16 Sectional view showing internal construction of
an electromagnetic shaker head

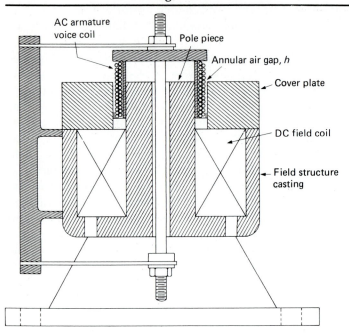

An electromagnetic shaker is rated according to its vector force capacity, which in turn is limited by the current-carrying ability of the voice coil. Temperature limitations of the insulation basically determine the shaker force capacity. The driving force is commonly simple harmonic (complex waveforms are also used) and may be thought of as a rotating vector in the manner of harmonic displacements discussed in Section 4.2. The force used for the rating is the vector force exerted between the voice and field coils.

Rated force, however, is never completely available for driving the test item. It is the force developed within the system, from which must be subtracted the force required by the moving portion of the shaker system proper. It may be expressed as

$$F_n = F_t - F_a, \qquad (17.17)$$

in which

F_n = the net usable force available to shake the test item,

F_t = the manufacturer's rated capacity, or total force provided by the magnetic interaction of the voice and field coils,

F_a = the force required to accelerate the moving parts of the shaker system, including the voice coil, table, and appropriate portions of the voice coil flexure beams.

Table 17.2   Specifications for typical electromagnetic exciter systems

Maximum Rated Force, lbf	Frequency Range, Hz	Maximum Double Amplitude, in.	Weight of Moving Armature, lbm
50	0–5,000	$\frac{1}{2}$	$\frac{3}{4}$
100	0–10,000	1	$1\frac{1}{4}$
300	0–5,000	1	$2\frac{3}{4}$
1,500	0–4,000	$1\frac{1}{4}$	20
3,200	5–3,000	$\frac{1}{2}$	25
20,000	5–3,000	1	95
40,000	2–2,000	1	250

In practice, it is often convenient to think in terms of the total vector force, F_t, and to simply add the weight of the shaker's moving parts to that of the test item and any required accessories such as mounting brackets, etc. Table 17.2 lists the specifications for typical commercially available electromagnetic shaker systems.

17.16.2   Mechanical-Type Exciters

There are two basic types of mechanical shakers: the directly driven and the inertia. The directly driven shaker consists simply of a test table that is caused to reciprocate by some form of mechanical linkage. Crank and connecting rod mechanisms, Scotch yokes, or cams may be used for this purpose.

Another mechanical type uses counterrotating masses to apply the driving force. Force adjustment is provided by relative offset of the weights and the counterrotation cancels shaking forces in one direction, say the x-direction, while supplementing the y-force. Frequency is controlled by a variable-speed motor.

There are two primary advantages in such inertia systems. In the first place, high force capacities are not difficult to obtain, and second, the shaking amplitude of the system remains unchanged by frequency cycling. Therefore, if a system is set up to provide a 0.05-in. amplitude at 20 Hz, changing the frequency to 50 Hz will not alter the amplitude. The reason will be understood if it is remembered that both the *available* exciting force and the *required* accelerating force are harmonic functions of the *square* of the exciting frequency; hence as the requirement changes with frequency, so too does the available force.

17.16.3   Hydraulic and Pneumatic Systems

Important disadvantages of the electromagnetic and mechanical shaker systems are limited load capacity and limited frequency, respectively. As a result, the search for other sources of controllable excitation has led to investigation in the areas of hydraulics and pneumatics.

Figure 17.17   Block diagram of a hydraulically operated shaker

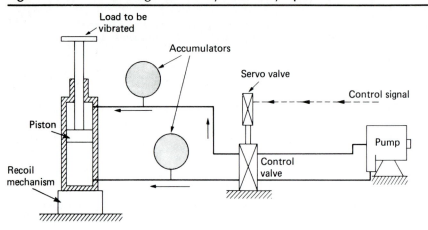

Figure 17.17 illustrates, in block form, a hydraulic system used for vibration testing [16]. In this arrangement an electrically actuated servo valve operates a main control valve, in turn regulating flow to each end of a main driving cylinder. Large capacities (to 500,000 lbf) and relatively high frequencies (to 400 Hz), with amplitudes as great as 18 in., have been attained. Of course, the maximum values cannot be attained simultaneously. As would be expected, a primary problem in designing a satisfactory system of this sort has been in developing valving with sufficient capacity and response to operate at the required speeds.

17.16.4   Relative Merits and Limitations of Each System

Frequency Range
The upper frequency ranges are available only through use of the electromagnetic shaker. In general, the larger the force capacity of the electromagnetic exciter, the lower its upper frequency will be. However, even the 40,000-lbf shaker listed in Table 17.2 boasts an upper useful frequency of 2000 Hz. To attain this value with a mechanical exciter would require rotative speeds of 120,000 rpm. The maximum frequency available from the smaller mechanical units is limited to approximately 120 Hz (7200 rpm) and for the larger machines to 60 Hz (3600 rpm). Hydraulic units are presently limited to about 2000 Hz.

Force Limitations
Electromagnetic shakers have been built with maximum vector force ratings of 40,000 lbf. Variable-frequency power sources for shakers of this type and size are very expensive. Within the frequency limitations of mechanical and hydraulic systems, corresponding or higher force capacities may be obtained at

lower costs by hydraulic shakers. Careful design of mechanical and hydraulic types is required, however, or maintenance costs become an important factor. Mechanical shakers are particularly susceptible to bearing and gear failures, whereas valve and packing problems are inherent in the hydraulic ones.

Maximum Excursion

One inch, or slightly more, may be considered the upper limit of peak-to-peak displacement for the electromagnetic exciter. Mechanical types may provide displacements as great as 5 or 6 in.: however, total excursions as great as 18 in. have been provided by the hydraulic-type exciter.

Magnetic Fields

Because the electromagnetic shaker requires a relatively intense fixed magnetic field, special precautions are sometimes required in testing certain items such as solenoids or relays, or any device in which induced voltages may be a problem. Although the flux is rather completely restricted to the magnetic field structure, relatively high stray flux is nevertheless present in the immediate vicinity of the shaker. Operation of items sensitive to magnetic fields may therefore be affected. Degaussing coils are sometimes used around the table to reduce flux level.

Nonsinusoidal Excitation

Shaker head motions may be sinusoidal or complex, periodic or completely random. Although sinusoidal motion is by far the most common, other waveforms and random motions are sometimes specified [17]. In this area, the electromagnetic shaker enjoys almost exclusive franchise. Although the hydraulic type may produce nonharmonic motion, precise control of a complex waveform is not easy. Here again, future development of valving may alter the situation.

The voice coil of the ordinary loudspeaker normally produces a complex random motion, depending on the sound to be reproduced. Complex random shaker head motions are obtained in essentially the same manner. Instead of using a fixed-frequency harmonic oscillator as the signal source, either a strictly random or a predetermined random signal source is used. Electronic *noise* sources are available, or a record of the motion of the actual end use of the device may be recorded on magnetic tape and used as the signal source for driving the shaker. As an example, electronic gear may be subjected to combat-vehicle motions by first tape-recording the output of motion transducers, then using the record to drive a shaker. In this manner, controlled repetition of an identical program is possible.

17.17   Vibration Test Methods

Two basic methods are used in applying a sinusoidal force to the test item: the *brute-force method* and the *resonance method* [18]. In the first case the item is attached or mounted on the shaker table, and the shaker supplies sufficient force to literally drive the item back and forth through its motion. The second method makes use of a mechanical spring-mass fixture having the desired natural frequency. The test item is mounted as a part of the system that is excited by the shaker. The shaker simply supplies the energy dissipated by damping.

17.17.1   The Brute-Force Method

Brute-force testing requires that the exciter supply all the accelerating force to drive the item through the prescribed motion. Such motion is generally sinusoidal, although complex waveforms may be used. The problems inherent in an arrangement of this sort are shown by the following example.

Suppose a vibration test specification calls for sinusoidally shaking a 10-kg test item at 100 Hz with a displacement amplitude of 2 mm (double amplitude = 4 mm). What force amplitude will be required?

Maximum force will correspond to maximum acceleration, and maximum acceleration may be calculated as follows:

$$\text{Circular frequency} = \Omega = 2\pi \times 100 = 628 \text{ rad/s},$$

$$\text{Maximum acceleration} = \text{the displacement amplitude} \times (\Omega)^2$$

$$= \left(\frac{2}{1000}\right)(628)^2 = 789 \text{ m/s}^2,$$

$$\text{Maximum force} = \frac{ma}{g_c} = \frac{(10 \times 789)}{1 \,(\text{kg} \cdot \text{m/s}^2)(\text{N} \cdot \text{s}^2/\text{kg} \cdot \text{m})}$$

$$= 7890 \text{ N, or about } 1770 \text{ lbf.}$$

This, of course, is the force amplitude required to shake the test item only. If support fixtures are required, they too must be shaken along with the moving coil of the shaker itself. Suppose these items (the fixture and voice-coil assembly) have a mass of 5 kg; then an additional vector force of 3945 N is required. The rated capacity of the shaker must therefore be a minimum of about 12,000 N.

17.17.2   The Resonance Method

The resonant system uses some form of spring-mass fixture to which the test item is attached. Figure 17.18 illustrates a small experimental setup whose characteristics will be described. The test item weighed 7.7 lbm and the test specifications required a sinusoidal vibration at 50 Hz with an amplitude of

Figure 17.18   Exciter-driven table for horizontal motion

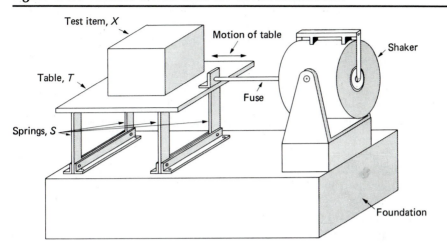

$\pm\frac{1}{8}$ in. A spring-supported table was designed, as shown in the figure. As initially tested, the resonance frequency was 58 Hz. Addition of a small mass fine-tuned the frequency to 50 Hz. The final weight distribution was as follows:

Test item	7.7
Table with mounting accessories	4.75
Moving weight of exciter	0.35
$\frac{1}{3}$ weight of leaf springs	0.50
Total	13.3 lbm

As a test of the maximum capacity of the system at 50 Hz, it was found that the 5-lbf shaker could actually move the table with test load through an amplitude of ±0.17 in. at 50 Hz. The force required to accomplish this may be calculated as follows:

$$\text{Maximum acceleration} = S_0\Omega^2 = (0.17)(2\pi \times 50)^2 = 16{,}750 \text{ in./s}^2,$$

$$\text{Necessary accelerating force} = \frac{ma}{g_c} = \left(\frac{16{,}750}{386}\right) \times 13.3 = 577 \text{ lbf.}$$

Obviously the 5-lbf shaker did not supply the force: The necessary accelerating forces were supplied almost in their entirety by the springs.

It will be readily realized that a resonant system of this type is limited to *one frequency*. Although a limited range of application might be designed into such a system through use of adjustable springs and masses, in general the system must be designed for the problem at hand and for that only.

Other forms of mechanically resonant systems include vertical spring-mass arrangements, the free-free beam [19, 20], tuning-fork systems [19], etc.

17.18   Shock Testing

Mechanical engineers are called on to design machinery to operate at higher and higher speeds. As speed goes up, accelerations increase, for the most part, not in direct proportion, but as the square of the speed. Both the magnitude of acceleration and acceleration gradients are increased. Resulting body loads often become much greater than applied loading, therefore becoming very significant factors in the design. The complexity of many problems has led to an area of investigation generally referred to as *shock testing* [21].

Actually, shock testing is only one of two phases of a broader classification that might better be called *acceleration testing*. Acceleration testing includes any test wherein acceleration loading is of primary significance. Included would be tests involving static or relatively slowly changing accelerations of any magnitude. Shock testing, on the other hand, is usually thought of as involving acceleration transients of moderate to high magnitude. In both cases the basic problem is to determine the ability of the test item to continue functioning properly either during or after application of such loading.

The more passive type of acceleration testing involves constant or relatively slowly changing accelerations, which, however, may be of high magnitude. It involves the use of centrifuges, rocket sleds, maneuvering aircraft, and the like, for the purpose of testing the capabilities of system components, including the human body, to withstand sustained or slowly changing high-level accelerations. Such tests are usually of quite specialized nature, generally applied to the study of performance in high-speed aircraft and missiles. Therefore, we shall only note this phase in passing and shall devote our primary attention to the first type of acceleration testing—namely, shock testing.

Most military apparatus must satisfactorily pass specified shock tests before acceptance. Equipment aboard ship, for example, is subject to shock from the ship's own armament, noncontact mine explosions, and the like. Aircraft equipment must withstand sharp maneuvering and landing loads, and artillery and communication equipment are subject to severe handling in crossing rough terrain. In addition, many items of industrial and civilian application are also subject to shock, often simply caused by normal handling during distribution, such as railroad-car humping, mail chuting, etc.

As a result, shock testing has become accepted as a necessary step in determining the usefulness of many items. It is becoming generally recognized, however, that to be meaningful, considerably more than magnitude of acceleration must be considered. In addition to magnitude, the rate and duration of application, along with the dynamic characteristics of the test item, must all be studied in setting up a useful shock test.

In general terms, shock testing may be divided into two broad categories: *low-energy* and *high-energy* (Fig. 17.19). Low-energy testing corresponds to the application of high accelerations over short time intervals. The terms sharp, intense, violent, and abrupt might be applied. However, the resulting

Figure 17.19   Typical characteristics of mechanical shock-producing machines of low, medium, and high energy

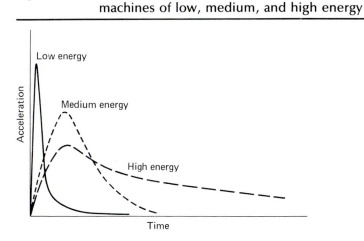

velocities (hence energies) may not be great. On the other hand, high-energy shock is applied for lengths of time permitting the buildup (or deterioration) of relatively high velocities. Acceleration magnitudes accompanying high-energy shock are commonly relatively low. This type of shock might be referred to as *impulsive, dynamic,* etc. Sometimes the latter is referred to as *energy loading* as opposed to *impact loading.* The severity of a shock test is very subjective: For certain items, high-energy shock is more severe; for others, the opposite may be true.

17.19   Shock Rigs

Several different methods are used for producing the necessary motion for shock testing. The approach generally taken is to store the required energy in some form of potential energy until needed, then to release it at a rate supplying the desired acceleration-time relation. Methods for doing this include the use of compressed air or hydraulic fluid, loaded springs, and the acceleration of gravity. The latter is the most commonly used.

17.19.1   Air Gun Shock-Producing Devices

Basically, the air gun system uses a piston, which moves within a tube or barrel under the action of high-pressure air applied to one face of the piston. Energy is stored by pressurizing the air in an accumulator. The high pressure is applied to the piston while it is restrained by a mechanical latching mechanism. When released, the piston with test item attached is sharply accelerated. Air trapped in the downstream portion of the cylinder serves to decelerate the piston, finally bringing it to rest. Machines based on this principle of operation have been made with energy capacities of more than 10^6 ft · lbf [22].

17.19.2   Spring-Loaded Test Rigs

As the name indicates, these machines use some form of mechanical spring for storing the energy required for acceleration. One machine designed to provide vertical accelerations uses helical tension springs attached at one end to a test carriage and at the other end to anchors that may be moved to put various initial tensions in the springs. With the springs initially tensioned, the test carriage (with test item) is released by a mechanical triggering mechanism and is accelerated suddenly upward. After the carriage has traveled a predetermined distance, the carriage ends of the spring strike stationary hooks. The spring's working stroke is thereby limited; however, the carriage continues upward until stopped by gravity.

17.19.3   A Hydraulic-Pneumatic Rig

The essentials for one hydraulic-pneumatic machine are illustrated in Fig. 17.20. In operation, an initial pneumatic pressure is introduced to chamber B. This pressure acts over the larger piston area. Hydraulic pressure is then increased in chamber A until the hydraulic pressure over the small area overcomes the pneumatic pressure acting over the larger area. At the instant the piston is lifted, the hydraulic pressure is suddenly applied to the larger area, producing a sharp impulsive-type upward motion. The form of the pulse is controlled by the shape of the metering pin [23, 24].

Figure 17.20   A form of hydropneumatic shock-producing device

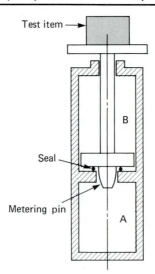

17.19.4   Gravity Rigs

There are two commonly used gravity-type shock rigs: the drop type and the hammer type. The hammer type is often referred to as a high-*g* machine and normally provides higher values of acceleration than the drop machine does.

Basically the drop machine consists of a platform to which the test item is attached, an elevating system for raising the platform, a releasing device that allows the platform to drop, and an impact pad or arrester against which the platform strikes. Guides are provided for controlling the fall (Fig. 17.21).

Acceleration-time relations are adjusted by controlling the height of drop and type of arrester pad. Pad selection is of great importance in determining the exact shock characteristics. If the pad is very rigid, for example, an acceleration pulse of very short duration results. On the other hand, a more flexible pad provides a longer time base, *Magnitude* of peak acceleration is controlled by adjusting the height of drop.

Rubber pads shaped to provide the desired acceleration-time pulse may be used. Sand pits have also been used in conjunction with variously shaped impacting surfaces. As another example, the test platform may be equipped with shaped pins or punches that strike and penetrate blocks of lead: The shape of the pins is designed to provide the desired pulse form.

Figure 17.22 illustrates the operation of a hammer-type shock-producing device that has been used to study the effects of head injury resulting from an impact. Strain propagation throughout the skull structure resulting from controlled frontal, rear, or side impacts were evaluated. In addition, extensive

Figure 17.21   A drop-type mechanical shock-producing machine

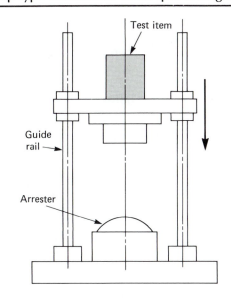

Figure 17.22   An application of a pendulum-type shock-loading
device to a research project

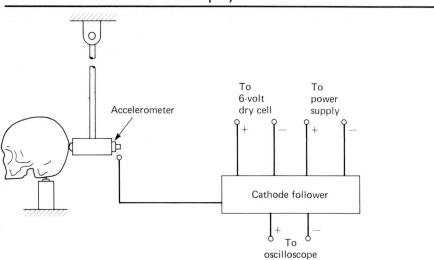

studies have been conducted on pressure wave propagation through simulated
brain tissue [25].

17.19.5   Relative Merits and Limitations of Each Shock Rig

Each of the shock-testing machines discussed in this section possessed certain
distinctive characteristics. The air-gun type produces what may be called a
high-energy shock. Generally speaking, high energy is synonymous with *high
velocity,* and to reach a high velocity, considerable displacement of the test
item is required. High velocity can be acquired only by relatively large
accelerations, or relatively long time intervals, or a combination of the two. In
either case, the test item will be displaced a considerable distance.

On the other hand, the drop and hammer machines are of the low-energy
category. High acceleration levels are possible, but only for short time
intervals. This results in comparatively low test-item velocities and hence low
energies. The hydraulic-pneumatic machine would be classified as a medium-
energy machine.

17.20   Practical Shock Testing

Some of the parameters pertinent to shock testing are illustrated in the
following example. Figure 17.23(a) illustrates an arrangement for a laboratory
experiment in dynamic-stress analysis. Pendulum *OA* is caused to swing freely

Figure 17.23   (a) Laboratory setup for demonstrating analysis of
mechanical shock, (b) typical experimental result
obtained from apparatus shown in (a)

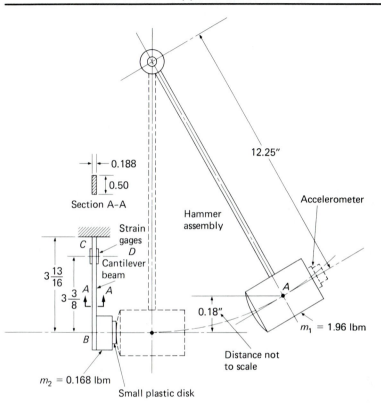

(a)

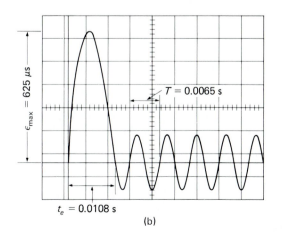

(b)

and strike a steel cantilever beam BC, on the end of which is mounted a small mass. A small plastic disk is placed at the point of impact to promote inelastic impact. Strain gages on each side of the beam are placed at D. Gage output is appropriately amplified and fed to a CRO for readout. Figure 17.23(b) shows the CRO trace obtained when the pendulum swings from rest through an arc equivalent to $h = 0.18$ in. Calibration-based readout values are shown in the figure.

Example 17.1

Applying theoretical relationships, compare analytical and experimental values of (a) maximum strain, (b) time of contact between pendulum tup and beam, and (c) the period of free vibration of the beam. Pertinent dimensions and masses are shown in the sketch. Equivalent mass, m_1, of the pendulum is taken as the mass of the tup plus one-third of the mass of the arm. Equivalent mass, m_2, is taken as the sum of the small mass at B plus one-third of the mass of the beam.

Note that the initial potential energy of the pendulum will be completely converted into kinetic energy at impact (assuming negligible losses). On the basis of the inelastic conditions between masses m_1 and m_2, the law of conservation of momentum will apply, from which the impacting energy loss may be evaluated. The remaining energy must be absorbed by the beam.

Time of contact may be estimated using the assumption that the strain-time relationship is a half sine wave corresponding to the free vibration of the beam as though both m_1 and m_2 are rigidly attached to it.

Solution.

a.   The velocity of initial contact of hammer and beam is

$$V_1 = \sqrt{2gh} = \sqrt{2 \times 386 \times 0.18} = 11.8 \text{ in./s.}$$

The velocity immediately after initial contact may be calculated using conservation of momentum, or,

$$V_2 = \left(\frac{m_1}{m_1 + m_2}\right) \quad V_1 = \left(\frac{1.96}{1.96 + 0.168}\right) \times 11.8 = 10.9 \text{ in./s.}$$

Therefore, energy to be absorbed by the beam is

$$U_2 = \frac{\frac{1}{2}(m_1 + m_2)V_2^2}{g_c} = \frac{\frac{1}{2}(1.96 + 0.168) \times (10.9)^2}{386} = 0.327 \text{ in.} \cdot \text{lbf.}$$

For the beam,

$$I = \frac{bd^3}{12} = \frac{0.5 \times (0.188)^3}{12} = 2.769 \times 10^{-4}\,\text{in}^4.$$

Work required to deflect the beam is

$$U_2 = \int_0^L \frac{M_1^2\,dx}{2EI} = \frac{F^2 L^3}{6EI},$$

where

$$M_1 = Fx \qquad (0 \le x \le L).$$

Substituting and solving for F_{max},

$$F_{max} = \sqrt{\frac{6EIU_2}{L^3}} = \sqrt{\frac{6 \times 30 \times 10^6 \times 2.769 \times 0.327 \times 10^{-4}}{(3.8125)^3}}$$

$$= 17.15\,\text{lbf}.$$

At the strain gages, $M_2 = 3.375 \times F_{max} = 57.88\,\text{in.} \cdot \text{lbf}.$

$$\varepsilon = \frac{M_2 c}{EI} \qquad (\text{where } c = 0.188/2)$$

$$= \frac{57.88 \times 0.094}{30 \times 10^6 \times 2.769 \times 10^{-4}}$$

$$= 655\,\mu\text{-strain}.$$

b. To estimate the time of contact between pendulum tup and beam, assume that for one-half cycle of vibration the tup is rigidly attached to the end of the beam; therefore,

$$\text{Period of vibration} = 2\pi \sqrt{\frac{m_1 + m_2}{kg_c}}$$

$$= 2\pi \sqrt{\frac{1.96 + 0.168}{449.7 \times 386}}$$

$$= 0.022\,\text{s}.$$

Time for one-half cycle = time of contact = $t_e = 0.022/2 = 0.011\,\text{s}.$

c. Calculating the period of the free beam (with m_2 at its end).
$$T = 2\pi\sqrt{m_2/kg_c} = 2\pi\sqrt{0.168/449.7 \times 386} = 0.0062\,\text{s}.$$

17.21   Elementary Shock-Testing Theory

An impact or shock test may be simplified as shown in Fig. 17.24. This diagram may be thought of as representing a drop machine of the spring-retarding type. Here M_1 represents the table mass; k_1 is the modulus of the retarding spring; and M_2 is the test item, or a portion thereof, supported by a linear spring of modulus, k_2. A viscous damping coefficient, c_2, may be assumed acting between the table and the test mass. If we assume c_2 to be comparatively small, we may write

$$\frac{M_1}{g_c}\frac{d^2S_1}{dt^2} + k_1S_1 - k_2(S_2 - S_1) = 0 \tag{17.18}$$

and

$$\frac{M_2}{g_c}\frac{d^2S_2}{dt^2} + k_2(S_2 - S_1) = 0 \tag{17.18a}$$

Rewriting Eq. (17.18) as

$$\frac{d^2S_1}{dt^2} + g_c\frac{k_1S_1}{M_1} - g_c\frac{k_2}{M_1}(S_2 - S_1) = 0 \tag{17.19}$$

and noting that if

$$k_1 \gg k_2 \quad \text{and} \quad M_1 \gg M_2,$$

we have

$$\frac{k_2}{M_1} \ll \frac{k_1}{M_1} \quad \text{and} \quad \frac{k_2}{M_1} \ll \frac{k_2}{M_2}.$$

Figure 17.24   Idealized shock situation

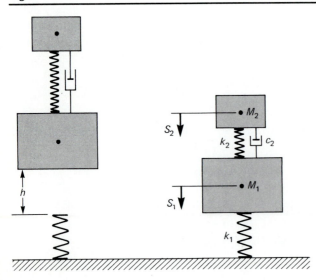

Thus Eqs. (17.18) and (17.18a) may be written as

$$\frac{d^2S_1}{dt^2} + \omega_1^2 S_1 = 0 \qquad \left(0 \le t \le \frac{\pi}{\omega_1}\right), \tag{17.20}$$

$$\frac{d^2S_1}{dt^2} = 0 \qquad \left(t \ge \frac{\pi}{\omega_1}\right), \tag{17.20a}$$

and

$$\frac{d^2S_2}{dt^2} + \omega_2^2(S_2 - S_1) = 0 \qquad (t \ge 0), \tag{17.20b}$$

where

$$\omega_1^2 = \frac{k_1 g_c}{M_1} \quad \text{and} \quad \omega_2^2 = \frac{k_2 g_c}{M_2}.$$

(Note that although the subscript n is not used, ω_1 and ω_2 represent undamped natural frequencies.) The solution to Eq. (17.20), subject to the initial conditions

$$S_1 = 0 \quad \text{and} \quad \frac{dS_1}{dt} = v_1 = \sqrt{2gh} \qquad \text{at } t = 0,$$

can be expressed as

$$S_1(t) = \frac{v_1}{\omega_1} \sin \omega_1 t \qquad \left(0 \le t \le \frac{\pi}{\omega_1}\right). \tag{17.21}$$

Defining the relative displacement between the masses as

$$S = S_2 - S_1,$$

we can write Eq. (17.20b) as

$$\frac{d^2S}{dt^2} + \omega_2^2 S = -\frac{d^2S_1}{dt^2}, \tag{17.22}$$

which gives

$$\frac{d^2S}{dt^2} + \omega_2^2 S = v_1 \omega_1 \sin \omega_1 t \qquad \left(0 \le t \le \frac{\pi}{\omega_1}\right), \tag{17.23}$$

$$\frac{d^2S}{dt^2} + \omega_2^2 S = 0 \qquad \left(t > \frac{\pi}{\omega_1}\right). \tag{17.23a}$$

Considering the initial conditions

$$S = \frac{dS}{dt} = 0 \qquad \text{at } t = 0,$$

we can write the solutions to Eqs. (17.23) and (17.23a) as

$$S = \frac{v_1\omega_1}{\omega_1^2 - \omega_2^2}\left(\frac{\omega_1}{\omega_2}\sin\omega_2 t - \sin\omega_1 t\right) \qquad \text{for } 0 \le t \le \frac{\pi}{\omega_1},$$

and

$$S = \frac{2v_1\omega_1^2\cos(\pi\omega_2/2\omega_1)}{\omega_2(\omega_1^2 - \omega_2^2)}\sin\omega_2\left(t - \frac{\pi}{2\omega_1}\right) \qquad \text{for } t \ge \frac{\pi}{\omega_1}.$$

Maximum displacements are as follows:

$$S_{max} = \frac{v_1}{\omega_2[(\omega_2/\omega_1) - 1]}\sin\frac{2n\pi}{(\omega_2/\omega_1) + 1} \qquad \text{for } 0 \le t \le \frac{\pi}{\omega_1},$$

and

$$S_{max} = \frac{2v_1\omega_1^2\cos(\pi\omega_2/2\omega_1)}{\omega_2(\omega_1^2 - \omega_2^2)} \qquad \text{for } t \ge \frac{\pi}{\omega_1}.$$

Here n is a positive integer, chosen to make the sine term as large as possible, while

$$\frac{2n}{(\omega_2/\omega_1) + 1} < 1.$$

If the gradient of acceleration (d^2S_1/dt^2) is very low so that the peak value $v_1\omega_1$ is reached very slowly, the maximum disturbance experienced by M_2 will be very nearly proportional to (d^2S_1/dt^2). In other words, under such conditions secondary vibrations would not be excited in the supported system. Under these conditions an equivalent static displacement, S_{st}, may be obtained from Eq. (17.23) by dropping the time-dependent terms, or

$$S_{st} = \frac{v_1\omega_1}{\omega_2^2}. \qquad\qquad (17.24)$$

Mindlin, Stubner, and Cooper [27] call the ratio (S_{max}/S_{st}) the *amplification factor*, A. Solving for A, we obtain

$$A = \frac{\omega_2/\omega_1}{(\omega_2/\omega_1) - 1}\sin\left[\frac{2n\pi}{(\omega_2/\omega_1) + 1}\right] \qquad \text{for } 0 \le t \le \frac{\pi}{\omega_1}, \qquad (17.25)$$

and

$$A = \frac{2(\omega_2/\omega_1)\cos(\pi\omega_2/2\omega_1)}{1 - (\omega_2/\omega_1)^2} \qquad \text{for } t \ge \frac{\pi}{\omega_1}.$$

It will be observed that the amplification factor is dependent only on the ratio ω_2/ω_1. By plotting Eqs. (17.25), we obtain the upper curve of Fig. 17.25. Damping ratios, $\beta_1 = C_2/2\sqrt{M_2 k_2}$, reduce the magnitude of the amplification factor as shown.

Figure 17.25 Amplification factors for linear undamped cushioning with perfect rebound

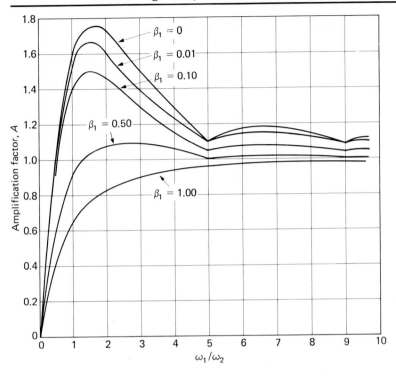

The foregoing theoretical consideration of the shock problem is quite elementary in so far as a very simple pulse excitation was assumed and a simple single-degree-of-freedom test situation was considered. In the field, and also when most shock test machines are used, the exciting pulse or disturbance consists of a complex acceleration-time relation including many frequency components of different amplitude, and the test item will normally possess many degrees of freedom, hence many modes of vibration. The theoretical results are quite useful, however, in providing information on the limiting maxima that may be expected.

Readers interested in pursuing the theoretical aspects of this problem further are directed particularly to Ref. [27].

Suggested Readings

Bloss, R. L., and M. J. Orloski (eds.). *Precision Measurement and Calibration* (selected NBS papers on mechanics), NBS Special Publ. 300, vol. 8. Washington, D.C.: U.S. Government Printing Office, 1972.

Booth, S. F. (ed.). *Precision Measurement and Calibration* (selected papers on heat and mechanics), NBS Hndbk. 77, vol. 2. Washington, D.C.: U.S. Government Printing Office, 1961.

Den Hartog, J. P. *Mechanical Vibrations.* 4th ed. New York: Dover, 1985.

Kistler, W. P. *Precision Calibration of Accelerometers for Shock and Vibration.* Test Engineering, May 1966, p. 16.

Weaver, W. Jr., S. P. Timoshenko, and D. H. Young. *Vibration Problems in Engineering.* 5th ed. New York: John Wiley, 1990.

Problems

17.1 A simple frequency meter is described in the final paragraph of Section 17.3. The undamped first-mode frequency of a uniformly sectioned cantilever beam may be calculated from the expression [28]

$$f = 3.52 \sqrt{\frac{EIg_c}{mL^4}} \ \text{Hz}$$

where

E = Young's modulus for the material of the beam,

m = the mass of the beam per unit length,

I = the area moment of inertia of the beam section,

L = the length of the beam.

For a steel wire of $1\frac{1}{2}$-mm (0.0039-in.) diameter, plot the resonance frequencies over a range of $L = 10$ to 25 cm (3.94 to 9.84 in.). (Use density of steel = 7900 kg/m³.)

17.2 Referring to Problem 17.1, we may design an instrument of wider range by providing a series of small masses to be attached to the outer end of the beam. In this case,

$$f = \frac{1}{2\pi} \sqrt{\frac{3EIg_c}{(M_1 + M_B)L^3}} \ \text{Hz}$$

where

M_1 = the mass of the attachment,

M_B = one-third the mass of the beam.

Design a system to cover the range of $50 < f < 2000$ Hz.

17.3 Obtain an inexpensive crystal-type phonograph pickup designed for replaceable phonograph needles. The least expensive will be quite satisfactory. Place a small overhanging mass in the stylus chuck, connect the pickup output to an oscilloscope, and use it to investigate various sources of mechanical vibration. (*Note:* The device should be quite useful as a "frequency" pickup; however, it will not be adequate for meaningful amplitude readout.)

17.4 The waveform from a mechanical vibration is sensed by a velocity-sensitive vibrometer. The CRO trace indicates that the motion is essentially simple harmonic. A 1-kHz oscillator is used for time calibration and 4 cycles of the vibration are found to correspond to 24 cycles from the oscillator. Calibrated vibrometer output indicates a velocity amplitude ($\frac{1}{2}$ peak-to-peak) of 3.8 mm/s. Determine (a) the displacement amplitude in mm and (b) the acceleration amplitude in standard g's.

17.5 A vibrometer is used to measure the time-dependent displacement of a machine vibrating with the motion

$$y = 0.5 \sin(3\pi t) + 0.8 \sin(10\pi t),$$

where y is in cm and t is in s. If the vibrometer has an undamped natural frequency of 1 Hz and a critical damping ratio of 0.65, determine the vibrometer time-dependent output and explain any discrepancies between the machine vibration and the vibrometer readings.

17.6 An accelerometer is used for measuring the amplitude of a mechanical vibration. The following data are obtained:

> Waveform: simple sinusoidal;
>
> Period of vibration = 0.0023 s;
>
> Output voltage from accelerometer = 0.213 V rms;
>
> Accelerometer calibration = 0.187 V/standard g.

What vibrational displacement (amplitude) is sensed by the accelerometer? Express your answer (a) in millimeters; (b) in inches.

17.7 An accelerometer is designed to have a maximum practical inherent error of 4% for measurements having frequencies in the range of 0 to 10,000 Hz. If the damping constant is 50 N · s/m, determine the spring constant and suspended mass.

17.8 A seismic-type vibrometer is characteristically a relatively fragile instrument, whereas an accelerometer is relatively rugged. Explain why this is so.

17.9 Refer to most any introductory vibrations text and review the material on viscous damping including that on the logarithmic decrement. Use such equipment as may be available to obtain a displacement-time record for a decaying damped free vibration, such as that shown in Fig. 17.26(a). The vibrating element may be a simple cantilever beam with concentrated end-mass, as shown in Fig. 17.26(b), etc. The phono-pickup vibration sensor of Problem 17.3 should suffice as a transducer and an oscilloscope or oscillograph as the terminating device. Using data as defined in the figure, determine the damped frequency of vibration and the viscous damping coefficient

$$c = \frac{M\omega_{nd}\delta}{\pi g_c},$$

where

$$\omega_{nd} = \text{the circular frequency of free vibration,}$$

$$\delta = \text{the logarithmic decrement}$$

$$= \frac{1}{q} \ln\left(\frac{S_m}{S_{m+q}}\right).$$

Figure 17.26   (a) Displacement-time record and
                   (b) equipment setup for Problem 17.9

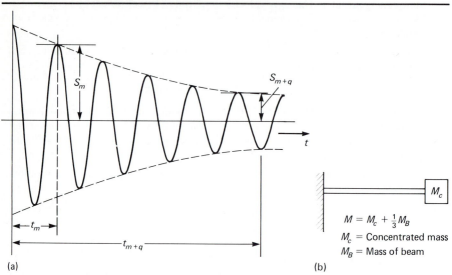

(a)   (b)

$M = M_c + \frac{1}{3} M_B$

M_c = Concentrated mass

M_B = Mass of beam

17.10 An accelerometer is being calibrated by the procedure illustrated in Fig. 17.9. However, instead of observing straight lines, as shown in Fig. 17.10(b), each of the points of light traces a path similar to that shown in the upper center illustration of Fig. 10.12. This trace demonstrates an unsatisfactory condition that should be corrected if a good calibration is to be obtained. What is your analysis of the problem?

17.11 An accelerometer is to be calibrated by the high-g method shown in Fig. 17.11. If the effective pendulum length is 1 m, the initial release angle measured from the vertical is 60°, and the maximum rebound angle is 50°, determine the calibration factor, K, given by Eq. (17.15a). Assume the acceleration waveform is a half sine wave with an amplitude of 500 mV and a pulse duration of 0.005 s.

17.12 An electromagnetic-type sinusoidal vibration exciter having a rated force capacity of 25 N (5.62 lbf) is to be used to excite a test item weighing 3 kg (6.61 lbm). If the moving parts of the shaker have a mass of 0.75 kg (1.65 lbm) and the amplitude of the vibration is 0.15 mm (0.0059 in.) (double amplitude = 0.30 mm), determine the maximum excitation frequency that can be applied.

17.13 At full rated load of 5000 N, a load cell deflects 0.10 mm. If it is used to measure the thrust of a small (200-kg) jet engine, what maximum frequency component of the thrust may be accurately measured if the inherent error is limited to 2%? Damping is negligible. What is the basis that you use for determining the limiting frequency?

References

1. Fibikar, R. J., "Touch and vibration sensitivity," *Prod. Eng.* 27: November 1956, p. 177.

2. Hudson, D. E. and O. D. Terrell, A preloaded spring accelerometer for shock and impact measurements. *SESA Proc.* 9(1): 1, 1951.

3. Pennington, D. *Piezoelectric Accelerometer Manual.* Pasadena: Endevco Corporation, 1965.

4. Kistler, W. P. Precision calibration of accelerometers for shock and vibration. *Test Eng.*: 16, May 1966.

5. Edelman, S. Additional thoughts on precision calibration of accelerometers. *Test Eng.*: 17, November 1966.

6. Lewis, R. C., Electro-dynamic calibration for vibration pickups. *Prod. Eng.* 22: September 1951.

7. Unholtz, K. The calibration of vibration pickups to 2000 CPS. *ISA Proc.* 7: 325, 1952.

8. Easily made device calibrates accelerometer. *NBS Tech. News Bull.*: 94, June 1966.

9. Wildhack, W. A. and R. O. Smith, *A Basic Method of Determining the Dynamic Characteristics of Accelerometers by Rotation.* ISA Paper No. 54-40-3, 1954.

10. Conrad, R. W. and I. Vigness. Calibration of accelerometers by impact techniques, *ISA Proc.* 8: 166, 1953.

11. Perls, T. A. and C. W. Kissinger, *High-g Accelerometer Calibration by Impact Methods with Ballistic Pendulum, Air Gun, and Inclined Trough.* ISA Paper No. 54-40-2, 1954.

12. Weiss, D. E. Design and application of accelerometers. *SESA Proc.* 4(2): 1947.

13. Burns, J. and G. Rosa. Calibration and test of accelerometers, *Instrument Notes 6.* Los Angeles: Statham Laboratories, December 1948.

14. Levy, S. and W. D. Kroll, Response of accelerometers to transient accelerations. *NBS J. Res.* 45: 4, October 1950.

15. Welch, W. P. A proposed new shock-measuring instrument, *SESA Proc.* 5(1): 39, 1947.

16. Adler, J. A. Hydraulic shakers. *Test Eng.*: April 1963.

17. Crandall, S. H. *Random Vibration.* New York: John Wiley, 1959.

18. Unholtz, K. Factors to consider in setting up vibration test specifications, *Mach. Des.* 28: 6, March 22, 1956.

19. Wozney, G. P. Resonant vibration fatigue testing. *Exp. Mech.*: January 1962.

20. Application and design formulae for free-free resonant beams. *MB Vibration Note-book.* New Haven, MB Manufacturing Co., March 1955.

21. Lazarus, M. Shock testing: A design guide. *Mach. Des.*: October 12, 1967.

22. Armstrong, J. H., Shock-testing technology at the Naval Ordnance Laboratory. *SESA Proc.* 6(1): 55, 1948.

23. Brown, J. Selection factors for mechanical buffers. *Prod. Eng.* 21: 156, November 1950.

24. Brown, J. Further principles of buffer design, *Prod. Eng.* 21: 125, December 1950.

25. Marangoni, R. D., C. A. Saez, D. A. Weyel, and R. A. Polosky. Impact stresses in human head-neck model. *J. Eng. Mech. Div., ASCE* 104 (EM1): 1978.

26. Mindlin, R. D., F. W. Stuber, and H. L. Cooper. Response of damped elastic systems to transient disturbances, *SESA Proc.*, 5(2): 69, 1947.

27. Den Hartog, J. P., *Mechanical Vibrations*. 4th ed. New York: McGraw-Hill, 1956, p. 153.

CHAPTER 18

Acoustical Measurements

18.1 Introduction

Sound may be described on the basis of two considerably different points of view: (a) from the standpoint of the physical phenomenon itself, or (b) in terms of the "psychoacoustical" effect sensed through the human process of hearing. It is very important that these basically different aspects be kept continually in mind. To measure the particular physical parameters associated with a specific sound, either of simple or complex waveform, is a much simpler assignment than to attempt to evaluate the effects of the parameters as sensed by human hearing.

Occasionally, the reasons for measuring a sound may not be associated with hearing. For example, sound pressure variations accompanying high-thrust rocket motor or jet engine operation may be of sufficient magnitude to endanger the structural integrity of the missile or aircraft [1]. Structural fatigue failures have been induced by sound excitation. In such cases measurement of the parameters that are involved does not directly include the psychoacoustical relationship. But in the great majority of cases the effect of the measured sound *is* directly related to human hearing and this added complication is therefore unavoidable.

Noise may be defined as unwanted sound. Noise affects human activities in many ways. Excessive noise may make communication by direct speech difficult or impossible. Noise may be a factor in marketing appliances or other equipment. Prolonged ambient noise levels may eventually cause permanent damage to hearing or, of course, it may simply impair efficiency of workers because of the annoyance factor. All these aspects of sound are unavoidably coupled with human hearing. Seldom are mechanical engineers concerned with the production of *pleasing* sounds. Almost always they are concerned with noise, its abatement, and its control.

Physically, airborne sound is, within a certain range of frequencies, a periodic variation in air pressure about the atmospheric mean. The air particles oscillate along the direction of propagation, and for this reason the waveform is said to be longitudinal. For a single tone or frequency (as opposed to a sound of complex form), the oscillation is simple harmonic and may be expressed as [2]:

$$S = S_0 \cos\left[\frac{2\pi}{\lambda}(x \pm ct)\right], \tag{18.1}$$

where

S = the displacement of a particle of the transmitting medium,

S_0 = the displacement amplitude,

λ = the wavelength = $\dfrac{c}{f}$,

x = the distance from some origin (e.g., the source), in the direction of propagation,

c = the velocity of propagation,

t = time, and

f = frequency.

For right-running wave in a gaseous medium [2]:

$$p = -B\frac{\partial S}{\partial x} = -B\frac{2\pi}{\lambda}S_0 \sin\left[\frac{2\pi}{\lambda}(ct - x)\right], \tag{18.2}$$

where

p = the pressure variation about an ambient pressure, P_{amb}, and

B = the adiabatic bulk modulus.

18.2   Characterization of Sound (Noise)

In its simplest form, sound is a *pure tone*; that is, it is of one frequency. Such a source is extremely difficult to produce. In addition to the inherent purity of the signal source and its coupling to the air, it requires the elimination of all reflections from surrounding objects. Such reflections or reverberations produce standing waves that at the point of observation can produce distortions caused by interaction between the directly incident wave and the returning reflections. *Free-field* conditions may be approximated in an *anechoic (no-echo) chamber.** The walls of such a chamber are lined with sound-absorbing materials formed in wedges such that small reflections as may result from the

---

* An imperfect but inexpensive substitute for an anechoic chamber is an open field in as quiet a location as possible. Preferably the site should be on a slight hill or hummock.

initial encounter with the wall are directed again and again into the absorbent materials until essentially all the energy has been absorbed. In such an environment, sound simply travels outward and away from the source, with no return. Even so, the mere placement of a transducer into the space will cause unwanted distortions.

Random noise, rather than a pure tone, is by far the more common subject of investigation. Random noise is produced from a number of discrete sources whose outputs combine to form the whole. The sounds reaching the transducer (or the ear) are commonly from more than the initiating sources alone. The basic originating sources of sound energy may be called *primary* sources. In mechanical engineering these sources typically result from interacting machine parts such as gear teeth or bearings; from vibrating members of a wide variety, such as housing panels, shafts, and supports; or from hydraulic, pneumatic, or combustion sources. Sound waves traveling outward from a primary source are intercepted by surrounding objects such as ceilings, walls, and other machinery. A part of such incident waves is absorbed and the remainder is reflected (Fig. 18.1). Each point of reflection then becomes a *secondary source,* "heard" by any pickup device and thereby becoming confused with the initial or primary sounds. Combinations of primary and secondary sounds from *different* reflecting sources produce *standing waves,* resulting in reinforcements and nulls throughout the environment. In the most common situation the primary sounds are not pure tones but possess certain randomnesses in both amplitudes and frequencies. As a result, the standing waves are not necessarily fixed in space but may be thought of as dancing about throughout the environment. Beats may result (Sections 4.4 and 10.8). It is obvious, then, that evaluation of the parameters associated with a given sound—e.g., from an internal combus-

Figure 18.1 Sketch illustrating primary and secondary sources of sound (noise)

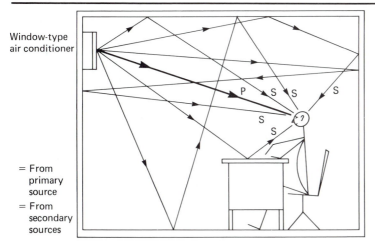

tion engine or an air compressor—becomes very difficult indeed. Under normal circumstances, separation of subject and environment becomes relatively impossible. In fact, one may question the real value of such a separation. Although anechoic-chamber testing may help in understanding or treating certain specific noise sources, to produce meaningful, true-to-life results must necessarily include interaction between both the primary and the secondary sources. The environment is an unavoidable adjunct to any analysis.

18.3 Basic Acoustical Parameters

18.3.1 Sound Pressure

In the presence of a sound wave, the instantaneous difference in air pressure and the average air pressure at a point is called *sound pressure*. The unit of measurement is the pascal (Pa), or newtons per square meter, or, more commonly, the micronewton per square meter (μN/m^2). The microbar (μbar) is also used. It is equal to one dyne per square centimeter.*

As noted previously, sound pressure is the difference between instantaneous absolute pressure and the ambient pressure. We now define the mean-square sound pressure as:

$$p_{\text{rms}}^2 = \frac{1}{T} \int_0^T p^2 \, dt \tag{18.3}$$

where

$$p = p(t) = \text{instantaneous sound pressure at a point (Pa)},$$
$$p_{\text{rms}} = \text{root-mean-square sound pressure (Pa)},$$
$$T = \text{averaging measurement time (sec)}.$$

For a pure-tone sound wave described by

$$p = P \sin \omega \left(\frac{x}{c} - t \right) \tag{18.3a}$$

where

$$\omega = 2\pi f$$

the mean-square sound pressure at any point x is given by

$$p_{\text{rms}}^2 = \frac{P^2}{T} \int_0^T \sin^2 \omega \left(\frac{x}{c} - t \right) dt. \tag{18.3b}$$

---

* 1 μN/m^2 = 0.00001 μbar.

If we choose $T = 1/f = 2\pi/\omega$,

$$p_{rms} = \frac{P}{\sqrt{2}}$$ (18.3c)

where P is the amplitude of the pure tone sound wave.

In typical measurement situations, sound pressure is averaged over a time interval large enough to include several periods of all frequencies of interest. Thus the contribution of partial periods does not significantly affect the rms value.

18.3.2 Sound Pressure Level

The ratio between the greatest sound pressure that a person with normal hearing may tolerate without pain and that of the softest discernible sound is roughly 10 million to 1 (Fig. 18.2). This tremendous range suggests the use of some form of logarithmic scale. Recall that the decibel (Section 7.12) provides such a scale and it is on this basis that most sound or noise measurements are

Figure 18.2 Typical sound pressures and sound pressure sources

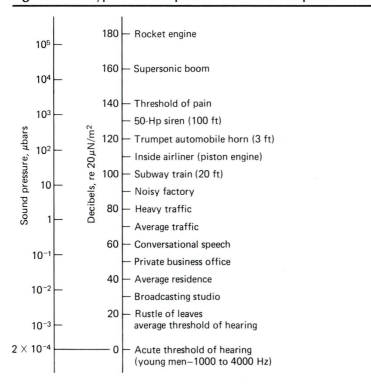

made. Basically, the decibel is a measure of *power ratio*:

$$dB = 10 \log_{10} \left(\frac{\text{power}_1}{\text{power}_2} \right). \tag{18.4}$$

Because sound power is proportional to the *square* of sound pressure [3], Eq. (18.4) may be written as

$$dB = 10 \log_{10} \frac{p_1^2}{p_0^2} = 20 \log_{10} \left(\frac{p_1}{p_0} \right). \tag{18.5}$$

We see that the decibel is not an absolute quantity but a comparative one, which, however, can be used *in the manner of* an absolute quantity if referred to some generally accepted base. This is done and the *rms* value,

$$p_0 = 0.00002 \, \text{N/m}^2 = 20 \, \mu\text{N/m}^2,$$

is widely accepted as the standard reference for sound pressure level. This value takes on added significance when we note that it corresponds quite closely to the acute threshold of hearing (Fig. 18.2). Therefore,

$$\text{SPL} = 20 \log \left(\frac{p}{0.00002} \right) \qquad \text{re: } 20 \, \mu\text{N/m}^2, \tag{18.6}$$

where

$$\text{SPL} = \text{the sound pressure level, in dB}$$

$$p = \text{the rms pressure from a sound source, in N/m}^2.$$

Note: Inasmuch as all sound level measurements use base 10 logarithms, the reference will be omitted throughout the remainder of the chapter. In addition, the appended "re: $20 \, \mu\text{N/m}^2$" will be omitted with the understanding that this is the standard of reference.

Example 18.1
What is the sound pressure level corresponding to an rms sound pressure of $1 \, \text{N/m}^2$?

Solution.

$$\text{SPL} = 20 \log \left(\frac{1}{0.00002} \right) \approx 94 \, \text{dB}$$

Special attention should be directed to the use of the word *level*. Various terms yet to be discussed use the word: sound level, loudness level, noise level, etc. Use of the word level implies a logarithmic scale of measurement expressed in decibels. Remembering this fact should help in keeping the units straight.

18.3.3 Power, Intensity, and Power Level

As suggested by Eq. (18.4), sound involves energy whose magnitude can be expressed in terms of power. The common unit is the watt, W. As the power radiates outward from an ideal point source, it will be continually spread over larger and larger areas of space. *Sound intensity* at any location is expressed in terms of watts per unit area. For a plane or spherical wave, the intensity I in the direction of propagation is [3] (see Fig. 18.3)

$$I = \frac{(p_{\rm rms})^2}{\rho_0 c} = \frac{W}{A},\qquad (18.7)$$

where

ρ_0 = the average mass density of the medium,

c = the speed of sound in the medium,

W = acoustic power in watts,

A = area.

A further discussion of the application of vector sound intensity measurement is given in Section 18.7.3.

Sound power level (PWL) is expressed in decibels and therefore must be given in terms of a reference level that is usually taken as 10^{-12} W. Power level is therefore defined as

$$\text{PWL} = 10 \log (W/10^{-12})\,\text{dB} \qquad \text{re: } 10^{-12}\,\text{W}. \qquad (18.8)$$

There is no instrument for measuring power level directly. However, the quantity can be calculated from sound pressure level measurements [4].

Figure 18.3 Relation between sound power and intensity for a spherical wave

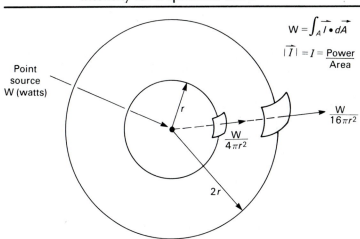

18.3.4   Combination of Sound Pressure Levels

When two pure-tone sounds occur at the same time, the combined effect depends on the sound pressure amplitudes, frequencies, and phase relationship at the receiver. Consider two sound waves at a point in space described by

$$p_1 = P_1 \cos(\omega_1 t + \phi_1), \tag{18.9}$$

$$p_2 = P_2 \cos(\omega_2 t + \phi_2), \tag{18.9a}$$

where

$$p = \text{instantaneous sound pressure},$$

$$P = \text{sound pressure amplitude},$$

$$\omega = 2\pi f = \text{circular frequency},$$

$$\phi = \text{phase angle}.$$

The instantaneous sound pressure due to the two waves is the sum of the two instantaneous sound pressures. The mean-square sound pressure of the combined pure tones is given by

$$p_{\text{rms}}^2 = \frac{1}{T} \int_0^T (p_1 + p_2)\, dt \tag{18.9b}$$

Substituting Eqs. (18.9) and (18.9a) into Eq. (18.9b) and integrating, we obtain [5]

$$p_{\text{rms}}^2 = \begin{cases} \dfrac{P_1^2 + P_2^2}{2} = p_{\text{rms1}}^2 + p_{\text{rms2}}^2 & \omega_1 \neq \omega_2 \\[2ex] \dfrac{P_1^2 + P_2^2}{2} + P_1 P_2 \cos(\phi_1 - \phi_2) & \omega_1 = \omega_2 \end{cases} \tag{18.9c}$$

where the averaging time, T, satisfies the relation

$$T \gg \frac{1}{f_{\text{min}}}$$

and f_{min} is the lowest frequency. Thus if two pure tones of equal amplitudes, frequency, and zero relative phase are combined, the resulting mean-square sound pressure is

$$p_{\text{rms}}^2 = P_1^2 + P_1^2 \cos 0 = 2P_1^2 = 4p_{\text{rms1}}^2$$

and the resulting increase in sound pressure level over the sound pressure level of one pure tone is

$$\text{SPL}_{\text{COMB}} - \text{SPL}_1 = 20 \log_{10} \left(\frac{2p_{\text{rms1}}}{p_0} \right) - 20 \log_{10} \frac{p_{\text{rms1}}}{p_0}$$

$$= 20 \log_{10} 2 = 6.02,$$

which is about a 6-dB increase.

Figure 18.4 Diagram for adding or subtracting decibels
(Courtesy: GenRad Inc., Concord, MA)

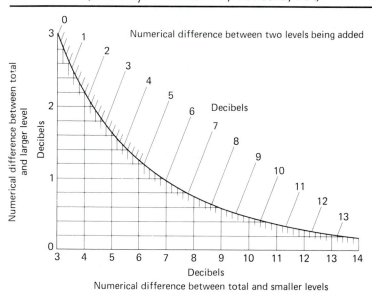

If we now determine the effect of adding two pure tones of equal sound pressure magnitudes but different frequencies, the first expression of Eq. (18.9c) should be used. Thus the difference between the combined sound pressure level and the sound pressure level of one pure tone is

$$\text{SPL}_{\text{COMB}} - \text{SPL}_1 = 20 \log_{10} \frac{\sqrt{2}\, p_{\text{rms1}}}{p_0} - 20 \log_{10} \frac{p_{\text{rms1}}}{p_0}$$

$$= 20 \log_{10} \sqrt{2} = 3.01$$

or about 3 dB. Most industrial and community noise problems that we experience represent a combination of many *uncorrelated* sources. There may be random variations in both amplitude and frequency. For *uncorrelated* noise sources, the first expression in Eq. (18.9c) applies. Thus the total mean-square sound pressure is simply the sum of the component source mean square pressures. In order to simplify the combination of *uncorrelated* noise sources, Fig. 18.4 can be used to add or subtract noise sources.

Example 18.2
A business machine is added to an office. The original ambient SPL was 68 dB. After the machine was added, the sound pressure level rose to 72 dB. What SPL was contributed by the machine?

Solution. Using Fig. 18.4, the difference between the total and the smaller levels is 4 dB. The 4 dB vertical line intersects the difference-between-decibel-levels line at 1.8 dB. Thus the machine decibel level is $68 + 1.8 = 69.8$ dB.

18.3.5   Attenuation with Distance

In a *lossless, free space* there is a 6 dB decrease in sound pressure level (SPL) for each doubling of distance [6]. This may be shown as follows. From Eq. (18.5), we have

$$\mathrm{SPL}_1 = 20 \log\left(\frac{p_1}{p_0}\right) \quad \text{and} \quad \mathrm{SPL}_2 = 20 \log\left(\frac{p_2}{p_0}\right);$$

therefore

$$\mathrm{SPL}_2 - \mathrm{SPL}_1 = 20[\log(p_2) - \log(p_1)] = 20 \log\left(\frac{p_2}{p_1}\right). \tag{18.10}$$

From Eq. (18.7) we see that for a spherical wave

$$\frac{p^2}{\rho c} = \frac{W}{(4\pi r^2)};$$

hence

$$\frac{p_2}{p_1} = \frac{r_1}{r_2}. \tag{18.11}$$

Combining Eqs. (18.10) and (18.11) gives us

$$\mathrm{SPL}_2 - \mathrm{SPL}_1 = 20 \log\left(\frac{r_1}{r_2}\right). \tag{18.12}$$

Question

In a free field, what will be the difference in sound pressure levels between points A and B if point B is twice as far from the source as is A?

$$\mathrm{SPL}_A - \mathrm{SPL}_B = 20 \log\frac{r_B}{r_A} = 20 \log 2 = 6.02.$$

Caution. The relation expressed by Eq. (18.12) holds only for free-field conditions. In certain instances the relation may be used to confirm the existence of a free field. (See Section 18.8.)

18.4 Psychoacoustic Relationships

As mentioned earlier, in only a relatively few situations is human hearing disassociated from sound measurements. Measuring systems and techniques are therefore unavoidably greatly influenced by the physiological and psychological makeup of the human ear as a transducer and the brain as the final evaluator. In terms of input magnitudes and frequencies, the human hearing system is quite nonlinear. Figure 18.5 shows average thresholds of hearing and tolerances for young persons. It will be noted that the greatest sensitivity occurs at about 4000 Hz and that a considerably greater SPL is required for equal reception at both lower and higher frequencies. Figure 18.6 shows the free-field *equal-loudness* contours for pure *tones* as determined by Robinson and Dadson at the National Physical Laboratories, Teddington, England.

Loudness is a measure of relative sound magnitudes or strengths *as judged by the listener.* It is a subjective quantity depending on both the physical waveform emanating from the source and on the *average* of many human hearing systems (persons) as receptors. The quantity is measured in *loudness level* and the unit is called the *phon* (pronounced "fon," to rhyme with "up*on*"). The loudness level in phons is numerically equal to the sound pressure level in dB at the frequency of 1000 Hz. Note that for each contour in Fig. 18.6, the loudness level and the sound pressure level are equal at only one point, namely, at 1000 Hz. It is quite important to keep in mind that the

Figure 18.5 Thresholds of hearing and tolerance for young people with good hearing (Courtesy: GenRad Inc., Concord, MA)

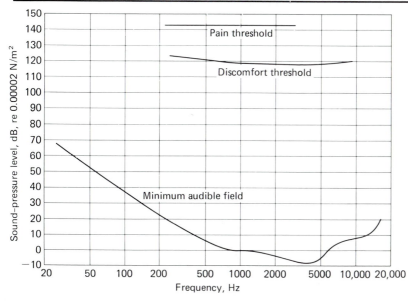

Figure 18.6   Free-field equal-loudness contours for pure tones
(Courtesy: GenRad Inc., Concord, MA)

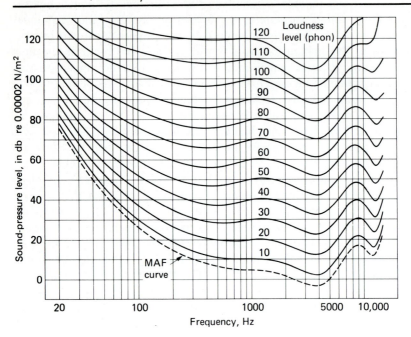

Figure 18.7   Relationship between loudness in sones and SPL

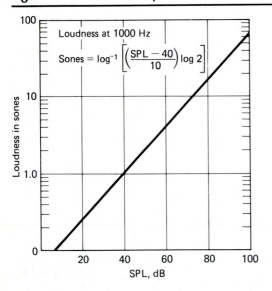

equal-loudness contours of Fig. 18.6 are based on *pure tones*. Loudness levels of complex sounds are considered in Section 18.8.

To further understand the curves in Fig. 18.6, first note that the SPL of 30 dB at 1000 Hz corresponds to 30 phons. On average, for a person to sense a loudness of 30 phons at 100 Hz would require an SPL of 44 dB; at 9000 Hz, 40 dB; and so on.

It is clear that loudness level in terms of the phon is a logarithmic quantity. Although this is quite useful, still another measure of strength is employed. *Loudness* (note the absence of the word *level*) is measured with a linear unit, the *sone* (rhymes with zone). One sone is the loudness of a 1000-Hz tone with an SPL of 40 dB (note that this also corresponds to 40 phons). A tone that sounds n times as loud has a loudness of n sones, etc.

We see that the sone is tied to the SPL at the one common pure tone of 1000 Hz and 40 dB. Figure 18.7 shows values at other sound pressure levels.

18.5 Sound-Measuring Apparatus and Techniques

Measurement of the parameters associated with sound use a basic system made up of a detector-transducer (the microphone), intermediate modifying devices (amplifiers and filtering systems), and readout means (a meter, CRO, or recording apparatus). Most sound-measuring systems are used to obtain psychoacoustically related information. It is therefore necessary to build into the apparatus nonlinearities approximating those of the average human ear. Elaborate filtering networks also provide the basis for analyzers, devices for separating and identifying the various frequency components or ranges of components forming a complex sound.

18.5.1 Microphones

Most microphones incorporate a thin diaphragm as the primary transducer, which is moved by the air acting against it. The mechanical movement of the diaphragm is converted to an electrical output by means of some form of secondary transducer that provides an analogous electrical signal.

Common microphones may be classified on the basis of the secondary transducer, as follows:

1. Capacitor or condenser
2. Crystal
3. Electrodynamic (moving coil or ribbon)
4. Carbon

The *capacitor* or *condenser microphone* is probably the most respected microphone for sound measurement purposes. It is arranged with a diaphragm forming one plate of an air-dielectric capacitor (Fig. 18.8). Movement of the

Figure 18.8   Schematic representation of the condenser-type microphone

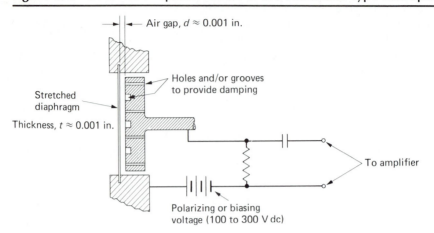

diaphragm caused by impingement of sound pressure results in an output voltage [7]:

$$E(t) = E_{\text{bias}} \frac{d'(t)}{d_0} \qquad (18.13)$$

where

$E(t)$ = the output voltage,

E_{bias} = the polarizing voltage,

d_0 = the original separation of the plates,

$d'(t)$ = the change in plate separation caused by sound pressure fluctuations.

The capacitive microphone is widely used as the primary transducer for sound measurement purposes.

The *electret microphone* is a special form of the condenser type. Whereas the common condenser type requires an external polarizing voltage, the electret type is self-polarizing. The diaphragm is constructed of a plastic sheet that has a conductive coating on one side. The coating serves as one side of the capacitor.

The *crystal microphone* uses a piezoelectric-type element (Section 6.14), generally activated by bending. For greatest sensitivity, a cantilevered element is mechanically linked to the diaphragm. Other constructions use direct contact between diaphragm and element, either by cementing (element placed in bending) or by direct bearing (element in compression). Crystal microphones are extensively used for serious sound measurement.

The *electrodynamic microphone* uses the principle of the moving conductor in a magnetic field. The field is commonly provided by a permanent magnet, thereby placing the transducer in the variable-reluctance category (Section 6.12). As the diaphragm is moved, the voltage induced is proportional to the *velocity* of the coil relative to the magnetic field, thereby providing an

Figure 18.9   Schematic diagram of the electrodynamic microphone

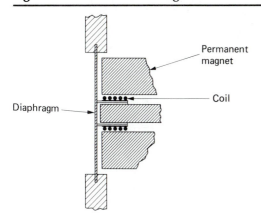

analogous electrical output. Two different constructions are used, the *moving coil* (Fig. 18.9) and the *ribbon* type. The inductive member of the latter type consists of a single element in the form of a ribbon that serves the dual purpose of "coil" and diaphragm.

The secondary transducer of the *carbon microphone* consists of a capsule of carbon granules, the resistance of which varies with change in sound pressure sensed by a diaphragm. Its limited high- and low-frequency response precludes its use for serious sound measurements and it is mentioned here merely to round out the list. The limited-frequency characteristics of the carbon microphone, coupled with its ruggedness, make it ideal for use as the transmitter in the ordinary telephone handset.

Microphone Selection Factors

An ideal microphone used for measurement would have the following characteristics:

1. Flat frequency response over the audible range
2. Nondirectivity
3. Predictable, repeatable sensitivity over the complete dynamic range
4. At the lowest sound level to be measured, output signal that is several times the system's internal noise level
5. Minimum dimensions and weight
6. Output that is unaffected by all environmental conditions except sound pressure

The capacitor-type microphone undoubtedly enjoys the top position for sound measurement use, although the crystal type runs a close second. As with all measurement, the presence of the sensor (microphone) unavoidably alters (loads) the signal to be measured. In this application the microphone should be

Table 18.1 Summary of microphone characteristics

Type of Microphone	Principle of Operation	Relative Impedance	Linearity	Advantages	Disadvantages
Capacitor	Capacitive	Very high	Excellent	Stable: holds calibration. Low sensitivity to vibration. Wide range.	Sensitive to temperature and pressure variations. Relatively fragile. Requires high polarizing voltage. Requires impedance-coupling device near microphone.
Crystal	Piezoelectric	High	Good to excellent	Self-generating. May be hermetically sealed. Relatively rugged. Relatively inexpensive.	Requires impedance-matching device. Relatively sensitive to vibration.
Electrodynamic	Reluctive	Low	Good	Self-generating.	Physically large.
Carbon	Resistive	Moderate	Poor	Very rugged. Inexpensive.	Severely limited frequency range.

as physically small as possible. It is obvious, however, that size, particularly diaphragm diameter, must have an important influence on both sensitivity and response. Microphones are therefore available in a range of sizes, and final selection must be based on a balance of the requirements for the specific application. Table 18.1 summarizes microphone characteristics.

18.5.2 The Sound Level Meter

The basic sound level meter is a measuring system that senses the input sound pressure and provides a meter readout yielding a measure of the sound magnitude. The sound may be wideband, it may have random frequency distribution, or it may contain discrete tones. Each of these factors will, of course, affect the readout.

Generally the system includes weighting networks (filters) that roughly match the instrument's response to that of human hearing. The readout, therefore, includes a psychoacoustical factor and provides a number ranking the sound magnitude in terms of the ability of the human measuring system.

A block diagram of the basic sound level meter is shown in Fig. 18.10, and Fig. 18.11 shows a commercially available system. When used in conjunction with a readout, the system is commonly referred to as a *sound level recorder*.

Figure 18.12 displays the internationally standardized weighting characteristics selectable by panel switch. We can see that the filter responses selectively discriminate against low and high frequencies, much as the human ear does. It is customary to use characteristics A for sound levels below 55 dB, B for sounds between 55 and 85 dB, and C for levels above 85 dB, as shown in Fig. 18.12. Certain broad generalities as to the frequency makeup of a sound may be made by taking separate readings with each network.

18.5.3 Frequency Spectrum Analysis

Although determination of a value of sound pressure or sound level provides a measure of sound intensity, it yields no indication of frequency distribution. For noise abatement purposes, for example, it is very desirable to know the

Figure 18.10 Block diagram of a typical sound level meter, or sound level recorder

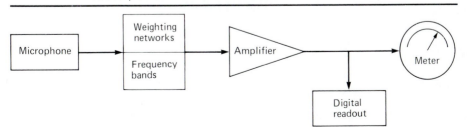

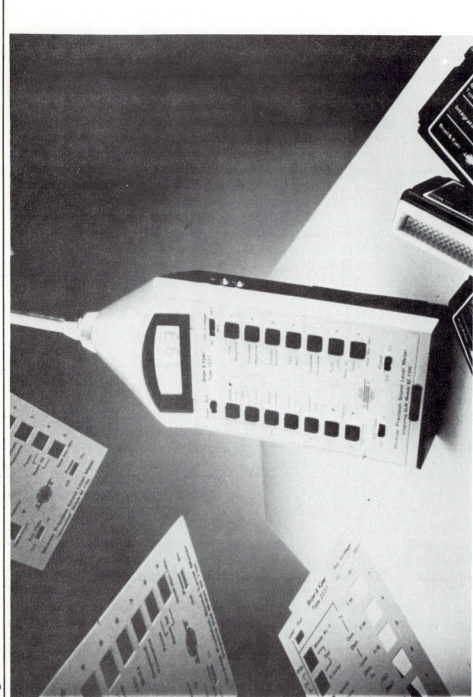

Figure 18.11 Sound level meter, Type 2231 (Courtesy of Bruel & Kjaer Instruments, Inc., Marlborough, MA)

Figure 18.12 Standard weighting characteristics for sound level meters

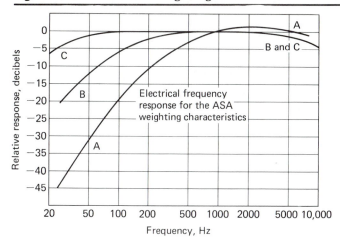

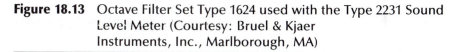

Figure 18.13 Octave Filter Set Type 1624 used with the Type 2231 Sound Level Meter (Courtesy: Bruel & Kjaer Instruments, Inc., Marlborough, MA)

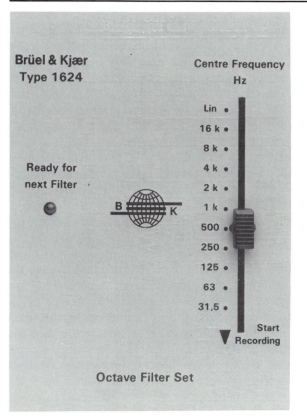

Figure 18.14 Typical overlapping filter characteristics
of band-type sound analyzers

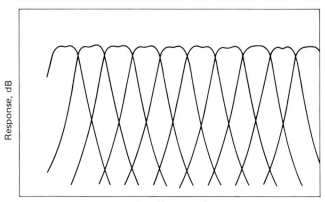

Response, dB

Log (frequency)

predominant frequencies involved. This can often point directly to the prominent noise sources.

Determination of intensities versus frequency is referred to as *spectrum analysis* and is accomplished through the use of band-pass filters (Section 7.20). Various combinations of filters may be used, determined by their relative band-pass widths. Probably the most commonly used are the "full-octave" filters having center frequencies as follows: 31.5, 63, 125, 250, 500, 1000, 2000, 4000, 8000, and 16,000 Hz. Figure 18.13 shows such an instrument, and Fig. 18.14 depicts typical frequency characteristics of the filters. "Lin" refers to flat response over the complete frequency range.

In addition to full-octave spectrum division, $\frac{1}{2}$, $\frac{1}{3}$, $\frac{1}{10}$, and other fractional octave divisions are also used. Although the ear may be capable of distinguishing pure tones in the presence of other tones, it does tend to integrate complex sounds over roughly $\frac{1}{3}$-octave intervals [8], thus lending some additional importance to $\frac{1}{3}$-octave analyzers. Table 18.2 lists the octave and $\frac{1}{3}$-octave band center frequencies and lower and upper band limits.

Simple analyzers are used by taking a separate reading for each pass band. It is seen therefore that an appreciable period of time is required to scan the range. Not only does the bulk of data increase with reduced bandwidth, but the constancy of the sound source also becomes of greater importance as the necessary time for measurement increases. A partial solution to the latter problem is to use a tape recorder for sampling and then analyze the recorded sound at leisure. It is obvious that to be useful for faithfully recording a sound source, the tape recorder must be of highest quality. Commonly available recorders for speech and music are not usable; their frequency response requirements are purposely made nonlinear over the required range of frequencies.

Table 18.2 Octave band and $\frac{1}{3}$-octave band limits

Octave Band

Center Frequency	Lower Limit	Upper Limit
31.5 Hz	22.4 Hz	45 Hz
63	45	90
125	90	180
250	180	355
500	355	710
1 kHz	710	1.4 kHz
2	1.4 kHz	2.8
4	2.8	5.6
8	5.6	11.2
16	11.2	22.4

$\frac{1}{3}$-Octave Band

Center Frequency	Lower Limit	Upper Limit
20.0 Hz	18.0 Hz	22.4 Hz
25.0	22.4	28.0
31.5	28.0	35.5
40	35.5	45
50	45	56
63	56	71
80	71	90
100	90	112
125	112	140
160	140	180
200	180	224
250	224	280
315	280	355
400	355	450
500	450	560
630	560	710
800	710	900
1,000	900	1,120
1,250	1,120	1,400
1,600	1,400	1,800
2,000	1,800	2,240
2,500	2,240	2,800
3,150	2,800	3,550
4,000	3,550	4,500
5,000	4,500	5,600
6,300	5,600	7,100
8,000	7,100	9,000
10,000	9,000	11,200
12,500	11,200	14,000
16,000	14,000	18,000
20,000	18,000	22,400

18.5.4   The Discrete Fourier Transform

Application of the DFT is of particular usefulness in acoustical work. Recall (Section 4.6) that this calculation produces an amplitude-vs.-frequency display. It provides a very convenient means for determining the contributions of the various harmonic components making up a complex noise signal.

18.6   Applied Spectrum Analysis

Harmonic or Fourier analysis of complex waveforms is discussed in Section 4.6 and is further mentioned in Appendix B. From those discussions it should be clear that even for relatively simple waveshapes, the number of required numerical manipulations may easily make the procedure prohibitive from a time-benefit standpoint. This limitation becomes especially important when the variety of nonrepetitive conditions needing analysis taxes the capacity of even the larger computers. For this reason an accelerated procedure referred to as the *fast Fourier transform,* or *FFT,* has been developed. If N represents the number of harmonic coefficients to be determined, ordinary harmonic analysis requires roughly N^2 separate computations, whereas the FFT requires approximately $N \log_2 N$—a marked reduction. It has been found that many waveforms encountered in acoustics, mechanical vibrations, and most electrical quantities permit practical application of the FFT.

FFT procedures may be hand-calculated using equations similar to those of discrete Fourier analysis discussed in Section 4.6.1. Alternatively, data may be taken from a signal display—from a CRO screen, for instance—and analyzed with a computer by using an appropriate program and inputting the data by hand. Better still, the data may be directly recorded through a computer's A/D board. To obtain greatest usefulness from the method, however, an integrated instrumentation package, the FFT analyzing system, can be employed. This system permits real-time processing with nearly immediate on-the-scene results.

Figure 18.15 is a schematic diagram of an FFT analyzing system. As shown, the signal is sensed by a microphone, vibration pickup, and so on, and is then conditioned by amplification or filtering, as may be required. A sample of the conditioned input is then taken over a time interval, T. Then the time base is divided into an integral number of equal increments. The number is usually a power of 2.* For *each time increment,* Δt, the A/D converter then outputs an amplitude in digital form for analysis by the logic section. The function of the logic section is to transform (convert) the sample from the time domain to the frequency domain for final display or numerical listing.

The output from the FFT calculation usually consists of the amplitude of a sine and cosine pair for each integral multiple of the fundamental frequency.†

---

* Note that $2^8 = 256_{10}$ is commonly used. Counting zero as the initial value, this corresponds to 0 to 255_{10} or 0 to FF_{16}, or 1 byte of information.

† The fundamental frequency here is directly related to the time duration of the waveform being analyzed [See Eq. (4.20)].

Figure 18.15 Block diagram of an FFT analyzing system

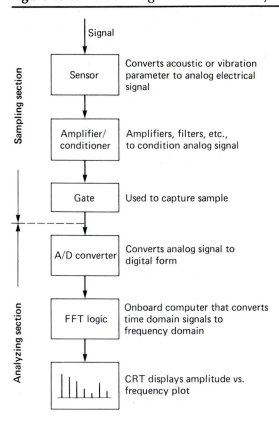

The number of pairs obtained is just equal to the number of original data points. Each sine and cosine pair can be considered as a vector with real and imaginary parts, which can be combined into a vector with absolute magnitude and a phase angle (see Section 4.4). For many acoustical analyses the phase angle is ignored and the vector magnitude at each integral multiple of the fundamental frequency is used as the result of the analysis.

The squared value of each vector magnitude is sometimes called the *autospectral value* and the set of the squared values, the *power spectrum,* even though these values may not represent power at all.

The sampling time interval, T, is often referred to as the *data window duration.* The experimenter sets this value by adjustment of the gating time, hence the sample's length. When obviously repetitive waveforms are encountered, the logical value is clear: the period of the cycle. But when the input is complex and nonrepetitive, the selection of a proper sampling time interval is in doubt. In such a case many primary sources of signal plus secondary sources may be present. In mechanical applications these sources may consist of gear and bearing noises, elastic resonances, combustion sounds, hydraulic noises, and the like. In the case of acoustical measurements the environment contributes also. In many situations there are identifiable frequency sources such as rpm's, rate of gear tooth meshings, etc. Because of its analyzing speed, the FFT system becomes extremely useful in cases of this sort. A variety of sampling time intervals may be tried and, by comparison of readouts, a "best" result selected. In certain instances, it may be decided that no significant, constant harmonic relationships are present. Of course, this result also represents pertinent information.

Some final considerations in the proper use of the FFT analysis include (1) the "shape" of the data window chosen in addition to its time duration and (2) the A/D converter's sampling frequency rate in relation to the highest frequency component desired from the analysis. With regard to (1), the shape of the data window indicates how the data points will be weighted. For example, a rectangular window indicates that all data points will be treated as equally important, whereas a "raised cosine" or "hanning" window gives greater emphasis to data points in the interior of the interval and lesser importance to those at the ends of the interval. (Reference [9] contains a more detailed discussion of data windows.) For consideration of (2), the simple rule of thumb is that the data sampling rate, $1/\Delta t$, must be more than twice the highest frequency component desired from the analysis (See Section 4.6.2).

18.7 Measurement and Interpretation of Industrial and Environmental Noise

18.7.1 Equivalent Sound Level, L_{eq}

For studying long-term trends in industrial or environmental noise, it is convenient to use a single number descriptor to define the noise history. The descriptor most often used is L_{eq}, that is, the continuous dB level that would have produced the same sound energy in the same time (T) as the actual noise history.

Equivalent sound level is obtained by averaging the mean-squared sound pressure over the desired time interval and converting back to decibels. Thus

$$L_{eq} = 10 \log_{10} \left(\frac{\overline{p_{rms}^2}}{p_0^2} \right) \tag{18.14}$$

where $\overline{p_{rms}^2}$ = time average of mean-square sound pressure. Since

$$SPL = 10 \log_{10} \frac{p_{rms}^2}{p_0^2}$$

and

$$\frac{p_{rms}^2}{p_0^2} = 10^{SPL/10} \tag{18.15}$$

we can write Eq. (18.14) as

$$L_{eq} = 10 \log_{10} \left(\frac{1}{T} \int_0^T 10^{SPL/10} \, dt \right) \qquad (18.16)$$

where

$$L_{eq} = \text{equivalent sound level, in dB,}$$
$$SPL = \text{rms sound pressure level,}$$
$$T = \text{averaging time, which is specified.}$$

Note that equivalent sound level is an energy average and will, in general, differ from the arithmetic average of the sound levels and the median level. High SPL readings tend to dominate the equivalent sound level. Readings of 20 dB or more below the peak level make a small contribution to the equivalent sound level. Most integrating sound level meters, such as the one shown in Fig. 18.11, calculate the L_{eq} electronically for any user-specified averaging time. Typical averaging times can range from minutes to days, depending upon the specific situation [5]. Figure 18.16 illustrates the difference between a time varying SPL reading and its L_{eq} value. Although the L_{eq} value may be determined for any weighted scale, it is most commonly used with the A-weighted scale.

Figure 18.16 Comparison of SPL and L_{eq} values

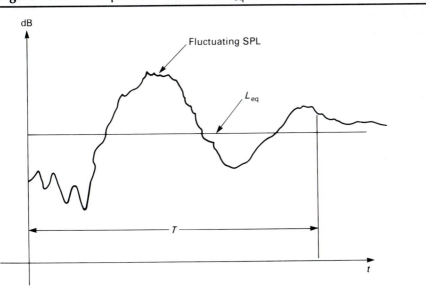

18.7.2   Sound Exposure Level (SEL)

When noise is of a transient nature, such as an automobile passing by an observer, an airplane passing overhead, or impact noise caused by a forging operation, the usual acoustic measurement of SPL is not only difficult to obtain but also is more difficult to interpret. Some sound level meters have a peak-hold capability, whereby the maximum SPL value measured may be captured and stored. The question arises as to how important this value is, especially when comparing one measurement with a second in which some noise attenuation fixes have been performed.

For this situation the L_{eq} measurement as described in Section 18.7.1 is performed, except that the averaging time T is taken as 1.0 s. The sound exposure level is thus defined as

$$\text{SEL} = 10 \log_{10}\left(\int_0^T 10^{\text{SPL}/10}\, dt\right) \tag{18.17}$$

where T is 1 s. If we compare Eq. (18.16) with Eq. (18.17), we observe that

$$\text{SEL} = L_{eq} + 10 \log_{10} T \tag{18.18}$$

and the SEL and L_{eq} value differ significantly only for large values of averaging time T. Thus the SEL reading, since it is a measure of acoustic energy, can be used to compare unrelated noise events because it is normalized to 1.0 s.

18.7.3   Sound Intensity Measurement

When a particle of air is displaced from its mean position, there is a temporary increase in pressure. The pressure increase acts in two ways: to restore the particle to its original position and to pass on the disturbance to the next particle. The cycles of pressure increases and decreases propagate as a sound wave. There are two important parameters in this process. The pressure and the velocity of the air particles oscillate about a fixed position. Sound intensity is the product of pressure and particle velocity [10].

$$\text{Sound intensity} = \text{pressure} \times \text{particle velocity}$$

$$= \frac{\text{force}}{\text{area}} \cdot \frac{\text{distance}}{\text{time}} = \frac{\text{energy}}{\text{area} \times \text{time}} = \frac{\text{power}}{\text{area}}$$

In an active field, the pressure and velocity vary simultaneously and the pressure and particle velocity are in phase. Only for this case is the time-averaged value of the intensity not equal to zero. Sound intensity may be defined as

$$I_r = \frac{1}{T}\int_0^T p u_r\, dt \tag{18.19}$$

where

$$p = \text{instantaneous pressure at a point,}$$
$$u_r = \text{air-particle velocity in the } r \text{ direction.}$$

The air-particle velocity at a point can be expressed in terms of the pressure gradient at a point [5]:

$$u_r = -\frac{1}{\rho} \int_{-\infty}^{t} \frac{\partial p}{\partial r} \, dt \approx -\frac{1}{\rho} \int_{0}^{t} \frac{(p_B - p_A)}{\Delta r} \, dt,$$

where the partial derivative has been replaced by a finite-difference approximation. If the pressure, p, is replaced by the average pressure, then the sound intensity determined by Eq. (18.19) becomes

$$I_r = -\frac{1}{T} \int_{0}^{T} \frac{(p_A + p_B)}{2\rho} \left(\int_{0}^{t} \frac{(p_B - p_A)}{\Delta r} \, dt \right) dt, \qquad (18.20)$$

where

$$\rho = \text{mass density of air,}$$
$$\Delta r = \text{spacing between points } A \text{ and } B,$$
$$p_A \text{ and } p_B = \text{instantaneous pressures at points } A \text{ and } B, \text{ respectively,}$$
$$T = \text{averaging time.}$$

Sound intensity is determined by a two-microphone method, where the two microphones are placed face to face and separated by a distance Δr by a spacer, as shown in Fig. 18.17a. Figure 18.17(b) shows the directivity

Figure 18.17   Sound intensity measuring systems: (a) microphone spacing and pressure gradient approximation, (b) sound intensity component measured

(a)

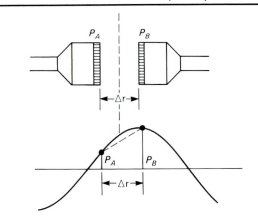

continued

Figure 18.17 *continued*

(b)

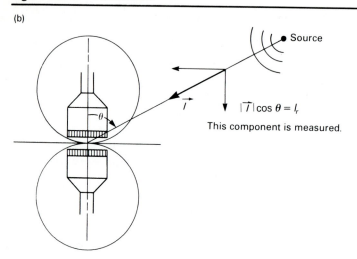

characteristic for the sound intensity measuring system. Note that when the angle θ is 90°, the sound intensity component is zero, since there is no difference in the pressure signals being measured. This feature makes this measurement useful in locating the sources of noise in complex sound fields.

A sound intensity analyzing system consists of a two-microphone probe system and an analyzer. The microphone probe system measures the two pressures p_A and p_B and the analyzer does the integration to find the sound intensity. A typical commercial system is the Bruel and Kjaer Type 3360 Sound Intensity Analysis System.

18.8   Notes on Some Practical Aspects of Sound Measurements

With all measurements, the act of measuring disrupts the process being evaluated. The sound pressure existing at the microphone diaphragm is *not* the sound pressure that would exist at that location were the microphone not present. *True* free-field conditions cannot be measured. The diaphragm stiffness characteristics, the housing properties, and so on do not correspond to the properties of the air that they displace. For wavelengths smaller than the diaphragm dimensions (high frequencies), the diaphragm has the characteristics of an infinite wall and for sounds arriving perpendicular to the diaphragm face, the pressures approach twice the value were the diaphragm not present. For wavelengths several times the diaphragm diameter, this effect is negligible.

Figure 18.18 Variations in microphone response depending on angle of incidence of sound wave

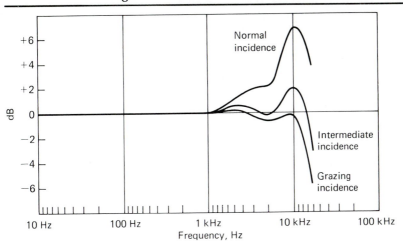

In addition, for other angles on incidence* the effect is reduced. Figure 18.18 illustrates typical response curves for various angles of incidence.

We see, therefore, that as a general practice, it is advisable to avoid "pointing" the microphone at the sound source. One should use an angle approaching 90° (grazing) incidence. One must remember, however, that in a reverberant field there are multiple sound sources from reflective walls, etc., so one must exercise judgement and carefully consider the individual situation. One reference suggests an angle of incidence of approximately 70°.

An exception to the preceding occurs when one is attempting to isolate discrete sound inputs, perhaps in conjunction with a spectrum analyzer or band analyzer. In such a case it may be desirable to point the microphone to suspected areas and also to place the microphone much closer to the source than would otherwise be done.

At what distance from the source should the microphone be placed when a general measurement is being made? Figure 18.19 shows a condition typical of many sound fields, produced by such sources as machinery: sources made up of many individual noisemakers such as gears, bearings, housings, and so on. The cross-sectional areas in the figure depict distances over which *transverse* movement of the microphone produces variations in SPL readings for a presumably constant input. In the near-distance interval the discrete sources may be identifiable. In the far distances the field is reverberant, and reflective sources cause SPL variations.

---

* Zero degrees incidence occurs when the microphone axis is parallel to the direction of sound propagation; i.e., the wavefront is parallel to the microphone diaphragm.

Figure 18.19 Typical SPL variations dependent
upon microphone placement

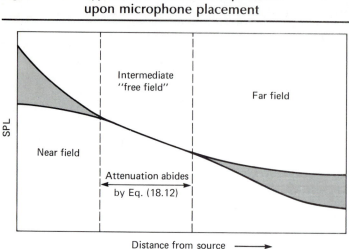

In the free-field interval, transverse movement of the microphone should
make little or no difference in readings. In addition, movement toward or away
from the source should provide readings approximating those predicted by Eq.
(18.12). This latter fact may be helpful in establishing that free-field conditions
do or do not prevail. It is important to note that in many cases the free-field
interval is not present: Near- and far-field conditions overlap.

18.9   Calibration Methods

As with most electromechanical measuring systems, sound measurement
involves a detector-transducer (the microphone) followed by intermediate
electrical/electronic signal conditioning and some sort of readout stage.
Calibration of the complete system involves introducing a known sound
pressure variation and comparing the known with the readout. In practice such
a calibration may be made; more commonly, the various component stages
may be calibrated separately. For example, sound level meters often provide a
simple check of the electrical and readout stages by substituting a carefully
controlled and known voltage for the microphone output and adjusting the
readout to a predetermined value for the particular instrument. This procedure
ignores the microphone altogether; however, it does provide a convenient
method for checking the remainder of the system.

Comparative methods may also be used whereby a microphone or a
complete system is directly compared with a "standard." Of course, as with all
comparative methods, it is imperative that great care be exercised to ensure
that a true comparison is made. It is necessary that both the standard and the

test microphones "hear" the identical sound. Our introductory discussion at the beginning of this chapter should indicate the difficulties in achieving this result.

Standard sound sources involve mechanical loudspeaker-type drivers for producing the necessary pressure fluctuations. One system* uses two battery-driven pistons moving in opposition in a cylinder at 250 Hz. A precalibrated pressure level of 124 dB $\pm$ 0.2 dB is produced. A somewhat similar system† uses a small, rugged loudspeaker as the driver. In both cases proper calibration is maintained only if the design coupling cavity is employed between driver and microphone. This restricts this type of calibrator to compatible microphones: those from the same company.

A more basic calibration method is known as *reciprocity* calibration. In addition to the microphone under test, a *reversible,* linear transducer and a sound source with a proper coupling cavity are required. Calibration procedure is divided into two steps. First, both the test microphone and the reversible transducer are subjected to a common sound source and the two outputs; hence their ratio is determined. Second, the reversible transducer is used as a sound source with known input (current), and the output (voltage) of the test microphone is measured. It can be shown [11, 12] that this step provides a relationship that is a function of the product of the two sensitivities: that of the test microphone and that of the reversible transducer. Results of the two steps yield sufficient information to determine the absolute sensitivity of the tested microphone.

18.10 Final Remarks

Sound is a complex physical quantity. Its evaluation through measurement becomes doubly difficult in comparison with evaluation of other engineering quantities because of the necessary involvement of human hearing. It must, therefore, be recognized that the foregoing sections serve merely as an introduction to the subject. The student is directed to the appended list of Suggested Readings for further study.

Little distinction has been made throughout this chapter between sound and noise. The latter has been considered merely as "undesirable" sound. Noise, however, is becoming an increasingly important problem to the engineer due to the controls that are set up by ordinance and statute as well as company-union agreements and court decision intended to protect the employee. As a result, engineers are called on to design "quiet" machinery and processes. They will be increasingly criticized for their part in "silence pollution." (We do not use the common term "noise pollution" because it is not "noise" that is polluted.) It is quite necessary, therefore, that they be

---

* Pistonphone Type 4290, Bruel & Kjaer, Inc., Marlborough, MA.

† Type 4230 Sound-Level Calibrator, Bruel & Kjaer, Inc., Marlborough, MA.

knowledgeable not only in the field of noise measurement, but also in the theory and art of noise abatement.

Suggested Readings

Bell, L. *Industrial Noise Control: Fundamentals and Applications.* New York: Marcel Dekker, 1982.

Beranek, L. L. (ed.). *Noise Reduction.* New York: McGraw-Hill, 1960.

Beranek, L. L. *Acoustics.* New York: McGraw-Hill, 1954.

Bloss, R. L. and M. J. Orloski (eds.). *Precision Measurement and Calibration* (selected NBS papers on mechanics), NBS Spcl. Publ. 300, vol. 8. Washington, D.C.: U.S. Government Printing Office, 1972.

Booth, S. F. (ed.). *Precision Measurement and Calibration,* vol. 2 (selected NBS technical papers on heat and mechanics), NBS Handbook 77, Washington, D.C.: U.S. Government Printing Office, 1961.

Fath, J. M. *Standards of Noise, Rating Schemes, and Definitions: A Compilation.* NBS Spcl. Publ. 386. Washington, D.C.: U.S. Government Printing Office, 1973.

Hassall, J. R. and K. Zaveri, *Acoustic Noise Measurements.* 4th ed. Marlborough, Mass.: Bruel & Kjaer, 1979.

Lord, H. W., W. S. Gatley, and H. A. Evensen. *Noise Control for Engineers.* New York: McGraw-Hill, 1980.

Peterson, A. P. G. and E. E. Gross, Jr. *Handbook of Noise Measurement.* 8th ed. Concord, Mass.: GenRad, 1980.

Pierce, A. D. *Acoustics: An Introduction to Its Physical Principles and Applications.* New York: McGraw-Hill, 1981.

Rayleigh, J. W. S. *The Theory of Sound,* vols. 1 and 2. New York: Dover, 1945.

Problems

18.1 Prepare a spreadsheet template for adding sound pressure levels from multiple sources. Provide for up to eight entries. Check the template against the results for the examples in Section 18.3.4.

18.2 Pumps are being selected for a fluids-handling system that is being designed. The upper capacity of the system is 10,000 gal/min. Among many parameters considered is that of noise. Combinations from a range of six pumps are being considered. The pump capacities along with the individual sound pressure levels that they generate are listed as follows:

Capacity (gal/min)	SPL (dB)
2000	89
2500	91
4000	93
5000	95
6000	96
7500	98

Determine the combination of pumps that will yield the lowest combined sound pressure level in dB (*Note:* The advantages of spreadsheet templates become obvious when working this and the following problem.)

18.3 The results from Problem 18.2 may be further refined by recognizing that more involved piping will be required as the number of pumps increases, thereby resulting in an added sound source. Use the following estimates of pipe noise to the various combinations of pumps and recalculate the total sound pressure levels.

Number of Pumps	Add the Following dB
2	80
3	83
4	86
5	89

18.4 The noise level at the factory property line caused by 12 identical air compressors running simultaneously is 60 dB. If the maximum SPL permitted at this location is 57 dB, how many compressors may be run simultaneously?

18.5 Calculate the SPL and intensity at a distance of 10 m from a uniformly radiating source of 2.0 W.

18.6 A sound pressure level meter used to measure the SPL of an engine without a muffler gave a reading of 120 dB. After attaching the muffler, the same meter gave a reading of 90 dB. Determine (a) the rms sound pressure before the muffler was attached, and (b) the percentage reduction in rms sound pressure amplitude when the muffler was used.

18.7 The SPL of a single rocket engine on its test stand and at a distance of $\frac{1}{2}$ mi is 108 dB. What SPL, at the same distance, would result if a cluster of five such engines were tested together? If two of the five were shut down, what drop in SPL would be expected?

18.8 Data from two lawnmower manufacturers regarding the noise produced by each as measured on the factory floor resulted in the following:

Mower	SPL (A)	Distance Directly Above Mower
A	100 dB	30 ft
B	85 dB	50 ft

Estimate the SPL (A) when these mowers are used in grass cutting at a distance of 100 ft at an ambient temperature of 25°C.

18.9 The following data were obtained from an octave band frequency analysis of a steam boiler based on "flat," or "linear," frequency weighting.
  a.   Construct a decibel-versus-frequency histogram of the data.
  b.   Superimpose on this histogram the following data corrected for A-weighting.

Octave Band Center Frequency, Hz	SPL, dB
31.5	79
63.0	77
125.0	76
250.0	81
500.0	80
1,000.0	82
2,000.0	82
4,000.0	72
8,000.0	67
16,000.0	60

18.10 In order to determine whether free-field conditions are approximated, an engineer finds the SPL at 10 m from the source to be 89 dB. What must be the SPL at 7.5 m in order for free-field conditions to exist?

18.11 Figure 18.20 illustrates a simple experimental setup that may be used to demonstrate a variety of basic principles discussed throughout this book. A small transistor-type loudspeaker (available from most electronic parts dealers) is cemented to a length of plastic pipe [Fig. 18.20(a)]. A signal generator is used to drive the speaker with a sinusoidal waveform [Fig. 18.20(b)]. The tube-speaker combination possesses two fundamental resonance frequencies, that of the "organ pipe" (Section 14.12.1) and the acoustical/electrical resonance frequency of the

Figure 18.20 Setup for Problem 18.11

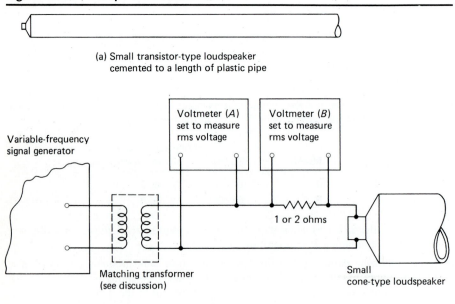

(a) Small transistor-type loudspeaker
cemented to a length of plastic pipe

Voltmeter (A) set to measure rms voltage

Voltmeter (B) set to measure rms voltage

Variable-frequency signal generator

1 or 2 ohms

Matching transformer
(see discussion)

Small
cone-type loudspeaker

(b) Electrical circuitry

speaker element. Voltmeter *B,* along with the 1 or 2 Ω resistor, is used to monitor the current to the speaker, and voltmeter *A* monitors the voltage. If voltage measured by *A* is held constant as a range of frequencies is swept by the signal generator, a sharp drop in current will be found at the frequency corresponding to resonance of the "organ pipe." This is the frequency at which oscillation may be maintained with minimum *power* expenditure. If the measuring system is sensitive enough, other resonances may also be determined, that of the speaker itself and the higher-mode frequencies of the pipe. The minimum-power principle is often quite useful in vibration testing employing shaker-driven systems (Chapter 17). Increased power transfer may be had by insertion of a matching transformer (Section 7.24). The turns ratio is not too critical; however, many signal generators have a 600 Ω output impedance and the input impedance of the speaker may be about 16 Ω. Care must be exercised to prevent overdriving the speaker and destroying its voice coil.

References

1. Skilling, D. C. Acoustical testing at Northrup Aircraft, *SESA Proc.* 16(2): 121, 1959.

2. Randall, R. H. *An Introduction to Acoustics.* Reading, MA: Addison-Wesley, 1951.

3. Beranek, L. L. *Acoustics.* New York: McGraw-Hill, 1954, p. 12.

4. Peterson, A. P. G. *Handbook of Noise Measurement,* 9th ed. Concord, MA: GenRad, Inc., 1980.

5. Wilson, C. E. *Noise Control.* New York: Harper and Row, 1989.

6. Beranek, L. L. (ed.). *Noise Reduction.* New York: McGraw-Hill, 1960, p. 186.

7. Keast, D. N. *Measurements in Mechanical Dynamics.* New York: McGraw-Hill, 1967.

8. Ranz, J. R. Noise measurement methods, *Mach. Des.:* November 10, 1966.

9. Peterson, A. P. G. *Handbook of Noise Measurement.* 9th ed. Concord, MA: GenRad, Inc., 1980, p. 133.

10. *Sound Intensity,* Technical Note. Brüel and Kjaer, 1986. 2850 Naerum Denmark.

11. Various papers, Handbook 77, Vol. II, *Precision Measurement and Calibration, Heat and Mechanics,* National Bureau of Standards, 1961.

12. *American National Standard Method for Calibration of Microphones,* S1.10–1966. New York: ANSI, 1966.

APPENDIX A

Standards and Conversion Equations

Standards

Gravity acceleration	9.80665 m/s^2 (round to 9.81) 32.174 ft/s^2 (round to 32.17)
Standard atmospheric pressure	1.01325 E + 05 Pa (round to 1.013 E + 05) 14.696 psia (round to 14.7)
Dimensional constants	$g_c = 1\,\text{kg} \cdot \text{m/N} \cdot \text{s}^2$ $g_c = 32.174\,\text{lbm} \cdot \text{ft/lbf} \cdot \text{s}^2$

Conversion Equations

Note: The relationships that follow are in the form of equations, solved for the pertinent SI unit. Consider, for example, the equation for length written in terms of meters and inches:

$$m = 2.540\,E - 02 \times in.$$

If we wish to find the number of meters corresponding to 36.00 in. (1 English yard), we would make the following substitutions,

$$m = \frac{2.540}{100} \times 36.00 = 0.9144;$$

that is, 36.00 in. or 1 English yard, is equal to 0.9144 m.

On the other hand, if we wish to find the number of inches that are the equivalent of 5.00 m, we would solve the equation for inches, as follows:

$$in. = m \times \frac{100}{2.54} = 5.00 \times \frac{100}{2.54} = \text{approx. } 197.$$

That is, 5.00 m and 197 in. represent approximately the identical length.

Note also that throughout the listing the computer printout convention for decimal place is used. For example, E + 03 = 10^3 = 1000. Likewise, E − 02 = 10^{-2} = 0.01.

Table A.1 Conversion factors

	SI				English or Other
Acceleration	m/s^2	=	3.048 000 E − 01	×	ft/s^2
	m/s^2	=	2.540 000 E − 02	×	in./s^2
Area	m^2	=	6.451 600 E − 04	×	in.^{2}
	m^2	=	9.290 304 E − 02	×	ft^2
Density (mass/vol)	kg/m^3	=	1.601 846 E + 01	×	lbm/ft^3
	kg/m^3	=	2.767 990 E + 04	×	lbm/in.^{3}
Energy (work)	J	=	1.055 040 E + 03	×	Btu
	J	=	1.355 818 E + 00	×	ft · lbf
	J	=	4.186 800 E + 00	×	calorie
	J	=	3.600 000 E + 03	×	W · h
Flow (vol/time)	m^3/s	=	2.831 685 E − 02	×	ft^3/s
	m^3/s	=	6.309 020 E − 05	×	U.S. liq gal/min
Force	N	=	1.000 000 E − 05	×	dyne
	N	=	4.448 222 E + 00	×	lbf
Length	m	=	1.000 000 E − 10	×	angstrom
	m	=	1.000 000 E − 06	×	micrometer (μm)
	m	=	2.540 000 E − 08	×	microinch (μin.)
	m	=	2.540 000 E − 02	×	in.
	m	=	3.048 000 E − 01	×	ft
	km	=	1.609 344 E + 00	×	statute mile
Mass	kg	=	4.535 924 E − 01	×	lbm
	kg	=	1.459 390 E + 01	×	slug
Moment (torque)	N · m	=	1.129 848 E − 01	×	lbf · in.
	N · m	=	1.355 818 E + 00	×	lbf · ft
Power	W	=	2.930 711 E − 01	×	Btu/h
	W	=	2.259 697 E − 02	×	ft · lbf/min
	W	=	7.456 999 E + 02	×	hp (550 ft · lbf/s)

Table A.1 *continued*

	SI				**English or Other**
Pressure (stress)	Pa	=	3.376 850 E + 03	×	in Hg (60°F)
	Pa	=	2.488 400 E + 02	×	in H$_2$O (60°F)
	Pa	=	4.788 026 E + 01	×	lbf/ft^2 (psf)
	Pa	=	9.806 380 E + 01	×	cm H$_2$O (4°C)
	Pa	=	1.333 220 E + 03	×	cm Hg (0°C)
	Pa	=	6.894 757 E + 03	×	lbf/in.^{2} (psi)
	Pa	=	1.000 000 E + 05	×	bar
	Pa	=	1.013 250 E + 05	×	atmosphere
Stress (see pressure)					
Temperature	K	=	°C + 273.15		
	K	=	(°F + 459.67)/1.8		
	K	=	°R/1.8		
	°C	=	(°F − 32)/1.8		
Torque (see moment)					
Velocity	m/s	=	3.048 000 E − 01	×	ft/s
	m/s	=	4.470 400 E − 01	×	mph
Viscosity	Pa · s	=	4.788 026 E + 01	×	lbf · s/ft^2
	Pa · s	=	1.000 000 E − 03	×	centipoise
Volume	m^3	=	3.785 412 E − 03	×	U.S. liq gal
	m^3	=	1.638 706 E − 05	×	in.^{3}
	m^3	=	2.831 685 E − 02	×	ft^3
	m^3	=	1.000 000 E − 03	×	liter
Vol/time (see flow)					

Some Additional Conversion Factors

	ft · lbf	=	7.782 E + 02	×	Btu
	gallons	=	4.329 E − 03	×	in.^{3}
	mph	=	6.214 E − 01	×	km/h
	knots	=	5.396 E − 01	×	km/h
	miles	=	6.214 E − 04	×	m
	°R	=	459.67	+	°F

APPENDIX B

Theoretical Basis for Fourier Analysis

The theoretical basis for the harmonic-analysis procedure may be described as follows: Any single-valued function $y(x)$ that is continuous (except for a finite number of finite discontinuities) in the interval $-\pi$ to π, and which has only a finite number of maxima and minima in that interval, can be represented by a series in the form

$$y(x) = \frac{A_0}{2} + A_1 \cos x + A_2 \cos 2x + \cdots + A_n \cos nx + \cdots$$

$$+ B_1 \sin x + B_2 \sin 2x + \cdots + B_n \sin nx + \cdots. \qquad \textbf{(B.1)}$$

If each term in Eq. (B.1) is multiplied by dx and integrated over any interval of 2π length, all sine and cosine terms will drop out, leaving

$$\int_a^{2\pi+a} y(x)\, dx = \int_a^{2\pi+a} \frac{A_0}{2}\, dx = A_0 \pi$$

or

$$A_0 = \frac{1}{\pi} \int_a^{2\pi+a} y(x)\, dx. \qquad \textbf{(B.2)}$$

The factor A_n may be determined if we multiply both sides of Eq. (B.1) by $\cos mx\, dx$ and integrate each term over the interval of 2π. In general there are the following terms:

$$\int_a^{2\pi+a} \sin nx \cos mx\, dx = 0$$

and

$$\int_a^{2\pi+a} \cos nx \cos mx\, dx = 0, \qquad \text{except for } m = n.$$

For the special case $m = n$,

$$\int_a^{2\pi+a} \cos^2 nx \, dx = \frac{1}{2n} [nx + \sin nx \cos nx]_{-\pi}^{\pi};$$

$$= \pi;$$

hence

$$\int_a^{2\pi+a} y(x) \cos nx \, dx = A_n \pi$$

or

$$A_n = \frac{1}{\pi} \int_a^{2\pi+a} y(x) \cos nx \, dx. \tag{B.3}$$

[*Note:* For $n = 0$, Eq. (B.3) reduces to Eq. (B.2).]

   In like manner, if we multiply both sides of Eq. (B.1) by $\sin mx \, dx$ and integrate term by term over the interval 2π, we may obtain

$$B_n = \frac{1}{\pi} \int_a^{2\pi+a} y(x) \sin nx \, dx. \tag{B.4}$$

Calculation of Fourier Coefficients for Special Periodic Waveforms

Square Wave

Considering the square wave shown in Fig. 4.9(a) we have for one full period

$$y(x) = A \qquad (0 \le x \le \pi)$$
$$y(x) = -A \qquad (\pi \le x \le 2\pi), \tag{B.5}$$

where $x = \omega t$. Applying Eq. (B.2) to Eq. (B.5), where we choose $a = 0$ for convenience, we have

$$A_0 = \frac{1}{\pi} \int_0^\pi A \, dx - \frac{1}{\pi} \int_\pi^{2\pi} A \, dx,$$

$$A_0 = 0. \tag{B.6}$$

From Eq. (B.3) we have

$$A_n = \frac{1}{\pi} \int_0^\pi A \cos nx \, dx - \frac{1}{\pi} \int_\pi^{2\pi} A \cos nx \, dx,$$

or

$$A_n = 0. \tag{B.7}$$

Similarly, from Eq. (B.4),

$$B_n = \frac{1}{\pi} \int_0^\pi A \sin nx \, dx - \frac{1}{\pi} \int_\pi^{2\pi} A \sin nx \, dx,$$

or

$$B_n = \frac{2A}{n\pi} [1 - \cos n\pi]$$

and, finally,

$$B_n = \begin{cases} \dfrac{4A}{n\pi} & \text{for } n \text{ odd,} \\ 0 & \text{for } n \text{ even.} \end{cases} \tag{B.8}$$

Substituting the results from Eqs. (B.6), (B.7), and (B.8) into Eq. (B.1) we obtain

$$y(x) = y(\omega t) = \frac{4A}{\pi} \sum_{n=1}^\infty \frac{\sin(2n-1)\omega t}{(2n-1)} \tag{B.9}$$

Sawtooth Wave

For the sawtooth wave shown in Fig. 4.9(d) we have

$$y(x) = \begin{cases} \dfrac{A}{\pi} x & (0 \le x \le \pi), \\ 2A - \dfrac{A}{\pi} x & (\pi \le x \le 2\pi). \end{cases} \tag{B.10}$$

Applying Eqs. (B.2) and (B.3)

$$A_0 = \frac{1}{\pi} \int_0^\pi \frac{Ax}{\pi} \, dx + \frac{1}{\pi} \int_\pi^{2\pi} \left(2A - \frac{A}{\pi} x \right) dx = A \tag{B.11}$$

and

$$A_n = \frac{1}{\pi} \int_0^\pi \frac{Ax}{\pi} \cos x \, dx + \frac{1}{\pi} \int_\pi^{2\pi} \left(2A - \frac{A}{\pi} x \right) \cos nx \, dx,$$

or

$$A_n = \frac{2A}{n^2\pi^2} [\cos n\pi - 1].$$

Considering various integer values for n we obtain

$$A_n = \begin{cases} 0 & \text{for } n \text{ even,} \\ \dfrac{-4A}{n^2\pi^2} & \text{for } n \text{ odd.} \end{cases} \tag{B.12}$$

Finally, from Eq. (B.4),

$$B_n = \frac{1}{\pi} \int_0^\pi \frac{Ax}{\pi} \sin nx \, dx + \frac{1}{\pi} \int_\pi^{2\pi} \left(2A - \frac{Ax}{\pi}\right) \sin nx \, dx,$$

and therefore

$$B_n = 0. \tag{B.13}$$

Using the results of Eqs. (B.11), (B.12), and (B.13) in Eq. (B.1) the Fourier series for the sawtooth wave becomes

$$y(x) = y(\omega t) = \frac{A}{2} - \frac{4A}{\pi^2} \sum_{n=1}^{\infty} \frac{\cos(2n - 1)\omega t}{(2n - 1)^2}. \tag{B.14}$$

Fourier series for some other waveforms are given in Table 4.1.

Discrete Fourier Analysis

When an analog waveform is sampled at N points separated by time increments of Δt, the apparent waveform has a period $T = N \Delta t$ and a fundamental frequency $\Delta f \equiv 1/T$, and it is defined at the sampling times $t_r = r \Delta t$, for $r = 1, \ldots, N$. In a manner analogous to the continuous Fourier series, a discretely sampled waveform can be represented as a finite sum of frequency components having frequencies $0, \Delta f, 2\Delta f, \ldots, (N/2)\Delta f$:

$$y(t_r) = \frac{A_0}{2} + \sum_{n=1}^{N/2-1} [A_n \cos(2\pi n \, \Delta f t_r) + B_n \sin(2\pi n \, \Delta f t_r)]$$

$$+ \frac{A_{N/2}}{2} \cos\left(2\pi \frac{N}{2} \Delta f t_r\right) \tag{B.15}$$

for N an even number. Only a finite number of frequencies are present in this sum, because frequencies greater than $(N/2) \Delta f$ (the *Nyquist frequency*; Section 4.6.2) vary too rapidly to be resolved with the points taken.

The coefficients of the discrete Fourier series are found in essentially the same way as those of the continuous Fourier series, except that we must sum rather than integrate. Hence, we multiply Eq. (B.15) by the mth frequency component and sum over time:

$$\sum_{r=1}^{N} y(t_r) \cos(2\pi m \, \Delta f t_r) = \frac{A_0}{2} \sum_{r=1}^{N} \cos(2\pi m \, \Delta f t_r)$$

$$+ \sum_{n=1}^{N/2-1} \sum_{r=1}^{N} \cos(2\pi m \, \Delta f t_r)$$

$$\times [A_n \cos(2\pi n \, \Delta f t_r) + B_n \sin(2\pi n \, \Delta f t_r)]$$

$$+ \frac{A_{N/2}}{2} \sum_{r=1}^{N} \cos(2\pi m \, \Delta f t_r) \cos\left(2\pi \frac{N}{2} \Delta f t_r\right) \tag{B.16}$$

It turns out that

$$
\sum_{r=1}^{N} \cos\left(2\pi m \, \Delta f t_r\right) \cos\left(2\pi n \, \Delta f t_r\right) =
\begin{cases}
0 & \text{for } m \neq n \\
\dfrac{N}{2} & \text{for } m = n \\
N & \text{for } m = n = 0 \text{ or } m = n = \dfrac{N}{2}
\end{cases}
$$

Likewise

$$
\sum_{r=1}^{N} \cos\left(2\pi m \, \Delta f t_r\right) \sin\left(2\pi n \, \Delta f t_r\right) \equiv 0.
$$

Thus, Eq. (B.16) reduces to the statement that

$$
\sum_{r=1}^{N} y(t_r) \cos\left(2\pi m \, \Delta f t_r\right) = \frac{N}{2} A_m \qquad m = 0, 1, \ldots, \frac{N}{2}
$$

A similar calculation fixes B_n. From this, the harmonic coefficients of the discrete Fourier series are

$$
A_n = \frac{2}{N} \sum_{r=1}^{N} y(r \, \Delta t) \cos\left(\frac{2\pi r n}{N}\right) \qquad n = 0, 1, \ldots, \frac{N}{2} \tag{B.17}
$$

$$
B_n = \frac{2}{N} \sum_{r=1}^{N} y(r \, \Delta t) \sin\left(\frac{2\pi r n}{N}\right) \qquad n = 1, 2, \ldots, \frac{N}{2} - 1 \tag{B.18}
$$

APPENDIX C

Number Systems

In general:

$$N_b = \sum a_m b^n = \cdots + a_n b^n + a_{(n-1)} b^{(n-1)} + a_{(n-2)} b^{(n-2)} + \cdots$$

where

b = base or radix of the particular system

    = the number of distinct character types required to express a quantity or magnitude,

N_b = a number in the system,

a = coefficient,

n = positional value or power.

The Decimal System

The decimal system is based on ten digits, 0 through 9; the base or radix is 10. Hence

$$N_{10} = \sum a_m 10^n.$$

For example:

$$524.3_{10} = (5)(10)^2 + (2)(10)^1 + (4)(10)^0 + (3)(10)^{-1}.$$

The digits 5, 2, 4, and 3 are the coefficients corresponding to the respective *positions* or powers 2, 1, 0, and -1. In the decimal system the coefficients may have integral values ranging from 0 through 9, inclusive.

The Binary System

The binary system employs two digits only, namely, 1 and 0, and

$$N_2 = \sum a_m 2^n.$$

For example, the representation 1101.1_2 is interpreted as

$$(1)(2)^3 + (1)(2)^2 + (0)(2)^1 + (1)(2)^0 + (1)(2)^{-1}$$
$$= 8 + 4 + 0 + 1 + \tfrac{1}{2} = 13.5_{10}.$$

This example demonstrates the procedure for *converting* a binary number to the equivalent decimal number. The two numbers 1101.1_2 and 13.5_{10} each represent an identical quantity.

The Octal System

This system is based on eight digits, 0 through 7, and

$$N_8 = \sum a_m 8^n.$$

The coefficients a_m do not include the decimal digits 8 and 9.
     For example,

$$375.3_8 = (3)(8)^2 + (7)(8)^1 + (5)(8)^0 + (3)(8)^{-1}$$
$$= 192_{10} + 56_{10} + 5_{10} + 0.375_{10} = 253.375_{10}.$$

The Hexadecimal System

This system employs 16 as the base or radix. Additional digital symbols are required. Those used are 0, 1, 2, 3, 4, 5, 6, 7, 8, 9, A, B, C, D, E, and F. These 16 characters correspond to the base-10 numbers 0 through 15, respectively.

$$N_{16} = \sum a_m (16)^n.$$

For example,

$$D8B.2_{16} = (D)(16)^2 + (8)(16)^1 + (B)(16)^0 + (2)(16)^{-1}$$
$$= (13)(16)_{10}^2 + (8)(16)_{10}^1 + (11)(16)_{10}^0 + (2)(16)_{10}^{-1}$$
$$= 3467.125_{10}.$$

Octal and Hexadecimal Formatted Binary

A binary number may be arranged (grouped) in ways that make it easy to convert it to either octal or hexadecimal equivalents.

Consider 110111001100_2, which we will rewrite as

$$110 \quad 111 \quad 001 \quad 100_2$$

If we consider each subgroup as a single octal digit, we obtain

$$6 \quad 7 \quad 1 \quad 4 \quad \text{or} \quad 6714_8$$

In like manner we can rearrange the same binary number into subgroups of four, as follows:

$$1101 \quad 1100 \quad 1100_2$$

We see that this is equal to DCC_{16}.

We can confirm the legitimacy of all of this by converting each number, the binary, the octal, and the hexadecimal to the equivalent decimal number, which we find is equal to 3532_{10}.

Why Need We Be Concerned with the Various Number Systems?

1. The decimal system is common in everyday usage but it is not a convenient system around which to build or use a computer. A computer would have to distinguish between ten different states, as opposed to only two when binary is used.

2. Although binary requires a much longer string of symbols to define a given magnitude, the advantage of the simple Yes/No operation more than offsets the use of longer numbers. Machine language employs binary arithmetic.

3. *Why octal or hexadecimal?* The fundamental computer language is binary, but for a human that system would be extremely awkward and error prone. Hexadecimal and/or octal can be thought of as simply a crutch used by humans to communicate with the computer in the computer's own tongue—machine or assembly language.

Converting a Base-10 Number to One of a Different Radix

In the preceding paragraphs we have established procedures for converting numbers of various bases to equivalent magnitudes of base-10. We shall now demonstrate a method* for performing the reverse: converting a base 10 number to equivalent binary, octal, and hexadecimal numbers. We will use Tables C1, C2, and C3.

---

* There are other methods than the one demonstrated.

Table C.1   Values of 2^n

n		2^n
	Etc.	
−1	↑	0.5
0		1
1		2
2		4
3		8
4		16
5		32
6		64
7		128
8		256
9		512
10		1024
11		2048
12	↓	4096
	Etc.	

Decimal to Binary

Let's convert the number 713_{10} to the equivalent binary number. We will do this by successively subtracting the largest values of 2^n that each remainder will permit. The procedure is demonstrated as follows (refer to Table C.1):

$$
\begin{array}{rl}
713_{10} & \\
\underline{-512} & = 2^9 \\
201_{10} & \\
\underline{-128} & = 2^7 \\
73_{10} & \\
\underline{-64} & = 2^6 \\
9_{10} & \\
\underline{-8} & = 2^3 \\
1_{10} & \\
\underline{-1} & = 2^0 \\
0 &
\end{array}
$$

Recalling that the powers of the radix correspond to the positional orders, we may write:

$$1011001001_2 = 713_{10}.$$

Decimal to Octal

Employ Table C.2 to convert the number 713_{10} to the equivalent octal number. A procedure similar to the one used for the previous example may be used. In this case, however, we select the "largest" components in terms of both powers of the radix, 8, and also the required coefficients.

$$
\begin{array}{ll}
713_{10} & \\
\underline{-512} & = 1 \times 8^3 \\
201_{10} & \\
\underline{-192} & = 3 \times 8^2 \\
9_{10} & \\
\underline{-8} & = 1 \times 8^1 \\
1_{10} & \\
\underline{-1} & = 1 \times 8^0 \\
0 &
\end{array}
$$

From this we determine $713_{10} = 1311_8$.

Decimal to Hexadecimal

In a similar manner, convert 713_{10} to the equivalent hexadecimal number using Table C.3.

$$
\begin{array}{ll}
713_{10} & \\
\underline{-512} & = 1 \times 8^3 \\
201_{10} & \\
\underline{-192} & = C \times 16^1 \\
9_{10} & \\
\underline{-9} & = 9 \times 16^0 \\
0 &
\end{array}
$$

From this calculation we may write $713_{10} = 2C9_{16}$.

Table C.2 Values of $(a)(8)^n$

| Coefficients | Powers, n | | | | | | |
a	6	5	4	3	2	1	0
1	262 144	32 768	4 096	512	64	8	1
2	524 288	65 536	8 192	1 024	128	16	2
3	786 432	98 304	12 288	1 536	192	24	3
4	1 048 576	131 072	16 384	2 048	256	32	4
5	1 310 720	163 840	20 480	2 560	320	40	5
6	1 572 864	196 608	24 576	3 072	384	48	6
7	1 835 008	229 376	28 672	3 584	448	56	7

Table C.3　Values of $(a)(16)^n$

Coefficients a	Powers, n						
	6	5	4	3	2	1	0
1	16 777 216	1 048 576	65 536	4 096	256	16	1
2	33 554 432	2 097 152	131 072	8 192	512	32	2
3	50 331 648	3 145 728	196 608	12 288	768	48	3
4	67 108 864	4 194 304	262 144	16 384	1 024	64	4
5	83 886 080	5 242 880	327 680	20 480	1 280	80	5
6	100 663 296	6 291 456	393 216	24 576	1 536	96	6
7	117 440 512	7 340 032	458 752	28 672	1 792	112	7
8	134 217 728	8 388 608	524 288	32 768	2 048	128	8
9	150 994 944	9 437 184	589 824	36 864	2 304	144	9
A	167 772 160	10 485 760	655 360	40 960	2 560	160	10
B	184 549 376	11 534 336	720 896	45 056	2 816	176	11
C	201 326 592	12 582 912	786 432	49 152	3 072	192	12
D	218 103 808	13 631 488	851 968	53 248	3 328	208	13
E	234 881 024	14 680 064	917 504	57 344	3 584	224	14
F	251 658 240	15 728 640	983 040	61 440	3 840	240	15

APPENDIX D

Some Useful Data

Table D.1 Properties of water: SI system

Temperature Degrees C (F)	Absolute Viscosity Pa · s	Density kg/m^3
5 (41)	15.188×10^{-4}	1000
10 (50)	13.077	999.7
15 (59)	11.404	999.1
20 (68)	10.050	998.2
25 (77)	8.937	997.0
30 (86)	8.007	995.7
35 (95)	7.225	994.1
40 (104)	6.560	992.2
45 (113)	5.988	990.3
50 (122)	5.494	988.1

Table D.2 Properties of water: English system

Temperature Degrees F (C)	Absolute Viscosity lbf · s/ft^2	Density lbm/ft^3
40 (4.44)	3.23×10^{-5}	62.42
50 (10.0)	2.72	62.41
60 (15.56)	2.33	62.37
70 (21.11)	2.02	62.30
80 (26.67)	1.77	62.22
90 (32.22)	1.58	62.11
100 (37.78)	1.43	61.99
110 (43.33)	1.30	61.86
120 (48.89)	1.15	61.71

Table D.3   Properties of dry air at atmospheric pressure: SI system

Temperature Degrees C (F)	Absolute Viscosity* Pa · s	Density† kg/m³
0 (32)	1.68×10^{-5}	1.26
10 (50)	1.73	1.22
20 (68)	1.80	1.18
30 (86)	1.85	1.14
40 (104)	1.91	1.10
50 (122)	1.97	1.07
60 (140)	2.03	1.04
70 (158)	2.09	1.00
80 (176)	2.15	0.97
90 (194)	2.22	0.94
100 (212)	2.28	0.924

* Over the range from atmospheric pressure to about 7000 kPa ($\approx$1000 psia) the viscosity of dry air increases at a rate of approximately 1% for each 700-kPa (100-psi) increase in pressure.

† For pressure other than atmospheric, use $\rho/\rho_{atmos} = P/P_{atmos}$.

Table D.4   Properties of dry air at atmospheric pressure: English System

Temperature Degrees F (C)	Absolute Viscosity* lbf · s/ft²	Density† lbm/ft³
40 (4.44)	0.362×10^{-6}	0.0794
50 (10.0)	0.368	0.0779
60 (15.6)	0.374	0.0764
70 (21.1)	0.379	0.0749
80 (26.7)	0.385	0.0735
90 (32.2)	0.390	0.0722
100 (37.8)	0.396	0.0709
110 (43.3)	0.401	0.0697
120 (48.9)	0.407	0.0685

* Over the range from atmospheric pressure to about 7000 kPa ($\approx$1000 psia) the viscosity of dry air increases at a rate of approximately 1% for each 700-kPa (100-psi) increase in pressure.

† For pressures other than atmospheric, use $\rho/\rho_{atmos} = P/P_{atmos}$.

Table D.5 Some values of gravitational acceleration

	m/s^2	ft/s^2
Standard	9.806 65	32.174
Location		
Ft. Egbert, Alaska	9.821 83	32.224
Key West, Florida	9.789 70	32.118
Batavia, Java	9.781 78	32.092
Karajak Glacier, Greenland	9.825 34	32.235
Pittsburgh, Pennsylvania	9.801 05	32.156
Latitude 40°, 26′, 40″		
Longitude 79°, 57′, 13″ W		
Elevation 908.35 ft		
Moon	1.67	5.48
Planet Mercury	3.92	12.86
Planet Jupiter	26.46	87.07

Table D.6 Specific gravities* of selected materials

Material	Specific Gravity
Mercury	13.596 @ 0°C
	13.546 @ 20°C
	13.690 @ −38.8°C (liquid at freezing point)
	14.193 @ −38.8°C (solid at freezing point)
Gasoline	0.66 to 0.69
Kerosene	0.82
Seawater	1.025
Oil (Meriam Red—a common manometer oil)	0.823
Carbon tetrachloride	1.60
Tetrabromo-ethane	2.96
Ethyl alcohol /water mixture at 20°C—% alcohol by weight	
0	1.00
20	0.97
40	0.94
60	0.89
80	0.84
100	0.79

* Specific gravity is the ratio of the mass of a body to that of an equal volume of water at 4°C, or at some other specified temperature.

Table D.7   Some additional material properties

Temperature °F	Medium Heating Oil		Heavy Heating Oil	
	Specific Gravity	Viscosity: lbf · s/ft²	Specific Gravity	Viscosity: lbf · s/ft²
40	0.865	10.82×10^{-5}	0.918	789×10^{-5}
60	0.858	7.85	0.912	390
80	0.851	6.03	0.905	200
100	0.843	4.59	0.899	109

Table D.8   Kerosene viscosity

Temperature, °C	Viscosity, Pa · s
0	28.7×10^{-4}
20	19.2
40	13.4
60	9.6
80	7.7
100	6.7

Table D.9   Young's modulus

For steel	$E \approx 29.5 \times 10^6$ lbf/in.² $\approx 20.3 \times 10^{10}$ Pa
For aluminum	$E \approx 10 \times 10^6$ lbf/in.² $\approx 6.9 \times 10^{10}$ Pa

APPENDIX E

║║║

Stress and Strain Relationships

E.1 The General Plane Stress Situation

Suppose an element dx wide by dy high is selected from a general plane stress situation in equilibrium, as shown in Fig. E.1. Assume that the element is of uniform thickness, t, normal to the paper.

From strength of materials it will be remembered that, for equilibrium, τ_{xy} must be equal to τ_{yx}. We will therefore employ the symbol τ_{xy} for both. We will also assume that not only must the complete element be in equilibrium, but so also must be all its parts. Therefore, if the element is bisected by a diagonal ds long, each half of the element must also be in equilibrium.

As shown in Fig. E.2, there must be a normal stress σ_θ and a shear stress τ_θ acting on the diagonal area. If the various stresses shown on the partial element, Fig. E.2, are multiplied by the areas over which they act, forces are obtained as shown in Fig. E.3. Let all the directions be considered positive, as shown. We note that $dy/ds = \cos\theta$ and $dx/ds = \sin\theta$. Summing forces normal to the diagonal plane and solving for σ_θ we obtain

$$\sigma_\theta = \sigma_x \cos^2\theta + \sigma_y \sin^2\theta + 2\tau_{xy}\sin\theta\cos\theta.$$

A more convenient form of this equation may be had in terms of double angles. By substituting trigonometric equivalents,

$$\sigma_\theta = \tfrac{1}{2}(\sigma_x + \sigma_y) + \tfrac{1}{2}(\sigma_x - \sigma_y)\cos 2\theta + \tau_{xy}\sin 2\theta. \tag{E.1}$$

Using this equation, the stress on any plane may be determined if values of σ_x, σ_y, and τ_{xy} are known.

Figure E.1 Element subject to plane stresses

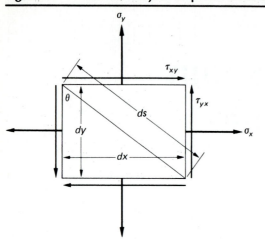

Figure E.2 Element used to define positive stress directions

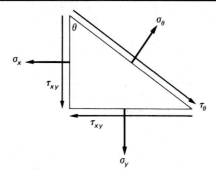

Figure E.3 Element illustrating the requirement for force equilibrium

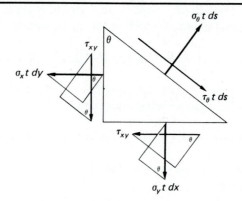

Figure E.4   Stresses acting on an element on the outer surface
of a shaft subject to torsion and bending

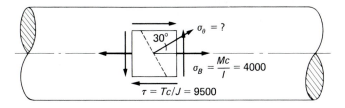

Example E.1

A shaft is subject to a torque, T, which results in a shear stress, $Tc/J = 9500\,\text{lbf/in.}^2$ (Fig. E.4), and at the same time and at the same point a bending moment due to gear loads causes an outer fiber stress, $Mc/I = 4000\,\text{lbf/in.}^2$. What will be the normal stress on the outer surface in a direction $30°$ to the shaft centerline?

Solution.

$$\sigma_{30°} = \tfrac{1}{2}(\sigma_x + \sigma_y) + \tfrac{1}{2}(\sigma_x - \sigma_y)\cos 2\theta + \tau_{xy}\sin 2\theta$$
$$= \tfrac{1}{2}(4000 + 0) + \tfrac{1}{2}(4000 - 0)\cos 60° + 9500\sin 60°$$
$$= 11{,}240\,\text{lbf/in.}^2.$$

Of course, the normal stress, $11{,}240\,\text{lbf/in.}^2$, is not necessarily the maximum normal stress, because the angle $30°$ was chosen at random; undoubtedly some other angle may result in a larger normal stress.

E.2   Direction and Magnitudes of Principal Stresses

To calculate the maximum normal stress, the particular angle θ_1 determining the plane over which it will act must be found. This angle may be found by differentiating Eq. (E.1) with respect to θ, setting the derivative equal to zero, and solving for the angle θ_1. This should also give us the plane over which the normal stress is a minimum.

$$\frac{d\sigma_\theta}{d\theta} = -(\sigma_x - \sigma_y)\sin 2\theta + 2\tau_{xy}\cos 2\theta = 0$$

or

$$\tan 2\theta_{1,2} = \frac{\pm 2\tau_{xy}}{\pm(\sigma_x - \sigma_y)}. \tag{E.2}$$

Two angles, $2\theta_{1,2}$, are determined by Eq. (E.2), and consideration of the trigonometry involved shows that the two angles are 180° apart. This result means, then, that the two angles $\theta_{1,2}$ are 90° apart, and leads to a very important fact: *The planes of maximum and minimum normal stress are always at right angles to each other.*

The maximum and minimum normal stresses are called the *principal stresses,* and the planes over which they act are called the *principal planes.* We have just found, therefore, that the principal planes are at right angles to each other. If we know the direction of the maximum normal stress, we automatically know the direction of the minimum normal stress.

Now we would like to find an expression for the principal stresses. From Eq. (E.2) we may write

$$\sin 2\theta_{1,2} = \frac{2\tau_{xy}}{\sqrt{(\sigma_x - \sigma_y)^2 + (2\tau_{xy})^2}},$$

$$\cos 2\theta_{1,2} = \frac{(\sigma_x - \sigma_y)}{\sqrt{(\sigma_x - \sigma_y)^2 + (2\tau_{xy})^2}}.$$

(E.3)

Substituting these values in Eq. (E.1) gives us the principal stresses, which we shall designate σ_1 and σ_2, and

$$\sigma_{\theta max} = \sigma_1 = \tfrac{1}{2}(\sigma_x + \sigma_y) + \tfrac{1}{2}\sqrt{(\sigma_x - \sigma_y)^2 + (2\tau_{xy})^2},$$

$$\sigma_{\theta min} = \sigma_2 = \tfrac{1}{2}(\sigma_x + \sigma_y) - \tfrac{1}{2}\sqrt{(\sigma_x - \sigma_y)^2 + (2\tau_{xy})^2}.$$

(E.4)

Example E.2

Referring to Example E.1, determine the magnitudes of the principal stresses and the positions of the principal planes relative to the shaft centerline.

Solution. Using Eq. (E.4),

$$\sigma_1 = \tfrac{1}{2}(4000 + 0) + \tfrac{1}{2}\sqrt{(4000 + 0)^2 + (2 \times 9500)^2}$$

$$= 11{,}708 \text{ lbf/in.}^2$$

and

$$\sigma_2 = \tfrac{1}{2}(4000 + 0) - \tfrac{1}{2}\sqrt{(4000 + 0)^2 + (2 \times 9500)^2}$$

$$= -7708 \text{ lbf/in.}^2.$$

From Eq. (E.2),

$$\tan 2\theta = \frac{2 \times 9500}{4000 - 0} = \frac{19{,}000}{4000} = 4.75,$$

$$2\theta = 78.1°,$$

$$\theta = 39.05°.$$

The orientation is as shown in Fig. E.5.

Figure E.5 The principal stresses corresponding to the
situation shown in Fig. E.4

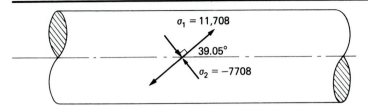

$\sigma_1 = 11{,}708$

$39.05°$

$\sigma_2 = -7708$

E.3   Variation in Shear Stress with Direction

Following the same procedure used for normal stresses, and again referring to
Fig. E.3, if the forces parallel to the diagonal plane are summed, the following
shear relations are obtained. The equation for the shear stress on any plane in
terms of σ_x, σ_y, and θ is

$$\tau_\theta = \tfrac{1}{2}(\sigma_x - \sigma_y)\sin 2\theta - \tau_{xy}\cos 2\theta. \qquad (E.5)$$

The angle determining the planes over which the shear stresses are maximum
and minimum may be determined by the relation

$$\tan 2\theta_s = \frac{\mp(\sigma_x - \sigma_y)}{\pm 2\tau_{xy}}. \qquad (E.6)$$

By substituting the angles determined by Eq. (E.6) in Eq. (E.5), relations for
maximum and minimum shear stress may be obtained.

$$\text{Maximum shear stress} = \tau_{\theta max} = \tfrac{1}{2}\sqrt{(\sigma_x - \sigma_y)^2 + (2\tau_{xy})^2},$$

$$\text{Minimum shear stress} = \tau_{\theta min} = -\tfrac{1}{2}\sqrt{(\sigma_x - \sigma_y)^2 + (2\tau_{xy})^2}. \qquad (E.7)$$

We must be careful, however, because the shear stress extremes given by
Eq. (E.7) account for only the x- and y-directions. Consideration of the
three-dimensional condition often shows the greatest shear stress to occur on
yet another plane. See Section E.6.

E.4   Shear Stress on Principal Planes

Equations (E.7) allow us to determine the shear stress on any plane defined by
θ. Therefore, let us substitute the expressions for $\theta_{1,2}$, Eq. (E.2), in Eq. (E.5)
and thereby determine the shear stresses acting over the principal planes.
Doing this gives

$$\tau_{1,2} = \frac{-\tfrac{1}{2}(\sigma_x - \sigma_y)(2\tau_{xy}) + \tau_{xy}(\sigma_x - \sigma_y)}{\sqrt{(\sigma_x - \sigma_y)^2 + (2\tau_{xy})^2}}$$

$$= 0.$$

This proves a very important fact about any plane stress situation: *The shear stresses on the principal planes are zero.* This in itself often provides the necessary clue to determine the orientation of the principal planes by inspection. In any case where it can be said, "there can be no shear on this plane," then the fact we have just established tells us that the plane we are referring to is a principal plane. Or, often just as important, if it can be said that shear stresses *do* exist on a plane, we know the plane *cannot* be one of the principal planes.

In strain-gage applications, knowing the directions of the principal planes at the point of interest provides a very decided advantage. With this information, gages may be aligned in the principal directions, and usually only two gages are required. More important, however, is the fact that the calculations become much simpler and less time consuming (Section 12.15.2).

E.5   General Stress Equations in Terms of Principal Stresses

Checking back on our original assumptions in Section E.1, we see that we assumed a simple element subject to two orthogonal normal stresses, σ_x and σ_y, and shear stresses, τ_{xy}. Using the information since developed, namely, that the principal stresses are at right angles to each other and that the shear stresses are zero on the principal planes, we may now rewrite certain of our equations in terms of principal stresses σ_1 and σ_2.

By selecting just the right element orientation, i.e., aligning it with the principal planes, our basic element could be made to appear as it does in Fig. E.6. σ_1 and σ_2 are orthogonal stresses, and we know also that there will be no shear on the planes over which they act. Therefore, any of our equations

Figure E.6   An element subject to principal stresses

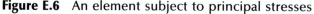

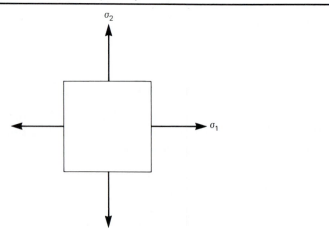

written so far may be modified by substituting σ_1 and σ_2 for σ_x and σ_y, respectively, and making the shear stress equal to zero.

Whereas substitution in Eqs. (E.1) and (E.5) yields particularly useful relations, substitution in most of the others simply confirms our definitions.

Substituting in Eqs. (E.1) and (E.5) gives

$$\sigma_\theta = \tfrac{1}{2}(\sigma_1 + \sigma_2) + \tfrac{1}{2}(\sigma_1 - \sigma_2)\cos 2\theta, \tag{E.8}$$

$$\tau_\theta = \tfrac{1}{2}(\sigma_1 - \sigma_2)\sin 2\theta. \tag{E.8a}$$

These equations are particularly useful in helping us visualize the overall stress condition as shown in the following section.

E.6   Mohr's Circle for Stress

Let us establish a coordinate system with σ_θ plotted as the abscissa and τ_θ as the ordinate (Fig. E.7). The shear stress corresponding to the principal stresses is zero; hence σ_1 and σ_2 will be plotted along the σ_θ-axis.

If a circle is drawn passing through σ_1 and σ_2 points and having its center on the σ_θ-axis, the construction shown in Fig. E.7 will result. It will be noted that for any point on the circle the distance along the abscissa represents σ_θ, and the ordinate distance represents τ_θ. This construction, which is very useful in helping to visualize stress situations, is known as Mohr's stress circle.

At this point we should consider the third or z-direction. In general, three orthogonal stresses σ_x, σ_y, and σ_z, along with corresponding shear stresses τ_{xy},

Figure E.7   Mohr's circle for plane stresses

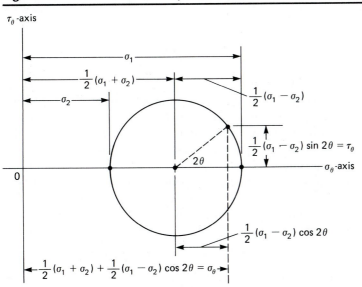

Figure E.8   (a) An element subjected to normal and shear stresses
on three orthogonal planes, (b) Mohr's circles for
stress for the element shown in (a)

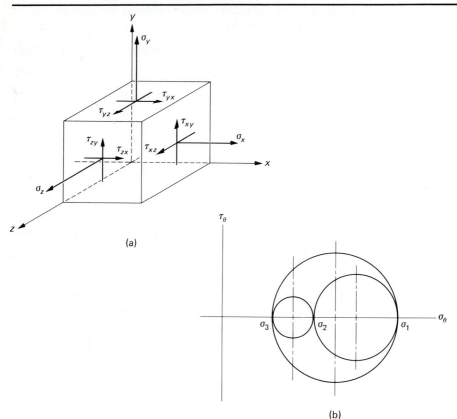

(a)

(b)

τ_{yz}, and σ_{zx}, will occur on an element, as shown in Fig. E.8(a). In this case a third principal plane exists, over which, as for the two-dimensional case, *shear is zero*. Also, it can be shown* that the three principal planes, along with the three principal stresses σ_1, σ_2, and σ_3, are at right angles to one another. By considering the three directions in combinations of two, we may reduce the problem to three related two-dimensional situations. The resulting combined Mohr's diagrams are illustrated in Fig. E.8(b).

In the majority of cases, strain gages are applied to free, unloaded surfaces and the condition is thought of as being two-dimensional. It is well, however, to consider every condition in terms of three dimensions, even though the third stress may be zero, and to plot or merely sketch the three-circle Mohr's

---

* For example, in most intermediate solid mechanics books.

diagram. This procedure often reveals a maximum shear that might otherwise be overlooked.

A few examples will demonstrate the power of the Mohr diagram.

Example E.3

Figure E.9(a) shows a simple tension member. We know there is no shear stress on a transverse section; hence we know that this must be a principal plane. Since the other principal plane must be normal to the first principal plane and hence be aligned with the axis of the specimen, the normal stress on this plane must be zero. Therefore,

$$\sigma_1 = \frac{F}{A} \quad \text{and} \quad \sigma_2 = \sigma_3 = 0.$$

Plotting Mohr's circle for this situation gives us Fig. E.9(b). One of the Mohr circles degenerates to a point in this case.

By inspection we see that the maximum shear stress is equal to $F/2A$

Figure E.9   Mohr's circle of stresses for a simple tension member

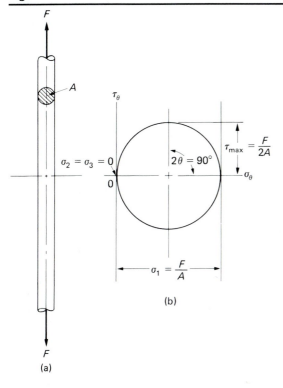

(b)

(a)

Figure E.10   Mohr's circle of stresses for the free surface of a cylindrical
pressure vessel; P = pressure, D = shell
diameter, and t = shell-wall thickness

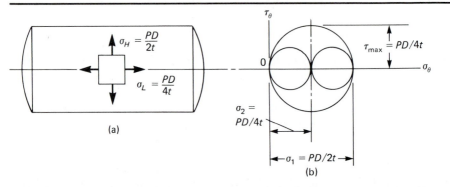

(a)

(b)

at an angle $2\theta = 90°$ or $\theta = 45°$ measured relative to the axis of the
specimen. This confirms our previous knowledge of the stress condition
for this simple situation.

Example E.4

Figure E.10(a) shows a thin-walled cylindrical pressure vessel. From
elementary theory, the hoop, longitudinal, and normal stresses may be
calculated by the following relations:

$$\sigma_H = \frac{PD}{2t}, \qquad \sigma_L = \frac{PD}{4t}, \quad \text{and} \quad \sigma_N = 0.$$

Consideration of the nature of the stress field makes it difficult to
imagine shear stresses on planes parallel to the hoop, longitudinal, or
normal directions. Assuming this to be correct, then the circumferential,
the longitudinal, and the normal directions must be the principal direc-
tions and σ_H, σ_L, and σ_N must be the principal stresses. Mohr's circle for
this situation is shown in Fig. E.10(b). The maximum shear is seen to be
$PD/4t$, over a plane inclined 45° to the circumferential and normal
directions.

Example E.5

Figure E.11(a) shows a shaft in simple torsion. From courses in strength
of materials we know that the shear stress on the outer fiber acting on a
plane normal to the shaft centerline is equal to Tc/J. The fact that shear
stress exists on this plane eliminates it from consideration as a principal

Figure E.11 Mohr's circle for a shaft subject to "pure" torsion; $T =$ torque, $J =$ polar moment of inertia of section, and $C =$ distance from neutral axis to fiber of interest

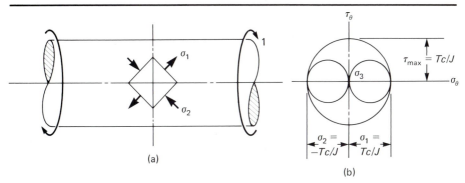

(a)

(b)

plane. Since principal planes must be normal to each other, we immediately think of the one other symmetrical possibility—the two planes inclined 45° to the shaft centerline. Careful consideration of the stresses that may exist on these two planes leads us to conclude that tension would exist on one and compression on the other. The wringing of a wet towel is often used as an example of this situation. In the one 45° direction, the threads of the towel are obviously in tension, while in the other 45° direction, compression is employed to squeeze out the water.

Because of symmetry, we are led to the conclusion that the magnitudes of the two stresses are equal. Plotting equal tensile and compressive principal stresses, using Mohr's circle construction, gives us Fig. E.11(b).

The third principal direction is normal to the shaft surface and we see that $\sigma_N = 0$. Although the preceding discussion can hardly be considered rigorous proof, Fig. E.11(b) does represent the actual stress situation for a shaft subject to pure torsion. We know that maximum shear stress is equal to Tc/J. Therefore, inspection shows us that the principal stresses σ_1 and σ_2 must also have the sample magnitude, Tc/J, tension and compression, respectively.

Example E.6

Figure E.12(a) shows a thin-wall spherical pressure vessel, for which elementary theory shows that the stress in the wall abides by the following relation:

$$\sigma = \frac{PD}{4t}.$$

Figure E.12   Mohr's circle for the free surface of a
                   spherical pressure vessel

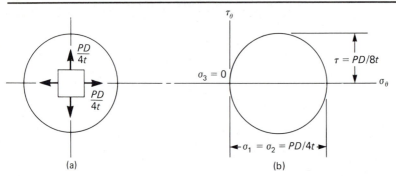

(a)   (b)

In this case it is difficult to see how direction has significance. At any
point on the outside of the shell, the normal stresses must be equal in all
directions, simply because of symmetry. We must therefore conclude that

$$\sigma_1 = \sigma_2 = \frac{PD}{4t} \quad \text{and} \quad \sigma_3 = \sigma_N = 0.$$

Mohr's diagram for this condition is shown in Fig. E.12(b), from which it
is seen that $\tau_{max} = PD/8t$.

From the preceding discussion and consideration of Mohr's circle construction, we may now make the following general observations:

1. A stress state involving shear without normal stress is impossible.
2. Maximum shear stress always occurs on planes oriented 45° to the principal
   stresses and is equal to one-half the algebraic difference of the principal
   stresses.
3. The shear stresses on any mutually perpendicular planes are of equal
   magnitude.
4. The sum of the normal stresses on any mutually perpendicular planes is a
   constant.
5. The maximum ratio of shear stress to principal stress occurs when the
   principal stresses are of equal magnitude but opposite sign.

E.7   Strain at a Point

Through use of Hooke's law and the stress relations developed in the
preceding pages, the following relations for strain at a point may be derived:

$$\varepsilon_\theta = \tfrac{1}{2}(\varepsilon_x + \varepsilon_y) + \tfrac{1}{2}(\varepsilon_x - \varepsilon_y)\cos 2\theta + \frac{\gamma_{xy}}{2}\sin 2\theta, \tag{E.9}$$

$$\frac{\gamma_\theta}{2} = \tfrac{1}{2}(\varepsilon_x - \varepsilon_y) \sin 2\theta - \frac{\gamma_{xy}}{2} \cos 2\theta. \tag{E.9a}$$

Comparison of the above two equations with Eqs. (E.1) and (E.5), respectively, indicates that with a minor exception (the shear strains γ are divided by 2, whereas their counterparts are not), the stress and the strain relations at a point are functionally alike. It follows, therefore, that we can draw a Mohr's diagram for strain, provided the ordinate is made $\gamma_\theta/2$. This is sometimes useful in treating strain-rosette data.

Example E.7

Power piping is subject to a combination of loading whose complexity will serve as an interesting example of a combined stress situation. In addition to pressure loading, differential expansion between the hot and cold conditions may superimpose bending, torsional, and axial loading.

Of course, the primary problem involved in piping design is the determination of the loading brought about by pipe expansion, end movements, and movement-limiting stops. In the simple situations good estimates of these loads may be determined analytically, through the use of computer programs.

In this example it will be assumed such preliminary work has been finished and the critically stressed location found. The remaining problem, then, is to combine the stress components and to determine the net stress condition. The problem is as follows:

Pipe Data (14-in, Schedule 100)

Outside diameter = 14 in.,

Wall thickness = 0.937 in.,

Inside diameter = 12.125 in.,

Bending moment of inertia = 825 in.^{4},

Bending section modulus = 117.9 in.^{3},

Torsional moment of inertia = 1650 in.^{4},

Torsional section modulus = 235.8 in.^{3},

Cross-sectional area = 38.47 in.^{2},

Young's modulus = 23×10^6 lbf/in.^{2},

Poisson's ratio = 0.29.

Loading Data

Internal pressure = 620 lbf/in.^{2},

Bending moment = 700,000 in. · lbf,

Torsional moment = 480,000 in. · lbf,

Axial load = 35,000 lbf tension.

Problem. For the outer surface of the pipe, calculate (a) the maximum shear stress, (b) the principal stresses, (c) the direction of the stress σ_1 relative to the axis of the pipe, (d) the axial and circumferential unit strains, and (e) the principal strains. Also, (f) sketch Mohr's diagrams for stress and for strain.

Solution. The stress components are found as follows:

$$\text{The ratio } \frac{\text{I.D.}}{\text{O.D.}} = \frac{12.125}{14} = 0.87,$$

which is the range usually termed *thin wall*. Hence,

$$\sigma_H = \text{hoop stress} = \frac{PD}{2t} = \frac{620 \times 12.125}{2 \times 0.937} = 4011 \text{ lbf/in.}^2,$$

$$\sigma_L = \text{longitudinal stress} = \frac{PD}{4t} = \frac{1}{2}\sigma_H = 2005 \text{ lbf/in.}^2,$$

$$\sigma_B = \text{bending stress} = \frac{Mc}{I} = \frac{700,000}{117.9} = \pm 5937 \text{ lbf/in.}^2,$$

$$\tau = \text{torsional stress} = \frac{Tc}{J} = \frac{480,000}{235.8} = 2035 \text{ lbf/in.}^2,$$

$$\sigma_A = \text{axial stress} = \frac{F}{A} = \frac{35,000}{38.47} = 910 \text{ lbf/in.}^2.$$

These conditions are illustrated in Fig. E.13.

a.   Using Eq. (E.7), we obtain

$$\tau = \tfrac{1}{2}\sqrt{(8852 - 4011)^2 + (2 \times 2035)^2} = 3162 \text{ lbf/in.}^2.$$

From Fig. E.14 we see, however, that the true maximum shear stress is

$$\tau_{max} = \left(\frac{9593}{2}\right) = 4796 \text{ lbf/in.}^2.$$

Figure E.13   Axial and hoop stresses acting in the pipe of Example E.7

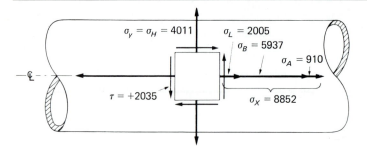

Figure E.14   Principal stresses and principal stress directions
                for the pipe in Example E.7

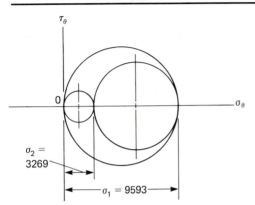

b.   Using Eq. (E.4) (see also Fig. E.15), we get

$$\sigma_1 = \tfrac{1}{2}(8852 + 4011) + 3162 = 9593 \text{ lbf/in.}^2,$$
$$\sigma_2 = \tfrac{1}{2}(8852 + 4011) - 3162 = 3269 \text{ lbf/in.}^2.$$

Also,

$$\sigma_3 = \sigma_N = 0.$$

c.   Using Eq. (E.2), we obtain

$$\tan 2\theta_{\sigma_1} = \frac{2 \times 2035}{(8852 - 4011)} = \frac{4072}{4841} = 0.8407,$$

$$2\theta_{\sigma_1} = 40°4', \qquad \theta_{\sigma_1} = 20°2'.$$

d.   Using Eq. (12.5), we obtain

$$\varepsilon_x = \frac{1}{23 \times 10^6}[8852 - 0.29(4011 + 0)] = 334 \ \mu\text{-strain},$$

Figure E.15   Principal stress element for Example E.7

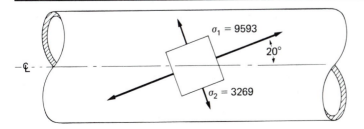

Figure E.16   Mohr's strain diagram (outer surface) for Example E.7

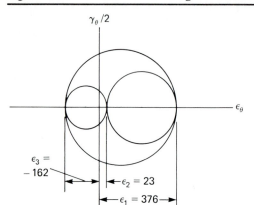

$$\varepsilon_y = \frac{1}{23 \times 10^6}[4011 - 0.29(0 + 8852)] = 63 \ \mu\text{-strain},$$

$$\varepsilon_z = \frac{1}{23 \times 10^6}[0 - 0.29(8852 + 4011)] = -162 \ \mu\text{-strain}.$$

e.   Again, from Eq. (12.5),

$$\varepsilon_1 = \frac{1}{23 \times 10^6}[9593 - 0.29(3269 + 0)] = 376 \ \mu\text{-strain},$$

$$\varepsilon_2 = \frac{1}{23 \times 10^6}[3269 - 0.29(0 + 9539)] = 22 \ \mu\text{-strain},$$

$$\varepsilon_3 = \frac{1}{23 \times 10^6}[0 - 0.29(9593 + 3269)] = -162 \ \mu\text{-strain}.$$

f.   Mohr's diagrams for stress and for strain are shown in Figs. E.14 and E.16, respectively.

[*Student assignment:* Modify the preceding calculations and diagrams for conditions on the inner pipe surface ($\sigma_3 = \sigma_N = -620 \ \text{lbf/in.}^2$).]

‖‖

Pseudorandom Normally Distributed Numbers

Table F.1 lists 500 pseudorandom, pseudonormal numbers. They were generated using a Macintosh computer and MicroSoft Quick BASIC. The simple program, Listing A, averages uniformly distributed random numbers producing approximately normally distributed random numbers.

Listing A

```
140   REM—A program for generating pseudonormally
160   REM—distributed random numbers.
180   PRINT
200   RANDOMIZE TIMER: REM- set random number seed
220   N1 = 25:   REM- Number rand/norm. values generated
240   N2 = 3:   REM- Number uniform rand. values averaged
260     FOR J = 1 TO N1
280       FOR K = 1 TO N2
300       X = X + RND
320       NEXT K
340     REM—Average N2 random values
360     R = X/N2: Rem—R is resulting pseudorandom number
380     REM—Adjust R to three decimal places
400     PRINT INT(1000*R + 0.5)/1000
420     X = 0:REM—reset X for next iteration
440     NEXT J
460   END
```

The values in Table F.1 may be construed as a "population" and parts thereof as "samples." The values in a column provide a sample of 50 items. Values in a row yield a sample of 10 items. Columns and/or rows may be combined to provide larger samples. In addition the values may be modified

Table F.1 Five hundred random normal numbers

	A	B	C	D	E	F	G	H	I	J
1	0.293	0.339	0.33	0.558	0.571	0.235	0.329	0.341	0.264	0.622
2	0.52	0.425	0.468	0.297	0.225	0.347	0.58	0.625	0.468	0.641
3	0.49	0.338	0.65	0.493	0.595	0.759	0.765	0.453	0.785	0.374
4	0.381	0.221	0.571	0.222	0.393	0.431	0.542	0.639	0.758	0.783
5	0.331	0.358	0.249	0.824	0.402	0.551	0.183	0.241	0.459	0.643
6	0.232	0.502	0.46	0.763	0.565	0.91	0.445	0.691	0.499	0.315
7	0.459	0.423	0.518	0.328	0.448	0.182	0.614	0.591	0.224	0.507
8	0.514	0.45	0.776	0.431	0.482	0.458	0.647	0.464	0.648	0.523
9	0.669	0.349	0.552	0.492	0.273	0.252	0.664	0.733	0.8	0.574
10	0.325	0.494	0.641	0.664	0.44	0.681	0.226	0.688	0.837	0.678
11	0.175	0.177	0.52	0.779	0.393	0.538	0.48	0.456	0.465	0.401
12	0.394	0.758	0.889	0.322	0.44	0.747	0.413	0.534	0.236	0.708
13	0.176	0.191	0.299	0.635	0.732	0.551	0.602	0.564	0.381	0.678
14	0.602	0.439	0.724	0.5	0.607	0.69	0.406	0.436	0.537	0.501
15	0.695	0.564	0.593	0.439	0.449	0.692	0.796	0.505	0.445	0.336
16	0.551	0.753	0.433	0.536	0.766	0.499	0.532	0.643	0.583	0.752
17	0.568	0.617	0.415	0.466	0.576	0.508	0.276	0.679	0.538	0.353
18	0.291	0.584	0.721	0.349	0.64	0.315	0.718	0.715	0.725	0.466
19	0.395	0.558	0.667	0.229	0.492	0.728	0.622	0.522	0.319	0.254
20	0.871	0.297	0.784	0.497	0.394	0.35	0.427	0.77	0.257	0.515
21	0.498	0.848	0.509	0.778	0.469	0.418	0.583	0.707	0.608	0.581
22	0.646	0.369	0.528	0.585	0.815	0.686	0.5	0.528	0.53	0.55
23	0.186	0.574	0.774	0.469	0.496	0.452	0.276	0.394	0.67	0.495
24	0.507	0.114	0.547	0.582	0.7	0.732	0.235	0.794	0.232	0.479
25	0.561	0.471	0.272	0.639	0.459	0.77	0.461	0.363	0.6	0.618
26	0.657	0.58	0.532	0.494	0.494	0.395	0.434	0.414	0.483	0.507
27	0.779	0.317	0.424	0.653	0.724	0.132	0.199	0.383	0.746	0.77
28	0.165	0.887	0.156	0.612	0.234	0.295	0.511	0.565	0.08	0.183
29	0.604	0.757	0.445	0.437	0.38	0.408	0.684	0.631	0.83	0.661
30	0.157	0.281	0.695	0.813	0.289	0.397	0.376	0.452	0.545	0.553
31	0.444	0.857	0.495	0.575	0.748	0.402	0.444	0.489	0.407	0.521
32	0.475	0.286	0.514	0.41	0.309	0.482	0.419	0.592	0.607	0.393
33	0.421	0.78	0.417	0.338	0.24	0.547	0.35	0.04	0.579	0.659
34	0.358	0.545	0.451	0.708	0.49	0.383	0.589	0.499	0.455	0.472
35	0.59	0.576	0.415	0.335	0.171	0.605	0.306	0.375	0.577	0.505
36	0.612	0.334	0.461	0.597	0.278	0.397	0.506	0.451	0.494	0.453
37	0.473	0.606	0.762	0.274	0.484	0.658	0.368	0.57	0.616	0.185
38	0.549	0.711	0.692	0.104	0.393	0.588	0.686	0.499	0.637	0.274
39	0.823	0.281	0.674	0.659	0.736	0.402	0.473	0.38	0.175	0.343
40	0.586	0.524	0.615	0.825	0.706	0.409	0.686	0.717	0.185	0.362

Table F.1 *continued*

	A	B	C	D	E	F	G	H	I	J
41	0.372	0.396	0.54	0.508	0.436	0.584	0.79	0.68	0.26	0.402
42	0.296	0.495	0.44	0.674	0.791	0.314	0.874	0.614	0.612	0.203
43	0.565	0.291	0.253	0.439	0.765	0.266	0.61	0.647	0.532	0.34
44	0.638	0.104	0.611	0.363	0.522	0.293	0.145	0.246	0.457	0.739
45	0.888	0.631	0.662	0.721	0.247	0.577	0.296	0.543	0.365	0.677
46	0.326	0.641	0.435	0.219	0.688	0.501	0.456	0.414	0.462	0.659
47	0.625	0.53	0.404	0.462	0.563	0.489	0.385	0.588	0.533	0.588
48	0.573	0.682	0.149	0.595	0.522	0.752	0.642	0.33	0.351	0.536
49	0.803	0.658	0.289	0.366	0.263	0.761	0.51	0.299	0.763	0.65
50	0.319	0.486	0.641	0.297	0.594	0.542	0.258	0.321	0.54	0.37

through use of multipliers and/or additives. Of course if this is done, population and sample must be treated alike in order to maintain comparability.

Listing B is presented here as a basis for study and perhaps modification to suit particular needs.*

Listing B

```
100  REM  HISTOGRAM  17 JULY 1977  JOHN M. NEVISON
110    RANDOMIZE TIMER
120  REM  PRINT A HISTOGRAM OF THE DISTRIBUTION
130  REM  OF N9 RANDOM NUMBERS.
140
150  REM  VARIABLES:
160  REM    (H)...THE LENGTH OF EACH HISTOGRAM BAR
170  REM    I.....THE HISTOGRAM INTERVAL
180  REM    J,K...INDEX VARIABLES
190  REM    M.....THE MAXIMUM H()
200  REM    X.....A RANDOM NUMBER
210
220  REM  CONSTANTS:
230    LET H9 = 20          'NUMBER OF HISTOGRAM BARS
240    LET L9 = 35          'LENGTH OF THE LARGEST BAR
250  REM                    '  IN CHARACTERS ACROSS THE PAGE
260    LET N9 = 300         'NUMBER OF RANDOM NUMBERS
270    LET R9 = 3           'NUMBER OF RND'S IN EACH X
280
290  REM  DIMENSIONS:
300    DIM H(20)
310
```

---

* This listing is adapted from John M. Nevison, *The Little Book of Basic Style,* Addison-Wesley, Reading, MA, 1978, p. 105. Reproduced by permission.

```
320   REM   FUNCTIONS:
330
340   REM   FNL CONVERTS LENGTH TO LENGTH IN CHARACTERS SO
350   REM   THAT FNL(M) WILL BE L9 CHARACTERS LONG
360
370     DEF FNL(L) = INT(L/M * (L9-1) + 1.5)
380
390   REM   MAIN PROGRAM
400
410   REM   GENERATE X, SORT IT INTO THE RIGHT HISTOGRAM BAR,
420   REM   H(K), CHECK FOR A NEW MAXIMUM. WHEN DONE,
430   REM   PRINT THE HISTOGRAM.
440
450   LET I = R9/H9
460   LET M = 0
470
480   FOR J = 1 TO N9
490
500     LET X = 0
510     FOR K = 1 TO R9
520       LET X = X + RND
530     NEXT K
540
550     LET K = INT(H9*X/R9) + 1
560     LET H(K) = H(K) + 1
570     IF H(K) <= M THEN 590
580       LET M = H(K)
590   NEXT J
600   NEXT J
610
620   PRINT "FROM 0 TO "; R9; " IN INTERVALS OF "; I; "."
630   PRINT "MAXIMUM HEIGHT IS "; M; "POINTS."
640   FOR J = 1 TO H9
650     PRINT "I";
660     FOR K = 1 TO FNL(H(J))
670       PRINT "*";
680     NEXT K
690     PRINT
700   NEXT J
710
730   END
```

Suggested Reading

Hanson, A. G. Simulating the normal distribution, *Byte*, p. 137, October 1985.

APPENDIX G

||||||||||||||||||||||||||||||||||||||||||||||||||||||||

Statistical Tests of Least-Squares Fits

The method of least squares is described in Section 3.15.1. Here, we consider some statistical tests of least squares results. Recall that least squares is essentially a method of averaging out the y *precision error* in data that satisfy an underlying straight-line relationship, $y = a + bx$.

As in Section 3.15.1, for the various measured values of x_i, y_i is the experimentally determined ordinate and $y(x_i) = a + bx_i$ is the corresponding value calculated from the fitted line; n is the number of experimental observations used. Least squares minimizes the sum, S^2, of squared vertical deviations from the fitted line:

$$S^2 = \sum_{i=1}^{n} [y_i - y(x_i)]^2$$

to obtain

$$a = \frac{\sum y_i \sum x_i^2 - \sum x_i \sum x_i y_i}{n \sum x_i^2 - (\sum x_i)^2} \tag{G.1}$$

and

$$b = \frac{n \sum x_i y_i - \sum x_i \sum y_i}{n \sum x_i^2 - (\sum x_i)^2} \tag{G.2}$$

Our objective in this appendix is to characterize the quality of the least squares fit.

The simplest test of the fit, the correlation coefficient r^2, is described in Section 3.15.1. An algebraic form of r^2 convenient for hand calculation is

$$r^2 = b \cdot \frac{n \sum x_i y_i - \sum x_i \sum y_i}{n \sum y_i^2 - (\sum y_i)^2} \tag{G.3}$$

Figure G.1   Precision error in least squares fitting

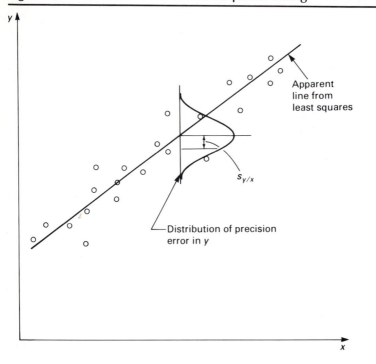

When $r^2 \to 1$, the precision error goes to zero, yielding a "perfect" fit. However, as noted before, one usually obtains $|r| > 0.9$ whenever the data appear to fall on a straight line; the correlation coefficient is not a very sensitive indicator of the precision of the data.

Instead, we may consider the *standard error* of the y-data about the fit, $s_{y/x}$:

$$s_{y/x} = \left(\frac{1}{(n-2)} \sum_{i=1}^{n} [y_i - y(x_i)]^2 \right)^{1/2} = \sqrt{\frac{S^2}{n-2}} \qquad \textbf{(G.4)}$$

This quantity approximates the standard deviation of the precision error in y_i (Fig. G.1); if the precision error is small, so will be $s_{y/x}$.

We can make $s_{y/x}$ more revealing by introducing *total* squared variation of the data set (which includes both the precision error and the straight-line variation of y with x),

$$S_{yy}^2 = \sum_{i=1}^{n} (y_i - y_m)^2,$$

where y_m is the mean measured y_i:

$$y_m = \frac{1}{n} \sum_{i=1}^{n} y_i$$

Letting $s_{yy}^2 = S_{yy}^2/(n - 1)$ be the *mean* total variation and performing some algebra, we find

$$\frac{S_{y/x}}{S_{yy}} = \left(\frac{n - 1}{n - 2}\right)^{1/2} (1 - r^2)^{1/2} \tag{G.5}$$

The fraction on the left is the ratio of the standard deviation of the data about the line to the mean total variation of the data; thus $(1 - r^2)^{1/2}$ is a better gauge of the precision error in the data than is r itself [1]. For pocket-calculator work, the following formula is useful in calculating $s_{y/x}$ from Eq. (G.5):

$$s_{yy}^2 = \frac{\sum\limits_{i=1}^{n} y_i^2 - \frac{1}{n}\left(\sum\limits_{i=1}^{n} y_i\right)^2}{n - 1}. \tag{G.6}$$

Most calculators provide the various sums as part of the least squares calculation.

When least squares is justifiable, confidence intervals for the slope and intercept can be calculated under the assumption that the precision error in y_i satisfies the normal distribution [1, 2]. In this case, the true slope lies within the $c\%$ confidence interval

$$b \pm t_{\alpha/2, v} \frac{S_{y/x}}{S_{xx}} \qquad (c\%) \tag{G.7}$$

and the true intercept is within

$$a \pm t_{\alpha/2, v} S_{y/x} \sqrt{\frac{1}{n} + \frac{x_m^2}{S_{xx}^2}} \qquad (c\%) \tag{G.8}$$

where $t_{\alpha/2, v}$ is the t-statistic with $v = n - 2$ degrees of freedom at an $\alpha = (1 - c)$ level of significance, and S_{xx}^2 is

$$S_{xx}^2 = \sum\limits_{i=1}^{n} (x_i - x_m)^2 = \sum\limits_{i=1}^{n} x_i^2 - \frac{1}{n}\left(\sum\limits_{i=1}^{n} x_i\right)^2 \tag{G.9}$$

for $x_m = (1/n) \sum x_i$. Like other statistical confidence intervals, those just given may become unacceptably broad when only a few data points have been fitted; an eyeball estimate of the uncertainty may again yield a more realistic result.

Finally, statistical tests for outliers can also be applied, under the assumption of normally distributed precision error in y_i. For example, if $s_{y/x}$ approximates the vertical standard deviation of the data, then only one point in 370 will have more than $3s_{y/x}$ vertical deviation from the line; for a small data set, points beyond $3s_{y/x}$ may be considered so unlikely as to be discarded.

Example G.1

In Example 3.16, a least squares fit was used to find the stiffness of a cantilever beam. Quantify the vertical precision error in the data, and use a statistical confidence interval to estimate the uncertainty in the beam stiffness.

Solution. In Example 3.16, we found

$$n = 9$$

$$\sum x = 1801$$

$$\sum x^2 = 5.109 \times 10^5$$

$$\sum y = 33.50$$

$$\sum y^2 = 179.3$$

$$\sum xy = 9959$$

The data and the line-fit are shown in Fig. 3.24.

From Eq. (G.5), we calculate $s_{y/x}/s_{yy} = 0.0970$, which indicates that the data's vertical standard deviation about the fitted line is only 9.7% of the mean vertical variation of the data. Thus most of the vertical variation of the data is explained by the straight-line relation between y and x rather than precision error.

The confidence interval for the slope is calculated from Eq. (G.7) (at 95% certainty, say). With Eq. (G.6) and (G.9) and Table 3.6, we obtain:

$$s_{yy} = 2.612$$

$$S_{xx} = 387.8$$

$$s_{y/x} = 0.2534$$

$$t_{0.025,7} = 2.365$$

from which the uncertainty in the slope is

$$\pm t_{\alpha/2,v} \frac{s_{y/x}}{S_{xx}} = \pm 0.001543 \qquad (95\%)$$

Hence, with 95% certainty, $0.0175 < b < 0.0205$ and so $478\,\text{N/m} < K < 560\,\text{N/m}$. The 95% uncertainty in K is about $\pm 8\%$.

The statistically determined uncertainty is about the same as that obtained from the eyeball estimated confidence limits on the fitted line (Fig. 3.24). Which approach is easier to use?

References

1. McClintock, F. A. *Statistical Estimation: Linear Regression and the Single Variable*. Research Memo 274, Fatigue and Plasticity Laboratory, February 14, 1987. Cambridge: Massachusetts Institute of Technology.

2. Miller, I. R., J. E. Freund, and R. Johnson. *Probability and Statistics for Engineers*. 4th ed. Englewood Cliffs, N.J.: Prentice Hall, 1990.

ANSWERS TO SELECTED PROBLEMS

Chapter 2

2.2 **(b)** 310.93°K

2.3 **(b)** 1,680.23°R

2.5 .01124 lbf

2.8 $1\dfrac{\text{gal}}{\text{min}} = 63.09\,\dfrac{\text{cm}^3}{\text{s}}$

Chapter 3

3.1 183.2 to 216.8

3.3 **(a)** 10.2 kΩ ± 0.0863 kΩ

3.4 2.24%

3.7 7.45%

3.10 1.256%

3.12 ≈ ±10%

3.15 ≈65 marbles

3.18 ≈93

3.19 Packing is significant

3.20 Significant difference at 99% confidence level

3.23 $1.239 < \mu < 1.279$

3.24 No significant difference

3.29 No difference

3.30 Coin not fair

3.32 Normally distributed at 95% confidence level

3.35 Significant variation from standard

3.38 Force = 28.8 + 30.7 × Deflection

Chapter 4

4.1 **(a)** $15/2\pi$ rad; **(b)** $y(t) = 100 + 109.77\cos(15t - 1.046)$

4.8 $V(t) \approx 6.7\cos(2{,}026.8t) + 2.9\sin(3{,}040.3t) - 0.9\cos(4{,}053.7t)$

4.12 **(d)** $V(t) \approx 10\sin\omega t + 4\cos\omega t - 2\sin 6\,\omega t$
$\omega \approx 2\pi/.036$ rad/s

4.13 **(a)** $f_{NO} = 2048\,\text{Hz}$ **(b)** $2\,\text{Hz}$ **(c)** $1000\,\text{Hz},\ 1500\,\text{Hz},\ 2000\,\text{Hz}$

4.16 $f_s > 4000\,\text{Hz}$, $N = 8000$ pts. sampled

Chapter 5

5.3 $1.57°\text{F}$

5.5 $f \approx 160\,\text{Hz}$

5.6 $35\,\text{Hz}$

5.11 $\tau = V_o/\dot{m}$

5.16 $f \approx 1{,}124\,\text{Hz}$

5.19 Error ≈ 277; $\phi \approx 24°$

5.21 **(a)** $2.0\,\text{s}$ **(b)** $E_R = 0.8\,\text{V}$

Chapter 6

6.4 $U_K/K \approx \pm 3\%$

6.11 $U_K/K \approx \pm 5.7\%$

6.13 $K = 498\,\text{N/m}$

Chapter 7

7.4 $e_0/e_i = \dfrac{kR_T}{R_B}\left[\dfrac{kR_T}{R_M} + \dfrac{kR_T}{R_B} + 1\right]$

7.6 $R_2 = 63\,\Omega$

7.8 $R_4 = 141.9\,\Omega$

7.13 For $100\,\text{N}$, $\Delta e_0 = e_0 = 0.24\,\text{V}$

7.24 $R_3 = 1\,\text{k}\Omega$

7.28 $e_{rms} = 6.0\,\text{V}$

7.29 **(a)** $fc = 4.82\,\text{kHz}$

Chapter 8

8.1 **(b)** 10

8.5 **(a)** $\varepsilon_V/\Delta V_{fs} = \dfrac{1}{2^8}$

Chapter 9

9.4 $0.55\,E_o$

9.5 0.50

9.6 10.0

Chapter 11
11.5 $d = 3.147246$ in

11.8 $L = 3.147179$ in

11.14 $N = 54.3$

11.19 $d = 1.761703$ in

Chapter 12
12.1 $V = 0.3$

12.2 **(a)** $\sigma = 12,750$ psi

12.8 **(a)** $\sigma_a = 9,300$ psi

12.11 $F = 1.68$

12.20 $\Delta e_o / e_i = (F/4)(2\varepsilon_T)$

12.23 $P_{max} = 53,800$ watts

Chapter 13
13.5 $F = 0.17$ lbf

13.7 $F_e = 293$ N

13.13 $U_K / K = 3.11\%$

Chapter 14
14.9 $h = 0.651$ m

14.13 $P_2 - P_1 = 16.65$ kPa

14.15 $O = 9.6°$

14.17 $M = 2.56$

14.23 $y_{max} = 3.4 \times 10^{-3}$ in

14.27 $\varepsilon_H = 322\ \mu$ strain; $\varepsilon_L = 76\ \mu$ strain

14.33 $f = 3400$ Hz

14.34 $f = 310$ Hz

Chapter 15
15.3 $Re_D = 6.9 \times 10^5$

15.7 $\Delta P = 39.35$ kPa

15.14 $Q = 1660$ gal/hr

15.17 $Q = 4.5\ \text{m}^3/\text{s}$

15.25 $\Delta P = 530$ psi

15.29 $V = 1.085$ m/s

15.34 $V_2 = 21.7$ m/s

Chapter 16
16.2 $T = 70.5°C$

16.6 $\Delta R = 0.33\ \Omega/°C$

16.27 $T_p = 179°C$

16.28 $T_p = 400°C$

16.29 $T_{max} = 276°F$

Chapter 17

17.4 $a_o = 0.41\,g$

17.7 $k = 13.4 \times 10^6\,N/m$

17.12 $f = 34\,Hz$

Chapter 18

18.2 Five 2000 gpm pumps

18.5 $I = W/A = 0.0016\,W/m^2$

18.7 **(a)** $SPL_5 = 115\,dB$; **(b)** $SPL_3 = 112.8\,dB$ Decrease $= 2.2\,dB$

18.10 $SPL = 91.5\,dB$

INDEX